INSECT ECOLOGY

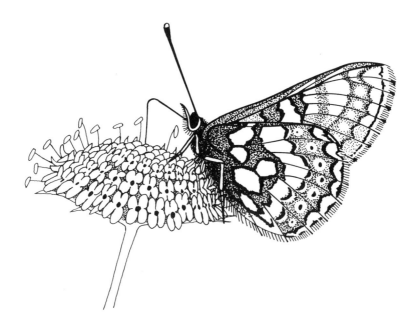

A male marsh fritillary butterfly, *Euphydryas aurinia,* feeds on an inflorescence of devil's-bit scabious, *Succisa pratensis.*
Drawing by Steve Price.

INSECT ECOLOGY

Third Edition

Peter W. Price
Northern Arizona University

John Wiley & Sons, Inc.
New York • Chichester • Weinheim • Brisbane • Singapore • Toronto

This text is printed on acid-free paper.

Library of Congress Cataloging-in-Publication Data
Price, Peter W.
    Insect ecology / Peter W. Price.—3rd ed.
        p.   cm.
    Includes bibliographical references (p.   ) and indexes.
    ISBN 0-471-16184-5 (cloth : alk. paper)
    1. Insects—Ecology. I. Title.
    QL496.4.P75   1997
    595.717—dc21
                                                              96-47739

Printed in the United States of America
10 9 8 7 6 5 4 3 2 1

To my collegues, friends,
teachers, students, and loved ones,
all of whom have enriched my experience
and my fascination with the
mystery of nature

Contents

Preface

My motivation for preparing this third edition of *Insect Ecology* has been the teachers, researchers, and students around the world who have used earlier editions, for a more up-to-date treatment of the field is overdue. I have continued to keep the book appropriate for advanced undergraduates, graduate students, and researchers, and my general approaches to the subject have remained the same as in the second edition.

In this revision I have added four new chapters, and considerable new material to some others, without deleting the previous content—hence the book is longer. However, with new chapters on the importance of insect ecology, the development of theory, hypotheses on plant and herbivore interactions, and an effort to develop the beginnings of a synthesis in population dynamics, I believe the reader will find a more cohesive end product.

Another change in this edition includes subheadings in chapters so that chapter contents may be found and comprehended more readily. Also, many figures have been added, which I hope will contribute to an aesthetically pleasing experience for the reader. Some figures have been redrawn with improved clarity. However, many features of the former editions of this book have remained the same or have been developed further.

Organization—The book covers the major conceptual themes in ecology and these are related to insect ecology. I have retained a close link between ideas and relevant empirical studies. The chapters usually have a strong historical basis, for it remains important to acknowledge the roots of our science and those scientists who contributed to the development of the discipline, whether or not they studied insects.

References—I have used citations to the literature extensively for several important reasons: (1) we should give credit where it is due; (2) sources can provide the reader with greater insights and many leads into the literature; (3) references provide information on the types of sources available to the insect ecologist; and (4) a student may learn something about the assimilation of knowledge from the types and diversity of sources cited.

Figures and tables—I believe in providing clear figures that embellish the text and a few simply to provide pleasure for the reader. Therefore, I have spent much time on figure design and finding suitable material for illustration. Figures are not necessarily easy to understand unless they are studied in detail, but they are a fundamental means of communicating quantitative and qualitative information. Some colleagues have noted my love of complex flow diagrams, yet for me they capture a large idea in a concise form that is, in my opinion, simpler to grasp than extended text. Tables also provide large bodies of information in compact form. They enable summarization and synthesis of diverse studies and present the weight of evidence for a particular argument developed in the text.

Debate—Some topics in ecology are in a state of flux. This is natural, healthy, and desirable in an actively developing field. In covering such issues, my approach is to present the studies relevant to the debate, usually in historical order, and to leave the reader with the latest developments. Thus the role of competition in insect communities is an unresolved issue, and both sides of the evidence are discussed. Similarly, there are generally strongly held feelings about the concept of the vacant niche, and the orderly or disorderly nature of trophic interactions in communities. The swing of ideas from one position to another is a part of the development of science, and I have found it illuminating to compare coverage of certain topics in the first and second, and in this third edition of *Insect Ecology*, spanning over 20 years of scientific development.

Synthesis—I believe that more focus is needed on synthesis in ecology and the development of empirically and factually based theory. We have performed superbly in collecting the details of intricate interactions, such as those among three trophic levels or other plant–herbivore interactions, but synthesis and theory are slow in developing. What are the themes that enable synthesis? How do we develop concepts that encapsulate a large body of knowledge? After a decade of concern for these issues I have taken a tentative step toward synthesis in population dynamics in this edition, for the field provides a theme encompassing essentially many areas of ecology. Such a synthesis will be controversial, no doubt, but I hope it stimulates debate and alternative approaches, and the students of today will, after all, develop the field in the future. Being aware of the need for synthesis and having some themes to criticize will perhaps provide an interesting educational experience for the reader, as it has for me.

Peter W. Price
Flagstaff, Arizona

May, 1997

Acknowledgments

My privilege over more than three decades has been to meet with the majority of authors cited in this book. Others have corresponded or sent reprints. All these contacts have enriched my scientific experiences on a personal and professional level. I hope that I have done justice to the many fine contributions from my colleagues. Several colleagues and their students have pointed to the need for more clarification or corrections and I thank them all. In particular, Jan Giliomee, head of entomology at the University of Stellenbosch, and his students provided a detailed set of corrections and suggestions.

This book is the work of many skilled and committed professionals whose care, concern, and knowledge are evident on every page. I am grateful to them all. I thank Dr. Philip Manor, senior editor at John Wiley & Sons; the production editor, Millie Torres; and many others at Wiley for their commitment to this edition. Louella Holter, with speed and accuracy, performed all word processing for this edition. Art and illustration so important for this book were provided by my brother, Steve Price; Ron Redsteer at the Bilby Research Center on campus; Lana Johnson of Koepke Illustration in Lincoln, Nebraska; and Tad Theimer, a former graduate student in our department and now a faculty member. Daniel Boone photographed many figures. I am deeply indebted to them for their finesse in advancing the information and pleasure for the readers and for me. Many others who contributed to former editions of the book have left their mark on this edition. I have used many figures from other sources and I am grateful for permission for their use. This is acknowledged specifically after the captions for individual figures.

I remain indebted to those who have provided me with a unique education, without which I would not have written this book. Richard Root was my doctoral dissertation advisor at Cornell University, providing a rich milieu for science, learning and debate in his ecological research group. I was also fortunate to be instructed by the late William Brown in evolution and William Dilger in behavior. This blend of ecology, evolution, and behavior at

Cornell has enriched this book and my own interest and involvement with science. The graduate and undergraduate students in my research group over the years have also contributed significantly to my education, and many are cited in this book.

Over the last few weeks of book preparation, working on indexes, I reflected on the richness of personal interactions and the pleasure in meeting so many colleagues, of the many research and lecturing trips around the country and the world, and the opportunities others have provided. I owe so much to many colleagues. Finally, my debt of gratitude to my wife, Maureen, must be acknowledged for providing a stable and nurturing home environment for almost 30 years, in which I have found the tranquility and freedom to write three editions of *Insect Ecology*.

P. W. P.

INSECT ECOLOGY

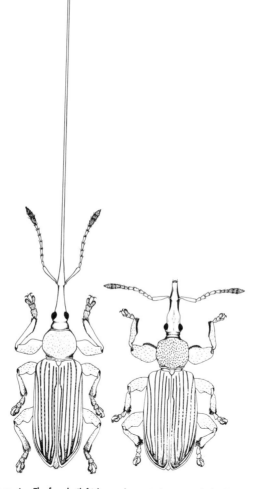

Cycad weevil, *Antliarhinus zamiae*. The female (left) has a phenominal rostrum which allows her to drill a hole between adjacent sporophylls in female *Encephalartos* (Zamiaceae) cones and into the gametophyte, after which the female uses a highly extended ovipositor to lay several eggs in the gametophyte, on which the larvae feed (Donaldson, 1992). From Scholtz and Holm (1985). Drawing by M. Johnson.

Introduction

Insect ecology is an important field in the sciences because of its breadth, depth, and impact on human existence (Chapter 1). The subject of ecology covers a wide range of topics, from molecular properties such as enzyme systems adapted to function well under certain environmental conditions, to organismal, population, community, and ecosystem structure and function. Two of the introductory chapters of this book deal with the ends of this continuum of relationships, discussing major components and processes in ecosystems (Chapter 2), and how small organisms such as insects relate individually to an environment that inevitably constrains their design and activity in nature (Chapter 3).

Much of ecology is devoted to identifying and understanding patterns in nature. Ecology could be defined as the science devoted to the development and study of theories of the natural history of organisms. Such theories can relate to cellular processes in relation to environmental factors, all the way up to global patterns. A framework for considering theories and concepts in ecology is provided by discussing major processes in ecosystems. But the nature of scientific theory and how theory is developed requires some consideration, as discussed in Chapter 4.

Ecology can be defined also as the science of relationships of organisms to their environment. How they relate depends on their structure. Thus a chapter dealing with aspects of insect design seems fundamental to understanding their ecological relationships. Ideally, one would enter the field of insect ecology with a detailed knowledge of insect body form and function obtained, perhaps, in a course on insect physiology, including environmental physiology, or in a course on the physiological ecology of insects. To cover such material in this book would double its size, making it unreasonably large. Therefore, in Chapter 3 only the link is provided between central aspects of ecology and those of physiology.

These chapters set the stage for considering intermediate levels of interaction involving trophic relationships (Part II), populations (Part III), and communities (Part IV).

Importance of Insect Ecology

> The most beautiful and most profound emotion one can experience is the sensation of the mystical . . . it is the source of all true science.
>
> Albert Einstein

> Few things indeed have I known in the way of emotion or appetite, ambition or achievement, that could surpass in richness and strength the excitement of entomological exploration.
>
> Vladimir Nabokov

NUMBER OF INSECT SPECIES

Insects are indeed mystical creatures, for the more we learn about them the more interesting the questions become. Even the number of insect species on earth has provided a continuing debate that probably will never be resolved in a definitive manner. But it is the sheer numbers of species, the range of their activities, and their ubiquity that make them important in ecosystems and to human existence. "Like it or not, insects are a part of where we have come from, what we are now, and what we will be" (Berenbaum, 1995, p. xiii).

Of the 1.82 million described species of plants and animals, insects are by far the richest group, with almost 60% of species (Fig. 1.1) (Stork, 1988). Even when all described entities from viruses and bacteria to chordates are included, and something like 950,000 described species of insects is accepted, the insects still represent about 60% of the biota, according to Samways (1994).

3

Two species from the extinct order Paleodictyoptera: (left) *Homaloneura lehmani,* from the Upper Carboniferous, feeding on a Cordaitales cone (based on Kukalová-Peck, 1991; by permission of Cornell University Press). (right) a *Goldenbergia* species from the lower Permian feeding on a megaspore (based on Brodsky, 1994; by permission of Oxford University Press). Drawing by Tad Theimer.

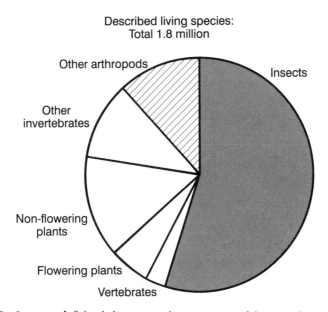

FIGURE 1.1   Percentage of all described species in each major taxon out of the estimated 1.8 million species described. From Stork (1988); with permission from the Linnean Society of London.

The total number of insect species both described and undescribed, of course, is difficult to estimate. Estimates have ranged from 3–5 million (May, 1986) to 30 million (Erwin, 1982), 50 million (Erwin, 1988), and even 80 million species (Stork, 1988). There seems to be a general move in the 1990s for estimates around 5–10 million species (Gaston, 1991, 1992). Hammond (1992) has used 8 million species as a general estimate in relation to the inventory of global diversity.

Whatever the total number of insect species on this earth, each has its own ecology, and each has an impact on many other species. The richness of ecological interactions is bewildering, but important.

ANTIQUITY OF INSECTS

Not only are the numbers of insect species staggering, but they are incredibly diverse, stemming from a very long evolutionary history. Colonization of land by arthropods probably preceded the evolution of land plants in the Silurian and Devonian periods, more than 400 million years ago. Centipedes, millipedes, trigonotarbids, scorpions, spiders, Collembola, and apterygote insects had all left their record in the Devonian, and flying insects are recorded from the Carboniferous (Fig. 1.2) (Kukalová-Peck, 1991; Gray and Shear, 1992;

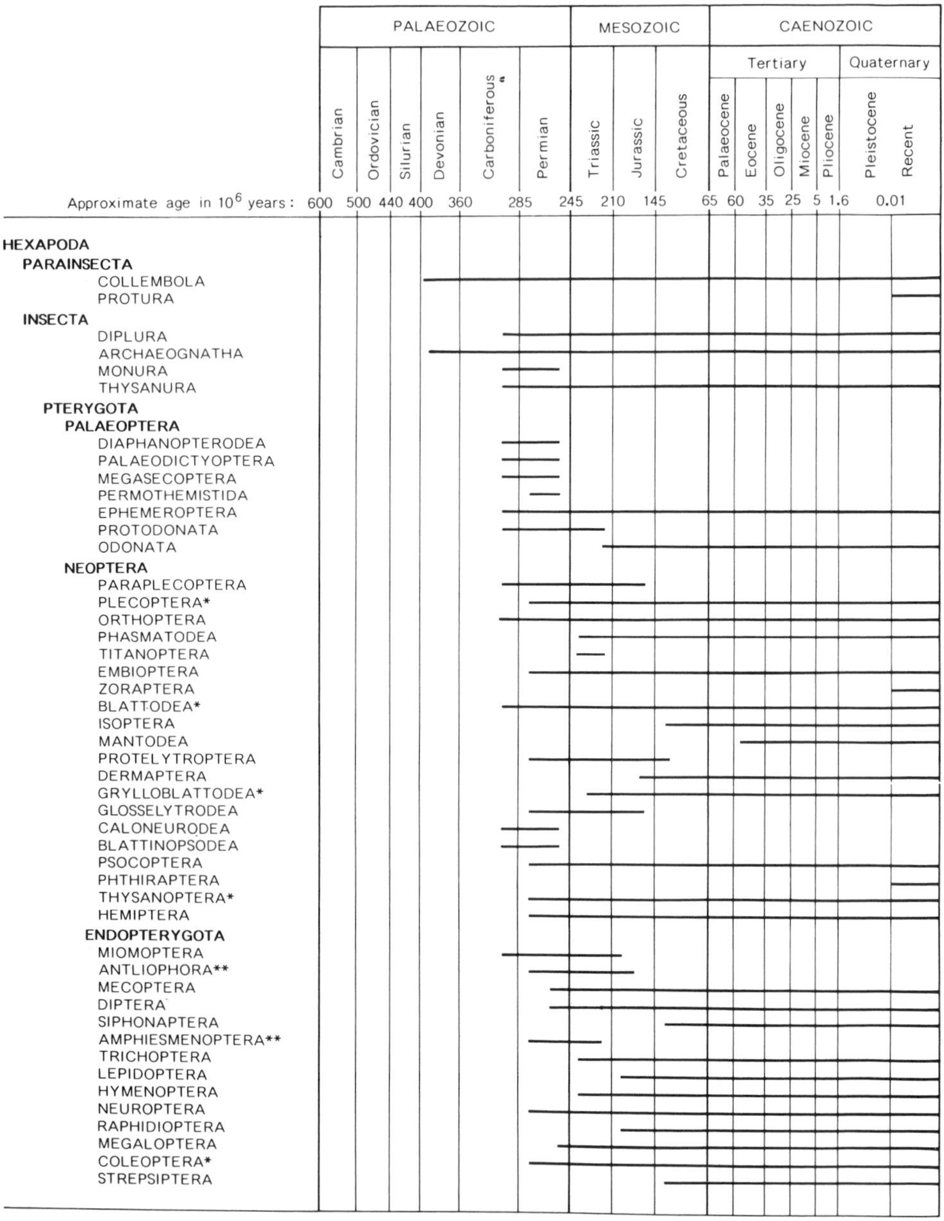

	PALAEOZOIC						MESOZOIC			CAENOZOIC						
											Tertiary				Quaternary	
	Cambrian	Ordovician	Silurian	Devonian	Carboniferous	Permian	Triassic	Jurassic	Cretaceous	Palaeocene	Eocene	Oligocene	Miocene	Pliocene	Pleistocene	Recent
Approximate age in 10^6 years:	600	500	440	400	360	285	245	210	145	65	60	35	25	5	1.6	0.01

HEXAPODA
  PARAINSECTA
     COLLEMBOLA
     PROTURA
  INSECTA
     DIPLURA
     ARCHAEOGNATHA
     MONURA
     THYSANURA
  PTERYGOTA
    PALAEOPTERA
     DIAPHANOPTERODEA
     PALAEODICTYOPTERA
     MEGASECOPTERA
     PERMOTHEMISTIDA
     EPHEMEROPTERA
     PROTODONATA
     ODONATA
    NEOPTERA
     PARAPLECOPTERA
     PLECOPTERA*
     ORTHOPTERA
     PHASMATODEA
     TITANOPTERA
     EMBIOPTERA
     ZORAPTERA
     BLATTODEA*
     ISOPTERA
     MANTODEA
     PROTELYTROPTERA
     DERMAPTERA
     GRYLLOBLATTODEA*
     GLOSSELYTRODEA
     CALONEURODEA
     BLATTINOPSODEA
     PSOCOPTERA
     PHTHIRAPTERA
     THYSANOPTERA*
     HEMIPTERA
  ENDOPTERYGOTA
     MIOMOPTERA
     ANTLIOPHORA**
     MECOPTERA
     DIPTERA
     SIPHONAPTERA
     AMPHIESMENOPTERA**
     TRICHOPTERA
     LEPIDOPTERA
     HYMENOPTERA
     NEUROPTERA
     RAPHIDIOPTERA
     MEGALOPTERA
     COLEOPTERA*
     STREPSIPTERA

FIGURE 1.2   Geological history of hexapods, by Kukalová-Peck (1991). Double asterisks indicate a stem group; single asterisks mean that the stem group is included in the fossil record. A stem group is one belonging to an unresolved basal lineage. Reprinted from I. D. Naumann (ed.), *The insects of Australia: A textbook for students and research workers*, 2nd ed., p. 142. Copyright © 1991 by the Commonwealth Scientific and Industrial Research Organisation (Division of Entomology). Used by permission of the publisher, Cornell University Press.

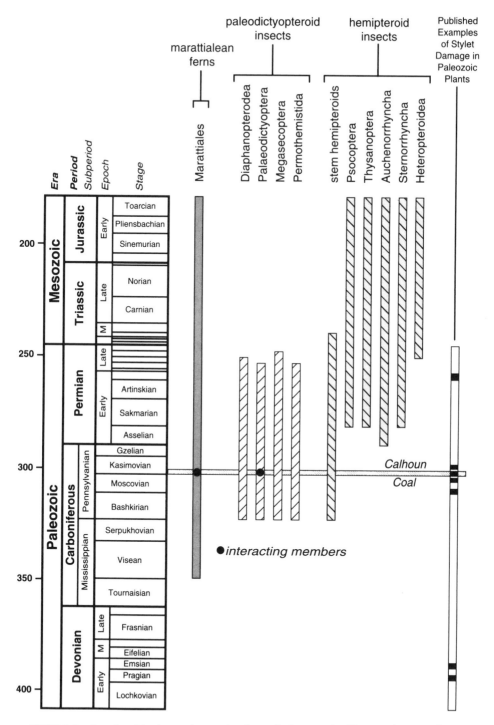

FIGURE 1.3 Chronological distributions of marattialean ferns, paleodictyopteroid and hemipteroid insects, in the Paleozoic and Mesozoic eras, and records of stylet damage in Paleozoic plants. From Labandeira and Phillips (1996a); reprinted by permission of the Entomological Society of America.

7

Gray and Boucot, 1994). Many orders are represented in Upper Carboniferous deposits.

There is also good evidence of insect herbivores, probably members of the Palaeodictyoptera with piercing-and-sucking mouthparts, feeding on tree ferns about 302 million years ago, and equivalent plant damage has been found as far back as the Early Devonian, up to 395 million years ago (Fig. 1.3) (Labandeira and Phillips, 1996a). Even insect galls have been found on *Psaronius* tree-fern fronds from 302 million years ago in Kasimovian coal deposits (Fig. 1.4) (Labandeira and Phillips, 1996b). Evidence suggests a possible sawfly gall-former, an association between insect and plant still seen today in sawfly gall-formers on fern stems and petioles.

These ancient ecological relationships pose many challenges for understanding the insects: their origins from a marine arthropod stock, the origins of flight, the phylogenetic relationships among orders, the radiation of insect groups on plants as herbivores and pollinators, and so on. The tremendous range of form and function in the insects and their intricate interactions with each other and other taxa make their ecology much richer than sheer numbers would suggest.

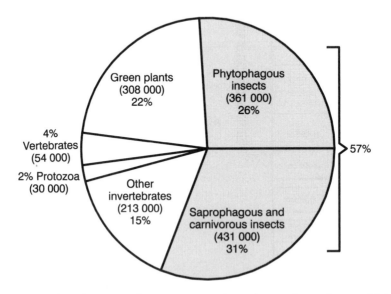

FIGURE 1.4   Number and percentages of species in major taxa showing the preponderance of insects and their distribution among phytophagous, and saprophagous plus carnivorous groups. Slightly modified from Strong et al. (1984a).

ROLE OF INSECTS IN FOOD CHAINS AND FOOD WEBS

Ornithologists frequently consider insects only as bird food and hardly worthy of their interest. But this attitude does emphasize the point that insects are of prime importance in moving energy through food chains and food webs (cf. Chapter 2). Among the major taxa, about 26% are estimated to be phytophagous insects, converting plant biomass to animal energy (Fig. 1.4) (Strong et al., 1984a). Another 31% are saprophagous and predaceous insects, forming important components of terrestrial and freshwater food webs (e.g., Figs. 22.14–22.17). Fully 57% of terrestrial species are involved as intermediaries in food webs, usually at the second and third trophic levels based on plants and at many different levels in food webs of detritivores.

"Ants are everywhere, but only occasionally noticed. They run much of the terrestrial world as the premier soil turners, chanellers of energy, dominatrices of the insect fauna—yet receive only passing mention in textbooks on ecology" (Hölldobler and Wilson, 1990, p. 1). Even in the cool north temperate a single population of a single *Formica* species consisted of 73 nests and about 12 million workers in 10 acres of land, and weighed 100 kg! Termites reach similar densities (Wilson, 1971). The social insects are our ubiquitous neighbors, moving enormous volumes of material, housing a large diversity of inquiline species, and passing energy in a multitude of directions in food webs. Jones et al. (1994, p. 373) emphasized that many species, including ants and termites, act as ecosystem engineers. "Ecosystem engineers are organisms that directly or indirectly modulate the availability of resources to other species, by causing physical state changes in biotic or abiotic materials. In so doing they modify, maintain and create habitats." As engineers, ants modify soils and vegetation and enrich local sites (Elmes, 1991), and termites have similar effects.

Some ecologists have argued that there is no such subject as insect ecology—there is simply the subject of ecology with insects playing a prominent role. This is a valid perspective, but whichever point of view we favor, it is the sheer number of species, the richness of their interactions, and their economic impact that justify focused attention.

BENEFICIAL INSECTS

Insects as pollinators, predators, parasitoids, and human food play a prime role in the regulation of populations. We use them in the biological control of insect pests and weeds. They inform us about our dead, with the field of forensic entomology providing some of the first studies of ecological succession (e.g., Mégnin, 1894; Smith, 1986; Haskell and Catts, 1990) (Table 1.1). Insects till our soil and recycle the corpses of plants and animals. They

TABLE 1.1   Ecological Succession of Insects on Human Cadavers in Europe as Described Largely by Mégnin in 1894.

Fauna	State of Corpse	Approx. Age of Corpse
A. Exposed Corpses		
First wave *Calliphora vicina* (Dipt., Calliphoridae)	"Fresh" (variable with season)	
C. vomitoria (Dipt., Calliphoridae)		
Lucilia spp. (Dipt., Calliphoridae)		First 3
Musca domestica (Dipt., Muscidae)		months
M. autumnalis (Dipt., Muscidae)		
Muscina stabulans (Dipt., Muscidae		
Second wave *Sarcophaga* spp. (Dipt., Sarcophagidae [may occur in first wave]	Odor developed	
Cynomya spp. (Dipt., Calliphoridae)		
Third wave *Dermestes* (Col., Dermestidae)	Fats rancid	
Aglossa (Lep., Pyralidae		
Fourth wave *Piophila casei* (Dipt., Piophilidae)	After butyric fermentation	
Madiza glabra (Dipt., Piophilidae)	protein of "caseic"	3–6 months
Fannia (Dipt., Fanniidae)	fermentation	
Drosophilidae (Dipt.)		
Sepsidae (Dipt.)		
Sphaeroceridae (Dipt.)		
Eristalis (Dipt., Syrphidae)		
Teichomyza fusca (Dipt., Ephydridae)		
Corynetes, Necrobia (Col., Cleridae)		
Fifth wave *Ophyra* (Dipt., Muscidae)	Ammoniacal fermentation	
Phoridae (Dipt.)		
Thyreophoridae (Dipt.)	Evaporation of	
Nicrophorus (Col., Silphidae)	sanious fluids	4–8 months
Silpha (Col., Silphidae)		
Hister (Col., Histeridae)	Remaining body	
Saprinus (Col., Histeridae)	fluids now absorbed	
Sixth wave Acari		6–12 months
Seventh wave *Attagenus pellio* (Col., Dermestidae)	Completely dry	
Anthrenus museorum (Col., Dermestidae)		
Dermestes maculatus (Col., Dermestidae)		
Tineola bisfor elliella (Lep., Tineidae)		1–3 years
T. pellionella (Lep., Tineidae)		

TABLE 1.1   (Continued)

Fauna	State of Corpse	Approx. Age of Corpse
	A. Exposed Corpses	
	Monopis rusticella (Lep., Tineidae)	
Eighth Wave	*Ptinus brunneus* (Col., Ptinidae)	3+ years
	Tenebrio obscurus (Col., Tenebrionidae)	
	B. Buried Corpses	
First wave	*Calliphora and Muscina stabulans*	
Second wave	*Ophyra*	
Third wave	Phoridae (*Conicera* may appear on surface	1 year
Fourth wave	*Rhizophagus parallelocollis* (Col., Rhizophagidae)	
	Philonthus (Col., Staphylinidae)	2 years

SOURCE: K. G. V. Smith, *A manual of forensic entomology.* Ithaca, N.Y.: Cornell University Press, 1986.

provide food for many species we like to eat, such as freshwater fish, gallinaceous birds, and many other items in the more exotic cuisines of the Orient. As models for the dry-fly fisherman's lures, they have filled many a pleasant hour in the creation and use of insect mimics (e.g., Halford, 1897; Leonard, 1950; Usinger, 1956). Insect products include honey, wax, silk, shellac, and cochineal, and their galls provide tanning acids and the basis for inks (Metcalf and Metcalf, 1993). The rich ecology of insects as they interact, and as a result benefit humans, covers many fields of research and enters into human ecology itself.

INSECT PESTS

There can be no doubt that destructive insects have provided ecology with major challenges, enlivened the field, and provided the incentive and funding for a massive research program. Much of insect ecology, and ecology in general, has been driven by a need to understand the insects that threaten our crops, our livestock, our timber supply, our ornamental plants, our health, and our homes. The term *applied entomology* is often used to cover these areas, which misses the point that basic ecological knowledge is fundamental in the understanding of insect pests in natural environments and those modified by humans. Most insect ecology courses are probably motivated by the need for solutions to the vexing challenges that insects pose for human existence. Agricultural, forest, horticultural, veterinary, and medical entomology are all nested within the field of insect ecology. Even the fields of toxicology and pesticides have, of necessity, become ecological in scope because toxins enter food webs and travel over the landscape (cf. Chapter 2). Integrated pest management involves the ecologically sound planning of landscapes and their manipulation for sustained productivity and ecosystem health.

Arthropod-borne diseases, such as malaria and typhus, are some of the most lethal ailments for humans, other vertebrates, and the vectors themselves (e.g., Ewald, 1983). They have influenced the results of more wars than have the generals (Zinsser, 1935) and will play a major role in human population regulation in the future, as human populations soar (Garrett, 1994). The complexity of the transmission dynamics of these vector-borne diseases, involving pathogenic organisms, insect and other arthropod vectors, wild animals, and humans, remains as a major ecological and epidemiological challenge for entomologists.

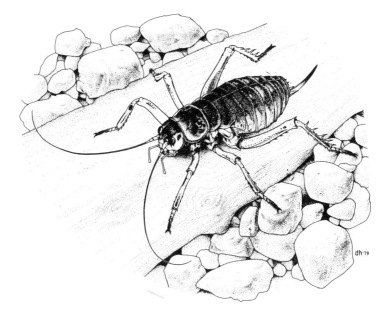

FIGURE 1.5   The Stephens Island weta (*Deinacrida rugosa*) is confined to Stephens and Mana islands in Cook Strait between the North and South islands of New Zealand. It was introduced to Maud Island in 1978, where it has flourished, but was once probably widespread on many islands and the mainland of New Zealand. From Howarth and Ramsay (1991); with permission. Drawing by D. H. Helmore, Landcare Research.

BIOLOGICAL CONSERVATION

With insects so numerous and important in any habitat, accompanied by rapid habitat destruction and fragmentation, concern regarding the preservation of insect species has developed rapidly (e.g., Collins and Thomas, 1991; Gaston et al., 1993; Samways, 1994). The field has evolved in parallel with concern over the preservation of biodiversity in general (e.g., Wilson, 1988, 1992), and the need to understand the nature of rare species (Kunin and Gaston, 1993; Gaston, 1994).

Relatively large insects on islands are particularly vulnerable when exposed to novel predators and habitat destruction. High endemicity, flightlessness, large size, and poorly developed defensive mechanisms all contribute to the endangered status of many species (cf. Howarth and Ramsay, 1991). The wetas of New Zealand are now restricted to small islands, mostly free of placental predators such as rats and mice. They are large and long lived, and may have filled the "mouse niche" in New Zealand (Fig. 1.5). The large Wellington speargrass weevil (Fig. 1.6) is long lived and affected significantly by habitat destruction and rat predation on New Zealand's north and south islands. Invading organisms of many kinds threaten indigenous, especially locally endemic organisms (e.g., Drake et al., 1989).

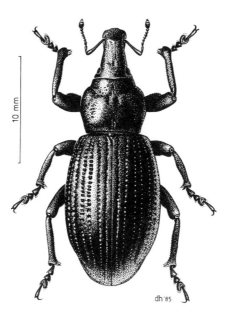

10 mm

dh '85

FIGURE 1.6   The Wellington speargrass weevil (*Lyperobius huttoni*) is almost 2.5 cm long, with populations strongly reduced by habitat destruction and predation by rats. From Howarth and Ramsay (1991); with permission. Drawing by D. H. Helmore, Landcare Research.

The remarkable high species richness of insects in tropical rain forest, and the rapid destruction of large areas, make conservation efforts particularly pressing (e.g., Erwin, 1988; Wilson, 1992). Yet more important is dry tropical forest, of which perhaps only 2% remains in Central America, with only 0.08% in preserves. The importance of a strongly ecological perspective is illustrated by the restoration of the landscape undertaken by Janzen and associates in what has become Guanacaste National Park in northwestern Costa Rica (e.g., Janzen, 1986a, 1988; Sun, 1988). The park now encompasses restored lowland forest and uplands essential for the many insect species that migrate during the dry season into the wetter montane vegetations, such as the evergreen forest and cloud forest above 500 m in elevation.

Much ecological research is needed to preserve insects and their habitats, for the essential characteristics of the habitat that enable species to persist indefinitely are usually unknown. Extinction, conservation, and nature reserves are discussed in Chapter 24.

ART, LITERATURE, AND SPIRITUALITY

Through most of human existence, both prehistoric and historic, we lived with insects in a most intimate way (Busvine, 1976). Human and insect ecology were interlocked. "Cleanliness was viewed with abhorrence. Lice we called 'pearls of God' and were a mark of saintliness" (Russell, 1946). After Thomas à Becket was murdered in Canterbury Cathedral in 1170, his nine layers of clothing were removed, innermost being a tight-fitting suit of coarse hair cloth. "The innumerable vermin which had infested the dead prelate were stimulated to such activity by the cold that his hair cloth garment, in the words of the chronicler, 'boiled over with them like water simmering in a cauldron' and the onlookers 'burst with alternate fits of weeping and laughter between the sorrow of having lost such a head and the joy of having found such a saint'" (MacArthur, 1927, p. 487).

Such intimate relations with insects stimulated both art and literature, as most engagingly described by Busvine (1976) (Fig. 1.7). Among the many poems and songs that Busvine discusses, Mephistopheles's "Song of the Flea" from *Faust,* by Goethe, illustrates a lovely kinship between humans and fleas, even if sung in jest.

> A king there was, be't noted,
> Who had a lusty flea,
> And on this flea he doted
> And loved him tenderly.
> A message to the tailor goes
> Swift came the man of stitches
> Ho, measure this youngster here for clothes
> And measure him for breeches.

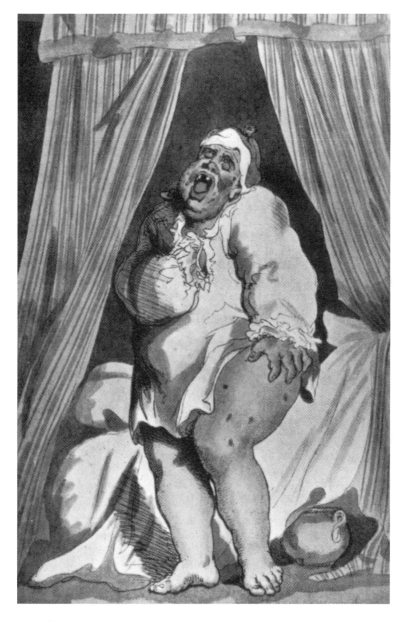

FIGURE 1.7 This yawning, lovely maiden, waking from sleep, captures the reality of the interactions among insects and humans not so long ago. The title of this work of art is "A tit-bit for the bugs." It was created by Rowlandson in 1793, and reproduced in Busvine (1976).

In silks and satins of the best
Soon was the flea arrayed there
   Ribbons he had his breast
Likewise a star displayed there.
Prime minister he grew anon
With star of large dimensions
Got title, rank and pensions.
And lords and ladies, high and faire
Were grievously tormented
Sore bitten the queen and her maidens were
But they did not dare resent it.
They were afraid to scratch
Howe'er our friend might sack them
But we without a scruple catch
And when we catch, we crack them.

Not all creations were complimentary to the entomologist, for village idiots have been cast in ballets, such as *La Fille mal gardée,* as butterfly collectors equipped with a net and performing *grands jetés* across the stage in pursuit of their insectan quarry.

While Kellert (1993) found a majority of people today holding strongly negative views on insects and other invertebrates, it remains impressive how great an impact insects have had on human culture, cuisine, and spirituality. A series of articles in the *American Entomologist* has covered much of this field: "Beetle gods of ancient Egypt" (Kritsky, 1991a); "Insect jewelry" (Akre et al., 1991); "Insects in the mythology of Native Americans" (Cherry, 1993); "Insect cartoons: When do they appear in newspapers and magazines?" (DeJong, 1994); "Insects: Old food in new Japan" (Pemberton and Yamasaki, 1995); "Insect tattoos on humans: A 'dermagraphic' study" (Pearson, 1996).

Insects have affected or interacted with human ecology in most cultures. Harvest time for wild-bee honey in tropical Africa generated a festival spirit for the Pygmies (Turnbull, 1961, 1962). Hopi Indian stories and legends are full of insect characters: the cicada, the pipe player; the robber fly with its primal creativity; the fly who was allowed to feast on roasted lamb and in return discovered a lost boy and rejuvenated him (Fig. 1.8) (Malotki, 1978). Among the Aborigines of Australia, insects were used as food, medicines, and inspiration for works of art (Waterhouse, 1991).

It may be stretching the point to claim that insect ecology includes the fields of art, literature, and spirituality. Yet there is such a close linkage of the insects with human pleasure and wonder that we should at least acknowledge the impact of insects on our environment, both biological and cultural, and therefore on our ecology. Many of the great naturalist explorers of the nineteenth century were dismayed and thrilled by the spectacle of insects. Darwin (1860), Belt (1888), and Spruce (1908) were all amazed at the density of "snowing butterflies" in the tropics. Bates (1962) noted that the ecology of

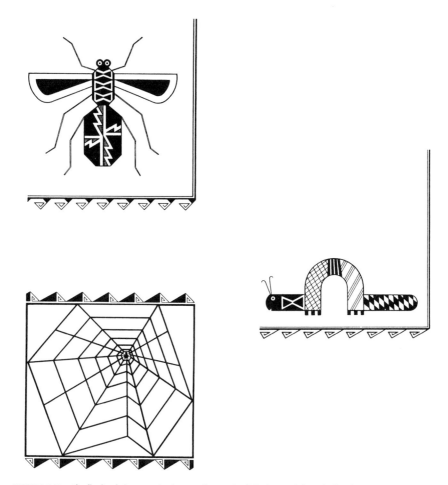

FIGURE 1.8 The fly that fed on roast lamb, a wood worm that hid a boy, and the web of Spider Woman, all characters from Hopi tales. The illustrations are based on typical abstract and geometric art by Pueblo artists and their ancestors, and are by Anne-Marie Malotki. From Malotki (1978); with permission.

butterflies would become an important part of science. Wallace (1869) was captivated by the large tropical beetles and butterflies and was able to collect in Borneo 24 new species of beetle every day, including large wood-boring brentids, longhorns, and weevils; while in Nicaragua, Belt (1888) found the leaf-cutting ants a plague in his garden.

Generally, today, insect ecologists share with the naturalist explorers such as Darwin, Wallace, Bates, Belt, and Spruce a fascination with the natural history of insects and the scientific approach to converting our interest and enthusiasm into general hypotheses and scientific theory.

Major Components and Processes in Ecosystems

ECOSYSTEMS

Ecosystem is a term coined by the phytosociologist Tansley (1935, 1939) for an area that includes all organisms therein and their physical environment. He emphasized the need to study such "whole systems" because no natural phenomenon can be adequately understood in isolation. Thus the ecosystem is a basic unit in nature. "There is a constant interchange of the most various kinds within each system, not only between the organisms but between the organic and the inorganic" (Tansley, 1935, p. 299).

The actual concept of the ecosystem is much older than the term itself. Forbes (1887) wrote in his paper "The lake as a microcosm" that a factor affecting one species must have an influence on all the species present, and he emphasized the need for making a comprehensive survey of the entire system in order to understand satisfactorily any part.

The term *ecosystem* was soon widely adopted by ecologists (e.g., Lindeman, 1942; Odum, 1953; Evans, 1956). Evans (1956) pointed out the many processes involved in ecosystem dynamics, many of which are dealt with in more detail in subsequent chapters: photosynthesis, decomposition, herbivory, predation, parasitism, and the symbiotic activities that influence the transport and storage of matter and energy. The interactions between organisms provide the pathways along which matter and energy are moved, as

Food for insects and its acquisition runs as a major theme through insect ecology. Here a carabid beetle in the genus *Scaphynotus* attacks a snail. The long narrow head and thorax are adapted for entering and eating snails in their shell. Drawing by Alison Partridge.

in food chains and food webs. In the abiotic part of the ecosystem, circulation of matter and energy is complemented by such processes as evaporation and precipitation and erosion and deposition.

COMPONENTS OF ECOSYSTEMS

The ecosystem is composed of many biologically meaningful units. Breaking it down into smaller and smaller parts, we may recognize the **community** as the organisms that interact in a given area. Thus the ecosystem can be defined as "the biotic community plus its abiotic environment" (Lindeman, 1942). **Compound communities** are composed of merging and interacting **component communities,** which are assemblages of species associated with some microenvironment or resource, such as a food plant, a tree hole, or leaf litter (Root, 1973). The component community may be formed of several **guilds** of species, being groups of species that exploit the same resource in a similar manner (Root, 1967; see also Jaksic, 1981). For example, of the leaf-eating insects, some suck plant juices, some chew off strips of leaf, and some chew small holes in leaves—the sap-feeding guild, the strip-feeding guild, and the pit-feeding guild, respectively (Root, 1973; and Fig. 2.1). Guilds are composed of **species,** which are groups of organisms that can interbreed but are reproductively isolated from other such groups (Mayr, 1963). Species are composed of **populations,** which are groups of actually or potentially interbreeding individuals at a given locality (Mayr, 1963), and populations are composed of **individual organisms.** Each individual is likely to be unique, having its own **genotype,** or genetic makeup, and its own **phenotype,** or the expressed characteristics as modified by the interaction of genotype and environment (Mayr, 1963).

The ecosystem is not the ultimate unit that can be recognized. **Biomes** are composed of ecosystems of similar vegetation type, such as tropical rain forest, desert, temperate forest, tundra, and steppe. All biomes form the **biosphere,** which includes all parts of the earth where life exists.

BIOTIC INTERACTIONS IN ECOSYSTEMS

An essential part of the ecosystem concept and ecosystem function is that elements of the system interact, and in ecology we have categorized organisms by the way in which they interact. Those species that fix the sun's energy and utilize inorganic chemicals to form complex organic molecules are essentially self-nourishing, or **autotrophs.** They produce organic matter, on which all life depends, and are thus also termed **producers.** The major autotrophs or producers in any ecosystem are green plants, either micro- or macroscopic, utiliz-

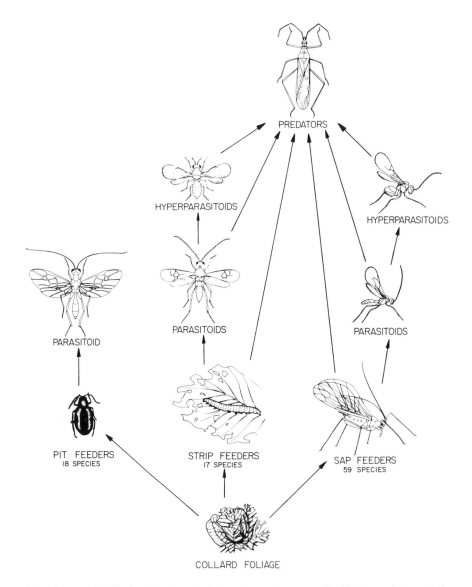

FIGURE 2.1 Simplified food web based on collard plants showing three major guilds of herbivores, their parasitoids, hyperparasitoids, and predators. Pit feeders include flea beetles. Strip feeders are caterpillars. Sap feeders are sucking insects such as aphids. Based on Root (1973). From Price (1984c); with permission.

ing photosynthesis as a source of energy for synthesis of organic molecules. Organisms that feed on autotrophs are naturally **consumers** and are nourished by others—hence the term **heterotrophic.**

Those heterotrophic species that feed on dead plants or animals play an important role in recycling nutrients in the ecosystem and are termed **saprophages** or **decomposers.** Heterotrophic species feeding on living organisms are called **herbivores** if they feed on plants and **carnivores** if animals are consumed. **Predators** kill and eat their food or **prey; parasites** sap energy and nutrients from their living **host** and live in or on this host. The term **parasitoid** denotes a species of insect that requires and eats only one animal in its life span by living parasitically as a larva on a host, but the adult is free-living and may ultimately kill many hosts by leaving eggs or larvae near or on the host that eventually consume the host. **Grazers** and **browsers** pluck plant parts without usually killing the plant and without being small enough to live in or on the plant. **Mutualists** benefit the species with which they are closely associated.

This variety of interactions results in a complex mesh of relationships referred to as the **food web,** composed of interlocking linear feeding links called **food chains** by Elton (1927). Elton used the example of the insect community based on pine trees in England (Fig. 2.2), where the food chains such as pine tree → caterpillar → ichneumonid wasp → spiders → spider wasp and pine tree → aphid → lady beetle → spiders → spider wasp were linked into a food web by the same spiders feeding on both ichneumonid wasps and the predators of aphids. Another food web was illustrated in Fig. 2.1. In aquatic systems we also see the great importance of insects in making connections in energy flow and movement of matter in ecosystems (Fig. 2.3). Schoenly et al. (1991) provide a broad spectrum of food webs involving the trophic relationships of insects. Although most food webs illustrate the negative impact of interactions such as herbivory, predation, and parasitism, of equal importance are the beneficial effects of mutualists, soil conditioners, and the enemies of herbivores, in terms of benefit to the plant. Some of this great variety of interactions is illustrated in Fig. 2.4 which also shows many of the abiotic influences on this complex component community.

MAJOR PROCESSES IN ECOSYSTEMS

Although the complexity of ecosystems is already apparent, and the number of species and interactions may be very high, there are relatively few major processes at work. These can be classified into energy flow, biogeochemical cycling, ecological succession, and the evolution of species, which will be treated in order.

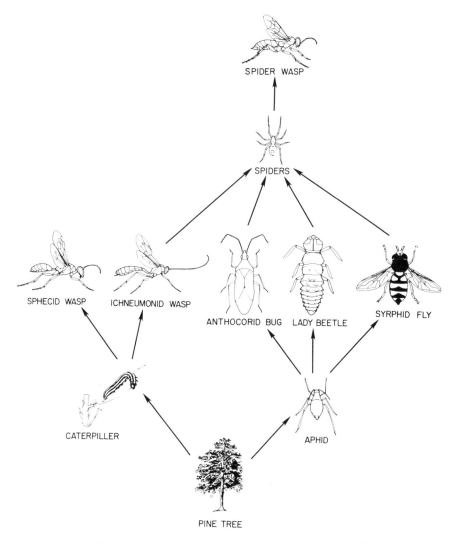

FIGURE 2.2   Food web based on pine trees growing in Surrey, England. From Price (1984c); with permission.

Energy Flow

Energy flows through an ecosystem. It does not cycle as chemicals do in biogeochemical cycles. This is explained by the laws of thermodynamics: The
first law states that energy can be neither created nor destroyed, and the second law states that in every energy transformation potential energy is reduced

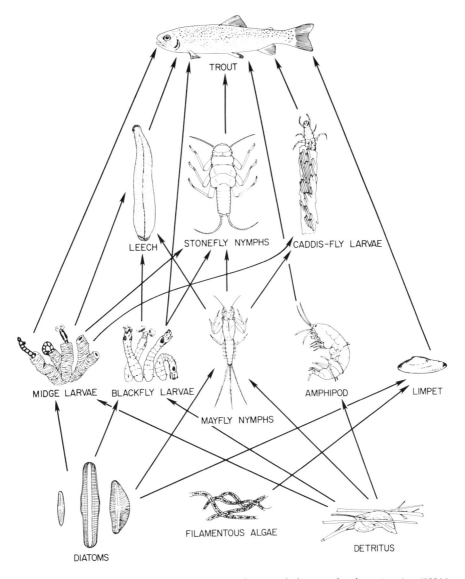

FIGURE 2.3   Aquatic food web illustrating the importance of insects as herbivores and predators. From Price (1984c); with permission.

because heat energy is lost to the system in the process. Thus, as food passes from one organism to another, potential energy contained in the food supply is reduced step by step until all the energy in the system becomes dissipated as heat (Fig. 2.5). Therefore, there is a unidirectional flow of energy through a system, with no possibility for recycling of energy.

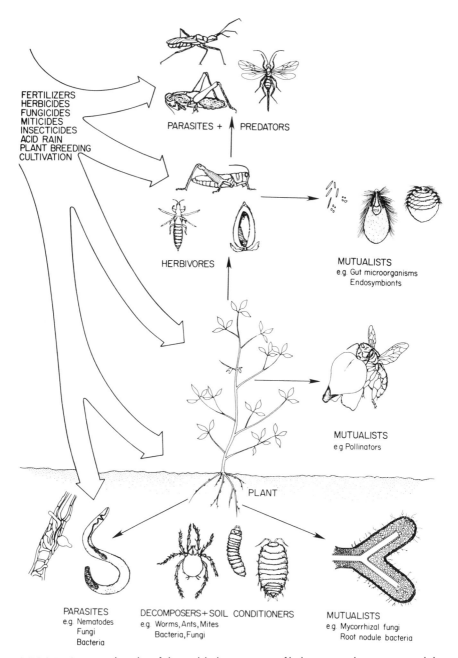

FERTILIZERS
HERBICIDES
FUNGICIDES
MITICIDES
INSECTICIDES
ACID RAIN
PLANT BREEDING
CULTIVATION

PARASITES + PREDATORS

HERBIVORES

MUTUALISTS
e.g. Gut microorganisms
    Endosymbionts

MUTUALISTS
e.g Pollinators

PLANT

PARASITES
e.g. Nematodes
    Fungi
    Bacteria

DECOMPOSERS + SOIL CONDITIONERS
e.g. Worms, Ants, Mites
    Bacteria, Fungi

MUTUALISTS
e.g. Mycorrhizal fungi
    Root nodule bacteria

FIGURE 2.4   Some interrelationships of plants with herbivores, enemies of herbivores, mutualistic organisms, and abiotic factors, such as may be found in a soybean field. From Price (1984c); with permission.

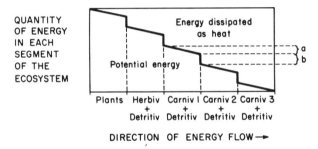

FIGURE 2.5 Use of energy as it flows through an ecosystem. It assumes that all potential energy is consumed either while alive or when dead; that is, none is stored. At each trophic level, maintenance energy is lost as heat (a), and energy is lost as heat in each transformation from one trophic level to the next (b). Ultimately, all energy in the system is dissipated as heat. Thus the system can be maintained only by an outside supply of energy. Herbiv, herbivore; Carniv, carnivore; Detritiv, detritivore.

A real example of energy flow, studied by H. T. Odum (1957) at Silver Springs, Florida (Fig. 2.6), illustrates the large amount of energy that is dissipated as heat at each trophic level; the rapid decline in the amount of energy in the food chain from producers (see net production) to herbivores, carnivores, and top carnivores; and the importance of decomposers in the system. This subject is treated in more detail in Chapter 12.

Biogeochemical Cycling

By contrast to energy flow, chemicals remain in the ecosystem indefinitely and are not dissipated (Fig. 2.7) unless erosion or movement out of the system occurs. Chemicals are constantly cycling in the ecosystem, and they move between the biological and the geological components of the system, from rocks and soil to plant, to herbivore, and up the food chain, when decomposers return them to the soil—hence the term **biogeochemical cycling.**

The nitrogen cycle is an example of a largely complete chemical cycle in natural ecosystems (Delwiche, 1970; and Fig. 2.8) with little leaching out of the system. In agricultural systems large inputs of nitrogen fertilizer may result in considerable leaching and unidirectional flow of nitrogen from agricultural fields into aquatic systems which become polluted with excessive nitrogen.

Pesticides that may be only slightly soluble in water may also be leached from agricultural fields and enter aquatic ecosystems. Small organisms feeding in the water selectively retain chlorinated hydrocarbons such as DDT in fatty tissues. Larger organisms feed on many small organisms and again selectively retain the pesticide. At each trophic level in the food web the toxic substances become more concentrated—termed **biological concentration** (Fig. 2.9).

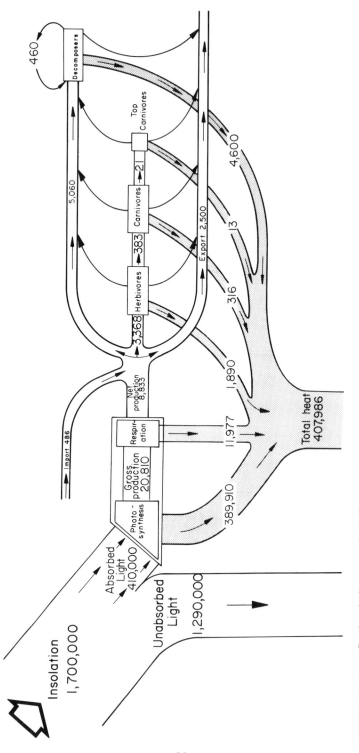

FIGURE 2.6 Energy flow through Silver Springs, Florida, illustrating the rapid loss of heat energy in the trophic system. After Odum (1957) and Phillipson (1966). From Price (1984c); with permission.

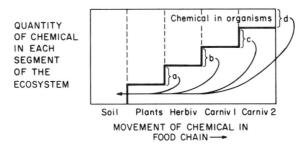

FIGURE 2.7   Cycling of a chemical in the ecosystem assuming no erosion. Chemicals in organisms not consumed by the higher trophic level [(a) to (d)] slowly return to the soil as they are released through decay.

Ultimately, concentrations of toxins become so high in the top predators that reproductive ability is impaired and populations start to decline. When birds die, the insecticide is recycled.

Model ecosystems have been designed to investigate the movement of many manufactured chemicals, including insecticides (Metcalf, 1977), coal products (Lu et al., 1978), carcinogens (Lu et al., 1977), and drugs (Coats et al., 1976).

Ecological Succession

One of the most obvious attributes of ecosystems is that they change, particularly in vegetational attributes, but also in the kinds of animals that reside in the ecosystem. The change is usually predictable and directional, so a pattern of change can be observed. We call this **ecological succession** because of the succession of species that come and go on any patch of land.

Succession occurs at any site that becomes available for colonization because species arrive at different times, and they alter conditions, which benefit some species while being deleterious to others. Ecological succession was first recognized in plants on sand dunes by Cowles (1899) because the changes are displayed in space as sand piles up against a lake or seashore, making new ground available for colonization proximal to the water, followed by all subsequent stages more and more distal from the water. Many other studies on the vegetation of sand dunes and their ecological succession have been made (Tansley, 1939; Shelford, 1963; Chevin, 1966; Bakker et al., 1974), and frequently, insects and other arthropods have been shown to change with ecological succession—tiger beetles (Shelford, 1907), spiders (van der Aart, 1974), pseudoscorpions (Weygoldt, 1969), and entire insect communities (Callan, 1964; Chevin, 1966).

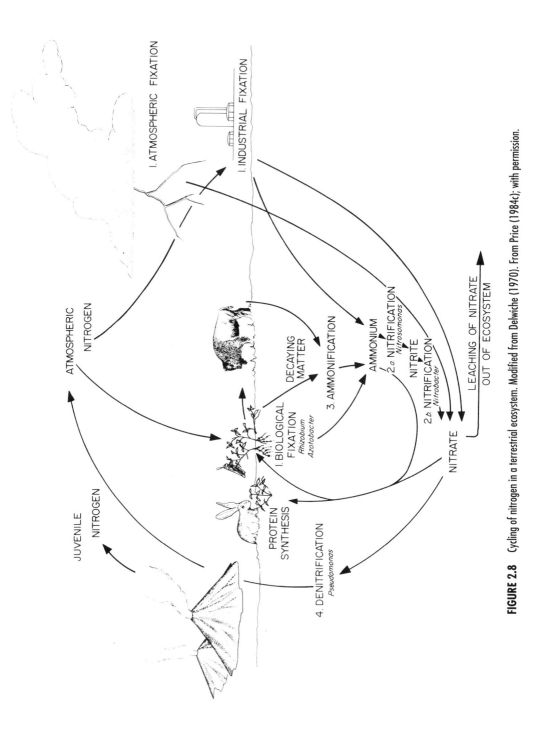

FIGURE 2.8 Cycling of nitrogen in a terrestrial ecosystem. Modified from Delwiche (1970). From Price (1984c); with permission.

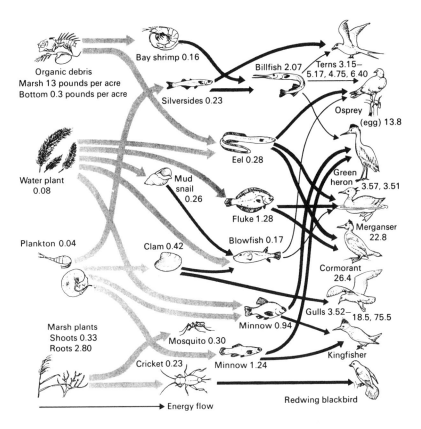

FIGURE 2.9 Movement of DDT in part of the food web in a Long Island estuary, New York, illustrating biological concentration of the insecticide. Energy and DDT flow in the direction of arrows and DDT concentrations are expressed in parts per million (see also Woodwell et al., 1967). From G. M. Woodwell. Toxic substances and ecological cycles, *Sci. Am.,* 1967, **216**(3):24–31. Copyright © 1967 by Scientific American, Inc. All rights reserved.

Succession has been observed repeatedly in other insect communities also. On living plants the insect community tends to change with the age of the plant, observed on broom by Waloff (1968a,b) and on pines by Martin (1966). In stored grain there is a change in insects, mites, and fungi as the grain is broken down (van Bronswijk and Sinha, 1971; Coombs and Woodroffe, 1973). Insects in rotting logs also follow successional patterns (Blackman and Stage, 1924), as do the flies attacking wounds in sheep (Haddow and Thomson, 1937; MacLeod, 1937) and carrion communities in general (Schoenly and Reid, 1987). Parasitoids may follow successional trends as conditions in the host population change (Van den Bosch et al., 1964; Price, 1973b), and carabids, diplopods, and isopods

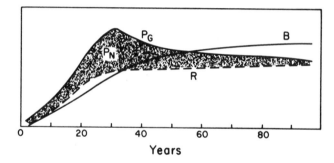

Years

FIGURE 2.10 Trends in succession leading to climax forest: R, total community respiration; B, total biomass; P_G, gross production due to photosynthesis; P_N, net production. From Odum (1969); with permission.

have been shown to change with time on spoil heaps of coal mines (Neumann, 1971).

Successional trends are so pervasive in nature that many trends have been recognized. Odum (1969) has defined these patterns in very broad strokes (Fig. 2.10 and Table 2.1). When a plant community becomes established, the young stages of succession must have a photosynthesis (P)/respiration (R) ratio greater than 1, while in older stages this ratio approaches 1. The energy fixed tends to be balanced by the energy cost of maintenance as later stages of succession are reached, and therefore the P/R ratio should be a good functional index of the relative maturity of the system. As long as P exceeds R, organic matter and biomass (B) will accumulate in the system and B/P will tend to increase and B/R and B/E (energy utilization) will increase. Theoretically, then, the amount of standing crop biomass supported by the available energy flow increases to a maximum in the mature stage of vegetation development. As a consequence, the net community production in an annual cycle is large in young systems and small or zero in mature systems.

Many of the trends listed in Table 2.1 are discussed later in the book. In Chapter 12 it will be seen that productivity increases as the woody component in the community increases. Early stages in succession are composed of short-lived organisms, whereas later stages contain long-lived species of plants and insects. Also detritivores are extremely important in mature deciduous forest. In Chapter 14 on life histories of insects the concepts of r- and K-selection on life history traits are considered and related to niche specialization (Chapter 20) and competition (Chapter 21). Aspects of diversity and stability in succession are treated in Chapter 23. Thus ecological succession provides an important concept that links many areas of ecology. Golley (1977) provides a valuable overview of concepts and patterns in ecological succession (see also Bazzaz, 1990).

TABLE 2.1   Summary of Trends in Ecological Succession.

Ecosystem Attribute	Developmental Stage	Mature Stage
Community Energetics		
1. Gross production/community respiration (*P/R* ratio)	Greater or less than 1	Approaches 1
2. Gross production/standing crop biomass (*P/B* ratio)	High	Low
3. Biomass supported/unit energy flow (*B/E* ratio)	Low	High
4. Net community production (yield)	High	Low
5. Food chains	Linear, predominantly grazing	Weblike, predominantly detritus
Community Structure		
6. Total organic matter	Small	Large
7. Inorganic nutrients	Extrabiotic	Intrabiotic
8. Species diversity—variety component	Low	High
9. Species diversity—equitability component	Low	High
10. Biochemical diversity	Low	High
11. Stratification and spatial heterogeneity (pattern diversity)	Poorly organized	Well-organized
Life History		
12. Niche specialization	Broad	Narrow
13. Size of organism	Small	Large
14. Life cycles	Short, simple	Long, complex
Nutrient Cycling		
15. Mineral cycles	Open	Closed
16. Nutrient exchange rate between organisms and environment	Rapid	Slow
17. Role of detritus in nutrient regeneration	Unimportant	Important
Selection Pressure		
18. Growth form	For rapid growth (*r*-selection)	For feedback control (*K*-selection)
19. Production	Quantity	Quality
Overall Homeostasis		
20. Internal symbiosis	Undeveloped	Developed
21. Nutrient conservation	Poor	Good
22. Stability (resistance to external perturbations)	Poor	Good
23. Entropy	High	Low
24. Information	Low	High

SOURCE: Reprinted with permission from Odum (1969).

Evolution of Populations and Species

Although the sciences of ecology and evolution developed independently to a certain extent (e.g., Orians, 1962), they are so intimately related that we must be constantly aware of evolutionary processes in ecological systems. This is because as soon as a population or a species evolves, its relationships with other organisms and the abiotic environment change, so its ecology changes. In turn, when environmental change occurs, a population or species is likely to evolve in response to this change. Thus the title of Hutchinson's (1965) book *The ecological theater and the evolutionary play* is very apt. Darwin (1859) stressed repeatedly in *The origin of species* that is was the interaction between organisms that resulted in powerful natural selection and evolution of species. And, of course, the study of interaction between organisms is a major preoccupation among ecologists. Therefore, in a preceding edition of this book (Price, 1975c), I defined ecology as the study of environments for evolutionary processes. I stressed that this definition focuses attention on the dynamics within a given environment, the functional relationships between organisms, and the analysis of control mechanisms in the ecosystem. Dynamics, functional relationships, and control mechanisms are key words in an evolutionary ecology permitting, in addition to the questions on what exists, the far more intriguing questions of why and how.

Evolutionary change occurs because a large variety of genotypes are produced in a population through mutation, leading to new genes, and recombination, resulting in novel combinations of genes. Environmental factors also play a role in determining the development of these genotypes into phenotypes and the natural selection among phenotypes. As the environment changes, some types will be selected for and others will be selected against, resulting in "a differential perpetuation of genotypes" (Mayr, 1963) which is characteristic of the evolutionary process.

Since individuals make up populations, and every individual is likely to have a unique genotype, the genotype and phenotype of the individuals will profoundly influence such population phenomena as population size, rate of change, persistence, evolution, and extinction (Fig. 2.11). These interacting factors have been called the **life system** of an organism by Clark et al. (1967).

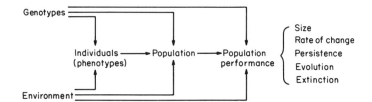

FIGURE 2.11   Interacting factors in the life system that influence population phenomena. After Clark et al. (1967).

Kettlewell (1959) called the evolution of **industrial melanism** in the peppered moth, *Biston betularia,* in Britain, "Darwin's missing evidence" because it occurred during Darwin's lifetime and it illustrates so well the evolution of a species in response to environmental change and the agents involved in natural selection. As industrialization intensified during the Industrial Revolution, pollutants settled on trees, killing the light-colored lichens and darkening trunks and branches with soot. The peppered moths settled on tree bark during the day and remained cryptic on a background of lichens because of the mottled black-on-white markings. Rare black mutants were easily detected by birds and eaten. With the darkening of tree trunks, the typical peppered moths became less and less cryptic and the rare mutants became better concealed. Natural selection imposed by predatory birds selected against the typical form and the black mutants survived better. In industrial areas the frequency of the melanic morph increased from less than 1% in the population to 90% in less than 100 years. But the evolutionary processes have not stagnated, because the environment continues to change. With the introduction of "smokeless" zones in many areas of Britain, pollution has been reduced and the frequency of melanic morphs has declined noticeably (Bishop and Cook, 1975). More recent reviews provide an ongoing picture of evolutionary change (Bishop and Cook, 1979, 1980), and this example is discussed in more detail in Chapter 16.

The very rapid evolution of insecticide resistance in arthropods provides a good index of the immense evolutionary potential in members of this phylum (Georghiou, 1972; Plapp, 1976; Brown, 1978). Hundreds of cases are now known where evolutionary response to selection from insecticides has enabled populations to persist at epidemic levels; 139 cases in arthropod species of medical importance and 225 cases in agricultural systems (Georghiou and Taylor, 1977). The evolutionary pace in the development of resistance is so great that during the period 1967–1975, the number of known resistant species increased by 62.5%! (See also Metcalf, 1989, 1994; Roush and Tabashnik, 1990; Metcalf and Metcalf, 1993.)

The enormous evolutionary potential of insects in ecosystems must be kept in mind in any ecological studies, a point that we shall return to in later chapters (c.f. Gould 1991, 1994; Gould et al. 1991).

The World of the Insect: Size and Scaling in Moderately Small Organisms

DIFFERENT WORLDS

> Life has a range of magnitude narrow indeed compared to that with which physical science deals; but it is wide enough to include three such discrepant conditions as those in which a man, an insect and a bacillus have their being and play their several roles. Man is ruled by gravitation, and rests on mother earth. A water-beetle finds the surface of a pool a matter of life and death, a perilous entanglement or an indispensable support. In a third world, where the bacillus lives, gravitation is forgotten, and the viscosity of the liquid, the resistance defined by Stokes's law, the molecular shocks of the Brownian movement, doubtless also the electric charges of the ionized medium, make up the physical environment and have their potent and immediate influence on the organism. The predominant factors are no longer those of our scale; we have come to the edge of a world of which we have no experience, and where all our preconceptions must be recast. (Thompson, 1942, p. 77)

Thompson (1942) had a perspective of life all too rare among biologists, for even now we are quite ignorant of the real environment in which small organisms live and the way in which they relate individually to physical factors and forces, not only through form and function but also through behavior. Van Valen (1973) remarked that body size has received surprisingly little study in biology. Gould (1974), understandably a great admirer of

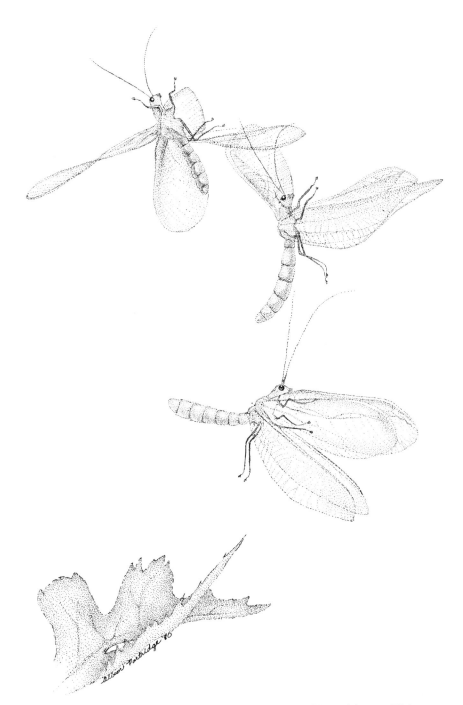

A green lacewing flies off from a leaf with marvelous ease and great maneuverability. Such features of flight in insects relate directly to scale effects associated with their small size. Drawn with permission from a multiflash photograph by Stephen Dalton (1975). Drawing by Alison Partridge.

Thompson, has reiterated the gulf between experiences felt by large and small organisms:

> We are prisoners of the perceptions of our size, and rarely recognize how different the world must appear to small animals. Since our relative surface area is so small at our large size, we are ruled by gravitational forces acting upon our weight. But gravity is negligible to very small animals with high surface to volume ratios; they live in a world dominated by surface forces and judge the pleasures and dangers of their surroundings in ways foreign to our experience. (Gould, 1974, p. 21, with permission from *Natural History,* Vol. 83, No. 1; copyright the American Museum of Natural History, 1974)

These quotations open up a massive area of interest for entomologists and ecologists which must, of necessity, be dealt with superficially in this chapter.

RANGE IN SIZE OF ORGANISMS

Thompson (1942) provides the range in length of organisms so that we can better understand the sizes and magnitudes of difference we are to discuss (Table 3.1). Living things span eight or nine orders of magnitude in their linear dimensions (L). But if these organisms were built on the same scale, and we compared surface areas, which vary as L^2, or volumes and weights of organisms, which vary as L^3, the range in magnitudes would be even greater, in fact 21 orders of magnitude (Schmidt-Nielsen, 1975; and Table 3.1). Thus different kinds of forces will have very different effects on organisms, according to their length, surface area, volume, weight, surface-to-volume ratio, or effective diameter; according to whether they live on land or in water, be it fresh or saline; and according to whether they walk, run, swim, or fly. But in all cases the form of the organism and the changes in form as it moves and grows will be under the influence of various forces, so that an organism is a diagram of forces, from which can be judged the most important forces acting upon it (Thompson, 1942).

Considering specific cases, let us draw some distinctions between large and relatively small organisms, such as humans and insects. Then we may better empathize with insects and understand the strictures on their design and their ecological implications.

Went (1968) discusses human size in relation to that of another highly social animal, the ant. The human is three orders of magnitude larger on a liner scale and eight orders of magnitude on a scale of weight. The ants have been social for millions of years, yet technology has not developed. Although our genus, *Homo,* is only about 2 million years old, and we have been highly social, living in villages for 10,000 years, technology is highly developed. Went contends that our size has been a dominating factor in the development of technology, just as an ant species is constrained by its size from making use of technology.

TABLE 3.1   Size Range in Terms of Length and Weight of Organisms in Relation to Molecules.

Length (m)		Organisms and Other Structures in Relation to:		Weight (g)
		Length	**Weight**	
			Blue whale	$>10^8$
10^7	10,000 km	Quadrant of earth's circumference		10^7
10^6	1 000 km			10^6
10^5			Human	10^5
10^4				10^4
		Mt. Everest		
10^3	1 km			10^3
10^2		Giant trees: *Sequoia*	Hamster	10^2
		Large whale		
10^1		Basking shark		10^1
		Elephant, ostrich, and human		
10^0	1 m			10^0
		Dog, rat, and eagle		
10^{-1}			Bee	10^{-1}
		Large insects, small birds, and mammals		
10^{-2}	1 cm			10^{-2}
		Small insects and very small fish		
10^{-3}	1 mm			10^{-3}
		Minute insects		
10^{-4}			Large amoeba	10^{-4}
		Protozoa and pollen grains		
10^{-5}				10^{-5}
		Large bacteria and human erythrocytes		
10^{-6}	1 μm			10^{-6}
		Minute bacteria		
10^{-7}			*Tetrahymena*	10^{-7}
		Viruses		
10^{-8}		Large molecules		10^{-8}
		Starch molecule		
10^{-9}	1 nm			10^{-9}
		Water molecule		
10^{-10}	1 Å		Malaria parasite	10^{-10}
				10^{-11}
				10^{-12}
			PPLO *Mycoplasma*	$<10^{-13}$

SOURCE: Data from Thompson (1942) and Schmidt-Nielsen (1975).

It can only be regarded as fortuitous that humans happened to be the right size for handling physically determined environmental factors. Went provides an example on the dimensional limitations of fire. A flame cannot be smaller than a few millimeters since the ignition point of gases is high and there must be enough heat in a flame to counteract the cooling of the local environment, the flow of cool air into the flame, and to volatilize the fuel to be burned. Solid fuels such as wood or coal require higher ignition temperatures and therefore more mass of fuel to keep the fire alight. An alcohol burner can be much smaller and cooler than the smallest wood fire. This critical mass of wood needed for a fire happens to be just right for warming human beings, both in terms of size and heat generated, but much too large and hot for ants to manipulate or tolerate. Thus humans have used fire to an advantage, whereas ants will never be able to benefit from fire.

FORCES ACTING ON ORGANISMS

Another factor in human technological supremacy is a person's size in relation to the kinetic energy the person can develop while manipulating tools such as a pick or axe (Went, 1968). Kinetic energy increases as length raised to the fifth power (L^5). Thus if humans were only half as small, they could mine coal and ores with only 1/32 of the energy. This accounts for Gould's (1974) sympathy for the dwarfs in Wagner's "Das Rheingold," who were expected by Albericht to excavate precious minerals with mining picks. The comforting corollary is that children only half our size fall with only 1/32 of the impact, and thus rarely hurt themselves. But also, a man twice as tall as normal would fall with 32 times more force, with no doubt disabling consequences. To be this tall an animal must rely on locomotion on all fours. For the insect, kinetic energy decreases with decreased size so rapidly that it is no longer a force that can be manipulated easily to advantage, but also one that is no longer a threat should a beetle lose its balance. At this size, cohesion and adhesion are becoming dominant forces in the way of life.

Since gravity acts on weight and not linear dimensions, the pull of gravity increases on L^3. Therefore, large organisms are in the grip of gravity, but small organisms are under slight pull. For those that live in close proximity to other surfaces (as insects do), the force of cohesion or adhesion becomes as important (Went, 1968; and Fig. 3.1). Molecular attraction increases as L^2 and decreases as the square root of the distance between surfaces. For example, a 1-mm cube with a flat surface will adhere to another surface since adhesion balances the force of gravity. But a 1-cm cube will fall as gravity exerts a force 50 times greater than adhesion, and a 0.1-mm cube is under 50 times greater adhesive force than gravity.

It is clear from Fig. 3.1 that organisms a little above and below 1 mm operate in a world where gravity and molecular forces are within the same order of magnitude, and it is easy to defy gravitational force. Insects can easily

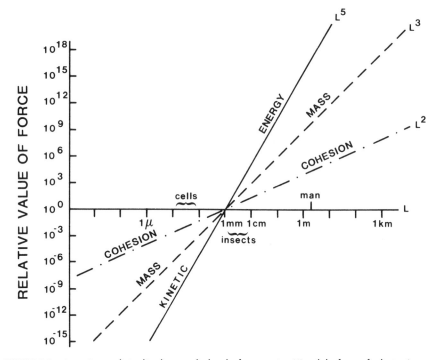

FIGURE 3.1   Approximate relationships between the length of an organism (L) and the forces of cohesion (proportional to L^2), mass or weight (proportional to L^3), and kinetic energy (proportional to L^5). The size range of 1 mm to 1 cm is occupied by many insects where the forces of cohesion and kinetic energy have similar relative values. Modified from Went (1968).

walk up vertical surfaces to colonize tall structures such as trees and dwellings. The smaller insects can readily walk under horizontal surfaces as long as they have "flat feet," as does a fly, so that adhesion is very effective. But the reduced influence of gravity also results in less traction for small organisms, so claws, hooks, suckers, and glandular hairs are used to ensure a foothold. Also, the surface tension of water may trap an insect in a drop of water or render its wings useless for flight. "A man coming wet from his bath carries a few ounces of water, and is perhaps 1 percent heavier than before; but a wet fly weighs twice as much as a dry one, and becomes a helpless thing" (Thompson, 1942, p. 51).

Went (1968) identifies two different physical worlds for organisms: the macroworld or Newtonian world where gravitational force rules design and movement; and the microworld or molecular world, where molecular forces predominate. Thompson (1942) chose to subdivide the molecular world of Went into that of the diving beetle and the bacillus just as Bidder (1931) had divided aquatic environments into the Brownian realm, the Stokesian realm, and the Archimedean realm (Hutchinson, 1971). (Stokes law states that the force required to move a sphere through a given viscous fluid at a low uniform velocity is directly proportional to the velocity and radius of the sphere.)

Failure to recognize the fundamental and nonlinear differences between small organisms and large organisms is common in both scientific and fictional literature, as Thompson (1942) and Gould (1974) have said. It is ludicrous to wonder at the strength of an ant if it were to become the size of a man, or the jump of the flea, or the volume of sound produced by a cricket. Science fiction in book and film, as in fairy tale, usually assumes that forces do not vary with size. Yet we know that large weighty animals have a very different design from small light ones, that large birds look very different from small birds, and even within a species, that differently shaped people excel at different physical enterprises. Forgotten is the fact that all these differences are under the constraints of physical laws. Once we understand how these laws operate, the patterns among animals begin to emerge and can be understood in turn.

SMALL ANIMALS AND EXOSKELETONS

One of these basic patterns is that many small animals have exoskeletons, whereas large ones without exception have endoskeletons. The full explanation for greater mechanical efficiency of the exoskeleton is attributable to Galilei (1638), who wrote near the end of his discourses of the second day:

> But, in order to bring our daily conference to an end, I wish to discuss the strength of hollow solids, which are employed in art—and still oftener in nature—in a thousand operations for the purpose of greatly increasing strength without adding to weight; examples of these are seen in the bones of birds and in the many kinds of reeds which are light and highly resistant both to bending and breaking. For if a stem or straw which carried a head of wheat heavier than the entire stalk were made up of the same amount of material in solid form it would offer less resistance to bending and breaking. This is an experience which has been verified and confirmed in practice where it is found that a hollow lance or a tube of wood or metal is much stronger than would be a solid one of the same length and weight, one which would necessarily be thinner; men have discovered, therefore, that in order to make lances strong as well as light they must make them hollow.

More readily available is Currey's (1970) explanation followed here, for we need to compare the properties of solid cylinders (as in endoskeletons) with those of hollow cylinders (as in exoskeletons). These represent the two extremes in design seen in neither insects nor vertebrates, but convenient as simple models.

When a beam or cylinder is bent downward, the upper surface becomes concave and thus compressed; the lower surface becomes convex and stretched. The center remains unstressed. Thus hollow cylinders are universally more efficient than solid ones. Let I be the second moment of area, which is proportional to rigidity and so provides a useful index of rigidity. For a hollow cylinder, I can be calculated from the formula

$$I = \frac{1}{4} \pi (R^4 = r^4)$$

where R and r are the external and internal radii, respectively. I/R is proportional to strength, providing another useful index.

We must assume that an animal can afford a certain amount of tissue for a skeleton and then test how this will be best used. It is clear that a hollow cylinder design will be advantageous, but what ratios of $r:R$ are most economical? The same amount of tissue will be used if the cross-sectional area (A) of the hollow cylinder remains the same. Then

$$I = \frac{A}{2} \left(R^2 - \frac{A}{2\pi} \right)$$

so that we can see how strength and rigidity are likely to change as R increases while keeping the amount of tissue devoted to the cylinder constant (Table 3.2).

Rigidity increases very rapidly with increasing R and strength also increases, although in a less spectacular manner. Therefore, with a certain amount of tissue available for support, an exoskeleton is theoretically always more economical than an endoskeleton.

But as Currey (1970) says, all optimal situations are a compromise, so we must understand also the disadvantages of external, hollow skeletons. These disadvantages are:

1. As R increases, with a given amount of tissue, the walls of the cylinder must become thinner. Beyond a certain rather small body weight, given that about 10% of the body weight is devoted to skeleton, the skeleton will become so thin that it collapses under the influence of elastic buckling (McMahon, 1973).

TABLE 3.2   Relationships in Cylinders between Radius, Rigidity, and Strength.

External radius R	1	2	4	10
I (proportional to rigidity)	1	7	31	199
I/R (proportional to strength)	1	3.5	7.8	19.9

SOURCE: Currey (1970).

2. An exoskeleton must necessarily be stiff and thus in order to grow, the animal must molt. Small animals when molted will be held in shape by cohesion of molecules and by hydrostatic pressure, for gravity will misshape the animal little. But large animals when unsupported would collapse under the force of gravity, and the newly secreted exoskeleton would "harden to produce an animal like a great tough pancake" (Currey, 1970, p. 8).

3. Exoskeletons are very sensitive to impact, and more so in large creatures than small since the stress of impact is proportional to length.

4. Any scratches or notches in the exoskeleton greatly weaken it, and large animals exert more force on their own exoskeleton when moving than small ones do, and thus would tend to damage it very much more readily than would small ones.

Thus the exoskeleton of an insect is an efficient use of material, but it constrains the insects from being large. However, since the insect must remain small, its surface-to-volume ratio is high and transpiration would be excessive if its integument were as porous as ours; the hardened waterproof exoskeleton becomes an important barrier to water loss. In fact, the majority of other small terrestrial animals have some sort of hardened covering: snails (Mollusca), isopods, millipedes, centipedes, and arachnids (Arthropoda). The integument of small terrestrial organisms has a profound influence on the sorts of environments that can be exploited and the whole physiological ecology of the species (Chapman, 1969; Wigglesworth, 1972; Edney, 1977; Heinrich, 1981; Hepburn, 1985).

OXYGEN CONSUMPTION

The presence of an exoskeleton has also permitted the evolution of many invaginations of the integument to form tracheae, along which oxygen can pass rapidly to every part of the body. Some permeability constants for oxygen are: through muscle, 8×10^{-14} $cm^2/atm \cdot h$; through water, 2×10^{-13} $cm^2/atm \cdot h$; through air, 660 $cm^2/atm \cdot h$ (Alexander, 1971). Thus diffusion of oxygen down tracheae can proceed 800,000 times more rapidly than through tissues. Even if only 1% of a cross section of an insect contained tracheae, the rate would still be 8000 times the rate if no tracheae were present. Thus, for a given size, insects can be much more active than most other animals, or they can develop to a considerably larger diameter while retaining the same potential for activity as smaller animals (Fig. 3.2). For example, for the same oxygen consumption in the tissues (say, 0.3 cm^3/cm^3 tissue·h), a flatworm would have only a maximum radius of 0.05 cm and an earthworm with oxygen distributed by blood a radius of 0.3 cm, but an insect could have a radius of 0.9 cm. This is three times larger than a worm and 18 times larger than a flatworm.

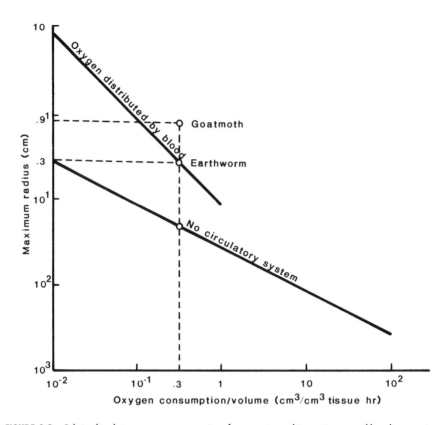

FIGURE 3.2   Relationships between oxygen consumption of an organism and its maximum possible radius assuming that organisms are cylindrical. The regression lines are for animals with no circulatory system and animals with a circulatory system but no specialized respiratory organs (e.g., an earthworm). The point above the earthworm represents the maximum possible radius for a caterpillar, such as a goatmoth larva with an extensive tracheal system. Modified from Alexander (1971).

Since insects are constrained in their size by an exoskeleton, they need little oxygen (Fig. 3.3), although as organisms get smaller generally they need more oxygen per gram (Fig. 3.4). This is because they need a higher metabolic rate with greater energy consumption. In general, oxygen consumption tends to increase as weight raised to the 0.75 power ($W^{0.75}$), and therefore oxygen consumption per unit weight decreases as weight raised to the negative 0.25 power ($W^{-0.25}$). The figures for insects are for adults ranging in size from fruit flies to locusts, and even the difference between resting and active insects is between one and two orders of magnitude. But there is no reason why the immature stages of holometabolous insects should need more oxygen per gram than do flatworms or perhaps earthworms, and thus the range of oxy-

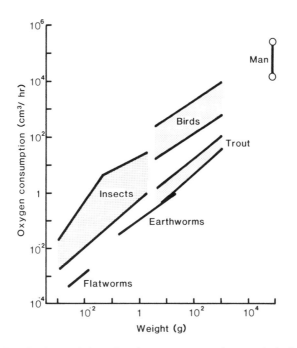

FIGURE 3.3   Relationships between body weight and oxygen consumption for various kinds of animals. For insects, the body weights represent sizes from fruit flies to locusts and the range for any weight shows minimum consumption while at rest and maximum consumption while flying. From R. McNeil Alexander, *Size and shape,* Edward Arnold (Publishers) Ltd., London, 1971. Reprinted with permission.

gen requirements of a single holometabolous insect over its life cycle may be enormous, almost three orders of magnitude. This is certainly part of the reason why larvae (immature holometabolous insects) can exploit highly secluded or deeply buried sites with low oxygen concentrations, for example, in decaying organic matter, logs, litter, and dung, or deep inside standing trees, or under bark.

SMALL ANIMALS AND FLIGHT

Another basic pattern seen among animals is that many moderate-sized to small organisms can fly (e.g., insects, birds, and bats), whereas no very large animals manage to stay aloft for long. Scaling effects are at work in the flapping flight mechanism (as opposed to gliding flight). In the simplest of

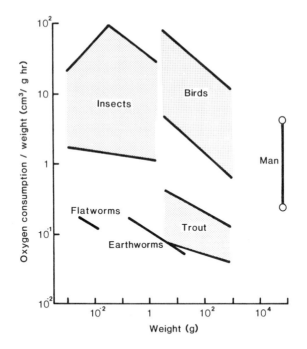

FIGURE 3.4   Relationships between body weight and oxygen consumption per unit of body weight using the same data as in Fig. 3.3. From R. McNeil Alexander, *Size and shape,* Edward Arnold (Publishers) Ltd., London, 1971. Reprinted with permission.

relationships, since weight (W) is proportional to L^3, and wing area (A) is proportional to L^2, wing loading (W/A) will vary directly with length (Pennycuick, 1972):

$$\frac{W}{A} \propto \frac{L^3}{L^2} = L$$

The stalling speed of a flying animal is given by the square root of wing loading and is thus proportional to $L^{1/2}$. Thus, for geometrically similar birds, if bird A has n times the wing span of species B, it should fly $\sqrt{n}$ times as fast. Large animals will fly faster than small animals and their stalling speeds must be higher. For example, large birds must gain more speed for takeoff than small ones, but the wing loading increases with weight (Table 3.3). Thus, two factors work against large flying organisms, and we see some large birds taking off from cliffs, while others, such as swans and some ducks, use considerable distances as their watery runways. As Thompson (1942) points out, a

TABLE 3.3   Wing Loadings in Some Flying Animals Expressed as the Weight in Newtons Carried Per Square Meter of Sustaining Wing Area. 1 N/m² = 0.0102 g-force/cm².

Organism	N/m²
Bats—small[a]	10–20
Birds	
Medium-sized and large	30–170
Small passerines	20–50
Hummingbirds[a]	20–30
Swifts, swallows, and bee eaters[a]	13–25
Insects	
Coleoptera, large—Lamellicornia[a]	12–40
Hymenoptera, large—Vespoidea and Apoidea[a]	8–44
Diptera, large—Brachycera and Cyclorrhapha[a]	5–20
Lepidoptera, large—Sphingidae[a]	4–12
Lepidoptera, medium-sized—Noctuidae[a]	3–6
Coleoptera, small[a]	1–6
Syrphinae—true hoverflies[b]	3–11
Odonata—dragonflies[b]	1–6
Drosophila virilis (wing length 3 mm)[b]	3–4
Encarsia formosa (wing length 0.6 mm)[b]	1.2
Lepidoptera Rhopalocera—butterflies[b]	0.4–2

[a] Most species show normal hovering.
[b] Slow forward flight and hovering involve unusual aerodynamic mechanisms.
SOURCE: Reprinted with permission from Weis-Fogh (1976).

sparrow may have a minimum speed of 20 mph, but an ostrich, if it could, would have to fly at 100 mph to keep aloft at its weight. In contrast, Johnson (1969) lists some speeds of insects that range from 0.2 mph in a mosquito to aphids at 1.2–1.5 mph; scolytid beetle, 3.5–5.1 mph; noctuid and sphingid moths, 10–15 mph; migratory locust, 11–18 mph; and tabanid flies, 9–40 mph.

In fact, the muscle power required to fly (P_r) at any given speed is proportional to $W^{1.17}$ (Pennycuick, 1972). Bird A, which weighs twice as much as bird B, requires $2^{1.17}$ or 2.25 times as much power to fly at its minimum-power speed. Thus larger birds must produce more power per gram of muscle than smaller birds because they must fly faster as a result of higher wing loading.

It turns out that the muscle power available for flight (P_a) increases with weight at a slower rate than the muscle power required, and thus there is an upper limit to the weight of flying animals. This is because mass of flight muscles is proportional to weight, or L^3, and the flapping frequency (f) is proportional to L^{-1} for mechanical and aerodynamical reasons (Pennycuick,

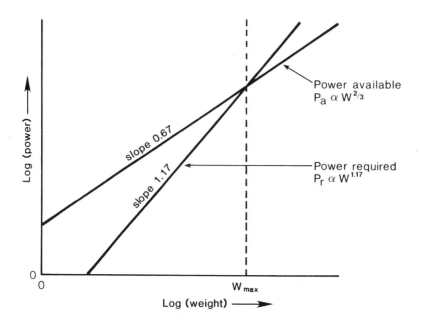

FIGURE 3.5   Relationship between the power required to fly and the power available for flight. Modified from Penny-cuick (1972).

1972). Combining these two proportionalities and using weight instead of length,

$$P_a \alpha W^{2/3}$$

(because $P_a \alpha L^3 \times L^{-1} = L^2 \alpha W^{2/3}$). Thus, if an animal A weighs twice as much as species B, it needs 2.25 times as much power to fly but will have only $2^{2/3}$ or 1.59 times as much power produced from its muscles. Thus the power available relative to that required must decrease as weight increases (Fig. 3.5). The notable result for the entomologist is that for such small organisms as insects, P_a greatly exceeds P_r, flight must seem relatively effortless, and performance can be spectacular (see Ellington, 1991).

As birds get smaller their wing-beat frequency increases until the smallest hummingbirds have frequencies of about 100 beats/second. All flight muscle in birds is neurogenic, in which each contraction is initiated by an action potential, and each muscle must pass through its contraction and relaxation phase, the whole process probably taking a minimum of 10 ms. Thus there is a physiological lower limit to the size of birds since they cannot have a higher wing beat than 100 beats/second (Pennycuick, 1972).

Although the heaviest flying insect, the goliath beetle (*Goliathus goliathus*), weighs about 40 g, the majority do not exceed 5 g, whereas few flying vertebrates get below this weight. The lightest birds and bats are about 2 g. The question must then be asked: If birds and bats are limited in size by the physiology of wing beat, how can even smaller animals fly? With the exception of the butterflies, most insects fly at 50–2000 beats/second, and many have therefore avoided the wing-beat frequency barrier. In the Lepidoptera and Odonata, neurogenic control is present, whereas in many other groups, myogenic or fibrillar muscles are present in which wing-beat frequencies can be much higher. This is because nerve impulses are not synchronized with muscle contractions and occur at a much lower frequency. Contractions are stimulated by mechanical changes within the muscle itself at a frequency defined by the mechanical resonance of the wings and thorax (Pennycuick, 1972). With this mechanism, unique to insects, many small flying species have confounded the effects of scale. In addition, elastic storage of energy by the flight system may confer increased mechanical advantages to insects (Ellington, 1985).

Another broad generalization about locomotion is that flying animals can cover enormous distances in a lifetime, whereas walking and running animals of equivalent size move relatively locally. Tucker (1969) compares the two modes of locomotion in terms of speed and endurance. He points out that birds are clearly superior to ground animals in both characteristics; even small birds fly for hours at speeds over 20 mph, and larger birds such as ducks can cruise at 40–50 mph. Some birds are able to fly for 10 hours without feeding and for 1000 miles or more. Similar feats of speed and endurance are seen also in insects, but not in any ground-dwelling animals.

These differences between flying and surface locomotion make an enormous difference between the ways in which animals can exploit their environment and the way they become physiologically adapted. Seasonally abundant foods such as insects in temperate regions can be exploited by insectivores without the need of adaptations to tolerate winter temperatures. Migration can become a way of life for both birds and insects. The migrations of monarch butterflies and milkweed bugs are well known. The massive dispersal flights of locusts and other grasshoppers, cicadellids, and other insects are equally impressive (Johnson, 1969; Rainey, 1976; Rabb and Kennedy, 1979; Showler, 1995). The soaring of aphids (Pennycuick, 1972) must have a profound influence on the distance to which aphids are carried by prevailing winds (Johnson, 1969). Long-distance movement becomes a way of life for some species in the world of flying animals. Swimming is an even more efficient form of locomotion (Fig. 3.6), and again we see long-distance movement to be common, as in salmon, eels, turtles, and whales. Even the

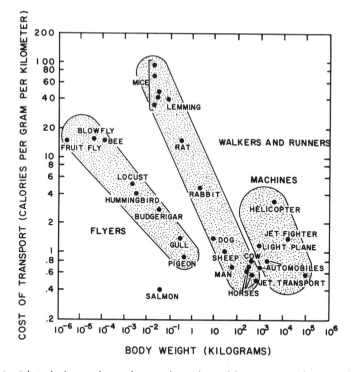

FIGURE 3.6 Relationship between the size of an animal or machine and the energetic cost of movement by means of walking or running, flying, or swimming (salmon only). From V. A. Tucker, The energetics of bird flight, *Sci. Am.*, 1969, **220**(5):70–78. Copyright © 1969 by Scientific American, Inc. All rights reserved.

extensive migrations of large herbivores such as zebra and wildebeeste are relatively minor compared to organisms three to seven orders of magnitude smaller.

ENERGETIC COSTS OF WALKING, FLYING, AND SWIMMING

What sort of scaling factors are involved with these differences between ground, air, and water transport among animals? Clearly, they must be many and complex, and they are not fully understood (Schmidt-Nielsen, 1972). But it can be shown empirically that the cost of locomotion per unit weight is lower for larger animals, as with oxygen consumption. However, the slopes of regressions of cost of transport per unit weight against body weight, both on logarithmic scales, are lower for flying animals than for walking animals (Fig. 3.6) and slightly lower for swimming animals than for flying animals

(Schmidt-Nielsen, 1972). Thus flying animals can move very economically and we can begin to understand why they can fly such impressive distances. Some direct comparisons can be made in Fig. 3.6. A gull, about the same body weight as a rat, can move 10 times more efficiently, and a locust, 1000 times smaller than a rabbit, can move almost as efficiently (see Pedley, 1977, for further information).

Insects have been limited in size by an exoskeleton that becomes less and less efficient with increasing size, and a respiratory system using movement of air down tracheae that becomes slower and slower with size. With small size goes the very high cost of locomotion by walking, and we can begin to understand the tremendous selective pressure on insects to become efficient fliers. The least change in morphology resulting in increased performance in the air would be very strongly reinforced or, perhaps, the initial selective advantage of small extensions of the thorax was thermo-regulation (Kingsolver and Koehl, 1985). There seems to be a set of interacting factors that promoted the probability of flight evolving in insects, which can be derived from three basic insect characteristics (Fig. 3.7). This hypothetical set of interacting forces on the evolution of insects merely shows that flight was likely. It does not conflict with any of the present theories on the evolution of flight (see Alexander and Brown, 1963; Wigglesworth, 1976; Kukalová-Peck, 1978, 1991; Brodsky, 1994), nor does it explain why other terrestrial arthropods, or immature insects, have no wings. It must be realized, however, that the evolution of flight has been an exceedingly rare event. Only four groups of organisms have been involved. A great diversity of insects probably had wings 340 million years ago or even earlier (Kukalová-Peck, 1991), and had the air to themselves for at least 150 million years. Pterodactyls were abundant 70–150 million years ago, birds 130–180 million years ago, and bats about 60 million years ago. It is significant that small organisms have shown the most dramatic adaptive radiation ever seen on this earth for exploitation of airspace.

SCALING AND INSECT ECOLOGY

Some rather obvious conclusions are important relative to the ecology of animals. First, since an animal is constrained in size by the physics of its body design, as an insect is, the evolution of wings avoids excessive costs in locomotion that would be incurred in running. Using the data in Fig. 3.6, we can estimate that running for a locust is 40 times more expensive than flying, and for a bee 47 times more expensive. For very small insects inaccuracies in measurement may be important, but according to the graph in Fig. 3.6, running for a fruit fly would be 470 times as expensive as flying. Tiny insects such as mymarid and trichogrammatid wasps may even be "swimming" through air (Weis-Fogh, 1976), with presumably more efficient use of energy than in other forms of flying.

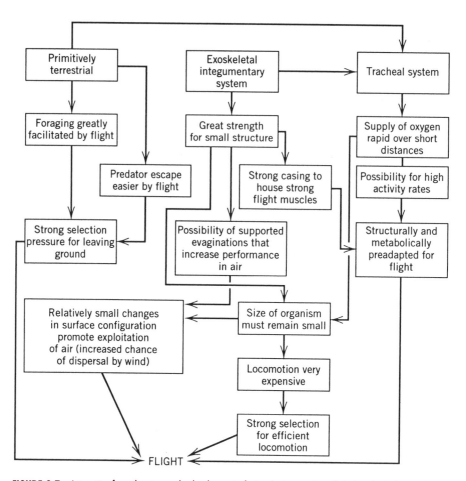

FIGURE 3.7   Interacting forces bearing on the development of wings in insects, given their three basic characteristics: their primitively terrestrial evolution, an integument of skeletal tissue, and a tracheal system (actually dependent on the exoskeleton and the insects' origin on land).

Second, flying organisms can become much more specialized in feeding and the utilization of other aspects of their environment if resources are generally low per unit area. Flower constancy and traplining would be impossible in bees and butterflies (see Chapter 11); such resources as dung and carrion, tree holes, and slime fluxes would have depauperate insect faunas, for colonization by foot would be too difficult. The diversity of organisms would be greatly reduced on these resources. The low oxygen consumption of immature insects also contributes importantly to their ability to utilize these resources.

Third, very small ground-dwelling animals, such as flightless insects, must remain close to their source of food. Perhaps the only answers to the high cost of locomotion are to live in or on the food, or to be a generalist, such as a predator, or to specialize on a very common resource.

Fourth, during harsh environmental spells, many more small favorable sites exist than large ones if they can be discovered. Searching ability, greatly facilitated by flight, becomes very important during harsh periods and in harsh environments. As resources become depleted, small patches could be discovered by flight and sustain a small organism through, say, a drought that may be devastating to larger organisms. As Van Valen (1973) observed, after examining global patterns of recent flowering plants, birds, and mammals, as body size becomes smaller there are more species in each taxon, a pattern also illustrated by May (1978c).

Fifth, flying organisms can be relatively rare per unit area. For example, cecropia moths *(Hyalophora cecropia)* and other saturniids are seldom abundant but they are able to fly great distances in response to the pheromone of the opposite sex, with males known to fly 5 miles per night (G. P. Waldbauer, personal communication).

In conclusion, size and scaling, and the physical forces that influence them, can help us considerably in an understanding of the limitations of particular designs in animals and the potentialities inherent in given sizes and forms. These considerations provide insights into foraging patterns of animals, performance, population structure, and large-scale distribution. We can begin to see some of the forces that have resulted in the diversity of life on this earth.

Chapter 4

Development of Theory in Insect Ecology

Acentral aspect of the development of science is discourse and debate. For effective debate we need to know where the science has been, its current status, and what the goals should be. Without such clear perspective it is tempting to follow our noses rather than our minds into every fascinating nook and cranny of nature into which insects are likely to lead us. Therefore, if we wish to build a vigorous and effective science of insect ecology, it is worth devoting a chapter to the way scientists work, the manner in which science progresses, and the ultimate goals that we wish to achieve. "The basic goals of ecology are seldom stated clearly by ecologists, if, indeed, most ecologists even have opinions about them" (Orians, 1962, p. 257) remains a valid comment, so some corrective measures are in order.

To start with, we can learn from two debates in the literature that cover fascinating plant and animal interactions; they involve extinct animals making the hypotheses hard to test, and they open up an arena for debate that could continue for decades. One interchange involves the tambalacoque tree and the dodo, and the other concerns the fruits the gomphotheres ate.

THE TAMBALACOQUE TREE AND THE DODO

Temple (1977) wrote a fascinating article about the dodo as the presumed disperser of a tree's seeds, and how extinction of the dodo now threatened

Conceptual view of the life cycle of an aggressive species of bark beetle that breeds in the cambium of killed trees and attacks living trees *en masse,* causing their death. Based on the concept illustrated by Leticia Arango Caballero in Tovar et al. (1995). Drawing by Tad Theimer.

extinction of the tree species. The argument became cited in many textbooks of ecology. The main points Temple made are as follows.

Among the many endemic species on the island of Mauritius in the western Indian Ocean were the tambalacoque tree, *Calvaria major* (Sapotaceae), a large primary forest timber tree, and the dodo, *Raphus cucullatus* (Raphidae), a large flightless bird weighing about 12 kg (or 26 lb). Temple argued that coevolution had occurred between the tree and the dodo. The trees produced large 5-cm-diameter fruits with a hard, woody, thick-walled endocarp—a stone pit. To germinate, the seeds needed to be eaten by dodos, swallowed, abraded in the dodo's gizzard, and then excreted. The gizzard contained large stones for crushing tough food. From the fruit the dodo acquired a pulpy succulent mesocarp and a food source, and the tree acquired a seed dispersal agent and seed preparation for germination. A mutualism evolved that was obligate for the tree. Coevolution occurred because thin-walled seeds would be digested and killed by dodos, so in an adaptive response to this seed predation the *Calvaria* population gradually evolved tougher and tougher endocarps.

The dodo was hunted to extinction by 1681. Hence, according to Temple, no seeds of the tambalacoque tree had germinated since that time. As a result, the tree was almost extinct, with only 13 old, overmature, and dying trees known in a remnant of native forest, with each tree estimated at over 300 years old. No younger trees were known to exist, although trees produce apparently fertile seeds. No seeds germinate naturally and there has been no rejuvenation for 296 years at least. No seeds germinate when planted under nursery conditions. Temple conducted an experiment, force-feeding 17 *Calvaria* pits to turkeys. Some pits were retained for 6 days by turkeys, 7 pits were crushed in the gizzards of turkeys, but 10 seeds were regurgitated or passed out with feces. These 10 seeds were planted in a nursery and three germinated. "These may well have been the first *Calvaria* seeds to germinate in more than 300 years" (Temple, 1977, p. 886). This argument was widely read and cited.

Horn (1978), Owadally (1979), and Witmer and Cheke (1991) opened up the debate on the validity of this interaction, and some of the issues are covered briefly here. Temple's experiment was inadequate, lacking a control and with low sample size. Germination of *Calvaria* seeds in a nursery had been reported in 1946, with success ranging from 2.5 to 20% seeds. The germination of seeds had been described and illustrated in 1941, involving a weak cap on the endocarp which splits off in the absence of abrasion. Abrasion did not improve seed germination. A survey of trees published in 1941 reported 33 trees in a 1-ha plot with three trees in the size class 10–14 cm diameter thought to represent trees 30–50 years old. There was an even-aged distribution of plants, suggesting long-term germination in the past centuries. Current estimates of tree populations are in the range of several hundred trees.

Gould (1980) reviewed the story of the dodo and the tambalacoque tree in his article "Nature's odd couples" in *The panda's thumb*. His concluding re-marks are relevant to our perspective on science. "This exchange highlights a disturbing issue in the transmission of news about science to the public. Many sources cited Temple's original story. I did not find a single mention of the subsequent doubts. Most "good" stories turn out to be false or at least overextended, but debunking doesn't match the fascination of a clever hy-pothesis. Most of the "classic" stories of natural history are wrong, but noth-ing is so resistant to expurgation as textbook dogma (Gould, 1980, p. 288). Gould noted that Thomas Huxley had said that a beautiful theory can be "killed by a nasty, ugly little fact."

Nevertheless, there is much that is positive about Temple's original re-search and the ensuing debate. The paper presented an interesting idea, gen-erating much debate and advancing scientific understanding. Novel ideas are essential to the progress of science, even if they are wrong. As Einstein noted, it is the sensation of the mystical that is the source of all true science. The generation and testing of hypotheses is essential. What one reads or hears may turn out to be false, but if it is interesting and heuristic, it still has merit. Most advances in science emerge from a mixture of ideas, many of which are wrong: the phlogiston theory, the flat earth theory, and so on. And nobody has established that dodo seed predation did not select for the thick and hard endocarp of the *Calvaria* seed. These may well be elements worth more study and expansion. We should beware of throwing out the baby with the bathwa-ter, a subject we discuss later in this chapter.

THE FRUITS THE GOMPHOTHERES ATE

Another novel perspective on fruit and animal interactions was developed by Janzen and Martin (1982). They noted that many fruits in the dry tropical forests of Costa Rica would remain uneaten were it not for introduced large mammals such as horses and cattle. They proposed that many species of tree produced fruits adapted to dispersal by members of the extinct megafauna from Pleistocene times, including the long-jawed mastodonts or gompho-theres.

Janzen and Martin identified a megafaunal dispersal syndrome involving fruits and seeds adapted to dispersal by the now-extinct large animals of Cen-tral America, including horses, gomphotheres, ground sloths, camels, and others. Some elements of the syndrome are that fruits are large and indehis-cent, they contain a pulp rich in sugar, oil, or nitrogen, and seeds are not dis-persed by wind or explosive release from the fruit. Such fruits are similar to those in Africa presently eaten by large mammals. Seeds are protected by an endocarp resistant to rupture by large mammal teeth and the digestive tract, or they are very small. Many fruits are abscised, fall to the ground and be-come available to large nonarboreal mammals, and they are unattractive to arboreal or flying frugivores such as monkeys, birds, and bats. Many fruits rot on the ground in the absence of large mammals. If present, peccaries, tapirs, agoutis, and small rodents usually perform the role of seed dispersers or predators, but fruits are favored by introduced Old World animals such as cattle, horses, pigs, and elephants.

Thus, in the absence of introduced Pleistocene-like large mammals, many tree species in dry deciduous tropical forest, such as in Guanacaste National Park, Costa Rica, seem to have had no effective seed dispersers for 10,000 years. "Modern habitats are Pleistocene anachronisms" (Janzen and Martin, 1982, p. 26). Pratt (1982) suggested that similar relationships may have evolved in Australia.

Howe (1985) was not convinced of the megafaunal dispersal syndrome hypothesis (see also Howe and Westley, 1988). He argued that the syndrome is not adequately defined and no clear criteria are identified, such as in the wind dispersal syndrome or the ant dispersal syndrome. He noted that many plant species in the list provided by Janzen and Martin are effectively dispersed by living mammals and birds, and that survival of tree species for 10,000 years or perhaps 200 generations is most unlikely without dispersal agents. Also, large herbivores are of necessity very general feeders, taking fruits only opportunistically, and seed dispersal mechanisms are rather generally adapted to a wide variety of agents. Therefore, it is hard to accept that a specific megafaunal dispersal syndrome would evolve.

DEBATE AND PROGRESS

These two interesting cases on plants and frugivores illustrate some important points about the nature of science, debate, and progress in understanding ecology. Both subjects were novel and interesting, and they attracted attention and stimulated debate. Science benefited from the dialogue even though a definitive conclusion has not been reached. *Calvaria* seed design may have been influenced by dodos in evolutionary time, and gomphotheres might have been important dispersers of some large fruits in Central America.

The papers and debates also illustrate our common tendency to invoke evolutionary adaptations as a way of explaining relationships without considering the phylogenetic background from which the interacting species are derived. How much of a species' form and function can be attributed to specific adaptations and how much to phylogenetic constraints? Gould and Lewontin (1979) argued that too much of the biological literature fails to place adaptations and species in their phylogenetic context and the constraints of morphology in the history of the lineage. Did the fruit design of *Calvaria* trees diverge significantly from the tree's progenitors when exposed to dodo predation on Mauritius? And do species of trees with the megafaunal dispersal syndrome differ in fruit design and display substantially from their

relatives or forebears? Such questions require broadly comparative approaches for which the methods are well developed (e.g., Harvey and Pagel,
1991). But without setting a study in a phylogenetic context, we may end up
with many idiosyncratic and somewhat anecdotal reports that attempt to explain how the real world is working, without the substance and credibility
that ensure enduring importance.

Especially in the study by Temple, an idiosyncratic relationship was fea-
tured, unset in the broad scope of developing theory and/or hypotheses on
seed dispersal mechanisms. This is typical of much of the ecological literature, which remains unrelated to the development of general ecological theory. This raises a question as to the nature of a scientific theory.

SCIENTIFIC THEORIES AND HYPOTHESES

We tend to use the words *theory* and *hypothesis* interchangeably. This is un-
fortunate because we really need distinctive words for conjectural statements
and well-established factual explanations. After all, the theory of gravity, the
theory of relativity, and the theory of evolution are far stronger than conjec-
ture or hypotheses. They are factually based, broad explanations of the natural world. They are worthy of the term *scientific theory*. Therefore, in this
book I retain a clear distinction between theory and hypothesis. A **scientific
theory** can be defined as the factual explanation of a pattern in nature. The
explanations and patterns are well established and widely accepted in the academic community, after repeated attempts to falsify the theory. The term **hypothesis** can then be reserved for a conjectural or tentative statement about
nature, inadequately tested and in need of much more study before it can be
accepted as fact (Moore, 1993; P. W. Price, 1991a, 1996b).

This use of the terms *theory* and *hypothesis* clarifies the process of science
because we naturally work from hypotheses to theory, and the development
of theory should be a driving force and focus for many scientists most of the
time. But terms such as *theoretical ecologist* are well entrenched even though
most deal principally with hypotheses, so we cannot expect universal acceptance of the clear dichotomy proposed here.

Recently, authors have differed in how they regard the terms. Moore
(1993) uses hypothesis and theory as proposed here, while Booth et al.
(1995) in *The craft of research* prefer not to use the terms at all, but use, instead, such words as *question* and *thesis*. Lumley and Benjamin (1994) think,
evidently, that theory is not an important part of research. Pickett et al.
(1994) make the important point that theories change and develop, and ultimately **mature theory** is generated. "The most useful theories will incorporate
explicit assumptions, clear domain, clear concepts and definitions, a body of
fact, confirmed generalizations, laws, models, a framework with translation
modes, and hypotheses. Not only must some large proportion of the compo-

nents of theory be present, but the individual components must be well developed for a theory to be maximally useful. Development refers to exactness, empirical certainty, applicability to observation, and derivativeness of complex components" (Pickett et al., 1994, p. 100).

Did Darwin (1859) develop a theory of evolution? Using the term *theory* as advocated here, he did not. Some of his facts were incorrect, such as those on the origin of variation in populations and the mechanisms of heredity (cf. Price, 1996a). Do we have a theory of evolution now, or a mature theory of evolution? Certainly (cf. Mayr, 1982; Mayr and Provine, 1980; Moore, 1993). Is the theory of evolution still developing? Naturally!

If the reader accepts that scientists should work toward the development of theory, we should broach the questions of why theory is important and how it is developed.

IMPORTANCE OF THEORY

Theory is central to science because of its explanatory and predictive power. Theory explains in a mechanistic way how driving forces in nature result in broad patterns and in our ability to reach broad generalizations. Darwin (1859) explained the whole of the unity and diversity of life with what he called five "laws" of nature. "These laws, taken in the largest sense, being Growth with Reproduction; Inheritance which is almost implied by reproduction; Variability from the indirect and direct action of the external conditions of life, and from use and disuse; a Ratio of Increase so high as to lead to a Struggle for Life, and as a consequence to Natural Selection, entailing Divergence of Character and the Extinction of less improved forms" (Darwin, 1859, pp. 489–490). These processes explained how life on Earth had developed and predicted the continued evolutionary responses under the influence of natural selection.

Such broad and factual explanations of nature are, of course, important both for those interested in fundamental questions about nature and for scientists investigating practical problems. The dichotomy of basic and applied science becomes unwarranted when we appreciate that the solution to practical problems may well contribute to the development of mature theory as much as fundamental inquiry. And focus on the development of general theory advances science much more rapidly than if studies are undertaken in isolation.

Are there five or more "laws" that underlie a general theory of ecology? Does the field of ecology or insect ecology encompass broad, factually based theory? Is there rapid development of theory in ecology? Any kind of response to these questions would be widely debated and controversial, no doubt, but the reader may reach a conclusion based on the number of times in this book the words *hypothesis* and *theory* are appended to conceptual developments in ecology.

DARWINIAN METHODOLOGY

Darwin (1859, 1871, etc.) was so successful in establishing the basis for a theory of evolution that we may well learn from the methodology he apparently used. My personal interpretation of Darwin's methodology (Price, 1991a) is no doubt debatable, but it features many of the elements in a research program that may generally be appreciated.

Darwin was a naturalist. He collected beetles, rocks, and shells and enjoyed hunting. He observed the results of the agrarian revolution firsthand, and collected factual information as to pigeon and rabbit breeding. He was an empiricist first, and the patterns he saw in nature were based on factual, firsthand evidence: cultivated crops never fail to show variation, many traits in parents are inherited by their progeny, artificial selection can result in profound change in a lineage, and so on.

Such empirical evidence of real patterns in nature formed the factual basis for Darwin's *Origin of species* and the beginnings of a theory of evolution. In his autobiography Darwin wrote, "I worked on true Baconian principles, and without any theory collected facts on a wholesale scale, more especially with respect to domesticated productions, by printed enquiries, by conversation with skilful breeders and gardeners, and by extensive reading" (Darwin, 1892, p. 42). This factual basis enabled a convincing thesis to be developed, even though some of the hypotheses on mechanisms were incorrect.

Based on empirical observations, Darwin was able to detect broad patterns in nature. No doubt Lyell's (1830) interpretations of how pattern developed in geology had a profound impact, and Darwin quickly adopted Lyell's point of view while on the *Beagle* voyage. He saw pattern in relationships among Pleistocene fossils and living species, pattern in morphology among the finches on the Galápagos Islands, pattern in the emotions of humans and wild relatives, and pattern in morphology, embryology, and heredity. Once broad patterns are detected, the hypotheses on mechanisms that result in pattern can be generated and tested relatively easily. The hypotheses adopted may be right or wrong, but if the factually based pattern is sound, the substance of the thesis can remain intact.

Once broad patterns are detected, mechanistic explanations for the pattern form the basis for the development of theory. Such explanations would start as hypotheses and then be tested over time, as for example, with Darwin's hypotheses on natural selection, the origin of variation in nature, and the mechanism of heredity. Testing hypotheses over decades may well result in refutation of some and the erection of alternatives. Other hypotheses would stand the tests of time and eventually become generally accepted. Thus a theory develops from empirical facts, to pattern detection, to hypotheses, continued testing and the erection of alternatives, and the ultimate general acceptance of a body of knowledge that accounts for broad patterns in nature (Moore, 1993; Pickett et al., 1994).

Darwin was an expert critic of his developing theory and was able to weigh the evidence in an effective manner. He pondered the difficulties with his hypotheses on how the well-developed wings of a bird or bat could evolve by gradual natural selection, and how ant societies could evolve and function when most females had lost the ability to reproduce. Darwin collected massive amounts of information and weighed its validity in relation to his thesis. He took a balanced view on a broad array of evidence and eventually created a convincing case for mechanisms resulting in the transmutation of species. As we collect information at increasing rates it becomes increasingly difficult to sort and weigh the facts. We tend to reject hypotheses based on a single exception and we often forget to set out clearly the weight of evidence in favor of a particular hypothesis.

Darwin's example illustrates the importance of many aspects of science relevant to ecologists: being a naturalist able to observe the empirical aspects of the natural world, the detection of pattern among all the details of nature, the use of hypotheses to formulate mechanistic explanations of pattern, the testing of pattern and mechanism and weighing the evidence. With long-term devotion to science, perseverance against formidable odds, a fascination for nature, and a methodical and analytical mind, Darwin laid the foundation for biologists' most comprehensive theory.

EMPIRICISM, MECHANISM, AND THEORY

The scientific debate on the development of theory may seem chaotic, with a strongly negative element, and certainly lacking in objectivity and impartiality. Sometimes, progress of one step is followed by regress of two. Nevertheless, there is a formidable power in the combination of logic, debate, and factual knowledge over protracted periods of time. The advance of knowledge is in fits and starts. A brief example illustrates some aspects of this debate.

We focus on the interchange between Tilman (1991a,b), Shipley and Peters, (1990, 1991), and Poorter and Lambers (1991). Tilman (1988) had brought much evidence to bear on hypotheses he advanced on the dynamics and structure of plant communities, based on resource competition among species. Shipley and Peters (1990) performed experiments to obtain empirical data for testing Tilman's model and hypotheses. They noted (p. 136) that "this model is general and powerful, and, if correct, it would provide an important unifying framework for plant ecology. The theory's importance as a motivator of experiment and discussion is already established."

However, Shipley and Peters rejected Tilman's model based on their empirical studies, finding no significant trends for predicted correlations and a significant correlation opposite to a prediction by Tilman. This paper was followed by a response from Poorter and Lambers (1991), who noted a flawed experimental approach, making the argument by Shipley and Peters inconclusive. Poorter and Lambers also included evidence undermining Tilman's hypotheses. Tilman (1991a) added his own concerns about Shipley and Peters's results: (1) they misinterpreted parts of the theory and so could not refute it; (2) their data run counter to several published reports on the relationships tested; and (3) habitat specificity of the plant species studied influenced important aspects of the study.

Naturally, a rebuttal was to be expected from Shipley and Peters (1991)—"the seduction by mechanism," ending with an important discussion on how mechanistic models can have several purposes and how falsification of mechanistic hypotheses poses serious problems for the researcher. For example, not all traits in a mechanistic model are experimentally manipulable, such as evolved life history characteristics. "The *appearance* [my emphasis] of success forms a substantial part of the allure of mechanism" (Shipley and Peters, 1991, p. 1281).

In a fine retort, Tilman (1991b, pp. 1283–1284) emphasized "the schism between theory and ardent empiricism."

> Ecological research, like all science, is most effective if it is based on the continual interplay of observation, hypothesis generation (theory), and experimentation. Empiricism is clearly a part of this process, as is theory. However, Shipley and Peters suggest that the falsification of one prediction of a mechanistic model indicates that the model is "wrong" and thus not useful in explaining other patterns. This is an extreme, absolute interpretation that sees a model as the mathematical embodiment of ecological truth. In contrast, mathematical ecologists view models as abstractions . . . simplifications that, in the words of May (1973a, p. 12) are "caricatures of reality, and thus have both the truth and falsity of caricatures." All models are caricatures, even the models of chemistry and physics, which can be precise predictors for some phenomena but wrong for others.
>
> I often call my models "mechanistic." Shipley and Peters interpreted this to mean that they must incorporate the irrefutable underlying mechanisms that cause a process. In contrast, I call my models "mechanistic" to distinguish them from classical models that are more phenomenological [i.e., involving description and classification of phenomena, not mechanism]. This distinction was elaborated in Tilman (1989, p. 94): "Clearly, phenomenology and mechanism are not absolute entities but idealized ends of a spectrum. Any theory that explicitly includes environmental constraints and organismal trade-offs will be more mechanistic than most current theory. It is likely that, along the spectrum from phenomenological to mechanistic theory, there will be a point that is optimal for explaining any given ecological pattern. It is very possible to produce theory that is too mechanistic, that loses generality without gaining significant

predictive power. The optimal point will be found only through the usual trial-and-error process of science. It is always possible to produce a theory that is more or less mechanistic than a given theory. I do not propose mechanistic theory as an absolute good, but rather suggest that many present theories may lack predictive power because they are not sufficiently mechanistic; that is, they do not explicitly deal with environmental constraints and organismal trade-offs."

Thus, there is a level of mechanistic detail that best fits the type of question being asked. The search for mechanism and predictive ability is a search for the appropriate level of abstraction. This search is seductive not because one gets lost in a morass of mechanistic detail (as Shipley and Peters suggested) but because it ultimately allows a better understanding of the simple forces that have shaped nature. (Tilman, 1991b, p. 1285)

SEARCH FOR PATTERN

The discovery of pattern in nature is one of the great challenges for the practicing scientist in ecology. "Study major, broad, repeatable patterns" (Tilman, 1989, p. 90). "Without patterns in nature to guide us, we risk being overwhelmed by detail," wrote Lawton (1991, p. 84), citing Southwood (1988). "A compendium of case histories is unenlightening if recurrent patterns are not sought and interpreted, and a generality is of little use if the exceptions far outweigh the cases that approximate the expected" (Colwell, 1984, p. 394). How does the ecologist find pattern in nature?

There are bound to be ecological patterns associated with the phylogenetic relationships among species. We will see these types of patterns in Chapter 14 when we look at life histories and reproductive strategies, and in Chapter 24, on paleoecology, biogeography, and biodiversity. As species diverge from a common stock into new ecological environments, we can expect to see evolutionary responses, although the ground plan of the lineage is retained. If there are macroevolutionary differences in life histories, we may well expect large ecological differences, or macroecological differences, in population dynamics (cf. Chapter 19). Orians (1962) was a strong proponent for an evolutionary perspective in any enterprise involving the understanding of ecological interactions.

Another basis for pattern concerns the many kinds of gradients over a landscape, or on a global scale: elevational gradients, moisture gradients, and latitudinal gradients all result in strong environmental changes and therefore impose pattern on utilization by species. Louda (1982) won the Mercer Award from the Ecological Society of America for her paper, "Distribution ecology: Variation in plant recruitment over a gradient in relation to insect seed predation," using an elevational gradient in coast sage brush vegetation in California. Many of Louda's papers (1983, 1989; Louda et al., 1987a,b) have used gradients, such as climate, moisture, and complexes of factors, for understanding ecological relationships. Productivity and other gradients in lakes have proved very useful in the detection of pattern in aquatic systems (e.g., Schindler, 1978; Carpenter, 1988; Carpenter and Kitchell, 1988; Cole et al., 1994). Latitudinal gradients have been a focus of attention among ecologists for decades in terms of species richness, as discussed in Chapter 23. We may even employ the gradients in plant architecture, such as plant size or shoot length, or leaf size, to search for pattern in utilization by insects (e.g., Lawton, 1983; Price et al., 1995a).

Time also acts as a gradient in ecology, imposing patterns of many kinds. Ecological succession is a process in time, whether it be vegetational changes and associated changes in the fauna (cf. Chapters 2 and 23) or changes in the insect fauna as a log or a corpse decays (cf. Chapter 1). Aging on a landscape takes other forms in terms of plant age, with associated gradients of growth rate, reproductive status, and other factors. Leaves age and change, creating gradients in moisture content, sugar, chemical defenses, and toughness (cf. Chapters 5 and 12). Insects age and die and populations through the generations change in character such that carnivores evolve different strategies to utilize different stages in the life cycle (cf. Chapter 14). Seasonality and phenology are time associated and clearly impose strong patterns on the landscape and the associated organisms.

Strongly comparative studies over the many kinds of gradients in nature have provided ecologists with important insights to organization in nature. Gradients may not be displayed clearly in space or time, but for comparative purposes we can organize data onto such gradients, as was necessary for studies on the effects of lake nutrients and productivity. Such comparative techniques are illustrated many times in this book, but there is much room for increased emphasis on broad comparative studies among insect ecologists, for example, in ecological genetics (Chapter 16), population dynamics (Chapter 19), competition (Chapter 21), and community or assemblage structure (Chapter 22).

Oksanen (1991) has been critical about progress in ecology, claiming that we keep doing the same kind of thing: reinventing the wheel. But insect ecologists have such an incredibly rich and diverse fauna to work with, and therefore such powerful comparative opportunities, that real progress can be made rapidly if we concentrate on the development of scientific theory. The challenges are considerable, but our rigorous science has generated a very large number of studies that can now be assimilated in an effort to discover pattern in nature. This is not to say that this book concentrates on broad patterns in nature and their underlying mechanisms, for we remain in many areas at a largely phenomenological level, with many hypotheses extant relevant to a particular interaction, such as plant and herbivore interactions (Chapter 6). Time is needed to order the relative importances of the hypotheses in terms of the generation of pattern in nature. For insect herbivore populations, for example, do plants as food or carnivores play a stronger role in dynamics, or is the interplay between the first and third trophic levels critical? (See, e.g., Hunter and Price, 1992; Power, 1992; Strong, 1992; Menge, 1992.) Where are the general hypotheses on population dynamics that may ultimately form the basis for theory? (See Chapters 17–19.) And how are we progressing with the mechanistic understanding of insect species richness on islands? (See Chapters 22 and 24.)

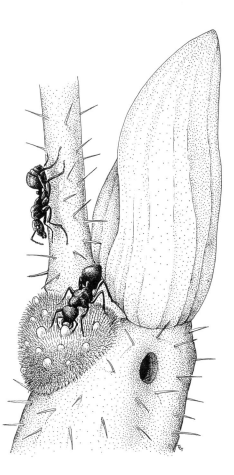

An *Azteca* ant takes a Müllerian body from the trichilium, or site of Müllerian body production, at the petiole base of a *Cecropia* tree in the neotropics. The opened prostoma, or entry site, directly below the bud, provides access to the hollow stem in which the ants nest. Drawing by Tad Theimer.

Trophic Relationships

Since food is essential to all organisms, feeding relationships have been a central focus among ecologists, and a theme that runs throughout the science. Ecologists who looked for patterns in nature regarded food as an organizing factor. They saw that types of food and feeding led to predictable relationships between trophic levels: in food webs (Part I), in interactions between plants and herbivores (Chapters 5 and 6), between prey and predators (Chapters 7 and 8), between hosts and parasites (Chapter 9), between organisms with mutually beneficial effects (Chapters 10 and 11), and in the ways in which energy and nutrients move in ecosystems (Part I and Chapter 12). In many communities insects play key roles as herbivores and predators and, in turn, form an important source of food for other organisms. The interactions between insects and their food plants (Chapters 5 and 6) and insects and their predators (Chapters 7 and 8) thus influence profoundly the amount of energy that passes from one trophic level to another (Chapter 12). The roles of parasites (Chapter 9) and mutualists (Chapter 10) are more difficult to determine, although it is clear that such trophic relationships have been exceedingly successful in the insects.

Food is a recurring theme throughout the other parts of this book. Food may limit population growth; it influences dispersion of the population and the timing and extent of dispersal (Part III). Feeding on different items leads to divergence of ecological niches, the coexistence of species, structure in the community, and the distribution and abundance of species (Part IV).

Chapter 5

Plant and Insect Herbivore Relationships

At the end of their paper, "Butterflies and Plants: A Study in Coevolution," Ehrlich and Raven (1964) said: "Probably our most important overall conclusion is that the importance of reciprocal selective responses between ecologically closely linked organisms has been vastly underrated in considerations of the origins of organic diversity. Indeed, the plant–herbivore interface may be the major zone of interaction responsible for generating terrestrial organic diversity" (see also Ehrlich and Raven, 1967). Brues (1924) had gone a considerable way to reaching this conclusion 40 years earlier.

DEFENSE AND ATTACK

There is a constant warfare. Acquiring sufficient energy to enable reproduction is the essential preoccupation of every member of a species. An individual would certainly be maladapted if it maintained a trait that permitted easy access by other species to the energy store represented by the protoplasm of

A copper butterfly adult (Lycaenidae) feeds on the nectar of a daisy, but its larvae eat plants and are tended and protected by ants. Drawing by Tad Theimer.

its body. The only exception to this statement can be seen in relationships for mutual benefit where the energy store, or usually prescribed parts of that store, are left available for a species in the higher trophic level, provided that the organism receives a significant gain by doing so. This aspect of plant–herbivore interactions is covered in various parts of this chapter and in Chapters 10 and 11. Individuals must defend their collected energy against all would-be consumers. If another organism higher in the trophic system can break down the defense, it has access to a lucrative energy source, and furthermore, this food will not have to be shared among other species until others are also able to break down the defense mechanism. The winner of this arms race receives a double advantage. As plants are the producers, these are the first that must defend themselves—against the herbivores. An attack by a herbivore selects an adaptation for defense in the plant, or even a counterattack. This cycle of attack and counterattack between organisms of adjacent trophic levels is the essence of the **coevolutionary process.** Plants and insects have been engaged in it for millions of years, so some very refined results may be expected in this age-old interaction, although how much mutually induced evolution has been involved is hard to determine (see Chapter 6).

The basic components required by a plant, and manufactured by the plant, can be classified into three biochemical groups: (1) those that catalyze biochemical reactions, (2) those that participate in building processes in the plant organism, and (3) those that supply energy. All plants require inorganic ions and must produce enzymes, hormones, carbohydrates, lipids, proteins, and phosphorus compounds for energy transfer. These compounds are all intimately involved in the growth and reproduction of the plant.

PRIMARY AND SECONDARY METABOLIC PRODUCTS

However, the plant kingdom contains a vastly greater variety of chemicals than those involved in these primary functions. Presumably the other chemicals are by-products in the synthesis of primary metabolic products, and lacking a means of excreting them, the plants store them in any convenient place within the plant structure. Some plants manufacture a particular chemical and others do not; this phenomenon suggests that this chemical is not an essential ingredient in plant metabolism. These kinds of substances have been called **secondary metabolic products.**

The fact that these products are very common in plants is supported by the large volume of literature on the chemical taxonomy of plants, which is based primarily on secondary metabolic products. Closely related plants are likely to share similar metabolic pathways for primary products, with the simultaneous production of similar secondary metabolic products. A single gene mutation, which may change a single enzyme, may be reflected in a change in the type of secondary metabolic product produced. Therefore, chemical taxonomy tends to be a more precise analytical test than morphology for determining phylogeny in evolution, with the pleiotropic effects that may occur in the phenotype as a result of gene mutation.

HOST SELECTION BY INSECTS

Insects keyed into this system of chemical relationships between related plant species very early in their evolution. For example, several families of plants contain glucosinolates, the precursors of mustard oils—Capparidaceae, Brassicaceae or Cruciferae, Tropaeolaceae, and Limnanthaceae, and certain insects will feed only on members of these families, for example, the flea beetles *Phyllotreta cruciferae* and *Phyllotreta striolata* (Feeny et al., 1970; Louda and Mole, 1991; Robinson, 1991). Mustard oils are irritants, capable of causing serious injury to animal tissue, and are also among the most potent antibiotics known from higher plants. For proof of their potency a teaspoonful of horseradish or mustard should be tasted.

The family Solanaceae is well known for the alkaloids members contain (Bernays and Chapman, 1994). For example, green parts of the potato, *Solanum tuberosum,* contain solanine; tobacco, *Nicotiana* spp., contains nicotine, and the deadly nightshade, *Atropa belladonna,* produces atropine. Some insects are restricted to members of this family in their feeding habits. Examples are the Colorado potato beetle, *Leptinotarsa decemlineata,* certain flea beetles, and tobacco and tomato hornworms, *Manduca sexta* and *M. quinquemaculata.*

Two closely related families of plants, Asclepiadaceae and Apocynaceae, include members that have a milky sap and cardiac glycosides, or cardenolides (Malcolm, 1991). Each family has two specific insect herbivores from the arctiid genera, *Euchaetias* and *Cycnia,* as follows:

Genus	Asclepiadaceae (milkweeds)	Apocynaceae (dogbanes)
Euchaetias	egle	*oregonensis*
Cycnia	*inopinatus*	*tenera*

Closely related plants support closely related insects. In the evolution of specific differences in insects, slight but permanent changes in digestive or other enzymes permitted the exploitation of closely related plants. This permitted the rapid speciation within a group once the early members had become tolerant to, or able to break down, the toxic substances in these plants. This is the gist of Ehrlich and Raven's (1964) paper. Fraenkel (1959, 1969) has emphasized the important role of secondary plant substances in host selection of herbivorous insects.

Characteristics of secondary metabolic products may be summarized as follows (mostly from Whittaker, 1970a):

1. They are not essential to the basic protoplasmic metabolism of the plant.
2. If the plant is considered by itself without reference to other organisms, there is no evident reason why the plant should produce them at all.
3. They are of irregular or sporadic occurrence, appearing in some plants or plant families and not in others; this fact reinforces the view that they are not essential to plant metabolism.
4. The occurrence of the same or related secondary compounds in related plant species makes these compounds important concerns of chemical taxonomy of plants, although some compounds have often been independently evolved.
5. Many of these products are produced in large quantity and are metabolically expensive; they must be serving a valuable purpose.
6. Many of these products are toxic to animals and other plants, or at least repellent. Whittaker (1970a) states: "The view has developed through observations of Dethier (1954), Fraenkel (1959), Ehrlich and Raven (1964) and others, that the secondary plant substances have their primary meaning as defenses against the plant's enemies." Enemies include pathogens, herbivores, or competitors, and this biochemical inhibition of feeding by animals or plants is called **allelopathy.**

CHEMICAL DEFENSE IN PLANTS

Allelopathic chemicals affecting plant–plant interactions have been reviewed by Rice (1974) with a brief section on allelopathy between plants, insects, and other animals. But chemicals similar to those used in plant–plant interactions are at least of equal importance in the plant's defense against herbivores. The production of nicotine by the tobacco plant is an obvious example; the toxin is so effective in reducing nonspecialized insect attacks on the plant that it has been employed as an insecticide by humans. The pyrethrins in pyrethrum flowers have been used for the same purpose. Sinigrin, a mustard oil glucoside in Brassicaceae or Cruciferae, protects plants from a large array of potential herbivores (e.g., Erickson and Feeny, 1974). Indeed, herbivores may have exerted the primary selective pressure for the production of toxic chemicals that subsequently became effective in plant–plant interactions (L. E. Gilbert, personal communication). An ovipositing female insect can locate the majority of progeny of a reproducing plant, which leads to heavy mortality of seeds or seedlings and strong selection for defense. Allelochemicals that reduce competition are usually effective only after the plant is well established, with correspondingly weaker reinforcement for production of the chemicals.

Feeny (1968) tested the effect of tannins produced in oak leaves on the growth rate and pupal weight of the winter moth, *Operophtera brumata,* which is a serious pest of oak in Europe and eastern Canada. He extracted condensed tannins (not hydrolysable) from old (September) oak leaves where the tannin content was 2.4% of the fresh weight of the leaf. The tannin was mixed with casein in an artificial diet so the protein became tanned, just as it would when a caterpillar chewed up an oak leaf. He used third and fourth instar larvae to start the experiment, and the results are shown in Fig. 5.1. One may imagine how the tannins in oak leaves reduce the fecundity of the resulting adults, if the larvae grow to about half the size of those without tannin in their diet. Also, the larvae eat less of the leaf because they are much smaller, and the populations are reduced because of lower fecundity (see also Hagerman and Butler, 1991). A similar role to tannins may be played by endopeptidase inhibitors in legume seeds that reduce digestion of nitrogen compounds (Applebaum, 1964).

At the problem-solving level the difficulties with manipulating secondary metabolic products in food plants become complex. This type of chemical is frequently responsible for the attractive taste of the plant (e.g., mustard oils in crucifers), but the same chemical may act as a feeding or oviposition cue to a specific pest species and an antibiotic to nonspecific pests. This dilemma was found in the cucumber by DaCosta and Jones (1971). The bitter gene, *Bi,* is dominant and is responsible for production of the antibiotic cucurbitacins and the bitter taste. These tetracyclic triterpenoids stimulate feeding by the specific cucumber beetle (Metcalf and Metcalf, 1992), so plants homozygous for the recessive gene, *bi,* lacking cucurbitacins, would offer a

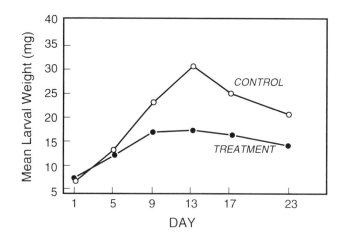

FIGURE 5.1   Change in mean larval fresh weights of winter moth, *Operophtera brumata,* with time, for larvae fed on a casein diet (control) and a casein complexed with tannin diet (treatment). The mean larval fresh weight increments (initial to peak weight) were 25.2 mg on the control diet and only 11.7 mg on the treatment diet. From Feeny (1968).

means of beetle control. However, *bibi* gene plants lack resistance to general pests such as two-spotted mites. The stepwise coevolutionary process is seen clearly here (Fig. 5.2).

There are so many plants, in such an array of taxonomic groups that produce chemicals with insect hormone activity, that the possibility of chance occurrence is almost certainly ruled out (Bowers, 1991). This situation must surely indicate convergence of armament strategies in the coevolutionary warfare, although caution is warranted when the term *coevolution* is invoked (e.g., Futuyma and Keese, 1992). The hormone mimics may be of the juvenile hormone type that maintain the immature condition, or of the moulting hormone type that synchronize moulting activity, both vital processes in insect development.

Juvenile hormone activity has been found in the firs *(Abies)* and Douglas fir *(Pseudotsuga)* (Sláma, 1969). Interest was generated in these chemicals by the discovery that certain North American paper products contained juvenile hormone activity for the European bug, *Pyrrhocoris apterus* (see Sláma and Williams, 1965, 1966; Sláma, 1969, 1979). When reared in contact with these papers the bugs did not mature but developed into giant nymphs. The hormone mimic, now named *juvabione,* was found to be present in the wood of balsam fir, *Abies balsamea,* a major constituent of North American paper pulp (Bowers, 1991).

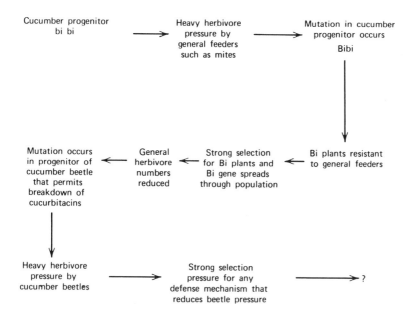

FIGURE 5.2   Possible stages in stepwise coevolution between the cucumber and its herbivores. From information in DaCosta and Jones (1971).

Moulting hormone activity has been found in *Podocarpus, Polypodium,* and *Vitex* (Sláma, 1969, 1979). Screening of plant species in Japan revealed that 40 provided active extracts or 4% of the 1056 species tested (Williams and Robbins, 1968). Bracken fern is known for its relatively high yield of ecdysone derivatives, and these are very active when injected into locusts. However, Jones, and Firn (1978) found only trace amounts, not effective in herbivore defense, and locusts can eat the foliage with impunity as they break down the hormone (Carlisle and Ellis, 1968), no doubt a counterattack evolved after the ecdysonelike chemicals in bracken had been effective defensive chemicals for a few thousand years. These ecdysterols are also found in members of the family Amaranthaceae, although activity differs greatly from one insect species to another and from one ecdysterol to another. Some are certainly insecticidal when applied topically, and others do inhibit normal development (Hikino and Takemoto, 1972).

ALLOCATION OF PROTECTIVE CHEMICALS

It is to be expected that plants put more energy into protecting reproductive organs than into vegetative parts (cf. McKey, 1974). Reproductive organs are usually smaller and so more easily damaged; they are richer in nutrients and so a greater prize for the herbivore; and indeed, they are the basis for the entire existence of the plant. For example, the psychoactive ingredient in marijuana, *Cannabis sativa,* is concentrated in flowers and seeds of the plant. Hypericin, a toxic agent in *Hypericum* species, of which the Klamath weed is one, is present at a concentration of 30 μg/g wet weight in the lower stem, 70 μg/g wet weight in the upper stem, and 500 μg/g in the flower (Rees, 1969). The pyrethrum plant *Chrysanthemum cinerariaefolium* produces pyrethrins known to have a high insecticidal activity. These compounds produce the startling "knock-down" effect characteristic of many household insecticides, and are concentrated in the flowerheads, with most chemical activity in the seeds. The very rapid action of these chemicals minimize damage to plant parts that could be very rapidly destroyed but for this protection. Many members of the family Rosaceae produce seeds with cyanogenic activity (e.g., apple, almond, and peach).

In the same vein, long-lived plants should be expected to have better chemical protection than more ephemeral species have. Cates and Orians (1975) found this expectation to be justified in a large sample of plants, a pattern discussed in more detail later in this chapter.

Herbivore pressure may also result in polymorphism in plant populations or differential allocation of chemical defense between populations (Berenbaum and Zangerl, 1992; Fritz and Simms, 1992). Cates (1975) showed that wild ginger, *Asarum caudatum,* has two morphs. Where numbers of general herbivores such as slugs are very low, a morph that allocates much energy to growth rate and seed production is common. However, where slugs are abundant, a morph with reduced growth rate and seed production exists, but more energy is channeled into a potent chemical defense. Jones (1962) had found similar results in acyanogenic and cyanogenic morphs of *Lotus corniculatus,* where the acyanogenic clones produced more seed and grew and cloned more rapidly where herbivore pressure was low. The costs and benefits of defense against herbivores are clearly portrayed in these studies. In his further extensive work on polymorphism in populations of *Lotus corniculatus,* evidence for maintenance by selective herbivory from slugs and snails has mounted (e.g., Jones 1966, 1972; Ellis et al., 1977a,b). However, this is not the complete explanation for the existence of polymorphisms at all sites, so caution in interpretation is required (Ellis et al., 1976; Jones et al., 1978).

PRIMARY ROLES FOR TOXIC CHEMICALS

Indeed, alternative hypotheses to explain the presence of secondary metabolic products in plants have usually been investigated inadequately. The copious literature indicating that the plants' enemies are major selective agents for the production of toxic chemicals is convincing. However, these chemicals may be used in primary processes in the plant. Many toxic chemicals exist in a state of dynamic equilibrium, with a half-life of a few hours in some cases (e.g., Loomis, 1967; Burbott and Loomis, 1969; Croteau et al., 1972). When C- and N-labeled nicotines were fed to tobacco plants, the label appeared in amino acids, sugars, and organic acids (Tso and Jeffrey, 1959, 1961), all of primary importance to the plant. Although about 12% of fixed carbon per day in a tobacco plant is used in nicotine biosynthesis, almost 40% of the nicotine in a plant is degraded in a 10-hour photoperiod (Robinson, 1974; see also Robinson, 1979). These findings, which are representative of many others, should prompt the search for adaptive roles of toxic chemicals that are independent of selection pressure from the plant's enemies. They may act as storage for essential chemical ingredients, or as regulators in biochemical processes (Seigler and Price, 1976; Swain, 1976; Seigler, 1977; Jones, 1979). Jones et al. (1978) cite studies indicating that the major selective pressures on polymorphism in cyanogenesis at some locations involve soil water stress, and temperature at others, while temperature also influences the expression of cyanogenesis. Clearly, more studies involving the many alternative hypotheses to account for the presence of secondary metabolic compounds are desirable, and populations of plants polymorphic for these offer one potent means of analysis (Jones, 1971).

RESPONSE OF INSECTS TO CHEMICAL DEFENSE

The insect herbivores can clearly withstand the many chemical defenses of plants or they would not be so successful in terms of numbers of species and population sizes. Vertebrates contain enzymes in the liver, their detoxifying center, that promote a metabolic attack on a variety of drugs and pesticides (Krieger et al., 1971; see also Freeland and Janzen, 1974; Brattsten, 1992). Similar enzymes have been found in invertebrates, and in lepidopterous larvae they are concentrated in the midgut tissues. These microsomal mixed-function oxidases catalyze a huge diversity of reactions (Brattsten, 1979, 1992) and are often instrumental in the development of insecticide resistance in insect populations. Thus although the pyrethrins in *Chrysanthemum* were once probably a potent defense, and although they still protect the flowers from many herbivores, they are unable to counter those insects with high microsomal oxidase production. An almost incredible twist to this coevolutionary race, as Whittaker and Feeny (1971) mention, is that perhaps *Chrysanthemum* has delivered a wining blow, a temporary one at least, by manufacturing sesamin, an inhibitor of mixed-function oxidases! And so the step-by-step coevolutionary process continues.

A fascinating example illustrating the complex adaptive biochemical traits acquired by an insect permitting it to use a toxic chemical concerns the bruchid beetle, *Caryedes brasiliensis,* which is a specialist on the neotropical legume, *Dioclea megacarpa,* studied in detail by Rosenthal (1983). Larvae of this weevil feed in the large seeds, which contain the nonprotein amino acid canavanine in high concentrations: up to 13% of the seed dry weight and 55% of all seed nitrogen. Canavanine is highly toxic to the vast majority of organisms because, as an analog of the protein amino acid arginine, it becomes incorporated into polypeptide chains and adversely affects the normal function of all proteins formed of canavanine-containing polypeptide chains. The chemical is therefore a very effective defense against all herbivores because the seed's nitrogen store is locked into a toxic chemical. The escalation of canavanine content in seeds may well have been under the selective pressure from granivores. However, *C. brasiliensis* broke through this chemical barrier and is the only species known to feed on the seeds of *D. megacarpa.*

The major biochemical adaptations acquired by the beetle include the following.

1. Transfer RNA in *C. brasiliensis* discriminates between arginine and canavanine, so that even in the presence of abundant canavanine it is not incorporated into polypeptide chains.
2. Bruchid beetles and other insects produce arginase, involved with hydrolysis of arginine, but the resulting urea, rich in nitrogen, cannot be used as a nitrogen source in most insects because urease, which could convert urea to usable ammonia, is usually absent. In *C. brasiliensis* arginase converts canavanine to canaline and urea, urease converts the urea to carbon dioxide and ammonia, making canavanine a rich nitrogen source for the synthesis of protein amino acids.
3. Arginase and other enzymes are normally strongly inhibited by canavanine and canaline, but in this bruchid beetle such inhibition is not apparent.

4. Canaline is also highly toxic to most insects, so canavanine utilization also necessitates the presence of an enzyme, which *C. brasiliensis* has, that breaks down the canaline to homoserine and ammonia, providing more nitrogen for protein biosynthesis. Whether this enzyme is novel to the beetle or a broad-spectrum enzyme is not yet known, but given the common insecticidal property of canaline, a rare enzyme is implicated.

5. Most terrestrial insects excrete nitrogenous wastes as uric acid, whereas *C. brasiliensis* excretes only small amounts of uric acid but large amounts of ammonia and urea, necessitating further adaptations for dealing with the highly toxic ammonia.

Rosenthal (1983) explains the advantages to the beetle of evolving to exploit *D. megacarpa* seeds.

1. A very rich nitrogen source is acquired.
2. Complete absence of competition from other metazoan granivores results.
3. By becoming a specialist on these seeds the range of toxic chemicals to which the species was exposed declined to a small number that were predictably present. The beetle needed only a minimal detoxification system.
4. Young beetle larvae have safe living quarters because the nutritious walls around them are toxic to most other organisms.
5. Sequestration of canavanine by larvae may well account for the lack of parasitoids attacking *C. brasiliensis*.

The puzzle in this story is how the beetle could acquire so much largely novel genetic information enabling it to exploit such toxic seeds. Rather than a stepwise accrual of genes it seems that a quantum jump in novel genes was required if canavanine was to be used at all, since all five major biochemical adaptations were needed simultaneously. Rosenthal (1983) suggests that the beetle acquired a microbial symbiont already with much of the biochemistry intact, for the microbe would be found typically associated with decaying *D. megacarpa* seeds. By sustaining this mutualist in the gut, a quantum leap in biochemical wizardry was possible. Many bacteria have been isolated which are capable of utilizing canavanine and/or canaline. Another possibility considered by Rosenthal is that the genomic information required by the beetle to exploit these seeds was acquired either from the plant or an adapted bacterium though transfer via a virus, an intriguing possibility that might increase in credibility if amino acid sequencing showed homologies unique to *C. brasiliensis* and *D. megacarpa*.

In addition to the very instructive detail contained in Rosenthal's studies synthesized in his 1983 paper, there are three important points: (1) toxic chemicals are used in primary metabolic pathways by plants, acting as important storage compounds; (2) toxic chemicals in plants have repercussions up the trophic system and may significantly effect the enemies of herbivores, as discussed later in this chapter; and (3) the acquisition of mutualists may be a major step leading to new evolutionary opportunities for both species involved (see Chapter 10).

TEMPORAL AVOIDANCE OF CHEMICALS

Rather than do battle with plant toxins, herbivores may also avoid them. There exists a paradox in an earlier statement about caterpillars feeding on

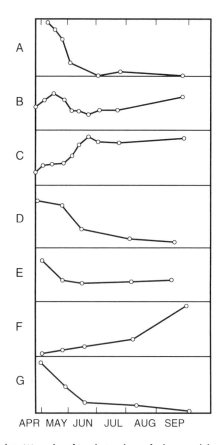

FIGURE 5.3 General trends in (A) number of Lepidoptera larvae feeding on oak leaves (including winter moth larvae), (B) reducing sugar content of oak leaves (percent dry weight as glucose), (C) sucrose content, (D) water content, (E) protein content, (F) tannin content, and (G) protein/tannin ratio of oak leaves. The two variables that correlate well with decline of winter moth numbers are water content and protein content. Feeny suggests that since some Lepidoptera larvae feed on oak leaves late in the season, sufficient water is available for insect growth. Protein availability is reduced by the quantity of nonhydrolysable tannin in the leaf. The protein/tannin ratio gives an estimate of the available protein and just as available protein becomes limiting so the feeding stage of the winter moth is completed. After Feeny (1970).

oak leaves. The winter moth is a serious pest on oaks, yet the tannins in the leaves are very effective in reducing individual growth rates, fecundity, and population growth. The answer to this paradox lies in the temporal avoidance by the winter moth caterpillars of the condensed tannins in oak leaves. Feeny (1970) actually set out to find why winter moth larvae hatch from their eggs so early in the season, sometimes even before leaves are available to feed upon, and with tremendous mortality resulting in the population. Later there is an apparent abundance of food for a period that would allow two generations of winter moth to succeed, although there is only one generation per year. He studied many different factors, and the results, among others, are given in Fig. 5.3. Feeding early in the season makes best use of the protein available, and the selection pressure to feed early is sufficiently strong to modify the entire life history of the moth. These studies explain a general pattern observed by Varley (1967) that all very abundant lepidoptera on oak feed very rapidly on early leaves, and the uncommon species feed on leaves rich in tannins and grow slowly (Fig. 5.4). This pattern of early feeding has

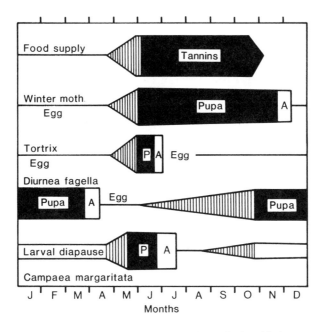

FIGURE 5.4 Relationships between food supply, tannin content in leaves of oak, and feeding patterns of Lepidoptera species on oak. The winter moth and *Tortrix* moth develop rapidly on young leaves low in tannins. *Diurnea fagella* develops slowly on old leaves high in tannin, and *Campaea margaritata* develops slowly in the fall and rapidly in the spring. Larval growth periods are shown in vertical shading. Months are designated by their first letter, and spring in this north temperate climate occurs in April (A) and May (M). From Varley (1967); with permission.

also been noticed by Drooz (1971) in the elm spanworm, and by Futuyma and Gould (1979) and Futuyma and Wasserman (1980) in other species.

Avoidance of toxins is also possible for other insects. Numerous insects such as some aphids (e.g., Way and Cammell, 1970) feed on dying foliage and so avoid toxic compounds (as well as for nutritional reasons, cf. Fig. 17.5) or other defensive strategies on the part of the plant. The very fine piercing mouthparts of the true bugs, Hemiptera and Homoptera, enable them to feed between pockets or ducts of toxin in the host plant and thus achieve a spatial avoidance.

USE OF SECONDARY METABOLIC PRODUCTS BY INSECTS

Secondary metabolic products can be utilized by insects as cues to identify the plant for feeding and breeding purposes once the insect species has evolved a mechanism for tolerating or detoxifying the plant's defense (Fraenkel, 1959, 1969). If an insect species is able to utilize a toxic plant for food, by some physiological adjustment, the advantages are at least fourfold, as pointed out for the earlier example on *C. brasiliensis*:

1. The herbivore gains a source of food that cannot be utilized by any other herbivore, or very few other herbivores, and competition for food is minimized.
2. This food is very easily recognized by its secondary compound label (Fig. 5.5).
3. Feeding on this food may also impart a toxic or unpalatable characteristic to the herbivore (e.g., Brower and Brower, 1964; Malcolm, 1991). Predation pressure on this herbivore will therefore be reduced.
4. The antibiotic properties of many toxic chemicals may protect the herbivore against pathogens. (For example, Frings et al., 1948, demonstrated the antibacterial action in the blood of the large milkweed bug, which ingests heart poisons in its diet.)

Eisner et al. (1962) suggested the possibility that a predaceous cerambycid beetle has taken this process one step further, so that the predator gains at least temporary protection by eating an "unpalatable" prey—unpalatable, that is, to most other predators. The tachinid parasite, *Zenillia adamsoni,* of monarch butterfly larvae, which feeds on poisonous milkweeds, also contains active poisons in its body (Reichstein et al., 1968), as does *Zenillia longicauda,* which feeds on lepidopterous larvae in the genus *Zygaena* which feed on cyanogenic plants such as *Lotus corniculatus* (Jones et al., 1962). Other parasitoids, such as *Apanteles zygaenaeum,* on *Zygaena* contained the enzyme rhodanese active in the detoxification of cyanogenic compounds (Jones, 1966).

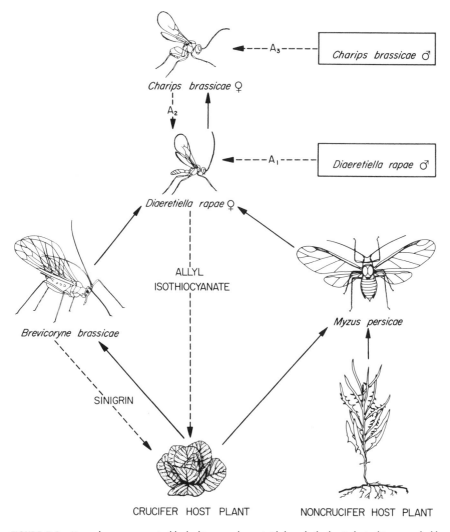

Charips brassicae ♂

Charips brassicae ♀

A₂

Diaeretiella rapae ♂

Diaeretiella rapae ♀

ALLYL
ISOTHIOCYANATE

Brevicoryne brassicae

Myzus persicae

SINIGRIN

CRUCIFER HOST PLANT   NONCRUCIFER HOST PLANT

FIGURE 5.5   Many plants are recognized by herbivores and parasitoids by volatile chemicals. In this case studied by Read et al. (1970) the integrity of the community is maintained by chemical links. Sinigrin from the host plant attracts the specialized aphid *Brevicoryne brassicae*, whereas a closely related compound, allyl isothiocyanate, attracts the primary parasitoid *Diaeretiella rapae*. From Price (1984c); with permission.

Some plant products are merely eaten by the herbivore, stored, and regurgitated in defense against predators. These are not really secretions but sequestered substances. Eisner et al. (1974) show that a diprionid sawfly uses the resin it obtains from pine needles, stores it in diverticula of the gut, and regurgitates it as a sticky blob at the mouth. This is distasteful to vertebrate predators such as chickadees and finches and gums up the mouthparts of invertebrate predators and the ovipositors of hymenopterous parasites.

Eisner et al. (1971) discovered that incorporated in the defensive secretion of the grasshopper, *Romalea microptera,* there is a repellent herbicide derivative presumably obtained by feeding on plants sprayed with herbicide. This surely indicates that insects may use to their advantage any chemical that is available, manufactured or natural.

These examples illustrate that plant qualities frequently do have impact right up the food chain (Price et al., 1980; Price, 1981a; Rowell-Rahier and Pasteels, 1992). The complexity of chemically mediated interactions in a community is illustrated in Fig. 5.6, which summarizes the ecology of body odor

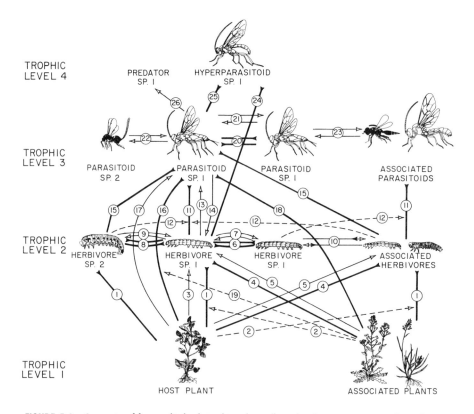

FIGURE 5.6   Community of four trophic levels involving chemically mediated interactions—semiochemicals. Arrows are placed against the responding organism. Thick solid lines and solid arrows illustrate attraction to a stimulus (e.g., 1, 4, 11, and 24). Thin solid lines and open arrows illustrate repulsion (e.g., 3, 13, 17, and 26). Thin dashed lines show indirect effects such as interference with another response (e.g., 2, 12, and 19). All interactions are discussed in Price (1981a). From Price (1981a); with permission.

for terrestrial insects. Body odor of plants and insects plays an important role in nature.

In interaction 1 (Fig. 5.6) herbivores are attracted to the host plant by volatile chemicals, and parasitoids may respond to similar compounds (interaction 16), as in Fig. 5.5. But toxic chemicals acquired from the host plant by the herbivore may be effective as defense against enemies (interaction 13). For example, Campbell and Duffey (1979) showed that increased tomatine levels in tomato plants were more deleterious to the parasitoid *Hyposoter exiguae* than to its host, *Heleothis zea,* feeding on tomato (interaction 3 was less potent than interaction 13). But when toxic chemicals are sequestered by parasitoids, as in the *Zenillia* species mentioned above, they become effective defenses against the fourth trophic level (interaction 26). As we will see later (Chapter 22), associated plants may interfere with host finding by herbivores (interaction 2) and by enemies of herbivores (interaction 19). All interactions in Fig. 5.6 are explained fully with examples in Price (1981a).

Physical characteristics of plants also mediate the herbivore–enemy interaction (Price et al., 1980). For example, trichomes on leaves may slow the searching rate of predators and parasitoids to the point where enemies become ineffective. Galls that grow relatively large provide more protection against the herbivores' enemies than do smaller galls, and galls with extrafloral nectaries attract foraging ants which interfere with attack by parasitic wasps (Washburn, 1984).

MONOPHAGY AND POLYPHAGY

If it is so advantageous to specialize in feeding on toxic plants, why are some species **monophagous** (feeding on one species of food) and others **polyphagous** (feeding on many species of food)? Also, why do some species feed on apparently innocuous hosts? There are many solutions to any given problem in nature and the evolutionary process ensures that many possible avenues are tested (Bernays and Chapman, 1994). Some species have never had the evolutionary chance to crack potent plant defenses because the gene pool has not produced the necessary combination of enzymes. Others have evolved to specialize by dealing with a few potent plant chemicals. Yet others evolved to generalize by utilization of many less toxic plants. These seem to be the two basic alternatives and both require quite an expensive metabolic commitment on the part of the insect. The monophagous species must produce large quantities of an enzyme to detoxify their food, or they must evolve storage mechanisms, as in the case of diprionid sawfly larvae. Conversely, Krieger et al. (1971) have shown that the polyphagous species of insects usually produce much more of the microsomal mixed-function oxidases in their midguts to deal with the very diverse array of plant chemicals in their potential diet.

Regarding the search for food, the monophagous species will have to search harder, but it can become highly specialized in its search and it has those chemical markers to cue in on. The polyphagous species has a more abundant food source, but its chemical cues are less distinctive, or produced in lesser amounts, and the insect cannot afford to specialize on any one of the variety of chemicals in its diet. The monophagous species will usually have to share its food source with few other herbivores, whereas the polyphagous species may be faced with many potential competitors. There are no outright winners in nature and every avenue for adaptive gain sets rather precise limits on how far this advantage can be exploited.

It becomes clear that if either the monophagous or polyphagous feeding strategies are metabolically costly, the cost of gaining this energy must reduce the amount of energy that can be channeled into reproduction. That is, even the food source of herbivores contains an environmental resistance factor in the form of toxic chemicals. Presumably, insects fed on a bland (nontoxic) diet would show their real biotic potential and those fed on natural foods would show their biotic potential minus the environmental resistance contained in food (rather like Feeny did with winter moths on artificial diets). In the comparison of the growth parameters r and r_{max}, this factor should be calculated (see Chapter 13).

If several evolutionary lines of insects adapt to feeding on a toxic plant, and use these toxins as cues for their behavior, as frequently happens, the result will be a unique insect fauna on each set of plants that are chemically related, a well-defined component community. This makes the study of the insect fauna of poisonous plant families very interesting and valuable for understanding community organizations (see Chapter 22).

APOSEMATIC COLORATION

The family Asclepiadaceae, the milkweeds, contains some species that manufacture cardiac glycosides which are emetic when ingested by vertebrate predators at half their lethal dosage (e.g., Brower, 1969). This is an important point for the herbivorous insect that sequesters these glycosides because if the predator is killed, there is no chance of the predator population learning to avoid feeding on an **aposematic species** (one that advertises its distasteful nature by being brightly colored; cf. Guilford, 1990). There is only a small group of insects in four insect orders that feed consistently on *Asclepias* in temperate North America, and they are found almost exclusively on milkweed (Table 5.1 and Fig. 5.7). All but one are aposematically colored, suggesting that they all are chemically protected. Contrary to this expectation, Brower (1969) has claimed that monarch butterflies that have fed in northeastern North America are palatable because their food plants lack cardiac glycosides. However, Duffey (1970) showed that the species of milkweed mentioned by Brower do contain glycosides.

TABLE 5.1 Herbivorous Insect Community on Milkweeds (Asclepiadaceae) in Temperate North America which are Mostly Aposematic Species. The One Exception is Placed in Parentheses (Personal Observations). Other Insects Feed Sporadically on Milkweeds but are Not Found Predominantly on These Plants.

Order	Family	Species	Coloring
Coleoptera	Cerambycidae	*Tetraopes tetraophthalmus*	Red with black spots
		Tetraopes femoratus	Red with black spots
		Tetraopes quinquemaculatus	Red with black spots
	Chrysomelidea	*Labidomera clivicollis*	Red and black pattern
	(Curculionidae	*Rhyssomatus lineaticollis*	Black)
Lepidoptera	Danaidae	*Danaus plexippus*	Adult—orange and black
			Larva—black, yellow, and white stripes
	Arctiidae	*Euchaetias egle*	Adult and larva—white, yellow, and black
		Cycnia inopinatus	Adult—white, yellow, and black
			Larva—orange with grey hairs
Hemiptera	Lygaeidae	*Oncopeltus fasciatus*	Red and black
		Lygaeus kalmii	Red and black
Homoptera	Aphididae	*Aphis nerii*	Yellow and black

PALATABILITY SPECTRUM

It now appears that there is a great variation in glycoside content in monarch butterfly populations, from zero to concentrations sufficient to cause emesis (Brower et al., 1972). Thus there exists a **palatability spectrum** for the butterfly's predators (Brower et al., 1968; Malcolm, 1991). Scudder and Duffey (1972) (see also Feir and Suen, 1971) found cardiac glycosides in aposematic lygaeids feeding on milkweeds, and these chemicals were also incorporated into the insects' defensive secretion. Beetles, such as *Tetraopes* spp., that are specific to milkweeds also contain glycosides (Duffey and Scudder, 1972). We may infer from the studies above that the majority of milkweeds of the genus *Asclepias* contain cardiac glycosides and that most insects that feed on these plants are protected by ingesting and sequestering these toxic chemicals.

COUNTERATTACK BY PLANTS

The production of sesamin in the pyrethrum flower that inhibits mixed-function oxidase activity has been mentioned. A more extensive set of

FIGURE 5.7 Two members of the milkweed fauna: (*a*) large milkweed bug, *Oncopeltus fasciatus;* (*b*) milkweed long-horn beetle, *Tetraopes femoratus,* on swamp milkweed, *Asclepias incarnata.* Both species are red on the lighter areas of the body and black on the remainder. Thus the insects contrast strikingly with the green leaves of the host plant and they sit prominently on upper foliage, a behavior that differs considerably from that of cryptic species. Drawing by Alice Prickett.

counter-responses may be seen in leguminous plants (Fabaceae) presumably in response to herbivore pressure by bruchid (pea and bean) weevils that oviposit on legume pods. A summary provided by Janzen (1969a) and Center and Johnson (1974) covers physical and chemical defense:

1. Some species produce gum when the seed pod is first penetrated by a newly hatched larva—this may push off the egg mass, or drown the larvae, or hamper their movements.
2. Pods may dehisce, fragment, or explode, scattering seeds to escape from larvae coming through the pod walls and from ovipositing females.
3. Some species have pods free of surface cracks, as some bruchids cannot glue eggs on a smooth surface.
4. Some species have indehiscent pods and thus exclude those species that oviposit only on exposed seeds.
5. Some species have a layer of material on the seed surface that swells when the pod opens and detaches the attached eggs.
6. Many species have poisonous or hallucinogenic compounds such as alkaloids, saponins, pentose sugars, and free amino acids (primary response).
7. Some are rich in endopeptidase inhibitors, making digestion of the bean by the bruchid very difficult.
8. Some have a flaking pod surface that may remove eggs laid on it.
9. In *Acacia* spp. the immature seeds remain small throughout the year and abruptly grow to maturity just before being dispersed.
10. In a species of *Cassia* the seeds are too thin to allow a bruchid to mature.
11. In many wild herbaceous legumes the seeds are so small that bruchids cannot mature in them. Janzen (1969a), Mitchell (1977), Johnson and Slobodchikoff (1979), and Johnson (1981) consider further examples.

As seen in the list above, one method of avoiding predation is to disperse seeds or fruits rapidly and extensively, with the resultant greater difficulty in discovering food for species that are seed specialists. The multitude of seed-dispersal mechanisms among plants obviously accomplish this aim. Adaptations for wind dispersal and for attaching to animals take many forms. Animals carry fruits and seeds in the mouth or gut and feed on the palatable part of the fruit or seed coat, and the seed is dispersed and may be buried by small mammals, harvester ants, and wood ants, or seed-caching beetles (see Sudd, 1967; Harper et al., 1970; Manley, 1971; Kirk, 1972, 1973; Alcock, 1973a). Many seeds have oily coats or appendages (caruncles and elaiosomes) that are eaten by ants and other insects (Sudd, 1967; Davidson and Morton, 1981; Buckley, 1982; Beattie, 1985; Huxley and Cutler, 1991), while the seed remains intact and is dispersed. Some seeds improved germination after passing through the gut of animals (Krefting and Roe, 1949; Janzen and Martin,

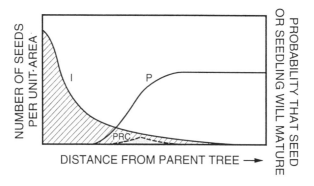

FIGURE 5.8   Relationship between density of seeds (*I*) and probability of escaping from a seed or seedling predator (*P*) relative to the distance of seeds from the parent tree. The product of *I* and *P* values gives the population recruitment curve (*PRC*), where progeny are most likely to survive at some distance from the parent. From D. H. Janzen, Herbivores and the number of tree species in tropical forests, *Am. Nat.*, 1970, **104**:501–528. Published by the University of Chicago Press. Copyright © 1970 by the University of Chicago. All rights reserved.

1982; Janzen, 1986b). This advantage can be obtained at the cost of producing a palatable fruit.

The further the seeds are dispersed, the less rewarding will predation be on these seeds, and a pattern such as that proposed by Janzen (1970) may be observed that relates seed density to the probability of survival of seed (Fig. 5.8). There is a minimum distance between parent and progeny where survival is likely. The population recruitment curve is clearly the product of seed density and probability of survival. Thus if a species does not evolve an effective dispersal mechanism to make seeds unprofitable, the only alternatives are to make them unpalatable, or very small.

PLANT APPARENCY HYPOTHESIS

Two basically different kinds of plant defense by chemicals have been mentioned in this chapter and have been recognized by Feeny (1975, 1976, 1992) and Rhoades and Cates (1976) (Rhoades, 1979). Feeny's (1976) terminology and arguments are used here. Oak leaves were seen to contain 2.4% tannins in fresh weight when old and the percentage may reach even higher levels of 5% or more. The defensive effect of tannins was dosage dependent, so as concentration of tannins increased, inhibition of larval growth increased: hence the relatively high percentage of digestibility-reducing chemicals in old oak leaves. Tannins act as **quantitative** (dosage-dependent) **barriers** to insects and are effective on insects that habitually feed on plants rich in tannins.

By contrast, the major chemical defenses of crucifers are mustard oils, which occur at relatively low concentrations. Allylglucosinolate in *Brassica nigra* represented only 0.18–0.66% of leaf fresh weight, but this is enough to reduce the fitness of nonspecific herbivores significantly or to kill them (Erickson and Feeny, 1974). However, for herbivores specialized to crucifers the chemicals seem to have no inhibitory effect on growth and the effect is not dosage dependent (Slansky and Feeny, 1977). Whereas tannins seem to be effective against all insects, mustard oils are effective only against nonadapted herbivores, while some species have adapted to detoxify them. Chemicals such as mustard oils therefore act as **qualitative barriers** to herbivore attack.

These contrasting modes of defense appear to be typical of plants in different stages of ecological succession. **Quantitative defenses,** which are effective against all herbivores, will be effective in late successional plants, which tend to exist in pure stands, or stands of relatively low diversity, for long periods of time. These plants are "bound to be found" by herbivores in ecological time, or they are **apparent** to herbivores, so that relatively heavy investment in chemical defense, which is more or less universally effective, is highly adaptive. With long generation times the high costs of such defense are not prohibitive since energy for reproduction can be accrued over several growing seasons. Long generation times also seriously limit the value of qualitative defenses since specialists with very short generation times and very large population sizes relative to the host plant can counteradapt rapidly and so track closely new defenses in the host population. **Qualitative defenses** could not be effective against specialized herbivores, but such compounds are typical of plants in early stages of ecological succession. These plants, as many of the crucifers, are short-lived, colonizing species, adapted to fast growth, early maturity, and effective dispersal either in space or time. Thus the most effective defense against specialists appears to be escape in space and time. These plants are relatively hard to find or **unapparent** to specialists, but chemically protected against other herbivores that happen upon them. The relatively low investment in defense detracts little from rapid growth and reproduction, so important to early successional species, although examples are known in which morphs producing toxins have lower growth rates and reproductive capacity than morphs without (Jones, 1966; Cates, 1975) as mentioned in the section "Allocation of protective chemicals" earlier in this chapter. Generation times of host plant and herbivore are similar, so an adaptive gain in a host plant can spread rapidly in the population before herbivores adapt to cope with the new defense.

Feeny (1976) makes the point that the chemical defense pattern seen in ecological succession has important effects on the life histories of herbivores. In early succession much foliage is highly nutritious but qualitatively protected. Herbivores must therefore adapt with specific enzymes to detoxify the qualitative defenses, and sophisticated methods for host–plant location and identification, but once discovered such plants permit rapid herbivore growth and reproduction. The ephemeral plants thus impose, or permit, an ephemeral *r*-selected life history (see Chapter 14) on the insects that exploit them. Such brevity, patchiness, and unpredictability in plants and herbivores makes exploitation of herbivores by specialized predators and parasites very difficult.

In late succession, foliage is quantitatively defended and growth of herbivores is slow (Figs. 5.1 and 5.4), making them vulnerable to their enemies. Selective pressures for defense in herbivores will be high, making them more K-selected (see Chapter 14) than those in early stages of succession. Apparency of herbivores to their enemies is affected by plant apparency and the related defense mechanisms, a three-trophic-level interaction discussed earlier in this chapter. Feeny makes the comparison between butterflies: the tiger swallowtail, *Papilio glaucus,* caterpillars of which are tree-feeding and slow-growing, and *Papilio polyxenes,* the black swallowtail, whose larvae feed on herbs and develop rapidly. The former species shows several antienemy adaptations lacking in its relative:

1. Larvae mimic tree snakes, which predatory warblers may have learned to avoid while overwintering in Central or South America.
2. They construct a leaf roll into which they retreat.
3. They chew off petioles of partly consumed leaves, and thus remove potential body odor signals to predators or parasitoids, although the real adaptive value of the leaf-clipping behavior needs detailed investigation (Scriber, 1996).

Parasitoids and other carnivores use plant damage, mandibular gland secretions, and chemicals in frass for finding hosts (Vinson, 1975; Weseloch, 1981; Dicke et al., 1990c; Price, 1993) so removal of potential sources of kairomones can be highly adaptive.

ROLE OF TANNINS

The recognition of these patterns has been important in the development of understanding plant–herbivore interactions, but the situation is not as clear as it first seemed. Fox and Macauley (1977) found that *Eucalyptus* species in Australia were heavily attacked by insect foliovores even though condensed tannin concentration in leaves reached 4, 7, and 12% dry weight in three species, and tannin concentration had no effect on nitrogen use efficiency. Tannins were not acting as quantitative defenses in these climax forest species. Even young leaves contained very high condensed tannin concentrations, reaching 33% leaf dry weight in one tree of *Eucalyptus mannifera,* with considerable intraspecific variation (Macauley and Fox, 1980), so there was no temporal escape available to herbivores. Experiments using about 20% tannin in grasshopper diets (Bernays, 1978) also revealed some problems. In four species of grasshopper, 20% condensed tannin in their diet had no effect on survival, growth, consumption, digestion, and efficiency of conversion of digested food. Only in one species did 20% hydrolysable tannin have an adverse effect, causing damage to the epithelium of the midgut. Clearly, some herbivores can utilize food effectively with high concentrations of tannins, which cannot therefore be regarded as very general quantitative defenses. Bernays (1978) concluded that insects adapted to tannins early in their evolution and that tannins in plants exist primarily for other reasons (see also Swain, 1979). Bernays (1978, 1981) also found the action of tannic acid, a hydrolysable tannin, to be as a toxin rather than as a digestibility reducer, and pointed out that some hydrolyzable tannins act as feeding stimulants for some species. Using the argument that structure determines function, Zucker (1983) suggested that the very similar structures found in condensed tannins may well result in their function as general-purpose defenses, but hydrolysable tannins show enormous diversity in structure and are likely to be highly specific defenses, each tannin binding with a specific protein. Thus overall tannin content tells us nothing about their impact on any specified herbivore. In their review of insect nutritional ecology, Scriber and Slansky (1981) show that tree leaves generally have lower water and nitrogen content than herbaceous plants at the same time in the year, and that concentrations decline rapidly during the season. So water and nitrogen content of food may be as important as chemical defense. Nevertheless, elements of the general pattern recognized by Feeny (1976) and Rhoades and Cates (1976) appear to be robust, for the highest relative growth rates in insects have been found in forb-foliage-chewing Lepidoptera and the lowest in tree-foliage-chewing Coleoptera (Scriber and Slansky, 1981). The developing literature on this subject will be interesting to follow (cf. Tabashnik and Slansky, 1987; Mattson and Scriber, 1987).

The fact remains that plants show fundamentally different dispersion patterns and contain different kinds of chemicals, suggesting that herbivores need different kinds of host selection and foraging strategies. Kogan (1977) has categorized these into six general types, four of which are discussed here.

HOST DISCOVERY BY INSECTS

Early successional plants with qualitative defenses offer specific cues to herbivores which can search the chemical maze for a specific odor—the plants' body odor. Two of Kogan's models cover this situation. One is *Model IV: Host-finding for oviposition is directional.* Adult females fly upwind along odor gradients to a host plant. Contact chemoreceptors permit effective discrimination between plants, so eggs are laid on a suitable host plant. Larvae are less discriminating and may feed on many plants that lack feeding deterrents. Insects on the Solanaceae are typical of this pattern, such as *Leptinotarsa decemlineata, Manduca sexta,* and *Lema trilineata.* Ovipositing females are highly specific to plants in the Solanaceae or to groups within this family characterized by a certain type of alkaloid. *Lema trilineata* oviposits on species containing tropane alkaloids in the genera *Datura, Physalis, Atropa,* and others. However, larvae may feed on many plant families. First instar larvae of *Manduca sexta* will accept plants in the Scrophulariaceae, Brassicaceae or Cruciferae, Moraceae, Plantaginaceae, and Fabaceae or Leguminoseae (Yamamoto, 1974). Specificity of the species seems to be maintained solely by the importance of use by the female of a narrow band of chemical cues for finding and accepting suitable host plants.

Kogan's *Model V: Host-finding for oviposition and feeding are highly selective* relates to a similar situation. Here the biochemical basis for host–plant selection is the same for adults and larvae. Specific attractants to adults, oviposition excitants, and feeding excitants are present in the host plants. The herbivores of the Brassicaceae or Cruciferae respond to such cues, the mustard oils acting as attractants, oviposition excitants, and feeding stimulants to specific herbivores, but repellents to nonspecific insects. Kogan notes that success of this host-finding strategy is reflected by the wide variety of insects that are specialists on crucifers, representatives being in the Diptera, Lepidoptera, Coleoptera, and Homoptera.

Herbivores in climax communities will often lack these distinctive chemical cues for searching, and they may even be faced with a massive convergence of characteristics that, primarily or secondarily, play a role in defense: high tannin content, low nitrogen or other nutrients, low water, and high lignin and toughness. Models IV and V cannot be effective under these circumstances, but the searching problem is less severe. Suitable host plants are usually abundant where the adult emerges and the few alternative plants for oviposition may be rapidly tested by contact chemoreception within a short time span. If necessary, long-distance host location may be by visual cues such as vertical shapes or deep shade. Some chemicals may act as attractants at short range. This pattern seems to apply to the apple maggot, *Rhagoletis pomonella* (Prokopy, 1968; Prokopy et al., 1973; Prokopy and Owens, 1978), and possibly bark beetles.

An alternative host-searching strategy is covered by Kogan's model I, in which orientation to food plants is triggered by generalized mechanisms such as phototaxis and anemotaxis. Plant recognition is usually by contact chemoreception, and host acceptance results from the presence of stimulants common to plants such as sucrose and lipids and the absence of extremely potent deterrents. Such species tend to be polyphagous, as are some grasshoppers in the family Acrididae. Acrid grasshoppers may be found in early successional communities, but they reach their highest numbers in climax grassland of apparent plants. Nymphs or adults may move randomly or move upwind. On landing, an individual will test-bite a plant and accept or reject it on this basis.

It is interesting to note that where the herbivores in late succession must be quite selective in hosts, for example, as in many bark beetles which can be successful initially only in physiologically stressed hosts, searching again becomes difficult and the strategy shifts toward that in early successional herbivores. Kogan's model III describes this situation where adult searching is more or less random, but successful colonists emit an aggregation pheromone (often derived in part from host–plant chemicals) which results in mass attack. Host-plant searching becomes highly directional and the original attackers benefit from the reduced resistance of the host tree.

SPECIALIZATION BY HERBIVORES

The pattern of defenses in plants proposed by Feeny and the models of host-plant selection defined by Kogan suggest that early successional herbivores are likely to be more specialized, on average, than late successional herbivores (and perhaps their parasitoids will follow the same trends). This has been found by Futuyma (1976) for Lepidoptera in North America and the British Isles. From various sources he calculated the number of species in various lepidopteran taxa that fed only on members in one family of plants, and the number that fed on more than one family, both in relation to whether the hosts were woody or herbaceous. For each taxon studied he could therefore construct a 2×2 contingency table of "on one family" versus "on more than one family" against "woody plants" versus "herbaceous plants." A large majority of herbaceous plants were early successional, so the comparison is between the degree of specialization in early versus late successional herbivores. In the seven cases tested, specialists were more common than expected in early successional plants and less common than expected in late successional plants. Two such relationships are given in Table 5.2.

TABLE 5.2   Numbers of Lepidopteran Species Feeding on One or More Than One Family of Vascular Plants.

	Hosts			
	Woody		*Herbaceous*	
	Observed	*Expected*	*Observed*	*Expected*
A. Species of Butterflies—Eastern United States				
On one family	36	43	97	90
On greater than one family	17	10	13	20
Percent on one family	68%	82%	88%	82%
B. Species of Moths—British Isles				
On one family	119	130	96	85
On greater than one family	100	89	47	58
Percent on one family	54%	59%	69%	59%

SOURCE: After Futuyma (1976).

Futuyma (1976) suggested that the most probable explanation for this pattern was because chemical defenses were more diverse among early successional species than late successional species, just as Feeny (1976) has noted. A comparison of 13 plant families which were primarily woody and 13 primarily herbaceous showed that the eight major types of chemical defenses were represented more frequently in herbaceous families, except for the phenolics, which include the tannins. A list of defensive chemical types and the percent of families in the total 13 families that contained these chemicals for woody plants and herbaceous plants, in that order, are: phenolics 85%, 38%; quinones 38%, 69%; saponins 8%, 46%; alkaloids 23%, 77%; cyanogenic glycosides 23%, 38%; coumarin glycosides 23%, 38%; acetylenes 0%, 23%; and sulfur compounds 0%, 23%.

Although this pattern supports the general trends anticipated by the plant apparency hypothesis, there are clearly many species that are quite specialized on woody plants (68% in Table 5.2), so clearly there are factors selecting for a narrow diet even though tannins are a very general defense among woody species (Bernays and Chapman, 1994). These factors may involve both toxic chemicals, since as the list in the table shows, many woody plants contain toxins and high-specificity hydrolysable tannins, as suggested by Zucker (1983).

The implications of the plant apparency hypothesis for agriculture are important. We have tended to use in agriculture plants that are unapparent in their wild state, such as crucifers (Brassicaceae, e.g., cabbage), Solanaceae (e.g., potatoes and tobacco), Asteraceae or Compositae (e.g., lettuce), Apiaceae or Umbelliferae (e.g., carrots), and Poaceae or Graminae (e.g., wheat and corn). This is obviously because short life cycles yield rapid returns in food and the plants are nutritious, succulent, and tasty, just as they are for insects. Thus we grow unapparent plants in very apparent monocultures where quantitative defenses would be much more effective. Humans have compensated for the plant's inadequate defenses by application of immense doses of pesticides. Any cultural method that reduces crop patch size or increases species diversity in the crop would reduce plant apparency and thus the colonization rate of herbivores (e.g., Root, 1973; Risch, 1981; see also Chapter 22). Crop rotation increases diversity in time. Genetic diversity within a crop, each genotype with a different defense, would also reduce plant apparancy (e.g., Burdon, 1978, 1987; Fritz and Simms, 1992). Some quantitative defenses may also be selected for in agricultural crops. Varieties with low nutrient content, high tannins or silica, or high toughness may be selected for as long as the well-defended plant parts were not those consumed by humans or livestock. However, if nutritional imbalances were bred into crops, these would slow or halt development of pests, but we could correct the imbalance after harvest during processing. For example, the sodium content of a crop might be reduced without impairing production while making it a limiting factor for pest insects.

Implications for biological control and the basic understanding of predator and parasitoid biologies are also extensive. The plant defensive strategies make the quality of herbivores very different in early versus later succession, and thus predators and parasitoids are faced with basically different sorts of prey and hosts to exploit. For example, parasitoids in early successional communities are frequently faced with chemically protected hosts to which they must adapt, probably involving specialization to sequestered toxins such as alkaloids (see the section "Use of secondary metabolic products by insects" for examples and the text relating to Fig. 5.6). Parasitoids and predators of late successional herbivores are probably faced with a biochemically less diverse array of hosts and may be more generalized. This may also result in more species attacking each host in late succession than early succession, a trend that appears to be supported in the literature (Hawkins, 1988; Hawkins et al., 1990; Price, 1991a,b). Because of the characteristics of insects on unapparent plants, biological control may be hard to achieve, accounting in part for the relatively low success rate in agricultural systems (Price, 1981b). The augmentation of natural enemies (Ridgway and Vinson, 1977; Nordlund et al., 1981) and breeding plants to favor natural enemies offer compensating alternatives in cultivated areas.

SEMIOCHEMICALS

A multitude of important messages are communicated in the ecology of body odor. Any chemical involved in the chemical interaction between organisms is called a **semiochemical,** a term first proposed by Law and Regnier (1971), derived from the Greek *simeon,* meaning a mark or signal. (For a discussion of all terms used below, see Nordlund and Lewis, 1976; Nordlund, 1981). Semiochemicals are divided into **pheromones,** which are intraspecific communication chemicals, and **allelochemics,** which play a role in interspecific communication.

Allelochemics are further divided into **allomones,** which are largely defensive chemicals, and **kairomones,** which are emitted by one trophic level but utilized by a higher trophic level, usually for discovery of food items; it is favorable to the receiver but not to the emitter. Nordlund and Lewis (1976) also propose the terms *synomone* and *apneumone* for additional types of interaction. A **synomone** is adaptively advantageous to both emitter and receiver; for example, a plant emitting a chemical cue that is utilized by a predator or parasitoid to find its food on the plant also benefits by this behavior. An **apneumone** is a chemical released by nonliving substances that are beneficial to the receiver but detrimental to another organism in the substance. For example, a parasitoid such as *Biosteres longicaudatus* finds its fruit fly hosts by responding to fermentation products from rotting fruits, particularly acetaldehyde (Greany et al., 1977). Figure 5.6 illustrates most of these interaction types.

It is not clear if all these terms will become generally accepted by the scientific community. Blum (1977; see also Roitberg and Prokopy, 1987) argues that most chemicals are probably adaptive to the emitter in the form of pheromones and allomones, and by breaking the code the enemies have evolved to use them as kairomones. Therefore, Blum argues that chemicals should be named only according to their primary adaptive function for the emitter, while recognizing that reception may elicit a diverse array of responses according to which species decodes the message. However, primary adaptive functions are not always easy to determine, as discussed earlier in the section "Primary and secondary metabolic products."

SELECTIVE PRESSURE FOR PLANT DEFENSE

The selection pressures on plants to avoid herbivore attack may be difficult to observe, but sometimes plant populations being dramatically influenced by herbivores can be seen. These instances may be rare, as heavy herbivore pressure will produce the strong selection pressure that causes a new defense in the plant. A stable equilibrium may follow for many years, and we may be led to assume that the herbivore is innocuous. The plant may also lose vigor when developing the new defense, as during production of cyanogenic chemicals in *Lotus corniculatus*. A price must be paid for extra defense, and it would be valuable to know some quantitative estimates. A large percentage would go into resin production in pines or latex in milkweeds, but in these cases it is hard to know how to apportion the costs between protection and storage of useful chemicals and waste products. Many studies are needed to elucidate the costs of coevolutionary processes.

Some examples of heavy herbivore pressure may help appreciation of the possible brevity of the selection process. In 1966 a European flea beetle, *Psylloides napi,* that fed on yellow rocket, *Barbarea vulgaris,* was discovered new to New York State. Adults fed on leaves, while larvae burrowed in the petioles. By 1969 it had reached infestation proportions and was killing off entire local populations of its host and exerting a strong selective pressure. The episode would probably have gone unnoticed but for a detailed study on crucifer communities being undertaken by Root, Tahvanainen, and others at that time (Tahvanainen and Root, 1970). Now it appears that *P. napi* may act as a valuable biological control agent on yellow rocket.

Only a few reports of an apparently rare sawfly, *Monostegia abdominalis,* existed on its occurrence in North America before 1970. Then it was found in high numbers in remote areas of Quebec, killing relatively large areas of a loosestrife, *Lysimachia terrestris* (Price, 1970a).

Perhaps the most dramatic example of herbivore pressure has been witnessed in the control of Klamath weed by the introduced beetle, *Chrysolina quadrigemina.* (For an account of this, see Huffaker, 1967; Huffaker and

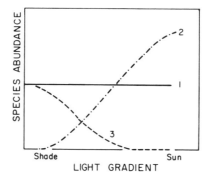

FIGURE 5.9 Schematic representation of abundance of plant host, *Hypericum,* and herbivore, *Chrysolina,* on a light-intensity gradient from shade to full sunlight: (1) *Hypericum* before introduction of *Chrysolina;* (2) *Chrysolina* on *Hypericum* soon after introduction; (3) resultant distribution of *Hypericum* with *Chrysolina* rare throughout the gradient.

Kennett, 1969.) Selection of plants by the beetles was not random but concentrated on plants in unshaded spots, so the pattern seen in Fig. 5.9 has emerged (see also Harper, 1969). *Chrysolina* populations were so high in exposed sites that *Hypericum* populations crashed, and *Chrysolina* numbers fell, too, because of lack of food. Because this leaf beetle cannot breed well in the shade, it might be concluded that (1) *Hypericum* is a shade-adapted species and (2) *Chrysolina* is a rare insect. Both conclusions are very misleading. The question should be posed, "Could the ecology of these species be properly understood if this interaction had not been witnessed?" and also, "How many other plant species have distributions dictated by herbivore pressure?" Increasing sensitivity in the ecological community to this question will no doubt yield more examples. For example, Parker and Root (1981) showed that distribution of the composite *Machaeranthera canescens* is limited by the grasshopper *Hesperotettix viridis* in New Mexico.

Where this sort of herbivore pressure acts on some populations and not on others, the process of evolution is accelerated tremendously, specialization may result, and both plant species diversity and herbivore diversity, and thus biotic diversity in general, must be strongly influenced by the coevolutionary process (see also Gillett, 1962; Ehrlich and Raven, 1964; Krieger et al., 1971).

MUTUALISTIC HERBIVORES

We have paid much less attention to herbivores that benefit plants than to the antagonistic relationship, a bias common in ecology which should be corrected as soon as possible (see Chapter 10). However, there is a growing literature asserting that at least some herbivores may well be beneficial to the plants they feed on. Owen and Wiegert (1976) point out that the activity of herbivores can increase conservation of nutrients by causing leaf fall over a prolonged season, and honeydew producers may increase nitrogen fixation beneath the plant. Owen (1980b) amplifies this theme and provides many suggestive references. Detling and Dyer (1981) cite several references indicating that herbivores contain plant growth regulators in relatively high concentrations. Although they found reduced biomass production after grasshopper feeding, Dyer and Bokhari (1976) had found enhanced plant regrowth, suggesting that growth regulation is variable depending on the plant and herbivore involved. Owen and Wiegert (1981, 1982) encourage the view that there exists a mutualism between grasses and grazers, although there is some dissension from this point (Herrara, 1982; Silvertown, 1982). Considerable interest is being focused now on insect herbivores as regulators in ecosystems, to be discussed in Chapter 12, influencing plant production and nutrient cycling. Few of these studies have measured impact on plant fitness, and beneficial impact on biomass production does not necessarily mean improved fitness and selection toward a mutualistic relationship between herbivore and plant. However, it is clear that some herbivores, such as yucca moths and ants on bull's-horn acacias, are mutualistic with their host plants (see Chapter 10).

The ecology of plant–herbivore interactions is a rapidly developing field. For further reference consult books by de Wilde and Schoonhoven (1969), Sondheimer and Simeone (1970), Harborne (1972, 1977, 1978), Van Emden (1972a), Gilbert and Raven (1975), Friend and Threlfall (1976), Wallace and Mansell (1976), Chapman and Bernays (1978), Rosenthal and Janzen (1979), Hedin (1983), Strong et al. (1984), Rosenthal and Berenbaum (1991, 1992), White (1993), and Bernays and Chapman (1994).

Hypotheses on Plant and Herbivore Interactions

With so many different kinds of plants and insect herbivores there are almost infinite kinds of interactions that may be studied. Each kind of plant provides many different resources for attack, and there are many ways that insects feed on plants (Fig. 6.1). Therefore, there are many opportunities for the study of ecological interactions.

KINDS OF HYPOTHESES AND APPROACHES

Hypotheses on the interactions between plants and herbivores constitute a wide diversity of approaches. These approaches cover very different scales, from variation within plants to differences across landscapes, from climatic effects over several years to momentary changes in food quality, from proximate mechanisms to ultimate evolutionary relationships. These approaches also differ in what aspect of the plant–herbivore interaction they emphasize: differences in plant chemical defense and patterns of difference, insect population dynamics and eruption of pest species, or the physiological constraints on plant metabolism and consequences for herbivores.

Therefore, this cake of hypotheses could be cut up in several different ways to provide some internal structure to covering the subject adequately. Suitable areas would be herbivore population dynamics, patterns in plant defensive

Gypsy moth, *Lymantria dispar,* showing a female ovipositing a large clutch of eggs on the trunk of a tree and a caterpillar on an oak leaf. A male is shown in flight, and a female is displayed with wings spread (top right). Based on Novák et al. (1976). Drawing by Tad Theimer.

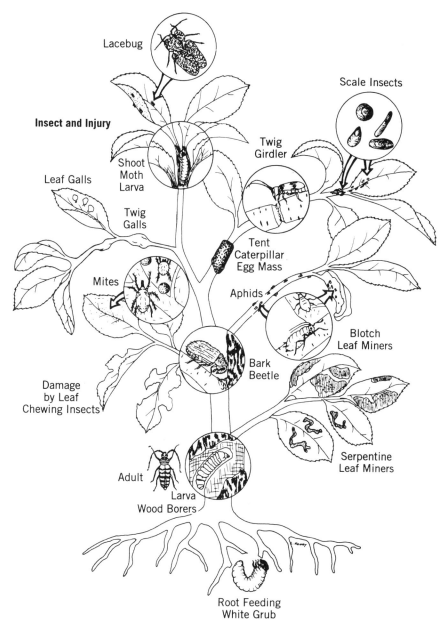

FIGURE 6.1 Examples of the plant parts attacked by insect herbivores and the different kinds of herbivores and modes of utilization. Reprinted from W. T. Johnson and H. H. Lyon, *Insects that feed on trees and shrubs,* 2nd ed., revised, p. 14. Copyright © 1991 by Cornell University. Used by permission of the publisher, Cornell University Press.

chemistry, and plant physiology and its effects on herbivores. These three areas would cover the majority of the large body of research. I choose to take an historical approach that covers all these approaches in the approximate order in which the hypotheses were generated. The historical approach is valuable, I think, because ideas tend to build on each other through time. One can evaluate a new contribution only by how it advances preceding knowledge. Emphasis on historical development also enables a clear view of the pace or momentum within a discipline. A clear indication of a healthy science is that the pace is quickening, a condition clearly seen in the area of plant–herbivore interactions.

In fact, this area is one of the most dynamic and rapidly developing in ecology today. It is the development of hypotheses that has stimulated the field of plant–herbivore interactions, coupled with enormous impetus from agriculture and forestry to understand pest herbivores and their population dynamics. Remarkably, this very vigorous field has been largely ignored in textbooks on ecology. This is despite the fact that herbivores are the central connection in trophic systems between autotrophic plants and higher trophic levels and that with 308,000 species of green plant in the world and 361,000 species of phytophagous insects, about half of the macroscopic species of organisms on earth are involved with plant–herbivore interactions (Fig. 1.4) (Strong et al., 1984). Ecologists and their textbooks have feasted attention on such interactions as predation and competition while largely ignoring an elementary interaction of probably far greater importance.

One hundred percent of individuals in a herbivore population and species, in each generation, must relate intimately to plants as food. Therefore, the ecology of plant food must be of paramount importance in the population biology and evolution of herbivores. The role of natural enemies is likely to be relatively less important because interactions do not affect every individual in every generation. This ordering of relative importance of plant food and natural enemies has not been a persistent perspective; for many years variation in plant quality was largely ignored, while emphasis was placed on the role of natural enemies in causing herbivore mortality. This shift in emphasis to the role of plant food quality and its variation will be discussed more when life table analysis is discussed in Chapter 13.

However, the area of plant–herbivore interactions has still to become a truly synthetic science with an integrated perspective of phytochemistry and physiology; herbivore physiology; population dynamics; the role of natural enemies; the interactions among the three trophic levels of plant, herbivore, and enemies of herbivores; and the abiotic forces of weather, soil quality, and other factors. This is a tremendous challenge for the future.

In the meantime the development of some ideas on plant–herbivore interactions is presented here. Emphasis is placed on population-level hypotheses involving variation in plant traits and consequences for insect populations and communities. Perhaps hypotheses on plant–herbivore interactions at the population level began with ideas developed during the late 1950s and early 1960s relating to the population dynamics of insect and mammal herbivores. These may be called the *climatic release hypothesis* and the *nutrient recovery hypothesis*.

CLIMATIC RELEASE HYPOTHESIS

Observations on insect pest outbreaks that follow periods of unusually warm dry weather are numerous and date back to the 1920s (Mattson and Haack, 1987). Mattson and Haack (1987) list over 20 examples, with half of the cases involving bark beetles in the family Scolytidae. Some of these outbreaks can be interpreted as a direct effect of increased temperature on insect survival because temperature is limiting to insect development in north temperate climates. The cases of the lodgepole needle miner, *Recurvaria starki* (Stark, 1959), and the fall webworm (Morris, 1964) probably fit this condition. But in other instances the interactions were more complex, as in the spruce budworm in eastern North America (Greenbank, 1956, 1963; Morris, 1963a). In addition to the direct effects of increased temperature, there were effects of plant quality: (1) mature stands of balsam fir showed signs of severe defoliation before stands of younger age classes, and (2) these mature trees flowered every year during a series of dry and sunny summers. High staminate flower production resulted in better conditions for spruce budworm larvae: (1) old staminate flower cups provide safe overwintering sites, so survival increases, and (2) after overwintering, larvae that feed on male flowers developed faster. Here the importance of plant quality variation became evident in the population dynamics of an herbivore, just as earlier studies by Blais (1952, 1953) and Miller (1957) had shown that foliage quality affected individual female fecundity.

These were important studies at a time when life table analysis was the principal approach to studying insect population dynamics, and plant quality hardly ever entered into a life table analysis. It was generally assumed that insects died because of bad weather, predation, parasitoids, disease, or a shortage of food. They did not die, according to life tables, because of poor nutrition or high chemical defense; nor did they "fail to be born" because of low fecundity imposed by a low-quality diet or a female's avoidance of low-quality resources (see Chapters 13 and 15 for details).

Even then, the population dynamics model developed by Morris (1963a,b) for the spruce budworm emphasized the roles of parasitism, predation, and a direct effect of temperature, because the model was for one location and change at this location from generation to generation. Therefore, differences in plant quality from locality to locality were not relevant to the model. More concerted focus on plant quality variation was developed by the *nutrient recovery hypothesis*.

NUTRIENT RECOVERY HYPOTHESIS

The nutrient recovery hypothesis was developed by Pitelka (1964) and Schultz (1964) to account for lemming cycles at Point Barrow, Alaska. Low densities of lemmings allow vegetation cover over the ground to increase, insulating the soil so that the depth to permafrost decreases. As a result, the concentration of nutrients in the rooting zone of plants increases, followed by an increased quality of foliage and an increased survival and reproductive rate of lemmings. An eruption of lemming numbers results. Large numbers of lemmings deplete the ground cover, and the ground heats up, lowering the level of permafrost, so the nutrient status of the soil declines because of leaching, or dilution of nutrients.

For the first time, apparently, qualitative change of plants was emphasized as the driving force in the population dynamics of an herbivore. The hypothesis has never been validated experimentally, but it received a lot of attention at the time and was probably influential in raising the consciousness of researchers about the possibilities for herbivore dynamics to be driven from below, from the resource base, rather than from above, by natural enemies.

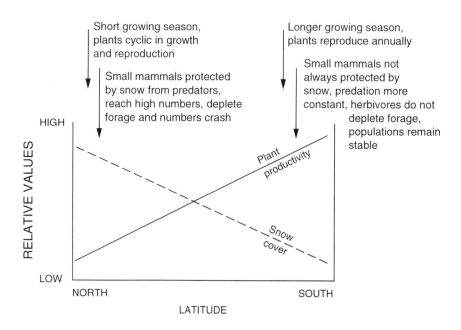

FIGURE 6.2   Hypothetical relationships between latitude, relative persistence of snow cover in winter, plant productivity per season, and small mammal population dynamics. North is equivalent to northern Finland and south to central Europe.

Even in 1983, almost 20 years after the nutrient recovery hypothesis had been proposed, Batzli (1983) wrote that experimental data on the effects of nutritional changes in forage on rodent populations was available only for temperate, not arctic, microtine rodents, although he did argue that forage quality could account for differences in densities of arctic rodents between habitats and seasons. But the role of forage quality in rodent cycles was not clear and may be coupled with changes in plant allelochemicals as well.

Laine and Henttonen (1983) added another perspective on microtine cycles by arguing that many plant species show synchronous cycles of growth and reproduction, independent of rodent population densities. The short growing seasons in northern Fennoscandia result in delayed reproduction while resources are accumulated by plants over several growing seasons. This cycle of accumulation of nutrients to a peak, allowing reproduction and subsequent depletion, drives the microtine cycles perhaps through nutrient quality and quantity changes.

As latitude declines toward the south, plant production increases, reproduction becomes more frequent and even annual, and the length of microtine cycles declines from 4–5 years in the north (e.g., 69°N) to 3 years in central and southern Finland, and populations are acyclic farther south. So plant growth, reproduction, and general quality may well be the independent variables that are the primary factors in microtine cycles. Predation is likely to be important in regulating numbers but probably plays a secondary role (Fig. 6.2).

Another important step came with White's repeated claims that plant stress drives insect herbivore population dynamics.

PLANT STRESS AND HERBIVORE SURVIVAL HYPOTHESIS

The link between insect eruptions and unusually warm dry weather, eruptions in mature or overmature stands of trees, and eruptions in stands on poor dry soils suggested that water stress may well play a role (Mattson and Haack, 1987). White was a strong advocate of this hypothesis beginning with his 1969 paper, and he provided a detailed mechanistic hypothesis.

White (1969) argued that plant stress could be estimated by the contrast between winter and summer availability of water, with high winter rainfall followed by low summer rainfall being physiologically stressful on plants, but low winter rains followed by low summer rains was less stressful. His *stress index* was based on the difference between the deviation from normal for winter rainfall and summer rainfall, and he concentrated studies on the psyllid, *Cardiaspina densitexta,* on the eucalypt tree, *Eucalyptus fasciculosa.* Psyllids suck plant sap and nymphs tend to be sedentary after settling as first-instar larvae. Psyllids on *Eucalyptus* are called lerp insects because they feed under a protective lerp formed from dried excrement.

White's method can be applied, for example, to the 1956 to 1963 outbreak of the psyllid, which coincided with the highest recorded stress index since 1910, for the second-longest duration of positive stress (Fig. 6.3). Other species of psyllids showed similar tends. White argued that the outbreaks were caused by physiological changes in the plant relating to nitrogen availability (Fig. 6.4). In a series of other papers, White (1974, 1976, 1978, 1984) expanded on this general theme to cover many other kinds of herbivorous insects, and nitrogen limitation for animals in general (White, 1993).

Some of the difficulties with these influential studies are symptomatic of the field in general, even today, and several mechanistic linkages need to be studied more carefully. As Mattson and Haack said in 1987 (p. 366), "Evidence to support the hypothesis that plant water deficits promote insect outbreaks is largely circumstantial" (see also Larsson, 1989).

Areas of concern in White's (1969) study are as follows:

1. In the earlier outbreak of *Cardiaspina densitexta* from 1914 to 1922, the outbreak started in a 7-year period of subnormal stress (Fig. 6.3).

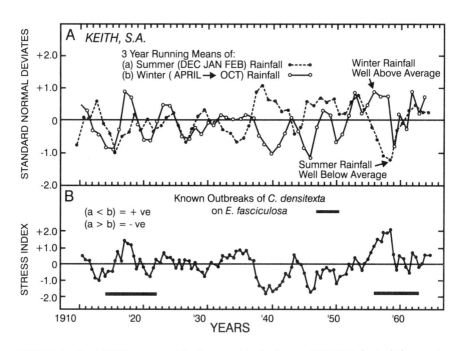

FIGURE 6.3 White's (1969) estimation of the plant stress index for the years 1910–1965, showing high stress coinciding with the outbreak of *Cardiaspina densitexta* from 1956 to 1963 at Keith, South Australia. From T. C. R. White, An index to measure weather-induced stress of trees associated with outbreaks of psyllids in Australia, *Ecology*, 1969, 50:905–909. Copyright © 1969 by the Ecological Society of America. Reprinted with permission.

2. No direct measure of plant stress was taken in terms of reduced plant performance relative to growth or reproduction under optimal conditions. Stress was inferred, not measured directly.

3. The physiological studies of change in stressed plants, cited by White, were performed on agricultural herbaceous plants and grasses, without an indication that the same kinds of processes were involved with large trees.

4. There was no direct link made between plant change under stress and individual insect survival and growth (i.e., effects on insect autecology and physiology need to be studied).

5. There was no direct link between plant change under stress and population change of the insect (i.e., effects on insect populations need to be studied), given complexities of compensatory factors such as competition and enemies. Many studies on plant stress have shown since that results are unpredictable and species-dependent when good experiments are performed (Larsson, 1989; Waring and Cobb, 1992).

6. White recognized that heavily stressed plants may frequently be deleterious to herbivores, so the degree of stress needs to be quantified on some scale from unstressed to moribund or dead (Larsson, 1989).

7. Increased stress may also reduce the effective defense of a plant, as White recognized, and reduced defense may become a more important factor. This is probably the case with bark beetle attack when oleoresin exudation pressure is lowered by water stress, with lowered resistance to attack (Mattson and Haack, 1987, p. 381; Berryman, 1982).

Despite these concerns, White did establish the *possibility* for a mechanistic relationship between changes in plant quality and insect population outbreaks. The precedent for ignoring plant quality was broken, and the potential of plant variation influencing insect herbivore ecology became well established.

Around this time, in the late 1960s, evolutionary ecology was becoming increasingly popular and interesting, with such questions being asked as "Why do many herbivorous insects feed so early in the season in temperate climates, when food becomes more abundant later in the season?" The question was answered by Feeny (1970), as discussed in Chapter 5, who showed that as oak *(Quercus robur)* leaves age, water content declines, protein content declines, and tannin content increases, such that available protein (that which probably remains untanned in the insect gut) declines rapidly in the first two months of a leaf's life. Thus, early feeding by the winter moth, *Operophtera brumata,* provides a temporal escape from plant defense, as well as synchrony with high leaf water content. The winter moth evolved a life history in response to a phytochemical problem of nutrition.

Feeny's paper should be regarded as a landmark in the development of plant–herbivore interactions. It united evolutionary ecology with chemical ecology in the study of plants and insects. The first attempt to synthesize this

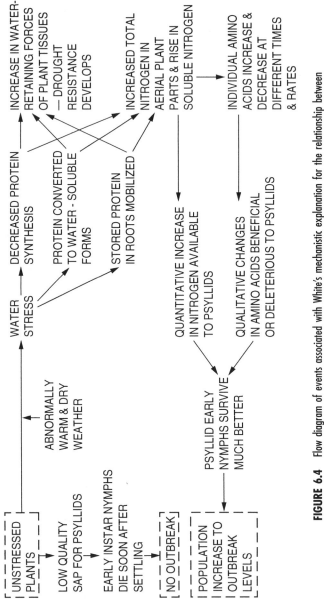

FIGURE 6.4 Flow diagram of events associated with White's mechanistic explanation for the relationship between plant stress and insect herbivore outbreaks. Starting with unstressed plants (top left) populations may remain low, as in sequence at left, or reach outbreak levels following the alternative sequence.

field came in 1970 with the publication of *Chemical ecology* (Sondheimer and Simeone, 1970), and the *Journal of Chemical Ecology* was started in 1975.

Feeny's paper also formed one of the bases for the next step in the development of hypotheses on plant–herbivore interactions, for the winter moth studies were to play a role in the plant apparency hypothesis (Feeny, 1975, 1976). The hypothesis was discussed in Chapter 5, but more detail is provided here. With this hypothesis we see a major shift in emphasis from understanding herbivore population dynamics to an approach using evolutionary ecology as a means to understand broad patterns in plant chemical defense against herbivores.

PLANT APPARENCY HYPOTHESIS

Feeny (1975, 1976) became intrigued by the large differences in investment of chemical defense in leaves of different plant species: from 2.4 to 5% synthetically expensive tannins in the fresh weight of old oak leaves to such low levels as 0.2–0.7% of synthetically cheaper mustard oils in the fresh weight of black mustard, *Brassica nigra*, leaves (Fig. 6.5). He noted fundamental differences between the action of these different kinds of chemical defenses, and together these observations provided the basis of his hypothesis. The hypothesis may best be summarized in table form (Table 6.1).

Rhoades and Cates (1976) expanded on this theme.

1. Quantitative defenses are generally *digestibility reducers:* they reduce the availability of plant nutrients (tannins, resins, cellulose, lignin.)
2. Qualitative defenses are generally *toxins,* which enter through the gut wall of the herbivore and interfere with biochemical processes. They have a pharmacological effect (many such chemicals are used as pharmaceuticals for humans: digitalin, atropine, nicotine).
3. *Digestibility reducers* tend to be *convergent forms of defense* for *predictable plant species* and *plant parts*—apparent plants and mature plant parts such as older leaves. They are costly defenses but are effective against both specialist and generalist herbivores.
4. Toxins are very diverse, from mustard oils, to alkaloids, to cardenolides, so they are *divergent* and cheap. They are therefore most useful in defense of *unpredictable plant species* and *plant parts*—unapparent plants and parts such as young leaves that are only briefly available to herbivores and may escape herbivore attack most of the time.
5. Thus even apparent, predictable plant species may have two forms of defense: one for early, ephemeral foliage, and one for later mature foliage. For example, in bracken fern the toxic cyanogenic glycosides and the digestibility-reducing tannins may both be found (Fig. 6.6). These

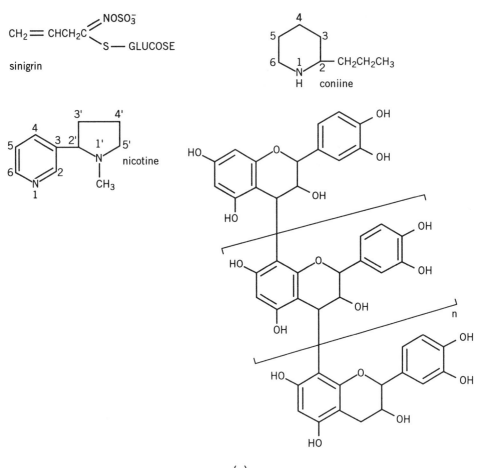

(a)

FIGURE 6.5 Some structures of chemicals involved as plant defenses against insects, showing the relatively simple structure of alkaloids and mustard oils or glucosinolates (from Robinson, 1991) when compared with tannins and lignins (from Hagerman and Butler, 1991; copyright Academic Press, Inc.). (a) Sinigrin is a mustard oil glucoside, or glucosinolate, from black mustard, *Brassica nigra*. Nicotine is the toxic alkaloid product of tobacco species in the genus *Nicotionia*. Coniine is the highly toxic alkaloid in the poison hemlock, *Conium maculatum*. Below right is a condensed tannin, with multiple units (*n*) of the flavonoid unit in parentheses, making it a polymer of flavonoid units. (b) Proposed structure of beech wood lignin.

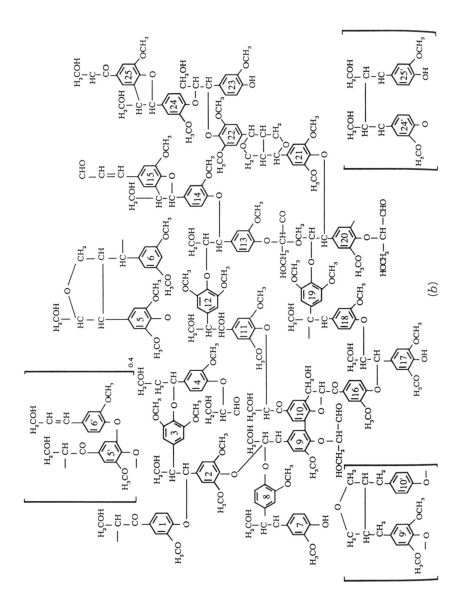

(b)

117

TABLE 6.1   Summary of the Plant Apparency Hypothesis.

Apparent Plants	*Unapparent Plants*
1. Plants are late successional.	1. Plants are early successional.
2. Plants exist frequently in pure stands.	2. Plants exist in high species richness mixtures; very patchy.
3. Plants are large (e.g., trees).	3. Plants are small (e.g., herbs).
4. Plants are long-lived (oaks, beech, fir, spruce, pines).	4. Plants are short-lived (weedy species, annuals).
5. Plants are "bound to be found" by herbivores (i.e., they are *apparent* to herbivores).	5. Plants are "hard to find" (i.e., they are *unapparent* to herbivores).
6. Heavy investment in defense is adaptive.	6. Most important defense is escape from discovery—little defense is needed.
7. Defense must be *general* and effective against generalist and specialist insects since all kinds will find apparent hosts.	7. Defense need be only against generalist herbivores that happen upon the plant. For this purpose, only highly toxic but low-concentration chemicals are needed.
8. *Quantitative defenses* are general defenses because they reduce the nutritional quality of food (e.g., condensed tannins bind with proteins and make them unavailable to herbivores).	8. Toxins, such as mustard oils, are repellent to nonspecialized insects but are not effective against specialists—*qualitative defenses*.
9. They are called *quantitative defenses* because they are dosage dependent: as concentration increases, negative effect increases.	9. *Qualitative defenses* have no inhibitory effect on the growth of specialists or their survival. Not dosage dependent.
10. Therefore, the higher the level of defense in a plant organ, the better it is protected. Leads to 2.4% tannins in oak leaves.	10. Therefore, low levels of defense are adequate for defense—less than 1% leaf fresh weight.
11. High defense investment can be tolerated in large, long-lived plants because energy and nutrients for reproduction can be stored and accumulated over long periods of time.	11. Low defensive investment is perhaps the only possible strategy for a small, short-lived plant that may depend on a single growing season for its reproductive effort.

trends are quite general across many taxa, with toxins declining with leaf age and digestibility reducers increasing with leaf age.

6. This emphasis on predictability of plant parts accounts for the apparent paradox that young leaves of creosote bush, *Larrea tridentata* and *L. cuneifolia,* are better defended with digestibility reducers, in the form of phenolic resins, than are mature leaves. The paradox is resolved by

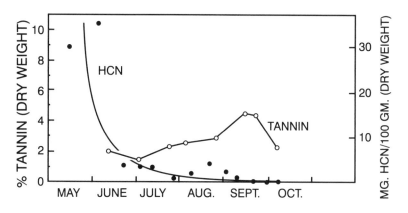

FIGURE 6.6   Chemical defenses of bracken fern in relation to the age of fronds which start emerging from the ground in May. Note the high level of cyanogenic glycosides (HCN) early in the season and the increasing levels of tannins later in the season. From Rhoades and Cates (1976); with permission.

Rhoades and Cates (1976) when it is seen that during the driest spells of the year, only the youngest leaves remain on plants. Young leaves are always present, and therefore apparent, or predictable (Fig. 6.7).

7. Therefore, the general picture that emerges from Rhoades and Cates's argument can be summarized as in Fig. 6.8.

Much later, Rhoades (1979) expanded on White's arguments by emphasizing concurrent changes in plant chemical defense during bouts of stress. He listed cases showing that, in general, under stress plants produce more toxins (e.g., alkaloids, cyanogenic compounds, glucosinolates) and decreased digestibility reducers (tannins, resins and essential oils). During stress a plant is assumed to reduce investment in high-cost defenses and increase allocation to low-cost toxins. The toxins are not effective against specialized herbivores, so the insects erupt to epidemic populations in response to improved plant nutritional status and reduced levels of digestibility reducers.

Emphasis on reduced defense in stressed plants is consistent with the long-held view among forest entomologists that bark beetle epidemics were associated with water stress, and as a result, reduced oleoresin exudation pressure.

Still, there are points of concern with these developments: (1) no clear relationships could be detected by Gershenzon (1984) between water stress and levels of phenolics in leaves, phenolics being one of the kinds of chemicals involved with digestibility reduction, and (2) many herbivores attack most, and survive best, on the most vigorous plant parts available and the least stressed plants in a population (see the section "Plant vigor hypothesis"). Therefore,

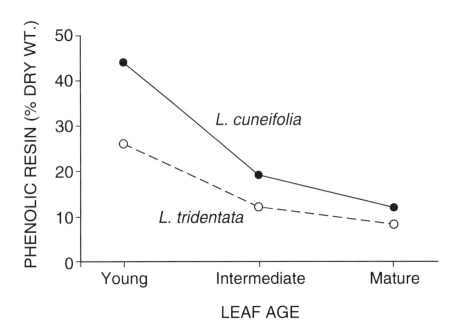

FIGURE 6.7   Levels of phenolic resins in two species of creosote bush, *Larrea cuneifolia* and *L. tridentata,* in relation to leaf age. Data from Rhoades and Cates (1976).

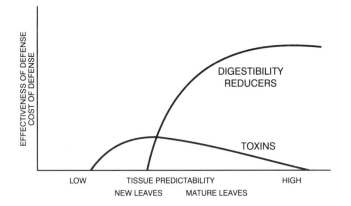

FIGURE 6.8   Hypothesis on the allocation of phytochemical defenses in terms of tissue predictability and the efficacy and cost of defenses. From Rhoades and Cates (1976); with permission.

we must begin to differentiate between kinds of herbivores, determined by very different responses to stressed and vigorous plants.

The hypothesis of plant apparency or plant predictability was a major stimulus on the development of studies on plant–herbivore interactions. It was the first hypothesis to clearly delineate patterns in phytochemical defense, coupled with mechanistic explanations for divergence of defense strategies. Some problems have been discussed in Chapter 5, but aspects of the hypothesis remain robust.

However, the emergence of general hypotheses at this time was complicated by concurrent developments on changing levels of chemical defense within individual plants in response to herbivory. Somehow herbivores changed defense levels in plants or induced chemical defense. In addition to the **constitutive defenses** of plants synthesized in the normal development of the plant, there were **induced defenses** synthesized after herbivore damage, or damage of any sort.

INDUCED DEFENSE HYPOTHESIS

In 1972, Green and Ryan reported experiments in which Colorado potato beetles, *Leptinotarsa decemlineata,* feeding on tomato plants, *Lycopersicon esculentum,* caused a rapid increase in proteinase inhibitors in leaves. Not only were damaged leaves affected, but undamaged leaves also rapidly accumulated proteinase inhibitors. Within 48 hours after severe damage to a single leaf, proteinase inhibitors reached over 2% of the soluble proteins in leaves throughout small plants. Insect damage had the same effect as that of any mechanical damage to the leaf. From the damaged plant part a **proteinase inhibitor inducing factor** (PIIF) is released that travels to other plant parts, inducing the defensive response. Proteinase inhibitors simply reduce the capacity of enzymes in the animal gut to digest protein, thereby reducing the nutritional value of plant tissue already low in nitrogen relative to the needs of animal herbivores.

One of the experiments reported in 1972 used young 8- to 10-cm tomato plants with two leaves. The lower leaf was wounded by crushing or punching holes with a paper punch and the upper leaf was assayed for inhibitor, using the terminal leaflet. The results showed a rapid increase in proteinase inhibitors within 48 hours: the more the damage, the more chemical defense was synthesized (Fig. 6.9). Two compounds, inhibitors I and II, have been isolated and characterized (Ryan, 1979). Wounding lower leaves of young tomato plants had a strong effect throughout the plant, while damage to an upper leaf had a much weaker systemic effect (Fig. 6.10) (Nelson et al., 1983).

Apparently, independent of this research by Green and Ryan, Haukioja and associates began studies on mountain birch *Betula pubescens* ssp. *tortuosa* and its major insect herbivores in northern Finland at the Kevo Subarctic

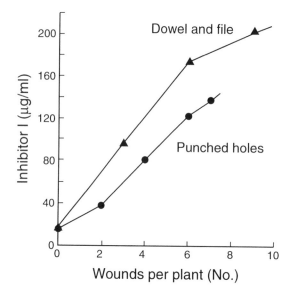

FIGURE 6.9   Experimental results from Green and Ryan (1973) showing the increase in proteinase inhibitor I in leaf liquids in an upper leaf when the lower leaf was damaged by crushing a leaf between a wooden dowel and a file, and when holes were punched in leaves. As the number of wounds per lower leaf increased the concentrations in the upper leaf increased. Reprinted with permission from T. R. Green and C. A. Ryan, Wound-induced proteinase inhibitor in plant leaves: A possible defense mechanism against insects, *Science*, 1972, **175:**776–777. Copyright 1972 by the American Association for the Advancement of Science.

Research Station. In 1976 and 1977 (Haukioja and Niemelä, 1976, 1977) they reported that when birch leaves were damaged and adjacent leaves were fed to caterpillars in the laboratory two days later, larval growth was retarded. The caterpillar studied was an eruptive species [*Oporinia* (=*Epirrita*) *autumnata* (Geometridae)], which caused heavy defoliation and tree death periodically in northern Finland. Later (Niemelä et al., 1979), they found that leaves adjacent to damaged leaves contained 9% more phenolics than controls and inhibited trypsin 40% better than controls, 2 days after treatment. Therefore, inhibition of protein digestion by the increased phenolics in leaves after damage may be the mechanistic explanation for delayed pupation times of larvae.

Later, Haukioja (1980, 1982) reported that the induced defenses were effective even in the year after defoliation, and full recovery of a tree to the predamaged state—the relaxation time for the induced defense—was 3–4 years at least. The effects in the year after defoliation were strong (Fig. 6.11). These findings, of a long-term inducible defensive response in birch after herbivore damage, immediately suggested the possibility that periodicity of out-

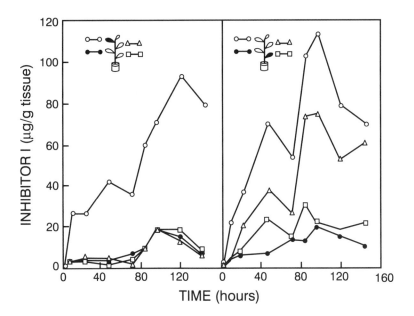

FIGURE 6.10   Experimental results showing levels of proteinase inhibitor I in terminal leaflets of young tomato plants when the uppermost leaf was wounded (left) and when the lowest leaf was wounded (right). Reprinted with permission from C. E. Nelson et al. in P. A. Hedin (ed.), *Plant resistance to insects,* American Chemical Society, Washington, D.C., 1983. Copyright 1983 by the American Chemical Society.

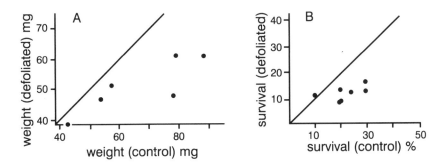

FIGURE 6.11   Comparisons on growth and survival of *Oporinia autumnata* larvae on foliage from undamaged control trees and from trees defoliated the year before. The line at 45° shows the trend expected if no effect of defoliation were evident, but almost all data points fall well below the line, showing much poorer performance on foliage from previously defoliated trees. From Haukioja (1980); with permission. Copyright © 1980 by OIKOS.

breaks of herbivores could be driven by induction (Haukioja, 1980), as predicted in the earlier hypothesis formulated by Haukioja and Hakala (1975).

The evidence now for induced defenses in plants is widespread and well documented (e.g., Kogan and Paxton, 1983; Karban and Carey, 1984; Edwards et al., 1985, 1986; Karban, 1986; Karban et al., 1987; Edwards and Wratten, 1987; Haukioja and Neuvonen, 1987; Tallamy and Raupp, 1991); although there have been some clear concerns developing (e.g., Fowler and Lawton, 1985; Hartley and Lawton, 1987).

Another form of induction results from regrowth of woody plants after heavy herbivore attack and the palatability of new stems, as winter browse, rather than leaf palatability during the summer as discussed above. Fox and Bryant (1984) focused on the snowshoe hare impact on woody plants in Alaska. When snowshoe hare populations reach a peak, many woody plants are overbrowsed or girdled, with subsequent regrowth of more juvenile shoots. These shoots are much better defended chemically than on more mature growth, and hare avoid more juvenile growth when they feed during the winter. Sprout palatability improves and volume increases with age and recovers in 2–3 years in earlier successional plants and in 4–10 years in mid to late successional plants. This unpalatability of juvenile growth is sufficient to cause the crash in snowshoe hare populations and subsequent recovery as growth matures, resulting in the 8- to 11-year cycles of hares in population density. However, since these woody tissues are utilized as winter browse by hares, but rapidly growing juvenile shoots will be available to insects in the summer, the results for hares and insects may be very different (Fig. 6.12). Note that damage to leaves and defoliation results in induced defenses in leaves available in summer to herbivores. However, pruning back stems reduces the number of leaves during regrowth, more nutrients are available per leaf, and leaf quality increases for herbivores. Therefore, in studies of herbivory it is most important to keep track of the type of damage and the season of feeding.

A remarkable development in the induced defense literature occurred in 1983, which has been referred to as *talking trees* (Baldwin and Schultz, 1983; Rhoades, 1983). The proponents argued that damaged plants released pheromones that stimulated increased defensive chemistry in nearby undamaged plants. Baldwin and Schultz (1983) used poplar *(Populus)* and maple *(Acer)*, and Rhoades (1983) used *Salix sitchensis*. No bioassays on the effects of chemical changes on insect growth and survival were performed by Baldwin and Schultz (1983), so the effects are potential, not realized effects. In 1981, Rhoades used a bioassay with the results shown in Fig. 6.13. The insect species was western tent caterpillar, *Malacosoma californicum pluviale*, in one of the tests. Ten test trees were loaded with about 600 caterpillars. After loading, growth of larvae was followed, on leaves from the test trees, on unattacked near-control trees about 3.3 m away and on far controls about 1.6 km from the test site. On leaves from the test trees larvae grew significantly slower than those on the far control on June 9 and 12, indicating in-

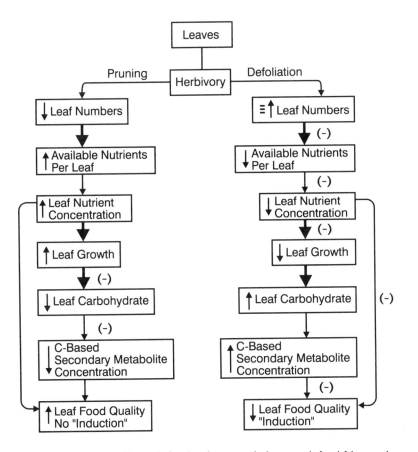

FIGURE 6.12 Flow diagram of effects on leaf quality after pruning back stems and after defoliation without stem pruning. Arrows in boxes indicate an increase (↑) or decrease (↓) in a variable. Arrows between boxes show a positive effect unless indicated by (−). The magnitude of the effect is suggested by the thickness of arrows. From J. P. Bryant et al. in D. W. Tallamy and M. J. Raupp (eds.), *Phytochemical induction by herbivores*. Copyright © 1991 by John Wiley & Sons, Inc. Reprinted by permission of John Wiley & Sons, Inc.

duced defenses on the trees attacked by larvae. However, there was reduced growth of larvae on the undamaged near-control trees on June 12, suggesting that defenses had been induced remotely from the test plants.

Rhoades (1983) did not consider these experiments to be proof of a pheromonal communication system between plants. For example, root connections between treatment and near-control plants were not found, but they could still exist. Mycorrhizal connections could play a role. Near controls oscillated a lot in their effects on larval growth and in four postloading data samples, differed significantly only on the last sample date. More sampling

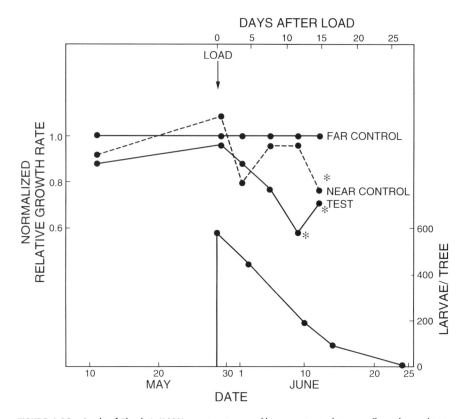

FIGURE 6.13   Results of Rhoades's (1983) experiment on possible communication between willows when a plant is damaged by insect herbivores. The date on which larvae were placed on test plants is indicated, and the mean number of larvae per tree is shown using the right-hand axis. All larval growth was normalized relative to growth on foliage from the far control trees, which is set at a constant of 1.0. Trends on the test and near control trees are shown, and significant differences from the far control are indicated with an asterisk. From D. F. Rhoades in P. A. Hedin (ed.), *Plant resistance to insects,* American Chemical Society, Washington, D.C., 1983. Copyright 1983 by the American Chemical Society.

was needed to make a convincing case. Disease in the treatments could not be ruled out (see also Fowler and Lawton, 1985).

If trees really do talk to each other, what is the adaptive significance? Induced defenses increase rapidly after damage, so why should a plant increase defense before damage when it may never be attacked? If damage to one plant has a population effect, all members of a population may be induced all the time, so herbivores would evolve to cope with this background level of induced defenses. Some answers may be contained in the results from many experiments conducted by Dicke and associates, for example on lima beans, spider mites, and predatory mites (e.g., Dicke, 1988; Dicke et al., 1990a–d). In closed experimental systems with controlled airflow, lima beans fed upon by spider mites release volatile chemicals such as terpenes and phenolics, which results in increased resistance of undamaged bean plants downwind (Fig. 6.14). Induced defenses on downwind plants may be involved, or adhesion of

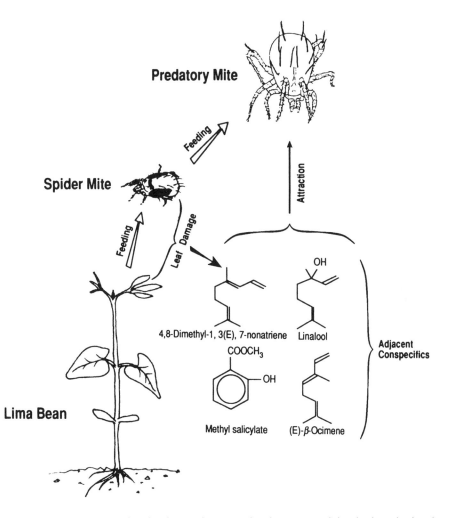

FIGURE 6.14   Interactions involving lima bean, spider mites, and predatory mites, including the chemicals released from damaged leaves that are attractive to the predatory mites and cause defense in plants downwind. Based on research by Dicke (1988) and Dicke et al. (1990a,b). From P. W. Price in T. A. van Beek and H. Breteler (eds.), *Phytochemistry and agriculture*, Oxford University Press, Oxford, 1993. Reprinted by permission of Oxford University Press.

the volatiles on leaves could be responsible. However, the volatiles also attract predatory mites to the damaged plant and its neighbors downwind. Since spider mites disperse on silk in wind, downwind plants are likely to be colonized, but predators may already be present. Takabayashi et al. (1991) found that plants contribute more to the volatiles than do the herbivores. Thus synomones are involved between plant and predatory mites, and perhaps pheromones are involved between plants (see Chapter 5 for definitions).

Another advance in the understanding of pheromonal communication between plants involves the actual chemicals that cause induction in nearby plants. Farmer and Ryan (1990, 1992) have shown that jasmonic acid and methyl jasmonate are powerful inducers of proteinase inhibitors, and that methyl jasmonate can induce proteinase inhibitors in adjacent plants. Methyl jasmonate is a common secondary compound in plants, such as in the sagebrush, *Artemesia tridentata,* and in the presence of this plant, induction in tomato plants occurred. Whether induction occurs between naturally co-occurring plant species is yet to be determined.

Studies on water stress, nutrient stress, and other forms of stress, including herbivore attack and induced defenses, converge in focusing attention on plant physiology and the constraints on phytochemical synthesis. A more plant-oriented focus has been important and stimulating. A major development in this area was the *carbon–nutrient balance hypothesis* developed by Bryant et al. (1983).

CARBON–NUTRIENT BALANCE HYPOTHESIS

Bryant et al. (1983) argued that limiting effects of nutrients are different from those of decreased carbon due to low light or carbon dioxide levels, summarized in Fig. 6.15. Low nutrients, especially nitrogen, would result in reduced photosynthesis because RuBP carboxylase used in photosynthesis is reduced when nitrogen is limiting. However, decline in growth is usually greater than the decline in photosynthesis (see strong arrows from reduced nutrient absorption box relative to those from the reduced photosynthesis box). Growth rate declines, but carbohydrate concentrations increase and become available for carbon-based defenses. Reduced nitrogen results directly in reduced nitrogen-based defenses, frequently toxins. Thus, for herbivores, a plant phenotype growing on poor soils will have low nutritional value and increased carbon-based defenses, likely to involve digestibility-reducing chemicals such as phenolics. If plant populations are continuously and predominantly exposed to low-nutrient conditions, they will evolve toward a strategy of growth with high constitutive carbon-based defenses effective against all or most herbivores (see Coley et al., 1985). For such plants, increased nutrients in the form of fertilizer or nutrient-rich sites will have the opposite set of effects, increasing nutritional value for herbivores and reducing carbon-based defenses.

On the other hand, a plant with a low carbon budget relative to nutrients shows a different and converse response (Fig. 6.15). In low-carbon environments, such as shady conditions, reduced photosynthesis has a strong effect on growth while nutrients are relatively high. If nitrogen is not used in growth, it becomes available for nitrogen-based defenses, which increase. Therefore, for herbivores adapted to and specialized on plants with nitrogen-based defenses such as alkaloids and cyanogenic glycosides, the reduced carbon-based defenses and increased nutritional status of the plant should be beneficial, remembering that increased chemical defenses in the form of toxins would have little or no effect on specialists. The same should be true for herbivores on the many plants with low or ephemeral nitrogen-based defenses which concentrate on carbon-based defenses such as phenolics.

This hypothesis helps to account for shifts in plant defense and differences in response by herbivores.

1. Heavy browsing of disturbance-adapted woody plants such as willow and birch, with high carbon reserves belowground, results in new juvenile shoots growing under relatively high carbon and low nutrient conditions and high carbon-based defenses. Juvenile twigs of *Betula papyrifera* ssp. *humilis* in Alaska have heavy resin accumulation on juvenile twigs, not present on mature twigs, and in winter are unpalatable to browsing mammals.
2. Fertilizer treatments will have different effects on herbivores, depending on the major defense of the plant species. If N-based defenses are major,

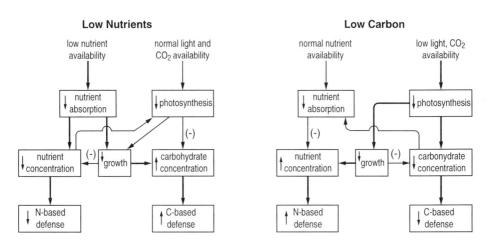

FIGURE 6.15 Carbon/nutrient balance hypothesis formulated by Bryant et al. (1983). Vertical arrows in boxes indicate an increase (↑) or decrease (↓) in a variable. Arrows between boxes show a positive effect unless a negative symbol (−) appears. The thickness of arrows indicates the relative strength of the effect. From Bryant et al. (1983); with permission. Copyright © 1983 by OIKOS.

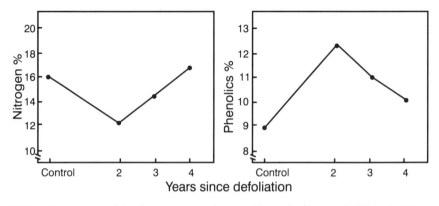

FIGURE 6.16   Nitrogen and phenolic content, expressed as percent dry weight of leaves, in birch leaves in relation to the defoliation history of the trees. Controls were undamaged trees, while the treatments were damaged trees in years 2, 3, and 4 after defoliation. From Tuomi et al. (1984); with permission. Copyright © 1984 by Springer-Verlag GmbH.

N fertilizer will increase defense. If C-based defenses are major, N fertilizer will reduce defenses.

3. Herbivores with slightly different responses to nutrient limitation and chemical defenses may respond differently to different plant treatments.
4. Herbivores on the same plant attacking early and late foliage may respond differently to plant treatments.

This hypothesis has received considerable support in the literature (e.g., Larsson et al., 1986; Bryant, 1987; Price et al., 1989). It may also be a hypothesis that accounts for long-lasting induced defenses in birches and other woody plants (e.g., Tuomi et al., 1984). Defoliation by herbivores of mountain birch, growing on nutrient-poor soils, for example, results in nutrient stress. Nitrogen in tissue declines and carbon is relatively abundant, so carbon-based phenolic defenses increase. This stress persists for 3–4 years (Fig. 6.16). The hypothesis has broad implications for understanding patterns in plant defenses and patterns in herbivore attack.

The emphasis on resource limitation resulted in another development in hypotheses on plant–herbivore interactions.

RESOURCE AVAILABILITY HYPOTHESIS

In the tropical rain forest on Barro Colorado Island, Panama, the pattern of herbivory studied by Coley (1983) did not fit the predictions of the plant apparency hypothesis. She studied 46 common tree species divided into two categories. *Pioneer species* were found only in light gaps, with high light availability after a disturbance such as a tree fall. The pioneers were early successional and very patchy in distribution, qualifying as unapparent species in Feeny's hypothesis. *Cecropia insignis* and *C. obtusifolia* were two common "weedy" species included in the study. The *persistent species* grew in the forest understory in shaded conditions. They grow much more slowly than the pioneer species, have denser wood, and are longer lived. Therefore, they persist into the mature forest canopy. Examples include *Calophyllum longifolium, Macrocnemum glabrescens,* and *Virola sebifera.* Saplings of each type were studied for leaf damage by herbivores and leaf characteristics.

Levels of herbivory did not differ between pioneer and persistent species *on young leaves.* The percent of young leaves grazed per day was 0.83 for pioneer species and 0.97 for persistent species, based on an annual average, a nonsignificant difference. This means that plants in early successional patchy sites had a probability of discovery by herbivores equal to that for persistent species. This contrasted with the plant apparency hypothesis.

For *mature leaves,* the pattern was different, with more grazing on pioneer species leaves than persistent species leaves (average of 0.24 and 0.04%/day, respectively). This difference correlated with higher values in mature leaves of persistent species of total phenols, tannins, fiber, cellulose, leaf toughness, and lower water content. Persistent species developed better-defended mature leaves than those of pioneer species.

These kinds of results led Coley (1983) to propose an alternative to the plant apparency hypothesis, based on resource availability of plants in light gaps and in the understory. High growth rates are possible in light gaps, with high light and relatively high nutrient availability. Compensation of loss from herbivory is rapid, so resources are used for rapid growth, with lower levels of defense. The pioneer species have rapid growth, relatively low defense by digestibility reducers, with rapid damage compensation being a major aspect of the relationship with herbivores. Large, short-lived leaves are commonly seen in these pioneer species, for example in *Cecropia.* In the shade of the understory, with high competition for nutrients, persistent species can grow only slowly and could not compensate for heavy damage of mature leaves. Heavy defense of long-lived leaves is the common theme. Thus habitat quality is the independent variable that causes selection for different defensive systems.

This hypothesis was extended to other vegetation types by Coley, et al. (1985). They argued that the differences between rapid- and slow-growing plants were persistent, even though apparency did not change. They cite 23 cases where herbivory is much higher on fast-growing than on slow-growing plant species, with examples from tropical forests, boreal forests, and arctic tundra. They also note that in the "throwaway economy" of fast growers, recyclable defenses should be used, such as alkaloids and other toxic chemicals. Conservation in the slow-growing species with long-lived leaves makes digestibility reducers a sounder investment in protection. Thus the plant apparency and resource availability hypotheses overlap considerably in the detected patterns, but the driving forces are recognized as different.

If a wide variety of herbivores, such as insects, microtine rodents, hares, beaver, moose, caribou, and monkeys, prefer rapidly growing plant species, in many cases to the practical exclusion of slow-growing species (10 cases of the 23; Coley et al., 1985), this does raise another question. If herbivores prefer the most vigorous plant species in a vegetation, what role does plant stress really have to play in herbivore population dynamics? Of course, the resource availability hypothesis relates to evolved and persistent differences between species, whereas the plant stress hypothesis relates to within-plant species variation in quality over space and time. But if there is such a general response by herbivores to rapidly growing plants, it does raise the question about herbivore responses to vigorous plants in a species relative to less vigorous plants of the same species in a local environment. The argument is valid that many herbivores attack more frequently the most vigorous members of a plant population or the most vigorous plant parts within a plant individual—the plant vigor hypothesis.

PLANT VIGOR HYPOTHESIS

The plant vigor hypothesis relates to within-plant-species variation in quality for herbivores (Price, 1991d). Foresters have long recognized that in some cases young vigorous trees are more susceptible than older plants to attack by herbivores. Some examples from W. L. Baker (1972), Furniss and Carolin (1977), and other sources are listed in Table 6.2. Much additional evidence can be found in the ecological literature. The pinyon pine cone and shoot borer, *Dioryctria albovitella,* on pinyon pine, *Pinus edulis,* attacks the most robust new shoots each year. Stout terminal shoots in the upper canopy are the preferred oviposition and feeding sites; these shoots are killed, causing a strong reduction of female cones on such shoots and a clipped appearance of susceptible trees (Whitham and Mopper, 1985; Mopper and Whitham, 1986). The leaf beetle *Chrysomela confluens* is 400-fold higher in density on juvenile narrowleaf cottonwood, *Populus angustifolia,* than on mature trees (Kearsley and Whitham, 1989). On narrowleaf cottonwood, *Pemphigus betae* selects the largest leaves in a population of leaves on the tree, although mature trees are attacked (Whitham, 1978) (see Chapter 15 for details).

Studies in the author's research group show repeatedly that gall-forming insects on shoots attack longer shoots more frequently than they do shorter shoots, and younger, vigorous plants more frequently than older, slower-growing plants. Larvae also survive better on longer shoots; the following species having been studied: *Euura lasiolepis* on *Salix lasiolepis* (Craig et al., 1986; Price and Clancy, 1986b), *E. mucronata* on *Salix cinerea* (Price et al., 1987a,b; Roininen et al., 1988), *E. exiguae* on *Salix exiguae* (Price, 1989), and *Diplolepis fusiformans* and *D. spinosa* on *Rosa arizonica* (Caouette and Price, 1989).

TABLE 6.2   Examples from the Forestry Literature that Indicate Utilization by Insect Herbivores of Vigorous Plants or Plant Modules more than Stressed Plants. Note that Species Oviposit Where the Larvae Will Feed (O/F Indicates a Link between Oviposition Site and Larval Feeding Site) and Feeding is Internal (W/P Indicates Feeding within Plant Parts).

Insect Taxon	*Host Plant Taxon*	*O/F*	*W/P*
Coleoptera			
1. White-pine weevil, *Pissodes strobi*	White pine, *Pinus strobus* Sitka spruce, *Picea sitchensis*	×	×
2. Pales weevil, *Hylobius pales*	Pines, *Pinus*	×	×
3. Pine root-collar weevil, *Hylobius radicis*	Pine	×	×
4. Pine weevil, *Pissodes approximatus*	Pine	×	×
5. Weevil, *Cylindrocopturus eatoni*	Pines, *Pinus ponderosa* and *P. jeffreyi*	×	×
6. Lodgepole terminal weevil, *Pissodes terminalis*	Pines, *Pinus*	×	×
Lepidoptera			
7. White-pine shoot borer, *Eucosma gloriola*	Jack pine, *Pinus banksiana*	×	×
8. Jack-pine shoot moth, *Eucosma sonomana*	Jack pine, *Pinus banksiana*	×	×
9. Cottonwood shoot borer, *Gypsonoma haimbachiana*	Eastern cottonwood, *Populus deltoides*	×	×
10. Pitch nodule maker, *Petrova albicapitana*	Jack pine, *Pinus banksiana*	×	×
11. European pine shoot moth, *Rhyacionia buoliana*	Pines, *Pinus*	×	×
12. Nantucket pine tip moth, *Rhyacionia frustrana*	Pines, *Pinus*	×	×
13. Aspen blotch miner, *Lithocolletis salicifoliella*	Aspen, *Populus tremuloides*	×	×
14. Ponderosa twig moth, *Dioryctria ponderosae*	Ponderosa pine, *Pinus ponderosa*	×	×
15. Pinyon pine cone and shoot borer, *Dioryctria albovitella*	Pinyon pine, *Pinus edulis*	×	×

SOURCE: Based on Price et al. (1990).

It is clear from our studies on *Euura* gall-forming sawflies that females have evolved to attack long shoots because larvae survive better there and stressed plants are very deleterious to young larvae. Shoot length on woody plants declines on average with plant age, and attack declines with plant or ramet age (Fig. 6.17) (e.g., Price et al., 1987a). On longer shoots larvae established in galls more successfully and survived better after establishment, with plant resistance factors being important and natural enemies playing no role

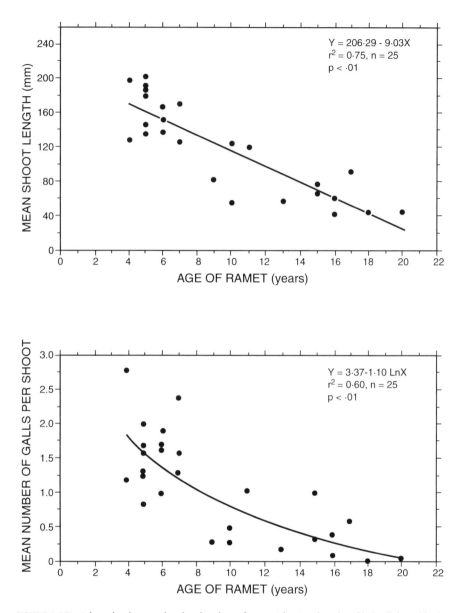

FIGURE 6.17   Relationships between shoot length and age of ramets (above) and number of bud galls formed by the sawfly, *Euura mucronata*, and age of ramet (below). The host plant is *Salix cinerea*, growing in Finland, which grows clonally producing many stems or ramets emerging at ground level. From Price et al. (1987a); with permission. Copyright © 1987 by Springer-Verlag GmbH.

(Price et al., 1987b). In experiments on water stress, survival was much higher for *Euura lasiolepis* on high-water treatment plants than on low-water treatment plants (Price and Clancy, 1986b). The experimental result mimicked survivorship curves on one wild willow clone after a wet winter providing good soil moisture, followed by a dry winter, willow stress, and reduced survivorship in the sawfly (Fig. 6.18).

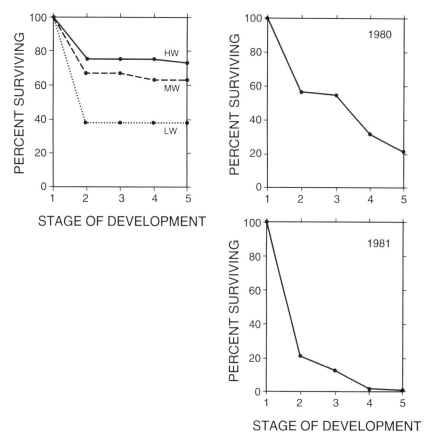

FIGURE 6.18 Comparison between an experiment on the effects of water availability on willow growth and attack and survival by the stem-galling sawfly, *Euura lasiolepis* (left), and survival in one wild clone after relatively high and low precipitation in 1980 and 1981, respectively (right). Note the similarity in low survivorship to stage 2 when water shortage reduces willow growth. Stages of development are: 1, egg; 2, early larva; 3, mid larva; 4, late larva; 5, larva in cocoon. Reduced survival on the wild clone to stages 4 and 5 relative to the experimental results is caused by higher parasitism on the wild clone. HW, high water treatment; MW, medium water treatment; LW, low water treatment. Based on Price and Clancy (1986b); with permission from the Ecological Society of Japan.

These results are similar to those for vertebrate herbivores also. Moose preferred trees clipped to simulate moose browsing five times more than control trees, for regrowth was more vigorous on the clipped trees (Danell et al., 1985). Moose-browsed birches were more heavily attacked than unbrowsed trees by mountain hare, psyllids, leaf gallers, leaf miners, and other insect herbivores (Danell and Huss-Danell, 1985).

The argument that many herbivores attack the most vigorous plant parts or plant individuals in a population is well documented. For insects it has been shown that larvae survive better in vigorous plant parts and that plant stress in wild and experimental plants is deleterious to herbivores. Many of the herbivores concerned are gallers or shoot borers, all of which attack actively growing plant parts. It has also been shown that protein and phenol levels in plants are not good predictors of plant resistance or susceptibility, while shoot length is an excellent predictor. Therefore, these kinds of herbivores may have circumvented the problem of low nutritional status and high defense in plants. Perhaps such rapidly growing tissues always have adequate or nonlimiting nutrients and low defense against specialists, as predicted by the plant apparency hypothesis.

Whatever the mechanisms allowing herbivores to attack vigorous plants, this category of herbivores must be recognized as distinct from those responding positively to plant stress, or aging plants. General theory must ultimately cope with why such differences exist and recognize that the plant vigor hypothesis is valid for many species. This hypothesis is thus a variation on the resource availability hypothesis, but it emphasizes variation within species and individuals rather than evolved differences between species.

DO HERBIVORES ACT AS SELECTIVE AGENTS ON PLANT DEFENSES?

Let us turn full circle and return to the beginning of Chapter 5, where plant and herbivore interactions were introduced with the concept of coevolution fostered by Ehrlich and Raven (1964). The concept of reciprocally induced evolutionary changes between plants and insects was illustrated by the authors using phytochemical defenses in plants and insect herbivores which evolve an ability to cope with those defenses in one way or another. The concept was set in general terms and invoked long spans of evolutionary time that result in major familial-level differences in phytochemical defenses, making coevolution very difficult to test. Cause-and-effect relationships have been difficult to establish unequivocally. Janzen (1980) stressed the need for reserving the term *coevolution* for species interactions that have actually produced the selective pressures for traits that are reciprocally induced.

While it is clear that insects have evolved in response to plant defenses as discussed at length in Chapter 5, it is by no means clear whether they have generally had a reciprocal impact on plants. The hypothesis of coevolution in varying formulations is appealing and widely employed (e.g., Gilbert and Raven, 1975; Futuyma and Slatkin, 1983; Nitecki, 1983; Thompson, 1982, 1994), but where is the sound evidence that insects have actually selected for the evolution of plant defenses? In his authoritative treatment of the *coevolutionary process,* Thompson (1994) provides no direct evidence of insects as selection agents for phytochemical defenses. He does document genetic aspects of defense against pathogens of the kind we discuss in Chapter 9 concerning a probable case of coevolution between plants and herbivores. But the chemical nature of the defense is not worked out and probably does not involve allelochemicals as constitutive defenses such as alkaloids, glucosinolates or coumarins. One example that Thompson provides is important, although the role of insects in evolutionary change is unknown. "Some populations may occur in environments free from enemies and therefore lose their defenses through natural selection, genetic drift, or a combination of the two. Hawaiian plants, which have been protected for much of their evolutionary history from most herbivores, are an extreme case: they are extraordinarily nonaromatic, nonprickly, and nontoxic. Carlquist (1970) has noted that although plants in the mint family are often strongly fragrant, the Hawaiian species are either odorless or nearly so. Almost all the plants catalogued in *Poisonous Plants of Hawaii* (Arnold, 1968) are plants introduced to the islands over the past several hundred years. The same applies to morphological defenses" (Thompson, 1994, pp. 164–1965).

Would the chemical nature of plants be essentially the same in the absence of insect herbivores? Although not using insects in their study, a promising approach to answering this question was taken by Bryant et al. (1989) using a biogeographic perspective, which may well be used by insect ecologists. As Thompson's example from Hawaii suggests, the geographic study of plant defenses where herbivores are common or rare will provide an avenue for answering a difficult question. Bryant and associates argued that the 10-year cycle of snowshoe hare, *Lepus americanus,* in Alaska and Siberia should have selected for highly defended juvenile shoots in winter because hare peaks are so devastating to birches and willows. In Finland, mountain hare, *L. timidus,* do not cycle as much, nor do they cause as much winter damage, selecting less for high plant defense. On Iceland, with no browsing mammals before Norse colonization, they expected even less defense. These predictions proved to be correct (Fig. 6.19). The order of palatability for birches was Icelandic highest, Finnish intermediate, and Siberian and Alaskan lowest. This ranking correlated well with the levels of defense in juvenile shoots in winter, based on internode resin and triterpene acids.

Also, snowshoe hare in Alaska would eat more of the more defended birches than the mountain hare from Finland, indicating a counter-response in snowshoe hares in Alaska to high plant chemical defense. This study suggests that true coevolution has occurred, with mutually induced responses in birches and hares in Alaska.

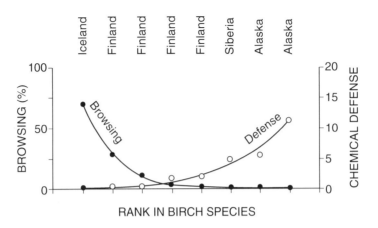

FIGURE 6.19   General results on the palatability of twigs in winter of young birches to hares in Alaska to Finland relative to the chemical defenses at the internodes, involving total resin and triterpene acid concentrations of juvenile birches (expressed as percent dry mass of internodes). From J. P. Bryant et al., Biogeographic evidence for the evolution of chemical defense against mammal browsing by boreal birch and willow, *Am. Nat.,* 1989, **136:**20–34. Published by the University of Chicago Press. Copyright © 1989 by the University of Chicago. All rights reserved.

Further studies in this vein are warranted, especially when plant populations can be found that have not been exposed to herbivores for a long time. Comparison between areas strongly affected by the Pleistocene glaciations, such as Finland, with areas not covered by ice, such as Alaska or south of the ice sheet, may be fruitful beyond the hare study by Bryant et al. (1989). But alternative hypotheses on the evolution of plant defenses are warranted, including the following: (1) the role of genetic drift as mentioned by Thompson (1994), (2) the adaptive value of phytochemicals in relation to the physical environment (e.g., temperature, salt, etc., as discussed briefly in Chapter 5), and (3) what are the plant responses as in carbon–nutrient balance and the resource availability hypotheses, for example to the very long summer days in the Arctic?

Clearly the pace in development of basic hypotheses on plant–herbivore interactions is quickening, and not all hypotheses have been discussed in Chapters 5 and 6. We should look for major advances in the future, and the prospect is exciting. But just how a synthesis of this broad field might be generated, and how theory is formulated, remain major challenges.

Chapter 7

Interactions Between Prey and Predator

In Chapters 5 and 6 some aspects of the interaction between the first and second trophic levels were covered. Moving up the feeding chain one step, we must now consider interactions between the second and third trophic levels. After the herbivores, all other members in the food chain of living organisms are either predators or parasites, so predation is an important subject in ecology. Added to the importance of general predators, many people regard the total consumption of any organism, plant or animal, as predation. Therefore, predation is a central issue in understanding how energy moves through the food chain and in understanding the adaptive strategies both for getting food and for the avoidance of becoming food. Since food is a critical resource for all organisms, predation must be considered to be at the heart of ecological thinking, and this explains why so many researchers have been fascinated by the subject. These factors justify the space devoted to the subject in this book.

PREDATION AS AN EXPLOITATION STRATEGY

The importance of predation as an exploitation strategy may be summarized under four main categories. First, predators play a prominent role in the flow

An underwing moth senses the ultrasonic chirps of an approaching bat and performs a protean display. Such displays frequently result in escape from the predator, discussed in this chapter. This is a composite of a painting by Walter Linsenmaier (1972) and of a photograph in Vaughan (1978). Drawing by Alison Partridge.

of energy through the community. Examples of predation can be seen at the feeding links between every trophic level. Second, predators and parasitoids have repeatedly been singled out as regulators of the animal populations upon which they feed. The biological control of insects, until recently, has relied on this notion, and in Chapter 17 it is seen how predation rates as a factor in regulation. It is certainly one of the most visible aspects of mortality. Third, predators play a role in maintaining fitness of the prey population. This statement is axiomatic when fitness is defined in terms of survival and the leaving of viable progeny. Those organisms eaten by predators are unfit. Those organisms best able to defend themselves or otherwise escape predation survive. Predation is therefore on old and decrepit individuals, those that are diseased or malformed, and those young whose parents are sufficiently unfit not to take adequate care of their progeny. This may range from a delinquent giraffe mother who does not defend her calf sufficiently to prevent lion predation, to a moth that lays its eggs in a position that makes them vulnerable to attack by parasitoids. Predation is one of the chief mortalities to which organisms are exposed and tends to maintain a healthy and vigorous breeding population of prey. Finally, predators act as selective agents in the evolution of their prey. Any severe mortality factor is likely to change the population permanently, and this is certainly true of predation. Some of the very diverse evolutionary trends are listed at the end of the chapter. Their diversity and commonness attest to the prevalence of predation as a selective factor in nature. Practically every insect species can be fitted into one or more of the categories listed, indicating that at some stage in its evolutionary development every insect species has been exposed to either heavy or prolonged predation. Perhaps predation pressure has produced one of the most visible evolutionary forces in the whole animal kingdom.

PARASITOIDS

In the insects there is a large group of species that parasitize others. In spite of Reuter (1913) making a clear distinction between these insects and true parasites, they are still commonly called parasitic insects, although since about 1960 the more precise term *parasitoid* has been used more frequently in the literature. A **predator** is defined as an organism that kills and consumes many animal-food items in its life span, although if plants are consumed completely, as are seeds or seedlings, predation may also cover these forms of feeding. In contrast, a **parasitoid** can be defined, as in Chapter 2, as an insect that requires and eats only one animal in its life span, but may ultimately be responsible for killing many. Here an egg is laid on or in a host (or prey), a larva hatches and consumes the host, becoming a free-living adult, which, if female, proceeds to oviposit on many other hosts that will nourish its progeny. This is where a single female may ultimately be responsible for the death of many individuals.

Some would argue that both types of food exploitation should be called predation. Certainly, the fact that both predator and parasitoid kill many food items makes them exactly comparable as far as effects on prey populations go. The only difference is that the predator kills food for itself and the parasitoid kills or paralyzes food for its progeny. However, predators such as insectivorous birds also collect prey for their young, and as far as population dynamics are concerned, the results are the same! Therefore, parasitoids must be considered under predation, and in fact a great deal of predation theory has been developed with parasitoids in mind. Only Lotka (1924), Nicholson (1933), and Nicholson and Bailey (1935) (see also Hassell, 1978) need be mentioned to show how important parasitoids have been in stimulating the development of ecological theory, but their work is discussed in Chapter 8, which considers predator populations rather than individuals.

Classic studies on predators of insects are considered in this chapter, for which the research itself was concentrated in the 1950s and 1960s. They have not been superseded in the literature and remain as enduring insights to the complex relationships among predators and their prey.

VERTEBRATE PREDATOR AND INSECT FOOD

By looking at a single species of predator, some general ideas may be obtained on how a predator exploits a given environment most efficiently, when prey species are most likely to be vulnerable to predation, and how they reduce this vulnerability. An example is the vertebrate predator, the great tit, *Parus major*. This bird is a large relative of the chickadees in North America: the black-capped, Carolina, and brown-capped chickadees. They are all insectivorous. By studying the great tit in some detail one may catch a glimpse of how this vertebrate predator views its insect food supply.

The great tit occurs throughout the Palearctic region and has been intensively studied in England, continental Europe, and Japan. In southern Holland, Luuk Tinbergen from the University of Groningen studied the great tit feeding its nestlings in a scots pine plantation called the Zwarte Berg. He wrote the manuscript of his 1960 paper during the winter of 1954 to 1955 and continued his field work in the summer of 1955. At the end of the year he died, so although the paper was published in 1960, the author could not have been aware of Holling's work in Canada, which began to appear in print in 1959 and which has also added significantly to the understanding of predation.

PROBABILITY OF ENCOUNTER HYPOTHESIS

Tinbergen argued in his **probability of encounter hypothesis** that if a very simple view of predation is adopted, and random movements of the predator and random dispersion of the prey are assumed, the number of a particular prey N_A captured by a predator can be expressed simply as

$$N_A = R_A D_A t$$

where N_A is the total number of species A prey captured by a predator in time interval t, D_A the population density of prey, and R_A the risk index of prey species A for a particular predator.

Prey Risk

Several factors influenced the risk index. For example, prey size was important. Very few small insects were collected and in the case of larvae of the pine noctuid moth *(Panolis flammea)*, none less than 18 mm were taken, although they represented 20% of the species available (Fig. 7.1). Each prey species also was subjected to a typical intensity of predation in relation to that of *Panolis* at any one time. Tinbergen took *Panolis* larvae (over 2 cm long) and gave them a risk value of 100 and then compared the relative risk of many other insects. For example, if species A was twice as abundant as *Panolis* in the environment but formed the same percentage of the food brought to the nest by the great tits, its relative risk would be 50% of *Panolis*. In this way Tinbergen could quantify the relative palatability, and ease in

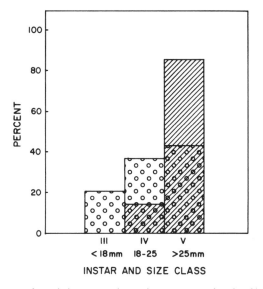

FIGURE 7.1   Percentage of *Panolis* larvae in each size class in twig samples (dotted) and in food of great tits (hatched). Data from Tinbergen (1960).

TABLE 7.1   Risk Index for Insect Species in the Zwarte Berg in Relation to Predation by the Great Tit, *Parus Major.*

Taxonomic Group or Common Name	Species	Risk Index	Important Characters in Relation to Risk Index
Lymantriid moth	*Lymantria monarcha,* 20 mm (larva)	225.0	Cryptic but large (6 cm)
Pine noctuid moth	*Panolis flammea,* 20 mm (larva)	100.0	Cryptic but large
Pine looper moth	*Bupalus piniarius* (adult)	95.0	Moths on ground but cryptic
Pamphiliid sawfly	*Acantholyda nemoralis*	49.0	Web easily seen, but larva smaller than *Panolis*
Tortricid moth	*Cacoecia piceana*	10.0	Same as for *Acantholyda,* difference not understood
Diprionid sawflies	*Diprion nemoralis* and *D. simile*	3.1	Solitary but conspicuous
Diprionid sawflies	*Diprion virens* and *D. frutetorum* (larvae)	2.2	Solitary and cryptic (green)
	Diprion virens (prepupae and pupae)	16.0	Solitary, pupae nonresinous
Diprionid sawflies	*Diprion pini* and *D. sertifer* (larvae)	0.1	Very conspicuous, colonial, resinous taste
	Diprion pini (prepupae and pupae)	22.0	Solitary, pupae nonresinous

SOURCE: Tinbergen (1960).

finding the important food items, in the great tits' diet (Table 7.1). In fact, Tinbergen found that R_A, the probability of a prey item being discovered and eaten by the predator, varies radically and depends on a number of factors: prey size, density of prey, availability of other food, developmental stage of prey (larva, pupa, or adult), and learning processes in the predator.

Density Effects and Palatability

It is clear that as far as the survival of the insect is concerned, its abundance, defensive, and palatability characteristics relative to the other prey species in the community will influence actual mortality enormously. One example from Tinbergen's work is given in Fig. 7.2. Even within a species, its density had a dramatic influence on the success of the predator in finding its food (Fig. 7.3). Here, at the lowest densities the percentage of a species in the food was relatively low. At higher densities the percentage showed a much steeper rise than in the expectation curve, and reached a high value at medium densities. At the highest densities the slope of the curve was much less steep than the slope

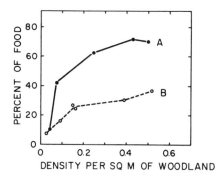

FIGURE 7.2   Percentage of total food taken by great tit represented by *Panolis* larvae in relation to their density in the trees and the availability of alternative prey: (A) during periods in which large *Panolis* caterpillars occurred together with prey of low risk index only; (B) when appreciable populations of species of high or moderate risk index were present (e.g., *Lymantria, Acantholyda,* or *Diprion* cocoons). After Tinbergen (1960).

of the expectation curve. Also, the palatability and other risk factors affected the quantity of species taken. For example, leveling off occurred at about 29% in *Panolis,* 25% in *Acantholyda,* and 12% in *Diprion sertifer.*

These points indicate the sorts of selective pressures under which prey populations change. They emphasize that to avoid predation, evolutionary change relative to the other species present may be just as important as absolute changes. These relative changes are difficult to evaluate, as it is hard to choose an adequate baseline on which to measure change.

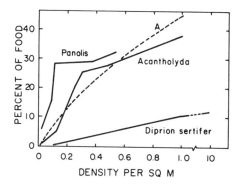

FIGURE 7.3   Percentage of a prey species in the great tit diet in relation to the density of the prey. Density figures refer to larvae over 20 mm long in *Panolis* and over 10 mm in the other species. Data for *Acantholyda* were collected in the presence of large *Panolis* larvae and for *Diprion sertifer* before the appearance of *Panolis.* (A) Expectation curve for *Acantholyda* derived from the probability of encounter hypothesis and the density of other species present. After Tinbergen (1960).

SEARCH IMAGE HYPOTHESIS

Tinbergen (1960) explains the change in risk of a species with density as follows. Tinbergen's own words are used to explain low risk at low densities, as these are the words that have created a controversy that is unlikely to be settled for a long time. His statement has strong supporters and vehement critics with very keen and observant biologists on both sides:

> The remarkably low risk at low densities is perhaps most easily explained by the hypothesis of searching images. We have assumed that specific searching images are adopted only when the species in question has exceeded a certain density. Accordingly the birds would make only a limited number of chance encounters when density is below the critical level, while predation would become more efficient at higher densities. The observations on *Acantholyda*, *Cacoecia* and *Panolis* (without alternative [food]) suggest that the critical density is rather sharply defined, for the increase in risk seems to be restricted to a rather narrow range of densities. The observations, however, are too few to warrant a definite conclusion on this point.

Tinbergen makes the point that **search images** for palatable prey are "accepted" at much lower densities than those for unpalatable prey (cf. Fig. 7.3). To quote further:

> Finally, we may ask why specific searching images are not used at low densities. From a functional point of view, this can be understood if we assume that the birds can adopt only a limited number of searching images. Under these circumstances it will be more profitable to search for abundant than for rare prey. Moreover, it may be supposed that a certain number of encounters is required for the acceptance and maintenance of a searching image.

Tinbergen was trying to assess the mental ability of the great tit, although he was aware of the importance of the profitability of search (cf. Royama's concept of profitability discussed later in this chapter). Tinbergen went on to explain the low risk at high density by saying that the great tit has a preference for a mixed diet. Therefore, an individual bird will not take more than 50–60% of its total diet in the form of a single species.

Tinbergen summarizes his hypothesis as follows: "the intensity of predation depends to a great extent on the use of specific searching images. This implies that the birds perform a highly selective sieving operation on the visual stimuli reaching their retina." This is "a kind of learning process." Dawkins (1971a,b) showed that chicks do learn to see cryptic food and thus supports Tinbergen's hypothesis. However, she argues cogently that the term *adopting a search image* has been used in conflicting ways and is not helpful. The term *learning to detect* may be preferable. Alcock (1973b) defined the concept as "hunting for certain visual cues associated with relatively cryptic food while ignoring or overlooking others."

CRITIQUE OF HYPOTHESIS

There is also something to learn from a brief summary of the criticisms of, and difficulties with, the hypothesis:

1. All prey species densities were not measured; therefore, the proportion of one prey in the total available could not be determined. The following ratio is needed:

$$\frac{\text{proportion of species } A \text{ present in total food available}}{\text{proportion of species } A \text{ taken by predator}}$$

2. The density of prey does not reflect the proportion of individuals that are unpalatable through disease and such. (Tinbergen recognized this important variability within a population.)
3. Tinbergen does not consider that within a species some individuals may be discovered more readily than others. If most favorable parts of niches are occupied first, and parts in which individuals are more vulnerable later, this could produce the same change in prey risk that Tinbergen documented. This applies to low risk at low density changing to high risk at moderate density. This would certainly be in accord with current concepts on predation covered in Chapter 8.
4. Within the bird's defended territory the density of prey may be very much reduced by predation, so that locally, profitability of searching for that prey and probability of finding it are very low, even though "average density" estimates indicate that only 30–40% of prey is taken. This may account for the apparent low risk at high density.
5. Tinbergen does not consider the distance of the prey from the nest, an important part of prey profitability.
6. Tinbergen makes three assumptions to explain the change in prey risk that are hard to test. First, he assumes that the bird uses a searching image. Second, he assumes that only a limited number of such images can be stored at one time. Third, he assumes a preference for a mixed diet.

In general, the problem of discovery of prey must be considered on a microenvironmental level and the situation is much more complex than it appears at first sight. These are all surmountable difficulties and Tinbergen's paper has stimulated a great deal of good research. Some is discussed later under the concept of profitability proposed by Royama in 1970.

COMPONENT ANALYSIS OF PREDATION

Clearly, a detailed study of the feeding rate of a predator in a simple environment was needed to avoid the difficulties that arise in interpreting Tinbergen's data. The essential components of a predator's life, and the act of predation, must be understood. Holling, then at the Sault St. Marie laboratory of the Canadian Forest Service, championed the component analysis approach to predator–prey interactions. He asked a simple question: "What are the parameters that are universal in the act of predation? By examining the components experimentally using various animals, he was able to develop models predicting results that are still proving to be true for real situations.

The simplicity in the concept of the component analysis approach can be seen by one of his early experiments (Holling, 1959a). He placed a blindfolded human subject in front of a table. On the 3-ft^2 table he thumbtacked sandpaper disks 4 cm in diameter. The subject was then asked to tap the table until she found a disk, remove the disk, and set it aside and then continue tapping, and so on. She tapped for 1 minute. For each density of disks eight replicates were made. The disk density ranged from 4 to 256 per 9 ft^2.

The results can be seen in Fig. 7.4. Why should there be a curvilinear response? Why should there not be a linear relationship in which the doubling of disk density doubled the rate of discovery? Here the number of disks picked up increased at a progressively decreasing rate as disk density rose. Holling pointed out that at low densities the majority of time was spent searching. Efficiency rose rapidly with increasing density until so many disks were found that most time was spent picking them up and laying them aside. The human "predator" could only handle a certain maximum number of "prey" disks per unit time.

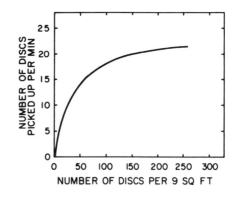

FIGURE 7.4   Graphical representation of Holling's disk equation showing the rate of discovery by a human "predator" in relation to the density of sandpaper disk "prey." After Holling (1959a).

DISK EQUATION AND FUNCTIONAL RESPONSES

Many of these curves have been described by Holling himself and subsequent workers for diverse real situations, and thus support Holling's **functional response** curve. The general **disk equation** is

$$N_A = \frac{aT_T N_o}{1 + aT_H N_o}$$

where N_A is the number of disks removed, N_o the density of disks, T_T the time interval available for searching, T_H the handling time, and a the rate of discovery (rate of search multiplied by probability of finding a given disk). It is taken for granted that the predator is hungry, but hunger is an additional important component of the functional response. The curve is representative of many cases of predation by invertebrates so far studied and fits many functional responses of parasitoids to host density. It was called a **type 2 response** curve by Holling. (Holling's type 1 curve represents a rather specialized example which is not dealt with in this book; see Holling, 1961, 1965.). But at the moment our concern is how a vertebrate predator such as the great tit might react to prey density.

Holling (1959b) found from his own experiments that an S-shaped functional response curve could be expected for vertebrates, and cited Tinbergen's data and that of a former student of Tinbergen, Mook (1963), which have been obtained in Canada, as other examples of the vertebrate functional response model (Fig. 7.5). Holling's explanation for this response is much simpler than Tinbergen's. Given the four essential components in the type 2 functional response, rate of successful search, time predator and prey are exposed,

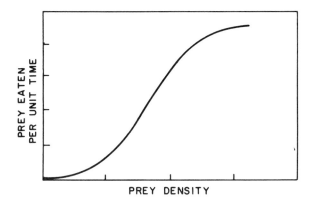

FIGURE 7.5 Sigmoid functional response, or type 3 response, showing the change in capture rate by an individual predator in terms of prey density.

handling time, and hunger (not incorporated into the disk equation, but see Holling, 1966), which must also be shared by vertebrate predators, only one new component need be added to describe adequately the **sigmoid response**— learning. The predator, with sufficiently frequent contact with particular items, will learn to find and recognize them and handle them more rapidly. At a given density its efficiency or rate of capture will increase rapidly. Ultimately, handling time will become the dominant component in the time available and the rise will decelerate to a plateau, just as Tinbergen had found. Since invertebrates may also show a sigmoid response [e.g, *Nemeritis canescens*, a parasitoid of moth larvae (Takahashi, 1968; Hassell, 1970, 1981)] it is best to refer to the two **functional responses** simply as **type 2** and **type 3 responses** without using the terms *invertebrate* and *vertebrate response*. Some vertebrates show a type 2 response (Holling, 1965). It is also clear now that learning is not the only factor that can result in a sigmoid functional response, but more active search in areas where prey are more abundant will yield the same response (Murdoch and Oaten, 1975; Hassell et al., 1977; Hassell, 1981).

Holling (1965) also found that the S-shaped functional response (or Holling's type 3 response) changed markedly with the strength of the stimulus from the prey just as Tinbergen had found (Fig. 7.6). Sawfly survival was much greater (30%) in the presence of a palatable prey—sunflower seeds. Tinbergen's comparative risk R_A can change radically as associated species change.

Concerning selection by avian predators, it is therefore maladaptive to be

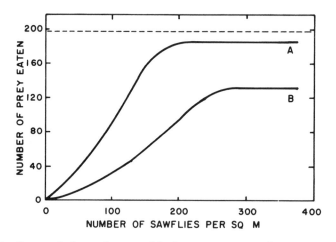

FIGURE 7.6   Change in the functional response of the deer mouse, *Peromyscus leucopus*, to the density of prey sawflies in cocoons in the presence of alternative food: (A) dog biscuits, a low-palatability alternate food; (B) sunflower seeds, a high-palatability alternate food. Dashed line indicates the total food eaten, sawflies plus alternative food. The number of dog biscuits eaten was expressed in "sawfly units" of approximately equivalent nutritive value. After Holling (1965).

associated in space and time with a less palatable species, as predation pressure is greatly increased. Some of the ways to avoid this situation are to be temporally and/or spatially separated or to mimic the unpalatable species. With such strong selective pressures as 30% advantage, evolution will be rapid and probably quite precise.

From the sawfly and dog-biscuit situation in the laboratory it is possible to extrapolate to the field where the sawfly coexists with another species of dog-biscuit-like palatability. Here the sawfly may evolve to look like its less palatable sympatriot, and thus form a **Batesian mimicry** complex where the model is relatively unpalatable and the mimic is palatable. [Batesian mimicry is defined as the result of evolutionary advergence in shape, pattern, color, or behavior of an edible "mimic" toward (close?) resemblance of a less palatable, and distantly related "model."] The levels of palatability have to change very little to make both species relatively unpalatable and provide an example of **Müllerian mimicry**—the result of evolutionary convergence in shape, patterning, and color of distantly related unpalatable species.

CRYPSIS AND DEFENSE

Other ways of avoiding predation are crypsis or defense. the effect of crypsis would be to alter the slope of the functional response, which has the same effect on survival as being in the presence of a more palatable species. If a prey species evolves a defensive mechanism, the shape of the response will change radically, as the species essentially becomes strongly repellent or distasteful. Holling's computer model predicted this change without any information on real situations. As prey density increases, the number taken increases initially and then the predator learns to avoid them (Fig. 7.7). Even with an invertebrate predator

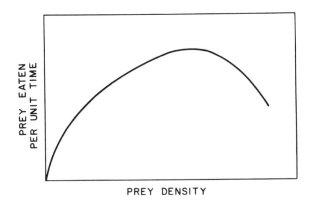

FIGURE 7.7   Functional response of a predator to a distasteful prey, illustrated by the change in rate of predation with increasing prey density. After Holling (1965).

feeding on a prey with a defensive behavior, as the prey density increases, the defense becomes more potent, which results in reduced predation.

Tostowaryk (1972) was the first to demonstrate this experimentally using the invertebrate predator *Podisus modestus* (Pentatomidae) on the colonial sawflies *Neodiprion swainei* and *Neodiprion pratti banksianae*. The defensive regurgitate of the prey larvae was described in Chapter 5. As the "colony" size increases, so predation decreases (Fig. 7.8). The normal colony ranges in size from 60 individuals at the beginning of the season to 20–40 at the end of the season. One of the determinants of colony size may well be the efficiency with which this collective defense can operate. These sawfly species are North American representatives of the diprionid family to which those that Tinbergen studied belong. Tostowaryk's studies help us to understand why the colonial diprionids had such a low prey risk in the presence of great tit predation.

In the light of Holling's work, another look at the great tit is worthwhile, this time through the camera lens and eye of Royama, who studied the species in Japan and at Wytham Wood near Oxford, England, while at Oxford University in the Edward Grey Institute. He set up cameras at the back of tit nesting boxes so that each time an adult brought prey to the nest a photograph could be taken automatically of the bird face and prey item. The precision of the photography was so great that each item could be identified, usually to species. A total of 29,000 photographs on 97 nestling days was collected in 3 years. At the same time, Varley and Gradwell, whose work is discussed in Chapters 17 and 18, supplied the insect abundance data from field sampling.

Royama (1970) found that the number of prey taken by great tits was not related to the number present in the tree canopy (Fig. 7.9). For example, with the winter moth, *Operophtera brumata* (Geometridae), the numbers of cater-

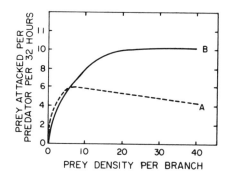

FIGURE 7.8   Functional response of third instar nymphs of the pentatomid bug, *Podisus modestus,* to (A) normally active second instar larvae of the sawfly, *Neodiprion pratti banksianae,* that showed a defensive behavior in response to predator attack, and (B) larvae of the same age and species immobilized by treatment in hot water and thus incapable of defense. Note that initially the active larvae are more heavily preyed upon as they are more visible to the predator, but as the prey density increases the colonial defensive behavior lowers the predation to less than 50% of that on the defenseless larvae. After Tostowaryk (1972).

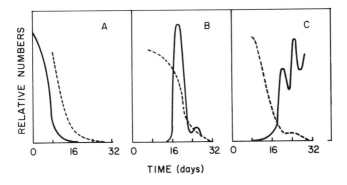

FIGURE 7.9 Abundance of some prey species in oak foliage (dashed line) and the number brought to the nest by great tits (solid line) during the spring in Wytham Wood. The vertical scales are not equivalent for the different sets of data. (A) Predation is reduced over 4 days before prey abundance is low; (B) predation increases dramatically as the prey larvae mature and pupate; (C) predation is greatest after the abundance of prey has declined. After Royama (1970).

pillars taken to the nest declined rapidly 4 days before their presence in the tree canopies (Fig. 7.9A). The reasons were unclear. One interpretation of the pattern in 7.9B for *Erannis leucophaearia* (Geometridae) is that larvae were not utilized while feeding in the leaves, but around the time the fully grown larvae were falling for pupation in the ground, they became exposed to predation. A similar explanation appears to be valid for *Orthosia stabilis* (Noctuidae) (Fig. 7.9C).

How could these rather unexpected results be explained? Royama proposed a new model based on (1) the action of natural selection, (2) the principle involved in the evolution of clutch size in birds (Lack, 1954, states that the clutch size in nidicolous species, i.e., that rear young in a nest, is adapted to the largest number of young for which the parents can provide enough food), and (3) the hunting efficiency of an individual predator in relation to the density of the prey (i.e., the disk equation of Holling, 1959a). These are three widely accepted ideas.

Royama also makes three assumptions:

1. Great tits have developed as efficient a method of hunting as their physiological capacity and their environmental conditions permit.
2. Composition of prey species in the habitat is constantly changing.
3. Different prey species occupy different niches.

CONCEPT OF PROFITABILITY

A predator would, therefore, have to search for different prey species in different niches and so must allocate its hunting time among them. As Royama assumed that the predator tries to maximize its hunting efficiency, the problem is to discover the most productive way for it to allocate its hunting time between the different prey species in different niches. To do this Royama proposed the **concept of profitability** of a prey species in its niche, where *profitability* may be measured as the biomass (or calorific or nutritional values) that the predator can collect in a given time in hunting this prey species.

Suppose that there are two alternative prey of the same biomass, and so on. Then the profitability of the prey species is N_A/T_T in Holling's disk equation terminology—the number caught per unit time. The profitability increases as the density of the prey species in its niche increases, but gradually flattens off (Fig. 7.10). Then suppose that the predator can move freely between two distinct niches A and B where prey species A and B live. In niche B the density of prey B is fixed as N_{oB} and the density of prey $A(N_{oA})$ is the only variable.

If N_{oA} is much smaller than N_{oB} (see $N_{oA} < N_{oB}$ in Fig. 7.10), there would be comparatively large differences between the profitability of A and B (cf. the difference between P_A and P_B). However, the difference would be much less, although niche A would now be the more profitable, if the density of prey A increased to $N_{oA'}$, at which density the profitability becomes only as high as $P_{A'}$ as compared with P_B. If the predator tries to increase its hunting efficiency, more time should be spent in niche B when $N_{oA} < N_{oB}$. If, however, N_{oA} increased to N_{oB} then P_A increases to P_B and the predator can take equal profits out of niches A and B. Allotment of time would be a matter of chance, and on average equal. If N_{oA} increased further to $N_{oA'}$ ($>N_{oB}$) and so does profitability P_A to $P_{A'}$, the difference between $P_{A'}$ and P_B is much smaller than that between P_A and P_B. Although more time should naturally be spent in niche A than in niche B, the total gain by doing so would be much smaller than it was when $N_{oA} < N_{oB}$, and more time was spent in niche B. For further

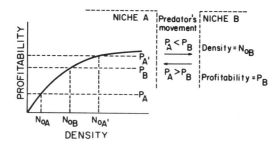

FIGURE 7.10 Profitability of niche A in relation to the density of prey species A and species B, and the predator's movements between insect niches. Note that the density of species A changes ($N_{0A} \rightarrow N_{0A}'$) in relation to the fixed density of species B (N_{0B}). The profitability changes, as N_A changes, according to Holling's disk equation. It is assumed that species A and B have the same biomass, palatability, visibility, and defensive reactions. After Royama (1970). Reprinted from *Journal of Animal Ecology* by permission of Blackwell Scientific Publications.

increase in N_{oA} there would be hardly any substantial increase in the profitability in niche A, and so a further increase in time spent in niche A would yield no increased profit. Clearly, from the predator's point of view, it is not the density of prey, but its profitability that is important (cf. Fig. 7.9).

As the composition of prey changes frequently, the predator has to pay attention to every potential niche to keep itself informed about profitable niches and to reject poor ones. It will stay a short time in poor niches and a longer time in good ones. However, as no substantial differences in profitability would be appreciated by a predator among niches where the prey density is above a certain level, no further increase in time spent hunting would be observed. Royama concludes that the time spent hunting in niche (T_T) will be an increasing function of prey density (N_o) but must flatten off (Fig. 7.11). From these curves the relationship of N_A to N_o is developed by substituting values of N_o and T_T in the disk equation, and the curves are all sigmoidal (Fig. 7.12).

If N_A is expressed as a percentage of the prey species concerned in the total number of prey taken by the predator, under Tinbergen's assumption that all other species are put in a single category whose density does not vary, it will yield the trend that Tinbergen in fact observed in his great tits. Therefore, the model proposed generates the trend that Tinbergen found and fits Royama's own data. If the **law of parsimony,** the principle that no more causes should be assumed than will account for the effect, is followed, then Royama's explanation of the sigmoidal response by vertebrate predators, being the more frugal in its assumptions, may seem more appealing than Tinbergen's. However, Alcock (1973b) showed that redwinged blackbirds can learn where they are likely to find food and what food they are likely to find, and these complementary processes support Royama's and Tinbergen's hypotheses, respectively. Even parasitoids may learn to use a similar pair of clues in searching for hosts (Taylor, 1974).

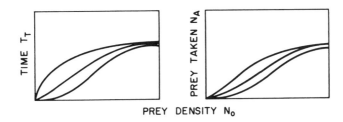

FIGURE 7.11   Time spent hunting by a predator in relation to prey density. The three examples are the possible shapes of curves representing T_T as an increasing function of N_0. After Royama (1970). Reprinted from *Journal of Animal Ecology* by permission of Blackwell Scientific Publications.

FIGURE 7.12   Relationship between the number of prey taken and prey density calculated from Fig. 7.11 using the disk equation. After Royama (1970). Reprinted from *Journal of Animal Ecology* by permission of Blackwell Scientific Publications.

AREA RESTRICTION OF SEARCH AND REWARD RATE

Croze also studied at Oxford University under the direction of Niko Tinbergen, Luuk Tinbergen's younger brother. Croze (1970) was critical of Royama's thesis and his comments supporting Tinbergen's search image hypothesis should be consulted. But Croze, who studied the searching behavior of the carrion crow (*Corvus corone*), provided some interesting additional information. He found that there were two important aspects to the crow's searching behavior: area restriction and reward rate. Once a crow found prey it tended to search more closely and in a more closely defined area, and as long as it was rewarded fairly frequently it would stay in the vicinity. Similar behavior has been observed by Smith (1974) in blackbirds and in thrushes searching for earthworms. This compares with Royama's ideas on niche searching and profitability, and the behavior will have an impact on prey populations. One response of the prey population is that a well-spaced dispersion pattern evolves (Fig. 7.13). As the prey population became more scattered, the intercatch distance increased disproportionately to the interprey distance. Increased distance between prey increases the searching time and so reduces the predator's functional response.

PREDATION ON POLYMORPHIC POPULATIONS

Then Croze asked what happens in terms of predator behavior and prey survival when the predator looks for more than one thing at a time? The question arose from the observation that many camouflaged animals are polymorphic for color and pattern (Sheppard, 1958) and that visual poly-

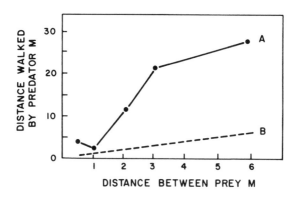

FIGURE 7.13 Relationship between the distance walked between prey items by a carrion crow (A) and the distance between the prey (B). As the prey distance increases the crow spends disproportionately more time searching for prey. After Croze (1970).

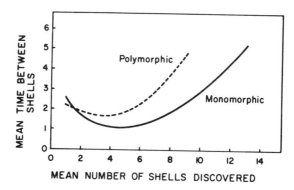

FIGURE 7.14 Mean time taken between the discovery of successive shells when a crow searched in a population of 27 shells, each shell set 12 m from the next. Each experiment was concluded when the crow gave up the search when it took more than 5 minutes to discover the prey item. After Croze (1970).

morphism may cause a reduction of mortality due to predation (see references for polymorphism in Table 7.2, and Chapter 16). Croze tested experimentally the idea that polymorphism increased survival of crow prey. Trimorphic populations suffered far less predation than monomorphic. A mean of 10.5 shells in polymorphic populations and only 3.5 shells in monomorphic populations were left by the crows searching among 27 regularly arranged shells; that is, a morph had a threefold selective advantage when occurring as one-third of a polymorphic population. In another experiment it took a crow 2.5 minutes to find the ninth shell in a monomorphic population and 5 minutes to find the ninth shell in a polymorphic population (Fig. 7.14). The crow gave up the search when it took longer than 5 minutes to discover a shell. This occurred after the thirteenth shell in the monomorphic population, and after the ninth shell in the polymorphic population.

Croze's conclusions on the ecological consequences of predation were as follows:

1. The first line of defense is camouflage.
2. Area-restricted search increases the selection pressure for a prey behavior that scatters their population.
3. The specificity of the searching behavior (search image) selects for:
   (a) Polymorphism within a camouflaged prey species.
   (b) Divergent coloration and behavior of sympatric species.

Direct evidence for a search image was reported by Pietrewicz and Kamil (1979), who studied blue jay responses to pictures with and without *Catocala* moth adults on tree trunks. The jays demonstrated an ability to improve detection of moths with successive encounters with a particular prey type. For experienced jays exposure to one prey type only enabled detection to improve from 75% of the time up to 100% of the time, indicating the extent to which the search image is important in predator–prey relationships.

MAGNITUDE OF SELECTION PRESSURE BY PREDATORS

There are some cases in which a good estimate of the adaptive advantage to protective devices can be obtained. In Tinbergen's study, *Panolis* filled 33% of the great tit's diet and *Diprion frutetorum* 3% at a certain time in the year. For every one *D. frutetorum* taken, 11 *Panolis* were taken; thus *Panolis* was 11 times more vulnerable to great tit predation than was the sawfly. If it is assumed that the populations of these two species were equal and that the great tit ate the equivalent of one complete population, it follows that 97% of the sawfly population survived and 67% of the *Panolis* population survived. Or for every 100 surviving sawflies about 70 *Panolis* survived, a 30% advantage to the sawfly. In the case of *Diprion pini* larvae, with colonial defense and distasteful nature, only 0.1% of the diet was filled by this species. For every one *D. pini* taken, 220 *Panolis* were taken, and conversely, using the same assumptions as for *D. frutetorum*, for every 100 *D. pini* that survived, 67 *Panolis* survived. In Croze's work an individual in a trimorphic population had a threefold advantage over one in a monomorphic population. For every 100 individuals that survived in a trimorphic population, 33 survived in a monomorphic population. In Holling's example given earlier, the same prey in the presence of a more palatable species (sunflower seeds) survived 30% better than in the presence of a less palatable species (dog biscuits).

Blest (1957a) provides an interesting example of how predators differ in their response to prey defense. The nymphalid butterfly, *Aglais urticae,* has the underside of the wings cryptic and the dorsal surface with a bright brown and orange pattern. At rest the butterflies sit with their wings closed, and when disturbed they suddenly open their wings, which elicits an escape response on the part of the predator. Blest gently removed the scales of some butterflies so that their pattern was gone and then tested the difference in predation. The normal butterflies had a 32% advantage over the treated butterflies when yellow buntings were the predator and a 63% advantage with reed buntings. The butterfly, *Nymphalis io,* also responds to predation by rapidly lowering its wings, but this species exposes four eyespots on its wings that mimic the vertebrate eye rather closely. Blest (1957a) removed spots from some of these individuals and tested the predation by yellow buntings. The number of overt escape responses by the predator was 31 for spotless and 128 for normal individuals, a 76% advantage to the spotted individuals.

For the melanic peppered moth *Biston betularia* in industrial areas, Kettlewell (1956) found that the advantage over the normal white moth with black spots was about 52%, and in the unpolluted country the survival of the nonmelanic was 66% better than that of the melanic form. These figures were calculated from release and recapture data. Where actual bird predation was observed in unpolluted areas, the typical form had a 6:1, or 86%, advantage over the melanic form.

Finally, Roeder and Treat (1961) found that noctuid moths detect the ultrasonic chirps of the searching bat and perform a protean display, so escaping predation. They studied 402 field encounters between bats and moths and recorded the survival of reactors and nonreactors to the bats' chirps. The advantage to evasive action proved to be 40% (for every 100 reacting moths that survived, there were only 60 surviving nonreactors).

POPULATION DISPERSION OF PREY

The frequency of very high selective advantages for defense in insects is noteworthy. This helps in understanding how so many incredibly intricate defenses have evolved. The speed of selection and adaptation can be particularly rapid in insects where fecundities are frequently high, with correspondingly high mortalities, and populations are large with high variability between individuals. One example on the evolution of crypsis should raise our expectations for the discovery of incredibly intricate modes of defense. The geometrid moth, *Nemoria arizonaria,* is bivoltine, each generation having caterpillars with completely different appearance and behavior (Greene, 1989). The spring generation feeds on oak catkins and is highly cryptic: the rich yellow color blends with the catkins and the rugose texture and projections make caterpillars excellent catkin mimics. This generation has small mandibles adapted for eating soft pollen grains and larvae always move from other substrates onto catkins. The summer generation feeds on leaves after catkins have withered and fallen. This morph is a twig mimic, with a greenish-gray integument, larger mandibles and head capsule for coping with tough leaves and larvae seek out twigs when placed on other plant parts. The morphs appear to be determined by a diet change from pollen to a high-tannin content in leaves.

In Table 7.2 are listed some of the responses to predation pressure that can be seen in insects. These are grouped according to how their defense is likely to influence population dispersion. Unpalatable or noxious species will benefit from a situation where the avoidance reaction of the predator is reinforced before enough time elapses for its learned response to be forgotten. Thus dense populations are likely to be an advantage. Conversely, as Croze points out, any palatable species should be dispersed widely to prevent a predator from forming a search image for these species or from increasing its efficiency in finding prey by any means at all. Similarly, if a species is palatable but has an intimidation display, the prey should also be well dispersed. Here there is no factor that will reinforce the predator's escape response, so that each defensive action must come to the predator as a completely unexpected stimulus. If the predator can learn to expect the response and learns the prey is palatable, the

TABLE 7.2 Some Responses to the High Selection Pressure of Predation, Particularly Vertebrate Predation. (References Provide Some Background for Each Type of Protection, Not Necessarily Support for This Ecological Classification of Protection.)

1. Protective strategies that result in selection for contagious distribution of prey—distasteful or otherwise noxious species:
   (*a*) Aposematic or warning coloration—advertising distasteful nature (Cott, 1940; Owen, 1980a; Guilford, 1990; Bowers, 1993).
   (*b*) Chemical defenses (Eisner, 1970; Blum, 1981; Bowers, 1992).
   (*c*) Müllerian mimics (Remington, 1963; Rettenmeyer, 1970; Owen, 1980a).
   (*d*) Intimidation displays and chemical defense, for example, colonial sawflies (Prop, 1960; Tostowaryk, 1972).
2. Protective strategies that result in selection for increased interprey distances—palatable species:
   (*a*) Crypsis and catalepsis (frozen posture usually with appendages retracted) (Thayer, 1909; Cott, 1940; de Ruiter, 1952, 1956; Hinton, 1955; Portmann, 1959; Keiper, 1969; Robinson, 1969; Sargent and Keiper, 1969; Owen, 1980a).
   (*b*) Intimidation displays (Blest, 1957a,b; Sargent, 1990).
3. Protective strategies that permit, but do not necessarily select for, contagious distribution of prey—palatable species:
   (*a*) Polymorphism (Sheppard, 1952; Cain and Sheppard, 1954a,b; Clarke, 1960, 1962; Wickler, 1968; Greene, 1989).
   (*b*) Batesian mimicry (Remington, 1963; Rettenmeyer, 1970; Waldbauer and Sternburg, 1975; Jeffords et al., 1979, 1980; Owen, 1980a).
   (*c*) Protean displays (Roeder, 1965; Humphries and Driver, 1970; Fullard, 1990; Sargent, 1990).
   (*d*) Phenological separation of prey from predator (Waldbauer and Sheldon, 1971; Evans, 1990).
   (*e*) Cellular defense reactions (against internal parasitoids only) (Salt, 1970; Stoltz and Vinson, 1979; Edson et al., 1981).

local population is doomed. A third group of defenses exists where selection may work in either direction, but here the direction does not depend on the type of defense the prey species had. Robinson (1969) proposed an alternative and valuable scheme of classification based on the primary defenses evolved by visually hunted prey. Heinrich (1979b, 1993) reviewed the ways in which caterpillars forage in relation to visually hunting predators such as birds, noting that palatable caterpillars behave in ways not so frequently seen among unpalatable species (see also Stamp and Casey, 1993). Palatable caterpillars show the following behaviors: they remain under leaves, feed only at night, move between feeding and resting sites, leave damaged but unfinished leaves and the associated indicators of their presence to feed on new leaves, and cut off partially eaten leaves. Herrebout (1969) cited a similar set of behaviors and other adaptations in caterpillars in relation to avoidance of parasitoids that may use both visual and chemical cues in searching, and Gross (1993) provided an excellent synthesis of insect defenses against parasitoids.

Thus the density of insect populations may well depend on the predation pressure they are exposed to and the defensive mechanism that has evolved in this interactive process. But this statement must be reexamined in Chapters 8, 17, 18, and 19, when the effect of predation on the regulation of prey populations is discussed.

There remains much interest in the act of predation and the relationships between predator and prey. General modeling approaches are reviewed by Hassell (1978, 1981, 1986), and foraging strategies are discussed for both terrestrial (Hassell, 1980; Griffiths, 1981; Janetos, 1982) and aquatic predators (Giller, 1980; Giller and McNeill, 1981; Pastorok, 1981). Foraging and the evolutionary aspects of insect prey species in relation to predation are treated in detail by Evans and Schmidt (1990) and Stamp and Casey (1993). Ambush predators, including ant lions (Griffiths, 1980, 1981) and orb-weaving spiders (Biere and Uetz, 1981), offer interesting subjects for research.

Predator and Prey Population Dynamics

As in Chapter 7, a historical perspective is developed in this chapter, concentrating on modeling aspects of predator and prey interactions. Studying the way in which early models developed provides insight to the thought processes involved and how the modeller attempts to extract the simplified essence of an interaction. Assumptions and relationships are defined clearly in the modeling process, providing a rigorous development of logic. But we should not forget that models are abstractions that simplify nature (Tilman, 1991b). "They are at best caricatures of reality, and thus have both the truth and the falsity of caricatures" (May, 1973a, p. 12). This does not lessen the importance of models, for simplification permits more general relationships to be explored, with the potential for the development of more general mechanistic theory.

CHARACTERISTICS OF A GENERAL PREDATOR

The characteristics of a general predator have been reviewed by Salt (1967), and a verbal description summarizes predator characteristics discussed in Chapter 7 and expands concepts about predation. Note that although Salt

An American robin feeds its young with insects, an abundant and high-protein resource. What is the role of avian predators in the population dynamics of insects? Drawing by Tad Theimer.

worked on *Woodruffia,* a protozoan predator, his review refers to predation in general:

1. Hunting is usually initiated by hunger and stopped by satiation.
2. Hunting site is determined primarily by inherited behavioral characteristics, although past experience on high concentrations of prey may modify these reactions.
3. Searching rate is a product of the rate of movement of the predator relative to that of the prey and the reactive perceptual field of the predator. Both values may be influenced by internal physiological states such as hunger, by external physical conditions, by prior experience of capture, and by prey characteristics such as size and color.
4. When selection of an individual prey animal is made from a group, the different or conspicuous individual is usually attacked. (One might wonder, then, how warning coloration can evolve; see Chapter 16 for a discussion. Contrary evidence to this generalization is given in Coppinger, 1970, and in Matthews, 1977.)
5. Predators do not attack all age or size classes of a prey species with equal frequency. Usually, some size or age classes of the prey are immune.
6. Although some predator species attack only one prey species, most predators subsist on several kinds of prey. Frequency of individuals in a diet is determined by the following factors, in order or importance:
   (*a*) Frequency of species in the environment.
   (*b*) Innate preferences of the predator.
   (*c*) Competition with other kinds of predators.
   (*d*) Profitability of prey, or the inverse of Tinbergen's prey risk.
7. Rate of capture by an individual predator is a product of the following:
   (*a*) All of the disk equation components (see Chapter 7).
   (*b*) Percentage of attacks that are successful. For most predators the degree of success is lower than usually supposed. One success in two attacks is a high value, and the average is about 1 in 10.
8. Responses by the predator to changes in prey density are determined by food habits of the predator:
   (*a*) If it is an obligate feeder on a single species of prey, the numbers of each population oscillate, with predator highs following prey highs. Emigration and immigration by the predator and exemption from attack of certain life history stages or size classes of the prey contribute to the stability in the oscillations. (See also the discussion of Nicholson and Bailey and the Leslie–Gower models later in this chapter.)
   (*b*) If the predator feeds on more than one species of prey, changes in density of one prey species bring about a shift in the percentage composition in the diet of the predator.
9. Predators react to changes in their own density in the same fashion as other animals by competing for a common requisite.

10. Competition between predators of two or more species for a common prey reduces the efficiency of both and may cause shifts in the percentage composition of the diet of both.

PREDATION MODELS

A consideration follows of how predator and prey populations interact with each other numerically. Royama's (1971) comparative study of predation models shows how the various models discussed here are related.

Lotka–Volterra Equations

Lotka (1924) and Volterra (1926) were the first to make any concrete predictions on predator–prey population interactions and reached the same conclusion independently. Their assumptions are as follows:

1. Animals move at random.
2. Every encounter with a prey results in a capture (cf. Salt, 1967, and above), and every prey that is captured is eaten. This is independent of densities of populations of predators and prey.
3. The populations of both predator and prey have all the qualities necessary for them to conform to the logistic theory—growth rate accelerates until a limiting factor causes deceleration.

The logistic equation is described in Chapter 13, but part of the theory uses the Malthusian exponential growth formula. The rate of increase of a single species of prey living by itself in unlimited space is

$$\frac{dN'}{dt} = r'_m N'$$

where r'_m is the maximum rate of increase per individual (birth rate) and N' is the density of prey population. In the absence of prey, predators would die of starvation. Therefore, the rate of decrease of a population by deaths—or the negative rate of increase—is

$$\frac{dN''}{dt} = d''N''$$

where d'' is the death rate, which is negative and N'' is the density of predator population. If both predator and prey are living together in a limited space, r'_m will be reduced by an amount that depends on the density of the population of predators [i.e., $(r'_m - ?)N'$]. Also, the negative coefficient d'' will be increased by an amount dependent on the density of the population of the prey species [i.e., $(d'' + ?)N'']$. Because of the assumptions made by Lotka and Volterra, the changes in r'_m will be proportional to N'' [i.e., $r'_m - ?N'')N'$] and those in d'' will be proportional to N' [i.e., $(d'' + ?N')N'']$. Therefore, the two equations become

$$\frac{dN'}{dt} = (r'_m - C'N'')N' \qquad \text{for prey populations}$$

and

$$\frac{dN''}{dt} = (d'' + C''N')N'' \qquad \text{for predator populations}$$

where C' is a constant [Volterra regarded it to be a measure of the ability of the prey to defend itself (cf. R'_A of Tinbergen)] and C'' is a second constant [Volterra regarded it as a measure of the effectiveness of offense of the predator (also included in R'_A of Tinbergen)]. The equations provide a *periodic solution,* in that the trends in densities of populations of prey and predator change direction in a systematic way (Fig. 8.1). Volterra called it the **law of periodic cycle.**

Quite small quality changes in animals result in the curve cutting either the abscissa or ordinate (in the phase diagram Fig. 8.1B) after several oscillations of increasing amplitude, that is, extermination of one species, leaving the other in complete possession of the field. If this happens to be the predator, it must switch food or die out also. Gause (1934) called this a *relaxation oscillation* to distinguish it from the classical oscillation, in which the closed loop indicates that both predator and prey live together indefinitely.

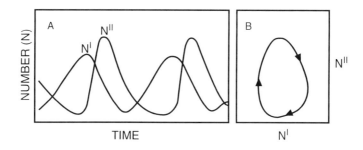

FIGURE 8.1 Population oscillations of prey (N') and predator (N'') predicted by the Lotka–Volterra equations. (A) Numbers plotted against time; (B) phase diagram indicating perpetual coexistence of prey and predator. From H. G. Andrewartha and L. C. Birch, *The distribution and abundance of animals.* Published by the University of Chicago Press. Copyright © 1954 by the University of Chicago Press. All rights reserved.

Nicholson–Bailey Equation

Nicholson and Bailey (1935) criticized the Lotka–Volterra equations on three major points:

1. Lotka and Volterra assumed that the "population reaction" when predator encounters prey is instantaneous. This is not so, and the time that elapses before the result of the "encounter" is evident may in some cases be as long as a generation of the predator. This is particularly true of parasitoids, where oviposition of an egg on the prey does not show a population effect—in terms of adult predators—until that egg has hatched and the larva has fed, pupated, and emerged as an adult. Therefore, there is an important lag effect between prey and predator populations.
2. They assumed that each individual in prey and predator populations is exactly equivalent to every other individual of the same species. No allowance was made for different age groups, for example, egg, larva, pupa, and adult.
3. Lotka and Volterra used calculus in their arguments that describe continuous changes in populations.

Nicholson and Bailey proposed a new model, with parasitoids in mind; that is, there is a 1:1 ratio between predator and prey numbers. Their assumptions should be compared with the qualities of predators and prey discussed in Chapter 7 and earlier in this chapter.

1. Prey is distributed uniformly in a uniform environment.
2. Ease with which prey can be found does not vary with density of population.
3. Predator population searches at random.
4. Appetites of predators (i.e., capacity for oviposition) are insatiable, independent of prey population density.
5. Predator has an "areal range" = constant = distance traveled during lifetime while searching multiplied by twice the distance from which it can perceive a prey. That is, the predator can receive visual or chemical stimuli from both sides of its body as it hunts; thus areal range is equivalent to the area searched in a lifetime.
6. Predator has an "area of discovery" = constant =

$$\frac{\text{number of prey found by predator during lifetime}}{\text{number of prey in "areal range" of the predator}}$$

This measures the proportion of prey found in the total area searched, or the efficiency of the predator in finding its prey.

With this information the number of prey in the next generation can be calculated using the formula

$$pa = \log_e \frac{u_i}{u_s}$$

where p is the parasitoid population density, a the area of discovery, u_i the host density in generation i, and u_s the host density in the next generation, or the number of hosts surviving.

For a parasitoid species with a given area of discovery there is a particular density at which it effectively searches a fraction of the environment equal to the fraction of hosts that are surplus, that is, the interest from the population, not the capital, and at this density exactly the surplus of hosts is destroyed. Also, there is a particular density of a host species (capital) that is just sufficient to maintain this density of parasitoids from generation to generation. When the densities of the interacting animals have these values, they must remain constant indefinitely in a constant environment—the steady state.

The slightest departure on either side of the steady state produces oscillations. When age distribution, or the lag in predator response, is ignored, oscillations continue indefinitely with a constant amplitude just as Volterra and Lotka predicted. When age distribution is taken into account, oscillations increase in amplitude with time as in a relaxation oscillation (Fig. 8.2). If the density of the host is slightly above the steady-state value and is being reduced by a parasitoid, when the host reaches its steady density, the density of the parasitoid is the result of the host density in the preceding generation, when it was above the steady state. This is where age distribution is important. There are more than sufficient parasitoids to destroy the surplus hosts, and the host density is still further reduced in the following generation. As a result, the densities of the interacting animals should oscillate about their steady values—with increasing amplitude.

This does not actually appear in nature. Nicholson and Bailey concluded that the probable ultimate effect of increasing oscillations is the breakup of the species population into numerous, small, widely separated groups that wax and wane and then disappear, to be replaced by new groups in previously unoccupied situations. Although this idea is currently an important one in evolutionary theory, Nicholson and Bailey could hardly justify this deduction from a mathematical model that assumes at the beginning that animals are distributed evenly over a uniform area.

Tinbergen and Klomp (1960) took a more pragmatic approach to the Nicholson and Bailey model by testing how variable factors would change the response of the populations from one of increasing oscillation amplitude. Some of their conclusions follow:

1. If it is assumed that the reproductive rate of the host is sensitive to density and decreases with density increase, naturally the oscillations can-

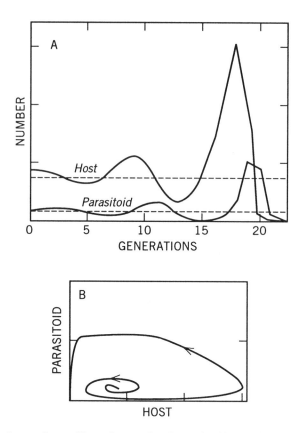

FIGURE 8.2   Population oscillations of host and parasitoid numbers predicted by the Nicholson–Bailey equation: (A) numbers plotted against time; (B) phase diagram indicating increasing oscillations. Steady densities are represented by dashed lines. After Nicholson (1933); by permission of Blackwell Scientific Publications.

not be so violent. The result is regular fluctuations or slight damping of oscillations, depending on the extent of the sensitivity of fecundity to density.

2. Egg production of parasitoids is likely to be finite. Therefore, the area of discovery is likely to decrease with host density at values above a critical one at which eggs cannot be produced rapidly enough. Therefore, the amplitude of oscillations will increase.

3. If mortality is dependent on the density of the population, the same result may occur in two ways:

   (a) If parasitoid mortality is dependent on parasitoid density and increases with density, oscillation amplitude is decreased.

(*b*) If host mortality is dependent on host density and increases with density, this also leads to damped oscillations.

4. Where percent predation increases with host density in a sigmoidal fashion, damped oscillations are generated, provided that at intermediate densities more than 25% of the prey is taken. Their major conclusion is that the type 3 functional response (although they did not use this term) can cause damping of the oscillations that are inherent in the host–parasitoid model of Nicholson and Bailey. Therefore, birds and small mammals appear to be important factors in the balance of herbivorous insect and parasitoid populations (for further discussion see the population regulation effect of Holling's type 3 functional response later in this chapter). It has been found since, however, that parasitoids may also show a sigmoid response as stated in Chapter 7 (see also Hassell et al., 1977).

Leslie–Gower model

The two models so far considered are deterministic, yet random processes obviously affect the predator–prey relationship in nature; thus a stochastic model may be more realistic. Leslie and Gower (1960) have developed such a model. They determined that given large populations of prey, although all prey individuals were accessible to predators, the two populations would fluctuate irregularly around the stable state once the equilibrium level had been reached. Extinction of either species was extremely unlikely (a mean of once in every 1.4×10^{11} units of time). As numbers were reduced, extinction became more likely, although if a certain proportion of the prey was unavailable to predation (optimally around 20%), the stability of the system was very much increased. Prey may become unavailable because of heterogeneity of the environment or because of predator characteristics (Salt, 1967, where *Woodruffia* encysts when prey reaches a certain low, unprofitable density).

Finally, if the rate of change of the predator population increases with increasing density, a very realistic possibility, so that predator populations respond more slowly at low prey populations than at high prey populations, using the Leslie–Gower model, a damping of oscillations toward equilibrium results (Fig. 8.3; see also Pielou, 1969). Holling and Ewing (1971) (see Holling, 1968, for similar discussion) have also presented models that predict damping of predator–prey oscillations where "a variety of behavioral mechanisms are included that tend to tune the system more to existing conditions than to the past. By so doing, the instability produced by the memory of the system is counteracted." That is, lag in predator response is compensated for by behavioral adjustments.

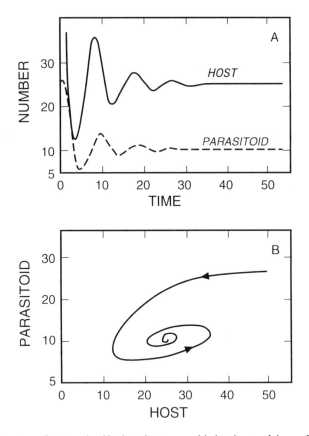

FIGURE 8.3 Damping oscillations predicted by the Leslie–Gower model when the rate of change of the parasitoid is density dependent (see under Hassell–Varley equation for definition): (A) numbers plotted against time; (B) phase diagram indicating decreasing oscillations. After Pielou (1969).

Hassell–Varley Equation

Hassell and Varley (1969) were familiar with the biology of parasitoids. From several published accounts of the efficiency of parasitoids in finding their hosts, they showed that Nicholson and Bailey's "area of discovery" (a) was far from constant for real parasitoid situations (Fig. 8.4), as Andrewartha and Birch (1954) had done before them. For several species and over several orders of magnitude, log a was linearly related to parasitoid density (p) in the following manner:

$$\log a = \log Q - (m \log p)$$

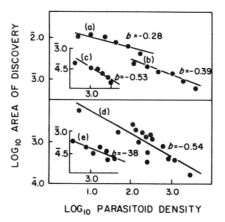

FIGURE 8.4   Relationships between searching efficiency (log$_{10}$ area of discovery) and density (log$_{10}$ parasitoid density) of five species of parasitoid. After Hassell and Varley (1969).

where a is the area of discovery, p the parasitoid density, Q the "quest constant," and m the "mutual interference constant." Both Q and m are calculated empirically from the regression line. Therefore, the factor $m \log p$ would increase with density and a would decline with density. Thus a in the Nicholson and Bailey model should be replaced by the density-sensitive Q/p^m. The new "parasite quest" theory model becomes, by substitution,

$$\log_e \frac{u_i}{u_s} = Q p^{1-m}$$

where u_i is the host density in generation i and u_s is the host density in the next generation.

In this new model the parasitoid population has a direct density-dependent factor built into its behavior. (A density-dependent factor is discussed in Chapter 17—it is any factor the effect of which increases as a percentage of the population as the population increases in density.) The greater the value of the mutual interference constant m, the greater will be the tendency for the host–parasitoid model to stabilize (Fig. 8.5). As m approaches Q, the parasitoid's behavior conforms more and more closely to the Nicholsonian system and the more unstable the interaction becomes. The more that is learned about parasitoids, the more it is realized how sensitive they are to the presence of others, but this topic is treated in Chapters 17 and 18.

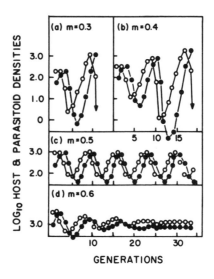

FIGURE 8.5   Population oscillations of host (open circles) and parasitoid (solid circles) predicted by the Hassell–Varley model with the mutual interference constant, *m*, increasing from 0.3 in (*a*) to 0.6 in (*d*). After Hassell and Varley (1969).

TESTS OF MODELS

Gause's Test

Gause (1934) was the first to test the Lotka–Volterra model experimentally. He used two protozoa: *Didinium nasutum*, the predator, and *Paramecium caudatum*, its prey. Reproduction is by binary fission; thus Gause made a clever choice, since the response of the predator to feeding is close to being instantaneous. Thus the food chain was bacteria → *Paramecium* → *Didinium*.

In his first experiment Gause used a homogeneous microcosm without immigration of prey or predator. The predator rapidly discovered all the prey and *Paramecium* and *Didinium* went extinct, but the bacteria remained abundant (Fig. 8.6A). The Lotka–Volterra model was not supported.

Next, Gause left a refuge present in the form of a sediment in the medium, where *Paramecium* were safe from attack by predators. Here the result was more unpredictable. In some microcosms all prey were devoured; in others the prey escaped from the predator and the predator went extinct (Fig. 8.6B); and in others the populations of predator and prey coexisted for several days. Gause then made a final effort to obtain oscillations in predator and prey populations. He did not use a refuge but artificial immigration, one of each species every third day. The result was oscillations that lasted variable lengths of time (Fig. 8.6C). But Gause pointed out that at the low population levels

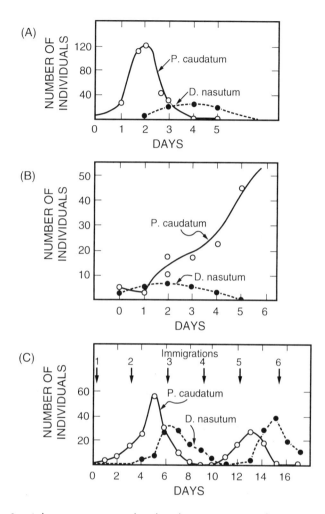

FIGURE 8.6   Gause's three experiments testing the Lotka–Volterra equation: (A) in a homogeneous microcosm; (B) in a microcosm with a refuge from predation; (C) with immigration of one of each species at times indicated. After Gause (1934).

between peaks, definite prediction on the outcome could not be made because in his words "multiplicity of causes acquires great significance. . . . As a result it turns out to be impossible to forecast exactly the development in every individual microcosm and we are again compelled to deal only with the probabilities of change." That is, a stochastic model is needed, not a deterministic model. Thus the Lotka–Volterra model was not supported by Gause's experiments. Luckinbill (1973) examined a similar system and obtained coexistence for about 100 generations of prey by maintaining food of prey as a limiting factor at peak prey densities.

DeBach and Smith's Test

Probably, DeBach and Smith, who published in 1941, were the first to test the Nicholson–Bailey model experimentally. (*Note:* Gause did not test a parasitoid–host situation, so that although Gause knew of Nicholson's 1933 paper, he was not testing the Nicholsonian model.) They used as subjects the house fly *Musca domestica* and a chalcid parasitoid *Nasonia* (= *Mormoniella*) *vitripennis*. They compared what was predicted by the Nicholson and Bailey model with what they found in their own cultures over seven generations (Fig. 8.7). The fit was remarkably good as far as it went, but it is not powerful evidence in support of the model. However, the experiment does show that oscillations appear to be an inherent component of the predator–prey interaction, although Wangersky and Cunningham (1957) predicted from the results that by the tenth generation the fluctuations would settle into a limit cycle rather than continue in oscillations of increasing magnitude (see also Andrewartha and Birch, 1954, for further discussion of this experiment). Burnett (1958) studied a similar system and observed increasing oscillations until extinction occurred in the twenty-first generation (see Varley et al., 1973, for discussion).

Huffaker's Test

Huffaker (1958) in the University of California at Berkeley also used laboratory studies to evaluate the basic relationships between predator and prey populations. He employed the six-spotted mite *Eotetranychus sexmaculatus*, which feeds on oranges, and its predator mite, *Typhlodromus occidentalis*. By

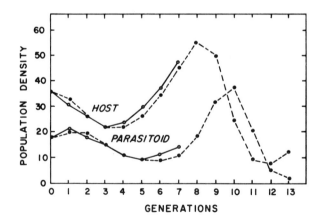

FIGURE 8.7   Predictions of the Nicholson–Bailey model (dashed lines) tested with populations of the host *Musca domestica* and the parasitoid *Nasonia vitripennis* (solid lines). After DeBach and Smith (1941).

setting out oranges and rubber balls on a tray he could alter the absolute amount of food and the dispersion of that food. In the simplest universe he had four large areas of food for prey (½-orange area each) grouped at adjacent, joined positions—a 2-orange feeding area on a 4-orange dispersion. Predators were introduced 11 days after the prey. The predator could discover the prey so easily that the prey were reduced in numbers to such a low level that the predators starved and became extinct, and the prey slowly increased (Fig. 8.8). Huffaker therefore increased the complexity of the environment by reducing the size of each food item and diluting their density in the universe. Even with small areas of food (⅒-orange area) for prey, alternating with 20 foodless positions, a 2-orange feeding area on a 20-orange dispersion, the predator and prey did not coexist. Finally, he used 120 oranges, each with ¹/₂₀-orange area available as food, a 6-orange feeding area on a 120-orange dispersion. Around the oranges he made a complex maze of petroleum-jelly partial barriers so there was no easy access from one orange to another. He introduced one six-spotted mite on each of 120 oranges and the 27 female predators were added 5 days later, one on each of 27 oranges dispersed evenly throughout the universe. He also added little posts from which prey would disperse by silken thread and a fan was supplied to produce a small air current. The predator was unable to disperse in this way.

Huffaker actually obtained three oscillations of predator and prey when finally the predator went extinct because of a shortage or prey (Fig. 8.9). Huffaker says, "However, by utilizing the large and more complex environment so as to make less likely the predator's contact with the prey at all positions at once . . ., it was possible to produce three waves or oscillations in density of predators and prey. That these waves represent a direct and reciprocal predator–prey dependence is obvious." Even in this elegant experiment, coexistence was not achieved simply by increasing environmental heterogeneity. The food quality of the prey deteriorated with increasing mite density, and acted in a density-dependent manner.

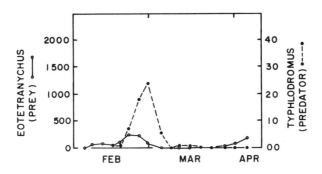

FIGURE 8.8   Results of Huffaker's test of the coexistence of a predator and a prey using a two-orange feeding area on a four-orange dispersion. After Huffaker (1958).

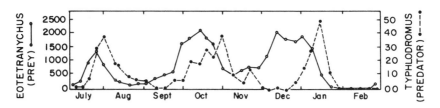

FIGURE 8.9   Results of Huffaker's final experiment when the prey and predator coexisted for three oscillations in a complex environment. After Huffaker (1958).

Luckinbill's (1973) studies identify food limitation of the prey as an important factor in producing limit cycles in predator and prey populations which may be achieved even in the absence of spatial heterogeneity or greater prey dispersal ability. However, Huffaker's work demonstrates the importance of four factors that may contribute to the stability of predator–prey interactions.

1. Environmental heterogeneity
2. Spatial relationships of predator and prey
3. Relative dispersal rates of predator and prey
4. Influence of the change in quality of prey food

NUMERICAL AND DEVELOPMENTAL RESPONSES

Type 2 and 3 functional responses were introduced in Chapter 7 and describe the response of a single predator to changes in prey density. However, this individual response also has a population effect. As prey density increases, each individual increases its consumption of food; food gathering becomes less expensive energetically and more energy can be channeled into reproduction. The reproductive rate increases and, therefore, the numbers in the population increase—a **numerical response**. Also, migrating predators will tend to stay longer in areas of high prey density, and this will result in population aggregations in relation to prey density—another aspect of the numerical response. In addition, Murdoch (1971) has argued that predators may show a **developmental response** because they eat more prey at higher densities and grow more as a result, and then kill more prey because of their larger size.

TOTAL RESPONSE

Therefore, the total response of predators to increasing prey density involves five factors: (1) individual functional response, (2) increased reproduction in the population, (3) increased immigration, (4) developmental response, and

(5) predator functional response to predator density. Evaluating the relative importance of each of these factors in field populations of predators and prey should prove to be challenging.

POPULATION REGULATION EFFECTS

The population regulation effects of Holling's type 2 and type 3 functional responses should now be considered (see Holling, 1965) to discover which type of predation is likely to produce the greatest prey population stability. The percent mortality necessary to stabilize the prey population (*NEC*) will vary considerably with prey density. It is equivalent to the potential for reproductive success of the population, or the mortality needed to counteract this potential. Reproduction will be very low when mates are too far apart for a high mating rate, and high at moderate densities—declining as shortages of resources increase (Fig. 8.10).

In Holling's type 2 response the percent mortality will constantly decline as prey density increases (cf. Fig. 8.11). Therefore, the result of the functional response can be superimposed on the reproductive potential curve (*NEC*) (Fig. 8.10A). An unstable equilibrium is developed at *EX* and *ES*. At prey densities below *EX*, predation will always be greater than the prey can

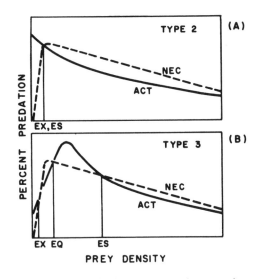

FIGURE 8.10   Population regulation effects of Holling's type 2 (A) and type 3 (B) functional responses. *NEC*, percent mortality necessary to stabilize populations (cf. Lewontin's, 1965, reproductive function in Chapter 13); *ACT*, actual percent predation; *EX*, threshold density for population extinction; *EQ*, equilibrium density; *ES*, threshold density for population escape. After Holling (1965).

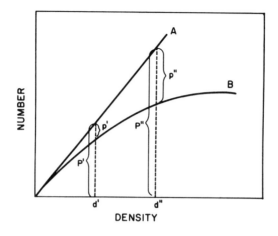

FIGURE 8.11   Decline in percent mortality resulting from the type 2 functional response. At low prey densities (d') the percentage of prey not taken ($p' \times 100)/P'$ is small, but as prey density increases (d'') the percentage not taken ($p'' \times 100)/P''$ increases. A, number of prey available; B, number of prey taken.

support and extinction results. At prey densities above *EX*, predation is never sufficient to stabilize the prey population, and the population can escape from the controlling influence of predation and can erupt. This is the escape equilibrium (*ES*). Here no feedback mechanism is likely to stabilize the population.

The situation is different, however, where the type 3 response is involved. The percent predation begins to increase as the prey density increases (Fig. 8.12). When this response is superimposed on the *NEC* curve (Fig. 8.10B) we see that a stable equilibrium (*EQ*) is developed. Below *EX*, the prey population goes to extinction because of the high predation rate, higher than that needed to stabilize the population. Above *ES*, actual predation is too low to stabilize the population, so that the prey escapes from the influence of predation and erupts. Above *EX*, however, the prey population will increase because predation pressure is lower than necessary—but it can increase to *EQ* only when predation becomes a controlling influence. Below *ES* the population can decline only as far as *EQ*. So the point *EQ* represents a stable equilibrium density. Below it, reproduction is greater than predation; above it, predation is greater than reproduction.

Therefore, it appears that the type 3 response is more likely to improve the stability of a prey population than is the type 2 response. Tinbergen and Klomp (1960) also concluded that vertebrate predation (i.e., birds), which exhibited the type 3 response in their example, could cause the damping of oscillations inherent in the host–parasitoid model of Nicholson and Bailey. More recently, avian predators of the fall webworm, *Hyphantria cunea* (R. F. Morris, 1972), and of the scolytid beetle, *Phleophthorus rhododactylus*

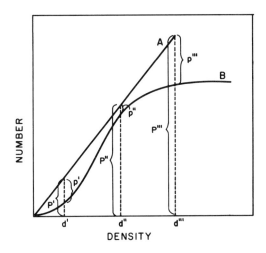

FIGURE 8.12 Change in percent mortality resulting from the type 3 functional response. Percent predation is low at low densities (d'), high at moderate densities (d''), and low at high densities (d'''). Notation as in Fig. 8.11.

(Waloff, 1968a), have been shown to respond as predicted by Holling's model.

EFFECT OF BATESIAN MIMICRY

Holling (1965) used the same type of reasoning to show the population effects of Batesian mimicry. Here he simplified the situation by having *NEC* (percent mortality necessary to stabilize populations) constant for the nonmimetic and mimetic population (Fig. 8.13). He assumed that if mimetic and nonmimetic populations were equal when predation was present, without predation the nonmimetic species must have a higher reproductive rate. Therefore, the percent mortality necessary to stabilize the population must be higher (cf. NEC_x and NEC_y in Fig. 8.13).

Holling's computer model showed that the development of mimicry lowers the proportion of prey destroyed and shifts the peak of the domed predation curve to a higher density (cf. P_x and P_y in Fig. 8.13). Holling concluded that despite the marked drop in the line of necessary mortality, the change is great enough to raise the equilibrium density of the mimics (EQ_y) significantly above that of the nonmimics (EQ_x). Therefore, the equilibrium density of a prey population can be increased by mimicking an unpalatable species.

Despite the apparent instability of the interaction between prey and a predator exhibiting a type 2 response, some factors may promote stability. Some predators (e.g., the snails *Thais emarginata* and *Acanthina spirata*) may

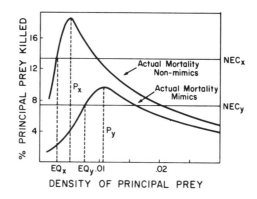

FIGURE 8.13   Effect of Batesian mimicry on population regulation predicted by use of the functional response model. After Holling (1965).

show a type 2 response in the presence of a single prey species. However, when several prey species exhibit a patchy dispersion, predators may become trained to the more abundant prey, which is attacked disproportionately by individuals and the predator population as a whole (Murdoch, 1969). Predators may then "switch" to another prey as relative population sizes change. This behavioral switching probably operates with a short time lag and has a potential stabilizing effect on the predator–prey interaction. Coccinellid beetle predators did not show this switching behavior when given a choice of aphid species (Murdoch and Marks, 1973), but *Notonecta* bugs did show switching (Lawton et al., 1974). Also, the type 3 response may be observed when refuges from predation exist, but as the prey population increases, a smaller percentage of it can occupy the refuge (Murdoch, 1973; see also criticism of Tinbergen's work in Chapter 7). Stability of the host–parasitoid interaction also increases when adult parasitoids interfere with each other's search and when they tend to aggregate at relatively high host densities (Hassell and Rogers, 1972; Hassell and May, 1973; Rogers and Hassell, 1974; Hassell, 1978).

In general, there are many stabilizing influences on the population interactions between predators and their prey. In this chapter we have considered the influence of logistic growth, functional and numerical responses of the predator, limitation of these responses in terms of appetite or egg laying, interference between predators and parasitoids at high densities, the role of refuges and relative dispersal rates of predator and prey, including habitat heterogeneity, and the importance of predators switching from one prey species to another. Excellent reviews on these mechanisms are provided by Murdoch and Oaten (1975) for predators in general, and by Hassell (1978) for arthropod predators.

The interaction between predators and prey and parasitoids and hosts have been studied following many avenues, including the importance of aggregation of predators (e.g., Hassell and May, 1973; Hassell, 1978; Jeffrey Waage, 1978), patchiness in host distributions (e.g., May, 1978b; Jeffrey Waage, 1979b; Nachman, 1981), optimal foraging strategies (e.g., Charnov, 1976a,b; Cook and Hubbard, 1977; Pyke et al., 1977; Hubbard and Cook, 1978), and the role of density dependence in stability (e.g., Wang and Gutierrez, 1980; May et al., 1981). More recent and extensive general developments in the field of predator and prey interactions are discussed in the following publications: Anderson et al. (1979), Berryman (1981, 1992), Roughgarden et al. (1989), Gilpin and Hanski (1991), Royama (1992), Cappuccino and Price (1995), Dwyer (1995), and Walde (1995).

Parasite and Host Interactions

Predators kill their prey and eat them or consume them alive. Parasites sap energy and nutrients from a host while it lives and may or may not eventually kill the host. This seemingly small difference between predators and parasites has very significant impact on the ecology and evolution of each group. If we compare predators and parasites in the Class Insecta, some generalizations can be made:

1. Predators tend to be larger, enabling them to kill by force; parasites tend to be small, at least relative to the host, so that as individuals they take only a small amount of nutrients from the host per day and can live in or on the host. Andrews (1991) explores extensively the ramifications of these apparently simple differences.
2. Predators tend to be generalized feeders because killing by force enables many species to be utilized as prey, while parasites tend to be specialized because living intimately with a host requires special adaptations that suit it only to one or a narrow range of hosts.
3. Because predators are generalized, speciation and adaptive radiation is limited by a relatively small number of ecological niches for organisms

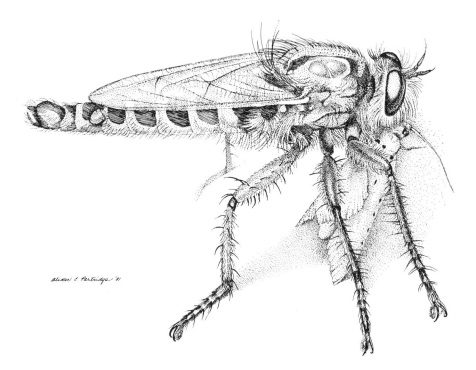

The act of predation is commonly observed, illustrated by this robber fly, *Promachus vagator* (Asilidae), eating a moth, but parasitism is frequently overlooked. The small parasitic mite feeding at the intersegmental membrane between abdominal segments 5 and 6 is relatively inconspicuous. After a photograph by Anthony Bannister, in Tweedie (1973). Used with permission. Drawing by Alison Partridge.

with a catholic diet, but for parasites specialization is so great that almost innumerable niches exist and speciation and adaptive radiation can be spectacular. These differences are discussed in more detail in the following paragraphs, after which we consider their ecological consequences.

RABBIT FLEA EXAMPLE

First, let us enter into the right frame of mind for thinking about parasites, using a rabbit flea's perspective. In the preceding two chapters we hardly discussed the lives of predators because they are so familiar, what they do when not acting as predators is not central, and how they kill prey is about as interesting as how humans eat. But all this changes when we consider parasites, for their life is so intimately associated with the host: each individual lives on a host and feeds on a host.

The host is habitat and food. The adult flea is a denizen of the rabbit's pelage while larvae live in the nest material. The rabbit flea (*Spilopsyllus cuniculi*) was studied extensively by Rothschild (1965) and Rothschild and Ford (1964, 1973). The tight linkage and synchronization between the host reproductive cycle and that of the flea are illustrated in Fig. 9.1. The reproductive cycle of the rabbit is illustrated, starting with an estrous doe, followed by mating, ovulation and 30 days of pregnancy, parturition, lactation, and the presence of young in the nest. Changes in hormone levels in the doe during this sequence are recorded within the circle.

The effect of rabbit reproduction on flea reproduction is given on the outside of the circle in Fig. 9.1. During copulation between rabbits, most fleas on the male transfer to the female. At ovulation adult fleas on the rabbit become more active, they are assembled at the principal site on the rabbit's ears, and after 20 days of rabbit pregnancy fleas begin to mature and ovarian development begins under the influence of rabbit hormones. Just before parturition adult fleas defecate rapidly, dropping blood into the nest on which larvae will feed when they hatch. Females become fully gravid and then male and female fleas move to feed on the newborn rabbits, and copulate. In addition to the hormonal influence of the host on the flea reproductive cycle, newborn rabbits produce an airborne factor that stimulates copulation—a kairomone, in fact. Eggs are laid in the nest, and larvae hatch and feed in the nesting material, followed by pupation and emergence of adults.

The adaptive advantages of this "hormone-bound" life cycle of the flea are clear enough. Fleas reproduce when nesting material is gathered in the spring, and the rabbits are not free-ranging. In fact, the female rabbit becomes quite sedentary, nesting away from the main rabbit warren, enabling large amounts of blood to be dropped by adult fleas into the nest, provisioning it for the fleas' young. The resources for the fleas during egg-laying suddenly increase from one adult rabbit to many nestlings, unable to groom and largely defenseless. Flea eggs and larvae develop in a warm nest with a plentiful supply

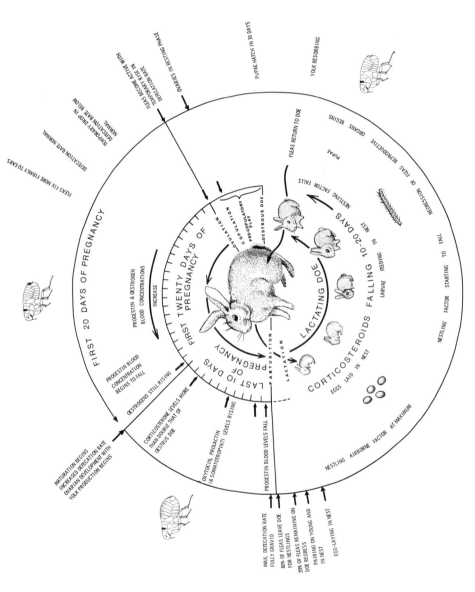

FIGURE 9.1 Synchronization of the breeding cycles of the host rabbit and the rabbit flea. From M. Rothschild and B. Ford, Factors influencing the breeding of the rabbit flea (*Spilopsyllus cuniculi*): A spring-time accelerator and a kairomone in nestling rabbit urine with notes on *Cediopsylla simplex*, another "hormone bound" species, *J. Zool. London*, 1973, **170**:87–137; by permission of Oxford University Press.

of dried blood and other debris, the parental adult fleas can return to the doe, and newly hatched fleas disperse with the rabbit nestlings or disperse on their own.

Such hormone-bound life cycles are also found in the Americas (Rothschild and Ford, 1972, 1973). In North America the flea, *Cediopsylla simplex,* breeds in a similar way to the rabbit flea described above, on the cottontail rabbit, *Silvilagus floridanus.* In Mexico, the Mexican rabbit, *Romerolagus diazi,* appears to have a parallel relationship with the flea, *Cediopsylla tepolita.* There is good reason to expect such close ties between parasites and their hosts to be of great antiquity.

PARASITE LIFE CYCLES

Such intimacy between host and parasite is the norm, frequently involving chemical signals of various kinds between the species, as we saw in Chapter 5. Both parasites on plants and animals also show remarkable convergence in life-cycle evolution, in which intermediate hosts become involved, and a complex sequence in the life history partitions activities among reproductive and dispersal or transmission stages between hosts (Fig. 9.2). Stretching out life cycles in a linear sequence starting with a sexually produced egg, the cycles of a liver fluke, a gall aphid, an adelgid, and a rust fungus look very similar, although the jargon used for each stage tends to obscure the convergence (Price, 1996a). We could add the life cycles of parasites vectored between hosts by insects, such as *Plasmodium* species, which cause malaria in humans and an unnamed disease in the vectoring mosquito.

For small parasites with widely scattered hosts a major obstacle in the life cycle is dispersal between hosts. The adaptive solution frequently involves massive production of individuals capable of dispersing or transmission to another host either by multiple cycles of asexual reproduction or by sexual reproduction (Fig. 9.2). Intermediate hosts or vectors may greatly increase the probability of recolonizing the primary host species. For larger species of insect with higher mobility, the nonparasitic adult female is capable of dispersing and finding hosts for its parasitic progeny, such as in butterflies, moths, and parasitoids. Nevertheless, the parasitic stage in a life cycle usually sets strict limits on the ecology of a species relative to that of predators.

LARGE PREDATORS AND SMALL PARASITES

Relatively speaking, predators are large and parasites are small. Among the insects, this generalization holds when common groups of predators such as Carabidae, which are insect and seed predators, Coccinellidae, and Vespidae

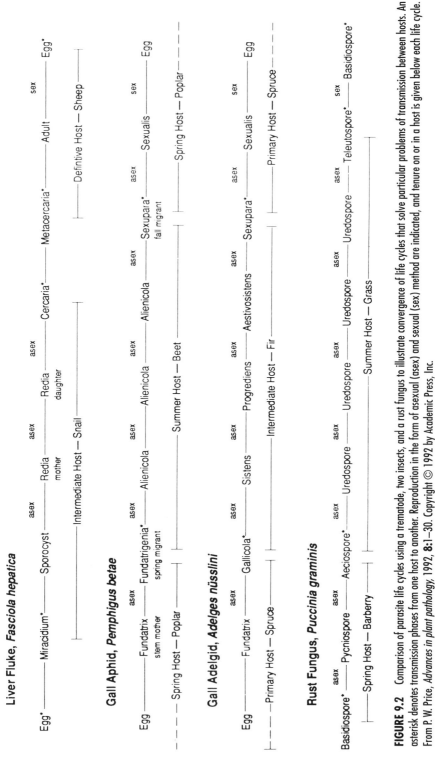

Liver Fluke, *Fasciola hepatica*

Egg* ———— Miracidium* ———— Sporocyst ———^{asex}——— Redia ———^{asex}——— Redia ———^{asex}——— Cercaria* ———— Metacercaria* ———— Adult ———^{sex}——— Egg*
                                                        mother   daughter
                                                    └——— Intermediate Host — Snail ———┘   └——— Definitive Host — Sheep ———┘

Gall Aphid, *Pemphigus betae*

Egg ———^{asex}——— Fundatrigenia* ———^{asex}——— Alienicola ———^{asex}——— Alienicola ———^{asex}——— Sexupara* ———^{asex}——— Sexualis ———^{sex}——— Egg
                    spring migrant   fall migrant
├—— Spring Host — Poplar ——┤   └—————— Summer Host — Beet ——————┘   └—————— Spring Host — Poplar —————— ┤ ├—

Gall Adelgid, *Adelges nüsslini*

Egg ———^{asex}——— Fundatrix ———^{asex}——— Gallicola* ———^{asex}——— Sistens ———^{asex}——— Progrediens ———^{asex}——— Aestivosistens ———^{asex}——— Sexupara* ———^{asex}——— Sexualis ———^{sex}——— Egg
├—— Primary Host — Spruce ——┤   └—————— Intermediate Host — Fir ——————┘   └—————— Primary Host — Spruce —————— ┤ ├—

Rust Fungus, *Puccinia graminis*

Basidiospore* ———^{asex}——— Pycniospore ———— Aeciospore* ———^{asex}——— Uredospore ———^{asex}——— Uredospore ———^{asex}——— Uredospore ———^{asex}——— Uredospore ———^{asex}——— Teleutospore* ———^{sex}——— Basidiospore*
├—— Spring Host — Barberry ——┤   └—————————————— Summer Host — Grass ——————————————┘

FIGURE 9.2 Comparison of parasite life cycles using a trematode, two insects, and a rust fungus to illustrate convergence of life cycles that solve particular problems of transmission between hosts. An asterisk denotes transmission phases from one host to another. Reproduction in the form of asexual (asex) and sexual (sex) method are indicated, and tenure on or in a host is given below each life cycle. From P. W. Price, *Advances in plant pathology.* 1992, **8:**1–30. Copyright © 1992 by Academic Press, Inc.

are compared with insect parasites on animals such as the Philopteridae, or bird lice, the Hystrichopsyllidae, the small mammal fleas, and the Cimicidae or bedbugs. These predators show no intimate living relationships with the prey they feed upon, whereas the bird lice spend their entire lives on a host, being wingless and able to transfer between hosts only during close contact of hosts while copulating or within the nest. The small mammal fleas and bedbugs have less intimate contacts with hosts, with inevitable consequences on their ecology and evolution discussed later in this chapter.

GENERALIST PREDATORS AND SPECIALIST PARASITES

Predators are usually more generalized in their diet, while parasites are more specialized. Carabids, vespids, coccinellids, and other predators feed on a wide variety of arthropods in terms of the number of species eaten, whereas bird lice and other parasites tend to be more specific to one host species or a few (see also Jeffrey Waage, 1979a). For example, bird lice have 87% of species known from only a single host in Israel (Table 9.1). It is hard to find lists of species eaten by insect predators because the emphasis is so frequently on which predators eat a certain prey species, and the serological technique used in such studies cannot easily be reversed to study all organisms in the predator's diet (e.g., Dempster, 1960; Young et al., 1964; Frank, 1967). Therefore, I have used for comparison the number of hosts used by parasitic insects compared with the number of species utilized by vertebrate predators and grazers (Table 9.1). The pattern is clear, showing that a concentration of parasitic species exploit only one, two, or three host species while predatory species tend to feed on 10 or more prey species. This dramatic difference would not change seriously if insect predators were included.

The families of parasitoids Ichneumonidae and Braconidae have been included under the parasite category. We saw in Chapter 8 that parsitoid adults act like predators in terms of population dynamics and impact on the host population. But the larvae live as parasites with the host and this intimacy leads to considerable specialization, as seen in Table 9.1. We will see in this chapter that much of the biology of parasitoids can be understood only in the light of their parasitic larval stage.

Another, perhaps surprising feature of Table 9.1 is the inclusion of herbivorous insects such as the leaf-mining flies in the family Agromyzidae, and the Gelechiidae in the Microlepidoptera. This is because insect herbivores that live in or on the plant and do damage to that plant while feeding on it fit the definition of a parasite perfectly (Price, 1975b, 1977, 1980). We commonly use the term *host plant* with the inference that the insect is a parasite on that plant. The important point is that insects do very similar things on plants and animals, and therefore come under the same kinds of ecological and evolutionary forces. Recognizing this fact greatly increases the potential for reach-

TABLE 9.1 Comparison Between Number of Host Species Utilized by Parasitic Insect Species (Given as the Percentage of Parasite Species in Each Host Number Class) and the Number of Species in the Diet of Predators and Grazers (Given as the Percentage of Species in Each Number Class).

	Parasites on:								Predators on:				
	Plants			Insects		Mammals and Birds			Seeds		Insects	Grazers	
Number of Hosts	Agromyzidae	Miridae	Gelechiidae	Ichneumonidae	Braconidae	Philopteridae	Hippoboscidae	Number of Species or Larger Taxa in Diet	Birds	Small Mammals	Bats	Small Mammals	Ungulates
1	57	22	53+	53	60	87	17	1	—	—	—	—	—
2	20	14	26+	18	14	9	9	2	—	—	—	—	10
3	7	13	6+	11	10	2	15	3	—	9	—	5	—
4	6	21	3+	4	7	1	13	4	—	9	—	—	—
5	4	8	—	5	2	1	7	5	—	—	—	—	—
6	3	6	—	3	1	—	4	6	3	—	—	5	—
7	<1	7	—	1	2	—	4	7	9	—	—	—	—
8	<1	1	—	1	<1	—	4	8	3	—	—	14	—
9	<1	4	—	1	1	—	2	9	4	17	—	—	—
10–19	2	3	—	3	2	—	4	10–19	43	35	25	40	10
20–29	<1	1	—	<1	—	—	7	20–29	20	26	25	31	20
30–39	—	—	—	<1	—	—	2	30–39	10	—	50	—	10
40–49	—	—	—	—	—	—	—	40–49	4	4	—	—	10
50–59	—	—	—	—	<1	—	2	50–59	1	—	—	—	10
60–69	—	—	—	—	—	—	—	60–69	3	—	—	—	10
70–79	—	—	—	—	—	—	—	70–79	—	—	—	5	—
80–89	—	—	—	<1	—	—	7	80–89	—	—	—	—	20
Number of species	267	72	112	514	214	122	46		70	23	4	22	10

a Frequently, only larger taxa are given, for example, the families or orders of insects eaten by bats.

SOURCE: Peter W. Price, *Evolutionary biology of parasites*. Copyright © 1980 by Princeton University Press. Reprinted by permission of Princeton University Press.

ing generalizations, and seeing patterns in nature, which are major preoccupations among ecologists.

Similarly, it is useful to remember that seed predators such as bruchid weevils (Bruchidae) and chalcid wasps (Megastigmidae) are exploiting seeds in precisely the same kind of way that parasitoids exploit insect hosts (Janzen, 1975; Price, 1975b). We need not call bruchids parasitoids, but the common mode of relating to resources should be recognized.

The degree to which insect parasites are specialized varies widely from taxon to taxon (Table 9.1). One of the patterns in this variation relates to the intimacy with which the life cycle of the parasite is associated with the host. In a comparison of insects utilizing mammals and birds as hosts we see a clear trend from the highly specific Philopteridae to the much less specific louse flies in the family Hippoboscidae (Table 9.2), with the bat flies (Streblidae), bot and warble flies (Oestridae), and the small mammal fleas (Hystrichopsyllidae) in intermediate positions. Accompanying this trend is a pattern of looser ties with the host.

As stated earlier, philopterid lice are wingless and the entire life cycle is spent on the host. Transfer between hosts is usually restricted to transfer between conspecific hosts, and specific adaptations to hosts are essential to

TABLE 9.2 Percentage of Species in Families of Parasitic Insects on Mammals and Birds in Each Class of Number of Hosts Attacked.

Number of Hosts	Philopteridae	Streblidae	Oestridae	Hystrichop- syllidae	Hippoboscidae
1	87	56	49	37	17
2	9	22	19	20	9
3	2	13	7	9	15
4	1	5	6	5	13
5	1	2	6	5	7
6		1	4	6	4
7		1	4	2	4
8			0	3	4
9			4	3	2
10–19			2	9	4
20–29				<1	7
30–39				<1	2
40–49					0
50–59					2
60–69					0
70–79					0
80–89					7
Number of parasites	122	135	53	172	46

SOURCE: Peter W. Price, *Evolutionary biology of parasites*. Copyright © 1980 by Princeton University Press. Reprinted by permission of Princeton University Press.

avoid the impact of frequent grooming and preening, which can decimate populations (Nelson and Murray, 1971). Other host characteristics may well be involved with host specialization such as the use of host cues in the synchronization of life cycles in the lice on the orange-crowned warbler (Foster, 1969).

The blood-sucking ectoparasitic flies on bats (Streblidae) are less specific than the bird lice. Females deposit fully developed larvae on surfaces in the bat roost where the larvae pupariate (i.e., form a puparium in which they pupate). Winged adults emerge but remain in close contact with the original host because of the confined space of bat roosts and the position of the puparium. Linkage between the two species remains quite intimate, accounting for relatively high specificity, but transfer between host species using the same roost reduces its level below that of the philopterids.

The bot and warble flies (Oestridae) utilize mammals as hosts, the subfamily Hypoderminae burrowing into host tissues and eventually lodging under the skin to form warbles, and the Oestrinae living in nasopharyngeal or orbital cavities of their hosts. Female flies are strong flyers and lay eggs on the host while larvae burrow into host tissue or cavities. Again the free-flying adult could oviposit on many hosts, but specialization is required, especially in the larval stage and during oviposition. Females in the Oestrinae may be considered as connoisseurs of noses. Where noses are rather similar, such as in sable antelope, roan antelope, and species of hartebeest, one parasite *Oestrus aureoargentatus* can attack them all. But animals with unusual noses have only one, specific species of oestrid: the giraffe is host to *Rhinoestrus giraffae*, the hippopotamus to *R. hippopatami,* and the African elephant to *Pharygobolus africanus* (see Zumpt, 1965).

Fleas on small mammals are usually less specific than oestrids. Adults are wingless ectoparasites and the larvae feed on debris in nest material. The adults are long-lived and have excellent jumping ability. Opportunities for transfer from one host species to another are considerable in burrow systems, and specificity is not selected for by the larvae living in intimate contact with the host. However, there is good evidence that there are close physiological ties between some fleas and hosts, resulting in synchronization of breeding as in the rabbit flea example given above (Rothschild, 1965; Rothschild and Ford, 1973).

The least specific parasites belong in the Hippoboscidae. The life cycle is similar to that in the Streblidae except that mainly birds are used as hosts, and the mature larva falls to the ground before pupariating in the soil. When the adult emerges it must fly about searching for hosts and the potential is great for attack of many different host species. Little specificity results, although even in this least specific group of parasites they are still much more restricted in the number of species utilized than predators (Table 9.1).

ADAPTIVE RADIATION IN PREDATORS AND PARASITES

Adaptive radiation in predators is minor compared to radiation among some parasite taxa. Askew (1971) estimated that 15% of all insects are parasitic on animals, Rothschild and Clay (1952) claimed that more than 50% of animal species were parasitic, and Arndt (1940) reckoned that 25% of animals in Germany were parasitic. The lower estimates reflect use of described species in the calculations, which clearly underestimates the number of parasites in the fauna. But the well-studied insect fauna of the British Isles, where even the most diminutive species seem to have been found and studied with loving care, permits a more accurate estimate of the relative abundance of predators and parasites based on checklists (Kloet and Hincks, 1945, 1964–1978). Most families can be classified into the categories predators, nonparasitic herbivores and carnivores, parasites on plants, parasites on animals, and saprophages (Table 9.3), covering 86% of the described species on the islands. Predators include dragonflies, lacewing flies, lady beetles, and vespid

TABLE 9.3 Feeding Habits of British Insects Based on Analysis of the Checklist of British Insects by Kloet and Hincks (1945).

Order	Predators	Nonparasitic Herbivores and Carnivores	Parasites On Plants	On Animals	Saprophages
Thysanura					23
Protura					17
Collembola					261
Orthoptera		39			
Psocoptera		70			
Phthiraptera				308	
Odonata	42				
Thysanoptera			183		
Hemiptera	123		283	5	
Homoptera			976		
Megaloptera	4				
Neuroptera	54				
Mecoptera	3				
Lepidoptera			2233		
Coleoptera	215	65	909	18	1637
Hymenoptera	170	241	435	5342	36
Diptera	54	231	922	311	1672
Siphonaptera				47	
Total	665	646	5941	6031	3646
Percent of insect fauna	3.9	3.8	35.1	35.6	21.5

SOURCE: Peter W. Price, *Evolutionary biology of parasites*. Copyright © 1980 by Princeton University Press. Reprinted by permission of Princeton University Press.

wasps. Nonparasitic herbivores and carnivores include pollinating insects such as bees, very general feeders (grazers) such as grasshoppers, and biting flies that do not live in or on a host but resemble more the grazing habit. Parasites on plants include thrips, cicadellid leaf hoppers, caterpillars, chrysomelid beetles, sawflies, and agromyzid leaf miners. Parasites on animals are represented by lice, bedbugs, parasitic Hymenoptera, parasitoid flies such as the Tachinidae, and the fleas.

Very distinctive differences exist between the numbers of insect predators and insect parasites: less than 4% of the fauna consists of predators, whereas over 70% of the fauna is parasitic, with about 35% parasitic on plants and 35% on animals. These impressive differences are not so much because there are more parasitic taxa at the family level or above, but because adaptive radiation has been much more extensive in parasitic taxa. The great diversity of butterflies and moths results as much from their parasitic habit as from the coevolutionary arms race stressed by Ehrlich and Raven (1964). As we shall see in the example of the hessian fly, the intimacy of the relationship between parasite and host forces a stepwise coevolutionary response in each interacting species.

The parasitic Hymenoptera in the British Isles number 5342 species and represent 26% of the insect fauna, illustrating several remarkable adaptive radiations in the families Ichneumonidae and Braconidae and several chalcidoid families. This radiation has resulted largely because of the parasitic aspects of the life history, rather than the predatory nature of adult behavior. If the 10 largest families in the British insect fauna in each of the categories predators, herbivorous parasites, and carnivorous parasites are compared, it is clear that the more intimate the relationship with the host, the greater the adaptive radiation can be (Table 9.4). The mean size for the 10 largest preda-

TABLE 9.4   Number of Species in the 10 Largest Families in the British Insect Fauna in the Categories of Predators, Herbivorous Parasites, Carnivorous Parasites Based on the Checklist by Kloet and Hincks (1945).

Predators		Herbivorous Parasites		Carnivorous Parasites	
Dytiscidae	110	Cecidomyiidae	629	Ichneumonidae	1938
Sphecidae	104	Curculionidae	509	Braconidae	891
Coccinellidae	45	Aphididae	365	Pteromalidae	649
Corixidae	32	Tenthredinidae	358	Eulophidae	485
Cucujidae	32	Noctuidae	298	Tachinidae	228
Hemerobiidae	29	Chrysomelidae	248	Philopteridae	176
Vespidae	27	Cicadellidae	242	Platygasteridae	147
Asilidae	26	Cynipidae	238	Encyrtidae	144
Anthocoridae	25	Olethreutidae	216	Diapriidae	125
Saldidae	20	Miridae	186	Ceraphronidae	108
Mean	45	Mean	329	Mean	489

SOURCE: Peter W. Price, *Evolutionary biology of parasites.* Copyright © 1980 by Princeton University Press. Reprinted by permission of Princeton University Press.

tory families is only 45 species. Among the herbivorous parasites we see an increasing number of species per family, with increasing intimacy of association from the ectoparasitic Miridae to the endoparasitic Curculionidae and the gall family Cecidomyiidae, and a mean family size of 329 species. In the carnivorous parasite group family size increases to a mean of 489 species, with very large numbers of species in the largest families. The 1938 species of ichneumonid represents 9.6% of the total insect fauna, and Townes (1969) estimates that there must exist 60,000 species in the world, although only 10% from the Neotropics and 35% from the Nearctic have been described. High specificity is selected for in these animal parasites because of (1) the host insect's ability to distinguish between self and nonself, resulting in encapsulation of nonadapted endoparasitic larvae (Salt, 1970; Edson et al., 1981; Godfray and Hassell, 1991; Godfray, 1994), (2) the relatively specific venoms used by ectoparasitic species for paralyzing hosts (Schmidt, 1982), and (3) the need for a reproductive strategy consonant with the survivorship curve of the host (Price, 1975a; see Chapter 14). Adaptive radiation has been spectacular.

More rigorous tests of the hypothesis that parasites have undergone greater adaptive radiation than predators have yielded mixed results. Cladistic analysis by Mitter et al. (1988) compared the number of species of phytophagous parasites in a taxon with the number in a sister taxon which did not include phytophages. In 11 of 13 sister-group comparisons the phytophagous group was larger than the nonphytophagous taxon. For example, the plant-feeding Lepidoptera contains about 140,000 species, while its sister group, the aquatic Trichoptera, numbers about 7000 species. The hypothesis was strongly supported, and the arguments are amplified in Mitter et al. (1991) and Farrell and Mitter (1993). A different result was obtained when carnivorous parasitic insect taxa were compared with predaceous or saprophagous sister groups. Wiegmann et al. (1993) found that in 15 comparisons the sister group of the parasitic taxon was larger in nine cases, and the hypothesis was not supported. Using flatworm parasites rather than insect groups, Brooks and McLennan (1993a,b) noted several difficulties in the identification and evaluation of the extents of adaptive radiation. They recognized the great potential of parasitic taxa for advancing understanding of adaptive radiation, without finding support for the hypothesis. Clearly, the intriguing questions about parasite ecology and evolution will remain perplexing, rich, and challenging.

GENETICS OF HOST–PARASITE RELATIONSHIPS

The intimacy of the relationship between host and parasite, and the coevolutionary nature of this interaction is best illustrated when the genetics are well understood (see Day, 1974, who concentrates on plant hosts and pathogenic microorganisms). The Hessian fly, *Mayetiola destructor* (Diptera: Cecidomyiidae), was a serious pest of wheat in North America in the 1920s. Hatchett and Gallun (1970) studied the heritability of virulence against wheat with two strains of the fly: GP, an avirulent strain; and E, a virulent strain. A peculiar feature in the genetics of cecidomyiids is that during spermatogenesis the paternal haploid genome is eliminated, so males reproduce as though homozygous for maternal traits.

The Mendelian genetics of this fly showed that the two races differed by a single gene and that the avirulence gene was dominant to the virulence gene (Fig. 9.3). If no segregation occurred, it was because of paternal genome elimination.

From these studies Hatchett and Gallun (1970) concluded that insect and host-plant genic systems were complementary, and that there was a gene-for-gene relationship between host-plant defense and parasitic insect attack. If a race of flies is to survive on a wheat variety it must have the recessive gene for virulence complementary to the dominant gene in the host plant for resistance.

Day (1974) indicates that this relationship between plant and insect parasite is essentially the same as those between plants and fungal pathogens described by Flor (e.g., 1956, 1971) who developed the **gene-for-gene concept.** Such systems can be most easily identified in pure-bred lines of agricultural crops, but they also must exist in nature.

HOST AND PARASITE VARIATION

Such close ties between parasite and host represent perfectly the coevolutionary process where it is demonstrated that change at the genetic level is selected for in pairwise interactions. A result of this coevolution is the great diversity of plant genotypes and parasite genotypes that can result from a small number of resistance genes in a population. In a plant population with n genes for resistance, each with a resistant or susceptible phenotype, the number of genotypes or races in the parasite population required to exploit all plants will be 2^n (Day, 1974). With five genes for resistance in the population, 32 races of parasite will be selected for, and with only 10 genes for resistance, 1024 races of parasites would be selected. Thus we should expect to find a great diversity of host suitabilities for insect parasites in natural populations.

This predicted variation among trees in a population has been known to exist for a long time in relation to insect herbivores, although the mechanisms behind this variation were not understood. As early as 1955 Morris cited five studies that supported his statement: "The fact that the variability in insect population between trees is so much greater than the variability within trees . . . may well prove to be a principle common to all forest insects" (p. 244). This kind of variation within tree populations has been studied more recently by Edmunds and Alstad (1978, 1981) and Whitham (1978, 1980, 1981).

Assume that A^a is a dominant gene conferring avirulence in the Hessian fly race. Race GP is A^aA^a. Assume that a^v is a recessive gene conferring virulence in the Hessian fly race. Race E is a^va^v.

Example 1. A cross between strains GP and E:

Race	GP♀ × E♂
Genotype	A^aA^a	$a^v\cancel{A^v}$ (paternal genome eliminated)
F$_1$	A^aa^v
Predicted	100% avirulent
Observed	100% avirulent

Example 2. Segregation in the F$_2$ generation when the male has derived its virulence gene from the paternal side:

F$_1$	(GP♀ × E♂)♀ × (GP♀ × E♂)♂
Genotype	A^aa^v	$A^v\cancel{A^v}$ (paternal genome eliminated)
F$_2$	A^aA^a	A^aa^v
Predicted	50% avirulent 50% avirulent
Observed	100% avirulent

Example 3. Segregation in the F$_2$ generation when the male has derived its virulence gene from the maternal side:

F$_1$	(GP♀ × E♂)♀ × (E♀ × GP♂)♂
Genotype	A^aa^v	$a^v\cancel{A^A}$ (paternal genome eliminated)
F$_2$	A^aa^v	a^va^v
Predicted	50% avirulent 50% virulent
Observed	51% avirulent 49% virulent

Example 4. Segregation in the F$_1$ backcross:

F$_1$ and P	(GP♀ × E♂)♀ × E♂
Genotype	A^aa^v	$a^v\cancel{A^v}$ (paternal genome eliminated)
	A^aa^v	a^va^v
Predicted	50% avirulent 50% virulent
Observed	40% avirulent 60% virulent

The similar backcross (E♀ × GP♂)♀ × E♂ gave a better ratio of 47% avirulent and 53% virulent.

FIGURE 9.3 Examples of the Mendelian genetics of Hessian fly races GP and E on Monon wheat, studied by Hatchett and Gallun (1970).

Edmunds and Alstad (1978, 1981) studied the black pineleaf scale (*Nuculaspis californica*) on ponderosa pine (*Pinus ponderosa*). They noticed that in heavy infestations of the scale some trees would be free of the insects. These trees, although colonized by crawlers, did not support persistent populations. They were resistant. Eventually, such trees may be colonized successfully by a strain of scale insects and a population would slowly build up to high densities. Using transfer experiments of scale insects from one tree to another, they demonstrated that trees differ very significantly in the survival success possible in the scales, that insects from different donor trees show different survival on a common recipient tree, and that insects survive much better when transferred within trees than between trees. The results point strongly to local insect population adaptation to individual trees defined by a genetic basis for host resistance and insect attack, and a tree population offers a complex genotypic and phenotypic mosaic to insect herbivores.

Whitham (1978, 1980, 1981) studied the gall-forming aphid *Pemphigus betae* on the narrowleaf cottonwood, *Populus angustifolia*. He demonstrated enormous differences in survival of aphids between trees (Table 9.5), between branches of the same tree, between leaves on the same branch, and even between positions on the same leaf (for details, see the section on foraging behavior in Chapter 15). He argued that an individual tree offers such a mosaic of varying susceptibilities to attack by aphids that somatic mutations may well play an important role in plant defense, generating a diversity of genotypes within a single plant (Whitham, 1981; Whitham and Slobodchikoff, 1981).

These very tight links between living resources and the parasites that exploit them should be expected. Understanding at the genetic level of interactions between hosts and their parasites will aid significantly in the development in the areas of coevolution, epidemiology, population dynamics, and community organization.

TABLE 9.5   Between-tree Variation in Suitability for *Pemphigus betae*. Shown are Densities and Survival Rates of Colonizing Stem Mothers on Different Host Trees of Narrowleaf Cottonwood (*Populus angustifolia*).

Tree (Mean Leaf Size, cm2)	Pemphigus betae	
	Galls/1000 Leaves	Percent Survival
4.7	7.7	24.2
5.1	0	—
6.1	13.8	66.0
8.8	105.7	78.5
15.9	183.8	87.0

SOURCE: Reprinted with permission from Whitham (1981).

ECOLOGICAL CONSEQUENCES

Some ecological consequences of the parasitic mode of life have already been mentioned in this chapter: specificity of host exploitation, opportunities for adaptive radiation, and close genetic ties with the host. In this section adaptive radiation will be examined more closely because it sheds light on other ecological consequences of parasitism, such as the equilibrium state of populations and communities and the importance of interspecific competition. In addition, effects of parasites on host biogeography will be discussed.

ADAPTIVE RADIATION

Can we obtain any estimates of the further potential in a taxon for adaptive radiation? This is as much an ecological question as an evolutionary question because we have to obtain some estimate of the number of ecological niches available to be exploited. With parasites we have perhaps a unique opportunity to estimate ecological niche availability because parasites are so specialized and they live in an easily defined niche space.

One estimate can be made by knowing the number of host species apparently available, the number of parasites exploiting these hosts, and estimating how many parasite species a host could support. In the British Isles there are 2177 species of nonaquatic plants that are available for colonization and 323 species of leaf-mining agromyzid flies. The mean number of agromyzid species per plant species is 0.15. Yet we know that some plants support eight agromyzid species and it is easy to calculate a possible 12–15 species coexisting on one plant species (Lawton and Price, 1979). Thus, conservatively, we could estimate that the present number of agromyzids represents less than 2% of the total potential radiation in this family, and perhaps only 1%.

Using a completely different approach, a similar result is evident for the parasitic wasp family Ichneumonidae (Price, 1980). Because fecundity of ichneumonids correlates with the probability of survival in the host defined by the host's survivorship curve (see Chapter 14 for examples), by identifying the range of host survivorship curves we can estimate the range of fecundities needed to fully exploit these hosts. This can be compared with the known range in the family. We also know that primitive ichneumonids attacked late stages in host development and the radiation in this family has been up the host survivorship curve, so to speak, to younger and younger host stages. We can therefore predict that ichneumonid species in early host stages will be more depauperate than on later stages. This is indeed the case. The range in fecundities needed to fully exploit pupal stage hosts is fully covered by existing species, but at all other host stages there is an increasing gap between fecundity needed and fecundity available (Fig. 9.4). Only 13% of available resources can be exploited by present-day ichneumonids attacking young

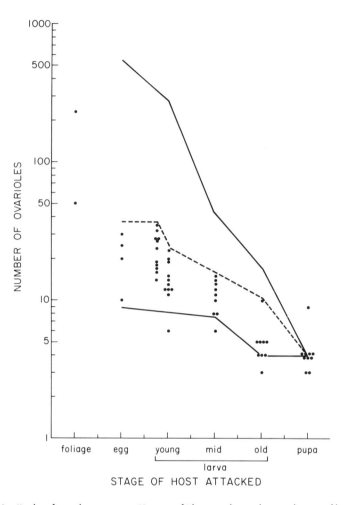

FIGURE 9.4   Number of ovarioles per ovary in 58 species of Ichneumonidae in relation to the stage of host attacked. Solid lines represent predicted maximum and minimum ovariole numbers required to exploit fully all host stages, with survivorship as in Fig. 13.9. The dashed line represents the upper limit in ovariole number reached so far in the adaptive radiation of the Ichneumonidae. From P. W. Price, *Evolutionary biology of parasites.* Copyright © 1980 by Princeton University Press. Reprinted by permission of Princeton University Press.

larvae and only 5% by those attacking eggs. Adaptive radiation can be much more extensive than it is at present even in what must be the largest family of metazoan animals on earth. Admittedly, members of other parasitoid taxa, such as the Chalcidoidea and Braconidae, may have preempted some of these resources, but not a large proportion of them.

Crude as they may be, these calculations illustrate the apparently inevitable consequences of a very specialized mode of exploitation: many ecological niches remain vacant, resources far exceed the species available to exploit them, and in any community of parasites many potential resources remain unutilized. These consequences of parasite adaptive radiation have obvious repercussions in the equilibrium state of populations and communities and the role of competition in communities to be discussed next.

EQUILIBRIUM IN POPULATIONS AND COMMUNITIES

The complexity of relationships in parasite life cycles appears to result in strong destabilizing forces acting on populations and communities. Resources are very patchy for a small, specialized organism and the patches are likely to be short-lived since they depend on a living host. For example, a host individual may die or the host population may go locally extinct, as discussed later. The net result is that populations of parasites in a patch increase and decline in very unpredictable ways without the development of the equilibrium condition, influenced profoundly by low probabilities of colonization of patches and high probabilities of extinction (Fig. 9.5). Here we are viewing patches as single oranges in Huffaker's orange arena for *Eotetranychus* and its predator, so that while area or global equilibrium may be evident, this is not experienced by the parasites in a single patch.

An example of this condition concerns the epidemiology of scrub typhus or chigger-borne rickettsiosis reviewed by Traub and Wisseman (1974). These authors note that scrub typhus has a "markedly patchy distribution of highly circumscribed foci or even microfoci of 'typhus islands' which may only be a few meters in diameter." Such patches may change rapidly from infective to noninfective within a few months. The factors resulting in high patchiness of

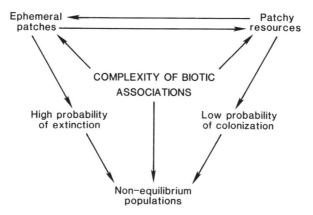

FIGURE 9.5   Patch dynamics in parasite populations—the general case. From P. W. Price, *Evolutionary biology of parasites.* Copyright © 1980 by Princeton University Press. Reprinted by permission of Princeton University Press.

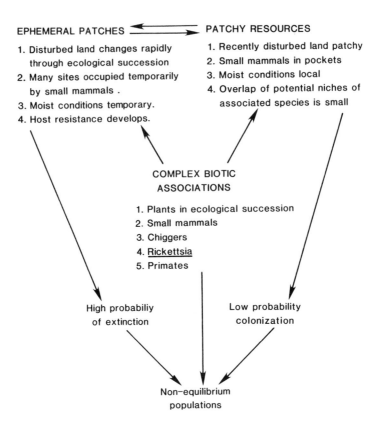

FIGURE 9.6 Patch dynamics in *Rickettsia* and chigger populations. From P. W. Price, *Evolutionary biology of parasites.* Copyright © 1980 by Princeton University Press. Reprinted by permission of Princeton University Press.

resources and ephemeral patches are summarized in Fig. 9.6, which adopts the basic scheme of Fig. 9.5. It seems unlikely that an equilibrium number of individuals in a patch will develop with so many and potent destabilizing forces at work.

This scenario can be illustrated in many parasite systems (see Price, 1980), with patchiness and ephemerality being common themes. Concerning the parasitic mite *Partnuniella thermalis* on its host, the brine fly, *Paracoenia turbida*, Collins (1975b, p. 250) notes "the pattern in space of resources for both parasitic and free-living stages of the mite will be a mosaic of high and low density patches with few patches of intermediate density." And the high costs associated with a low probability of colonizing host patches is stressed: "the 76%–91% mortality associated with *Partnuniella's* parasitic stage may be an irreducible cost of exploiting an unpredictable and extremely clumped distribution of hosts, combined with host defenses which effectively filter out adaptations that could improve host-finding ability" (Collins et al., 1976, p. 1229).

Where colonization probability is low and extinction probability high for most species in a patch, it is unlikely that an equilibrium number of species will develop, and nonequilibrium communities will be common. The equilibrium theory of island biogeography (discussed in Chapter 22 and 24) suggests that a highly predictable number of species should be expected in relation to island size. But it has been found repeatedly that if hosts are regarded as islands available for colonization by parasites, the prediction fails to account for much of the variation of number of species of parasites on area over which the host is distributed. Some examples include mites on small mammals (46% of variance accounted for) (Dritschilo et al., 1975), cynipid wasps on Atlantic oaks (41%) (Cornell and Washburn, 1979), agromyzid flies on umbelliferous plants (32%) (Lawton and Price, 1979), fungi on trees (28%) (Strong and Levin, 1975), arthropods on cushion plants (10–16%) (Tepedino and Stanton, 1976), and leaf hoppers on trees (0.04%) (Claridge and Wilson, 1976, 1978). Admittedly, other studies show better relationships (see Strong, 1979; Price, 1980), but there are almost as many studies accounting for less than 50% of the variance as those studies that explain more than half of the variance (see Table 24.2).

Where many resources remain unexploited and communities are depauperate relative to what could be supported, as we saw in the section on adaptive radiation, the forces leading to a steady and predictable number of species in the community are unlikely to be powerful. Particularly, interspecific competition may frequently be an unimportant organizing force, as discussed in the next section.

INTERSPECIFIC COMPETITION

Lawton and Strong (1981) reviewed the literature on phytophagous insects, mostly parasites, coming to the conclusion that interspecific competition is infrequent and usually not an important organizing force in communities. The importance of this conclusion cannot be fully appreciated until the conventional wisdom stressing the importance of competition is recognized (see Chapters 20 and 21). But there is convincing evidence in their favor.

Rathcke (1976a,b) studied stem-boring insect communities in tall-grass prairie plants with 13 common insect species present. She estimated that usually less than 10% of resources in a stem were utilized by a boring insect, usually less than 20% of stems were attacked, and that co-occurrence of species fitted a pattern generated assuming random distribution of species. She concluded that competition has not been a major organizing force within the guild of stem borers because resources of food and space were generally not limiting. Rathcke (1976a, p. 85) noted that the insects seldom existed in equilibrium populations—"a condition necessary for most predictions of competition theory."

Strong (1981, 1982) reached similar conclusions using experiments, an approach often ignored but obviously providing much more powerful analysis than the observation of species distributions. The hispine chrysomelid beetles that utilize rolled leaves of *Heliconia* plants in Costa Rica were used in experiments. Strong noted that only rarely is even 0.5% of rolled leaf eaten by all herbivores in the community, and no evidence exists that the remaining leaf tissue is unavailable. Strong found that rather than avoiding other species, these beetles tend to aggregate, and no antagonistic behavior was observed between species, suggesting that they normally feed on a superabundant resource.

For small specialized insects the time in which resources are available for colonization is usually very limited, so an individual unit of resource may pass through its available stage before it is discovered, even by a single species. The probability of all available species discovering this unit is very small, so that interspecific contact is greatly reduced. Leaves of the narrowleaf cottonwood, *Populus angustifolia,* are suitable for attack by the gall aphid, *Pemphigus betae,* for only 4–5 days immediately after bud break (Whitham, 1981). Any one umbel on wild parsnip used by the ovipositing parsnip webworm, *Depressaria pastinacella,* is available for about 2 days (Thompson and Price, 1977; Price, 1984b). For any gall-forming species that must find very active meristematic tissue in which to oviposit (Rohfritsch and Shorthouse, 1982), every resource unit is available only briefly. For specialized species that attack rapidly developing hosts, like plants, shortage of time may dominate the availability of resources, leaving many suitable locations uncolonized. Interspecific competition is unlikely to become important under such conditions.

BIOGEOGRAPHICAL AND LOCAL DISTRIBUTIONAL EFFECTS

The ecological effects of a specialized way of life play a role not only in the parasite's ecology but also in that of the host. Although many such effects can be recognized (Price, 1980), one of the most pervasive is the effect on host distribution, although this has not been widely recognized.

Insects, with their flight ability, play an important role as vectors of disease organisms between hosts, both vector and pathogen frequently being parasites. Wherever the pathogen has differential impact on two host species, it is likely to influence their distributions relative to each other.

Many native birds of Hawaii were probably driven to extinction by bird malaria (Warner, 1968). Before 1826 they were not attacked by malaria even though migrating shorebirds regularly carried the disease to the islands because a vector of the disease was not present. In 1826 the mosquito *Culex pipiens fatigans* was accidentally introduced and acted as a vector between the shorebirds and the indigenous birds of Hawaii, which were very susceptible to the disease, not having been exposed to it before. Only those indigenous bird species dwelling in mountain forest vegetation, remote from shores, remained unexposed to malaria and survived. The mosquito vector caused a distributional gap between the carriers of the disease and the nearest indigenous species. The extensive decrease in native birds of New Zealand such as the bellbird, saddleback, and kakapo is likely to have been caused, at least partially, by disease contracted from introduced species (Lack, 1954). Such gaps in distributions have been discussed by Barbehenn (1969) and Cornell (1974) with the suggestion that suitable habitat may remain unutilized by two species if cross infection of pathogens virulent in the receiver but not the normal host were caused by vectors.

Sylvatic plague is caused by the bacterium, *Yersinia pestis*, and is transmitted from host to host by fleas. In lava caves in California occupied by wood rats, *Neotoma cinerea*, plague epidemics are common (Nelson and Smith, 1980). It seems that the deer mouse, *Peromyscus maniculatus*, acts as a reservoir for the disease, being very resistant to its pathogenic effects. Because wood rats are very susceptible to plague, a mouse carrying infected fleas can start an epidemic in a cave, usually resulting in local extinction of the rats, thereby opening up very desirable nesting sites for the mice. The pack rat middens show successive occupation by rats and then mice, and the dynamics of small mammals in the lava caves could not be understood without recognition of the differential pathogenicity of the plague bacillus.

Insects are also very important as vectors of plant diseases (Carter 1973; Maramorosch and Harris, 1979). Many vectors, such as aphids, alternate between host species within a season so that transport of a disease between hosts is inevitable. With the likelihood of differential pathogenicity on hosts, a disease can have considerable effects on the distributional relationships between plant species. For example, western yellow blight of tomatoes can persist only when plants are grown close to sugar beets with curly top because the vector of the common pathogen (a virus) is the cicadellid, *Circulifer tenellus*, which breeds only on beet. If these were wild plants, the disease–vector–host plant relationships would probably result in spatial separation of the two plant hosts. Unfortunately, such relationships have been studied in detail only for agricultural crops, but similar effects are probably widespread in nature. Of 280 known arthropod-borne plant pathogens, 97% are transmitted by insects, mostly in the suborder Homoptera (Harris, 1979). Many more remain unidentified and an understanding of their role in natural vegetations remains an unexplored challenge to ecologists.

For visually hunting females searching for oviposition sites among plants, such as the butterflies in the genus *Heliconius*, Gilbert (1975, 1991) has suggested that selection on plants will be to diverge in visual similarity. The small number of discrete leaf shapes possibly may well set limits on the number of *Passiflora* species that can coexist because of visual searching by the butterflies. Thus community structure may be defined to a large extent by the herbivores. The remarkable diversity of leaf shapes in a community and the allopatric distribution of species with similar leaf shape make Gilbert's hypothesis very convincing (Fig. 9.7).

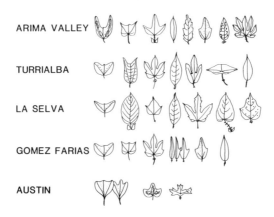

FIGURE 9.7   Leaf-shape variation among sympatric species of *Passiflora* at five localities: Arima Valley, Trinidad; Turrialba, Costa Rica; La Selva, Costa Rica; Gomez Farias, Mexico; and Austin, Texas. From L. E. Gilbert in L. E. Gilbert and P. H. Raven (eds.), *Coevolution of animals and plants.* Copyright © 1975, 1980. Reprinted by permission of the University of Texas Press.

Similar effects are probably imposed by parasitoids searching for hosts in galls, and an approach similar to Gilbert's would be rewarding in an analysis of gall communities. For example, Askew (1961), studying three coexisting galling species on oak, *Cynips divisa, C. longiventris,* and *C. quercus-folii,* found divergence in gall structure—*C. divisa* forms small hard galls, *C. longiventris* forms medium-sized hard galls, and *C. quercus-folii* forms large soft galls (Fig. 9.8). All occur on the underside of oak leaves. This divergence influences both the species of parasitoids on the *Cynips* species and the total number of species attacking: 14 species on the smaller gall, 11 on the medium-sized gall, and 10 on the largest.

These effects of parasites on biogeography and local distribution of their hosts have been inadequately studied, yet their potential for influencing population dynamics and community structure in plant and animal assemblages is great.

ADVANTAGES OF BEING PARASITIZED

It is not the conventional view to regard parasites as benefiting a host in any way. Yet, as we saw with herbivores in Chapter 5, the case can be made that enemies are not inevitably an unmitigated curse.

Any host to parasitic microorganisms may acquire new genetic material, with new potentials, because genes can be received from viruses, plasmids, viroids, and so on, or transferred from one host to another by such organisms (Anderson, 1970; Day, 1974). Many viruses replicate in vector hosts, with the

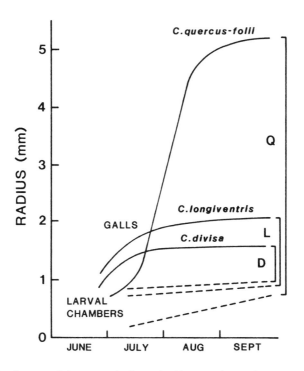

FIGURE 9.8   Development and ultimate size of galls stimulated by *Cynips divisa, C. longiventris,* and *C. quercus-folii.* Wall thickness is indicated on the right (D, L, and Q, respectively), being the difference between gall radius and radius of larval chamber (after Askew, 1961). From P. W. Price, *Evolutionary biology of parasites.* Copyright © 1980 by Princeton University Press. Reprinted by permission of Princeton University Press.

potential for leaving adaptive genetic material in host cells. As mentioned in Chapter 5, *Caryedes brasiliensis* may have acquired the capacity to utilize canavanine via a virus that transferred the necessary genome from the plant host or an adapted bacterium. Plasmids carried by the crown gall bacterium, *Agrobacterium tumefaciens,* regulate gall development by genetic material from the plasmid becoming incorporated into the host cell genome (Schell et al., 1979; Wullems et al., 1981). Viruses responsible for some plant diseases replicate in the vector host, such as wound tumor virus in *Agallia constricta* (Black, 1979). There is an interesting and suggestive correlation between the commonness of eriophyid mites as vectors of virus diseases of plants and the ability of these mites to initiate gall formation (Oldfield, 1970). Perhaps viruses regulate plant development into the gall (Cornell, 1983). The relationships between insects and microorganisms are incredibly diverse (e.g., Steinhaus 1946, 1963), making the potential high for acquisition of new genetic material from parasites. Anderson (1970) points out that many virus infections span large taxonomic distances. For example, the arthropod-borne viruses infect insects and vertebrates. Genetic material obtained from viruses is often inherited with the host genome, and Anderson argues that evolution depends heavily on transfer of viral DNA between organisms, even between plants and insects and insects and vertebrates.

Another possible advantage of being parasitized is that parasites may evolve into mutualists, leading to adaptive peaks and consequent adaptive radiation for both taxa involved. The pollinating yucca moths and fig wasps discussed in Chapter 10 were almost certainly parasitic initially.

Another advantage that works in ecological time is that parasites may modify competitive relationships. For an inferior competitor to carry parasites that are more deleterious to the dominant species has important consequences for the coexistence of the two species. In two plant associations involving *Eucalyptus pauciflora*, this species seems to persist by being less heavily attacked by parasites common to it and its associated congenor. Although *E. stellulata* grows more rapidly than *E. pauciflora* in the absence of insect parasites (Morrow, 1977; Morrow and La Marche, 1978), insects are more devastating on this species, causing death of 96% of shoots and consuming 50% of leaf area. On *E. pauciflora*, 76% of shoots were killed and 38% of leaf area destroyed. The combined effects of insects and fungi were also stronger on *E. dalrympleana* than on *E. pauciflora*, with 30 and 15% leaf area loss, respectively (Burdon and Chilvers, 1974). In young stands *E. dalrympleana* was the dominant species, but in the mature stand *E. pauciflora* was almost three times more abundant. Of course, the reverse situation can prevail where the competitive dominant can support herbivores more deleterious to an associated species. This appears to be the case with *Rumex obtusifolius*, the dominant, and *Rumex crispus* (Bentley and Whittaker, 1979; Bentley et al., 1980). The chrysomelid beetle, *Gastrophysa viridula*, shows a preference for *R. obtusifolius*, but when *R. crispus* is present this is more seriously injured. Being both competitively inferior and more susceptible to herbivory, *R. crispus* is likely to have a distribution noncoincident with that of *R. obtusifolius*, a tendency that Bentley and Whittaker (1979) observed. These kinds of interactions may well be common in nature, but they have been studied inadequately. Holmes (1979) gives further examples.

Parasites are known to affect the outcome of competition between insects also. In the absence of the sporozoan parasite *Adelina tribolii*, *Tribolium castaneum* usually won in competition with *T. confusum* in warm, moist environments, with the latter becoming extinct (Park, 1948). However, when *Adelina* was present in these laboratory experiments, *T. confusum* usually caused the extinction of *T. castaneum*. This kind of effect is also seen in nature. Feener (1981) studied attacks by the fire ant, *Solenopsis texana*, on nests of another ant, *Pheidole dentata*. When foraging groups of the fire ants intrude on *Pheidole* territory an enemy-specific alarm recruitment is initiated, causing major workers to defend the nest. In experiments using tuna fish bait Feener studied the effects of introducing fire ants at the bait when *Pheidole* was already foraging on it. He found that the presence of fire ants signifi-

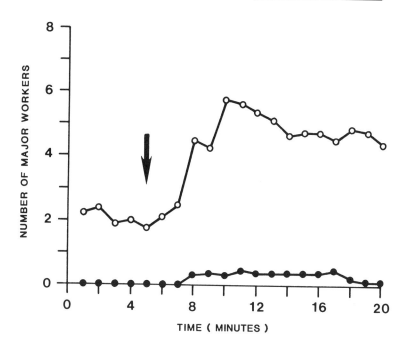

FIGURE 9.9 Responses of *Pheidole dentata* major workers to the addition of *Solenopsis texana*. The arrow indicates the time of addition of *S. texana;* open circles show the response in the absence of the phorid fly *Apocephalus;* and the solid circles show the response in the presence of the parasitic fly. From D. H. Feener, *Science,* 1981, **214:**815–817. Copyright © 1981 by the American Association for the Advancement of Science. Reprinted with permission.

cantly increased the number of major workers at the bait which were effective at combating the invaders, but this was only in the absence of the parasitic phorid fly, *Apocephalus feeneri,* which lays eggs only on major workers of *Pheidole* (Fig. 9.9). In the presence of these phorid flies almost all major workers remained under cover and were not effectively recruited to defend the bait against *Solenopsis.* In general, when the parasitic fly was absent, *Pheidole* dominated, and when it was present, *Solenopsis* dominated. Feener (1981) suggests that this kind of interference in aggressive interactions between ant species may be common in nature.

In this chapter we saw several cases of differential impact of a parasite on hosts: avian malaria in Hawaii and New Zealand, the plague bacterium in wood rats and deer mice, the virus on sugar beets and tomatoes, insects and fungi on *Eucalyptus,* and *Gastrophysa* on *Rumex.* Given that a parasite will rarely have equal effects on two host species, the potential for mediating interactions is enormous (Price et al., 1986). The interactions may well go beyond competitive interaction, for the effects of cross infection may be felt in the absence of competition. When resources are plentiful and with vectors carrying diseases, effects may be widespread. Such widespread interaction would best be classed under the heading *biological warfare,* which is certainly a very common but underappreciated phenomenon. Although the term is anthropomorphic, it is no less so than the term *competition.* Thus the potential for biological warfare by acquiring a parasite more deleterious to another species may be an important advantage to harboring parasites.

Davies et al. (1980) discuss at greater length the advantages of being parasitized, particularly in relation to the immune response in vertebrates [which can, of course, be stimulated by insects and other arthropods (Wikel, 1982)], while they also recognize disadvantages to parasitism.

Several general treatments provide further information on parasitic insects and their ecology and evolution: Askew (1971) and Price (1975d) on parasites and parasitoids, Marshall (1981) on ectoparasitic insects, Jeffrey Waage (1979a) on the evolution of insect–vertebrate associations, and Price (1980) and Toft et al. (1991) on the general evolutionary biology of parasites. Several books are devoted exclusively to parasitoid biology (e.g., Waage and Greathead, 1986; Godfray, 1994; Hawkins, 1994; Hawkins and Sheehan, 1994). Parasitic arthropods on birds are discussed in Loye and Zuk (1991), while human hygiene and disease are treated in many parasitology texts, Busvine's (1976) delightful treatise on insects, hygiene, and history, and the more alarming discussions of human disease epidemiology by Ewald (1994) and Garrett (1994).

In her compelling book *The coming plague,* Garrett (1994) argued that disruption in the ecology of humans, specifically their food production and general environment—including arthropod vectors—created serious imbalances fostering epidemic disease. As Louis Pasteur said, "The microbe is nothing; the terrain everything." Any blood-feeding arthropod, from ticks, lice, and fleas to sandflies, mosquitoes, tse-tse flies, black flies, and triatomid bugs, may exploit changing ecological relationships, vectoring pathogens among new sets of hosts or to new populations in novel biogeographic realms. Many more regional diseases are likely to become cosmopolitan, and new pathogens are likely to colonize humans as ecological change accelerates. For the insect ecologist there is a fascinating challenge, even though for humans in general the prospect is frightening. Of course, the arthropod vectors of human pathogens share our suffering!

Chapter 10

Mutualistic Associations

INTERSPECIFIC INTERACTION TYPES

A consideration of all possible interaction types between species can be made by comparing pairwise combinations of positive (+), negative (−), and no (0) impact on the interacting species (Table 10.1). Burkholder (1952) recognized nine such categories, filling the matrix, by distinguishing between impact of "strong" on "weak" organisms and then "weak" on "strong" organisms, with different names for each of the nine possibilities. But this complicates terminology unnecessarily. Others since then, reviewed by Starr (1975), have proposed a complex array of terms to cover the continuum of interaction types we see in nature, but the six given in Table 10.1 have usually been adequate. The terms *predation, parasitism,* and *mutualism* were discussed in Chapter 2. *Commensalism* denotes an interaction in which one species benefits and another species receives no impact (+0). In *neutralism* there is essentially no measurable interaction (00). *Amensalism* is the term used for an interaction in which one species suffers but there is no impact on the other (−0). *Competition* has a negative impact on both species (−−) because there is an active demand in members of each species for a resource that is in limited supply.

Much of conceptual ecology relates to the interactions with negative

Poorly studied relationship, possibly mutualistic, between an *Iridomyrmex* ant worker and the egg of the walkingstick insect, *Didymuria violescens*. In Australia, eggs are dropped from tall *Eucalyptus* trees and worker ants carry some into their nests. Drawing by Tad Theimer.

TABLE 10.1   Possible Interaction Types in Ecological Systems.

	−	0	+
+			++ Mutualism
0		00 Neutralism	0+ Commensalism
−	−− Competition	−0 Amensalism	−+ Predation Parasitism

impact, particularly predation and competition (King et al., 1975; Risch and Boucher, 1976; Vandermeer and Boucher, 1978; Travis and Post, 1979), perhaps resulting from this emphasis by Darwin (1859) in *The origin of species.* Much less emphasis has been placed on mutualism and commensalism, yet it can easily be demonstrated that these are very common interaction types in nature. Burkholder (1952) noted how mutualisms "are abundantly present in nature," and many books have devoted large sections to mutualism (e.g., Buchner, 1965; Hungate, 1966; Harley 1969; Faegri and van der Pijl, 1971; Proctor and Yeo, 1972; Jennings and Lee, 1975; Batra, 1979; and many others). The subjects range from pollination ecology, to mycorrhizae, to the gut fauna of animals eating a high-cellulose diet, to the evolution of eukaryote cells, and the many beneficial associations of insects with microorganisms. It is remarkable that despite this voluminous literature, ecologists have tended to regard mutualism as rather rare and/or unimportant in ecology. For example, Williamson (1972, p. 95) discusses mutualism in a single sentence saying that "its importance in populations in general is small," but spends seven chapters on other kinds of two-species interactions. Other authors have tended to emphasize the persistence of mutualism in the warm and humid tropics as if only there can such delicate relationships persist (Futuyma, 1973; May, 1976, 1981). This chapter is devoted to the argument that mutualism is one of the great forces in the ecology and evolution of species, of equal importance with predation, parasitism, and competition, at the population, community, and ecosystem level.

Fortunately, we have witnessed in the last decade a growing concern and appreciation for mutualism as a ubiquitous interaction in nature with relevance to the ecology and evolution of many species. The growing influence of the endosymbiotic theory of cell evolution fostered by Margulis (e.g., 1981, 1993; Margulis and Fester 1991) alerts us all to the broad ramifications of mutualism. Sapp (1994) provides a historical view on evolution by association, noting the chequered career of the science of mutualism. Boucher (1985) discusses the "renaissance" of interest relating to mutualism, and the topic has received increasing attention in evolutionary ecology (e.g., Beattie, 1985;

Margulis and Fester, 1991; Huxley and Cutler, 1991; Thompson, 1994; Price, 1996a).

Even an armchair approach to the commonness of interaction types in ecology should make us wary of earlier glib statements about the relative unimportance of mutualism. Recognizing that humans' feeble attempts at classifying nature grossly oversimplify its diversity, we can be sure that all intermediate stages between the categories in Table 10.1 exist; that a continuum across the matrix is real. This is so because each interaction exerts selective pressures on populations that tend to change the force of the interaction and ultimately perhaps its direction. Naturally, negative impact has a very strong selective influence to reduce it, so that $--$ competitive interactions will tend to become -0 amensal interactions, and then 00 neutral relation-

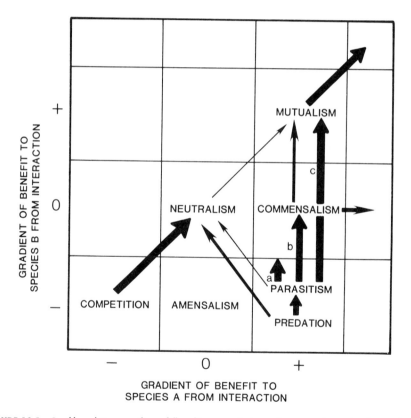

FIGURE 10.1   Possible evolutionary pathways followed in two-species interactive systems discussed in the text, based on the matrix of possible interactions given in Table 10.1. Three possible consequences of interaction between parasites and hosts are: (a) evolution of reduced negative impact; (b) "parasites" have no measurable negative impact on the host and act as commensals utilizing a parasitelike habit; (c) "parasites" become beneficial to their host.

ships. This is consistent with the theory on competition discussed in Chapter 21. Lawton and Hassell (1981, 1984) emphasized the commonness of amensalism in their review of competition studies on insects, noting that studies of insects in natural environments show that amensalism or near-amensalism "is the norm rather than the exception by a ratio of at least 2:1" (Lawton and Hassell, 1981, p. 793). The evolutionary impact of $--$ interactions will be to move species toward the right in Table 10.1, making interactions less negative (Fig. 10.1).

The same argument can be made for the $+-$ interactions of predation and parasitism. The power of the selective force of the species with negative impact was discussed at the end of Chapter 7, resulting in remarkable adaptations for reducing such impact. The evolutionary trend will be toward neutralism or commensalism. Highly virulent parasites also evolve toward a lesser negative impact because host resistance increases and transmission has a low probability of success if the host dies rapidly. This was nicely demonstrated in the myxoma virus in rabbits vectored by fleas and mosquitoes (Fenner and Ratcliffe, 1965). Many parasites seem to have virtually no impact on the host, representing more commensal relationships, and ultimately, an originally parasitic species may become a mutualist, as mentioned in Chapter 9. There are several evolutionary results to the parasite–host relationship tending to push the species toward a $++$ encounter, although selection for high virulence may persist in some cases (e.g., Ewald, 1983, 1994; Toft and Aeschlimann, 1991).

EVOLUTIONARY TRENDS IN INTERACTIONS

For mutualists, the selection pressures will be for improved benefit to each species in the relationship, although as we will see in Chapter 11, there are sometimes strict limits on how far this can proceed because there is a conflict between what is most adaptive for each species involved.

The general trend will be toward a reduction of negative impacts and a reinforcement of positive relationships (Fig. 10.1). A question then arises as to the limits preventing all relationships moving toward mutualism and all organisms rejoicing in May's (1981, p. 95) "orgy of mutual benefaction." The answer seems to be that many beneficial relationships are at the expense of another species. As Addicott (1981) points out, models of mutualism should include at least three species: the two mutualists and a third species, such as a predator or competitor or host, against which the mutualism is effective. Heithaus et al. (1980) have modeled such systems. Mutualisms frequently foster antagonistic relationships as we shall see later in the chapter. Thus the trends toward positive interactions between two species, depicted in Fig. 10.1, would change dramatically if a third species were added on a third dimension or if an even more realistic n-dimensional figure could be envisaged.

This armchair exercise would lead us to predict that mutualism should be at least as common as predation and competition in nature, in temperate regions as well as in the tropics. If we do not find this, it is probably wiser to look more closely than to dismiss it as unimportant. Fortunately, ecologists have begun to look more closely.

FREQUENCY OF MUTUALISM

To estimate the commonness of mutualism in insects we can return to the checklist by Kloet and Hincks (1964–1978), work through the orders listed in Table 9.3, and note the taxa in which mutualism is known to be involved (Table 10.2). This entails much extrapolation from the known examples, and

TABLE 10.2   Examples of Mutualism Involving Insects.

Order	Type of Food and Specific Taxon Involved	Mutualists	Number of Species in Britain[a]	References
Orthoptera	Plants high in cellulose	Microbes in gut?	30	
Isoptera	Wood high in cellulose	Protozoa and bacteria	0	Breznak (1975)
	Fungus	Fungus in gardens	0	Batra and Batra (1979)
Mallophaga	Dead skin	Endosymbionts	510	Buchner (1965)
Anoplura	Blood	Endosymbionts	25	Buchner (1965)
Thysanoptera	Plant sap	Endosymbionts	160	Buchner (1965)
Hemiptera	Plant sap or blood	Endosymbionts	530	Buchner (1965)
Homoptera	Plant sap	Endosymbionts	1,150	Buchner (1965)
		Ants	(300?)	Addicott (1978, 1979)
Lepidoptera	Plants high in cellulose (larvae)	Microbes in gut?	2,200	Jones et al. (1981a, b)
	Nectar (adult Macrolepidoptera)	Flowering plants	(850)	Faegri and van der Pijl (1971)
Coleoptera	Plants high in cellulose (larvae)	Endosymbionts and microbes in gut?	900	Buchner (1965)
	Scolytidae and Platypodidae	Fungi	(60)	Norris (1979), Kok (1979)
	Nectar and pollen (adults)	Flowering plants	?	Faegri and van der Pijl (1971)
	Seeds	Flowering plants	?	Kirk (1972, 1973), Manley (1971)

TABLE 10.2   (continued)

Order	Type of Food and Specific Taxon Involved	Mutualists	Number of Species in Britain[a]	References
Hymenoptera	Plants high in cellulose (larvae)	Endosymbionts and microbes in gut	430	Buchner (1965)
	Siricidae	Fungi	(10)	Madden and Coutts (1979)
	Haemolymph Braconidae and Ichneumonidae	Viruses	2,820	Stoltz and Vinson (1979), Edson et al. (1981)
	Nectar and pollen (Apoidea)	Flowering plants	250	Faegri and van der Pijl (1971), Heinrich (1979a)
	Seeds (ants)	Flowering plants	?	Sernander (1906), Heithaus et al. (1980)
	Extrafloral nectar (ants)	Flowering plants	?	Bentley (1977), Tilman (1978)
	Honeydew (ants)	Aphids	?	Addicott (1978, 1979)
	Fungus (ants)	Fungus in gardens	0	Weber (1979)
Diptera	Plants high in cellulose (larvae)	Endosymbionts and microbes in gut	900	Buchner (1965)
	Cecidomyiidae	Fungi and endosymbionts	(630)	Batra and Batra (1967), Buchner (1965)
	Nectar (adults)	Flowering plants	?	Faegri and van der Pijl (1971)
	Total mutualistic species		9,905	
	Total species in British Isles		22,000	
	Percent of species mutualistic		45%	

[a] Numbers are estimated from Kloet and Hincks (1964–1978). Numbers in parentheses are subsets of the numbers above.

only time will tell how reasonable this estimate is. Many of the mutualisms are probably quite loose and opportunistic, as in microbes in the gut of herbivorous insects. But five general types of mutualism can be observed:

1. Wherever food is rather pure, as in blood or plant sap, or has a high content of refractive chemicals, as in cellulose and keratin, microbes are usually involved in aiding in digestion and synthesizing micronutrients absent in the food
2. In the parasitoid families Ichneumonidae and Braconidae, viral particles are injected into the insect host which reduce host resistance to the parasitic larvae
3. Use of nectar and pollen by pollinators, which transfer pollen from one flower to another
4. Transport of seeds away from the parent plant, particularly by ants, which eat the small oily elaiosomes on the seed surface
5. The association of ants with insect species that excrete honeydew, as in the aphids, psyllids, and membracids, with ants providing protection for the mutualists against their enemies

Many other categories of looser mutualistic relationships could be listed, but these are harder to recognize and less common perhaps than the five major types.

This estimate of about 45% of insect species in the British Isles being involved in one mutualistic relationship or more illustrates the potential importance of mutualism in natural systems. Only a brief visit to these islands is enough to show that they are not tropical! Insects play an important role as consumers (Figs. 2.1–2.4 and Chapter 12), and the interaction between plant food and insect may be highly modified by the microbial gut fauna. Indeed, much of plant defense may be adapted to the microbes rather than to the insects (Jones et al., 1981a,b). The acquisition of microbes has resulted in impressive adaptive radiations onto foods deficient in nutrients or with nutrients locked away from enzyme systems typically found in the Metazoa, such as cellulose. The rise of the angiosperms, providing such conspicuous beauty for humans, was certainly due in part to the acquisition of reliable mutualists (Regal, 1977), with radiation of flowering plants, bees, butterflies, and moths probably fostering each other (see Chapter 11).

STABILITY OF MUTUALISMS

Many of these mutualistic relationships are very stable because transmission of the microbial organisms is usually ensured in special pouches or mycetangia, as in wood wasps and bark beetles (Batra and Batra 1967); by transovarial transmission, as in the sucking lice; from mother to larva in the tsetse fly and the Pupipara (Nycteribiidae, Streblidae, and Hippoboscidae) (Askew, 1971); and from mother to nymph in the termites (Breznak, 1975). Contact between the two species is never lost; they function as one organism except in the ways in which they modify the relation of other species to the environment, which are usually profound (see examples later in the chapter). Vance (1978) also noted the commonness of mutualism in temperate regions and suggested that the destabilizing tendencies observed in models is diminished by very tight coupling of the association. Of course, where each mutualistic species lives a more or less independent life, the relationship becomes less stable and the extent of the stability or instability of these systems is being debated in the literature.

MODELS

May's (1973b, 1976, 1978a, 1981) (Whittaker, 1975) models of mutualism, involving as an example a plant population and a pollinator population, illustrate that stable points exist when both populations are large, but extinction of one species is almost inevitable when populations become small (Fig. 10.2). Resources, in terms of food for the pollinator or transported pollen for the plant, simply become too scarce to support the population. May's argument is that a pollinator population Y, following the logistic equation (see

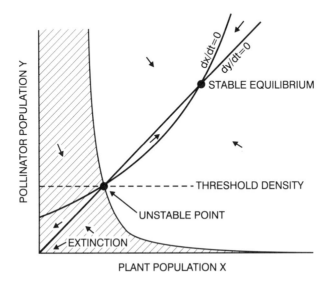

FIGURE 10.2   May's model of two mutualistically interacting populations X and Y, such as plant and pollinator populations. The threshold density of pollinators is the minimum number necessary to sustain the plant population at the unstable point. Arrows indicate the direction in which populations move. After May (1976).

Chapter 13), would have a carrying capacity proportional to the plant population size X. Therefore, equilibrium numbers of Y must lie along the line $dY/dt = 0$ (Fig. 10.2). The plant population will have equilibrium values that are concave because pollinators become inefficient at low plant densities and the plant population will decline, whereas at high plant densities the population will enter into intraspecific competition and reach a maximum density independent of pollinator abundance. According to May's model, a stable equilibrium position is reached when both lines of equilibrium numbers intersect and all points in the unshaded area converge toward this position. All points below the unstable point tend toward extinction of both species concerned. If such a system existed in an unstable environment, the populations would ultimately fall into the shaded zone and to extinction. Hence May's conclusion that mutualisms are likely to persist mostly where stability already prevails, as in the humid tropics. Vandermeer and Boucher (1978) develop a similar model for the plant-nitrogen-fixing bacterial (*Rhizobium*) interaction, and they consider a wider range of possible interactions, some of which are stable.

Goh (1979) also adopted the Lotka–Volterra equations used to study predation (Chapter 8) and competition (Chapter 21) to investigate the stability properties of mutualistic interactions between two or more species. He concluded that in two-species models the interaction is as stable as in similar competition models, but not as stable as in predator–prey models. In models using three or more species, those invoking mutualism were less stable in general than models of competition and predation. "The mathematical result suggests that in nature mutualism is less common than competition and predation" (Goh, 1979, p. 273).

Of course, it is a very weak argument to predict frequency of interaction types in nature by the stability properties of mathematical models. It ignores the important element of probability of encounter between interacting species, and as we have seen, such probabilities may be low for potentially competing parasites (Chapter 9), and mutualists obviously adapt to maximize encounter probabilities.

The models of May and Goh also emphasize the increase in equilibrium density of the mutualistic populations, as in other models by Vandermeer and Boucher (1978) and Travis and Post (1979). However, this is not a necessary result of mutualism, for survival and growth rate may improve without affecting the ultimate equilibrium density. Gilbert (1977) pointed this out in relation to *Heliconius* butterflies, which pollinate the cucurbits *Anguria* and *Gurania,* where the mutualism results in greater constancy of adult populations of all species involved. Stability of populations is increased, not necessarily the equilibrium density.

Modeling of mutualism should therefore include the study of effects on both equilibrium density and rate of growth of populations up to an unaltered equilibrium density. Addicott's (1981) models are designed to do this.

He examined the stability properties of two-species mutualisms using three basic models: model I, in which mutualism increases the equilibrium density of the species but does not affect the maximum possible growth rate, as in May's (1976, 1978a, 1981) models (Fig. 10.3); model II, in which both growth rate and equilibrium density are increased by mutualism; and model III, in which the two species affect the population growth rate of the other but not the equilibrium density. He compared results from these models with that of the null model based on the logistic equation (see Chapter 13), which represents the growth of one population in the absence of interaction with other species.

Addicott used computer simulation to study the time taken for equilibrium to be reached after a perturbation. He plotted the results on a polar coordinate system in which the angular axis is the initial angle of displacement from equilibrium and the radius represents the return time to equilibrium. The term α represents the positive value of the mutualistic interaction, equivalent to negative α values in predator–prey and competition models. Several values of α were used in the simulations.

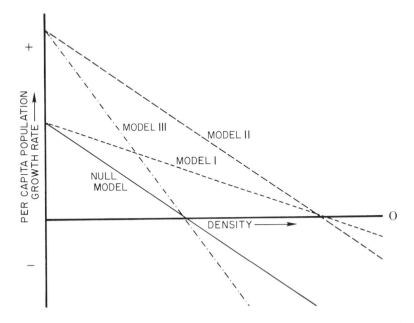

FIGURE 10.3 Addicott's models of *per capita* population growth rate of one species as a function of its own density, given a constant density of the second species, for the three basic models of mutualism and the null model. From Addicott (1981); with permission.

Model I predicted that as α becomes stronger the interaction becomes less stable and return times were usually longer than in the null model. Model II showed that return times were the same or less than in the null model, and the stronger α was, the shorter was the return time to equilibrium. In model III return times were invariably shorter than in the null model, and the stronger the mutualistic effect, α, the shorter was the return time. When all models are compared using an α value of 0.5, and the null model is included, the results can be summarized as in Fig. 10.4. Models II and III produce more stable relationships as the mutualism becomes stronger.

Heithaus et al. (1980) accepted the prediction of instability in mutualistic population models and examined the influence of a third species, such as a seed predator or plant competitor in the ant–seed dispersal mutualism. They concluded that seed predation increased stability while competition between plants decreased the probability of stable coexistence between the mutualists.

These results support the statement made earlier that mutualistic relationships are likely to evolve toward more interdependence, with the ultimate re-

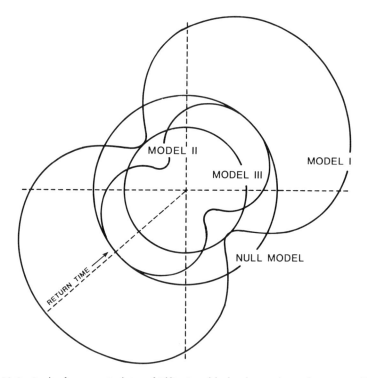

FIGURE 10.4 Results of computer simulations of Addicott's models plotted on a polar coordinate system. The return time to equilibrium after a perturbation for each model is represented by the distance to the center of the figure, and the extent of the perturbation is given by the angle of displacement. From Addicott (1981); with permission.

sults that they become obligatory, and the species become inseparable. This is a condition that May (1976, 1981) explicitly avoided because he was examining the dynamics of two-species systems in which each species could live apart from the other. But the empirical evidence, as summarized in Table 10.2, suggests that many, perhaps the majority, of mutualistic species live in close association. When microbes are involved, effective transmission from one generation of the insect species to the next is usually ensured.

So far, only models that consider the stability of mutualistic interactions have been considered, but two other kinds of models have been proposed. Keeler (1981) has developed a model addressing the question of when mutualism should evolve, based on the observation that closely related plants often differ in whether or not they have extrafloral nectaries that are visited by ants which tend to ward off other herbivores. She noted that the plant–ant mutualism was likely to evolve depending on several factors:

1. The low cost of the mutualistic relationship
2. The probability of contact between mutualists
3. The reward obtained
4. The presence or absence of alternative defenses in the plant

Roughgarden (1975) also proposed a model on when mutualism should evolve in marine systems. Another kind of model addresses the question of community-wide relationships between mutualists which influence the diversity of species interaction. King et al (1975) examine the species diversity of plant–pollinator systems, examining the relationship between rates of immigration and extinction of species in a community as in the theory of island biogeography (see Chapter 24).

Thus there are three kinds of models of mutualism. The best studied are those on population stability, but selection models and community models have not been examined extensively.

KINDS OF MUTUALISTIC RELATIONSHIPS

There is a very wide range in the kinds of relationships between mutualistic species, many of which involve insects (Table 10.2). The examples that follow will illustrate some of this range, the intimacy of the relationships, and the coevolutionary nature of many relationships. Some generalizations will be sought after consideration of the six examples provided.

1. Ants and plants

More than 90% of species of the genus *Acacia* in Central America are protected from herbivores by cyanogenic chemicals in the leaves (Rehr et al.,

1973). The remainder seem to have gained a more potent defense in the form of ants that live in close association with these plants. Belt (1874) discovered that some species, the bull's-horn acacias (e.g., *A. cornigera*), act as hosts to colonies of ants in the genus *Pseudomyrmex,* and the ants act as allelopathic agents for the plant (Brown, 1960; Janzen, 1966, 1967a,b). The ants gain protection from the plant by living in the swollen stipular thorns and food is provided by the plant—sugar is secreted by petiolar nectaries and protein is produced in small "Beltian bodies" growing at the tips of new leaves. The aggressive ants patrol the plant, ward off herbivores, and suppress potentially competitive plants by chewing the growing tips (Janzen, 1967b). Such suppression of plants around an occupied *Acacia* plant also make it much less vulnerable to fire that frequently sweeps through this dry-tropics vegetation (Janzen, 1967a).

Similar relationships exist between ants and *Cecropia* plants (Janzen, 1969b; Davidson and Fisher, 1991; Davidson et al., 1991) in which the precision of coevolution can be seen in the production of animal sugar (*glycogen*) in the extrafloral nectaries of the host plant, the only known case in the higher plants (Rickson, 1971) (see also Buckley, 1982).

Bentley (1977) and Bentley and Elias (1983) review the literature on extrafloral nectaries on plants and ants, and Hocking (1975) treats the ant–plant mutualism in terms of evolution and energy. The full array of ant and plant interactions is covered by Beattie (1985) and Huxley and Cutler (1991).

2. Figs and Fig Wasps

More than 900 species of figs (*Ficus*) have been recorded which are pollinated by very specific wasps in the family Agaonidae (Chalcidoidea)—the fig wasps. Many characters in both the figs and the wasps show closely coevolved properties, and the two taxa have obviously radiated together as a result of the initial development of a mutualistic relationship (Bronstein and McKey, 1989).

The fig is a false fruit formed by the enlarged receptacle of the inflorescence. The flask-shaped fruit encloses a large number of flowers (Fig. 10.5), and each inflorescence passes through the following stages (Galil and Eisikowitch, 1968; Ramirez, 1970; Wiebes, 1979): *phase A—prefemale,* in which the inflorescence is closed to entry by fig wasps (Fig. 10.6); *phase B—female,* in which the ostiolar scales loosen, female flowers ripen, and agaonid wasp females penetrate the inflorescence and oviposit into the ovaries; *phase C—interfloral,* where wasp larvae develop in developing galls formed from the fig ovaries, and unattacked fig embryos develop; *phase D—male,* where male flowers mature, wasps reach maturity and emerge from galls, and males inseminate females and bore holes in the receptacle, the females collect pollen

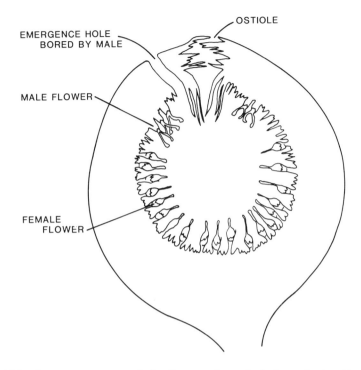

FIGURE 10.5 Diagrammatic cross section of a fig inflorescence showing the distribution of male and female flowers, drawn as if both were mature synchronously. After Galil and Eisikowitch (1968).

from the male flowers and emerge through holes bored by males and fly to fig inflorescences in phase B; and *phase E—postfloral,* in which seeds ripen and the receptacle ripens, becoming attractive to fruit-eating animals, which disperse the seeds.

The life cycle of the fig wasps is also complicated, and only a general pattern is depicted here, as there is considerable variation between species. Before females leave the inflorescence in phase D, they load up with pollen, packing it into special receptacles on or near the coxae of the front legs and the abdomen. Then they exit from the inflorescence through holes bored by males, fly to figs in phase B on another tree, and enter the ostiole. On entry they lose their wings and part of their antennae. Males are wingless and do not leave the fig in which they developed. In the lumen of the fig, females pierce with the long ovipositor the stigmas of the flowers and the length of the styles, ovipositing in the ovary (Fig. 10.7). After ovipositing in an ovary they pollinate the flower by scraping pollen out of a receptacle with their legs onto the stigmatic surface. They also pollinate flowers in which no wasp eggs are laid. Eggs hatch and larvae develop in the gall formed from ovary tissue.

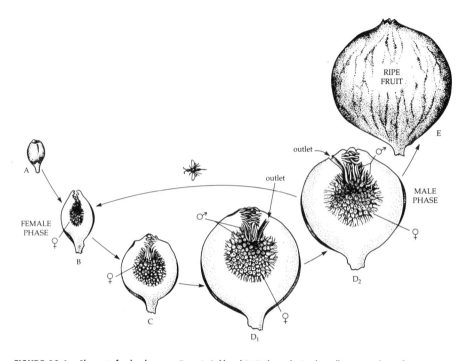

FIGURE 10.6 Phases in fig development. From J. Galil and D. Eisikowitch, On the pollination ecology of *Ficus sycomorus* in East Africa, *Ecology*, 1968, **49**:259–269. Copyright © 1968 by the Ecological Society of America. Reprinted with permission.

Males emerge before females, cut holes in the sides of ovaries, and inseminate the female inside. Females then collect pollen and leave the fig, completing the life cycle.

Some coevolved traits in the figs include:

1. The unique false fruit design, allowing only agaonids and a small number of closely related parasitic wasps to enter the inflorescence.
2. The extreme protogyny, with female flowers receptive several weeks before male flowers produce pollen, is clearly adapted specifically to the generation time of the fig wasp.
3. The inflorescence contains both stalked and unstalked flowers with short and long styles, respectively, making seeds more or less available to ovipositing fig wasps, some sacrificed for the mutualistic wasps and some reserved for seed production.

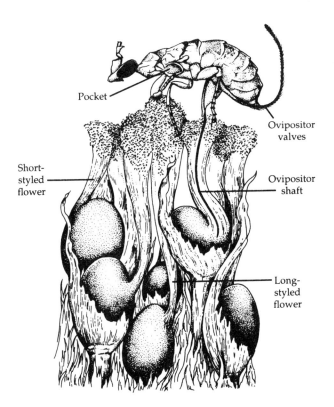

FIGURE 10.7   Female fig wasp, *Ceratosolen arabicus,* ovipositing in a short-styled flower of *Ficus sycomorus* and extracting pollen from a pouch to fertilize the flower. From Galil and Eisikowitch (1969); with permission.

Some coevolved features in the wasps include:

1. The specialized morphology of both male and female fig wasps (Fig. 10.8), involving:
   (*a*) The female body adapted for squeezing through the ostiole of the fig.
   (*b*) Pollen receptacles on the female.
   (*c*) A wingless male with a long abdomen for mating with a female in gall.
2. The specialized behaviors of loading and releasing pollen.
3. The specialized secretions in the female that promote gall formation.
4. The very specialized relationship, usually between one fig species and one fig wasp species, reminiscent of the highly specialized parasites discussed in Chapter 9.

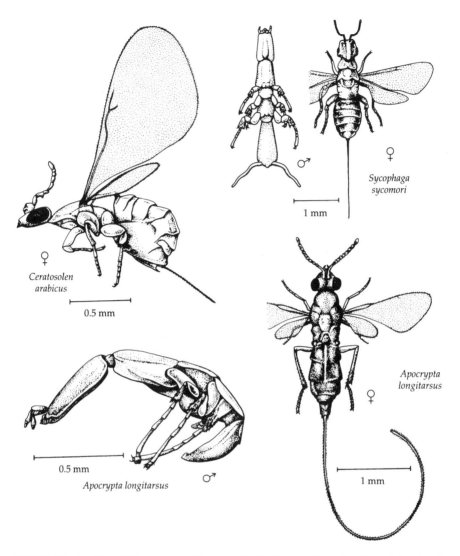

FIGURE 10.8 Inhabitants of *Ficus sycomorus* inflorescences. The female *Ceratosolen arabicus* (Agaonidae) is the pollinator and mutualist. *Sycophaga sycomori* (Agaonidae) is a parasite on the fig and a competitor with *C. arabicus* because it does not pollinate the fig but utilizes both long- and short-styled flowers as oviposition sites. *Apocrypta longitarsus* (Torymidae) is an inquiline utilizing galls formed by *Ceratosolen* and *Sycophaga*. From J. Galil and D. Eisikowitch, On the pollination ecology of *Ficus sycomorus* in East Africa, *Ecology*, 1968, **49:**259–269. Copyright © 1968 by the Ecological Society of America. Reprinted by permission.

Wiebes (1979), Bronstein and McKey (1989), and Janzen (1979), provide reviews on the fig–fig wasp relationship and show that there is more work to be done by ecologists on this fascinating system. Bronstein (1987; 1988a,b) and Bronstein and Hoffmann (1987) develop many intriguing aspects of the interaction.

3. Parasitoids and viruses

Viruses contained in the calyx of the ovary are injected into hosts when parasitoid females in the families Ichneumonidae and Braconidae oviposit internally (Stoltz et al., 1976; Edson et al., 1981). Edson et al. (1981) reared the ichneumonid *Campoletis sonorensis* on the tobacco budworm, *Heliothis virescens,* and with the injected egg they injected different components from the ovary: egg plus saline only, egg plus calyx fluid containing virus, and egg plus purified virus. As a control, hosts were parasitized naturally by *Campoletis* females. The results showed clearly that the virus was important in suppressing the cellular encapsulation response of the host that caused 100% encapsulation and mortality in the egg without virus treatment. The virus does not replicate in the parasitoid's host, only in the parasitoid's ovaries. Edson et al. (1981) assert that this is a mutualistic relationship between the ichneumonid and the virus and that it is the first example of obligate mutualism between a virus and a eukaryotic species. This subject was reviewed by Stoltz and Vinson (1979), who recorded eight species of ichneumonid and 23 species of braconid with viruses present in the calyx of the ovary.

The impressive adaptive radiations in the Ichneumonidae and Braconidae may well have been greatly stimulated by the acquisition of viral symbionts. In both taxa the most primitive species are ectoparasites, and the development of endoparasitism could have arisen by association with viruses that could suppress the encapsulation of foreign bodies in the host's haemolymph (see also Godfray, 1994). This is a case of biological warfare between species on different trophic levels. In Chapter 9 warfare between species on the same trophic level was discussed.

4. Termites and intestinal microorganisms

Termites and the closely related wood roach *Cryptocercus* are associated with protozoa in the intestine that have cellulolytic enzymes which enable the utilization of wood as food by the termites. Cleveland (1924, 1925) (Cleveland et al., 1934) established the mutualistic nature of this relationship, with the protozoa digesting wood and the termites supplying a constant supply of wood for the protozoa and an anaerobic chamber in which to house them. But it was Hungate (1936, 1938) (reviewed in 1955; see also Honigberg, 1967, 1970) who established that fermentation of cellulose in the protozoa provided nutrients for both the protozoa and the termites. Wood particles are engulfed by protozoa; the cellulose is fermented liberating energy for the protozoa. The acetic acid produced is absorbed through the hindgut wall of the termite and oxidized by cellular respiration, producing energy for the termite. About 70% of the cellulose in the diet is digested (Breznak, 1975).

In addition, with the protozoa many bacteria exist, some of which fix nitrogen, providing an important source of nitrogen for termites on such a low-nitrogen diet (Breznak, 1975). Young termites support bacteria with the highest rate of nitrogen fixation just when their requirements for protein synthesis are highest. When the bacteria are killed the protozoa die off within a few weeks, indicating that the bacteria are mutualistic to the protozoa as well as the termites. Methanogenic bacteria found in ruminants are also present in termites and may be partially responsible for digestion of the cellulose.

The close ties and coevolutionary aspects of this mutualistic relationship can be summarized as follows:

1. The radiation of the termites, involving about 2000 species, depended on the acquisition of protozoan mutualists with cellulolytic enzymes, which presumably originated in rotting wood.
2. The radiations or protozoa likewise depended on the mutualism, for unique genera and species of oxymonad, trichomonad, and hyper-mastigote flagellates are found only in termites and *Cryptocercus* (Breznak, 1975).
3. The division of labor in the digestion of cellulose represents well-attuned complementary systems.
4. Part of the hindgut of the termite is enlarged into the "paunch," which houses the protozoa.
5. Reinfection of protozoa and bacteria after molting and voiding of gut contents and infection of newly hatched nymphs is ensured by coprophagy. This specialized behavior has been suggested as the impetus to the evolution of eusociality in the termites (Cleveland et al., 1934; Wilson, 1971).

5. Fungi and Ambrosia Beetles

Three major groups of insects are agriculturalists nurturing fungi using specialized cultivation techniques to grow their food: the attine ants (e.g., Weber, 1979; Hölldobler and Wilson, 1990), the macrotermitine termites (e.g., Batra and Batra, 1979), and the ambrosia beetles (e.g., Kok, 1979; Norris, 1979). Many other insects are associated with fungi in a mutualistic manner (Batra and Batra, 1967; Batra, 1979; Wilding et al., 1989). All of these groups have stimulated fascinating studies on the nature of the mutualism, and the references above provide entries into the literature. Only the relationship between fungi and ambrosia beetles is discussed further here.

Ambrosia beetles in the families Scolytidae and Platypodidae bore into wood as adults forming galleries in the solid xylem. From the site in which they developed, they carry an inoculum of fungus in pockets called mycangia or mycetangia. These may be located at the base of each mandible as in *Xyleborus fornicatus,* or in an intersegmental pouch between the pro- and mesonotum in *Xylosandrus discolor,* or in sclerotized pouches in the base of the elytra as in *Xyleborus gracilis* (Norris, 1979). The fungi are nourished by the beetle while in the mycangia and multiply. During the boring of the tunnel it becomes inoculated with the fungus, which grows along the galleries and into the larval brood chambers, developing into a black mat of fungal hyphae on which adults and larvae feed (Fig. 10.9). The fungi digest woody cells in the tree, making nutrients available to the beetles, a food that is particularly rich in the amino acids lysine, methionine, arginine, and histidine. These nutrients derived from the mutualistic fungi are essential in the initiation of reproduction in female *Xyleborus* (Norris, 1979). Kok (1979) discusses the specialized metabolism of these beetles in relation to lipids in ambrosia fungi and dietary requirements of the beetles. Before leaving the gallery system after maturing, a beetle rocks back and forth in a gallery, and in so doing packs fungal spores into the mycangia for transport to a new host tree (Batra and Batra, 1967).

Norris (1979) summarizes the impact of associating mutualistically with fungi on the xyleborine scolytid beetles:

1. Sexual reproduction is dependent on the symbiotic fungus.
2. Progeny development and maturation is dependent on the fungus and its constant cultivation by the mother—presocial behavior is involved.
3. The male has become a short-lived, nonflying, nondispersing, inseminator of sisters or the mother.
4. Outbreeding seems to be absent. The whole social system and genetic structure of populations is influenced by the mutualistic relationship.

FIGURE 10.9   Ambrosia beetle in its tunnel lined with the ambrosia fungus it grazes and transports from one location to another. From S. W. T. Batra and L. R. Batra, The fungus gardens of insects, *Sci. Am.,* 1967, **217**(5):112–120. Copyright © 1967 by Scientific American, Inc. All rights reserved.

6. Ants and Aphids

The relationship between sap-sucking homopterans such as aphids, membracids, and psyllids, and ants is widespread and well known. Aphids excrete excessive sugar in their diet in the form of honeydew on which ants feed. Ants tend the aphids, protecting them from enemies to some extent, and sometimes transporting aphids to new plants. In fact, when some aphid species are attacked they release an alarm pheromone which stimulates a mass response in the aphid colony, and tending ants are also stimulated to defend the colony more aggressively (Nault et al., 1976). The beneficial effects of ants on aphids are very variable from one aphid species to another and in relation to the density of aphids present. For example, in five species of aphid on fireweed, *Epilobium angustifolium,* one species was unattended, *Macrosiphum valerianae,* and the ants had a negative impact on the persistence of colonies. Of the tended species, persistence of colonies of *Aphis salicariae* was not affected, while those of *Aphis varians* and *A. helianthi* were affected, positively at low densities and negatively or not at all at high densities (Addicott, 1979; Cushman and Addicott, 1991). It is important to note the density-dependent regulation in this mutualistic system and to note that there are self-limiting mechanisms in at least some of these mutualisms. Also, the species of ant affected the benefit to the aphid colony, with *Formica cinerea* and *F. fusca* improving colony persistence in *A. varians,* with *F. neorufibarbis* and *Tapinoma sessile* having no effect.

These very variable effects of the mutualism are also reflected in the literature on the impact of enemies on homopteran colonies tended and unattended by ants. Banks (1962) recorded reduced predation in tended aphid colonies, but increased attack by parasitoids. Burns (1973) found no influence of tending ants on rates of parasitism, while Bartlett (1961) found that some species of parasitoids were affected and others were not. Some parasitoids oviposited rapidly, thereby avoiding ant attack, while others tolerated the presence of ants.

As is to be expected with these looser kinds of mutualistic relationships, relying to a certain extent on the development of opportunistic associations, the results of the interaction are extremely variable. In some cases a purely negative impact results, presumably through predation; in others the aphids do not benefit but the ant gains honeydew, a commensalism; and in others a truly mutualistic interaction results.

GENERALIZATIONS

Some generalizations can be drawn from the previous six examples. Listing each example and the types of benefit derived from the mutualism for each species (Table 10.3), it becomes clear that most commonly, mutualists are im-

TABLE 10.3   Summary of the Six Examples of Mutualism with the Species Involved and the Benefits They Obtain from the Mutualism.

Example	Species 1	Species 2	Benefits to Species 1	Benefits to Species 2
1	Plants	Ants	Protection	Food and protection
2	Figs	Fig wasps	Pollination	Food and protection
3	Parasitoids	Viruses	Protection	Food, protection, and transmission
4	Termites	Protozoa and bacteria	Food	Food, protection, and transmission
5	Ambrosia beetles	Fungi	Food	Food and transmission
6	Aphids	Ants	Protection	Food

portant in supplying food and protection, while a third important role is transmission between sites suitable for reproduction, or in the case of pollination, transmission of gametes. In all the examples cited there is no obvious way in which the mutualists increase the carrying capacity of the environment, but the effects are on nutrition, survivorship, and colonization, all affecting the rate of population growth. Therefore, Addicott's (1981) model III appears to be a very realistic model for mutualism. It could be argued that for the microorganisms, the carrying capacity of the environment is increased over that possible in a free environment, but such freedom does not exist for any of the species involved, and the number of mutualistic hosts available is probably defined by factors other than the mutualism, as discussed in Chapter 17. Another generalization is that mutualisms are very common in temperate regions, and the tropical bias detected in the literature may well simply reflect the greater species richness in the tropics. Only two of the six examples, ants and plants and figs and fig wasps, are largely tropical, although many ants associate with temperate plants with extrafloral nectaries. There are many more tropical termites than temperate species, but ant–aphid relationships are probably more common in temperate regions because aphids are less abundant in the tropics (Eastop, 1972). A third generalization is that the biology of associating species is profoundly affected in terms of population structure and size, and mating system, leading to very different evolutionary circumstances. The evolution of eusociality in termites, the development of presocial behavior in ambrosia beetles, and the colonial habit of many aphids may all result from mutualistic relationships. The strong inbreeding and small effective population sizes seen in fig wasps, ambrosia beetles, and most of the microorganisms in these mutualisms contribute substantially to the potential for speciation and adaptive radiation (Price, 1980).

COMPLEXITY OF MUTUALISTIC SYSTEMS

One of the challenges in the future development of the ecology of mutualism is to learn how to cope with the great diversity of interactions involved. Even in an apparently simple system such as ants tending aphids on a plant, the interactions become very diverse and far reaching. Even if we ignore the many possibilities in the ant–aphid relationship, as discussed in example 6, and accept the density-dependent beneficial effect of ants on aphids, the web of interactions is complex (Fig. 10.10). Aphids have a negative impact on the host plant; they are parasites. Ants have a positive effect on aphids at low densities and a negative effect at high densities. Aphid species compete for ant mutualists (Addicott, 1978; Cushman and Addicott, 1989). Ants promote the development of aphid populations on plants and therefore have a negative impact on plant fitness. However, ants also act as predators on herbivores other than aphids, and therefore benefit the plant. Indeed, the ant–aphid mutualism may provide a net improvement in plant fitness because of effective reduction of leaf-eating herbivores. When stands of mountain birch (*Betula pubescens*) are heavily attacked by the geometrid moth, *Oporinia autumnata*, many trees are almost completely defoliated in Finland (Laine and Niemalä, 1980). But

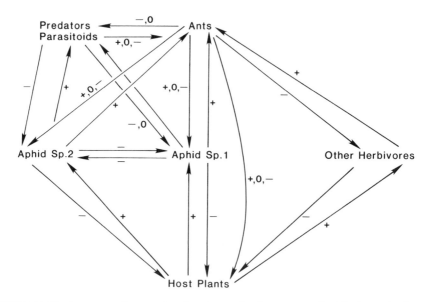

FIGURE 10.10 Interactions to be expected involving the mutualism between ants and aphids, showing that the sign of the interaction is likely to change for several interactions as relative densities of various components of the system change.

"green islands" of healthy birch 30–40 m across can be observed in a matrix of heavily damaged trees, and these are associated with nests of the ant, *Formica aquilonia*. Ants forage extensively in the tree canopies, and densities of ants are highly correlated with densities of the honeydew-producing aphid, *Symydobius oblongus*. There is little doubt that birches close to ant colonies gain a net increment in fitness over trees distant from ant colonies despite heavy aphid densities. In years of low incidence of *Oporinia* the trees with aphids may well suffer a net decrement in fitness. Thus all the interactions portrayed in Fig. 10.10 are in a very dynamic state, many changing sign with density changes in the component species, both in space and time. All forms of interaction can be observed: predation, competition, mutualism, parasitism, and presumably commensalism between periods when interactions are changing from positive to negative. Evaluating the costs and benefits to each species in this community would demand an extensive and detailed study. Gilbert (1977) illustrates a similar but more complex web of interactions, and Messina (1981) illustrates a case of plant protection resulting from an ant–membracid mutualism. Indeed, the conditional nature of mutualistic interactions and the highly dynamic shifts in relationships are receiving increasing attention, emphasized by Cushman and Whitham (1989) and Cushman and Addicott (1991).

During this discussion of mutualism I have avoided use of the word *symbiosis*—the living together in more or less intimate association of two dissimilar organisms. The classes of symbiosis are usually recognized as parasitism, commensalism, and mutualism. Many of the mutualisms discussed in this chapter are symbiotic, particularly parasitoids and viruses, termites and protozoa and bacteria, and ambrosia beetles and fungi, because individuals are almost never separated from their mutualist. Ants and acacias and fig wasps and figs are also closely associated, but ants and aphids share a less intimate relationship. A continuum exists between symbiotic mutualisms and nonsymbiotic mutualisms, so there is no great virtue in distinguishing between them as long as it is recognized that very different kinds of relationships are involved.

Mutualism is a fascinating subject deserving of more attention from ecologists. The theoretical basis for consideration of mutualism has developed very significantly since the early 1970s, so this should act as a stimulus for much more field work. Considering the commonness of mutualism among insects and the large variety of types of mutualisms represented, insect ecologists have the opportunity to make a major contribution to the development of this field. Much work has been done on the mutualism between pollinators and flowering plants, discussed in the next chapter.

Pollination Ecology

Insects, birds, and bats largely compensate for the inability in many plants to walk, run, and fly by acting as dispersal agents for pollen, seeds, and spores. Some plants do fly very well in the form of wind-dispersed seeds, and pollen is broadcast on the wind by many plant species (e.g., Niklas, 1985), but the majority depend on animal dispersal. Insects are very important pollen transporters (Proctor and Yeo, 1972; Faegri and van der Pijl, 1979). They transport spores of some mosses in the family Splanchnaceae (Bequaert, 1921; Erlanson, 1930) and spores of many fungi. They are important "pollinators" of rust fungi, carrying pycniospores from one pycnium to another (Alexopoulos, 1952), and they transport seeds of flowering plants (van der Pijl, 1972). All are mutualistic relationships.

RADIATION OF THE FLOWERING PLANTS

Insects are such vital mutualists with flowering plants that it is hard to imagine the existence of a large number of angiosperms without these animals. Regal (1977) makes a convincing argument that the dominance of the

A hawk moth forages for nectar from honeysuckle flowers. Moth is based on a painting by Walter Linsenmaier (1972). Drawing by Alison Partridge.

angiosperms depends largely on mutualistic relationships, with insects as pollinators and birds as seed dispersers. He substantiates the following hypotheses leading to the success of flowering plants:

1. Insect pollination permits a plant species to outcross with others of the species even though plant individuals are widely dispersed and very patchy in distribution, whereas wind pollination would be ineffective with such a population structure.
2. This dependence on insect pollination became important only when long-distance seed dispersal was made possible by birds, making likely colonization of widely dispersed patches of suitable terrain. Such long-distance dispersal greatly reduced seed predation (see Fig. 5.8).
3. With reliable long-distance dispersal of seed came the specialization of plants to specific microhabitats, and greatly reduced vulnerability to epidemic diseases.
4. The new ecological and genetic opportunities for flowering plants enabled impressive diversification in life form, growth strategies, and chemical defense, thus linking these hypotheses to those discussed in Chapter 5, where plant–herbivore interactions were regarded by Ehrlich and Raven (1964) as the driving force behind much of the adaptive radiation in the angiosperms.

Support for this scenario is also provided by Levin and Wilson (1976), who showed that evolutionary rates, measured in terms of chromosomal change, were highest in angiosperm herbs and declined in the order shrubs, hardwood trees, coniferous trees, and cycads. They argued that the small, semi-isolated populations in changeable habitats in which herbs frequently occur are conducive to rapid speciation. Such populations could persist only with effective pollination and dispersal over long distances.

INTEGRATION OF ECOLOGICAL CONCEPTS

Indeed, pollination ecology offers a very good example of how necessary it is to integrate many of the aspects of ecology treated in this book to evaluate ecological processes adequately. Obviously, many areas of ecology must be understood: the trophic structure of the community (Chapter 2); the coevolution between plant and pollinator (Chapter 5); energetic relationships (Chapter 12); demography of plant and pollinator (Chapter 13); reproductive strategies (Chapter 14); population dynamics (Chapter 17); niche segregation, competition, and species packing (Chapters 20 and 21); sociality, since many pollinators are social insects (Chapter 15); community ecology (Chapter 22); and finally, paleoecology and biogeography (Chapter 24), since we deal with isolation of populations and speciation in time and space. It should be clear in this chapter how the subject matter relates to all these areas of ecology.

Gilbert (1975, 1991) provides an excellent example of how the study of pollination ecology requires a detailed understanding of individual behavior of pollinators, population phenomena, and community-wide interactions. The incisive nature of this approach sets an example regarding how many other kinds of ecological studies should be undertaken. The brief account given here must make many shortcuts and will create a rather crude picture of the interactions involved. From detailed studies on the biology of the butterflies in the genus *Heliconius*, their larval food plants, *Passiflora*, and their adult food plants, *Anguria*, which provide pollen and nectar, Gilbert was able to determine how these coevolving species interacted to produce organization at the individual, population, and community levels (Fig. 11.1). *Heliconius* butterflies are long-lived, since they are able to digest pollen rich in amino acids (Gilbert, 1972). They have a highly developed visual system and learning ability. They roost gregariously and trap-line for the widely dispersed *Anguria* plants in flower. Young butterflies learn routes by following older members of the roost, and regular roosting at the same site enables the first plants on the trap-line to be found in the early morning, when pollen and nectar

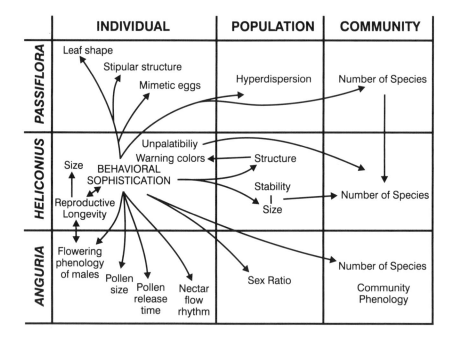

FIGURE 11.1 Summary of interactions between *Heliconius* butterflies, species of *Passiflora* (Passifloraceae) which are hosts to larvae, and the dioecious species of *Anguria* (Cucurbitaceae) which provide pollen and nectar to the butterflies. From L. E. Gilbert in L. E. Gilbert and P. H. Raven (eds.), *Coevolution of animals and plants.* Copyright © 1975, 1980. Reprinted by permission of the University of Texas Press.

supplies are high but when light conditions are very poor. Thus there is strong selection for male *Anguria* individuals to produce large inflorescences providing open flowers over a prolonged period, large amounts of pollen, and large pollen grains that can easily be utilized by *Heliconius* adults (Gilbert, 1972). Female flowering may be short, inflorescences small, and much energy is channeled into fruit and seed production. Because of heavy herbivore pressure, plants are widely dispersed and few species are found in the same community. In addition, in coexisting *Anguria* species, temporal differences in anther dehiscence and nectar flow are apparent, so few species can segregate adequately within a day. These factors and the inconspicuous nature of the flowers make flowers of this species a poor food source unless visual acuity and learning ability for trap-lining are highly developed, as in *Heliconius* butterflies. Since female flowers of *Anguria* are visited mainly "by mistake" by female *Heliconius,* a high proportion of females in the population would enable the highly intelligent pollinators to discriminate against them. The sex ratio (about $10:1$) is actually heavily skewed toward males, allowing the mimetic female flowers to receive sufficient visits to ensure pollination. Thus Gilbert explains factors involving the evolution of individual, population, and community characteristics of *Anguria* species as shown in Fig. 11.1.

THE FOSSIL RECORD AND COEVOLUTION

It has long been understood that pollination by animals involves a coevolutionary process, and this has been proceeding for perhaps 225 million years. The fossil record provides a time sequence of events, so that the steps in coevolution may be observed with clarity (Fig. 11.2). It provides us with few surprises. Winged insects were abundant in the Carboniferous long before flower-like structures were available as food sources. Even predatory groups (e.g., Neuroptera) had emerged before the Mesozoic. The first holometabolous insects became abundant in the late Carboniferous, Permian, and Triassic. Thus for the first time in history, many adults were seeking new food sources different from those of their larvae as the larvae became adapted to concealed feeding sites impossible for the adults to occupy. These were mandibulate adults and all were potential "mess-and-soil" pollinators (Smart and Hughes, 1973), and the Coleoptera were particularly numerous, although perhaps they preserve better and make better fossils. It is significant, then, that the first flowers evolved at this time—those of the gymnospermous Bennettitales—about 225 million years ago.

At first the flowers were small (5 mm) and had the appearance of a composite inflorescence (Leppik, 1960), which must have been insect pollinated. Flower size tended to increase with time, but rather suddenly about 140 million years ago flowers became 10–12 cm across. Smart and Hughes (1973) suggest that this large size may be correlated with the emergence of birds at

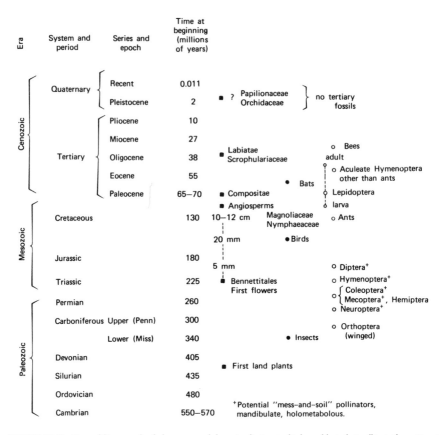

FIGURE 11.2 Time of first records of plant taxa and the animals, insects, birds, and bats that pollinate them. Time is not to scale. After Smart and Hughes (1973).

this time. No definite evolutionary link has been established between the Bennettitales and the angiosperms, but it is most likely that many insects were pollinators before other flying animals were extant.

Although the angiosperms emerged after birds had appeared, it is hard to imagine the possibility of such rapid and explosive radiation of this group in the absence of insects. Primitive angiosperms such as the Magnoliaceae and the Nymphaeaceae, which are still pollinated by insects today, appeared in the fossil record in the late Mesozoic. By the early Paleocene, Compositae (Fig. 11.3) were present that are evolved in inflorescence structure to provide a landing platform for large insects.

It is also significant that no highly evolved pollinators such as Lepidoptera and bees have been found before the angiosperms appeared. Although a head capsule of a lepidopteran caterpillar was found from the Cretaceous, adults

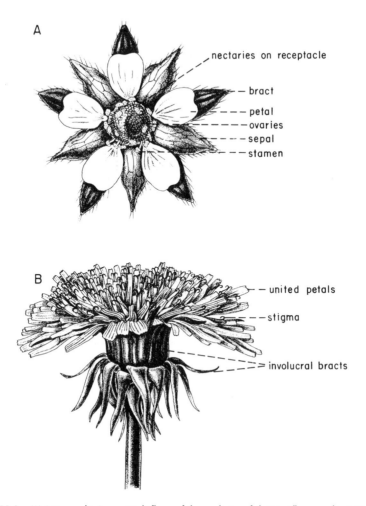

FIGURE 11.3 (A) Actinomorphic (symmetrical) flower of the rough cinquefoil, *Potentilla monspeliensis* (Rosaceae). The open, flattish arrangement of the flower permits many insects to reach nectar secreted on the receptacle and unspecialized "mess-and-soil" pollinators, such as beetles and flies, are common visitors. Reduction of floral parts has progressed to a certain extent: petals are reduced to five but stamens are still numerous (15—20 per flower). (B) Inflorescence of the dandelion. *Taraxacum officinale,* showing how the small flowers are grouped to form a landing platform for insects. Apparently, the flowers of the now extinct Bennettitales looked like this. Each flower is composed of five united petals which form a narrow tube, at the base of which nectar is secreted. Only insects with elongated mouthparts can reach nectar, although pollen is more easily obtained as stamens are arranged at the mouth of the tube.

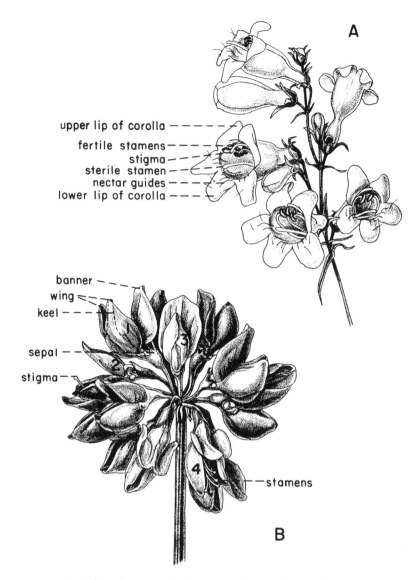

upper lip of corolla

fertile stamens

stigma

sterile stamen

nectar guides

lower lip of corolla

A

banner

wing

keel

sepal

stigma

stamens

B

FIGURE 11.4 (A) Part of an inflorescence of the foxglove beard-tongue. *Penstemon digitalis* (Scrophulariaceae). Flowers are zygomorphic (bilaterally symmetrical). The five petals are fused into a tube that narrows near the receptacle, preventing all but long-tongued insects such as bees from reaching the nectar secreted at the end of the tube. The lips of the three lower petals form a landing platform for bees, and nectar guides act as a visual cue for guiding the bee down the tubular corolla. There are four fertile stamens that dust the dorsum of a visiting bee. The stigma is similarly placed and this receives pollen when a pollen-carrying bee visits. A sterile stamen, the beard-tongue, lies ventrally in the corolla tube. In the Scrophulariaceae there is an evolutionary reduction of stamens from five to two. (B) Inflorescence of the crown vetch, *Coronilla varia* (Papilionaceae). Each flower is strongly zygomorphic with five petals fused at their bases. A banner petal is dorsal, and two wing petals are arranged either side of the two fused petals that form the keel (see flower 1). A heavy insect lands on the wings, depresses the wings and keel, and the stamens and stigma spring up in contact with the underside of the insects' abdomen (2). A pollinated flower is shown at 4. Thus nectar and pollen are protected in the closed flower (3) until a heavy bee visits the flower. Mess-and-soil pollinators are excluded.

were not found until the Oligocene, just before bees appeared in the fossil record, but it appears that the Lepidoptera must have evolved at the same time as the early angiosperms. It is notable also that bees appeared at the same time as the families Scrophulariaceae and Labiatae, with their highly specialized flowers. These families evolved zygomorphic arrangement of flower parts (bilaterally symmetrical as opposed to the actinomorphic or symmetrical flowers of more primitive angiosperms, e.g., Fig. 11.3A), reduced numbers of stamens, and a landing platform—obviously adaptations as bee flowers (Fig. 11.4). Finally, after bees had been evolving and diversifying for some 35 million years, those most beautifully bee-adapted families, the Papilionaceae and the Orchidaceae, left their stamp in the fossil record. These relationships are considered further by Crepet (1979).

MUTUALISM AND NECTAR CONTENT

The relationship between plant and flower pollinator is a mutualistic one. However, the extent of this mutualism is not yet fully appreciated. Baker and Baker (1973a,b, 1975, 1983a,b) provided evidence for an unsuspected and important aspect of the coevolutionary process. Nectar that may contain sucrose, glucose, and fructose has usually been considered as an energy source, while protein-making chemicals (amino acids) had to be found by pollinators elsewhere, by larval feeding in Lepidoptera, from pollen by bees, and from insects by vertebrates. But clearly if the plant could provide amino acids, foraging could be concentrated on the flowers, and both foraging and pollinating efficiency would improve. Therefore, the Bakers surveyed 266 species of flowering plants in California and discovered significant concentrations of amino acids in nectar (Table 11.1). This is no fortuitous array of amino acid sources,

TABLE 11.1   Relative Amounts of Amino Acids (Estimated on a Histidine Scale) of Various Types of Flowers. As Dependence on Nectar as a Source of Amino Acids Declines, the Amino Acid Level Declines.

Flower Type	Number of Species	Mean on Histidine Scale[a]
1. Fly (specialized for carrion flies)	8	9.25
2. Butterfly	41	6.68
3. Butterfly and bee	44	6.02
4. Moth	30	5.60
5. Bird	49	5.22
6. Bee	95	4.76
7. Fly (unspecialized)	34	3.77

[a] $1 \equiv 49\ \mu M$, $2 \equiv 98\ \mu M$, and $10 \equiv 25\ mM$.
SOURCE: Baker and Baker (1973a).

and these flowers provide an important source. Butterflies (except for *Helico-nius* butterflies, which digest pollen; see Gilbert, 1972) and moths have few other readily available sources of amino acids, and we see fairly high concentrations in the nectar of flowers adapted to pollination by Lepidoptera. However, birds use insects as a protein source and bees have a rich source in the form of pollen, and we see amino acid concentration reduced in flowers pollinated by these animals. Finally, in relatively unspecialized flowers both nectar

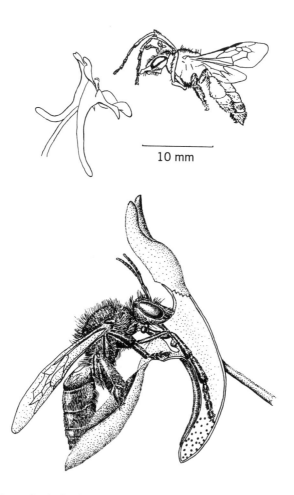

10 mm

FIGURE 11.5   (above) A female of *Rediviva longimanus* visits a twin-spurred *Diascia* flower. Note the long prothoracic legs of the bee, matched by the long spurs of the flower. From Whitehead et al. (1984); with permission. (below) A female *Rediviva neliana* collects oil secreted by trichomes in the spur of *Diascia capsularis*. The bee transfers the oil to the hind legs and carries it back to the nest. The oil becomes a component of larval food and may also be employed in nest construction. From Steiner and Whitehead (1990); with permission.

and pollen are freely available to pollinators, so that there is little need for an amino acid "supplement" to the nectar food source. The Bakers found amino acids in primitive angiosperm nectar, but with coevolution more and more has been produced, as can be seen in a comparison between the unspecialized fly flower to the tubular flowers adapted to pollination by butterflies and moths. (See Kevan and Baker, 1983, and Baker and Baker, 1983a,b, for additional developments).

Nectar and pollen are not the only rewards provided by flowers for pollination services. Stigmatic exudates; oils, resins, and gums; sexual attractants; and other products are produced by a wide variety of species (Simpson and Neff, 1983). Resins and gums may be used by female bees for nest construction and are produced by flowers of members of the Euphorbiaceae (e.g., Armbruster and Webster, 1979; Armbruster, 1983) and Clusiaceae (Skutch, 1971). Females of melittid bees in the genus *Rediviva* in South Africa collect oils from glands in the long spurs of *Diascia* flowers with their greatly elongated prothoracic legs (Fig. 11.5) (Whitehead et al., 1984; Steiner and Whitehead, 1988, 1990, 1991). Prothoracic leg lengths of females in one species tend to vary with flower spur length as flower species change over the geographical range of the bee, probably illustrating the importance of oil foraging for these *Rediviva* species.

FLOWER AND INFLORESCENCE DESIGN

The service required for the energy expended in nectar production is pollen transport and outcrossing. Such an important food source to animals as nectar and pollen can be manipulated by the evolutionary process so that the plant receives the best possible service from the pollinator. Models of optimal nectar allocation are discussed by Rathcke (1992), and Cruden et al. (1983) describe patterns.

Promotion of Outcrossing

The promotion of outcrossing can be achieved by several means, which are considered by Heinrich and Raven (1972):

1. Flower evolution can tend to restrict visits by a certain population of pollinators to flowers of a single species. This involves flower constancy, which is adaptive for the plant as well as the insect (see later).
2. The pollinator must not confine its visits to a single plant, and therefore the amount of nectar secreted per flower and per total plant at any one time must be precisely adjusted. Speaking in terms of evolution, the plant must control the caloric and amino acid reward presented synchronously. Agriculturalists and plant breeders select plants that must

grow at abnormally high densities. Foraging by bees in alfalfa, clover, and soybeans is extremely easy but may result in low seed set because not many flowers are visited. However, if plants with lower nectar secretions are selected for, the energy gain per unit effort would be balanced, and seed set would be improved. Heinrich and Raven (1972) explain how naturally reduced nectar supply has achieved this end in the red clover seed crops. Large numbers of pollinator visits produce greater crop yields. Short-tongued bumblebees take nectar by piercing the corolla without pollinating; but rather than reducing the seed crop, they actually lead to an increase in seed production. This is probably because the real pollinators (long-tongued bumblebees) visited more flowers when less nectar remained per flower after robbing.

3. The plant can only control the caloric reward in relation to a certain organism because each has a very different energy requirement. For example, a 100-mg bumblebee lands on a flower and uses about 0.08 cal/min while walking on the flower (see also Heinrich, 1973, 1975a, 1979a on bumblebee energetics), but 3-g sphinx moths and hummingbirds use energy at about 11 cal/min while hovering (Heinrich 1971a,b), a 140-fold difference. Hickman (1974) describes the need for secretion of small quantities of nectar in ant-pollinated flowers. He outlines many characteristics in flower design and location on the plant, plant morphology, plant population dispersion, and community characteristics that result in and reinforce this low-energy pollination system. However, many exceptions were noted by Beattie (1985), who also wondered why ants were not more commonly involved as pollinators. His own research with colleagues indicated a strong antibiotic effect by ants on pollen, equivalent to the bacteriocidal effects of secretions so important to ant hygiene. Except for the cases involving three genera of orchids, the production of viable seed after transport of pollen by ants has not been reported (Peakall et al., 1991).

Restriction of Visitors and Nectar Rewards

Therefore, plants evolve with flowers that restrict access to all but a few, or even one, pollinator species. Animals that require high levels of energy will not utilize species with low nectar production, and those plants with high nectar production must evolve protection against small organisms that may utilize nectar but not outcross the plants. A common evolutionary device is the development of a longer corolla tube so that only larger organisms (often with prolonged mouthparts, such as bees and moths) can reach the reward. Examples include many Scrophulariaceae (Fig. 11.4A), Labiatae, and Solanaceae, and an extreme case where nectar is secreted in the end of a spur 30 cm long in the Christmas Star orchid, *Angraecum sesquipedale,* and is perhaps pollinated by a magnificent moth, *Xanthopan morgani praedicta,* with a 15- to 20-cm wing span and a 30-cm proboscis (Kritsky, 1991b). Heinrich (1979a) illustrates a euglossine bee with a tongue almost twice the length of its body.

Shape, Color, and Scent

Shape, color, and scent of flowers may also be adapted for restricting pollinators. Flowers with much nectar may open and secrete nectar briefly at night, so that the flowers are available to bats and hawk moths but unavailable to many day-flying insects and birds. Bird-pollinated flowers are often scentless, red in color, and open during the day, and thus less conspicuous and less attractive to many insects (which do not perceive red light) and unavailable to nocturnal pollinators. Insect-pollinated flowers emit fragrances, while birds are poorly adapted for receiving such chemical stimuli. Bee flowers and many other insect-pollinated blooms responding to the insects' vision, which reaches into the ultraviolet, often show ultraviolet reflectance patterns (Eisner et al., 1969), which are presumably not visible to vertebrates. Detailed considerations of insect perception and the evolution of flower shape are provided by Leppik (1972 and references therein) and Meeuse (1973). Further details on ultraviolet reflectance patterns in flowers have been investigated by Jones and Buchman (1974), Brehm and Krell (1975). Thorp et al. (1975), and Frohlich (1976). Kevan (1978) gives an excellent review on coloration in flowers.

Clustering of Flowers and Blooming Times

Clustering of flowers will also influence the type of organism that can make a net energetic gain while foraging. If large inflorescences of small flowers open flowers synchronously, the energetic gain becomes more profitable for larger alighting pollinators than if flowers tend to open sequentially. We may see an evolutionary trend for greater flowering synchrony with increasing depth of nectar source where large pollinators exist in abundance. Outcrossing may also be promoted by evolution of inflorescences far larger than is optimal for attracting pollinators and achieving adequate pollination (Willson and Rathcke, 1974; Willson and Price, 1977; Willson and Bertin, 1979). Here plants may increase fitness by the low-cost strategy of being pollen donors, since seed and fruit production are relatively expensive in terms of energy and nutrients required. On a larger scale, synchronous blooming of one species in an area would reduce time and energy required by pollinators flying between plants, whereas brief blooming of individuals occurring at random over a long period would increase time and energy required to exploit the nectar source. In the latter case, either nectar reward must increase or pollinator size must decrease. Plant density also influences the distance of pollen-mediated gene dispersal in a density-dependent manner (Levin and Kerster, 1969a,b;

Antonovics and Levin, 1980). The more clumped the plants are, the shorter is the distance of pollinator movement, and the more probable is local differentiation of types within the plant species.

COMPETITION AMONG PLANTS AND AMONG POLLINATORS

In any one plant species the frequency of cross-pollination would be inversely related to the number of other species blooming at the same time if foragers visited flowers indiscriminately. Thus cross-pollination will be more effective and competition by plants for pollinators will be reduced if species bloom sequentially so that the same pollinators use a succession of blooming species (Rathcke, 1983). For example, Mosquin (1971) studied the activities of pollinating insects in relation to flowering phenology of plants near Banff, Alberta. From snow melt (early May) to the end of May, pollinating insects were abundant and competed for relatively scarce pollen and nectar. In early June, some "cornucopian species" (e.g., Willows, *Salix* spp., and dandelion, *Taraxacum offinale*) began flowering, offering virtually unlimited supplies of nectar and pollen. Insects left the spring flowers and foraged on the cornucopian species. In the presence of cornucopian species natural selection would therefore favor evolution of earlier or later flowering populations of competing species. The direction of movement in time will also depend on competition with other species. For example, Mosquin suggested that in the spring insects competed for flowers, but after the cornucopian species had completed flowering, flowers competed for pollinating insects (see also Linsley and MacSwain, 1947; Free, 1968: Frankie et al., 1974; Heinrich, 1975b). Several cases of character displacement in plants competing for pollinators have been suggested (Rathcke, 1983).

Cruden's (1972a) data suggest that there may be competitive displacement between hummingbirds and bees on an altitudinal gradient. Cruden found in Mexico that hummingbirds are more effective pollinators at high elevations (2300 m and above) during the rainy season because they remain active during cloudy and rainy weather. Also, nearly twice as many hummingbird species are reported in high-elevation communities as at middle elevations (1000–2300 m). Cruden considered that there was little or no difference in effectiveness between bees and birds as pollinators provided that flight conditions were favorable, as they were more frequently in middle elevations. However, the fitness of plants pollinated by bees at middle elevations was much higher (Table 11.2). Therefore, one might argue that bees are better pollinators. Since bees maintain their resources better than hummingbirds at middle elevations, this might be considered as a form of indirect competitive displacement. Of course, hummingbirds may occur at higher elevations for many reasons, but the possible interactions would be interesting to study.

TABLE 11.2   Pollination, Seed Set, and Fecundity in Labiates from Chiapas, Oaxaca, Mexico, and Durango.

Flower Type	Elevation	Pollination (%) (A)	Seed set (%) (Percent of Ovules that Develop Into Seeds) (B)	Fecundity (%) (A × B)
Bee	High	83	58	48
	Middle	93	91	85
Bird	High	92	87	80
	Middle	91	76	69

SOURCE: Cruden (1972a).

Cruden et al. (1976) also investigated moth pollination on elevational gradients and found reduced fruit set and fecundity with increasing altitude.

In deserts, blooming after rain leads to synchrony of many species and divergence is seen in a variety of ways: (1) daily time of blooming as in the *Oenothera* species that Linsley et al. (1963a,b, 1964) studied (e.g., see Fig. 11.6, (2) amount of caloric reward provided, (3) type of flower product (nectar, pollen, or both), and (4) structures affecting access to nectar or pollen. Thus many sympatric bee species specialize and diverge in foraging patterns and thus decrease competition for food, as discussed by Rathcke (1983) and Inoue and Kato (1992).

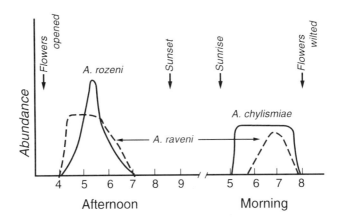

FIGURE 11.6   Pollen-gathering activity of three species of *Andrena* on *Oenothera clavaeformis,* a white evening primrose that opens in the late afternoon. Data from Linsley et al. (1963b).

FLOWER CONSTANCY

The term *flower constancy* relates to the foraging pattern of pollinators. Where flower constancy is high, foraging is restricted to one plant species during single trips, during a day, and even over several days. This high constancy may be seen in individuals and in colonies of bees. Flower constancy is adaptive for both the insect (frequently a bee) and the plant, and coevolution is likely to progress toward this end, particularly involving flower evolution to favor a certain type of very efficient pollinator. The adaptiveness for the plant is obvious; cross-pollination is ensured and nectar supply and pollen production can be adjusted efficiently to certain pollinating species and protected by floral parts to conserve this resource for the specialists. An early development in this evolution for constancy was the fusing of petals into a tube around the stamens. Then fewer insects can enter the flower, but those that can experience less competition and flower constancy is reinforced. The result is seen in the alpine flora of Europe, for example, where beetles, wasps, flies, and other unspecialized and inconstant insects that make up 86% of insect species on flowers with exposed nectar constitute only 37% of visitors to flowers with concealed nectar (Grant, 1950). Further evidence of this coevolution for flower constancy is seen in the number of taxonomic characters used to distinguish species in the same genus. Grant (1949) found that the percentage of these characters that were correlated with the pollination mechanism was very high in bird-pollinated plants (37%) and bee and long-tongued fly plants (40%), but low in plants that were pollinated indiscriminately (promiscuous) (15%) and in wind- and water-pollinated species (4%).

Flower constancy becomes more efficient for the bee as it learns to extract nectar and pollen while visiting the same flower type (Grant, 1950), and larvae can specialize metabolic processes for utilizing a constant source of nutri-

TABLE 11.3   Flower Constancy in Genera of Bees Indicated by the Percent of Individuals in a Sample That Contained Pollen from One Species of Plant. Note the High Frequency of Pure Pollen Loads in All Genera Except *Anthophora*. If Estimates were for Pollen Loads with 90% or More of One Pollen Type, the Percentage would be Much Higher.

Genus of Bee	Percent Pure Pollen Loads (from Various Sources)
Andrena	44, 64
Apis	62, 88, 93
Anthophora	20
Bombus	59, 49, 57
Halictus	84, 75
Megachile	75, 55

SOURCE: Grant (1950).

ents. Development may become more rapid and utilization more efficient. For honeybees, packing of pollen onto the corbicula, or pollen basket, seems to be more efficient when only one type of pollen is involved (Zahavi et al., 1984). Hence **pollen constancy** may be more accurate a term than flower constancy. Bees that visit flowers of a single plant species are called **monotropic** and those that collect a single species' pollen are **monolectic**. Those that visit a few related species are **oligotropic** and **oligolectic,** and those that visit many species are **polytropic** and **polylectic.** The constancy of bees can be estimated from examination of pollen types on the body after foraging, and we frequently see a high proportion of bees with a pure pollen load (Grant, 1950; and Table 11.3).

FORAGING

But bees in temperate regions (e.g., *Apis mellifera*) tend to forage in small ares of 10–20 m^2, and thus considerable inbreeding may result. The bees are constant to a particular species not only in single flights, as indicated by the pure pollen loads, but also over periods up to 10 or 11 days (Grant, 1950), and they are quite constant in a colony. We actually see a dynamic interaction between honeybee colonies that must influence the population structure of the plants immensely. This has been studied by Butler (1943, 1945) and Butler et al. (1943). The first flowers to bloom in the spring will have a colony of bees that utilize flowers in a certain area and will defend this area against intrusion by members of other colonies. Honeybees from other colonies arriving after local patches have been preempted will not be able to forage on a group-defended area and will wander over the field more or less at random, and these will cause much more extensive movement of pollen. But as a second plant species comes into bloom, the wanderers will become constant and defend areas. This second source does not attract the older bees, and new bees will act as wanderers until new sources become available. As a plant blooming period ends, competition for pollen ensues, a new wave of inconstancy results, and equilibrium is finally established when some groups forage in other areas.

This type of foraging dynamics may also be observed in the tropics on species that flower abundantly but intermittently. Roubik (1992) emphasized the lability of ecological niches in tropical bees and the flexibility of the mutualistic relationships involved in pollination systems over a landscape. Why are there so many tree species and so few bee species in the tropics?, he asked. The answer seems to be that highly social bee species that predominate in the tropics, such as in the Apidae (O'Toole and Raw, 1991), are generalists of necessity, with rapidly shifting foraging patterns and highly efficient utilization of plant nectar and pollen resources. Thus a relatively small number of bee species can provide services for a large number of plant species.

However, where food plants occur singly and far apart, a completely different foraging pattern has been observed among the pollinators and a different flowering pattern among the plants. For example, Janzen (1971) showed that euglossine bees forage for food over many kilometers of tropical forest and that females may visit the same set of plants each day. They behave as if they remember the location of food plants and act as if they are operating a trap-line for food. The same trap-line behavior has been observed for the butterfly, *Heliconius ethilla,* by Ehrlich and Gilbert (1973) and Gilbert (1975, 1991), one of the species discussed at the beginning of this chapter. Janzen (1971) described the significance of **trap-lining** to the host plants:

1. Outcrossing occurs at very low host-plant densities characteristic of tropical populations.
2. Only one or very few flowers need to be produced in any one day.
3. Much energy need not be stored for a large synchronized flowering.
4. Floral morphology can become specialized to the intelligent, effective pollinators, so that energy is not wasted on feeding species that do not carry pollen of that species over long distances.
5. Since only small amounts of energy are required for flowering, a woody plant can reproduce much earlier in its life history or under conditions of heavy competition for light or nutrients.
6. Floral visibility is less important and pollination can be effective in the heavily shaded forest understory.

Gilbert's (1975, 1991) studies demonstrate how these factors interact to produce organization of the host plants and the pollinators at the population and community levels. Many other fascinating aspects of pollination biology in the tropics are treated by Bawa and Hadley (1990).

SPECIATION OF PLANTS

Flower constancy and local foraging may lead to isolation of populations and eventually to speciation, with the strong possibility of sympatric speciation. Grant (1949) pointed out that bee plants may speciate more rapidly than do promiscuous species (being pollinated by many types of insect) because of this **ethological isolation** caused by the bees. Two processes may be involved. First, bees may cause an initial evolutionary divergence of flowers within a population as a result of selective pollination of mutant floral types. Sympatric speciation may be possible. (See Straw, 1955, for a supporting example of fixation of *Penstemon* hybrids by bees, wasps, and hummingbirds). Second, in the zone of secondary contact between populations that have been geographically isolated, but have become sympatric, bees may effectively isolate these populations, although only small differences in floral characters were acquired during isolation. Evidence for Grant's contention of more rapid speciation in bee plants is seen in the southern California flora, where there is an average of 5.94 species per genus in bee plants but only 3.38 species per genus in promiscuous insect-pollinated plants.

Dressler (1968), Dodson et al. (1969), Hills et al. (1972), and Dodson (1975) provide strong reason to believe that sympatric and allopatric speciation are highly likely in orchids pollinated by euglossine bees. Dressler uses the term **euglossine pollination** for the relationship in which orchid flowers are visited only by male euglossine bees that are attracted by odor and "brush" on the surface of the flower. This brushing action is apparently to obtain the fragrance, which is then transferred to the inflated hind tibiae (Evoy and Jones, 1971). It may be used in one of three ways (Dodson et al., 1969):

1. Male bees live longer than normal, but they die if deprived of the compounds. Thus the bees may metabolize the compounds to make up for a natural deficiency of their diet.
2. They may convert the compounds into sex attractants. Males fly in their territory leaving odor on leaves and produce a fine mist that may attract the females.
3. Males may use the compounds to attract other males of the same species so that several males congregate and may thus be more attractive to females, since the buzzing of the bees will be louder and the compounds more concentrated. Thus a lek system is involved (Kimsey, 1980).

Since only odor is involved in attraction, an odor-modifying mutuation may lead to pollination by another species of bee. A shift in size of bee may cause severe selection pressure and rapid morphological change, since flower morphology is crucial to pollination success (e.g., see Fig. 11.7), and sympatric speciation may result. Also, as an orchid species disperses, populations may come into contact with preadapted pollinators in new areas after long-distance dispersal, and again flower morphology would adapt to increase precision of the pollinating mechanism. Allopatric speciation may result. (See Williams, 1978, 1982, and Dressler, 1982, for other references).

Polymorphisms in plant populations may also be maintained by pollinators. In England, the wild radish *Raphanus raphanistrum* has two color morphs, white and yellow. The pollinating butterflies (*Pieris* spp.) and syrphid flies (*Eristalis* spp.) show a strong preference for yellow flowers, while some bumblebees (*Bombus* spp.) show a preference for white flowers (Kay, 1978). As long as both types of pollinator work in a population, the polymorphism will be maintained. Reproductive isolation could result in populations serviced only by pollinators showing such preferences. Mogford (1978) discusses other examples.

Conversely, insipient polymorphism may be suppressed by stabilizing selection on a monomorphic species, as seems to be the case in *Delphinium nelsoni* (Waser and Price, 1981). White mutants in this species suffer reduced

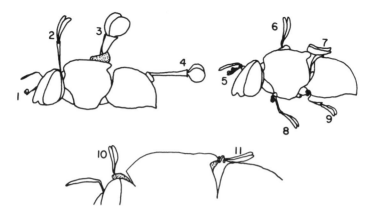

FIGURE 11.7   Outlines of male euglossine bees showing how the pollinia of 11 species of orchid are deposited in precise locations on the bee's body. This emphasizes the need for close coevolution between flower and pollinator and the probable changes in flower morphology with change in bee species and consequent behavioral and morphological changes. From Dressler (1968).

pollinator visitation and seed set because of visual difficulties in locating the nectar source. The normal blue flowers have a white target around the opening to the nectaries, which is lacking in the all-white flower.

SPECIATION OF BEES

In a similar vein, Cruden (1972b) discusses two ways in which plants may influence speciation in oligolectic bees. First, the host-plant population may be split geographically, which may result in two new species; oligolectic bees modify to the diverging plant populations; and we see closely related bee–plant pairs as we saw in coevolution of insects with toxic plants (Chapter 5), for example,

Andrena torulosa—Nemophila atomaria (Hydrophyllaceae)
Andrena crudeni—Nemophila menziesii

The second possibility depends on the local extinction of a host plant with a shift in the bee to a new pollen source and subsequent adaptation of the bee population to this new source. Adaptation will result as a response to changes in flower size, pollen size, and phenology of the new host plant (see Thorp, 1969). Thus the local populations will diverge from those where the original plant host survived and speciation may result.

POLLINATION ENERGETICS

The pollinating insect needs to make a net energetic gain while foraging. The cost of foraging is high in terms of calories used in flying, but the rewards must be higher. This relationship sets rather precise limits on which flowers can be pollinated, especially at low temperatures when it is energetically expensive to maintain body temperatures high enough for flying. A bumblebee cannot fly if its muscle temperature drops below 30°C or if it heats up to 44°C (Heinrich, 1979a). Precise mechanisms of thermoregulation are necessary in pollinating insects (May, 1979; Heinrich, 1981).

Some plants have specialized on bee pollination despite the relatively high cost of supporting the foraging of a relatively large species. For example, there are many species of gentian *(Gentiana)* that have a tightly pleated corolla tube that can be forced open only by strong bumblebees. This slows down foraging, but the reward is large because it is untapped by other species of pollinator, and a flower may provide 45 µL of nectar with 40% sugar in it (Heinrich, 1979a). This is 10 times more than other local plants, for 1 µL would contain 1.6 cal of sugar, and one flower would provide 72 cal, providing enough energy for a bumblebee to fly for 140 minutes! Thus gentian flowers may be quite rare and yet maintain bumblebees on a positive energy budget. Other bumblebee flowers, such as monkshood (*Aconitum* spp.), provide high nectar rewards in well-concealed nectaries.

Temperature has a profound effect on pollinators, particularly the poikilothermic insects (see Heinrich and Raven, 1972). For example, bumblebees can forage at 5°C or less by producing endogenous heat that elevates body temperature, but the energy cost is two or three times greater than at 26°C. Therefore, flowers pollinated at low temperatures should provide more caloric reward than those blooming at high temperatures, or they should be closer together so that they can be visited in rapid succession. Perhaps the clumping of spring flowers is adaptive in this way. It is not surprising, therefore, to see ecotypic variation in caloric reward, where more northerly ecotypes secrete more nectar (Heinrich and Raven, 1972). Similar trends may be seen on elevational gradients.

Also, cool temperatures in early morning in temperate regions can be utilized only by large insects capable of temperature regulation, and the flowers are usually large (e.g., *Oenothera* spp.) and we see a trend in decreasing size of pollinators as temperatures rise. *Bombus edwardsii* (0.12 g) had a thoracic temperature of 37°C while foraging at 2°C from a manzanita, *Arctostaphylos otayensis,* an energetically expensive operation. However, each flower contained 1.5 cal of sugar while the maintenance cost was about 0.8 cal/min, so that the bees could make an energetic gain while foraging at near-freezing temperatures. At noon when the flowers were visited by small insects each flower contained only 0.32 cal of sugar. Thus it may be energetically most profitable for bees to forage in the early morning when there is either little

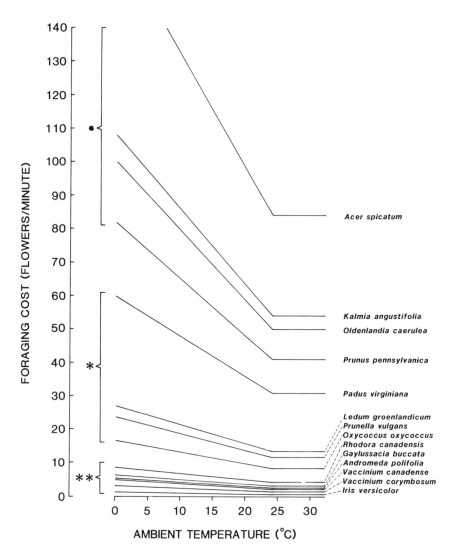

FIGURE 11.8   Estimated foraging cost in relation to nectar contents of flowers and temperature. Assuming that a queen bumblebee is 0.6 g in weight, it spends 50% of its time in flight and has a metabolic rate of 3.0 cal/min in flight. The plant species bracketed with two asterisks are those visited at both low and high temperatures; species bracketed with one asterisk are those visited frequently only at higher temperatures; and those bracketed with a solid spot are visited only occasionally at higher temperatures. From B. Heinrich in L. E. Gilbert and P. H. Raven (eds.), *Coevolution of animals and plants*. Copyright © 1975, 1980. Reprinted by permission of the University of Texas Press.

competition for nectar, a considerable accumulation of nectar, or the rate of nectar secretion is high (Heinrich and Raven, 1972). Conversely, low-energy sources such as golden rod, *Solidago* spp., are visited by bees at relatively high ambient temperatures.

A valuable perspective on what a bee selects to forage upon is obtained by measuring nectar contents of many flowers in a community and calculating how many flowers a bee would have to visit per minute to make a net energetic gain. This gain is reduced as temperatures decline, so the number of flowers visited must increase (Fig. 11.8). Even at 0°C only 10 flowers per minute need to be visited of the rhododendron, *Rhodora canadensis*, but 110 flowers per minute of lambkill, *Kalmia angustifolia*, would have to be visited for the same energetic gain (Heinrich, 1975c, 1979a). Clearly, there is a spectrum of flower types, some of which will be rewarding at any positive temperature, others that are rewarding only at the highest temperatures, and others that are seldom rewarding but may be visited occasionally.

Heinrich and Raven (1972) also discussed alternative strategies to the main theme. When temperatures get too cold, for example, during spring in temperate climates, or in northern latitudes, autogamy tends to increase, or nectar production increases for good competitive advantage, or very close clumping may achieve the same end. Where nectar rewards are very small, plants must be close and pollen dispersal will be very local, as is the case in Compositae, but wide dispersal of seed ensures constant mixing of gene pools and high genetic variability. In warm humid regions plants that are typically wind pollinated may become dependent on animals for pollen transfer. Grasses in tropical regions are visited by many insects (Bogdan, 1962; Soderstrom and Calderon, 1971), whereas elsewhere they are anemophilous.

DECEPTION IN POLLINATION SYSTEMS

Orchid plants are frequently widely separated but ensure outcrossing by the extreme precision in pollen transfer and reception (e.g., Dressler, 1968); pollen is not lost when flowers of different species are visited, and large numbers of seeds are set with a single pollination. Since flowers are widely separated in space and time, the pollinator cannot depend on flowers for its energy supply, no energy balance need be maintained, and thus about half of all orchid species offer no energy reward. Thus attraction is achieved frequently by deception: by mimicry of other flowers with nectar; by mimicking females of the pollinating male insects involving **pseudocopulation** (e.g., Coleman, 1933; Kullenberg, 1950, 1956a,b, 1961; Van der Pijl and Dodson, 1966; Steiner et al., 1994); by mimicry of potential hosts of parasitoids involving **pseudoparasitism** (van der Pijl and Dodson, 1966; Stoutamire, 1974); and by mimicry of other insects, which elicits territorial aggression by a bee and thus results in pollination, called **pseudoantagonism** (Dodson and Frymire, 1961; van der Pijl and Dodson, 1966). In the last case, orchid flowers of the genus

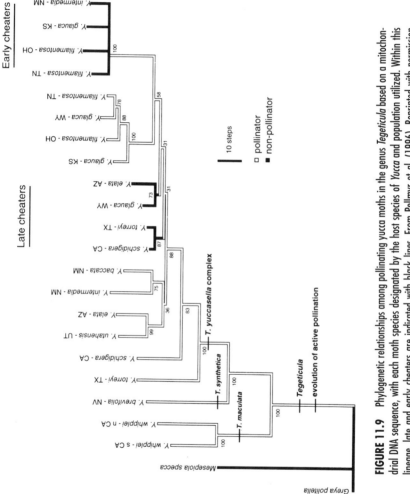

FIGURE 11.9 Phylogenetic relationships among pollinating yucca moths in the genus *Tegeticula* based on a mitochondrial DNA sequence, with each moth species designated by the host species of *Yucca* and population utilized. Within this lineage, late and early cheaters are indicated with black lines. From Pellmyr et al. (1996). Reprinted with permission from *Nature* **380**:155–156. Copyright 1996 by Macmillan Magazines Limited.

Oncidium mimic insects. The racemes of flowers are arched and the slightest breeze causes the flowers to dance. Male bees of the genus *Centris* attack the flowers as if they were flying insects, and pollination results. Deception in flowers other than orchids is discussed by Vogel (1978). Other bizarre methods of pollination involve figs and fig wasps, discussed in Chapter 10 (Baker, 1961; Galil and Eisikowitch, 1968, 1968; Ramirez, 1969, 1970; Bronstein and McKey, 1989); and yucca and the yucca moths (Riley, 1892, 1893; Holland, 1903; Powell and Mackie, 1966; Addicott, 1986; Addicott et al., 1990; Brown et al., 1994; Pellmyr and Huth, 1994) and globe flowers and their anthomyid pollinators and seed parasites (Pellmyer, 1992).

Of course, insects may also evolve the ability to cheat in pollination systems. Normally, yucca moth females use specialized tentacular mouth parts to collect pollen into a ball and transport it to the receptive stigmatic surface of the same or other flowers. Before pollination is completed she oviposts into the yucca ovary and her larvae eat developing yucca seeds. However, only a small proportion of the seeds are eaten, so the plant benefits from its specialized pollinator and the yucca moth benefits from the food and generally safe feeding location in the developing seed pods.

Only recently has it become clear that within the many highly specific yucca moths there are species that cheat (Pellmyr et al., 1996). Derived from the same lineage and genus, *Tegeticula,* some species have lost the ability to pollinate yuccas but oviposit into the fruits once flowers are pollinated by a noncheating species. One group of cheaters attacks fruits early in their development, and another group attacks fully developed fruits using a greatly elongated ovipositor (Fig. 11.9). When cheaters are present in a pod, all seeds may be destroyed.

Another form of cheating in pollination systems involves nectar robbing. Large carpenter bees in the genus *Xylocopa* are well known as nectar robbers, piercing the base of tubular flowers of species such as *Solanum* and ocotillo *(Fouquieria splendens)* and taking nectar without contacting stamens or stigma at the distal end of the corolla (e.g., Scott et al., 1993). However, in some cases pollen is gathered to provision the nest, in which case carpenter bees become mutualists with such species as ocotillo.

ADAPTIVE CONSTRAINTS

An important point, illustrated in this chapter, concerning many types of mutualism is that each associated species obtains a different advantage from the other. It is therefore unlikely that the mutualism will evolve to become an optimum relationship for both species. More likely, the species will exist in selective tension with each functioning at some suboptimal level because of these adaptive constraints.

An example of this conflict has been studied by Price and Waser (1979) and Waser and Price (1983). Clearly, an ideal situation for a pollinator is to

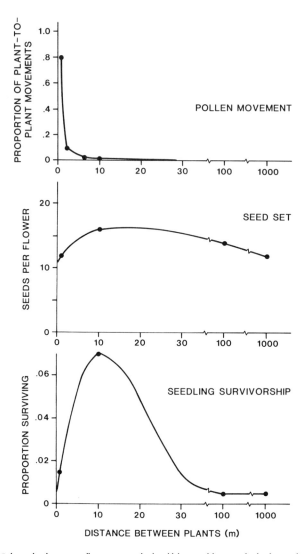

FIGURE 11.10 Relationship between pollen movement by bumblebees and hummingbirds; the number of seeds set per flower when receiving pollen from plants of different distances; and seedling survivorship to 2 years of age. Values at zero distance between plants relate to plants that received their own pollen. Plotted from data in Price and Wasser (1979).

be able to gather nectar very rapidly, so the closer the flowers are and the richer their reward, the better it is. But in many cases the plant gains when outcrossing is achieved and when the pollinator is forced to move greater distances than the ideal. The question that Price and Waser addressed concerned the best distance for pollen transfer from the point of plant fitness. They argued that cross-pollination between closely related individuals is likely to lead to inbreeding depression (Grant, 1975), whereas mating between genetically very different individuals will disrupt beneficial gene complexes and lead to outbreeding depression. Some intermediate level of pollen transport results in maximum plant fitness. Ideal distances are different for the pollinator and plant.

Price and Waser (1979) studied *Delphinium nelsoni* in western Colorado in the Rocky Mountains, which is pollinated there by bumblebee queens in the species *Bombus appositus* and *B. flavifrons* and two hummingbird species. Whereas the mean pollen-dispersal distance was only about 1 m, seed set of plants was highest when they received pollen from about 10 m distant, and seedling survivorship to 2 years of age was highest (Fig. 11.10). Waser and Price (1983) report similar results on further studies on *D. nelsoni* and *Ipomopsis aggregata*. The selective tension is apparent and is unlikely to be resolved by evolutionary change in either mutualist because differences in their requirements will persist.

Adaptive constraints are also likely when the distributions of plants and their pollinators do not overlap completely. "Most plants and pollinators move independently over the landscape, not in matched pairs" (Feinsinger, 1983, p. 307). Thus coevolution between plants and pollinators may be seen as broad patterns, but not so much at the individual species level, except in isolated cases such as in the Christmas star orchid. As Feinsinger (1983, p. 307) points out, "An obligate, exclusive, coevolving relationship does not necessarily increase the fitness of plant or animal; in fact, the opposite may often be true."

Pollinators may not overlap the blooming time of the plant population either. For example, hummingbirds migrate south before the end of the flowering season of scarlet gilia *(Ipomopsis aggregata)* in northern Arizona. Their preference is for red-flowered plants. But an alternative pollinator, the white-lined sphinx moth *(Hyles lineata)*, persists into the cooling temperatures of late summer. Hawkmoths forage after sunset from twilight into darkness and increasingly prefer white and light-pink flowers as light diminishes (Fig. 11.11). A remarkable discovery by Paige and Whitham (1985) is that individual plants shift from darker to lighter flower colors as the percentage of pollinators shifts increasingly to hawkmoth pollination. Thus plants can evolve to accommodate the pollinating species with different flower-color preferences and different phenologies.

Pollination ecology is a rapidly developing field (Faegri, 1978; Jones and Little, 1983). Access to literature on a wide variety of subjects is available through general treatments of pollination ecology (Proctor and Yeo, 1972; Faegri and van der Pijl, 1979; Jones and Little, 1983; Real, 1983; Buchmann and Nabhan, 1996), pollination by insects (Richards, 1978), pollination of agricultural crops (Free, 1970), and energetic aspects of pollination (Heinrich, 1975a). A very important aspect of pollination is the foraging behavior of pollinators, which is discussed in Chapter 15. Rathcke (1992) also demonstrates the central role of behavior in pollination biology and plant reproductive success. Further coverage is provided in broad treatments of plant and

FIGURE 11.11   The pollinating sphinx moth, *Hyles lineata*, visits a scarlet gilia inflorescence while a male rufous hummingbird, *Selaphorus rufus*, inspects opportunities. Based on Paige and Whitham (1985). Drawing by Tad Theimer.

animal interactions (Abrahamson, 1989; Howe and Westley, 1988). But the fascination with the reproductive biology of plants does not end with pollination, for many subsequent complexities are involved and are in need of further investigation, as revealed by Willson (1983) and Willson and Burley (1983).

Chapter 12

Energy Flow, Nutrients, and Ecosystem Function

Energy flow was treated in Chapter 2 as one of the major processes in ecosystems. Odum's (1957) study on Silver Springs, Florida, was used as an example of how a large energy input from the sun is used to fix energy by green plants, which is subsequently used by herbivores and carnivores, at each step much of the energy being dissipated as heat (Fig. 2.6). On returning to this subject a better idea may be obtained as to the details of energy flow, especially how insects play a role in this movement of energy.

SUN'S ENERGY

Since all energy (with very minor exceptions) is ultimately derived from the sun, the extent of this energy source must be known first. The quantity of solar energy entering the earth's atmosphere is approximately 15.3×10^8 cal/m^2 per year. A large part of this is scattered by dust particles or is used in the evaporation of water. The average amount of radiant energy actually available to plants varies with geographical location (Phillipson, 1966):

Georgia (latitude 33°N) 6.0×10^8 cal/m^2 per year
Michigan (latitude 42°N) 4.7×10^8 cal/m^2 per year
Britain (latitude 53°N) 2.5×10^8 cal/m^2 per year

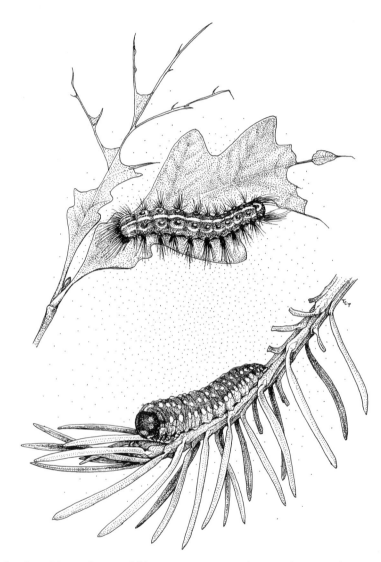

Caterpillars of two of the most destructive defoliating insects in eastern North America: the gypsy moth, *Lymantria dispar,* (above) and the eastern spruce budworm, *Choristoneura fumiferana* (below). Drawing by Tad Theimer.

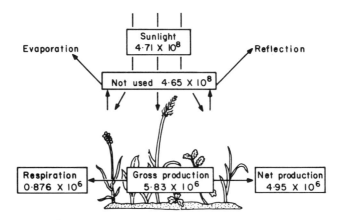

FIGURE 12.1   Energetic relationships at the primary producer level in a perennial grass and herb community of an old field in Michigan. Note that net primary production = gross primary production − respiration. Data from Golley (1960) and from Phillipson (1966).

More than 95% of this energy is immediately lost from the plants in the form of heat and heat of evaporation. The remaining 1–5% is used in photosynthesis and is transformed into plant tissues, with the energy stored in the chemicals in these tissues. The energetic relationships of the first trophic level are shown in Fig. 12.1. The energy values balance as far as methodology would allow, and the laws of thermodynamics are supported.

ECOLOGICAL EFFICIENCIES

At first sight it is remarkable that the photosynthetic efficiency (see Table 12.1) is so low. In the example given in Fig. 12.1 it is 4.95/4.71, or 1.05%. Other examples are given in Table 12.2, and the range appears to be 1.9–3.2% efficiency in agricultural crops and 2.2–3.5% in mature forest when estimated over a growing season (Hellmers, 1964).

However, these low figures grossly underestimate the ability of chloroplasts to fix energy. Individual plants synthesize organic matter by use of chemical energy, and this energy is released by oxidation (or respiration) of organic substances. Therefore, the gross production is reduced by respiration and the net primary production is usually 50–90% of the gross primary production. Although calculations may allow for cloud cover reducing available light, no correction is made for light absorbed by nonphotosynthesizing parts of the plant or for reflected light (e.g., Hellmers, 1964). Actually, the efficiency of chloroplasts to convert light energy into work has been estimated by Spanner (1963) to have a maximum value of 80% in bright light and 60% in low light intensity.

TABLE 12.1   Ecological Efficiencies of Energy Transfer Expressed as percentages (see Odum, 1971, Kozlovsky, 1968, for Additional Relationships).

Photosynthetic efficiency (e.g., Hellmers, 1964):

$$\frac{\text{net production} \times 100}{\text{visible light energy available}}$$

Assimilation efficiency (e.g., Odum, 1971):

$$\frac{\text{assimilation at trophic level } t \times 100}{\text{ingestion at trophic level } t}$$

Production (tissue growth) efficiency:

$$\frac{\text{production at trophic level } t \times 100}{\text{assimilation at trophic level } t}$$

Trophic level (Lindeman's efficiency (Lindeman, 1942):

$$\frac{\text{ingestion by trophic level } (t + 1) \times 100}{\text{ingestion at trophic level } t}$$

Individual heterotrophs do not assimilate all of the food they consume. Up to 90% of the total food intake may pass through the body and out as feces, giving an assimilation efficiency of 10% (see Table 12.1). At the other extreme some organisms may have an assimilation efficiency of 75% (Phillipson, 1966), with carnivores tending to have higher efficiencies than herbivores. It is not surprising then that trophic level (or Lindeman's) efficiency (see Table 12.1) is low, ranging from 5 to 20%.

Lindeman felt that there was evidence in 1942 that trophic level efficiency tends to increase as the food passes up the food chain. However, Slobodkin (1962) concluded that trophic level efficiency—what he called ecological efficiency—ranged from 5 to 20% and at that time there was no evidence to indicate a taxonomic, ecological, or geographic variation in ecological effi-

TABLE 12.2   Photosynthetic Efficiencies of Some Plants.

Plant	Location	Efficiency
Sugarcane	Java	1.9
Rice	Japan	2.2
Scots pine	Britain	2.2–2.6
Beech	Denmark	2.5

SOURCE: Hellmers (1964) and Ovington (1962).

ciency. However, the increase in our knowledge has been considerable since that time.

EFFICIENCY OF INDIVIDUAL INSECTS

Interest in the efficiency with which insects can convert food to their own biomass has developed rapidly over the past 20 years because of the compelling questions in need of an answer. Slansky and Rodriguez (1987a,b) have performed a wonderful service by compiling a large source list of literature and enunciating the points of view, questions to be answered, and methods available in the developing field of **nutritional ecology.**

The fundamental position in nutritional ecology, or the paradigm, is that "an organism has a genetically determined suite of performance value criteria [e.g., growth rate (or survival or pupal or adult mass, or total fecundity)] that prescribe its maximal fitness in a totally favorable, 'ideal' environment, but that natural environments, which are seldom ideal, place constraints on the consumption, utilization, and allocation of food, such that an individual seldom achieves its potential performance values. Because of selective pressures (e.g., variation in the quality and quantity of food; fluctuations in temperature, humidity, and other abiotic factors; and exposure to competitors, parasites, pathogens, and predators) imposed by natural environments, organisms have evolved means to evaluate their environment and respond to it in an adaptive manner, rather than merely being passively influenced by it" (Slansky and Rodriguez, 1987b, p. 5).

Using this approach, we can then ask many questions. How nutritious is the food provided by a plant in terms of growth rate of larvae and ultimate mass and fecundity? Does nutritional value change as leaves age, as Feeny (1970) asked for the winter moth (Chapter 5)? Does an ovipositing female select food plants that maximize the performance of her progeny—is ovipositional preference linked to larval performance? This question will be raised several times in Chapters 13 to 15, 17, and 19). We can ask if specialists are more efficient than generalists in converting food to insect mass, and a host of other questions. Thus nutritional ecology is central to much in the life of the insect and in insect ecology in general.

The research approach in nutritional ecology takes three basic steps after a question has been raised. First, individual performance is determined for as many of the life-cycle stages as possible under optimal conditions of the abiotic and biotic environment. Second, performance is evaluated under natural conditions to measure its reduction resulting from, for example, change in food quality, change in host plants utilized, increasing competition, suboptimal abiotic conditions, and so on. Third, the results in the former two steps can be used in the development of an understanding of the evolutionary constraints on an insect's lifestyle and the adaptive responses to these constraints. A strong comparative approach among related species of insects, or performance of one insect species on a related group of plants, or specialists versus generalists, provides a cogent approach to the discovery of pattern in nature and the mechanisms resulting in pattern. That is, nutritional ecology can play a critical role in the development of theory in ecology.

The methods used in nutritional ecology are well developed, starting with Waldbauer's (1968) landmark paper. Accurate measurements are made on consumption, utilization, and allocation of food using gravimetric methods. Great care is needed, as errors become magnified in the final nutritional indices calculated (Schmidt and Reese, 1986). Dry weight is often employed, which can be converted to energy budgets or nutrient budgets (Slansky and Rodriguez, 1987b). Inevitably some jargon and formulas are needed, but this technical language has become so much a part of the fabric of nutritional ecology that it is worth our efforts to explain and understand the essentials. Symbols are defined and formulas developed in Table 12.3. Digestibility and efficiencies are generally stated as percentage values. For example, the larval parasitoid, *Nemeritis canescens*, showed very high values of approximate di-

TABLE 12.3   Measurements, Symbols, Terminology, and Formulas Used in Nutritional Ecology.

Symbol	Definition	
C	Food consumption	} same process
I	Food ingestion	
F	Feces output	} same process
E	Egestion	
B	Biomass production	} same process
P	Biomass production	
G	Growth	
T	Time specified as any particular period	
GR	Growth rate (B/T)	
CR	Consumption rate (I/T)	
AD	Approximate digestibility	} same process ($I-F/I$)
AE	Assimilation efficiency	
ECD	Efficiency of conversion of digested food	} same process [$B/(I-F)$]
NGE	Net growth efficiency	
ECI	Efficiency of conversion of ingested food	} same process (B/I)
GGE	Gross growth efficiency	
$\overline{B}$	Mean body weight during T	
RCR	Relative consumption rate ($CR/\overline{B}$)	
RGR	Relative growth rate ($GR/\overline{B}$)	

Therefore, growth rate is calculated as follows:

$$GR = CR \times AD \times ECD$$

and efficiency of conversion of ingested food is calculated as

$$ECI = AD \times ECD$$

SOURCE: Based on Slansky and Rodriguez (1987b).

gestibility (AD) of 94%, efficiency of conversion of ingested food (ECI) of 61%, and efficiency of conversion of digested food (ECD) of 65% (Vinson and Barbosa, 1987). Rates are usually stated as milligrams per day, where dry weight of total mass is commonly used. Relative rates are used for comparing individuals of different size or age and are expressed as the absolute rate divided by mean body weight $(\overline{B})$. For example, relative growth rate (RGR) of the penultimate instar of the southern armyworm, *Spodoptera eridania,* fed on dill, lima bean, and cabbage, declined from about 0.46 mg/mg per day on dill to about 0.30 mg/mg per day on cabbage (Mattson and Scriber, 1987).

What are some of the patterns discovered by the approaches in nutritional ecology? First, food sources differ so much in essentials such as water content and nitrogen or protein content that we would expect some patterns to be driven from the bottom up, from primary producers to consumers. The extremes in ranges of total nitrogen are seen from wood with perhaps 0.25% dry weight to living animals and carrion with nitrogen up to 12% dry weight (Fig. 12.2). And water content ranges from seeds and dry wood as low as 2% up to phloem and xylem fluid at 98% or more. Ranges in contents of various foods for insects are given in Table 12.4. Noting the nitrogen and water contents of living insects in Fig. 12.2, it is clear that many insects must concentrate nitrogen and or water very effectively if they are to live on many substrates involving plant foods.

Plant foods also have the characteristic of changing in character through the season, as we saw in Chapter 5. Seasonal trends in water and nitrogen content of leaves show very large changes (Fig. 12.3), which influence consumption rates and efficiencies of insects feeding on leaves. For example, reducing the nitrogen content in the diet of the butterfly caterpillars, *Euphydryas chalcedona,* from 4% to 1.3% resulted in a 51% increase in consumption rate of food, which partially compensated for the reduced nitrogen content. Nevertheless, there was a decrease in growth rate of 20% (Slansky and Scriber, 1985). Therefore, we should expect to see differences in growth rate and total feeding time between feeders on young leaves versus those on old leaves. Because plants are so variable in time and among species, a generalized feeder such as the southern armyworm is likely to be exposed to an enormous range of food qualities, both for a single individual while "marching" from plant to plant, and within a population. Their capacity to perform well on this variety is remarkable (Fig. 12.4). Consumption rates *(RCR)* and efficiencies *(ECI)* vary considerably, as does the relative growth rate *(RGR)* in *Spodoptera eridania*. Relative growth rates may stay within the range 40–60% even though *ECI* values range from 15 to 60% by compensation in consumption rates from about 5 down to about 0.75 mg/mg per day, respectively (Fig. 12.4). The important process of compensation by feeding caterpillars is discussed in more detail in Chapter 17.

These kinds of variables in rates and efficiencies seen within species and between species, coupled with changes within individuals as they grow, result in wide ranges in nutritional indices within species and groups of species

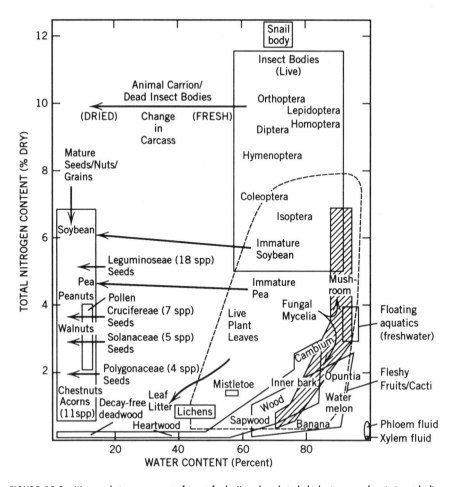

FIGURE 12.2   Water and nitrogen content of insect foods. Note the relatively high nitrogen values in insect bodies compared to most plant parts. Data are provided in Table 12.4. Nitrogen contents are percent dry weight except phloem and xylem fluid, which are given as percent fresh weight. From Slansky and Scriber (1985); with permission.

feeding on similar kinds of foods. This wide variation makes the discovery of patterns more difficult. Slansky and Scriber (1985) provide their own digestion of a large number of studies on herbivores, carnivores, and detritivores (Table 12.5). Ideally, one would compare, with a phylogenetic analysis, related groups with divergent feeding ecology, such as sucking insects that have diverged into herbivorous and carnivorous habits. However, in the absence of such analyses some general conclusions can be reached.

For herbivores, remarkably high values in digestibility *(AD)* and efficiencies of conversion *(ECD* and *ECI)* are seen for species feeding on forbs. The high values are generally as high as those for carnivores, making any ar-

TABLE 12.4   Nitrogen and Water Contents of Foods Consumed by Insects.

| | | Range in Contents | |
| | | | |
Food Category	Number of Species	Nitrogen (Percentage Dry Weight)	Water (Percentage Fresh Weight)
Foliage			
Aquatic			
Fresh water	17	0.6–3.5	—
Marine	12	0.5–4.5	67–94
Terrestrial			
Ferns	>5	—	71–77
Forbs	>98	1.5–9.7	50–93
Grasses and sedges	>33	1.2–4.5	44–80
Lichens	29	0.3–4.7	<50
Trees and shrubs	>255	0.6–6.9	36–87
Stems and shoots	—	0.5–3.8	52–94
Roots	—	0.3–2.5	50–88
Wood			
Cambium	—	1.0–5.0	89–91
Inner bark	—	0.1–2.2	52–88
Sapwood	—	0.06–0.3	50–70
Heartwood	—	0.04–0.22	25–49
Deadwood	—	—	2–30
Fungi	>21	0.1–8.2	75–91
Flowers	5	2.0–4.0	70–90
Fleshy Fruit	160	0.4–3.6	52–93
Pollen	2	2.5–4.0	11–12
Seeds, grains, and nuts	176	0.5–6.6	2–15
Phloem and xylem fluid	—	0.004–0.6	98–99.9
Detritus			
Freshly fallen	—	0.2–3.0	30–60
Enriched with fungi			
and bacteria	—	1.2–4.2	—
Invertebrates			
Insects	>15	6.6–12.0	60–89
Snails	—	12.4	79
Vertebrates			
Meat and fish	>12	5.0–15.0	50–81
Blood	3	—	80–83
Artificial diets	—	0–12.2	66–87

SOURCE: Slansky and Scriber (1985), with permission.

gument about trophic-level differences weak. However, at the low end of the range in values of *AD, ECD,* and ECI, herbivores can reach lower values than carnivores.

Among carnivores, parasitoids and predators show high levels of *AD* and *ECI,* but certainly not distinctly higher than the high end of the range in her-

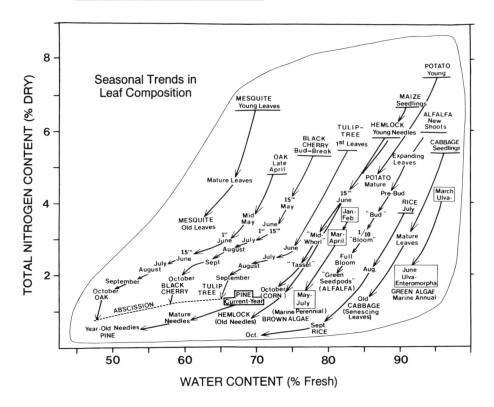

FIGURE 12.3   General seasonal trends in water and nitrogen contents of selected plants and plant parts, including annual and perennial forbs, grasses, deciduous and coniferous trees, and a marine alga. From Slansky and Scriber (1985); with permission.

bivores. In detritivores or saprophages, values of *AD* and *ECI* tend to be lower than for those insects feeding on living organisms, the biophages.

Other ecological questions can be examined with these kinds of studies, such as whether specialists are more efficient than generalists. The *feeding specialization hypothesis,* which suggests that specialists should show higher efficiency than generalists, has a long history, dating back at least to Dethier (1954). But even after much more data became available, Slansky and Scriber (1985, p. 131) noted that the degree of specialization of insects is a weak influence on nutritional indices relative to food quality. "In general, it appears that the nutritional quality of the plant accounts for the major portion of the variation in rates and efficiencies of food utilization and growth, mostly independent of the degree of specialization."

A final point on nutritional indices and efficiencies in food utilization is that efficiencies of dry weight utilization are generally comparable to those of energy utilization (Slansky and Scriber, 1985). Rarely do the indices differ by more than 10% (Table 12.6). Therefore, data generated in studies of nutri-

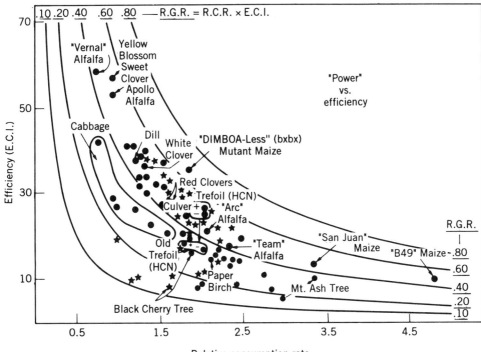

FIGURE 12.4   Relationships between RGR, RCR, and ECI for southern armyworms, *Spodoptera eridania,* in the last instar, fed many different plant species and varieties. Note that RGR values are moderately stable over large ranges of RCR and ECI because of the capacity of caterpillars to compensate RCR. High RCR ("power") results in low ECI values ("efficiency"). From Slansky and Scriber (1985); with permission.

tional ecology may readily be converted for studies on energetics and energy flow.

NUMBER OF TROPHIC LEVELS

The number of trophic levels may be limited by the amount of radiant energy reaching the biosphere or a part of the biosphere, the efficiency with which it is converted to organic material, and the efficiency with which each subsequent trophic level uses this energy. Since efficiencies are generally very low, we can expect very few links in the chain before all available energy is dissipated. Four or five trophic levels seem to be the maximum. As the energy source becomes smaller, more and more energy must be utilized to find this food until ultimately the energy is so sparse that more is expended in its

TABLE 12.5 Quantitative Food Utilization by Immature Arthropods.

Feeding Category	Taxon	AD (%)		ECD (%)		ECI (%)	
		Mean	Range	Mean	Range	Mean	Range
Herbivore							
Chewing							
Foliage							
Forb	Total	53	—	40	—	20	—
	Coleoptera	83	73–95	30	7–64	27	7–56
	Lepidoptera	53	16–97	40	2–87	20	1–78
	Orthoptera	50	25–86	38	6–67	17	3–34
Grass	Total	44	—	35	—	15	—
	Lepidoptera	45	24–68	43	18–77	19	9–34
	Orthoptera	44	19–82	32	8–71	13	4–37
Forb/grass mix	Orthoptera	35	16–62	41	23–63	24	14–45
Tree	Total	40	—	37	—	13	—
	Coleoptera	50	2–94	38	8–72	17	3–51
	Hymenoptera	26	12–77	39	12–77	9	4–17
	Lepidoptera	41	12–98	37	2–93	13	2–31
Root		—	—	49	18–63	—	—
Wood		54	3–93	65	—	8	2–14
Grain/seed		72	46–96	19	2–59	14	1–50
Nectar/pollen		88	—	53	—	47	—
Sucking							
Phloem, xylem, cells							
Forb		60	29–92	65	40–88	43	14–65
Grass		22	10–33	51	40–60	23	2–59
Tree		36	—	39	—	9	4–20
Seed		73	50–92	89	40–96	70	30–87

Carnivore						
Arthropod host/prey						
Total	81	—	37	—	35	—
Parasitoid	68	55–94	37	11–62	33	6–75
Predator	85	37–98	34	4–64	39	4–75
Arachnida	87	79–91	43	27–59	35	13–68
Vertebrate host	64	31–81	75	55–84	44	21–58
Detritivore						
Aquatic						
Total	40	—	50	—	13	—
	30	2–70	50	5–75	10	0.3–57
Crustacea	60	7–86	49	24–83	20	4–60
Terrestrial						
Total	32	—	29	—	2	—
Litter	41	7–70	50	—	9	8–10
Arachnida	17	14–20	—	—	—	—
Crustacea	39	5–82	18	5–32	6	—
Diplopoda	10	4–20	—	—	0.6	0.3–0.9
Dung	9	7–10	47	23–63	4	2–5
Ants	60	28–92	20	0.3–33	18	0.3–56
Other						
Wood/fur	51	38–66	20	6–37	10	3–18
Artificial diets						
Chewing						
Coleoptera	60	11–96	27	9–57	21	6–32
Lepidoptera	48	15–74	47	6–84	23	3–50
Orthoptera	51	19–95	39	16–70	25	3–45
Sucking						
Homoptera	71	30–85	27	11–87	18	8–35

SOURCE: Slansky and Scriber (1985), with permission.

TABLE 12.6 Comparison of Food Utilization Efficiencies for Dry Weight and for Energy.

Order and Species	Details	AD Dry Weight	AD Energy	ECD Dry Weight	ECD Energy	ECI Dry Weight	ECI Energy
Coleoptera							
Odontopus calceatus	Adult	54	51	19	22	10	11
Phytodecta pallidus[a]	L₂–L₄	50	52	54	55	25	29
Hemiptera							
Lygaeus kalmii	L₂–L₅						
	20°					60[a]	66
	30°					63[a]	69
Oncopeltus fasciatus	L₂–L₅						
	20°					76[a]	84
	30°					75[a]	83
Hymenoptera							
Diadromus pulchellus	Larva						
	15–25°					58	58
	20–30°					60	59
Megachile pacifica	Larva	88	88	53	67	47	59
Lepidoptera							
Papilio polyxenes	L₅, NY on S[b]	44	47	41	48	18	23
	L₅, CR on S[b]	53	57	28	33	14	18
	L₅, NY on D[c]	34	37	45	51	15	19
	L₅, CR on D[c]	31	30	51	65	15	18
Hyalophora cecropia	Larva	29	37	56	53	16	19
Orthoptera							
Acheta domesticus	Last instar	70	71	40	52	28	38
	First 10 days adult	73	73	35	46	25	34

[a] Values calculated from authors' data.
[b] New York (NY) vs. Costa Rica (CR) larvae on *Spananthe* (S).
[c] New York (NY) vs. Costa Rica (CR) larvae on *Daucus* (D).
SOURCE: Slansky and Scriber (1985), with permission.

exploitation than is gained from the food, so another trophic level cannot exist. Wiegert and Owen (1971) argue that basic differences in life histories of primary producers in terrestrial versus aquatic systems influence the length of food chains in these systems. Trees have a long generation time and a low biotic potential. The percentage of net production of a mature deciduous forest taken by herbivores is as low as 1.5–2.5 (see also Golley, 1972; Reichle et al., 1973; Van Hook and Dodson, 1974); thus little energy is passed up the food chain involving living organisms (biophages) and these chains are typically short, with three trophic levels. In consequence, much living material dies and becomes available to saprophages, making the saprophage-based food chains relatively important in deciduous forests. By contrast, in ocean water, phytoplankton have a high biotic potential and a short generation time. Regarding net primary production, 60–90% may pass to the primary biophages. A large proportion of fixed energy passes along the biophage food chain, which consists typically of four trophic levels, and the saprophage-based food chain is relatively insignificant.

Since radiant energy varies so much with latitude, some latitudinal differences should also be expected in the length of food chains and adaptations to exploit large and small energy supplies of organisms living in tropical rain forest and arctic tundra. However, Pimm and Lawton (1977) have countered this expectation, pointing out that food chains do not differ substantially in length between high-latitude and low-latitude systems, even though primary productivity ranges over at least four orders of magnitude in terrestrial ecosystems. They argue that the number of trophic levels is dictated by the stability of trophic relationships determined by the population dynamics in food webs. This position is developed in Chapter 23.

PRIMARY PRODUCTION

Photosynthetic production is the primary source of organic matter and potential energy, on which nearly all forms of life, including humans, are dependent. The only known exceptions to complete dependence on green plants are certain bacteria, which can synthesize organic matter using only chemical energy and inorganic materials, and humans who can obtain some of the energy needed from wind and water power, atomic energy, and direct solar energy conversions. The bacteria contribute a minute amount to the total supply of organic energy. Humans are completely dependent on photosynthesis for most nutritional requirements, and the energy obtained from fossil fuels, such as coal and oil, was produced by photosynthesis during earlier ages. Westlake (1963) points out that to understand community dynamics it is necessary to evaluate primary photosynthetic production and the factors that influence it, because this is one of the major factors regulating the growth and reproduction of all other organisms in the community.

TABLE 12.7   Net Primary Production (Biomass) on Fertile Sites at Different Latitudes (Metric Tons/Hectare Per Year).

Ecosystem	Climate	Organic Productivity
Desert	Arid	1
Deciduous forest	Temperate	12
Agriculture—annual plants	Temperate	22
Coniferous forest	Temperate	28
Agriculture—annual plants	Tropical	30
Rain forest	Tropical	50
Agriculture—perennial plants	Tropical	75

SOURCE: Westlake (1963).

Net primary production differs significantly between latitudes (Table 12.7) (see also Reichle et al., 1975; Lieth, 1978). Even at a given latitude the annual primary net productivity varies tremendously. Ovington et al. (1963) give figures for central Minnesota (Table 12.8). In natural vegetation productivity increases as the woody component increases (see also Fig. 2.10 and Table 2.1). Apparently, woody growth permits efficient utilization of time available for production because photosynthesis can start early in the spring and continue late into the fall. Oakwood production is very efficient, as no nutrients or cultivation energy are supplied by humans.

Viewing production in this way, one can make predictions as to the relative sizes of the herbivore biomass and the relative numbers of species of herbivores that can be supported by each ecosystem type. Herbivore biomass might be about 10% of the plant biomass, and since the latter are different by an order of magnitude, equivalent differences in the herbivores can be ex-

TABLE 12.8   Net Primary Production (Biomass) in Different Ecosystems in Central Minnesota (Metric Tons/Hectare Per Year).

Vegetation Sample	Ecosystem			
	Prairie	Savanna	Oakwood	Maize-corn
Herbaceous layer	0.920	1.886	0.182	9.456
Shrub layer	0.010	0.041	0.389	—
Tree layer				
Current year	—	2.833	4.046	—
Older	—	0.503	3.575	—
Total for aerial parts	0.930	5.263	8.192	9.456

SOURCE: Ovington et al. (1963).

pected. Even on the basis of biomass alone, more insect problems on corn than in the prairie should be anticipated, particularly since corn is a monoculture whereas biomass in the prairie is divided among 10–20 plant species. Corn production is high. Here a large part of the plant contains photosynthetic tissue, and much human effort goes into this production—cultivation and fertilizer. It would be interesting to compare the biomass of insects and other herbivores in corn that had no insecticide treatment with the biomass in an oakwood.

Not only will production change the character of the herbivore community, but the storage of energy by plants will also. In the savanna and oakwood ecosystems a large amount of energy is stored as woody tissue. Insects adapted to feed on this supply are not faced with an ephemeral food source and life histories can be prolonged. Prolonged life histories lead to slower population changes and increase the chances of stability of populations. The character of the community may be strongly influenced by energy storage. Some well-known examples of long-lived insect species that feed on woody tissues are the long-horn beetles (Cerambycidae), metallic wood-boring beetles (Buprestidae), and many other beetles, wood wasps (Siricidae), carpenter moths and leopard moths (Cossidae), some crane flies (Tipulidae), and the cicadas (Cicadidae), including the 13- and 17-year periodical cicadas. Typical prairie species that overwinter in the egg or adult stage and have an annual or multivoltine life cycle include grasshoppers (Acrididae), crickets (Gryllidae), leafhoppers (Cicadellidae or Jassidae), flea beetles (Chrysomelidae), and a great number of plant-feeding flies.

ENERGY UTILIZATION

Just as plants vary in their energy storage capacity, the herbivores have adapted many different ways of allocating assimilated energy. Allocation depends on the type of energy source they exploit for food (whether it is ephemeral or long-lasting), the physiological stresses to which they are exposed (weather, seasonal change), and the reproductive strategy evolved (many small progeny, few large; few large clutches of eggs, many small clutches). The possibilities that a herbivore or carnivore has available to adjust its energy utilization in evolutionary time and in a single life cycle are numerous (Fig. 12.5). Selection will normally work toward increasing the harvest, ingestion, and assimilation. But there is an obvious evolutionary choice between being sluggish and docile and expending little maintenance energy, and being highly mobile and utilizing much maintenance energy. At every ramification from here on there is an evolutionary choice between alternatives that affect the whole life of organisms and a choice within a life history according to the needs of each life stage. Insects are remarkable in showing diverse strategies in energy utilization from stage to stage. If the selection pressures that influence the apportionment of energy in animals and plants

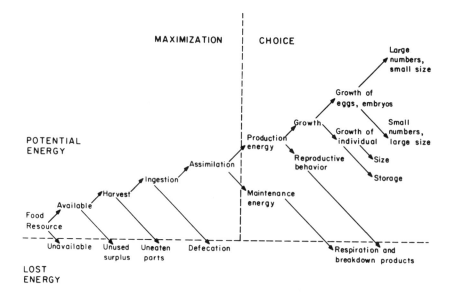

FIGURE 12.5   Energy flowchart for a population or individual of a consumer species. Adapted from R. B. Root (personal communication).

were fully understood, several major questions in ecology could be answered (e.g., see Chapter 14 on strategies in reproduction).

ENERGY FLOW IN INDIVIDUALS AND POPULATIONS

To better understand energy allocation, it helps to have a compartment model of energy flow in an organism to focus attention on the factors that need to be measured (Fig. 12.6). Batzli's (1974) model is conceptually rigorous and is used here. Energy does not become part of the organism until it is captured as reduced carbon. In heterotrophs this is when it has been assimilated through the gut wall. Egested material passes through the alimentary canal without entering the biochemical pool of the individual. Excretory products are lost from the biochemical pool of the organism and should be considered a cost of maintenance of that pool which is comparable to respiratory energy. Once energy (reduced carbon) has been fixed or assimilated by an organism, it becomes gross production and can travel through several pathways:

1. Respiratory metabolism, with production of ethanol, lactic acid, or CO_2.
2. Nitrogenous compounds excreted as waste products.
3. Organisms may perform work by moving a weight.

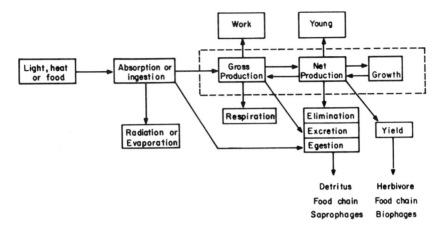

FIGURE 12.6 Model of energy flow through individual organisms (autotrophs or heterotrophs). Dashed line represents boundary of organism. Reproduced with permission from G. O. Batzli, Production, assimilation and accumulation of organic matter in ecosystems, *Journal of Theoretical Biology*, 1974, **45**:205–217. Copyright © 1974 by Academic Press Inc. (London) Ltd.

4. Reduced carbon may be incorporated into new molecules. When net production is positive, new tissues are produced more rapidly than old tissues are broken down.

From the model, gross production for autotrophs can be calculated by the formula

$$GPP = AR + NPP + AEX + AW$$

where GPP is the gross primary production, AR the autotrophic respiration, NPP the net primary production, AEX the autotrophic excretion, and AW the autotrophic work. When $AEX \simeq AW \simeq 0$, which is usually true, the equation simplifies greatly.

For heterotrophs,

$$GSP = HR + NSP + HEX + HW$$

where GSP is the gross secondary production, NSP is the net secondary production, and the other symbols are as for autotrophs, with H designating heterotrophic activity. Here HEX and HW will often be significantly greater than zero.

Tissue, or net production, may be lost in four ways:

1. Loss can be through reproductive propagules (young).
2. Portions of individuals may be sloughed off as dead material (elimination). This is very important in insects, as cast skins represent a consid-

erable loss in energy. The total dry weight of exuviae during the life of one aphid was 30% of the total dry weight of growth. In aphids this represents 14% of net production. A population of aphids on one tree may lose 300–360 kcal/year on exuviae (Llewellyn, 1972).

3. Another type of "sloughing" is in the form of secretions (elimination). Again this is an extremely important energy drain in insects when one considers the production of pheromones and defensive secretions.

4. Portions of an individual may be consumed by other organisms (yield). In this regard it is important to remember that biophages do affect the rate of production of their food; they essentially feed inside the organism box, whereas saprophages cannot affect the rate of production of their food and feed outside the organism box. This major difference must surely lead to fundamental differences in exploitation strategies of biophages and saprophages.

When an individual dies it becomes either elimination or yield, and all rates of energy flow become zero. Thus net production can be calculated by the formulas

$$\text{Autotrophs } (A): \quad NPP = AG + ARP + AEL + AY$$
$$\text{Heterotrophs } (H): \quad NSP = HG + HRP + HEL + HY$$

where G is the growth, RP the reproduction, EL the elimination, and Y the yield. Productivity of the population is obtained by summing GPP's and GSP's for individuals, and NPP and NSP is obtained by summing NPP's and NSP's for individuals, omitting the term for reproduction in the equation.

In Chapter 5 it was seen that a good deal of energy is put into reducing yield, which must reduce net production. This is done by manufacturing toxins and storing them, or by eliminating them at the appropriate time, growing spines or hairs for mechanical protection, or actively escaping an attack. All are drains from the energy pool of the organism. Then the choice between using energy for maintenance during care of young or for producing more young and taking less care of them can be made. It would be valuable to be able to partition the energy drains on the organism between the various elements of the total life strategy of the organism. Using such data comparatively for closely related organisms as well as for very different organisms would help tremendously in resolving debates on latitudinal gradients of clutch size, r- and K-selection, and reproductive strategies in general. However, more is needed than quantities for the compartments in the energy flow diagram. We need to subdivide respiration (R) into R for basic maintenance, and R for foraging, R for care of young, and R for competition. We would like to know how elimination is partitioned between caste skins in insects, sex pheromones, allelochemicals for competitive purposes, and defensive secretions. Thus to understand a reproductive strategy fully, many more details must be known than the model implies (see also Chapter 14 and Price, 1974b).

ROLE OF INSECTS IN ENERGY FLOW: BIOPHAGES

What is the role that insects play in energy flow, and how do they utilize their food energy? Odum et al. (1962) studied the population energy flow of the three major primary consumer groups living in early stages of old-field succession at the Atomic Energy Commission Savannah River area, South Carolina. These were the seed eaters, the Savannah sparrow, *Passerculus sandwichensis,* and the old-field mouse, *Peromyscus polionotus,* and the foliage-eating Orthoptera, grasshoppers, *Melanoplus femur-rubrum* and *M. biliteratus,* and a tree cricket, *Oecanthus nigricornis.* At average density levels energy flow was estimated to be 3.6 kcal/m^2 per year for sparrows, 6.7 kcal/m^2 per year for mice, and 25.6 kcal/m^2 per year for Orthoptera. Insects are very important members of the herbivore trophic level. Wiegert an Evans (1967) calculated that in an old field the Orthoptera and harvester ant populations together accounted for 81% of the total energy flow (assimilation) through the secondary producer trophic level.

Although the Orthoptera utilized a smaller part of the food available to them (2–7%) than the seed-eating sparrows and mice (10–50%), they channeled more into production than did the seed eaters. Maintenance costs of homeotherm vertebrates are much higher than for invertebrates, so actual production figures showed even greater differences than did the energy flow values (Table 12.9). Production by Orthoptera was 100 times greater than that by sparrows and 33 times greater than that by mice. Therefore, the insect herbivores not only consume more energy (in absolute terms) in the primary producer trophic level than did the vertebrates, but they also make much more energy available to the carnivore trophic level. They therefore play a major role in the energy flow of a community. If insects were not a part of the community, we would expect a significant drop in the possible number of trophic levels and a gross simplification of feeding links in the food web.

Wiegert and Evans (1967) have summarized energy flow data for old-field situations in Michigan and South Carolina (Table 12.10), and the following points are worthy of note. Ingestion rates in insects are very high in some cases. Even spittlebugs, which appear to be quite insignificant in the field, in-

TABLE 12.9   Production in an Old Field by Populations of Herbivores at Average Densities (kcal/m^2 per year).

Herbivore	Production	Percent of Assimilation Channeled into Production
Sparrows	0.04	1
Mice	0.12	2
Orthoptera	4.00	15

SOURCE: Odum et al. (1962).

TABLE 12.10   Summary of Annual Energy Utilization of Primary Consumers in Old-Field Communities in South Carolina (S.C.) and Southern Michigan (S.M.) (kcal/m^2 per year).

	Ingestion		Assimilation		Secondary Production	
	S.C.	S.M.	S.C.	S.M.	S.C.	S.M.
Sparrows	4.0	2.6	3.6	2.3	0.04	0.05
Mice	7.4	1.1	6.7	0.6	0.12	0.01
Orthoptera	76.9	3.7	25.6	1.4	4.00	0.51
Spittlebugs	—	1.5	—	0.9	—	0.08
Other herbivorous insects	7.7	0.7	2.6	0.3	0.40	0.10

SOURCE: Weigert and Evans (1967).

gest more than the mice in the same field. Golley and Gentry (1964) report that the seed-eating harvester ant, *Pogonomyrmex badius*, consumed more energy than did either sparrows or mice in the South Carolina old field. Assimilation efficiencies (Table 12.1) vary considerably, but the vertebrates are very efficient in assimilating their food. Well over 50% is usually assimilated, but seeds are a richer source of nutrients than are leaves. Insects are much less efficient in assimilation, as less than 50% of the ingested material is assimilated (except in the sedentary spittlebug). Even so, the unassimilated plant material that passes out as feces is considerable and is important to the decomposer-based trophic system. During an insect infestation frass becomes an important litter component. Cow dung may seem prominent in a meadow, but insect frass may be much more abundant! When secondary production is considered, insects are again seen to be a very important component of the community. In every case they produce more than vertebrates in the same area. For most people it is unexpected to find that spittlebugs in a field contribute more to secondary production than do sparrows or mice. When the large numbers of mature Orthoptera in a field late in the growing season are considered, perhaps it is less surprising that these insects contribute more than the vertebrates. For comparison, secondary production by elephants in Uganda was estimated to be 0.34 kcal/m^2 per year, less than that of Orthoptera in an old field in South Carolina and southern Michigan.

The great importance of insects in secondary production depends on their large populations and the efficiency with which they convert assimilated material into secondary production. Note the large differences between insects and vertebrates in production efficiency (see Table 12.1) given by Wiegert and Evans (1967) and Smith (1972) (Table 12.11). For the fairly sedentary mirid bug, *Leptoterna dolabrata*, efficiency ranged from 50 to 58% over a 5-year period (McNeill, 1971). Waldbauer (1968) has reviewed much literature on insect metabolism and efficiency. The reason for the extremely low value for

TABLE 12.11   Production/Assimilation Efficiencies of Insects and Mammals.

	Efficiency (%)	
Hemimetabolous insects		
Salt marsh grasshoppers	36.73	
Old-field grasshoppers	36.22	
Alfalfa field spittlebugs	41.45	

	Laboratory (%)	Field (%)
Holometabolous insects		
Lepidoptera		
Operophtera brumata	40.5	59
Hydriomena furcata	41.8	47
Erannis spp.	43.3	64
Cosmia trapezina	46.2	56
Harvester ant	0.7	
Mammals		
Old-field mouse	1.79	
Meadow vole	2.95	
Uganda cob	1.46	
African elephant	1.46	

SOURCE: Wiegert and Evans (1967) and Smith (1972).

harvester ants is not clear at present, but an investigation of the literature on biology and methods of study may provide some clues. It seems that the harvester ant converted most assimilated energy into maintenance and little into net production (Engelmann, 1966; see also Hadley, 1972).

The importance of aphids in community energetics is not generally realized, although the significance of honeydew as a food for numerous animals and fungi is better understood. The large number of specialized feeders on aphids—lacewings, syrphids, geocorids, braconids, chalcids, cynipids, and fungi—rival in diversity the many insects that feed on honeydew. Llewellyn (1972) studied the lime aphid, *Eucallipterus tiliae,* which feeds on lime trees in England. He found that these aphids consume 3672 kcal/m^2 per year. This rate is impressive when compared to that of other organisms. Beef cattle at 0.18/ha consume 730 kcal/m^2 per year, and oak tree caterpillars consume 154 kcal/m^2 per year. Of this energy consumed by aphids, only 5% went into production, but 90% was excreted as honeydew. In this light it is hardly surprising that honeydew is utilized by so many organisms and that ants domesticate them. Energy turnover equivalent to standing crop was 482 times per annum and the energy drain on a tree was 28,055 kcal/year.

Aphids do not destroy the photosynthetic machinery of the plant. They can therefore extract much larger amounts of energy per square meter of vegetation than do conventional grazers, without destroying their food source. Indeed, aphids may even increase fixation of carbon in the host plant by removing accumulated nutrients that if not utilized in this way, would depress the rate of photosynthesis (Way and Cammell, 1970). This parasitic habit is a great advantage to the Hemiptera and Homoptera and should stimulate more attention to this group in energetic studies and community organization in general (see also Wiegert, 1964a). In fairness to leaf cutter ants it must be stated that they may also increase production by their activities. Although a nest may reduce gross production of forest in Costa Rica by 1.76 kcal/m^2 per day, activity accelerated net production by at least 1.80 kcal/m^2 per day by returning material rich in nutrients to the forest floor (Lugo et al., 1973).

HERBIVORE IMPACT ON PLANTS

The impact that an insect population has on the primary producers depends on the type of damage they cause. If insects feed on leaves, then a 5–20% consumption concentrated on these leaves is usually insignificant as many crops can withstand a 30–40% defoliation without serious loss in harvest. However, if the 5–20% of plant production is taken by insects in the form of flowers and/or seeds, this is likely to be catastrophic to the plant population. There are a great many data on the effect of insects on crop production, or, conversely, how crop production increases with the application of insecticide. Unfortunately, this method of assessing the impact of insects has rarely been used to look at natural situations. A brief look at agricultural experiment station reports, such as that from the Rothamsted Experiment Station Annual Report for 1970 (C. G. Johnson, 1971), indicates what should be expected in agricultural situations. In the second year of insecticide treatment of an old pasture the production was 30% greater than in unsprayed areas. In newly sown rye grass in the second year of treatment, production was increased 25%. The production of a field bean crop increased from 1.07 metric tons/hectare to 1.61 metric tons/hectare when sprayed against bean aphids, an increase in production of 50%. When spray was applied to the roots of field beans, to kill the weevil, *Sitona* sp., production increased from 0.63 metric ton/hectare to 0.82 metric ton/hectare, a 30% improvement. Thus insects can exert a tremendous pressure on their plant food populations, particularly in agricultural situations.

In natural situations the calculation of herbivore impact is more difficult, as primary production and secondary production must be measured and, by detailed studies on consumption, assimilation, and production efficiencies of herbivores in the laboratory, the consumption of herbivores in the field can be estimated. Therefore, impact is usually underestimated in these situations, as there is no estimate of production of primary producers in the absence of herbivores. Some examples will aid in comparing agricultural and natural situations. Stockner (1971) found that the aquatic stratiomyid fly, *Hedriodiscus truquii,* the dominant herbivore on the algal mat that grows in thermal springs in Mount Rainier National Park, consumed only 0.5–1.0% of the primary production. Grasshoppers ingested 0.5% of net primary production in old-field vegetation but 2.5% in an alfalfa field (Wiegert, 1965). McNeill (1971) studied the mirid bug, *Leptoterna dolabrata,* in an unmown grassland and found that the resource utilization by this herbivore was only 0.06–0.23% of the primary production. Impact on the plants was increased by damage to plant tissue by the piercing mouthparts and the injected toxins from the salivary secretions. Andraejewska (1967) calculated that plant losses were 26–36% greater than plant consumption by the cicadellid, *Cicadella viridis,* in Poland. But even then the impact is very low compared to the agricultural situation. However, spittlebugs in alfalfa fields were calculated by Wiegert (1964b) to reduce photosynthetic fixation of energy by five times the total energy ingested by the insects. (Ingestion was 38.6 kcal/m^2 per year; cf. the value in Table 12.10 for spittlebugs in an old field.) This highly significant reduction, both for plant survival and crop yield, was due to removal of amino acids from the host plant.

The differences between the agricultural and natural settings raise the questions:

1. Why don't herbivores in natural environments eat more of the available resources?
2. What makes agricultural crops particularly vulnerable to herbivore pressure?
3. How do agricultural crops differ from natural communities?

They may be resolved by an understanding of plant–herbivore relationships (Chapter 5), community organization (Chapter 22), and diversity and stability (Chapter 23).

Another example that illustrates the importance of insects, and in this case arthropods and other invertebrates, involves the breakdown of leaf litter and the subsequent release of nutrients in forests. As already mentioned, Wiegert and Owen (1971) consider the saprophage-based food chains relatively important in forest ecosystems. For example, in an oak–pine forest, 2650 dry g/m^2 per year was fixed as gross primary production, and 1200 dry g/m^2 per year remained as net primary production. Of this, 360 dry g/m^2 per year resulted in litter available to decomposers, while only 30 dry g/m^2 per year was utilized by herbivores (Woodwell, 1970). This illustrates the greater importance of decomposers than of herbivores in temperate forest ecosystems if only energy is considered. This contrasts significantly with aquatic, grassland, and agricultural systems, where a much larger percentage of net primary production is utilized by herbivores. A comparison of many ecosystems has been made by Petrusewicz and Grodzinski (1975), where in all 13 cases from grassland and agricultural ecosystems consumption by herbivores is much higher than in any of the 16 forest ecosystems (Fig. 12.7).

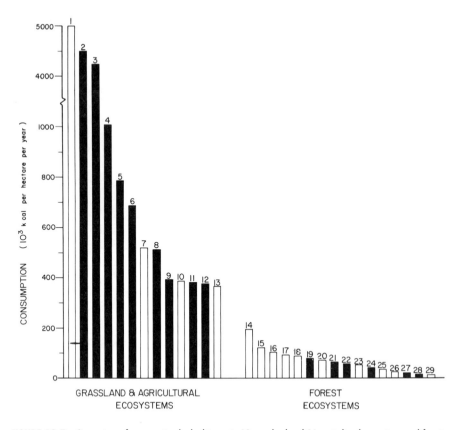

FIGURE 12.7   Comparison of consumption by herbivores in 13 grassland and 16 agricultural ecosystems and forest ecosystems. Insect examples are in black. After Petrusewicz and Grodzinski (1975). From Price (1984c); with permission.

Comparisons of herbivory between terrestrial and aquatic systems show large differences (Fig. 12.8) (Cyr and Pace, 1993). Almost 80% of primary production by aquatic algae is ingested by herbivores on average, and 30% of macrophytes. By contrast, a mean of 18% of aboveground terrestrial plants is ingested according to data reviewed by McNaughton et al. (1989, 1991). Such large differences have important implications for energy flow in ecosystems. In aquatic systems recycling of limiting nutrients is more rapid, and more energy will move up the food chain to herbivores and less to decomposers. Thus we may expect higher herbivore productivity and predator productivity in aquatic systems. In addition, the evolution of defenses against herbivores by algae in marine systems is likely to be well developed (cf. Hay et al., 1987; Hay and Fenical, 1988).

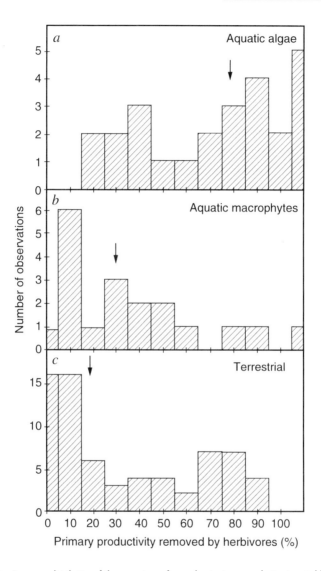

FIGURE 12.8 Frequency distribution of the percentage of annual net primary production ingested by herbivores in 44 aquatic sites for (*a*) aquatic algae, including phytoplankton and reef periphyton and (*b*) submerged and emergent macrophytes. (*c*) Herbivory in terrestrial systems with data from McNaughton et al. (1989, 1991). From Cyr and Pace (1993). Reprinted with permission from *Nature* **361:**148–150. Copyright 1993 by Macmillan Magazines Limited.

ROLE OF ARTHROPODS IN ENERGY FLOW: SAPROPHAGES

The role of arthropods in litter breakdown may be studied by placing fallen leaves in mesh bags, and after a period of time, extracting the arthropods and measuring the loss of litter weight (e.g., Crossley and Hoglund, 1962) or area (e.g., Edwards and Heath, 1963). Other methods are described by Edwards et al. (1970), who also provide much of the information discussed below. Collembola and oribatid mites are usually the most abundant arthropods in litter, although diplopods contribute significantly to decomposer metabolism. Nematodes and oligochaet worms are as important as these arthropods in litter breakdown. All these taxa are more abundant in less acid mull organic matter of oak litter than in mor derived from beech and coniferous litter.

Soil animals increase the speed of litter breakdown by disintegrating tissue, increasing surface area, changing the physical and chemical nature of the litter in other ways, and mixing it with inorganic matter (Edwards et al., 1970). All activities increase availability to bacteria and fungi. Much of the litter ingested by animals is egested as feces. In some soils most of the plant remains consist of feces. Larvae of a single species of terrestrial caddis fly, *Enoicyla pusilla*, consumed from 4.5 to 19% of litter produced in a stand of coppiced oak (Van der Drift and Witkamp, 1960). Of the material ingested, 7% was assimilated and 93% egested. The fecal pellets had a larger exposed surface, better aeration and water-holding capacity, and a higher pH than did the original litter. Microbial activity in the feces was much higher than in whole leaves, and on average was slightly higher than in mechanically ground leaves. Using litter bags of different sizes, Edwards and Heath (1963) showed that in 9 months 60% of leaf litter area was removed by earthworms; mites, collembolans, enchytraeid worms, and other small invertebrates removed an additional 30% of the leaf area, and when these were excluded, no visible breakdown of leaves occurred. Microarthropods have also been shown to be very important in desert ecosystems (Santos and Whitford, 1981; Santos et al., 1981; Zak and Whitford, 1988; Zak and Freckman, 1991). Witkamp (1971) has stated that litter breakdown, or fragmentation, is due to the litter fauna, whereas litter decomposition, or chemical deterioration, occurs through microbial action. As Edwards et al. (1970) conclude, soil invertebrates contribute significantly to leaf degradation, although total energy flow may be small compared to that through bacteria and fungi. Wallwork (1970) should be consulted for the role of specific taxa in the soil system, and Dickenson and Pugh (1974) review decomposition of various kinds of litter and the organisms involved. The growing emphasis on microbial ecology and complex interactions are captured in Edwards and Stinner (1988) and Gauge (1996).

The point made by Edwards et al. (1970) that soil invertebrates may contribute very little to total energy flow, although they are important in breaking down litter, suggests that factors other than energy must be studied to understand ecosystem function.

NUTRIENT LIMITATION

Although the acquisition and allocation of energy are important in ecology, energy need not be the most limiting factor in the food supply nor the most important organizing influence in the community. Nutrients are also essential to organisms, although many studies underemphasize their ecological importance. The limiting factor in the food of aphids is not energy but the level of amino nitrogen (Dixon, 1970), and aphids may excrete 90% of the energy they obtain from the plant host to obtain adequate nutrients (Llewellyn, 1972). Much of the energy in a plant may not be available to herbivores because microclimate is unsuitable: in the case of the sycamore aphids, temperatures in the upper canopy become too high in the heat of the summer, and leaves brushed by others in a breeze remain unoccupied (e.g., Dixon, 1970). Some parts cannot be reached because of plant and herbivore morphology (Way and Cammell, 1970) or because of chemical protection (e.g., Feeny, 1970), as discussed in Chapter 5. Chew (1974) has argued that although some animals may consume insignificant amounts of the primary production, their roles in the community are nevertheless important. Especially for herbivores, the nutrient status of the food is at least as important as its energy content (see also Paine, 1971; and Chapter 17). Carnivores and their prey have nutrient requirements that are more nearly similar, and carnivores tend to feed on a more diverse array of foods, possibly making the study of energy relationships in these species more valuable.

Chew (1974) notes that soil invertebrates are important as conditioners of litter for microbial decomposition, and as grazers of fungi which maintain vigor of fungul metabolism and, hence, rapid litter decay. Santos and Whitford (1981) and Santos et al. (1981) add to this list of important roles of soil arthropods based on their studies in the Chihuahuan desert. They argue that the microarthropods inoculate litter with fungal spores, thereby hastening decomposition; they graze on fungi; and predatory mites prey on nematodes feeding on bacteria and maintain a higher population of bacteria involved in decomposition.

INSECTS AS REGULATORS

Herbivores can also result in significant increases in plant productivity by prolonging plant growth and delaying plant senescence (Chew, 1974; Dyer and Bokhari, 1976; McNaughton, 1976, 1979). They can influence the diversity of plants in a community (Chew, 1974) and the quality of individuals in the community.

These kinds of realization have led to a major shift in thinking about insects in ecosystems from insects as consumers to insects as regulators. Mattson and Addy (1975) argue this point by noting that as host-plant conditions change, insects respond, providing feedback to host plants and other components in the ecosystem. As an example, they use the case of severe defoliation by insects in temperate forest trees, such as by the forest tent caterpillar, *Malacosoma disstria,* on aspen, *Populus tremuloides,* or by eastern spruce budworm, *Choristoneura fumiferana,* on balsam fir, *Abies balsamea.* Such defoliations have the following effects:

1. They change the host's physiological status.
2. They cause increased litter fall, including leaves, twigs, and branches, as well as frass and insect carcasses.
3. More nutrients reach the soil litter system through leaching from trees.
4. Many trees die in the classes of weakened, old, or suppressed trees, and this changes the distribution of light, heat, nutrients, and other abiotic factors for the survivors, enhancing their growth.
5. Activity of soil mircoorganisms is stimulated and numbers increase, accelerating nutrient cycling in the system. The insects therefore play an important role in maintaining relatively high primary production and nutrient cycling.

From this perspective Mattson and Addy (1975) mapped the relationship between the forest tent caterpillar and aspen in terms of all the possible relationships, from being mutualists (+ +) to having a negative impact on each other (− −) (see Table 10.1). They used the periodic coordinate system, where the coordinates 0°, 90°, 180°, and 270° represent interaction types + 0, 0+, − 0, and 0 −, respectively, and the angles midway between these coordinates represent + +, − +, − −, and + − interactions, respectively. Where there is no impact (0, 0) the relationship is recorded on the reference zero (RO) circle, while positive relationships become progressively distant from the circle outward, and negative interactions move inward as the strengths of interaction increase (Fig. 12.9). Total production with and without insects was used to evaluate the type of interaction involved. The interesting result of this analysis is that the relationship changes from one of commensalism in stands 26 and 27 years of age to one of parasitism at ages 28 and above, and the impact of the herbivores varies quite dramatically from year to year.

In a similar analysis, using spruce budworm on balsam fir (Fig. 12.10), even more significant changes take place, from commensalism (0, +) at ages 50–55, to parasitism (+, −) at ages 55–70, and to the opposite form of commensalism (+, 0) at ages 70 years and above. Ultimately, these older trees are killed, young vigorous trees dominate the forest, and the relationship between host trees and herbivores returns to the 90° position.

Mattson and Addy (1975) conclude that each forest system supports insects that vary in impact with seasonal and developmental changes in the plant. Some of these insect species respond dramatically to such changes, and in doing so act as regulators in the forest ecosystem. This phenomenon is one of the

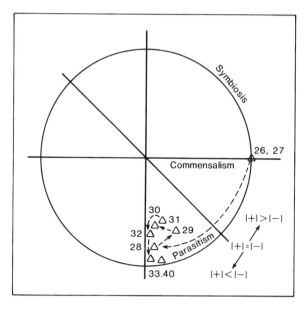

FIGURE 12.9   Relationships between forest tent caterpillar and aspen trees, plotted on the periodic coordinate system. The circle RO is reference zero when total production of insects plus vegetation is equal whether insects are present or absent. The open triangles present the position of the insect–plant relationship and the numbers show the age (in years) of the aspen stand. Dashed lines and arrows indicate the direction of movement of the relationship in the coordinate system. From W. J. Mattson and N. D. Addy, Phytophagous insects as regulators of forest primary production, *Science*, 1975, **190**:515–522. Copyright © 1975 by the American Association for the Advancement of Science. Reprinted with permission.

major kinds of perturbation in ecosystems, discussed by Loucks (1970), that influence plant succession, species diversity, and primary productivity.

Peterman (1978) adopts a view similar to that expressed by Mattson and Addy, arguing that the mountain pine beetle, *Dendroctonus ponderosae*, acts as a "manager" of lodgepole pine, *Pinus contorta* var. *latifolia*, rather than a pest of lodgepole pine. The beetle infestations naturally thin and "harvest" in stands of pine, creating fuel for fires that are essential in the reproduction of the tree, since seeds are released from the serotinous cones only after extreme heat from a fire. Such thinning also decreases the probability of very high density stands in the next generation, in which growth stagnates, because the seed bank does not increase long enough to produce high densities of seedlings. Peterman suggests that the age at which trees become less resistant to *Dendroctonus* may well have evolved in response to selection favoring a high probability of fire and consequent reproduction when the parental seed crop is adequate for effective regeneration, but before the crop would lead to

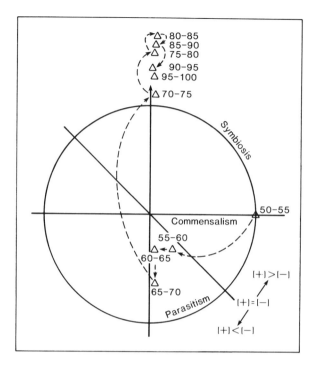

FIGURE 12.10   Relationships between spruce budworm and balsam fir plotted on the periodic coordinate system as in Fig. 12.9. From W. J. Mattson and N. D. Addy, Phytophagous insects as regulators of forest primary production, *Science*, 1975, **190:**515–522. Copyright © 1975 by the American Association for the Advancement of Science. Reprinted with permission.

intensive competition among its progeny. Progeny growing in open stands grow large crowns with high seed production. In addition, such "management" would provide more breeding sites for the beetle in the long run since stagnant stands of lodgepole pine have trees too small, with phloem too thin, to support beetle populations (see also Amman, 1977).

This concept, that consumers act as important regulators, is now well founded in the literature (Lee and Inman, 1975; O'Neill, 1976; Kitchell et al.; 1979; O'Neill and Reichle, 1980). Studies show that although energy movement may be small, nutrient movement may be very significant. Carlisle et al. (1966) found that small nonleafy materials such as insect frass, male flowers, and bud scales accounted for only 15% of the dry weight of the litter, but they contained 30% of the nitrogen, 40% of the phosphorus, and 25% of the potassium. They point out that most of these materials fall in the spring and early summer when soil microorganism activity is at a peak, thereby profoundly influencing the nutrient cycling in the forest. By causing damage to leaves, herbivores also promote leaching of nutrients from these leaves (Kimmins, 1972). We cannot judge adequately the role and importance of any organisms in an ecosystem by evaluating its role in energy flow. Pollinators move very small amounts of energy, relatively speaking, but their importance in many systems is obvious.

An important warning, issued by O'Neill (1976, p. 1244), is that "heterotrophs in the ecosystem may be capable of exerting effective control by changes in biomass that would be difficult to detect in field sampling." His analysis based on simple ecosystem design (Fig. 12.11) shows that only small changes in heterotroph biomass could reestablish ecosystem equilibrium after a perturbation. He noted that in a comparison of systems the greater the potential for rapid change, the larger was the standing crop of heterotroph regulators.

Cummins' (1973, 1974) conceptual model of stream ecosystem structure also illustrates the potential in this system of relatively small changes in one compartment having significant effects on many other compartments (Fig. 12.12). Some processes, such as leaching of nutrients from leaves falling into a stream and the colonization of coarse particulate organic matter such as twigs and leaves by microorganisms, occur very rapidly and profoundly influence the shredder compartment and ultimately the predators in the system. The importance of linkage between ecosystems is also illustrated, for the quantity and quality of leaves falling into a stream is defined by characteristics of the terrestrial system and the amount of leaching of leaves in that system. This understanding is especially critical where water quality is such a major issue in environmental biology (Cummins, 1974; Likens and Bormann, 1974).

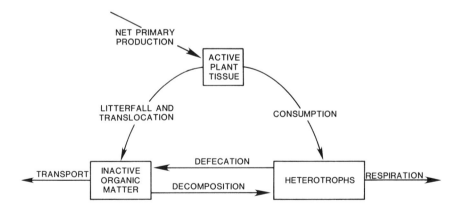

FIGURE 12.11 Simplified view of the ecosystem. The boxes indicate state variables and the arrows show transfers of energy between components of the system. From R. V. O'Neill, Ecosystem persistence and heterotrophic regulation, *Ecology*, 1976, **57**:1244–1253. Copyright © 1976 by the Ecological Society of America. Reprinted with permission.

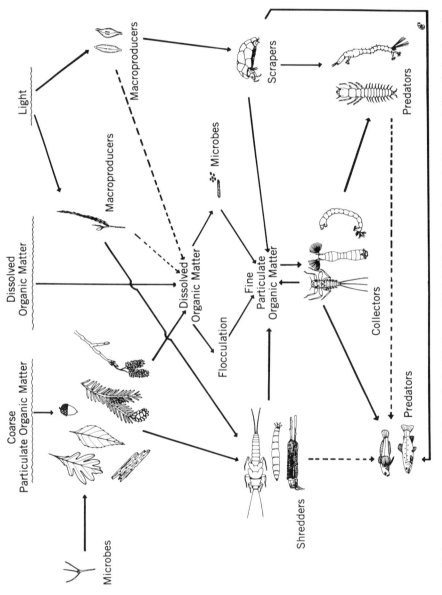

FIGURE 12.12 Model of stream ecosystem structure and function developed by Cummins (1973, 1974), illustrating the considerable potential effects of relatively small changes in quantity or quality of inputs to the aquatic system. Shredding insects include cranefly and caddisfly larvae and stonefly nymphs. Collecting insects include blackfly and midge larvae and mayfly nymphs. Scrapers include caddisfly larvae. Predators include both fish, sculpin and trout, and insects, fishflies and midge larvae. From K. W. Cummins, Structure and functions of stream ecosystems, *BioScience*, 1974, 24:631–641. Copyright © 1974 by the American Institute of Biological Sciences. Reprinted with permission.

Some books on ecosystem function that will provide further details on this subject include Wiegert (1976) on ecological energetics, Pomeroy (1974) on nutrient cycling, Lieth (1978) on primary production, Shugart and O'Neill (1979) on systems analysis, and Mattson (1977) on the role of arthropods in forest ecosystems. Through the 1980s and 1990s, excellent volumes in the "Ecosystems of the world" series have been published. The most recent volume is *River and stream ecosystems* (Cushing et al., 1995). Other recent volumes include the *Comparative analyses of ecosystems* (Cole et al., 1991), and *Linking species and ecosystems* Jones and Lawton (1995).

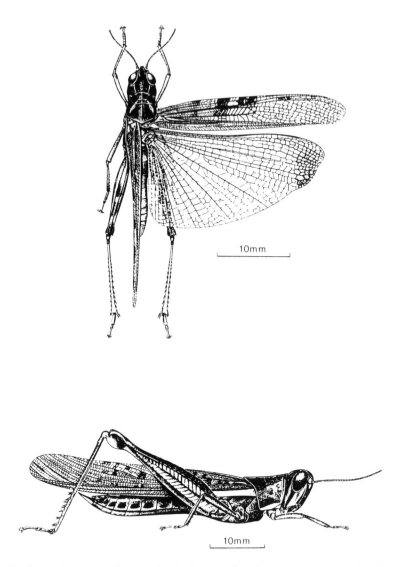

Two males of plague locust species from Australia, the Australian plague locust, *Chortiocetes terminifera* (Acridinae) (above), and the spur-throated locust, *Nomacridis guttulosa* (Cyrtacanthacridinae) (below). From Naumann (1991). Drawing by F. Nanninga.

III

Populations

Many qualities of individuals and interactions between them discussed in Part II are expressed ultimately as population phenomena. In addition, the quality and quantity of progeny a female produces and the timing of reproduction influence demography (Chapter 13) and life history characteristics (Chapter 14). The genetic constitution of individuals and populations (Chapter 16) and relative fitness under variable conditions affect demography, life history, and dynamics and persistence of populations. Foraging for food and finding mates are essential activities for survival and reproduction (Chapter 15). Factors discussed in Parts II and III contribute to the population dynamics of the species (Chapter 17) and attempts at modeling such population change (Chapter 18). An effort to synthesize several aspects of population biology is made in Chapter 19.

Population distribution, abundance, and persistence also depend on the quantity and location of resources available for exploitation and the manner in which individuals integrate into the trophic system. The relative availability of these attributes for coexisting species determines population abundance and the dominance and diversity characteristics of communities (Part IV). Ultimately, the genetic constitution and resultant persistence, size, and distribution of populations influence speciation and the geographic patterns seen in populations and species (Part IV).

Chapter 13

Demography: Population Growth and Life Tables

Demography is the quantitative analysis of characteristics of populations, particularly in relation to patterns of population growth, survivorship, and movement. The subject is therefore closely allied to population dynamics (Chapters 17 to 19) but the emphases have been rather different, with demography concentrating on the patterns of growth, life, death, and movement, while the causes and results of these phenomena are emphasized in population dynamics.

HUMAN POPULATION

Malthus in his famous essay of 1798 was explicit about the nature of the will in the human population for indefinite exponential growth: "I think I may fairly make two postulata. First, that food is necessary to the existence of man. Secondly, that the passion between the sexes is necessary and will remain nearly in its present state." Malthus pointed out that a population can increase in a geometric ratio, and subsistence increases only in an arithmetic ratio; thus population is bound to outgrow food supply. So Malthus

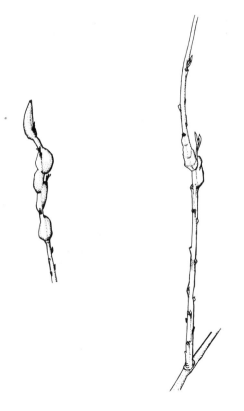

Galls of the sawfly, *Euura lasiolepis,* on the arroyo willow, *Salix lasiolepis,* from the western United States. From Price et al. (1995b). Drawing by Mary Bayless.

concludes: "and that the superior power of population cannot be checked without producing misery or vice, the ample portion of these too bitter ingredients in the cup of human life and the continuance of the physical causes that seem to have produced them bear too convincing a testimony."

EXPONENTIAL GROWTH

Geometric growth, or exponential growth, when a population increases by a constant factor per unit time, can be expressed by the equation

$$\frac{dN}{dt} = rN$$

where N is the number in a population and r is the rate of change per individual, or the instantaneous rate of population increase. This very rapid growth (Fig. 13.1a) is usually checked by factors that cause mortality—predators, parasites, severe weather, and aggression—or reduced fecundity, and ultimately by food shortage if all else fails. In a given environment a certain

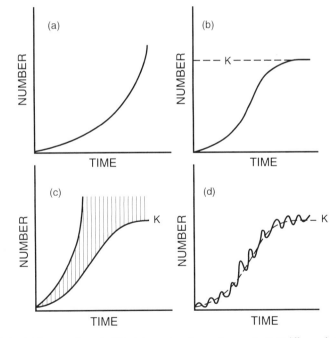

FIGURE 13.1   (a) Exponential growth; (b) logistic growth to the carrying capacity K; (c) difference between exponential growth and logistic growth caused by one environmental factor, competition for the resource that cannot support more than K individuals; (d) irregular growth of an insect population toward the carrying capacity.

amount of food is available to a population. If the population exploits less than this amount of food, it can continue to increase; if it exploits more than this amount of food, it is bound to decline; and if it eats precisely that amount of food, a steady state will be reached.

LOGISTIC GROWTH

Verhulst recognized in 1838 that there had to be a maximum amount of food available and therefore a limit to the human population. The closer the population came to this limit, the slower would its increase be, until there was no increase at all. The rate of population increase, r, must therefore be modified by a factor that decreases growth rate with increasing population density until a maximum population is reached that can be sustained by the resources. The maximum has been called the **carrying capacity,** K, and the factor can be written $(K - N)/K$, so the population growth formula becomes

$$\frac{dN}{dt} = rN \frac{K - N}{K}$$

which describes a **sigmoidal** or **logistic curve** (Fig. 13.1b). When $N = K$, $dN/dt = 0$, so unless K is depleted, the population levels off at the carrying capacity and a steady state is maintained. As N approaches K, $(K - N)/K$ will decrease and the rate of population growth will decline.

Other factors besides shortage of food are likely to be limiting, and these will reduce the maximum population below the carrying capacity determined by food availability alone. Thus the difference between the exponential growth and logistic growth represents the interaction between the **biotic potential** of the population and the food supply (Fig. 13.1c). Some authors have stated that the difference between these curves defines the **environmental resistance** or the effect of all factors that limit the biotic potential. But clearly the difference is the result of only one type of environmental factor—competition—either for food, space, or other resources in limited supply. Many other factors may reduce population growth and keep populations well below the carrying capacity and these are all a part of *environmental resistance.* This term and *biotic potential* were first used by Chapman in 1928. Since complete destruction of a food supply is rarely seen in nature, it seems that environmental resistance usually keeps the population below the carrying capacity. Pearl and Reed (1920) developed the same equation as Verhulst independently so the logistic growth model is frequently referred to as the Verhulst–Pearl equation.

As might be expected, there are several limitations to the logistic growth model because of the following assumptions:

1. Individuals are all equal in their reproductive potential. Clearly, immatures contribute nothing and matures will vary considerably in their productivity. The logistic model assumes an even age distribution so that the same proportion of individuals in a population are breeding all the time.
2. Reproduction is constant irrespective of climate and other variations.
3. Responses are instantaneous and there is no allowance for time lags.
4. The carrying capacity, K, is constant; K cannot be reduced by excessive feeding or changing influences on the trophic level below.
5. Environmental resistance through competition is a linear function of density.

SEASONAL REPRODUCTION

These assumptions may be acceptable for laboratory populations of insects under constant conditions, such as *Tribolium* flour beetles, where the generation time is also short and the food resources are maintained. In natural insect populations there is usually (1) a distinct breeding season—a season of population increase; (2) a feeding period during which the population decreases due to predation, disease, and such; and (3) unfavorable climatic conditions that interrupt activity and during which further mortality occurs. The very closest approximation to the Verhulst–Pearl model under these circumstances is an irregular climb in numbers to the carrying capacity (Fig. 13.1d). Thus the model is useful in formulating ideas about population regulation and stability but is not very helpful in predicting population change. The model has stimulated a great deal of research, but as is usual with the simplest and therefore most helpful models, this deterministic model has failed to describe many real situations. Mertz (1970) gives an excellent account of population growth and the use of models.

The exponential and logistic equations describe continuous population increase. However, population increase must be described in a different way when reproduction is seasonal, as it is in the great majority of insects and plants in temperate and desert conditions, and frequently also in the tropics. If the initial population size is N_0, and exactly 1 year, or one generation, later each individual has died and has been replaced on the average by λ offspring, the population size (N_1) after one season will be

$$N_1 = \lambda N_0$$

In the second (N_2) and subsequent $N_3, \ldots, N_t$) seasons it will be

$$N_2 = \lambda N_1 = \lambda(\lambda N_0) = \lambda^2 N_0$$
$$N_3 = \lambda^3 N_0$$
$$N_t = \lambda^t N_0$$

Geometric growth is again evident and if the population at any particular time in each generation is plotted, for example, the population at the start of each generation, an exponential growth curve results. If the entire population is figured through several generations, the pattern shown in Fig. 13.2 emerges. For seasonal breeders such as plants and insects, this growth model is more helpful. For example, even with a rate of increase of 2, there are large differences between population levels at N_3 and N_4. Considering that the fecundity of insects is frequently 100–200 eggs, these differences will be exaggerated, and it can be understood why in 1 year an insect species is quite inconspicuous and, apparently "all of a sudden," the following year, it is in epidemic proportions, doing considerable damage. No other explanation than this model for seasonal breeders needs to be invoked, although there may be many other factors that come into play.

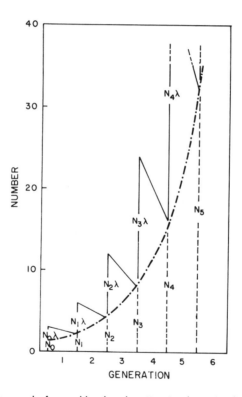

FIGURE 13.2 Population growth of seasonal breeders where $N_0 = 1$ and $\lambda = 2$ and all breeders die before the next breeding season. This model is therefore applicable to many insect populations. Note that if all points at a certain time in each generation are joined, they demonstrate exponential growth (see dashed-and-dotted line joining successive generations prior to breeding).

OVERLAPPING GENERATIONS

The description of population growth where generations overlap is more complex. We must know (1) how many offspring an average female will produce at each age interval in her life and (2) the number of individuals that are present at each age interval or in each age class. To do this the approach is simplified by following only the females in a population. When the female population size and the sex ratio are known, it is simple to calculate the male and total populations. Other assumptions remain the same as in the model for seasonal breeders without overlapping generations:

1. Each season the population acts in exactly the same way—no evolutionary or environmental changes occur.
2. Population birth and death rates do not vary in response to crowding.

LIFE TABLES

With overlapping generations the effects of aging on birth and death rates must be known. The life span of an insect cohort may be 5 months, which can be divided equally into 1-month periods. If x gives the age of individuals at the beginning of each period, and for each age the proportion of females that have survived to that age is designated as l_x—the **age-specific survivorship**—the statistics of the cohort can be tabulated as in Table 13.1.

Only at certain ages will females be capable of breeding and producing progeny. Much of the time they are immature or too old. By direct observation an estimate can be made of the number of daughters a female can be expected to leave each month, m_x—the **expected number of daughters** that will be produced at age x by a female who is still alive at age x—and each month can be considered as a breeding season (Table 13.2). The total of the m_x column gives the **gross reproductive rate**, that is, the expected total number of

TABLE 13.1   Age-specific Survivorship of a Cohort of Insects in a Hypothetical Population.

Age at Beginning of Interval, x	Probability at Birth of a Female Being Alive at Age x, l_x
0	1.0
1	0.8
2	0.5
3	0.3
4	0.1
5	0

TABLE 13.2   Hypothetical Life Table for an Insect Population in Which Maxiumum Longevity is 5 Months. R_0 = Net Replacement Rate.

Age at Beginning of Interval, x	Age Specific Survivorship, l_x	Expected Daughters, m_x	Reproductive Expectation, $l_x m_x$
0	1.0	0	0
1	0.8	0	0
2	0.5	3	1.5
3	0.3	2	0.6
4	0.1	0	0
5	0	0	0
	Gross reproductive rate = $\overline{5}$		$R_0 = \overline{2.1}$

births produced by a female who lives through all age groups. But population growth depends on the number of surviving females and their individual production of progeny, or the product of l_x and m_x. This gives the **reproductive expectation** of a female at age x for a female entering the population (Table 13.2). The total progeny left by an average female in the population in this example will be 2.1, designated R_0, the **net replacement rate,** defined as the number of daughters that replace an average female in the course of a generation. One female on average produced 2.1 females. Therefore, there was a population increase. A stable population would have an $R_0 = 1$. The simplest way to obtain this information is to select a group of newly born organisms and follow the life of this cohort until the last individual dies. However, population change can also be calculated by knowing the age distribution of a total population (see Fig. 13.3) and the parameters for each age interval as in Table 13.2. This is the method used for calculating human population growth and other populations that can be adequately censussed (see Poole, 1974, for further discussion).

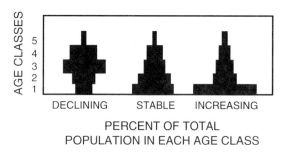

FIGURE 13.3   Age pyramids indicating the idealized pattern for declining, stable, and increasing populations. Note the differences between relative numbers of young in each population.

Once R_0 is known, the **instantaneous rate of population increase**, r, can be calculated by the approximation

$$r = \frac{\log_e R_0}{T}$$

or the number of progeny produced per unit time. The **mean generation time**, T, is the mean of the period over which progeny are produced and is estimated by the formula

$$T = \frac{\Sigma \, l_x m_x x}{\Sigma \, l_x m_x}$$

Birch (1948) provides a more accurate calculation of r.

Since R_0 is the net replacement rate per generation it is not a useful statistic for comparison between species, as generation times vary tremendously. Thus r is a much more useful statistic for comparing growth rates of populations of different species, or growth rates of populations of the same species under different environmental conditions. Under optimal conditions r is designated r_m or r_{max}, the **innate capacity for increase**, or the rate of increase in a population growing under optimal conditions (i.e., a population expressing its biotic potential). This value can be useful in comparing species strategies in reproduction. A very high r_m may mean that under natural conditions the populations suffer a high mortality (Smith, 1954). A low r_m may indicate low mortality in nature and would lead to a search for adaptive methods for avoiding mortality.

Because of each interval of time new members are added and present members disappear by death, natural populations with overlapping generations are a complex of individuals representing all possible age groups. It is therefore essential to know the age distribution in the population. This is normally presented as an **age pyramid,** and by the shape of the pyramid it is possible to identify young, stable, and senescent populations (Fig. 13.3).

Numerical changes in a population can be described completely by knowing the birth rate, the death rate, and the migration rate in that population. **Life tables** provide a way to tabulate births and deaths, and entomologists modified life tables so that migration could also be included. Thus the life table is really a summary statement on the life of a typical individual of a population or a cohort of individuals. From these data the expected life remaining to an individual can be calculated, whatever its age, and included in the table. This is the substance of an actuary's professional diet. Deevey (1947) states that a table showing the expectation of life at birth, at age 20, and at 5-year intervals after that was in use by the Romans in the third century. Life insurance rates depend on life expectation in the population, and by insuring many people a company can calculate its commitments from the **expectation of life** column, e_x, in a life table, which tabulates the mean life remaining to those attaining a certain age interval.

From the study of human populations we obtained a valuable method for looking at animal populations. The pioneers in this change were Pearl and Parker (1921); Pearl and Miner, who wrote their paper "The comparative mortality of certain lower organisms" in 1935; and Deevey in 1947. Deevey's paper provides many examples of the life table format and a method for the calculation of e_x.

To calculate e_x three steps are required. Step 1 calculates the mean number of individuals alive in each age interval. Since l_x is the number of females alive at the beginning of an age interval and l_{x+1} is the number alive in the next interval, the mean number present during age interval x will be $\frac{1}{2}(l_x + l_{x+1})$ and is designated L_x. Step 2 involves summing all the L_x values from the bottom of the table upwards giving values of T_x, providing the cumulative sum of the product of individuals and time units. Step 3 divides T_x by the number of individuals present, l_x, to obtain the average expectation of life in each age interval:

$$e_x = \frac{T_x}{l_x}$$

Note that l_x is used in this calculation, not L_x. By designating the cohort at the beginning of age interval 0 to 1000 individuals instead of the probability of 1.0, as in Tables 13.1 and 13.2, these calculations result in life expectation values as given in Table 13.3.

Coats (1976) used these estimates in her study of the parasitoid *Muscidifurax zaraptor*, which attacks the house fly, *Musca domestica*, and she discussed various approaches to the calculation of life history statistics.

TABLE 13.3 Example of the Calculation of L_x and T_x and the Life Expectation, e_x, Based on Data in Tables 13.1 and 13.2.

x	l_x	L_x	T_x	e_x
0	1000	900	2200	2.2
1	800	650	1300	1.6
2	500	400	650	1.3
3	300	200	250	0.8
4	100	50	50	0.5
5	0	0	0	0

SURVIVORSHIP CURVES

If l_x gives the number in a cohort, or a proportion in that cohort, which survives to age x, these values can be plotted to obtain a **survivorship curve.** Lotka (1924) gave some survivorship curves for humans (Fig. 13.4). Although maximum longevity was hardly increased, the mean longevity was greatest for the U.S. population. This difference was due largely to infant mortality, but even in middle age the mortality was higher in England. Survivorship may also be very different at the same time in different parts of the world. For example, survivorship curves for the United States and India in 1910 show practically the same characteristics as Fig. 13.4, with India similar to the Northampton population in 1780.

As Lotka pointed out, it is probably more informative to plot the survivors, l_x, on a logarithmic scale. Then a straight line indicates a constant rate of mortality throughout life, or no greater probability of dying at one age than another. Thus variation in slope measures what Lotka called the

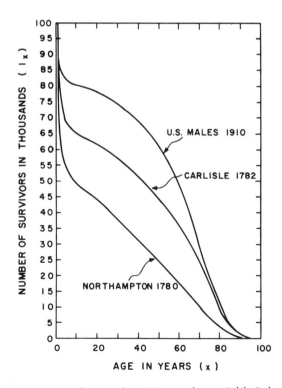

FIGURE 13.4 Survivorship curves for U.S. males in 1910; a population in Carlisle, England, in 1782; and another in Northampton, England, in 1780. Note that l_x is on an arithmetic scale. From Lotka (1924).

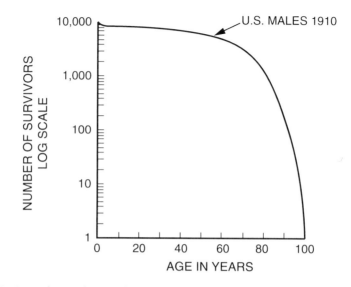

FIGURE 13.5   Survivorship curve for U.S. males in 1910 with l_x on a logarithmic scale. The slope of curve indicates directly the "force of mortality" at that age interval. From Lotka (1924).

force of mortality. The survivorship curve for males in the United States in 1910 is given in Fig. 13.5. At first the slope is very steep, it decreases at about the fourth year of life, when it reaches a minimum, and then increases continuously to the end of life. In fact, different organisms show different survivorship curve characteristics, and it is possible to determine the vulnerable stages in a life history by comparing survivorship curves. There seem to be three general shapes to the curve (Fig. 13.6). The type I curve shows a very low death rate of young and high mortality in old age. Seldom do natural populations actually show this type of curve exactly, but some are close to it, for example, the present human population, some *Drosophila* populations, the rotifer *Proales,* and Dall mountain sheep. The type II survivorship seems to apply to *Hydra* and many species of birds after the end of their first summer, although juvenile mortality must be high. Examples are the blackbird, song thrush, American robin, starling, and lapwing (Deevey, 1947, but see Botkin and Miller, 1974). The type III curve is less often observed, but if they were studied, many parasitic organisms would show this survivorship, where huge mortalities occur in early stages. Some insects show this pattern in less extreme form (Figs. 13.8 and 13.9). Other examples are some fish populations and organisms with pelagic eggs and larvae such as oysters. Many examples of survivorship curves show characteristics of two or more types combined during a life cycle, for example, the large milkweed bug, *Oncopeltus fasciatus,* and its tropical congener, *O. unifasciatellus* (Fig. 13.7).

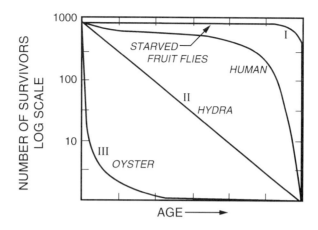

FIGURE 13.6 Generalized survivorship curves of types I, II, and III plus the curve for a human population represented by U.S. males in 1910. Reprinted from A. S. Boughey, *Ecology of populations,* by permission of Macmillan Publishing Co., Inc. Copyright © 1968 by Arthur S. Boughey.

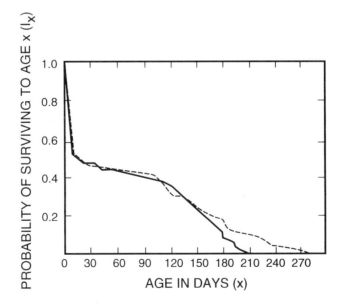

FIGURE 13.7 Survivorship curves for *Oncopeltus fasciatus* (solid line) and *O. unifasciatellus* (dashed line), showing lack of fit to any of the generalized trends. From Landahl and Root (1969).

The typical life table has the following columns (see Deevey, 1947):

x = age of cohort

d_x = number dying in age interval (the usual basic data from which the life table is constructed), the number in a cohort is usually 1000 individuals

l_x = number surviving at beginning of age interval (usually out of 1000 born)

$1000q_x$ = mortality rate (per thousand) of those alive at beginning of age interval, or d_x as a percentage of l_x

e_x = expectation of life, or mean lifetime remaining to those attaining age interval

LIFE TABLES FOR NATURAL INSECT POPULATIONS

Morris and Miller (1954) were the first to adapt the life table format for the study of natural insect populations. They were not so interested in the life expectancy of an individual as they were in the cause of mortality at a particular age interval. By quantifying these causes the population dynamics of the insect could be better understood. They studied the spruce budworm, *Choristoneura fumiferana,* in New Brunswick, Canada. Not only could births (i.e., eggs) be counted, and deaths due to parasites, predators, disease, and winter weather, but also dispersal to and from the site could be estimated. So all the important determinants of population size were quantified (see Table 13.4).

The age interval x was not divided up into equal lengths of time but into life stages: eggs, instar I, hibernacula, instar II, instars III–VI, pupae, and adults, and an additional column d_xF was added, the factor responsible for d_x. This d_xF column listed all the mortality factors that could be quantified, and the d_x and $100q_x$ columns contained actual values for each factor. The effect of an insecticide such as DDT on the population could also be included (Table 13.5).

Since their development for natural populations, life tables have been used widely in insect population studies (see Harcourt, 1969). There are now sufficient life tables from natural populations of insect herbivores (see Table 13.6) to make some tentative generalizations about the shape of survivorship curves. From the 22 life tables examined, it appears that there are two basic types, although intermediates occur (Fig. 13.8). Type A shows very high mortality in early stages with more than 70% occurring by the midlarval stage. An extreme case is seen in the homopteran *Lepidosaphes ulmi,* in which the crawler stage suffers heavy mortality (Samarasinghe and LeRoux, 1966). Many of the species showing heavy early mortality are free living and exposed, although some are apparently protected by buds or webs. In the latter case, establishment of early larvae is probably difficult. Members of type B have a convex survivorship curve, with 40% or less mortality by the midlar-

TABLE 13.4   Life Table for the 1952–1953 Generation in a Relatively Low Population of Spruce Budworm in the Green River Watershed, New Brunswick. Note That the Cohort Studied (l_x) is the Mean Number in an Egg Mass, and the Age Intervals (x) are Not Equal but Are Defined by the Developmental Stages of the Insect. The Original Publication Should be Consulted for Details of the Calculations Involved.

Age Interval, x	Number[a] Alive at Beginning of x, l_x	Factor Responsible for d_x, d_xF	Number[a] Dying during x, d_x	d_x as Percentage of l_x, $100q_x$
Eggs	174	Parasites	3	2[b]
		Predators	15	9[b]
		Other	1	1[b]
		Total	19	11[b]
Instar I	155	Dispersion, etc.	74.40	48
Hibernacula	80.60	Winter	13.70	17
Instar II	66.90	Dispersion, etc.	42.20	63
Instars III–VI	24.70	Parasites	8.92	36
		Disease	0.54	2
		Birds	3.39	14
		Other–inter.[c]	10.57	43
		Total	23.42	95
Pupae	1.28	Parasites	0.10	8
		Predators	0.13	10
		Others	0.23	18
		Total	0.46	36
Moths (SR = 50:50)	0.82	Sex	0	0
Females × 2	0.82	Size	0	0
		Other	0	0
		Total	0	0
"Normal" females × 2	0.82	—	—	—
Generation	—	—	173.18	99.53
Expected eggs	62	Moth migration, etc.	−513	−827
Actual eggs	575			

Index of population trend: Expected   36%
Actual   330%

[a] Number per 10 square feet of branch surface.
[b] Use of whole numbers results in inaccurate summation for total $100q_x$.
[c] Other factors minus mutual interference among all factors.
SOURCE: Morris and Miller (1954).

val stage. The majority of species are apparently protected by their burrowing habit or colonial defense behavior. The only exception is *Pieris rapae* on cultivated cabbages (Harcourt, 1966), where natural mortality factors may have been ameliorated. Survivorship curves for both *Coleophora serratella* and *Leucoptera spartifoliella* suggest that types A and B will prove to be extremes in a continuum of types between them, as indicated in Fig. 13.9, summarizing 20 survivorship curves for species of Lepidoptera (see Table 13.7 for species names and sources).

TABLE 13.5   Life Table for the 1952–1953 Generation in a High Population of Spruce Budworm in the Green River Watershed, New Brunswick. Starvation Was a Serious Mortality Factor, Due to Death of New Foliage Caused by Late Frosts, and an Aerial Application of an Insecticide, DDT, Inflicted Additional, but Relatively Minor, Mortality.

Age Interval, x	Number[a] Alive at Beginning of x, l_x	Factor Responsible for d_x, d_xF	Number[a] Dying During x, d_x	d_x as Percentage of l_x, $100q_x$
Eggs	2176	Parasites	1	<1[b]
		Predators	174	8[b]
		Other	21	1[b]
		Total	196	9[b]
Instar I	1980	Dispersion, etc.	1148	58
Hibernacula	832	Winter	141	17
Instar II	691	Dispersion, etc.	484	70
Instar III–VI	207	Parasites	2.90	1
		Disease	0.30	<1
		Birds	1.70	1
		Starvation	165.30	80
		DDT	8.30	4
		Other – inter.[c]	26.70	13
		Total	205.20	99
Pupae	1.80	Parasites	0.13	7
		Predators	0.11	6
		Other	0.27	15
		Total	0.51	28
Moths (SR = 54:46)	1.29	Sex	0.10	8
Females × 2	1.19	Size	0.57	48
		Other	0.00	0
		Total	0.57	48
"Normal" females × 2	0.62	—	—	—
Generation	—	—	2175.38	99.97
Expected eggs	47	Moth migration, −199 etc.		−423
Actual eggs	246			
	Index of population trend: Expected 2%			
	Actual 11%			

[a] Number per 10 square feet of branch surface.
[b] Use of whole numbers results in inaccurate summation for total $100q_x$.
[c] Other factors minus mutual interference among all factors.
SOURCE: Morris and Miller (1954).

A much larger sample of life tables, 530 in all, on 124 species of holometabolous herbivorous insects provides a more general perspective than that accomplished in Figs. 13.8 and 13.9. Cornell and Hawkins (1995) discovered that survivorship curves type A and B can be found in an array of species within a single general feeding type: for example, in the native exo-

TABLE 13.6   Sources for Life Tables of Natural Populations of Insect Herbivores Which Provide Survivorship Data Used in Construction of Fig. 13.8 (except those in group C).

Group	Insect Species	Insect Order	Habit	Reference
A: 70% or more of mortality by midlarval stage:				
	1. *Choristoneura fumiferana*	Lepidoptera	In bud or web	Morris and Miller (1954)
	2. *Bupalus piniarius*	Lepidoptera	On foliage	Klomp (1966)
	3. *Plutella maculipennis*	Lepidoptera	On foliage	Harcourt (1963)
	4. *Spilonota ocellana*	Lepidoptera	In bud	LeRoux et al. (1963)
	5. *Archips argyrospilus*	Lepidoptera	In leaf roll	LeRoux et al. (1963)
	6. *Rhyacionia buoliana*	Lepidoptera	In bud and shoot	W. E. Miller (1967)
	7. *Operophtera brumata*	Lepidoptera	On foliage	Embree (1965)
	8. *Recurvaria starki*	Lepidoptera	In needle mine	Stark (1959)
	9. *Sitona regensteinensis*	Coleoptera	On root nodules	Waloff (1968b)
	10. *Lepidosaphes ulmi*	Homoptera	On branches	Samarasinghe and LeRoux (1966)
	11. *Arytaina genistae*	Homoptera	On foliage	Waloff (1968b)
B: 40% or less of mortality by midlarval stage:				
	1. *Ostrinia nubilalis*	Lepidoptera	In stems	LeRoux et al. (1963)
	2. *Coleophora serratella*	Lepidoptera	In case	LeRoux et al. (1963)
	3. *Lithocolletis blancardella*	Lepidoptera	In leaf mine	Pottinger and LeRoux (1971)
	4. *Pieris rapae*	Lepidoptera	On foliage (cultured environment)	Harcourt (1966)
	5. *Neodiprion swainei*	Hymenoptera	On foliage, colonial	McLeod (1972)
	6. *Scolytus scolytus*	Coleoptera	Under bark	Beaver (1966)
	7. *Scolytus ventralis*	Coleoptera	In twig mines	Berryman (1973)
Between A and B:				
	Leucoptera spartifoliella	Lepidoptera	In twig mines	Waloff (1968b)
C: Probably in group A, but uncertain, as no data are provided for midlarval stage:				
	1. *Bruchidius ater*	Coleoptera	In pod	Parnell (1966)
	2. *Apion fuscirostre*	Coleoptera	In pod	Parnell (1966)
	3. *Phytodecta olivacea*	Coleoptera	On foliage	Richards and Waloff (1961)

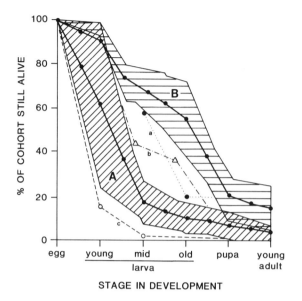

STAGE IN DEVELOPMENT

FIGURE 13.8 General trends in survivorship curves for 19 species of herbivorous insects listed in Table 13.6. Shaded areas indicate the zones that include all survivorship curves in the group indicated, the species of which are listed in Table 13.6. Dots are midpoints between limits of the shaded areas, on the vertical scale, and the heavy lines joining these indicate the general characteristics of type A and B survivorship curves. Part of the survivorship curves for three species not included in the shaded areas are given separately: (a) *Coleophora seratella;* (b) *Leucoptera spartifoliella;* (c) *Lepidosaphes ulmi.* Reprinted from P. W. Price (ed.), *Evolutionary strategies of parasitic insects and mites.* Published by Plenum Publishing Company Limited, 1975.

phytic group, the leaf rollers or webbers group, and groupings of leaf miners, borers, and gallers. However, some interesting patterns did emerge, even though much variation in survivorship curves remains to be explained. These patterns have heuristic value and should stimulate more detailed comparative studies among phylogenetically related groups:

1. The external feeders, or exophytics, had a 5–10% greater risk of mortality before the adult stage than endophytics such as borers and gallers.
2. Natural enemies were the main cause of death, although plant factors have probably been underestimated in the preparation of life tables.
3. Plant factors are strongest in early larval stages and cause death of endophytic species more often than exophytic species (15% vs. 2% of records).
4. Natural enemies have most impact on later life stages and cause death of free feeders more frequently than concealed feeders (51% vs. 44% of records).

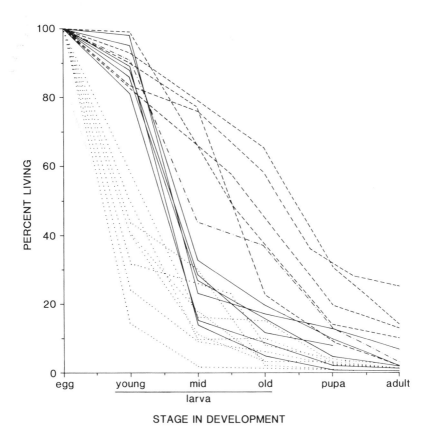

FIGURE 13.9 Survivorship curves for 20 species of Lepidoptera published in the literature, plotted as the percent of the initial cohort alive at each stage of development. Species and sources are provided in Table 13.7. From P. W. Price, *Evolutionary biology of parasites.* Copyright © 1980 by Princeton University Press. Reprinted by permission of Princeton University Press.

5. The importance of competition and natural enemies did not vary with the successional status of the habitat or with latitude.
6. Herbivore survival did not differ significantly in relation to successional status, natural or cultivated habitats, or whether herbivore species were indigenous or exotic.

As is so frequently the case, the enormous diversity of insect species, life histories, and habitats utilized makes detection of pattern a real challenge. That any pattern is detected at all in a sample of 124 species is a cause for optimism that life history traits have such strong effects that they may be used further in the examination of survivorship curves.

TABLE 13.7 Sources for 20 Survivorship Curves for Species of Lepidoptera Given in Fig. 13.9, Classified According to Type A and Type B Survivorship Curves as in Fig. 13.8 and Table 13.6.

Group	Insect Species	Percent Living at Midlarval Stage	Family	Habit	References
A: 70% or more of mortality by midlarval stages:					
	1. *Dendrolimus spectabilis*	2	Lasiocampidae	On foliage	Kokubo (1965)
	2. *Recurvaria starki*	9	Gelechiidae	In needle mine	Stark (1959)
	3. *Dendrolimus pini*	10	Lasiocampidae	On foliage	Schwerdtfeger (1968)
	4. *Papilio xuthus*	11	Papilionidae	On foliage	Watanabe (1976)
	5. *Operophtera brumata*	14	Geometridae	On foliage	Embree (1965)
	6. *Choristoneura fumiferana*	15	Tortricidae	In bud and web	Morris and Miller (1954), Miller (1963)
	7. *Bupalus piniarius*	16	Geometridae	On foliage	Klomp (1966)
	8. *Rhyacionia buoliana*	18	Olethreutidae	In bud and shoot	W. E. Miller (1967)
	9. *Plutella maculipennis*	23	Plutellidae	On foliage	Harcourt (1963)
	10. *Spilonota ocellana*	26	Olethreutidae	In bud	LeRoux et al. (1963)
	11. *Archips argyrospilus*	27	Tortricidae	In leaf roll	LeRoux et al. (1963), Paradis and LeRoux (1965)
	12. *Pieris rapae*	29	Pieridae	On foliage	Dempster (1967)
	13. *Hyphantria cunea*	30	Arctiidae	Colonial in web on stems and leaves	Itô and Miyashita (1968)
Between A and B:					
	14. *Tyria jacobaeae*	33	Arctiidae	On foliage and flowers	Dempster (1971, 1975)
	15. *Leucoptera spartifoliella*	44	Lyonetiidae	In twig mines	Waloff (1968b)
B: 40% or less mortality by midlarval stage:					
	16. *Pieris rapae*	66	Pieridae	On foliage	Harcourt (1966)
	17. *Coleophora serratella*	66	Coleophoridae	In bud and case	LeRoux et al. (1963)
	18. *Ostrinia nubilalis*	76	Pyralidae	In stems	LeRoux et al. (1963)
	19. *Panolis flammea*	77	Noctuidae	On foliage	Schwerdtfeger (1934, 1968)
	20. *Lithocolletis blancardella*	79	Gracillariidae	In leaf mines	Pottinger and LeRoux (1971)

The shapes of insect survivorship curves are of interest in understanding the reproductive strategies of parasitoids that attack these insects (see Chapter 14) and in applied control. Control methods should aim at reducing survival to less than two progeny per female on average rather than obtaining a certain percentage kill without regard to the fecundity of the females. For example, even a 98% mortality after insecticide application, in a species where females lay 200 eggs, would still result in a doubling of the population in the generation. The survivorship curve indicates the vulnerable stages of each species and may lead to emphasis of control efforts on this stage. For example, application of insecticide before the midlarval stage on group A species may frequently be very effective, since high natural mortality indicates exposed and vulnerable immatures, on which insecticides could be effective.

METHODS OF STUDY

If the age at which an animal dies can be observed directly, it is a relatively simple matter to construct a life table. A classic case of this was studied by Murie (1944) in his monograph. "The Wolves of Mount McKinley." Murie was interested in the mortality of Dall Mountain sheep caused by wolf predation. He collected skulls of sheep that had accumulated over a period of years and determined the sheep's age at death from the annual rings on the horns. Thus Deevey (1947) was able to construct a good life table for the sheep.

For insects, this method is not available and we are forced (1) to observe a cohort very closely, usually therefore in unnatural conditions, or (2) to take successive population samples to see how the population declines through a generation. Morris has probably been the most influential worker in the development of insect sampling techniques with a view to developing life tables (e.g., Morris, 1955). Points mentioned here refer to the spruce budworm, but the principles can readily be applied to other species, especially herbivores. First, the detailed life history must be known so that all stages of the species can readily be found in the trees. This includes eggs, mining larvae, larvae in webbed foliage, and pupae that are about 2 cm long. Good life history data also enable accurate timing of samples of the various life stages. To obtain this information, a full-scale study is usually necessary, as life histories of so few insects are adequately known, and phenology changes from place to place.

Second, a sampling unit must be selected so that it is easily collected in the field and subject to as little variation as possible, which might be caused by defoliation in the case of herbivore populations. The sampling unit should also be reasonably small, so that many can be collected, and be quantifiable in terms of numbers per tree and numbers per unit area. For herbivorous insects a part of the host plant usually serves as the sample unit. Morris (1955) selected a lateral half of a whole branch of balsam fir so that insects could be expressed per unit area of branch surface. The balsam fir has a triangular growth pattern on branches, with all side branches more or less in one plane, so the area of a sample could be calculated quite accurately from two measurements: length and maximum width. By measuring the diameter at breast height of the tree from which the sample was taken, tables could be consulted to find the total branch area per unit area of ground. Thus populations could be expressed in absolute terms.

Third, one must determine the best location of the sample. Since budworm density varies tremendously with height in the crown, sampling could become very complicated and laborious. For survey work when rapid sampling was necessary, only branches from the midcrown, representing mean population levels, were taken.

Fourth, midcrown in tall stands may be 40–50 feet above ground, so ladders and pole pruners, or trucks equipped with hydraulically operated platforms as for telephone line repairs, are essential. So the technical difficulties are gradually surmounted. The technical details of sample collection, the difficulty with creating statistically rigorous techniques, and the length of time required before meaningful data are accumulated have all conspired to limit to a relatively small number the studies on forest insect populations (see Chapter 17).

Relative abundances are easier to obtain by means of trapping techniques (see Southwood, 1966, 1979), and for studies of insects as food for vertebrates these methods provide a much greater return per unit effort.

RETROSPECTIVE AND FUTURE USE OF LIFE TABLES

In retrospect, it is now clear that the methodology of life table construction more or less dictates the kinds of answers we obtain. Life tables on insects have tended to emphasize mortality factors impinging on a cohort of eggs, and subsequent stages. Perhaps they should be called "death tables"? Except for a few examples, they have left largely unexplored the effects of natality measured in a direct way. If plant resources are unsuitable for larval survival, perhaps an ovipositing female will choose not to lay eggs, but rather, disperse into another area. Perhaps plant resistance to endophytic species is strong enough to obscure any evidence of oviposition, as with hypersensitive reactions to attack (e.g., Fernandes, 1990; Cappuccino, 1992). If natality and mortality, bottom-up effects, top-down effects, and lateral effects are not evaluated adequately, biases will result in life table construction. One can make a valid case for overemphasis on mortality factors, especially those caused by carnivores, in the majority of life table studies undertaken to date.

Plant Resistance and Natality

A particularly illuminating case of a life table that has captured the bottom-up effects of plant resistance, and its impact on female ovipositional behavior and fecundity, concerns the fir engraver beetle, *Scolytus ventralis.* Berryman (1973) studied this bark beetle on grand fir, *Abies grandis,* for many years and estimated in each generation the role of plant resistance on natality. This case and other bark beetles are explored further in Chapter 19 in relation to their population dynamics.

Bark beetles bore through the bark of a tree, down to the cambial layer, where a nuptial chamber is constructed, from which maternal galleries are directed along the cambium, engraving both the xylem and inner bark with a characteristic pattern for each beetle species (e.g., Fig. 15.14). The cambial layer of the tree trunk can be considered as the "heart" of the tree and is inevitably well defended against attack. In coniferous host trees, such as grand fir, resin canals in the bark prove to be a formidable barrier to attack by bark beetles. Once ruptured by boring activity, resin canals release a flow of resin into the wound which can immobilize and kill beetles: resinosis or "pitching out." If entrance to the cambium is gained successfully, for example if the tree's defenses are reduced by drought stress, a female constructs a maternal gallery and oviposits in egg niches along its length. Larvae hatch and bore in the cambial layer away from the maternal gallery, fanning out until feeding is completed. They then pupate and emerge through the bark by cutting an individual emergence hole.

Thus each female bark beetle and her progeny, practically speaking, writes its own life table, from the attempted entry into the tree, successful or unsuccessful, to the emergence of her progeny, or their death caused by competition for cambial space, predators, parasitoids, or fungi. By knowing the mean fecundity of females in a population, Berryman (1983) could calculate the reduction in natality in a cohort resulting from resinosis and plant defenses. He could also evaluate the effects of competition among females for gallery space and competition among larvae for foraging space in the cambium.

Therefore, Berryman's (1973) life tables start with the number of eggs that could have been laid by the number of females actually attacking a tree per 1 ft^2 of bark surface. The original cohort at the top of the life table is "number of potential eggs." Then losses in natality due to resinosis and female death are calculated. For scolytid beetles, then, we have an unusual opportunity to readily assess the roles of plant resistance and natality, relative to those factors commonly estimated in life tables: weather, predation, parasitism, and so on. Some results are provided in Table 13.8.

In 3 out of 5 years, reduced natality was the most important factor in loss from the potential cohort of eggs, causing more loss than in any other life stage, including total larval life. In the other 2 years it was second to mortality of all larval stages combined.

TABLE 13.8 Reduced Natality Resulting from Plant Resistance in the Fir Engraver, *Scolytus ventralis*, Attacking Grand Fir, *Abies grandis.*

Generation	Density of Females	Percent Mortality of Females	Reduction in Egg Cohort Size	Number Dying	Rank Compared to Other Stages
1966–1967	12	44	719 → 403	316	Top
1967–1968	10	51	630 → 312	318	Top
1968–1969	18	54	832 → 380	452	Top
1969–1970	12	31	667 → 463	204	Second
1970–1971	20	37	1115 → 703	412	Second

SOURCE: Berryman (1973).

This is a very sobering result. Most life tables start with the cohort of eggs laid, without directly evaluating the number of eggs lost in a population of females, for various reasons. When this is estimated it proves to be of major importance.

Reduced Natality in Response to Plant Quality

A completely different kind of study illustrates how experimental approaches are essential in estimating the role of natality. In an excellent set of experiments, Ohgushi and Sawada (1985a,b; Ohgushi, 1986, 1992, 1995) showed in the herbivores lady beetle, *Henosepilachna niponica* (Fig. 13.10), feeding on the leaves of the thistle, *Cirsium kagamontanum*, that part of the population stability was imposed by females withholding eggs and resorbing them when plant damage became high. In cage experiments, once leaf damage

FIGURE 13.10 Herbivorous lady beetle, *Henosepilachna niponica* (now in the genus *Epilachna*). From Ohgushi (1995); with permission. Drawing by Akiko Fukui.

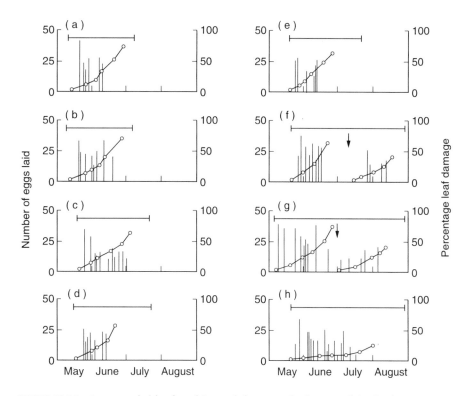

FIGURE 13.11   Oviposition schedules of caged *Henosepilachna niponica* beetles in controls (a–e) and in treatments in which females were transferred to new foliage at the date indicated by the arrow (f, g) and in which earwig egg predators were introduced (h). Vertical bars indicate the number of eggs deposited each day. Open circles show the percent of leaf damage and horizontal bars show the life span of each female. From Ohgushi and Sawada (1985a); with permission.

reached 30–50%, females stopped laying eggs and resorbed eggs in the ovary (Fig. 13.11). If females were transferred to new plants, they would resume oviposition. But if predatory earwigs, which ate eggs, were introduced to a cage, plant damage was reduced and beetles would lay eggs for much longer periods of time (Ohgushi and Sawada, 1985a).

Thus, while predation was commonly an important component of the stable population (Ohgushi and Sawada, 1985b; Ohgushi, 1986), the beetles had a self-regulatory capacity, resulting in close tracking of resource availability, even though the carrying capacity of leaves seemed to be higher than the self-regulated density of larvae. That is only 50% of leaf damage was enough to stop oviposition. (We might wonder if induced defenses in the plant were also involved?)

These studies show clearly how experimental work can reveal the capacity of female behavior to play a role in the population dynamics of a species. This kind of role is likely to have been missed completely in a typical sampling program to develop life tables.

TABLE 13.9   Oviposition and Survival of the Stem-galling Sawfly, *Euura lasiolepis,* on Uncaged Plants with High and Low Water Treatments.

	High Water	Low Water
Number of ovipositions per plant	7.8 (3.7× higher)	2.1
Percent survival	75 (1.9× higher)	39

SOURCE: Price and Clancy (1986b).

Caging females, it could be argued, does not provide a realistic evaluation of female behavior because a wild female would fly off and lay eggs elsewhere. But, in fact, oviposition on plants of different quality by uncaged female insects shows that they do refrain from laying eggs on lower-quality plants. Reduced natality on low-quality resources is real. In water treatment experiments using the stem-galling sawfly, *Euura lasiolepis,* oviposition was much lower on low- than on high-water-treatment plants (Table 13.9).

Hence, oviposition preferences resulted in an almost fourfold difference in sawfly density, and survival a virtual twofold difference, for an overall sevenfold difference in number of sawflies in the next generation. This is a large difference driven by the interaction of plant quality variation and female sawfly behavioral responses.

Similar but more detailed experiments were conducted using caged females, with similar but stronger effects (Preszler and Price, 1988). These results were in close agreement with survivorship curves on wild willows in wet and dry sites. On the experimental plants, a 10-fold difference in population size was observed resulting from the combined effects of female behavior and larval survival on high- and low-water-treatment plants. The number of eggs expected from two females per cage was estimated at 100, based on fecundity studies. This egg cohort, still within the female sawflies, was used to start the life table, enabling a full analysis of natality in response to host-plant variation (Table 13.10).

These are notable life tables relative to the data usually available for life table analysis, with results comparable to those for a bark beetle studied by Berryman. First, by far the largest difference in the two life tables is female response. Most of the lost potential for individuals in the next generation occurred because of maternal response: 27.5% of the 37.4% loss in the high-water treatment, and 85.1% out of 93.5% mortality in the low-water treatment. Second, when so much cohort loss occurs in the life table, this reduces the potential importance for death from natural enemies. In the low-water treatment, parasitoids, for example, were bound to play a less important role than maternal responses. Third, the effects of host plant resistance after the eggs were laid were also very strong in the low-water treatment, accounting for loss of 54.5% of the eggs in galls: the second most important factor in cohort loss. Fourth, the overall result is a very different kind of life

TABLE 13.10   Life Tables for the Stem-galling Sawfly, *Euura lasiolepis,* Starting with Eggs in Females Before Oviposition Commences, on Plants with High and Low Water Treatments.

Stage, x	Number Alive at Beginning of x, l_x		d_xF: Factor Responsible for d_x	Number Dying During x, d_x	
	High Water	Low Water		High Water	Low Water
Eggs in females	100	100	Failure to form gall	0	62.1
			Egg retention	27.5	23.0
			Total maternal response	27.5	85.1
Eggs in galls	72.5	14.9		0	0
First instar	72.5	14.9	Host effects— resistance	9.9 (13.6%)	8.1 (54.5%)
			Parasitism	0	0
Second instar	62.6	6.8	Host effects— resistance	0	0.3
Last instar	62.6	6.5			
			Total % mortality	37.4	93.5

SOURCE: Preszler and Price (1988).

table for an insect herbivore compared, say, to that for the spruce budworm developed by Morris and Miller (Tables 13.4 and 13.5). These kinds of differences may well be real, as discussed in Chapter 19 (Table 13.11). A comparison of simplified life tables for the galling sawfly and the spruce budworm reveals fundamental differences in the mechanisms involved in cohort reduction. Maternal responses and plant resistance in the *Euura* life table are of prime importance, while dispersal of larvae and larval mortality are major losses in *Choristoneura*.

TABLE 13.11   Comparison of Simplified Life Tables for *Euura lasiolepis* (in a dry site, cf. Table 13.10) and *Choristoneura fumiferana* (cf. Table 13.4).

	Percent of Individuals Lost Per Generation	
	Euura	Choristoneura
Female behavior, natality	85	0?
Plant resistance	8	0?
Dispersal of larvae	0	67
Winter mortality	0	8
Predation	0	11
Parasitism	0	7
Total mortality (%)	93	93

Sex Ratio and Maternal Effects

Even primary sex ratios may be biased in haplodiploid insects based on a female's response to plant quality (Craig and Mopper, 1993) and secondary sex ratios are generally biased in sawfly life tables in 47 species studied. With so many groups of arthropod herbivores with haplodiploid sex determination, such as the Hymenoptera, Homoptera (Iceryine coccids, Aleurodidae), Thysanoptera, and Acarina (Tarsonemidae, Tetranychidae) (White, 1973), more attention should be placed on the mechanisms resulting in biased sex ratios and their impact on population dynamics (Craig and Mopper, 1993). Sex ratio differences from $50:50$ can have strong impacts on natality, as seen in Dahlstein's (1967) study of *Neodiprion fulviceps* on ponderosa pine on Mt. Shasta, California. In one area only 23% of adults were females, resulting in a 54% loss of eggs in the next generation relative to a $50:50$ sex ratio. In other areas there were 29% and 33% females, resulting in 42% and 34% loss of eggs, respectively. Dahlstein interpreted these sex ratio biases as secondary factors, such as greater predation on the larger female cocoons while in the soil. However, the alternative hypothesis that differences resulted from primary sex ratio variation was not tested. Two studies indicate that females adjust sex ratio in relation to plant quality, one concerning a diprionid sawfly (Mopper and Whitham, 1992) and the other a tenthredinid sawfly (Craig et al., 1992).

Another factor that is inadequately studied in terms of life table construction is the effect of a female on the quality of progeny. This topic of maternal effects has been revitalized in a series of papers by Rossiter (1992, 1994, 1995). The maternal effects hypothesis was introduced in Chapter 6 and will be discussed again in relation to population dynamics in Chapters 17 and 19.

Several other factors may well be relevant to life table construction. Competition for ovipositional sites among females is evident in many species, but not in life tables. If Rhoades (1985) is correct that spacing and territoriality may be prevalent in stealthy herbivores, as discussed in Chapter 6, this may be a characteristic of fundamental importance in the understanding of populations with low-amplitude fluctuations. This subject is treated in Chapter 15. The question of the role of emigration or dispersal is perhaps the most vexing of all to study. Seldom is emigration from a study measured directly and effectively, a subject treated in Chapters 17 and 19.

With these retrospective views on life table construction it is clear that much more understanding can be derived when mechanistic studies capture a full range of factors from female behaviors during oviposition, parental effects, and competition among females for oviposition sites to the more generally studied aspects of larval competition, parasitism, and predation. Better studies on dispersal are needed, and certainly the genetics of populations and their evolution need incorporation. These are challenges for the future.

EFFECTS OF NATURAL SELECTION ON *r*

An evolutionary approach has been taken on the demographic consequence of life history variation. The instantaneous rate of increase, *r*, is not necessarily constant for a population or species over long periods of time. The possible effects of natural selection in changing *r* can be explored, and we may also inquire about how a certain *r* may be achieved by different means from one generation to another. How is population growth altered by changes in life history and differences in reproduction between populations and species? Do we have any insight into how selection works to define the most propitious time for reproduction? First, major differences in life history are examined. For example, some ideas in Cole's (1954b) classic paper "The population consequences of life history phenomena" are worth careful scrutiny.

One of the factors that Cole examined was how breeding once in a lifetime (**semelparity**) compared to breeding several times (**iteroparity**) influenced population growth (Fig. 13.12). This is a choice that the evolutionary process has made for organisms. From the figure the number of eggs a semelparous female would have to lay to achieve the same *r* as an iteroparous species can be calculated. For example, females of an iteroparous species may start to

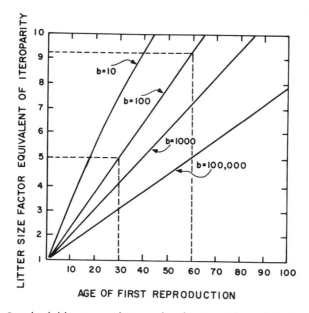

FIGURE 13.12 Factor by which litter size must be increased to achieve in a single reproduction the same intrinsic rate of increase, *r*, that would result from indefinite iteroparity; *b* is the brood size of an iteroparous species and the ordinate gives the factor by which this must be multiplied. From Cole (1954b).

reproduce at 30 days and lay 100 eggs in each clutch. How many eggs would a female of another species have to lay, if she laid one batch of eggs, to have the same population consequences? ($100 \times 5 = 500$ eggs.) If reproduction occurs at 60 days, then if an iteroparous species lays 100 eggs per clutch, a semelparous species must lay 925 eggs to produce the same population growth rate. Cole's paper should be consulted to see how selection is likely to work on these two basic differences in reproductive strategy.

With respect to many insect and plant species having annual life cycles and high fecundities, Cole argues that any selective pressure for iteroparity as a means of increasing r must be negligible. For example, an iteroparous insect that produces 30 eggs in each annual reproductive period would have to produce only 31 eggs in a single clutch if it were semelparous to achieve the same r. Even with only 30 eggs per clutch, the gain in becoming iteroparous is less than a 1% increase in r. With fecundities commonly ranging from 60 to 200 eggs per female insect, the iteroparous habit will usually be maladaptive. For the other extreme, consult Goodman (1974), who considers iteroparous species with very small clutch sizes.

Cole looked at another factor that influences population growth, the age at first reproduction, or the effect of delayed maturity on the rate of population increase (Fig. 13.13). For example, when the brood size is 10, the rate of increase is more than doubled by breeding at age 2 rather than at age 6 ($r = 1.29$ instead of 0.52). There must be a strong selection pressure to mature and breed early. What are the selective pressures working in the opposite direction? Cole's conclusions have been reexamined by Gadgil and Bossert (1970), Bryant (1971), and Charnov and Schaffer (1973).

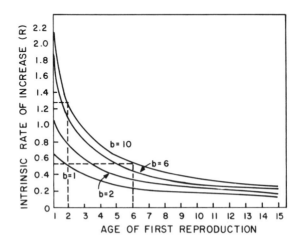

FIGURE 13.13 Effect of changing the time of first reproduction on the intrinsic rate of increase, r, of a population; b is the brood size for an indefinitely iteroparous species where a brood is produced at every time interval after first reproduction. From Cole (1954b).

Looking at reproduction in this way, it becomes clear that a human female that breeds at 20 years contributes much more to the population growth than one that breeds at 30 years. This should be taken into account when formulating policy on family size. Also, a female that produces females contributes more to subsequent population growth than does a female that produces males. We need only imagine two equal populations on different islands, where on one there are 99 males and one female and on the other there are 99 females and one male. The difference between the growth rates of these populations should be obvious.

SELECTION FOR COLONIZING ABILITY

Lewontin (1965) considered selection for colonizing ability, or slight shifts within a general strategy, not radically different evolutionary strategies such as iteroparity versus semelparity. Lewontin was interested in the age of first reproduction, the effect of timing of the peak of reproduction, and the population effect of the age at which the last offspring is produced. He was interested in small variation in these factors that could be selected for over a few generations because of changing conditions. He wanted to know also how the production of progeny by organisms that colonize unoccupied areas is likely to differ from progeny production in resident populations.

Lewontin asked what is a reasonable shape for the $l_x m_x$ curve and found that a triangular shape was realistic (Fig. 13.14), defined by the age at first reproduction (A), age at peak reproduction (T), and age at the end of reproduction (W). An example of this shape is provided by Coats (1976) for the parasitoid *Muscidifurax zaraptor*. Then by modifying the position of the reproductive function in time and changing its shape, he could compute how

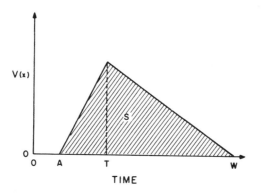

FIGURE 13.14   Generalized triangular reproductive function V(x) that represents the $l_x m_x$ curve of a cohort. A, age at first reproduction; T, peak reproduction; W, age at last reproduction; S, total number of offspring produced. From Lewontin (1965).

these changes affected r. He could ask questions such as: What is the relative advantage of increased fecundity of a certain amount as opposed to greater longevity? To increase r from 0.270 to 0.300, how much does fecundity have to be increased for different developmental rates? These questions are actually asked in nature, not through rational means but because innumerable strategies in life history are attempted by genotypes and their phenotypes produced by a process with a random element in it. Only the best are selected for.

Only two examples are given. First, what is the effect of changing only the age at reproduction? (See Fig. 13.15) What happens if natural selection reduces the time interval to first reproduction from 13 days to 10? If fecundity remains the same (at 600), the rate of increase changes from 0.270 to almost 0.330. To maintain the rate of increase at 0.270, a species with first reproduction at age 10 days would only have to produce 260 eggs instead of 600, a remarkable saving of energy as the result of a rather small adjustment in breeding time. The number of progeny that would have to be produced to increase r from 0.270 to 0.330 if reproduction started at day 13 is 2000. There is a considerable evolutionary choice in when to breed and how many progeny to produce.

The second example describes what happens when A and W remain constant and T changes (Fig. 13.16). How does the peak skewed to the left increase the rate of population increase? Here S is constant as the area of the reproductive function does not change. By moving T from day 28 to 16, r increases from 0.270 to 0.330.

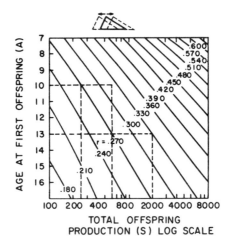

FIGURE 13.15 Effect of changing age at first reproduction, A, on the intrinsic rate of increase, r. Solid lines give equal r values for changing A and S. The change in the reproductive function is given above the graph; thus with the change in A, T and W also change. From Lewontin (1965).

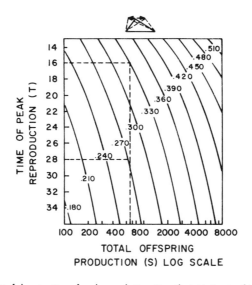

FIGURE 13.16 Effect of changing time of peak reproduction, T, on the intrinsic rate of increase, r. Solid lines give equal r values for changing T. S does not change. The change in shape of the reproductive function is given above the graph; its position in time does not change; that is, T changes but A and W remain unchanged. From Lewontin (1965).

SEXUAL DIFFERENCES IN SURVIVORSHIP

Another type of adjustment is related to differences between the sexes of the same species and differences between closely related species. For example, Blest (1963) found in saturniid moths that the postreproductive life of males and females differed from one species to another. He argued that selection should reduce the postreproductive life of cryptic and palatable species, as old individuals may improve the speed of learning of the predator. Conversely, selection should increase the postreproductive life of aposematic species, as old individuals will play a role in the predator's learning process without reducing the reproductive potential of the population, since neither males nor females feed. Kin selection must be involved here. Blest's reasoning is supported by data in Fig. 13.17 for males and Table 13.12 for females.

Landahl and Root (1969) found a similar situation with the aposematic milkweed bugs, *Oncopeltus fasciatus* and *O. unifasciatellus,* where males live longer than females. A high proportion of males in the population would reduce the probability of vertebrate predators taking females. Here males feed, but "the cost of sustaining the 'surplus' males may be outweighed by their value as decoys."

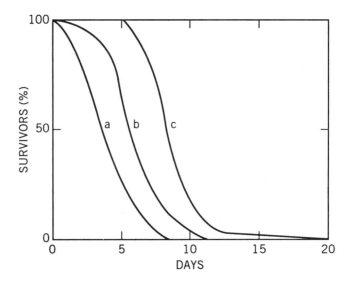

FIGURE 13.17   Difference in postreproductive survivorship curves of male saturniid moths (plotted as percent survivors): (a) for a cryptic leaf mimic that is palatable; (b) for a species with a forewing pattern that mimics a leaf, and the hind wings with eyespots used in an intimidation display but which is palatable; (c) for an aposematic species. After Blest (1963).

INDIVIDUAL ADJUSTMENTS

Even adjustments in reproductive effort can be made by individuals according to availability of food. Murdoch (1966a,b) found in carabids that at one time a female may breed early and thus reduce fecundity later, or another female of the same species at a different time may perform the reverse adjustment

TABLE 13.12   Postreproductive Survival of Female Saturniid Moths.

Days After End of Reproduction	Palatable Females	Unpalatable Females
1	10	0
2	3	13
3	0	6
4	0	12
5	0	7
6	0	4
7	0	1
8	0	0
9	0	2
Total	13	45

SOURCE: Blest (1963).

(see Chapter 14), rather like Lewontin's (1965) examples were T and W are manipulated.

Finally, as Frank (168) points out, it may be tautological to say that long-lived species have lower population growth rates and tend to live in stable environments. Many wood-feeding insects provide an example, as shown in Chapter 12. But differences in r in stable and unstable environments are discussed again in this book.

For the last section of this chapter the subject of demography has merged with that of reproductive strategies of organisms, an area of ecology that is expanded on in Chapter 14.

Life Histories and Reproductive Strategies

All individuals in all species are bound to maximize the number of progeny they leave in future generations. But the life history characteristics that result in maximizing survival of young are about as diverse as the number of insect species. Again, the ecologists' job is to attempt to identify patterns in this large variety of examples and to understand why such patterns exist. This is a difficult task because so many characteristics of the life cycle of insects may change, as we saw in Chapter 13: age at first reproduction, time of peak reproduction, time at end of reproduction, time of death, time of dispersal, fecundity, seasonality of reproduction, the care with which eggs are placed and protected, and the relationship between parents and young.

USE OF RELATED SPECIES

The interesting questions to be asked are many, but comparisons between closely related species, or life history variation in one species in different parts of its range, or polymorphism in life history traits, provide the most rigorous analysis because we know that life history variation has developed from a common phylogenetic stock and divergence of traits has probably resulted

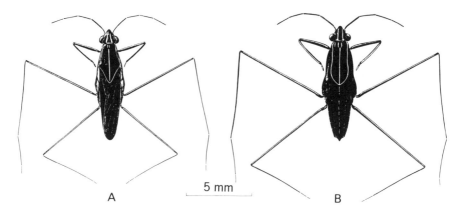

Polymorphism in the Australian water strider species, *Limnogonus luctuosus* (Gerridae), with the macropterous (A) and apterous (B) morphs. From Naumann (1991). Drawing by S. Monteith.

from differential natural selection because of differing ecological opportunities and constraints on life history evolution. Comparisons between vertebrates and insects, such as those made by Pianka (1970, 1974), cannot possibly help us to understand anything but the grossest kinds of differences in life history traits.

Interesting questions involving comparisons between related species, between intraspecific populations, and within populations would include the following:

1. Why do individuals live for different lengths of time?
2. Why do they produce different numbers of progeny?
3. Why does the schedule of progeny production differ?
4. Why do lengths of postreproductive life differ?
5. Why do some lay large eggs and some small?
6. Why do some breed sexually and others asexually?

To make these questions more tangible, some examples of differences in life histories and reproductive strategies will be given after the following cautionary note.

TACTICS AND STRATEGIES

The words **tactic** and **strategy** are used in ecology and evolutionary biology as a shorthand to depict the understanding that traits are under the influence of natural selection and undergo evolutionary change. No plan or intent is acknowledged in this usage, but the results are clearly solutions to environmental and phylogenetic constraints. Since tactics and strategies are adopted by humans to solve problems, we can then apply the words to evolutionary processes as long as we recognize the very important differences in the mechanisms involved.

The following examples will illustrate some of the many different approaches to understanding life history variation, the many kinds of differences that exist, and the different levels at which differences can be seen: intrapopulational, intraspecific, and interspecific. Examples are classified according to this last criterion, although some examples may serve to illustrate variation at more than one level.

INTRAPOPULATION VARIATION

The following examples illustrate **intrapopulational variation:**

1. *The pitcher-plant mosquito, Wyemyia smithii.* In north-temperate latitudes environmental conditions for insects are unfavorable during the winter for active development, and the onset of these conditions is unpredictable.

Insects usually pass this season in a nondeveloping, diapausal state, which necessitates physiological changes before the onset of harsh conditions. The interesting question can therefore be asked: How does selection work to produce high fitness for individuals in a population that exists in an environment with very unpredictable onset of harsh conditions?

Istock (1981 and references therein) addressed this question using the pitcher-plant mosquito, which lives in the water held in the pitchers of *Saracenea purpurea*. The population existed at Kennedy Bog in western New York State, near Rochester. In this population Istock discovered much variation in development time and diapause among individuals, and with breeding experiments established that the variation was under polygenic control. The two characters were genetically correlated such that a range in phenotypes was expressed from extreme fast-developing and diapause-resisting to extreme slow-developing and diapause-prone. The variation between these extremes is continuous, producing life history variation within the population from **univoltinism** (one generation per season) to **multivoltinism** (more than one generation per season) (Fig. 14.1). The figure depicts the genotype resistant to diapause with fast development, the genotype prone to diapause with slow development, and the intermediate, mixed, genotype produced by mating between the extremes.

The differences in reproductive potential between these genotypes is enormous (see potential net reproductive rate per season R_s in Fig. 14.1), and this is clearly a **mixed reproductive strategy** or **bet-hedging strategy** (Stearns,

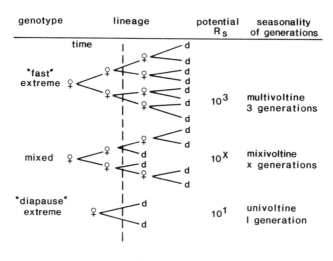

FIGURE 14.1   Lineages of *Wyeomyia smithii* from Kennedy Bog illustrating three types of lineage, from the "fast" extreme to an intermediate mixed genotype to the "diapause" extreme. The letter "d" shows the production of diapausing females in a lineage. From Istock (1981); with permission.

1976). The fast extreme genotype produces most progeny when the season is long and food is plentiful. The diapause extreme has high fitness when the season is short and food is scarce. The mixed genotype is adequately adapted to both extremes and best adapted to intermediate conditions, which presumably are most commonly encountered. We can consider this phenomenon as a genetic polymorphism with the mixed genotype as the heterozygote. The polymorphism is maintained in the population by the unpredictability of conditions in these phytotelmata, or plant-held waters. Understanding the genetic basis for this life history variation is the key, enabling life history studies on insects to become prominent and critical in the development of this field.

2. *The mushroom fly, Mycophila speyeri.*   Where food and microhabitats are patchy, and the patches are ephemeral, selection on small organisms that exploit these kinds of resources is likely to produce extremely high reproductive rates and peculiar modes of reproduction. Such conditions exist for many species of gall midge (Cecidomyiidae), which have selected for the evolution of **paedogenesis** in the group at least four times (Wyatt, 1967). This bizarre phenomenon involves reproduction by individuals with the form of an immature animal, and in the cecidomyiids involves viviparous parthenogenesis in larvae-form females (Gould, 1977). Larvae form within the body of the mother, feeding on her tissues, and emerge as larvae competent to reproduce, leaving few remains of the consumed matron. As Gould (p. 306) comments: "Greater love hath no woman."

In the mushroom fly three forms of paedogenetic females may be observed (Ulrich et al., 1972): female-producing, male-producing, and both male- and female-producing. Again we see a mixed strategy. As food availability declines, males and females are produced, and in the absence of food, they develop through the holometabolic pathway into insects with adult cecidomyiid form and the capacity to disperse.

But while food is plentiful, viviparous parthenogenesis has impressive results. Larvae of the mushroom fly molt once and reproduce as larvae, producing a mean of 19 young in 6 days and a maximum number of 38 young in 5 days, resulting in 20,000 reproductive larvae per square foot of mushroom bed in 5 weeks (Wyatt, 1964). In contrast, the sexual adults result from two larval and one pupal molt and require 2 weeks for development.

INTRASPECIFIC VARIATION

The following examples illustrate **intraspecific variation**:

1. *The flour beetle, Tribolium castaneum.*   A valuable experimental approach was taken by Mertz (1975) with laboratory populations of the flour beetle. The experiment was designed to study evolution of populations under different selective regimes:

(*a*) Selection against late reproduction by killing all adults after 10 days of adult life.
(*b*) Slightly less stringent selection against late reproduction, by killing all adults after 20 days of life.
(*c*) No selection on breeding time where some beetles lived more than 400 days

Selection was continued for 11 or 12 generations in treatments 1 and 2. Mertz found that there was a significant increase in early fecundity—about 15%—in treatment 1 but no such increase in treatment 2. No shifts in time of first reproduction or longevity were observed. Beetles in treatment 1 with higher early reproduction had lower later reproduction, so that total fecundity in a lifetime was equivalent in each treatment.

The importance of this experiment lies in the direct and simple demonstration that life history characteristics do change under different selective regimes and that such characteristics are under genetic control since they are heritable. Another important lesson is that some characteristics change much more rapidly than others; in the case of *T. castaneum* the fecundity schedule changed under strong selection, whereas time of first reproduction did not. It would be very interesting to do similar experiments with many organisms with short life cycles, such as *Drosophila* species, to see if rates of change of the different life history traits differ between populations or species—if the lability of characters differs in species living in rather disparate environments. Much more experimental work is needed relating to the possibilities discussed in Chapter 13 concerning Lewontin's (1965) studies.

2. *The fruit fly, Drosophila melanogaster.* Certainly, it would be valuable to study natural variation in life history traits of fruit flies, but experiments using artificial selection provide a cautionary tale. Adaptive shifts in these traits are complex and involve many characters that interact.

In a long series of selection regimes, Rose (e.g., 1984) and Service (e.g., 1987, 1993) studied the effects of changing the reproductive function (Fig. 13.14) by extending the age at last reproduction, *W*. They selected for postponed senescence. In control populations, flies were cultured such that selection was for reproductive success at age 14 days from oviposition, or for days 1–5 of adult life. The selection regime on populations for postponed senescence was on reproductive success at age 10–12 weeks from oviposition, with adults at about 60–75 days of adult life.

The selection experiment for postponed senescence was effective in producing increased longevity. This was associated with reduced fecundity in young females, increased resistance to starvation, desiccation, ethanol vapor, and a greater proportional lipid content. These flies also had lower activity levels and respiration rates when young (Service, 1987). There was a strong negative genetic correlation between high early fecundity and starvation resistance and postponed senescence.

Mating success involved yet another set of traits influenced by selection for postponed senescence (Service, 1993; Service and Fales, 1993). In competitive mating situations, males selected for postponed senescence were superior to controls only when they were older, but they were usually superior to controls in sperm competition. Also, in the period from 6 to 21 days after mating, males fathered more adult progeny than did controls (Service and Vossbrink, 1996).

These complexities in life history evolution may instruct us on mechanisms causing senescence (Service et al., 1988) and the age at last reproduction, W. Senescence may result from deleterious mutations effective in older individuals, the *mutation-accumulation mechanism*. An alternative mechanism may result from selection on traits that improve early fitness but which have negative effects later in life. In balance, improved early fitness would usually be favored, as in Fig. 13.14. This concept is known as the *antagonistic-pleiotropy mechanism* because genes with more than one effect would be involved.

The ingenious experimental test of these concepts involved reversed selection on the postponed senescence lineages, in which pleiotropic genes would allow trait change toward that of controls but mutated genes would be fixed and irreversible. Both mechanisms seem to have been involved. When selection for early-stage fitness was resumed, resistance to starvation declined and female early fecundity increased, supporting the antagonistic-pleiotropy mechanism. However, resistance to desiccation and ethanol vapor remained fixed, supporting the mutation-accumulation mechanism.

The complexities of these interactions illustrate the subtle nature of adaptive responses to changing ecological conditions. They are likely to be important in nature relative to changing conditions over a landscape or through time as populations change in density and distribution. Certainly, we should learn more about variation in these kinds of traits in nature (cf. Roff, 1992; Stearns, 1992).

3. *Water-striders in the genus Gerris.* These predatory bugs live on a wide range of water bodies in temperature climates: from habitats that freeze in winter, to temporary bodies of water, to permanent sites. Their wing shape and diapause have been studied by Vepsäläinen (1978 and examples therein) in Finland (60°N latitude) and Hungary (47°N latitude), shown in Table 14.1. This example actually involves life history differences at all levels: within populations, between populations, and between species.

Vepsäläinen (1978) uses the term *polymorphism* to mean the occurrence together in the same population of two (dimorphism) or more phenotypes belonging to the same stage in the life cycle, with the rarest phenotypes representing at least 1% of the population. This may involve a permanent genetically determined polymorphism, or a temporary environmentally induced change of morphs, called *polyphenism* or *seasonal polymorphism*. The patterns in Table 14.1 show that some species have long wings and are

TABLE 14.1   Wing-length Patterns of *Gerris* Species Studied in Southern Finland (About 60° N Latitude) and Hungary (About 47° N Latitude). *LW*, Long-winged; *SW*, Short-winged; ?, Not Known; —, Does Not Exist.

Wing-length Pattern	Winter = Diapause Generation (Univoltine Populations)	Summer = Nondiapause Generation of Multivoltine Populations		Additional Notes
		Finland	Hungary	
I. *LW* monomorphism				
G. rufoscutellatus	*LW*	*LW*	*LW*	*SW* known
II. Seasonal dimorphism (seasonal polyphenism)				
G. thoracicus	*LW*	*SW*	*LW*	Type I in Hungary
G. odontogaster	*LW*	*SW*	Dimorphic	
G. argentatus	*LW*	*SW*	Dimorphic	
G. paludum	*LW*	*SW*	Dimorphic	
III. Permanent dimorphism (genetic polymorphism)				
G. lacustris	Dimorphic	*SW*	Dimorphic	Partially type II
G. lateralis	Dimorphic	?	—	
G. asper	Dimorphic	—	?	
IV. *SW* monomorphism				
G. najas	*SW*	—	?	Dimorphic in Poland
G. sphagnetorum	*SW*	?	—	*LW* known

SOURCE: Reproduced with permission from Vepsäläinen (1978). © by Springer-Verlag, Inc., New York.

monomorphic—individuals are prepared for flight and dispersal throughout the season, and there is little intraspecific variation in this character. Other species show seasonal polyphenism, with short-winged (and an inability to fly) individuals present in the summer and long-winged individuals in the winter generation. In this type we see intraspecific variation with seasonal polyphenism in Finland involving only short-winged females but in Hungary a dimorphism in the summer generation. Still other species show permanent genetically regulated dimorphism, while two species are usually only in the short-winged form.

Once this impressive variation has been discovered, the question must be asked: What are the environmental conditions that select for such disparate life histories? But first we need to know how each wing morph is adaptive. The long-winged morph is clearly adapted to dispersal, but the cost of long wings involves a longer development and slower initiation of reproduction since the large flight muscles must develop before energy is shunted to maturation of the gonads. Short-winged individuals cannot disperse but reproduce more rapidly producing more generations in a favorable season. Therefore, "short wingedness is always the optimal within-site strategy, and large wingedness always the optimal between-sites strategy" (Vepsäläinen, 1978, p. 232). Which morphs exist and when are determined by the predictability of the patches of water and the isolation of these patches (Fig. 14.2). In conditions of high permanency of a patch and high isolation, populations will be monomorphic and short-winged, because sites will be too isolated for colonization except in very rare cases. Under conditions of low permanency and low isolation, the long-winged morph will predominate. In intermediate conditions both morphs will be selected for, and from Table 14.1 we can conclude that such conditions are by far the most commonly utilized by *Gerris* species in Finland and Hungary.

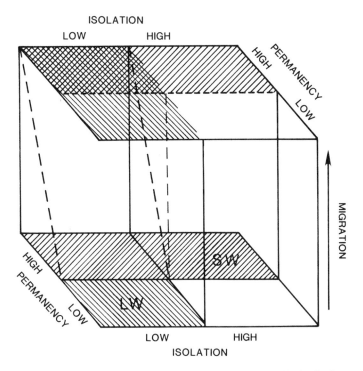

FIGURE 14.2 Relative predictability and isolation of habitats for *Gerris* species which select for the wing-length patterns shown in Table 14.1. For example, high permanency and high isolation select for short-winged (*SW*) monomorphic populations, and low permanency and low isolation select for long-winged (*LW*) monomorphic populations. The effect of migration (on the vertical axis) results in mixing of populations as in the crosshatched area at the upper left-hand corner. From Vepsäläinen (1978); reprinted with permission. Copyright © 1978 by Springer-Verlag, Inc., New York.

The intraspecific differences between populations in Finland and Hungary in *G. thoracicus, G. odontogaster, G. argentatus,* and *G. paludum* relate to the scenario depicted in Fig. 14.2. Hungary, being farther south than Finland, is hotter and drier, making water bodies less predictable and more temporary. Here the summer generation always preserves the dispersal option by having monomorphic long-winged individuals in *G. thoracicus* and dimorphism in the other three species (Table 14.1). In Finland even temporary waters are more persistent than in Hungary, and a generation of short-winged individuals has time to mature and breed before the long-winged morph is produced and disperses.

INTERSPECIFIC VARIATION

The following examples illustrate **interspecific variation:**

1. *Parasitoids in the families Ichneumonidae and Tachinidae.*   Among insects we observe a vast range in fecundities from species that lay one egg at a time, or produce one larva at a time as in the tsetse fly and in the Pupipara, to those that lay thousands of eggs or even millions of eggs in some social insects, and perhaps 30,000 eggs a day in *Macrotermes natalensis* (Wilson, 1971). The parasitoids do not show this vast range but a significant portion of it, with some species laying 20–30 eggs in a lifetime to those that lay about 5000 (Fig. 14.3). I have looked at what determines fecundity in these parasitoids and why such extreme variation exists among species (Price, 1973a, 1974a, 1975a).

In the Ichneumonidae and Tachinidae, fecundity correlates very well with the number of production lines for eggs, the ovarioles in each ovary (Fig. 14.3) making it easier to obtain data and compare a large number of species. When this is done for parasitoids on one host and plotted in comparison with the survivorship curve of the host, it becomes clear that the egg production of a parasitoid is closely tuned to when it attacks the host (Fig. 14.4). Since species attacking early host stages are internal parasites as larvae that emerge from the host cocoon, the larvae are subjected to the same survivorship curve as that of the host. At early host stages hosts are relatively plentiful and aggregated in this colonial sawfly, so many, but necessarily small, eggs can be laid quickly—hence the high fecundity of *Olesicampe*. *Euceros frigidus* lays eggs on foliage, and a planidial first instar larva climbs onto a sawfly larva as it passes by, but this parasitoid larva can only mature on another parasitoid that attacks this host, making the probability of finding the necessary host very low, and fecundity is very high to compensate for this. At the other extreme, parasitoids attacking cocoons hidden in the soil cannot find many hosts per unit time, but when they do, survivorship is quite high. Females produce a small number of large eggs. The differences in design of reproductive organs are very obvious, with large lateral oviducts for egg storage in the

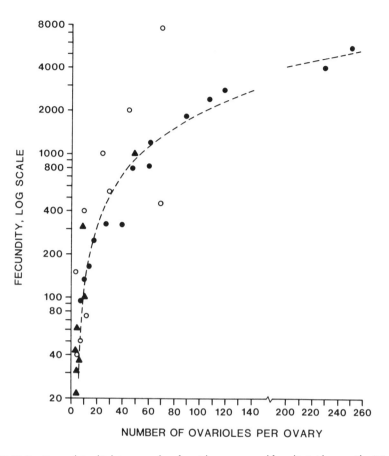

FIGURE 14.3   Linear relationship between number of ovarioles per ovary and fecundity in Ichneumonidae (triangles) and Tachinidae (circles). From Price (1975a); with permission.

high-fecundity species, and a short ovipositor for laying eggs into the host, contrasting with almost no lateral oviduct in the low-fecundity species, but a long ovipositor for piercing cocoons and reaching protected hosts and a large uterus gland for lubricating passage of the egg down the long and narrow ovipositor (Fig. 14.5).

When the ovariole number of many ichneumonids is plotted against the stage of host attacked (Fig. 14.6), the same pattern emerges as in Fig. 14.4, and the increased scatter of points results from differences in parasitoid biology. For example, the species with very low fecundity which attack eggs ("d" in Fig. 14.6) actually attack eggs in clusters, such as the egg cocoons of spiders, and the parasitoid life cycle is completed in the egg mass, so we should expect them to have ovariole numbers similar to parasitoids attacking host pupae (Price, 1975a). The species with higher than usual ovariole number for

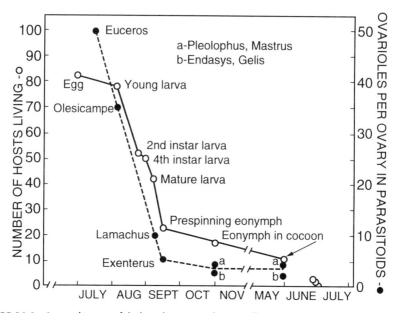

FIGURE 14.4 Survivorship curve of the host, the Swaine jack pine sawfly, in a typical generation (open circles and solid line; from McLeod, 1972), and the number of ovarioles per ovary in parasitoids, (solid circles and dashed line) plotted in synchrony with the time at which each species most commonly attacks the host. From Price (1974a).

attacking pupae ("h" in Fig. 14.6) is also unusual among the Ichneumonidae in being gregarious in the larval stage, so several eggs are laid per host, requiring a synchronous development of these eggs.

Primitive Ichneumonidae attacked late stage hosts, and the radiation of the family has been largely up the survivorship curve of the host to earlier stages (Price, 1974b). The inevitable increase in fecundity was permitted by ease of finding hosts, or provisioning young, and required by higher probability of death. During the phylogeny of the group there has been a tendency for egg production to balance mortality experienced in the host (see later discussion on the balanced mortality hypothesis).

A similar pattern in fecundity in relation to host characteristics is seen in the Tachinidae (Price, 1975a) (Figs. 14.3 and 14.5), where the probability of finding hosts seems to be the dominant factor in determining fecundity. These basic patterns in reproductive strategy are probably characteristic of the large family Ichneumonidae, with perhaps 60,000 species, the moderately large family Tachinidae, the Conopidae, and the parasitoid Sarcophagidae (Price, 1975a), illustrating the broad application of the constraints on fecundity for parasitoids.

2. *Milkweed bugs in the genus* Oncopeltus. Several species of tropical and temperate milkweed bugs have been studied, particularly in relation to diapause, migration, and egg production schedules (see Dingle, 1978a, 1981; Chaplin and Chaplin, 1981 and references therein).

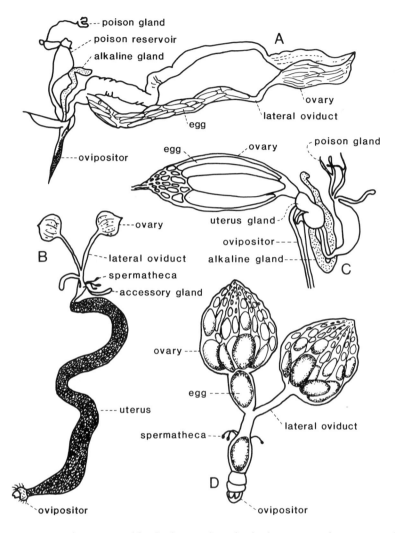

FIGURE 14.5   Reproductive organs of female Ichneumonidae and Tachinidae: (A) *Enicospilus americanus* (Ichneumonidae), a highly fecund species with a short ovipositor and large lateral oviducts; (B) *Leschenaultia exul* (Tachinidae), a highly fecund species that lays eggs on the host's food and are then eaten with food by the host; (C) *Trachysphyrus albatorius* (Ichneumonidae), with few ovarioles, short lateral oviducts, and a long ovipositor; (D) *Hyperecteina cinerea* (Tachinidae) with few (10) ovarioles per ovary, and storage for one egg in each of the lateral oviducts and the median oviduct, which lays eggs on adult hosts. From Price (1975a); with permission.

Concerning diapause, Dingle (1978a) predicted that in tropical locations, where food and climate were favorable throughout the year, no photoperiodically induced diapause would occur. This was supported in a study of five tropical species of *Oncopeltus* and the temperate species *Oncopeltus fasciatus*. Only in the last species was a diapause induced by shortening day lengths.

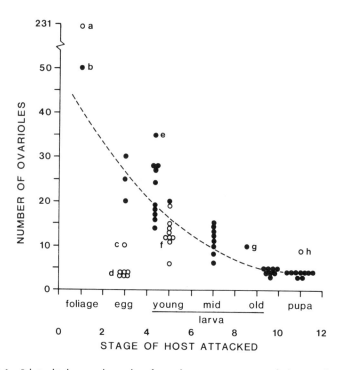

FIGURE 14.6   Relationship between the number of ovarioles per ovary in species of Ichneumonidae and the host stage attacked by each species. Open circles represent species that are exceptional: species at "d" and "h" are discussed in the text, others are explained in Price (1975a). From Price (1975a); with permission.

In relation to flight and migration, Dingle (1978a) predicted that tropical and island populations and species would contain a lower proportion of individuals showing long-duration flights than in temperate populations on the mainland. This was because temperate regions can be exploited by a species such as *O. fasciatus,* which cannot survive temperate winters, only by the evolution of extensive migration ability with overwintering in southerly latitudes. No such adaptation is required in the tropics, where only local dispersal is essential. Extensive dispersal in tropical islands is both unnecessary and deleterious since, once an island is left, the probability of finding another is low. This prediction was supported in both studies on different populations of *Oncopeltus fasciatus* and between species of *Oncopeltus* (Table 14.2). The temperate populations of *O. fasciatus* flew longer on average than tropical populations and a higher proportion of individuals in the population flew for 30 minutes or more during tethered flight. There was also an interesting difference between nondiapausing adults that were reproductively active and usually flying locally between milkweed patches and adults in reproductive diapause and ready to make the long migration to overwintering sites in the south. The percentage flying for 30 minutes or more was 24 in the nondia-

TABLE 14.2   Flight Characteristics of *Oncopeltus* Species.

Species	Population	Mean Flight Duration (Minutes)	Proportion Flying 30 Minutes or More
O. *fasciatus*	Iowa—diapause	46	0.70
	Michigan—nondiapause	39	0.24
	Florida	9	0.13
	Puerto Rico	1	0.03
	Guadeloupe	0.5	0.05
O. *unifasciatellus*		0.2	0
O. *sandarachatus*		1.5	0.03
O. *cingulifer*	Laboratory	0.5	0.02
	Trinidad	14.5	0.15

SOURCE: Data from Dingle (1978a).

pause population and 70 in the diapausing population. The tropical species all showed a low proportion of individuals flying for more than 30 minutes and most flying for a very short duration (Table 14.2). The maximum flight duration observed for O. *unifasciatellus* was only 5.2 minutes, whereas the longest flight for the Michigan population of O. *fasciatus* was 9.5 hours (see also Dingle, 1996).

Egg-production schedules also differed between temperate and tropical species (Landahl and Root, 1969). The temperate O. *fasciatus* starts reproducing earlier in life and has a higher age-specific fecundity than the tropical species, O. *unifasciatellus* (Fig. 14.7), thereby achieving a higher instantaneous rate of population increase, r. The clutch size of each species is similar, but clutches are laid more frequently in the temperate species. This is the kind of pattern to be expected in a temperate–tropical comparison, where the temperate species has a limited time in which to reproduce and more unpredictable conditions, leading to higher potential mortality of progeny (see the section "r- and K-selection" later in the chapter).

When local movement of *Oncopeltus* species in the Cauca Valley of Colombia were studied by Root and Chaplin (1976), differences were related to the kind of food required in order to mature. One species, O. *cingulifer*, was the more sedentary and could mature by feeding on vegetative tissues of milkweeds, whereas O. *unifasciatellus* was less sedentary and required milkweed seeds in order to mature. Milkweeds grow continuously in this area; patches flower and fruit asynchronously; and patches are subject to catastrophic setbacks, so that seed supplies are less predictable than vegetative parts, necessitating more local dispersal in the seed-dependent species. This difference contrasts with the differences in long-distance dispersal found by Dingle (1978a) (Table 14.2).

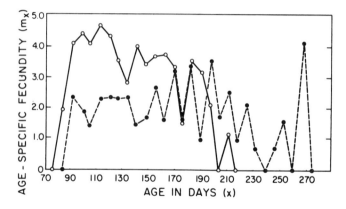

FIGURE 14.7   Change in egg-production (female eggs only) with age for *Oncopeltus fasciatus* (solid line) and *O. unifasciatellus* (dashed line). Since clutch sizes were similar, the higher rate of the former species early in reproduction was achieved by more frequent clutch production. From Landahl and Root (1969).

3. *Delphacid planthoppers.* As in the water-striders in the genus *Gerris,* delphacid species are commonly polymorphic for wing size, with macropterous individuals capable of dispersal and migration and brachypterous forms unable to fly. The frequency of these forms in species and populations indicates the extent to which they are adapted to the local habitat in terms of an ability to escape from it or to persist in it. Denno (1994; Denno et al., 1991) searched for broad patterns in wing polymorphism in relation to habitat persistence and found the following interesting results.

Habitat persistence was defined as the maximum number of generations that could be completed while the habitat persisted. Obviously, species occupying frequently disturbed sites, including agricultural crops, could complete only a few generations during the life of the habitat. Keeping vegetation type more or less constant at low-profile plants less than 1–2 m in height, the persistent habitats included bogs and marshlands. Data on 35 species of delphacids showed a very strong pattern of rapidly decreasing percent of macroptery in females and males as habitat persistence increased (Fig. 14.8). As in the gerrids, there is a trade-off between dispersal ability and reproduction. Macropterous forms have a longer preoviposition period and generally produce fewer eggs (Denno, 1994). Hence, in persistent habitats, brachypterous forms have an advantage over macropterous forms.

In polymorphic species, there is also a difference in the sensitivity to density in relation to the production of macropterous types. In temporary habitats, occupied by *Prokelesia marginata,* for example, males are almost always macropterous, and macroptery in females increases rapidly with density (Fig. 14.9). However, its congenor, *Prokelesia dolus,* occupying persistent habitats, has brachypterous and macropterous males and females at almost all densi-

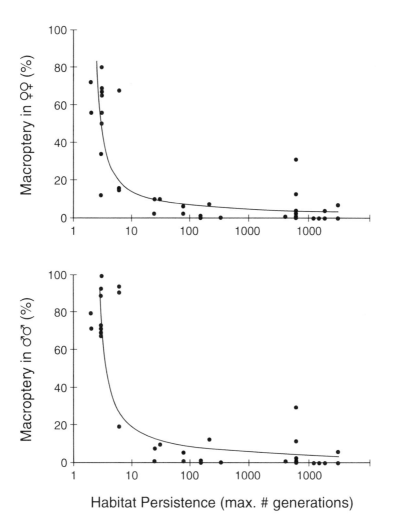

FIGURE 14.8   Natural frequencies of macroptery in females (top) and males (bottom) of 35 planthopper species in relation to habitat persistence. From R. F. Denno et al., Density-related migration in planthoppers (Homoptera: Delphacidae): The role of habitat persistence, *Am. Nat.*, 1991, **138**:1513–1541. Published by the University of Chicago Press. Copyright © 1991 by the University of Chicago. All rights reserved.

ties, and the frequency of macroptery increases, in relation to density increases, relatively slowly (Fig. 14.9).

Wing polymorphism in planthoppers and other groups provides an index for the discovery of broad patterns in life history variation and population dynamics, well exploited in the literature (e.g., Denno and Perfect, 1994;

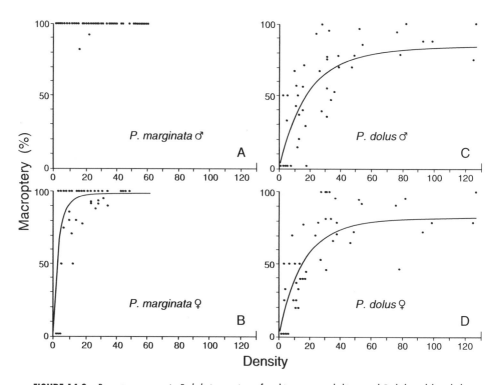

FIGURE 14.9   Percent macroptery in *Prokelesia marginata,* found in temporary habitats, and *Prokelesia dolus,* which typically occupies persistent habitats. As rearing density in cages increases, *P. marginata* shows a rapid increase in macroptery in females, while *P. dolus* shows a much less dramatic response in males and females. From R. F. Denno et al., Density-related migration in planthoppers (Homoptera: Delphacidae): The role of habitat persistence, *Am. Nat.,* 1991, **138:**1513–1541. Published by the University of Chicago Press. Copyright © 1991 by the University of Chicago. All rights reserved.

Denno and Peterson, 1995). As noted in Chapter 4, it is these kinds of broad patterns and their mechanistic understanding that contributes so importantly to the development of theory in insect ecology.

These examples illustrate many patterns of variation in life history traits involving diapause, fecundity and reproductive rates, timing of reproduction, wing polymorphism, and dispersal, together with the important evidence that these traits are under genetic regulation, although in certain examples there is a significant influence of environment on this expression of traits. Life histories are under the influence of natural selection.

Although these patterns represent variation over a considerable number of species, the question must inevitably be asked: Are there broader patterns in life history traits that cut across taxonomic boundaries to provide a more global picture? Two kinds of large-scale patterns have been discussed in the literature, classified as the **clutch size debate** and *r*- **and** *K*-**selection.** As with any effort to increase generality, questions of validity arise and controversy increases.

CLUTCH SIZE DEBATE

Lack's Hypothesis

Lack (1954) noted interspecific and intraspecific variation in clutch size in birds, with a trend toward decreasing clutch size with decreasing latitude. He argued that clutch size evolved through natural selection to correspond to the largest number of young for which the parents could, on average, find enough food. Itô (1959, 1980) has been a strong proponent of this view, generalizing to organisms other than birds by arguing that it is the procurability of food by the young that determines fecundity and clutch size, and parental care evolves, as in birds, where finding food is difficult. Cody (1966) agreed with Lack that in temperate regions, where periodic local catastrophes reduce populations and maintain populations below the carrying capacity of the habitat, K, natural selection will tend to maximize the reproductive rate, r. In the tropics, however, with a more climatically stable environment, where catastrophes are rare, populations will be at saturation densities, and any adaptive variation that will increase the carrying capacity will usually be favored by natural selection (MacArthur, 1962). Increasing K is equivalent to increasing the population density with the same resources. Here energy will be utilized in ways different from that in temperate regions. Selection will be for increased energy used in predator avoidance, intraspecific and interspecific competition, and more energy expended per individual progeny reared (see also Cody, 1971; Foster, 1974). Thus Cody used Lack's hypothesis but also extended the hypothesis to make it more generally valid.

Among the insects we do see examples of the trend for higher and earlier fecundity in life cycles in northern populations of insects compared to more southerly populations. This was seen in the comparison of milkweed bugs in the genus *Oncopeltus* (Fig. 14.7). Peschken (1972) also demonstrated this pattern in populations of the chrysomelid beetle, *Chrysolina quadrigemina*. After the beetle became established in California to control the Klamath weed, it was introduced into British Columbia for the same reason and there became adapted to local conditions. Peschken found that the northern beetles now lay more eggs than the California beetles and that these eggs are laid early in the beetle's life (Fig. 14.10), both traits contributing significantly to the increased reproductive rate r as clutch size theory would predict. In Lewontin's (1965) terminology (Fig. 13.14), both A and T have been shifted forward. However, when *Epilachna niponica* was moved 30 km south to a warmer climate in Japan, after 10 years females have evolved with a shorter

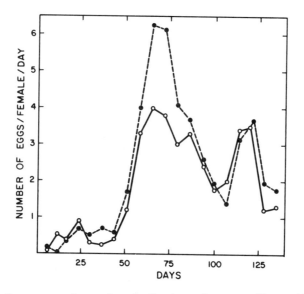

FIGURE 14.10 Change in egg production with age for *Chrysolina quadrigemina* in California (solid line) and British Columbia (dashed line), from day 1 to day 135 of their oviposition periods. From Peschken (1972).

life span and higher fecundity, opposite to the examples just provided (Oh-gushi and Sawada, 1996).

Another hypothesis relating to the clutch size of birds is Ricklefs's (1970) **counteradaptation hypothesis,** in which he invokes the importance of differential coevolutionary rates of predator and prey as a regulator of food intake, and consequent energetic allocation to egg production. Therefore, Ricklefs slightly modifies Lack's hypothesis but also incorporates some of Cody's general theory. Since evolutionary rates of predator and prey must be estimated, considerable difficulties exist in testing this hypothesis.

Balanced Mortality Hypothesis

An older hypothesis than Lack's on clutch size and fecundity may be called the **balanced mortality hypothesis** (e.g., Price, 1974a). Many ecologists have expressed the view that egg production is adapted to counter the relative hostility of the environment in which the organism lives (e.g., Rensch, 1938; Skutch, 1949, 1967; Smith, 1954). Cole (1954b) stated that the high fecundity frequently seen in parasites and marine organisms is an adaptation to ensure population survival when probability of death is high. Lack (1947a, 1949, 1954, 1966) has argued against this view, but it has gained support from many (e.g., Dobzhansky, 1950; Fretwell, 1969; Pianka, 1970). The supporters of this hypothesis would argue that the adjustment of egg production through the action of natural selection during evolutionary time results in a degree of balance between fecundity and mortality in the long term but cannot result in seasonal adjustment to mortality factors.

If the ichneumonid parasitoid example given earlier is examined (Fig. 14.4), the validity of the hypothesis can be tested. The relative egg-production capacity of females is known and the probability of survival of progeny can be estimated from the survivorship curve of the host (from McLeod, 1972), since immature parasitoids must suffer the same mortality as the host. Thus we know the actual probability of survival, derived from the host survivorship curve, and the predicted probability, derived from the balanced mortality hypothesis; that is, the probability of survival is the reciprocal of fecundity (or in this case relative fecundity estimated by ovariole number). The predicted and actual values are closely correlated, the regression line accounting for 93% of the variance (Fig. 14.11). Since the number of eggs laid is also dependent on the supply of larval food (hosts) available to the searching female, both Lack's hypothesis and the balanced mortality hypothesis are supported by this analysis.

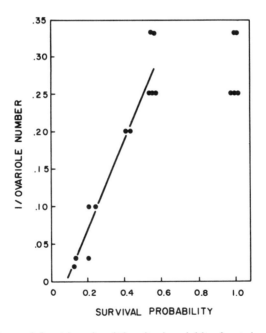

FIGURE 14.11   Reciprocal of ovariole number, which predicts the probability of survival to maturity of progeny, in relation to the actual probability of survival estimated from the host survivorship curve. The regression line accounts for 93% of the variance. The five points at a survival probability of 1.0 were not included in the regression, since egg production must be geared for the more stringent conditions represented by a survival probability of 0.6. At both probabilities the same five species were involved (cf. Fig. 14.4). From Price (1974a).

Williams (1966a) would argue that the hypothesis is back to front and that actually mortality balanced fecundity. Since populations over long periods of time have a zero growth rate, an increase in fecundity automatically increases the number of individuals that die. Also, a female with limited resources has two basic strategies available. She can put much energy into few progeny or little energy into many, with correspondingly predestined low and high mortality rates, respectively. However, the forces that cause increased fecundity must be understood. Another look at the parasitoid complex on *N. swainei*, and the family Ichneumonidae as a whole, throws light on the debate. If the phylogeny of the Ichneumonidae is examined, both the Tryphoninae and the Ephialtinae are considered to be primitive and they have relatively low fecundities. An evolutionary trend of changing hosts or time of attack in this primitive stock would thus start from a female with low fecundity, as pointed out above. If a female attacked an early and thus a vulnerable stage of a host in which mortality was high, she may well be unsuccessful in leaving viable progeny. Such a host could be exploited only with increased fecundity. Thus natural selection for increased fecundity results through the force of mortality. Mortality is the proximate selective factor that results, ultimately, in change in fecundity. Mortality is balanced by fecundity, in this case. Under other evolutionary conditions Williams may be correct. Willson (1971) also discusses conditions under which fecundity is balanced.

r- AND *K*-SELECTION

r- and *K*-selection have already been discussed under the clutch size debate. *r*-Selection operates to maximize reproductive rate in harsh, unstable environments where populations remain well below the carrying capacity of the environment and resources are not limiting except for time to reproduce successfully (MacArthur and Wilson, 1967). Populations will typically exist at the lower end of the logistic growth curve, before the inflection point induced by the onset of competition (see Fig. 14.1). At the other extreme of the continuum, in predictable, stable environments populations will frequently reach the carrying capacity of the environment and selection will operate by improving adaptations for living in these crowded, competitive conditions, where enemies such as predators and parasites will be very effective—*K*-selection. Populations will typically exist in the upper part of the logistic curve, close to *K*.

Under *r*-selection the reproductive rate will be maximized to counter harsh conditions, and early reproduction becomes an important component of *r* because of time shortage. To maximize *r* many progeny are produced per unit time, but little investment can therefore be placed in each. Under *K*-selection reproductive rate is reduced, but much investment is placed in each progeny in terms of protection from enemies, enabling progeny to compete successfully both inter- and intraspecifically.

The bare essentials of this hypothesis on *r-* and *K*-selection therefore involve five characteristics that must be integrated if the hypothesis is to be applied:

1. Relative stability of environmental conditions and population sizes
2. Reproductive rate of species, *r*
3. Carrying capacity of the environment relative to mean population size.
4. The frequency of intraspecific and interspecific competition
5. The importance of enemy attack

The hypothesis has been oversimplified too frequently to a comparison of fecundities only, without concern for the other necessary factors. This has tended to discredit the argument, and although exceptions can be given, as the following examples illustrate, the theory has provided an important broad conceptual framework and a theme with which to integrate many studies. In this book the concept of *r-* and *K*-selection is alluded to directly or indirectly many times. The literature on this subject may be entered by consulting MacArthur and Wilson (1967), Pianka (1970, 1972), Hairston et al. (1970), Thompson (1975), Stearns (1977, 1992), Calow (1978), Parry (1981), and Roff (1992).

Examples that support the theory of *r-* and *K*-selection have been given above (Figs. 14.7 and 14.10), but examples that do not support the theory also exist. Reproductive strategies on rich food sources and in harsh conditions may lead to convergent strategies, with parental care and high parental investment per progeny evolving in an apparently *r*-selecting environment as well as in a highly competitive environment. Wilson (1971) considered that presocial behavior in insects has evolved repeatedly and frequently in very different situations. Where physical conditions are harsh it is frequently necessary for the mother to protect her progeny, as does a ground-dwelling cricket, *Anurogryllus muticus,* from excessive moisture and fungal growth. Conversely, where rich food sources are exploited, competition is fierce and a parent may need to protect resources for her young. Such a case is seen in the *Necrophorous* beetles (e.g., *N. vespillio*), which build a nest and feed their young in a manner similar to birds. Another example concerns the phytophagous coccinellid, *Henosepilachna pustulosa,* which feeds on thistle, *Cirsium kagamontanum,* in central Japan. Ohgushi and Sawada (1981) studied populations on an elevational gradient from the bottom of a valley at 200 m above sea level to the top at 400 m, so the theory would predict a more *K*-selected strategy at the base of the valley and a more *r*-selected strategy at its head. In fact, in the downstream sites adult beetles lived for a shorter time and had a higher reproductive capacity than those of the upstream beetles. The authors found that thistle plant population stability was higher in the upstream sites, but predation pressure was higher in the downstream sites, indicating that factors were not correlated in the way predicted by the theory. Kambysellis and Heed (1971) also emphasized the importance of microenvi-

TABLE 14.3   **Some Representative Species of Hawaiian Drosophilidae Studied by Kambysellis and Heed (1971) in the Groups They Designate.**

Group	Species	Number of Ovarioles Per Fly	Number of Mature Eggs Per Ovariole	Ovipositional Substrate
I	*Scaptomyza mauiensis*	4	0.25	Morning glory flowers
II	*Antopocerus aduncus*	11	0.63	*Cheirodendron* leaves
IIIa	*Drosophila adiastola*	46	1.00	Leaves, fruits, stems, and roots of endemic plants
IIIb	*Drosophila sejuncta*	57	1–3	*Pisonia* bark

ronment for *Drosophila* larvae as the most important factor selecting on fecundity in Hawaii. The range in ovariole number per fly and the number of mature eggs per ovariole is considerable (Table 14.3), and a pattern is clear. Where resources are in very small packets, as is the pollen in morning glory flowers, and very ephemeral, ovariole number is low and only one egg is matured at a time (group I). Species utilizing small but longer-lasting substrates, such as decaying leaves (group II), have intermediate fecundities, while those utilizing decaying stems and similar substrates (group III) have the highest fecundity, with group IIIa usually depositing eggs singly and those in group IIIb laying clusters of eggs on the relatively large patches of resources found in decaying stems and branches. As the stability of the microhabitat increases, the rate of egg production of the species increases.

Idiosyncrasies of species may also influence egg production profoundly, as in the butterfly, *Heliconius charitonius*, which feeds on pollen when available, enabling it to produce eggs over a protracted period compared to females without pollen (Fig. 14.12 and Dunlap-Pianka et al., 1977), and in related species such as *Dryas julia* that do not feed on pollen.

These examples should act as a cautionary note, so that *r*- and *K*-selection are not invoked in a facile manner. Many factors impinge on the evolution of life histories, and many solutions result from the evolutionary process. It is clear that many solutions do not fit the hypothesis on *r*- and *K*-selection. Both Stearns (1992) and Roff (1992) are severely critical of the concepts of *r*- and *K*-selection, mostly because when they are invoked the necessary details for correct application of the terms are lacking. "The lack of a comprehensive theory of how life histories should evolve under density dependence is a minor scandal; responsible discussion of the strengths of *K*-selection is one approach to creating such a theory. Enough people have found it a useful framework in which to interpret their observations that it must contain an element of truth. The problem is to identify that element. Other papers use *r*- and *K*-selection as a convenient description of a pattern without having any idea of what caused it. That is a misleading waste of time . . . *r*- and *K*-selection

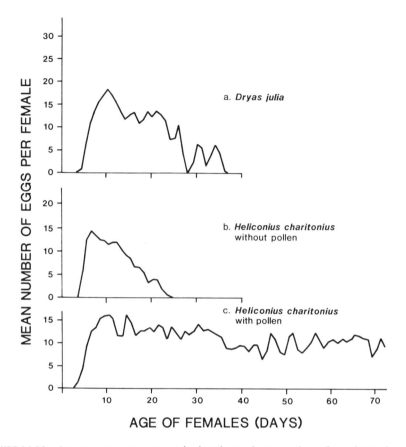

FIGURE 14.12   Oviposition patterns in (a) *Dryas julia,* (b) *Heliconius charitonius* without pollen, and (c) *H. charitonius* with pollen. From H. Dunlap-Pianka et al., Ovarian dynamics in heliconiine butterflies: Programmed senescence versus eternal youth, *Science,* 1977, **197:**487–490. Copyright © 1977 by the American Association for the Advancement of Science. Reprinted with permission.

was a helpful initial generalization with a part worth saving: the focus on density dependence" (Stearns, 1992, p. 207).

SURVIVORSHIP OF BREEDERS

Each reproductive strategy carries with it a cost of reproduction that influences the probability of survival of the parent. Williams (1966a) points out that it is convenient to recognize two categories of adaptation: (1) those that relate to the continued existence of the individual, provided that it retains its reproductive ability, and (2) those that relate to reproduction itself. The two are intimately related. Williams (p. 159) says, "Heart failure and mammary failure have exactly equivalent effects on the fitness of a female mammal (except in the one bottle-manufacturing species)." Anyway, continued existence of the individual and reproductive effort must be balanced. Williams (p. 171) asks: "What determines how much effort an organism should expend, and to how much danger it should expose itself, in its efforts to reproduce?" Parental sacrifices are sometimes enormous and sometimes slight. Williams (1966a, p. 172) says, "In dealing with this problem, I will take it as a basic axiom that selection will adjust the amount of immediate reproductive effort in such a way that the cost in physiological stress and personal hazard will be justified by the probability of success." Tinkle (1969) was stimulated by this focus of Williams (1966a,b) to look at the reproductive patterns in lizards. A large clutch size is a more risky proposition for a female lizard than a small one, and several clutches per year are riskier than a single one because (1) more energy must be channeled into eggs over a short period of time, (2) because the female is loaded down with the heavy eggs and therefore is less able to escape predators, and (3) because the female is exposed in an egg-laying position for a longer time per season. Therefore, it seems reasonable to anticipate that the species that have a larger fecundity per season will have a lower survivorship. This is the relationship that Tinkle found in 13 species of lizard.

Murdoch (1996a,b), working with ground beetles (Carabidae), found a similar relationship: "The survival of adult female Carabidae, from near the end of one breeding to the start of the next, is inversely proportional to the amount of reproduction done in that first breeding season." He also suggested that this is one of the compensatory mechanisms leading to population stability in Carabidae, which may also be widespread in other insects. Apparently, egg production was directly related to food supply (Murdoch, 1966a), so that when food supply was poor in the fall, reproduction was low, survival to the following spring was increased, and in fact the major effort in reproduction was delayed until more favorable conditions prevailed. Survival of the reproducing individual is subject to evolutionary change, even adaptive change within a generation in the case of carabid beetles.

SURVIVORSHIP OF PROGENY

What about survival of the progeny? In arthropods the number of progeny produced is subject to considerable variation between species as discussed already in the chapter. Mitchell's (1970) work on the analysis of dispersal in mites is also revealing. For, except for soil mites, habitats are discontinuous and transient. Many are predators and parasites, and their minute size prevents them from actively seeking new habitats with fresh food resources. The number of progeny that actually find a resource must be maximized and is dependent on the number of progeny produced by a female, and the probability of progeny finding food, which in turn depends on the accuracy in the aim of the dispersal mechanism. Thus there exists the simple relationship.

$$F = NP_d$$

where F is the number of founders, N the number of dispersants, and P_d the probability of discovery of a new resource. A founder must be a female and must have mated or be parthenogenetic. If the sexes disperse independently and unmated, the number of actual founders is greatly reduced because a male and female must arrive independently at the same place, thus reducing the probability of founding a colony. Actually, where mites disperse before mating, they tend to disperse in groups rather than as single individuals. There seem to be three patterns in mites identified so far:

1. *Tetranychid mites.* *Eotetranychus sexmaculatus*, the prey species that Huffaker used in his orange prey–predator experiments, is a member of this family. These tetranychids climb up stems, drop on a silken thread, and disperse in the wind. Dispersal as aerial plankton is potentially very wasteful of progeny, which may be compensated for in three ways according to Mitchell (1970). When resources begin to deteriorate, reproduction of females may be greatly increased so that there can be a significant increment to the standing crop of adult females within a week. Then the males may be smaller and less abundant than females, so that most of the standing crop biomass is females. Finally, mating usually occurs immediately after emergence of the female; thus almost all active females are mated and capable of founding a colony. Each of these ways seems to have been adopted by tetranychids. For example, one species, *Panonychus ulmi,* has males that weigh 2.8×10^{-6} g and make up 26% of the population. Females weigh 11.1×10^{-6} g and make up 74% of population. Thus the proportion of biomass of the population going into dispersants is $(0.74 \times 11.1)/[(0.26 \times 2.8) + (0.74 \times 11.1)] = 0.92$, or 92% of biomass! Similar and additional adaptations relative to dispersal have been described by Mitchell (1973) for *Tetranychus urticae*.

2. *Mesostigmatid mites.* One species that Springett (1968) studied is a predator of muscid fly eggs, which are laid on carrion. For dispersal they are phoretic and climb aboard carrion beetles (Silphidae) that have matured on the carrion; the beetles fly to fresh carrion and the mites reach new sources of fly eggs. This accounts for the large numbers of mites commonly seen on carrion beetles, such as *Necrophorus* and *Necrodes* spp. In fact, the relationship between the mites and the beetles is rather a beautiful one. If the beetles arrive at new carrion without mites, their progeny die in competition with fly larvae. If mites are transported to the carrion, they eradicate this competition, and the beetles on which the mites depend survive, a good example of mutualism. Apparently, because of this reliable dispersal method there seems to be

no budgeting of resources for increasing the numbers of dispersants. Males are only slightly smaller than females and there is a 50:50 sex ratio.

3. *Tarsonemid scolytid egg parasite.* This sort of mite, in the genus *Iponemus,* is commonly found on the elytral declivity of scolytid beetles (which is often hollowed) and on the thorax. Scolytid beetles burrow into bark and lay an egg in each egg niche along a gallery in the bark. The fertilized female mites ride a beetle to a new breeding site and drop off as the beetle lays an egg. The egg is immediately walled off by the parent so that a mite remains with a single egg. The mite matures and lays 60 eggs, which hatch, and the hatchlings mate. There are only about three males in a brood and these mate with the females produced from the same beetle egg. Up to three female mites can produce full broods on a single egg, the progeny do not feed on the egg, and males die without ever having fed. Females disperse on newly emerged adult beetles and do not feed until after they drop off on a freshly deposited egg. The mites retain a high N by reproducing so effectively on a single egg. Also, they maintain a high P_d, as only few beetle eggs are parasitized out of a total brood, so many beetle progeny can mature and carry the mites to new breeding sites. The efficiency of resource utilization for dispersants is high. Since males and females are the same size, the numbers yielded by a single female are adequate for calculation of the proportion of biomass going into dispersal. This is $57/(3 + 57) = 0.95$, or 95%. Here the principle of allocation is at work to promote the fitness of reproducing females (see later in this chapter).

In contrast to plants, moving organisms can adapt to placing their progeny in suitable situations for survival. In an interesting series of papers, Chandler (1967,1968a–c) describes how adults of several species of Syrphidae—flower or hover flies in which the larvae of some are predaceous on aphids and other small arthropods—are sensitive to the density of the aphids on a plant and lay eggs preferentially at certain densities. The species illustrate differing strategies in the placement of eggs. The obvious strategy is to lay most eggs where food for larvae is most abundant, a behavior shown by *Syrphus balteatus* (Fig. 14.13). This, however, may not always prove to be the best strategy, as aphid populations frequently increase rapidly and then decline just as rapidly. If a dense population happens to be on the decline, laying many eggs at that density is a wasteful activity. Another species, *Platycheirus manicatus,* shows the alternative behavior of laying eggs almost independently of density, and if host populations change rapidly, new ones may become available to progeny laid in the absence of hosts. Several other oviposition patterns are seen in these Syrphidae.

Another interesting variation on the theme of evolved patterns in egg allocation concerns the egg size of butterflies that feed on forbs and woody plants (Slansky and Scriber, 1985). Leaves of woody plants are generally tougher and less nutritious than those of forbs. Hence a caterpillar on a woody plant

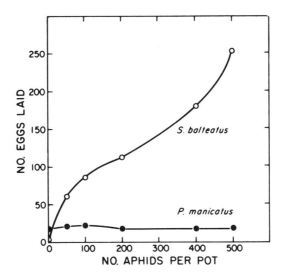

FIGURE 14.13   Differences in egg-deposition strategies between two syrphid flies, *Syrphus balteatus* and *Platycheirus manicatus,* on field beans in pots with various population sizes of *Aphis fabae.* Data from Chandler (1968b).

may feed longer and experience a higher probability of attack by carnivores than one on a forb, or the species may evolve to produce a larger egg to compensate for poorer nutrition. In the nymphalid butterflies this strategy seems to have evolved in many species (Table 14.4).

Clustering of eggs may also increase progeny survival if grouped larvae are more effective in defense than individuals. Among butterflies the subfamily Nymphalinae has a disproportionately large number of members that cluster

TABLE 14.4   Egg Sizes of Nymphalid Butterflies on Forbs and Woody Plants.

| Egg Size (mm³) | Number of Species | Percentage of Total Species in Each Egg Size Category that are: | |
		Forb-feeders	Shrub- and tree-feeders
0.100–0.199	10	80	20
0.200–0.299	6	83	17
0.300–0.399	3	67	33
0.400–0.499	9	56	44
≥0.500	10	20	80

SOURCE: Slansky and Scriber (1985), with permission.

their eggs (Labine, 1968). The larvae may have branched spines, synchronous head jerking, or a deterrent odor, all defenses that are enhanced by grouping (Ford, 1957). Many traits associated with group feeding are discussed by Bowers (1993) and Fitzgerald (1993).

PRINCIPLE OF ALLOCATION

We have seen repeatedly in this chapter, and in Chapter 12 (Fig. 12.5), that resources are allocated in different ways to growth, reproduction, survival, dispersal, few large eggs or many small eggs, and so on. Energy, nutrients, and time are utilized as efficiently as the process of natural selection permits by each species faced with the particular environmental conditions that prevail. Cody (1966) called this partitioning to promote fitness the **principle of allocation.** This principle implies that resources an organism has to partition between life history functions are limited and a compromise is always made in the allocation of these resources. A pitcher-plant mosquito may allocate some eggs to the nondiapause morph and others to the diapause morph, to spread the risk in an unpredictable environment. The mushroom fly can allocate much more per unit time to reproduction if it reproduces paedogenetically, but dispersal capacity is lost. If early egg production is increased in the flour beetle, later egg production declines. When delayed senescence was selected for in *Drosophila melanogaster,* several reductions in efficacy in early life were observed. Long-winged morphs of the water-strider spend more time developing and in the initiation of reproduction, but they maintain the ability to disperse. Similar relationships hold for delphacid planthoppers. Some parasitoid species produce many small eggs while others mature few large eggs, a pattern also seen in the Hawaiian Drosophilidae studied by Kambysellis and Heed (1971). When early egg production is high, longevity declines, as seen in carabid beetles and the phytophagous coccinellid *Henosepilachna pustulosa.* Almost every example used in this chapter illustrates the principle of allocation in one way or another.

The concept of allocation actually dates back much further than Cody (1966), in relation to the evolved form of species, under the concept of **material compensation.** Rensch (1959) traces the history of this concept back to ancient Chinese and Indian writers, and provides many examples from the insects where, for instance, wing size is inversely related to gonad size. Matsuda (1979) also gives examples, including the inverse relationship between wing size and hind-leg size in several insect orders, and Traub (1972) discusses the concept applied to many aspects of the structure of fleas. The study of such trade-offs in the evolution of life histories is well developed in the literature, as may be seen in Roff (1992) and Stearns (1992).

LIFE HISTORIES AND POPULATION DYNAMICS

The use of life history attributes of species to understand the evolutionary background to their population dynamics is one of the areas of synthesis in ecology receiving increasing attention. For example, Nothnagle and Schultz (1987) asked "What is a forest pest?" and analyzed many life history traits to find some answers. Wallner (1987) undertook a similar study comparing the biological attributes of endemic and epidemic insect species, otherwise known as *nonoutbreak* and *outbreak species,* respectively.

The themes that run through this literature include evolved characteristics of species and higher taxa which set limits on habitat use and population behavior, and the correlated nature of these traits. The evolutionary history of a genus or a species may well set **phylogenetic constraints** on the ecology of the species (Nothnagle and Schultz, 1987; Price et al., 1990). Traits may be correlated as sets of adaptations in response to certain environmental challenges, called **adaptive syndromes** by Root and Chaplin (1976; see also Root, 1975). "As organisms perfect a mode of life, their evolution is channeled so that a variety of adaptations are brought into harmony" (Root and Chaplin, 1976, p. 139). Eckhardt (1979, p. 130) defined the adaptive syndrome as "the coordinated set of characteristics associated with an adaptation or adaptations of overriding importance, e.g., the manner of resource utilization, predator defense, herbivore defense, etc." Eckhardt recognized that species faced with similar challenges in their evolutionary background are likely to evolve with similar adaptive syndromes. Therefore, groups of species with similar phylogenetic constraints may well show comparable adaptive syndromes. The search for these kinds of patterns among species and their mechanistic explanations creates an evolutionary basis for understanding aspects of population dynamics, as discussed in Chapter 19.

A fine example of this integrated perspective was discussed in Chapter 11 in relation to Gilbert's studies on *Heliconius* butterflies (Fig. 11.1). We saw how the evolved life history characters of species, such as reproductive longevity, size, and aposematic coloration, coupled with behavioral traits, in turn influenced population structure, size, and stability. Can similar approaches be used to reveal the evolutionary underpinnings of species with different kinds of population dynamics?

Wallner (1987) focused on the differences between K-selected rare species and r-selected eruptive species (Table 14.5). Several life history traits were considered, such as adult life span, mobility, fecundity, and degree of aggregation of eggs and larvae. He used the classification of Berryman and Stark (1985) and ranked types from K- to r-selected as rare, gradient, cyclic, and irruptive. Rare species are those that persist at low densities. Pest gradients occur when average density is determined entirely by site conditions and environmental disturbances such as drought. Pest cycles are when density is

TABLE 14.5 Biological Traits Associated with Different Kinds of Population Dynamics. The Relative Importance of a Trait Increases as the Number of Plus Signs Increases.

Biological Trait	K-selected ←———— Rare	Gradient	Cyclic	————→ r-selected Irruptive
Adult life span	+++	+++	+	+
Adult feeding	+++	+++	+	+
Habitat restriction	+++	+++	+	+
Response to plant defense	+++	+	+	++
Degree of host specificity	+++	+++	+	++
Flush–crash cycles	+	+++	+++	+++
Adult vagility	+	++	++	+++
Fecundity	+	++	+++	+++
Alteration of environment	+	+	++	+++
Degree of egg clumping	+	+	+++	+++
Degree of larval aggregation	+	+	+++	++
Response to weather	+	++	+++	+++
Importance of biological control	+	++	+++	+++
Utilization of foci or refuges	+	+	+++	+++
Incidence of polymorphism	+	+	+++	+++

SOURCE: Wallner (1987). Reproduced, with permission, from the *Annual Review of Entomology*, vol. 32, © 1987, by Annual Reviews Inc.

dependent on intrinsic factors yielding predictable cycles. Pest eruptions were defined as when populations remain very low but at irregular intervals increase to very high densities. The qualitative evaluation of each biological trait showed a clear pattern of concentration of important traits for rare and gradient species in the upper set of characters and the opposite for cyclic and irruptive species (Table 14.5). It would be valuable to know which of these factors are most basic in the evolution of a species and which are most derived. What are the phylogenetic constraints and the adaptive syndromes? Nevertheless, this study was one of the early attempts to bring to population dynamics an understanding based on life history evolution.

Nothnagle and Schultz (1987) used quantitative analysis of traits in the outbreaking species of forest Macrolepidoptera. They argued for differences in the adaptive syndromes of life history traits between eruptive and endemic species. Pest species were regarded as **risk concentrators** and nonpest species as **risk spreaders.** These terms are akin to those used by Rhoades (1985) of **opportunistic herbivores** and **stealthy herbivores,** respectively. As an example of the risks to which outbreak species are exposed, lymantriid and lasiocampid species, such as the gypsy moth and tent caterpillars, respectively, have the following adaptive syndrome. They overwinter as eggs and in a large mass, eggs hatch early in the spring, and larvae feed gregariously. The risks are high mortality if phenological overlap of egg hatch with flushing of new foliage is poor, larvae must find their own food often at considerable distance from the clutch of eggs, and aggregated larvae increase susceptibility to diseases. The variation in these factors may well form the basis for eruptive population dynamics.

Life history traits in eruptive and endemic forest Macrolepidoptera were also examined by Hunter (1991, 1995b). Again, aggregation of progeny was the most consistent difference between eruptive and endemic species, resulting from females with poor or no flight ability, eggs in masses, and gregarious feeding. Associated with this aggregation of progeny is limited dispersal. Females emerge with a full complement of eggs ready to lay—they are proovigenic—and being heavy with eggs, they are poor fliers or may not even fly at all, as in the winter moth and gypsy moth. The adaptive syndrome predisposes populations to increase locally under favorable conditions and reach eruptive status.

This topic of the relationship between life history traits and population dynamics will be discussed again in Chapter 19, when an attempt will be made to synthesize the many elements that contribute to an understanding of population dynamics.

Behavioral Ecology

Behavioral aspects of ecology cover a wide range of topics, for they involve every stage in the life cycle of an organism. Every day an organism must attempt to find enough food or a female must find suitable locations for laying eggs. Searching behavior for food is a vital activity for most organisms at one stage or another in the life cycle. In every breeding season a mature individual must search for a mate, court that mate, copulate, and perhaps engage in parental care. In social insects such as ants, termites, and bees, the colonies function as a unit largely because of the integrated behavior of the individual members.

This large subject is covered by many books that should be consulted for a broader and more detailed consideration than can be achieved in this chapter. A comprehensive treatment of insect behavior is provided by Matthews and Matthews (1978), and Alcock's (1979a) book on animal behavior considers many aspects of insect behavior. Behavioral ecology is treated specifically in Krebs and Davies (1978, 1981, 1991) and Morse (1980). Behavior in relation to reproduction of insects is the subject of a symposium volume edited by Blum and Blum (1979). Many aspects of social behavior are discussed by

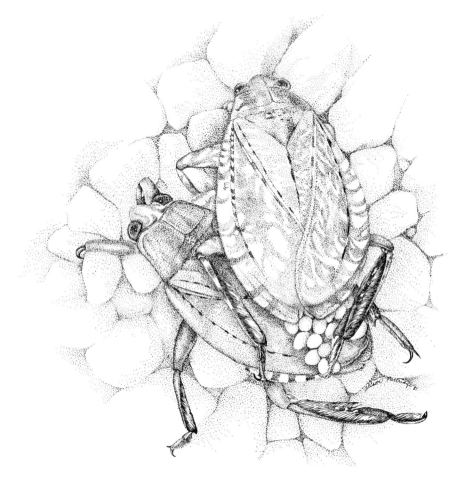

Paternal care of eggs or young among insects is rare. Here a male belostomatid bug (below), *Abedus herberti,* receives eggs on his back from a female (above) with which he has just mated. From an original photograph by Robert L. Smith. Drawing by Alison Partridge.

von Frisch (1967), Michener (1974), Wilson (1971, 1975), Hermann (1979–1982), and Hölldobler and Wilson (1990).

In this chapter we consider two major aspects of an insect's life: finding food and finding mates, and related topics. Both subjects come under the umbrella of foraging strategies. Since an individual must usually find food before it is ready to mate, the subjects are treated in that order.

FINDING FOOD

The question of how an insect finds food can be broken down into two major strategies: (1) the physiological aspects of food recognition and orientation to food, which is a complex subject in itself, dealt with briefly in Chapter 5 in relation to herbivorous insects [Dethier's (1976) fascinating book *The Hungry Fly* is devoted to the physiological basis of behavior in relation to fly feeding, a subject on which he spent much of his very active research effort, but this subject is not developed here]; and (2) the choice of the best food available and the allocation of time needed to exploit this food efficiently. How does a bee behave when foraging among flowers for nectar or pollen in order to make best use of its time? How does an aphid behave when selecting a site for rearing its brood? What are the important factors integrated together when such behavioral choices are made? These are the kinds of questions that are explored in this chapter.

HABITAT CHOICE BY AN APHID

An example will help to illustrate the value of this approach for understanding the ecology of insects. For a *Pemphigus betae* aphid it is important to find a suitable place at which to initiate gall formation, for in this gall she will rear all her progeny before she dies. In terms of the fitness of a female, this is a crucial part of her life cycle, for a poor choice of galling site can result in the production of no progeny, while a female at a good site may produce almost 200 progeny in the gall (Whitham, 1978, 1980). (See also Fig. 17.13 and associated text.)

A stem mother hatches from an egg on the bark of a narrowleaf cottonwood, *Populus angustifolia,* in early spring, and as a first instar nymph, searches for a very young leaf on which to initiate gall formation. Although the leaves destined to become the largest leaves on a tree are rare compared to smaller leaves, these are selected preferentially by stem mothers (Fig. 15.1). All the largest leaves were colonized, although they represented only 1.6% of the leaf population on a tree (Whitham, 1978). In fact, selection pressure for finding large leaves is so strong that these minute females, only about 0.6 mm

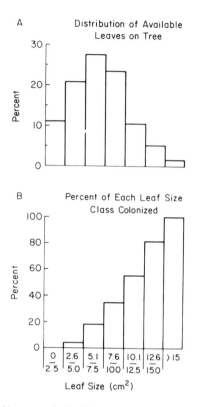

FIGURE 15.1   Leaf sizes available to (A) and utilized by (B) stem mothers of the galling aphid, *Pemphigus betae*. From T. G. Whitham, Habitat selection by *Pemphigus* aphids in response to resource limitation and competition, *Ecology*, 1978, **59:**1164–1176. Copyright © 1978 by the Ecological Society of America. Reprinted with permission.

long, will fight for hours over occupancy of the best leaves and the best position on those leaves (Figs. 15.2 and 15.3). Females are strongly territorial for a site about 3 mm long at the base of a large leaf in which they can produce many progeny, engaging in kicking and pushing contests to secure such favorable sites (Whitham, 1979).

FIGURE 15.2   Typical fighting posture of *Pemphigus betae* females, only about 0.6 mm long and in their first instar. The larger female usually wins the better territory, which is about 3 mm long at the base of the leaf. Drawing by Pam Lungé. From Whitham (1979). Reprinted by permission from *Nature*, **279:**324–325. Copyright © 1979 by Macmillan Journals Limited.

FIGURE 15.3 Positions of two *Pemphigus betae* females competing for the better territory at the base of the leaf. When the dominant female was removed at 17.10, the displaced female moved down to occupy the better position. From T. G. Whitham, The theory of habitat selection: Examined and extended using *Pemphigus* aphids, *Am. Nat.*, 1980, 115:449–466. Published by the University of Chicago Press. Copyright © 1980 by the University of Chicago. All rights reserved.

The selection pressure for an ability to find and utilize leaves that will become the largest in the population results from two related factors. First, galls on large leaves can be initiated easily with 100% success on the largest leaves, while galls on small leaves tend not to develop beyond a small swelling—they are abortive 80% of the attempts on the smallest class of leaves (Figs. 15.4 and 15.5). Second, females are much more fecund on large leaves than on small leaves, producing on average 189 progeny per gall on large leaves and 59 progeny per viable gall on small leaves (Whitham, 1978). The net effect of these two factors is that females on average produce almost six times more progeny on large leaves than on small leaves.

The problem of selecting an appropriate site for gall initiation is made more complex because a female has a choice between making a gall on an occupied leaf or initiating gall formation on a vacant leaf, meaning that the leaf is probably smaller. Even a third gall on a leaf may be developed successfully. The rules defining a correct decision are complex, for they involve an integration of leaf size and leaf occupancy, and the stakes are high. If a female climbs onto a large occupied leaf, she will probably contest the occupant with a kicking match, in which the larger aphid usually wins. The winner obtains the lowest and best position on the leaf and is likely to produce about 189 progeny. The loser must then make a decision on whether to initiate a second gall on that leaf or move to another leaf. If she stays, she faces a 20% chance of gall abortion, but if successful, she may produce 75 progeny. If she departs to a leaf that is smaller and also vacant, she faces a probability of 18%

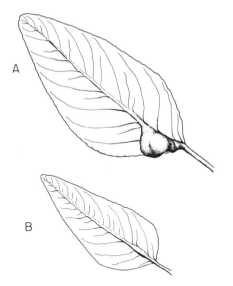

FIGURE 15.4 Successfully developed gall of *Pemphigus betae* (A) and aborted gall (B) on leaves of *Populus angustifolia*. From T. G. Whitham, Habitat selection by *Pemphigus* aphids in response to resource limitation and competition. *Ecology, 1978,* **59:**1164–1176. Copyright © 1978 by the Ecological Society of America. Reprinted with permission.

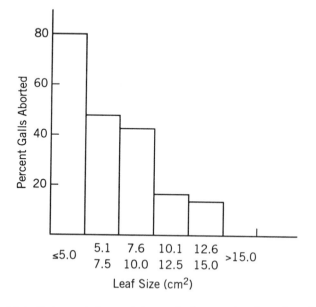

FIGURE 15.5 Relationship between leaf size and abortion of galls in *Pemphigus betae* when only one gall is attempted per leaf. From T. G. Whitham, The theory of habitat selection: Examined and extended using *Pemphigus* aphids, *Am. Nat.,* 1980, **115:**449–466. Published by the University of Chicago Press. Copyright © 1980 by the University of Chicago. All rights reserved.

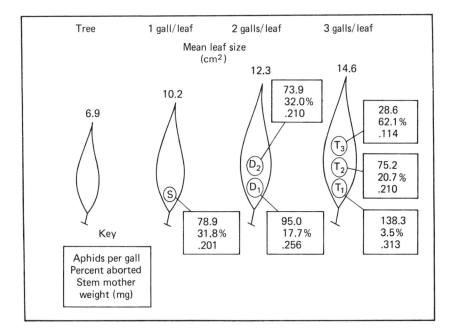

FIGURE 15.6   Effect of gall position on female fitness of *Pemphigus betae*. Leaves are drawn to scale and effects of position on leaf are given in boxes: (top) number of aphids per gall; (middle) percent aborted; (bottom) stem mother weight (mg). From T. G. Whitham, The theory of habitat selection: Examined and extended using *Pemphigus* aphids. *Am. Nat.*, 1980, **115**:449–466. Published by the University of Chicago Press. Copyright © 1980 by the University of Chicago. All rights reserved.

abortion and production of 95 progeny on a medium-sized leaf, and 32% abortion and 79 progeny on a small leaf. The full spectrum of choices and their consequences are defined clearly in Whitman's (1980) figure (Fig. 15.6). The remarkable fact is that for such minute insects searching in the large heterogeneous canopy of a cottonwood tree, they can achieve 84% of their potential fitness estimated when aphids are provided with unlimited search time, freedom to settle where they choose, and a perfect response to leaf quality and aphid density (Whitham, 1980).

How can aphids make such accurate choices? Studies by Zucker (1982) suggest that a possible cue used by a female is the concentration of phenolic glycosides. If she searched for the lowest concentrations that she could find available in a leaf population, this would result in correct decisions in terms of position of gall initiation on the leaf and choice of leaves. Individual leaves have a gradient of phenolic concentrations along the leaf with the lowest concentrations, usually at the base of the leaf. The largest leaf on a shoot has lower phenolic concentrations than those on the smallest leaf on a shoot on 95% of shoots analyzed, and when more than one gall is formed on a leaf, this leaf is consistently low in phenolic glycosides. Phenolic concentrations in leaves with 3, 2, 1, and 0 galls were 2.5, 2.9, 3.5, and 5.7 optical density units, respectively. Even trees with lower phenol levels in leaves are colonized more heavily than those with higher levels, although the wingless stem mothers are not able to choose between trees.

A general model of habitat choice formulated by Fretwell and Lucas (1970) and Fretwell (1972) was applied by Whitham (1980) to foraging by *Pemphigus betae*. The model assumes that the habitat that is most favorable to individuals will be occupied first, but if these individuals defend territories, the best sites will become fully occupied and other individuals will occupy less favorable habitats, with the poor habitats utilized last. Thus the fitness of an individual will depend on an interplay between habitat quality and the number of other individuals already present (Fig. 15.7). If the best locations

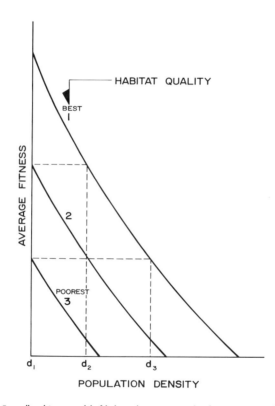

FIGURE 15.7   The Fretwell and Lucas model of habitat choice assumes that fitness is negatively correlated with density of competitors. As the best habitat is colonized and densities become high, poorer habitats are likely to be utilized. For example, the dashed lines indicate that with the same average fitness, populations may reach d_3 in habitat 1, d_2 in habitat 2, and d_1 in habitat 3. Therefore, an individual is likely to select the poorer habitat 2 if the population is below d_2 there, and at d_3 or above in the best habitat 1. From T. G. Whitham, The theory of habitat selection: Examined and extended using *Pemphigus* aphids. *Am. Nat.*, 1980, 115:449–466. Published by the University of Chicago Press. Copyright © 1980 by the University of Chicago. All rights reserved.

are occupied, an individual may benefit in terms of fitness by choosing lower-quality habitat, where there is less competition. The difficulty in assessing this model lies in measuring fitness of individuals in different-quality habitats.

The model can be tested very well, however, using the gall-forming aphids, because the progeny produced in the gall provide an accurate measure of fitness, and the gall is stationary, so position or habitat and number of galls per habitat can easily be evaluated. Whitham (1980) used the leaf as the habitat unit, and as already documented, leaf size provides a good estimate of habitat quality. Thus a family of fitness relationships could be developed for the range of leaf sizes utilized by aphids (Fig. 15.8). From these relationships it is

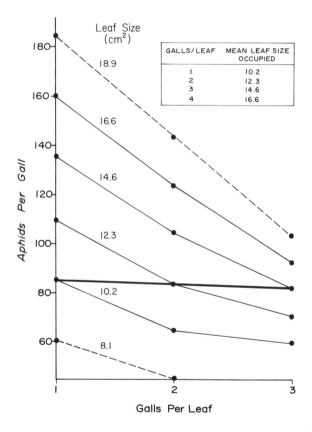

FIGURE 15.8 Fretwell and Lucas model applied to *Pemphigus betae* aphids showing habitats (leaves) of varying quality (size) and differences in population density per habitat (galls per leaf), in relation to an estimate of average fitness (aphids per gall). Lines show the expected fitness of a female selecting among leaf-size habitats and densities of galls per leaf. The bold solid horizontal line indicates average fitness for one, two, and three stem mothers per leaf, showing that fitness is evened out across densities, supporting the model's prediction. From T. G. Whitham, The theory of habitat selection: Examined and extended using *Pemphigus* aphids, *Am. Nat.*, 1980, **115**:449–466. Published by the University of Chicago Press. Copyright © 1980 by the University of Chicago. All rights reserved.

clear that when the best habitat is already occupied by one gall, a second aphid will leave more progeny by leaving to find a slightly smaller leaf that is unoccupied. The fitness of the second aphid would be about 150 progeny on a leaf of 18.9 cm^2, but 160 progeny when alone on a leaf of 16.6 cm^2. The alternative for the second aphid would be to enter into a territorial dispute with the original occupant, and should she be displaced, the second arrival could expect to leave over 180 progeny—hence the prolonged unladylike behavior for positions on the best leaves. The real test of the model comes when average fitness per habitat is calculated, for the model predicts that fitnesses should be evened out over habitats because habitat quality and competition are integrated. In fact, this is the case (Fig. 15.8), and again the wonder is that these minute aphids can make such excellent choices in a complex heterogeneous environment.

PREFERENCE–PERFORMANCE LINKAGE

Consequences of habitat choice for the aphid continue for two generations. Stem mother body weight increases with increasing leaf size, and as body weight increases, the number of aphid progeny per gall increases. The progeny produced are winged migrants which fly to their summer host, usually *Chenopodium album* (Chenopodiaceae) in the study area. Body weight of these winged migrants increases as stem mother body weight increases and the number of embryos carried by the migrants increases with body weight. Thus the original decisions made by the tiny stem mother on which leaf to attack and gall has dramatic consequences for her own reproductive success resulting from her own progeny. In addition, there are consequences for her progeny's reproductive success when they reach the summer host, just as in the case for a butterfly.

Whitham's elegant studies were ongoing independently from, but in parallel with, studies on butterflies. While small aphid stem mothers are constrained to walk in search of optimal sites for bearing young within a tree, a female butterfly has a considerable range over which she can make decisions on where best to lay her eggs. She may choose among habitats such as sunny or shady; she may select one plant species and not another, and she may even prefer one plant part relative to another in the same host-plant species. Whereas a butterfly female must make many decisions on where to lay eggs, for she may lay only one egg per host plant, the gall-forming aphid female makes only one decision. But the parallels between these two very different lifestyles are apparent enough.

In fact, we should expect that through natural selection, a female's decisions on where to lay her eggs, her preferences for oviposition, should relate to the welfare of her progeny, or in general terms their performance, as in the case studied by Whitham. An **ovipositional preference** and **progeny perform-**

ance linkage should be anticipated. The development of research on this concept reveals again the gradual refinement of concepts, working from a single perception of nature to a more pluralistic perspective on the range of linkages, from strong to absent, revealed by empirical research (Thompson and Pellmyr, 1991).

Singer (1986) sought a common, conceptually based lexicon relating to ideas on choosing, selecting, preferring, or discriminating among host plants and random attack. He also wished to focus on the essential consequences involving the fates of progeny, which may be measured as survival probability, growth rate, fertility and fecundity, or efficiency of conversation of host biomass to insect biomass (*ECI*; cf. Chapter 12 and Table 12.3). He chose the term **preference** as encompassing enough to cover the adult female's deviations from random behavior. Insect preference is indicated when plant species A and B are equally accessible to an ovipositing female, but more eggs are laid on host species A than B within a short period of time. More generally, preference may be detected when an ovipositing female has equal encounter rates on parts of an individual plant, such as leaves or stems, or long shoots and short shoots, or different plants within the same species, or different plant individuals in an array of species.

The consequences for progeny of an ovipositional preference can be evaluated and discussed using the single word **performance** (Singer, 1986), but a researcher may choose to measure the results for progeny of a mother's preference. Then we can ask: Does mother know best? (e.g., Courtney and Kibota, 1990). Does an ovipositing female make the best possible decisions during oviposition to maximize larval performance and her own fitness? Whitham's studies suggest that she does, but others indicate that mother does not know best (e.g., Thompson, 1988; Courtney and Kibota, 1990). In some cases either mother is sloppy or we do not understand the full range of selection factors working on a lineage relating to ovipositional decisions.

Thompson (1988, p. 4) defined the term *ovipositional preference* as "the hierarchical ordering of plant species by ovipositing females. In a choice trial in which plants of equal mass of several species are offered simultaneously, preference would be expressed as the proportion of eggs laid on each of the plant species." He used the term *performance* as a "composite term for survival at all immature stages (egg, larval, pupal), larval growth rate, efficiency as indicated by nutritional indices . . ., pupal mass, and resultant adult fecundity and longevity."

The fascinating possibility is that within a population of adults, there may be variation in preference and consequent performance. If some females in a population showed high preference and some low, this would indicate important variation on which natural selection could act. Perhaps host-plant shifting with consequent speciation might be involved (e.g., Price, 1996a). Or, as host-plant phenology changed in a season, low preference may enable a shift in oviposition from one plant species to another, with beneficial consequences for progeny (Wiklund, 1984). Or, perhaps, adult behavior is constraining in the evolutionary development of a lineage (Singer, 1994). Wiklund (1974, 1981, 1984) made important contributions by emphasizing variation in preference within a species, the swallowtail butterfly, *Papilio machaon*. He argued that a hierarchical preference of host plants would ensure oviposition on the best possible host plant for larvae when present, but that alternative hosts were within the ambit of successful larval utilization in other places or at other times. For such a wide-ranging species as *Papilio machaon*, such oviposition preferences optimize utilization of ideal hosts but allow for creative change in diet in novel habitats.

Although controversial, we may extrapolate the preference–performance concept to the population of insects in natural environments in which plant species, or plant modules, are not equally available as they are in choice experiments. But the laws of probability allow an expected attack rate based on frequency of the different categories of available resources and the observed frequency of attack. Care is needed to define what is available as resources at the time that females are ovipositing, such that expected attack and actual attack can be compared directly. In the case of Whitham's (1978, 1980) studies working on the aphid, *Pemphigus betae*, in a natural setting, we have already seen the validity of this extrapolation. Leaves on short shoots of poplar flush in a brief time and nymphs attack very young leaves. Hence what is available and what is utilized can be evaluated accurately. We lose some of the interesting variation within the population, but we gain broad ecological insight to the plant and herbivore interaction. This is particularly valuable and less restraining when we consider highly specialized insect species utilizing only one host species, but show a strong preference–performance linkage relating to plant module size variation. Such a view enables extension of the concept to the understanding of plant resources utilizable by herbivores and the effects on herbivore distribution, abundance, and population dynamics (see Chapter 13), especially Table 13.10).

Within this framework of the preference–performance linkage, many studies have evaluated its strength, using many different approaches (Table 15.1). One example is Rausher's (1982) test showing good correspondence between female ovipositional preference and larval relative growth rate (RGR), growth efficiency (ECI), and survival. Two populations of the checkerspot butterfly, *Euphydryas editha*, were compared. One population in Del Puerto Canyon (DP), in California, was composed of females that oviposited almost exclusively on *Pedicularis densiflora* (Scrophulariaceae) in the presence of host plants used by the butterfly species elsewhere, all in the same family as *Pedicularis*: *Castilleja foliolosa, C. affinis,* and *Collinsia bartsiaefolia*. Another population at Indian Flat (IF) was 120 km distant, in which females laid most eggs on *Collinsia tinctoria*, although other adequate food plants were available: *Plantago erecta, Penstemon brevifloris, Orthocarpus* sp., and *Castilleja* sp.

TABLE 15.1 Ten Examples of Tests of Preference–Performance Linkage in Insects.

Correspondence Between Female Preference and Larval Performance	Herbivore Species and Higher Taxa	Criterion for Evaluation of Performance	References
Good between species	*Euphydryas editha* (Lepidoptera: Nymphilidae)	Larval biomass gain over 4 days	Rausher (1982)
		Larval mass after 10 days	Singer et al. (1988)
		Survival of eggs and larvae to 10 days	Singer et al. (1994)
	Papilio machaon (Lepidorptera: Papilionidae)	Survival, development time and mass	Wiklund (1975, 1981)
Poor			
Poor host more apparent	*Anthocharis cardamines* (Lepidoptera: Pieridae)	Larval survival	Courtney (1981, 1982a,b)
Introduced host plants	*Pieris napi* and *Pieris occidentalis* (Lepidoptera: Pieridae)	Larval survival	Chew (1977)
Rare preferred host	*Euphydryas chalcedona* (Lepidoptera: Nymphalidae)	Larval preference	Williams (1983); Williams et al. (1983)
Introduced host plants	*Ostrinia nubilalis* (Lepidoptera: Pyralidae)	Larval establishment and survival	Legg et al. (1986)
Excellent within species	*Pemphigus betae* (Homoptera: Aphididae)	Survival to maturity and fecundity	Whitham (1978, 1980)
	Euura lasiolepis (Hymenoptera: Tenthredinidae)	Survival to adult stage	Craig et al. (1989)

387

In experiment 1, Rauscher used newly hatched larvae from each population and fed them on each of the major hosts, *Pedicularis densiflora* used in the DP population and *Collinsia tinctoria* used in the IF population. Larvae were weighed at the beginning of the experiment and after 4 days, and biomass gain was estimated. Survivorship was estimated over the same period of time. Performance was measured using only first instar larvae. In the DP population, larvae performed better on their habitual host than the IF population, and the IF population performed better on its habitual host than the DP population, both showing better performance than the other population on the host that is usually fed upon (Table 15.2). Survivorship differed in a similar way, although host species and population effects were not significant. Similar results were obtained in experiment 2, which measured growth rates, consumption rates, and growth efficiency of third instar larvae.

Of course, it would be preferable to measure performance over the full life of immatures in order to estimate performance. However, in the case of checkerspot butterfly larvae, they hatch from eggs in the spring, feed for only about 10 days, and then enter diapause for the summer. Hence it would be extremely difficult to follow cohorts of larvae through their full development. Another concern is that growth rate is usually assumed to result in larger adults with higher fecundity. But these relationships need to be established in each case, for they are by no means general (Leather, 1988, 1995).

Singer (1986) identified the protocols best suited to evaluate preference. He devised his manipulative technique specifically for the major species in Rauscher's studies: *Euphydryas editha* and the host plants in the genera *Pedicularis* and *Collinsia*. Some critical points are noted by Singer. First, the butterfly is docile and can be handled easily. It will oviposit on a suitable host readily if placed on it by hand, and therefore encounter rates can be controlled by the experimenter by exposing females at equal rates to common and rare species. Second, this kind of testing in the field ensures the use of plants in a natural state, uninfluenced by cutting, moving, growth in a greenhouse, or other artificial factors. Third, acceptance of a plant is shown by the butterfly curling the ab-

TABLE 15.2   Results of Performance Tests by Rauscher (1982) on First Instar Larvae of the Checkerspot Butterfly, *Euphydryas editha*.

Host Plant	Del Puerto Canyon Population (DP)		Indian Flat Population (IF)	
	Growth Rate[a]	Survivorship[b]	Growth Rate[a]	Survivorship[b]
Pedicularis densiflora	4.82[c]	0.88	3.65	0.65
Collinsia tinctoria	10.53	0.69	16.44[c]	0.81

[a] Growth rate is in mg/mg per 48 h.
[b] Survivorship is given as the proportion of larvae surviving.
[c] Growth rate on the preferred host plant is shown in italic.

domen and extension of the ovipositor for about 3 seconds. Fourth, the insect is removed from the host plant before oviposition, so the female is kept in a similar motivational state to oviposit. However, even with this technique, as minutes tick by this state changes—as motivation increases, the range of acceptable plant species broadens. Fifth, single females can be tested on one plant of, say, *Pedicularis* and then on a group of the smaller *Collinsia* plants of approximately equal mass. Replication, and changing order of plants offered at random, can yield a powerful experiment on female preference, the variation in preference in the population, and differences among populations. Sixth, rearing from egg to pupa, of species that have simpler life cycles than the checkerspot, on a variety of host-plant species tested for preference, would allow an evaluation of the preference–performance linkage.

Using this protocol, Singer (1983) first showed that there existed individual variation in host-plant preference in a population of *Euphyrdyas editha*. Later, this preference was shown to be heritable and was correlated with offspring performance measured as larval weight after 10 days of feeding on the host (Singer et al., 1988). Rapid evolution of host preference was also noted (Singer et al., 1992; Radtkey and Singer, 1995), some in response to human-induced changes in host plants in traditional sites caused by logging and cattle ranching (Singer et al., 1993). During rapid evolution of preference there was higher within-population variation in preference rank, but in stable populations, preference rank was invariant (Singer et al., 1994).

These studies highlight the important behavioral shifts in populations relating to the evolution of diet change, host-plant utilization, and the constraints and opportunities available to particular species (Thompson and Pellmyr, 1991; Singer, 1994). As we will see in the next example and in Chapter 19, the linkage between preference and performance is also informative on how closely related female ovipositional behavior is to variation in quality of host plants within plant species and how this may affect population dynamics.

Rather than study insect behavioral variation in relation to populations of plant species, we can employ the preference–performance approach to examine the response of insect populations to variation within host-plant populations. An example is provided by study of the gall-forming sawfly, *Euura lasiolepis*, which attacks only one species of willow, *Salix lasiolepis*, under natural conditions (Craig et al., 1989). The focus of interest was how a population of sawflies responds to willow shoot-length variation, and if a preference–performance linkage was evident. A field survey evaluated attack by females in relation to shoot length and larval survival. The relationship between adult female mass and fecundity was studied with caged females ovipositing on potted plants in the field. In a preference experiment, one female was placed in a cage with well-established potted willows in the field for 24 h. All shoots per plant were available to the female, and all had the same number of oviposition sites available, because females oviposit only through the petioles of the youngest leaves at the shoot tip.

With the field survey, Craig et al., established that larvae survived much better on long shoots than on short shoots, with shoot length correlated with ramet age and clonal genotype. Female mass was not well correlated with fecundity, so mass of larvae or pupae was not used as a measure of performance. Sawfly preference for long shoots was strong in the cages, with 64% of eggs laid on the longest 20% of shoots, and in the field the highest survival to the adult stage was recorded at about 85% on the longest shoots on the youngest ramets (number of shoots per length class was equal since all shoots were divided into five equal categories based on shoot length) (see also Table 13.10 for similar results on the same species).

This example makes a convincing case for very high preference–performance linkage, even though the linkage is within a plant species rather than between plant species as most studies on butterflies have been. This case and Whitham's studies may show the highest preference–performance associations reported to date. The tight linkage may well result from the close association of egg position with larval feeding site, within the gall for its entire life—a much tighter relationship than in butterflies, where larvae are free feeding and may travel among plant individuals in later instars, acting as grazers rather than parasites living on a single host (cf. Thompson, 1988; and Chapter 9).

Some reasons for the pattern found by Craig et al., (1989) for the galling sawfly are clear enough. Short shoots abscised much more frequently than do long shoots: over 80% in shorter categories and only about 10% of long shoots. Larvae die in abscised shoots. However, it is not understood why larvae in short shoots or plants in dry conditions die soon after eclosion from the egg.

Therefore, there is good reason to accept that it is possible to extend the study of preference–performance linkage to evaluating responses to plant quality and the construction of life tables (see also Chapter 13). However, care is needed in combining field observations with experiments. These behavioral studies are emerging as an important aspect in the understanding of the population dynamics of insect herbivores, so these linkages or their absence will be discussed again in Chapter 19.

OPTIMAL FORAGING MODELS

Many other models, in addition to the Fretwell and Lucas model, have dealt with the choices made by animals in allocating their effort in selecting (1) the type of food eaten, (2) the kind of patch they forage in, (3) the apportionment of time within and between patches, and (4) the rate and direction of travel while searching for food (Pyke et al., 1977). Starting with Emlen (1966) and MacArthur and Pianka (1966), the general reasoning has been that the more efficiently food or energy can be acquired, the higher will be the fitness of the animal. Thus natural selection should result in organisms that forage so that fitness is maximized. Since the models seek an optimum solution to the costs of foraging compared to the benefits of finding and consuming food, they are generally known as **optimal foraging models** and the theory is **optimal foraging theory** (Pyke et al., 1977; see also chapters in Krebs and Davies, 1978, 1991).

The four kinds of models recognized by Pyke et al. (1977) listed above all relate to insects either as food for predators or to an insect foraging itself. In both cases the models and their tests are relevant to insect ecology because predators' foraging tactics are likely to have selective impact on the insect population, which needs to be understood, and insect foraging itself, such as by parasitoids, is frequently for other insects as food. In the following paragraphs references relevant to insects and other arthropods are provided on each kind of model.

1. Optimal Diet: Models on Animal Choice of Food Types

The models relate to generalists that can utilize several kinds of food, such as predatory birds feeding on arthropods (e.g., Davies, 1977; Krebs, 1978; Craig et al., 1979), predatory insects (e.g., Charnov, 1976a; Davidson, 1978; Gittelman, 1978), and spiders (e.g., Morse, 1979), and pollinators that can choose between types of flower (e.g., Waddington and Holden, 1979; Marden and Waddington, 1981; Waddington and Heinrich, 1981; Waddington et al., 1981).

2. Optimal Patch Choice: Models on Animal Choice of Patch Type in Which to Feed

Many studies relate to use of patches by insectivorous vertebrates (e.g., Royama, 1970; see Chapter 7 in this book; Smith and Dawkins, 1971; Smith and Sweatman, 1974; Krebs, 1978; Heller and Milinski, 1979). But choice between plants as patches by ovipositing insects has been studied by Chew (1977) and Rausher (1979), and choice between leaves as patches within a plant by Whitham (1980), discussed in this chapter. Morse and Fritz (1982) studied the crab spider, *Misumena vatia,* and the consequences of patch choice (see also Iwasa et al., 1981).

3. Optimal Time Allocation: Models on the Time Spent in Different Patches

Several studies relate to insectivorous birds (e.g., Tullock, 1970; Krebs et al., 1974; Cowie, 1977), pollinators (e.g., Whitham, 1977; Hartling and Plowright, 1979; Pyke, 1981), insect predators (Cook and Cockrell, 1978), and parasitoid wasps (Cook and Hubbard, 1977; Hubbard and Cook, 1978;

Jeffrey Waage, 1979b; van Alphen and Galis, 1982). Much of this literature relates to decisions by foragers on when to leave one patch and move to another. It seems that an animal should leave a patch when the rate of food discovery, or host finding by parasitoids, falls to the average rate for the habitat. Such a patch would provide a marginal reward compared to other patches, and the predator, parasitoid, or pollinator will maintain a constant rate of reward over all patches in a habitat by allocating time efficiently. This *marginal value theorem* was developed by Charnov (1976b) and has been tested and supported by Cook and Hubbard (1977) using the parasitoid *Nemeritis canescens*.

4. Optimal Patterns and Rates of Movement

Much of this literature relates to the searching patterns of insect predators and parasitoids, partly because of its importance in modeling predator–prey interactions and its importance in biological control, and partly because it is relatively easy to follow the path of a walking insect either by tracing the path directly or using a video camera (e.g., Van Lenteren et al., 1976). Many references are cited by Pyke et al. (1977) and some more recent studies include Pyke (1978a,b), Waddington (1980), and Zimmerman (1979) on pollinating insects, and Bond (1980), Formanowicz (1982), and Carter and Dixon (1982) on insect predators.

TESTING MODELS

The value of optimal foraging models is that clear predictions can be derived from mathematical models and tested in experiments using foraging animals, so that predicted and observed patterns can be compared. The validity of two or more models can be compared with reference to a certain animal forging in clearly defined conditions. One example concerns models relating to foraging for food by animals that choose between food types. Pulliam (1974) developed a model using the following assumptions:

1. Foragers rank food types according to their net value.
2. Food types are therefore either always accepted or always rejected according to their rank.

Waddington and Holden (1979) constructed an alternative model with different assumptions:

1. Foragers do not always accept certain food types and reject others—they exhibit partial preferences.

2. Foragers are able to integrate the potential reward and cost of handling and distance to reward, such that items with low reward may be utilized if they are close to the forager.

Waddington and Holden (1979) tested these models using honeybees foraging in an artificial arena of flowerlike forms stamped on a flat surface, each flower supplied with sugar solution (= "nectar") or no solution. Two types of flower were arranged in patches, and the number and proportion of flowers with nectar and the caloric value of nectar were varied. Each flower patch was composed of two flower types, but the relative caloric value of each type was adjusted such that in patches A, B, and C they were the same and in patches D, E, F, and G, flowers of type 1 provided an increasingly larger reward than flowers of type 2 in that order. The visitation frequencies of honeybees to each flower type in each patch was observed. When rewards in flower types were equal (patches A, B, and C), the bees spent close to equal times foraging on both types (Fig. 15.9). As rewards became more disparate be-

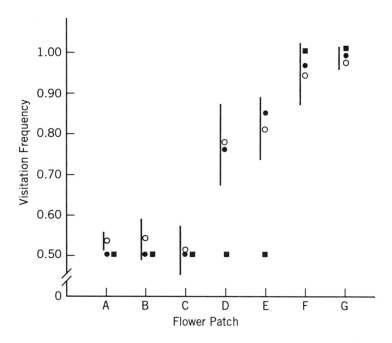

FIGURE 15.9   Visitation frequency of bees to type 1 flowers at each flower patch A to G. Open circles represent the mean visitation frequency and the vertical lines the 95% confidence interval around this mean. Closed squares show the predicted visitation frequencies using Pulliam's model, and closed circles show predictions from Waddington's and Holden's model. From K. D. Waddington and L. R. Holden, Optimal foraging: On flower selection by bees, *Am. Nat.,* 1979, **114**:179–196. Published by the University of Chicago Press. Copyright © 1979 by the University of Chicago. All rights reserved.

tween flower types, the bees showed a graded response spending about 75% of their time on type 1 flowers when the relative reward of type 2 flowers was almost half that of type 1 flowers (patch D) and 95% of their time on type 1 flowers when type 2 flowers provided only 20% of the reward supplied by type 1 flowers (patch F). This result comes closer to fitting the optimal foraging model of Waddington and Holden than the all-or-none choice model of Pulliam in this particular case.

DIFFICULTIES WITH OPTIMAL FORAGING

Difficulties with optimal foraging models involve several implicit assumptions.

1. They usually assume that only one factor at a time is optimized, usually rate of energy intake, without regard for other important factors involved with foraging, particularly the need to compromise foraging by care for avoiding predators and other dangers. Covich (1976) and Sih (1980) consider such compromises.
2. The currency that is optimized is seldom established as the critical currency for the organism under study. Energy is a commonly used currency, but for many animals nutrients may be much more limiting (see Chapter 17). However, energy may well be the critical resource for pollinators gathering nectar (see Chapter 11).
3. Most studies do not show a positive relationship between foraging efficiency and fitness, an essential ingredient of all foraging theory. Thus the best tests of models are those that come closest to a measure of fitness: studies on finding and ovipositing in hosts by parasitoids, and other studies, such as Whitham's (1980), when the number of progeny produced as a result of foraging is determined.
4. It is seldom established that time is so short for an animal that there will be measurable selection for optimizing use of time.
5. The presence of genetic variance in foraging ability on which natural selection can act is implicit in all models but has not been demonstrated in any. Maynard Smith (1978) reviews other problems with optimization theory.

However, whether these difficulties are explicitly stated or implicit in the optimal foraging models, such models can play an important role in the conceptual development of areas such as plant–herbivore relationships, predator–prey relationships, parasitoid–host relationships, and insect population dynamics. Considering that the first models are little over 25 years old, optimal foraging theory has provided a rigorous and creative research approach, and studies on insects have played a central role in the development of this body of theory. Insect ecologists can continue to play an important part in developing this theory and overcoming some of the difficulties associated with many of the models (see Kamil and Sargent, 1981).

FINDING MATES

Besides the search for food, the other major search that an insect must undertake is one for mates. Although this sounds like a simple subject to discuss, there are many alternative ways of finding mates, and the ecological and evolutionary consequences become quite complex. Also, females may be distributed in patches, and a male must then search among patches. Therefore, optimal foraging approaches may be used for mate searching (Parker, 1978a,b). In addition, there are the problems of persuading a mate to copulate, and preventing access by other males both during insemination and until the male's sperm is fully utilized. Sex ratio among progeny may then be regulated and the ideal ratio may change according to conditions (e.g., Werren, 1980; Charnov et al., 1981; Charnov, 1982). Thus there are well-developed conceptual links between finding food and reproductive activity.

Wilson (1975) stated that sex is an antisocial force in evolution—perhaps a surprising comment until one realizes that females usually profit most by mating with one male, who shares parental responsibilities fully with the female, while males usually profit most by inseminating many females without devoting energy to parental care. Offspring tend to overexploit their parents, while parents must frequently reject their young in order to breed again. There are strong conflicts within the family unit. In this section we explore how some of these conflicts are resolved and the consequences for each sex.

PARENTAL INVESTMENT

Trivers (1972) provided a general theory of parental investment that helps in understanding the diversity of sexual behavior. **Parental investment** is defined as any time or energy drain that increases the survival probability of progeny at the expense of the parent's ability to invest in other progeny. Usually, the female invests much more per offspring than a male invests. An egg costs more than a spermatozoan both in absolute terms and in terms of the proportion of energy an adult has which it can allocate to reproduction. Also, a female insect must find appropriate oviposition sites for eggs, whereas the male is usually uninvolved with the progeny once he has inseminated the female. Therefore, parental investment by a female is limited to relatively few progeny by the high-energy cost of eggs and oviposition, whereas a male can produce many progeny for the same energetic cost.

However, the reproductive success, or fitness, of both sexes is correlated directly with the number of progeny produced that survive to reproduce. Thus male reproductive success can be much greater than that of the female. Given this fact, we can expect male fitness within a species to be much more variable than that in females. For example, a male that copulates with three females is likely to leave three times as many progeny as a male that copulates with only one. Conversely, in many insect species a female will produce a similar number of progeny whether she is inseminated by one male or several. Thus strong competition between males for females is to be expected, whereas competition between females for males is unlikely to be acute. In general, we should expect to see adaptations in males for assuring insemination of several females developed to extremes not seen in the females in relation to competition for males. Darwin (1871) understood this clearly.

MALE PARENTAL INVESTMENT

However, such adaptations may be balanced by improved fitness in males that invest more per individual progeny than others. Fitness may be increased, for example, if a male offers a female some food, for several reasons (Thornhill, 1976; Thornhill and Alcock, 1983):

1. He may be preferred by a female compared to males that offer less or no food (e.g., in the mecopteran *Panorpa*; Thornhill, 1974).
2. He may supply the female with nutrients which increase the number of eggs or the quality of eggs she makes, or the rate of egg production (e.g., in *Heliconius* butterflies; Dunlap-Pianka et al., 1977).
3. If in a predaceous species, he may avoid being eaten by the female (e.g., in balloon or dance flies, Empididae; Kessel, 1955).

Thornhill (1976; Thornhill and Alcock, 1983) recognized three sources of food provided by male insects to females during courtship:

1. Glandular products of the male, such as from dorsal glands common in orthopteroid insects (see also Schal and Bell, 1982), salivary smears or globules used by some male flies and *Panorpa*, spermatophores in many orders of insects, and mating plugs.
2. Food collected by males—males in the Bittacidae (Mecoptera) and Empididae (Diptera) offer captured prey items. Males of a species of tiphiid wasp offer nectar to females, while males of *Stilbocoris natalensis* (Hemiptera: Lygaeidae) offer a fig seed injected with saliva as a nuptial gift.
3. The male himself may be eaten during, or immediately after, insemination, as observed in a carabid beetle, two species of ceratopogonid midges, and preying mantids. (In the last case only laboratory observa-

tions have been made.) This type of self-sacrifice may still increase male fitness if mates are very scarce, if competition for males is severe, or if females typically suffer from poor nutrition and thus egg production is greatly enhanced. If the male mates only once, he should be highly selective in choice of females to ensure many viable progeny, and he should be just as fastidious in selection as the female (Thornhill, 1976).

An extreme case of male parental investment involves species in which males take care of eggs or young exclusively—a rare phenomenon in animals in general, and insects are no exception. Of the thousands of known insect species, about 100 have reproductive behavior involving exclusive male parental care. All are restricted to the order Hemiptera, with known cases in the Belostomatidae, Gerridae, Reduviidae, Coreidae, and Aradidae (Melber and Schmidt, 1977; Ridley, 1979; Smith, 1980). A particularly well-studied case concerns the belostomatid *Abedus herberti,* investigated by Smith (1979a,b, 1980 and references therein). These bugs are quite common in shallow streams in Arizona. Females lay eggs on the backs of males and the males brood the eggs until hatching. Brooding consists of aerating eggs by exposing them to the atmosphere and rocking eggs below the water surface—brood pumping. These behaviors are essential to embryonic development and are extremely costly to the male for several reasons:

1. They are energetically demanding.
2. The hunting efficacy of the male is reduced, and during eclosion of nymphs the male does not feed at all.
3. There is an increased risk of predation while the male broods eggs.
4. The male is unable to mate until eggs are hatched or removed.

We should therefore anticipate that males are extremely selective about the females they mate with, as Thornhill (1976) predicts, and take considerable precautions to ensure that the eggs he broods are truly his kin. In courtship the roles are essentially reversed, with the females always initiating the interaction (Smith, 1979a). The courtship sequence is complex, involving sparring, embracing, clasping, and male pumping, all enabling the male and female opportunities for assessing the quality of their mate. Once oviposition commences, the male goes to extremes to ensure that he fertilizes every egg the female lays on his back, always copulating before egg laying starts, and copulating after every one, two, or at most three eggs are laid. The most "concerned" male copulated 100 times in 36 hours while he received 144 eggs from a female (Smith, 1979b). The evolution of paternal care in the belostomatids seems to involve a move from ovipositing in emergent vegetation, which may frequently be in short supply, to an aquatic site which is nevertheless well aerated, with the male's back providing the best possible substrate (Smith, 1980).

Thus parental investment may become very considerable, with great reduction in number of progeny sired, but nevertheless resulting in increased fitness. In his appendix, Thornhill (1976) and Thornhill and Alcock (1983) give an impressive list of insects in which males provide nourishment to females. The great potential importance of courtship feeding in insects may be glimpsed when Royama's (1966) data on courtship feeding of great tits are considered. He found that the energy contained in the food brought to the female by the male was very similar to the total energy in the clutch of eggs she laid. The implication is that without his help the female could not make a sufficient energetic gain to produce any eggs by herself. The important point is that the female can exert considerable selective pressure on males by demanding significantly more parental investment than a purely male strategy would dictate. The fascinating topic of nuptial gifts in insects is covered by Boggs (1995).

SEXUAL SELECTION

Those adaptations that are related to obtaining a mate are under the influence of what Darwin (1871) called **sexual selection.** It is a particular kind of natural selection. Sexual selection "depends on the advantage which certain individuals have over others of the same sex and species solely in respect to reproduction" (p. 568). Darwin goes on to make his meaning more explicit:

> When the two sexes follow exactly the same habits of life, and the male has the sensory or locomotive organs more highly developed than those of the female, it may be that the perfection of these is indispensable to the male for finding the female; but in the vast majority of cases, they serve only to give one male an advantage over another, for with sufficient time the less well-endowed males would succeed in pairing with the females; and judging from the structure of the female, they would be in all other respects equally well adapted for their ordinary habits of life. Since in such cases the males have acquired their present structures, not from being better fitted to survive in the struggle for existence, but from having gained an advantage over other males, and from having transmitted this advantage to their male offspring alone, sexual selection must have come into action. It was the importance of this distinction which led me to designate this form of selection as sexual selection. So again, if the chief service rendered to the male by his prehensile organs is to prevent the escape of the female before the arrival of other males, or when assaulted by them, these organs will have been perfected through sexual selection, that is by the advantage acquired by certain individuals over their rivals. (p. 569)

Darwin (1871) devoted two chapters in his book, *The descent of man and selection in relation to sex,* to secondary sexual characters of insects, many of which have evolved under the influence of sexual selection. For example, male appendages may be modified extensively for grasping a female, as in *Crabro cribrarius*. Males of stag beetles (*Lucanus*) and hercules beetles (*Dynastes*) and other scarabs are larger than females. The male stag beetles have greatly enlarged mandibles, and the male hercules beetles have huge horns on the prothorax and head. In each case males use their enlarged parts as weapons in jousting to gain access to a female. Males of many species attract females by sound (e.g., Diptera, Orthoptera, Homoptera, and Coleoptera), where sound quality and persistence must surely be under the influence of sexual selection. The dragonflies often show differences between the sexes in color and males may change color just before mating occurs. Male butterflies are aggressive toward members of the same sex, and Batesian mimicry in butterflies is frequently limited to the female sex. In the latter case one possible explanation is given by Darwin, who argues that males could be mimetic but they would be less attractive to females and thus eliminated through sexual selection.

Huxley (1938) made the distinction between two types of sexual selection: **epigamic selection,** which works on adaptations that influence the choices made between a male and a female, and **intrasexual selection,** which influences adaptations that improve intrasexual competition for mates, usually between males. As predicted earlier in the chapter, competition between males for mates, or intrasexual selection, will be strong in most cases, and it is in response to these selective pressures that a wide variety of strategies and tactics in breeding behavior and morphology have evolved.

The reproductive stakes are high for males, as Bateman's (1948) interesting experiment demonstrated. Into one cage he placed five males and five virgin females of *Drosophila melanogaster*. They could be distinguished as individuals by chromosomal markers, and the progeny produced could be attributed accurately to one or another parent. The experiment was replicated. All the females were courted by males and only 4% did not mate; those that did mate, mated only one or twice. For the males a very different picture emerged: 21% of males did not mate, males attempted to mate repeatedly, the most successful males produced almost three times more offspring than the most fecund females, and the reproductive success of males increased linearly with copulation frequency.

SPERM COMPETITION

The evolutionary problem for males is not only how to inseminate a female first, but how to prevent other males from inseminating her subsequently. The sperm from a second male may displace that of a previous male before the latter is used to fertilize any eggs. This problem of **sperm competition** is particularly serious in insects, for several reasons discussed by Parker (1970a). He describes the situation in which sperm competition should be maximal and argues that insects approach this theoretical maximum. It will occur when (1) females mate several times before fertilization of eggs, (2) the female stores all inseminated sperm, (3) only on death of the female do the stored sperm die, and (4) the sperm is used sparingly such that there is no wastage and one sperm is used to fertilize one egg. In the female insect, inseminated sperm are stored in the spermatheca, and sperm utilization is very economical in many species. Sperm are also kept viable in the female for many days or even months, during which time a female has opportunities for repeated insemination (see Parker, 1970a, for details). Therefore, we should expect strong intrasexual selection, which would produce adaptations in males to reduce sperm competition. It should not be forgotten, however, that the female also improves fitness by a reduction in sperm competition since a female that mates with a male with a sexual selective advantage will gain if this advantage is inherited by her male progeny (Parker, 1970a).

Sexually selected adaptations for reduced sperm competition are diverse and have been reviewed by Parker (1970a) (see also Walker, 1980; Thornhill and Alcock, 1983).

1. *Mating plugs.* Mating plugs are secretions of the male accessory glands which block the genital passage so that after insemination subsequent insemination by other males is prevented or impaired. They are common in Diptera (Culicidae and Drosophilidae), Hymenoptera, and Lepidoptera, and they have been observed in Orthoptera, Coleoptera, and Isoptera. The plug usually dissolves away in a few hours so that oviposition can proceed, but it seems to serve in preventing a second insemination until the female is physiologically unreceptive. In some cases a male secretion may even induce physiological unreceptivity in the females, as in the mosquito, *Aedes aegypti* (Craig, 1967).

2. *Prolonged copulation.* Prolonged copulation is very common in insects and this may serve the same function as mating plugs. Moths commonly mate for 24 hours. In the housefly copulation may last for an hour but insemination is complete in 10–15 minutes. In this species agents in the seminal fluid induce nonreceptivity in the female. They are not injected with the initial sperm, so prolonged copulation is adaptive, as the inductive agent is transmitted after the first 15 minutes of copulation and the male can prevent further insemination until the female is nonreceptive. This time spent in copulation is very costly to the male since he could transmit much more sperm per unit time to other females if copulation were brief. Evidently, the selective pressure for prolonged copulation is the stronger force. Where mating plugs are used, copulation can be brief, as in *Aedes* mosquitoes, which lasts 10–20 seconds.

3. *Passive phases.* A passive phase is a stage in the reproductive behavior of the male in which he stays attached to the female without genital contact (Parker, 1970a,b). In postcopulatory passive phases the male defends the female against insemination by other males, frequently while she lays eggs. Sperm competition is thereby reduced.

The elegant studies by Parker (see references in Parker, 1970a,b, 1974, 1978a, 1979) on the dung fly, *Scatophaga stercoraria,* have shown in a quantitative way the selective advantages of the postcopulatory passive phase. Around fresh dung in which females oviposit, the density of males is usually high. Males enter a passive phase after inseminating a female, remaining on the back of the female. The female then lays eggs while the male guards her from other males with specialized rejection responses preventing access to the female. Another male may "take over" a female if the original male is not aggressive in his defense. The female separates from the male when she has laid all her mature eggs. Parker (1970c) has shown that the last insemination is responsible for fertilization of 80% of the eggs subsequently laid, until another insemination.

Parker (1970a) recognizes several conditions as necessary for the passive phase to evolve in *Scatophaga:*

a. Mating occurs just before oviposition.
b. High male densities occur at the oviposition site, as is likely in such ephemeral, readily discovered, and concentrated resources as dung.
c. The second mate displaces sperm of a former ejaculate from the spermatheca.
d. Male-rejection reactions against others are effective in preventing a takeover.

A prolonged postcopulatory passive phase is seen in many Odonata, sometimes called the *tandem position.* They are also seen in the migratory locusts, *Locusta migratoria* and *Schistocerca gregaria,* and in most crickets (Alexander, 1961). Other interesting examples are seen in the Hemiptera (e.g., Odhiambo, 1959). The male reduviid, *Rhinocoris albopilosus,* enters a passive phase after mating until the female lays a clutch of eggs. The female then leaves and the male guards the eggs until after they have hatched. The female adds additional clutches of eggs to the original cluster and the male mates with her several times during this period. Other females may also contribute eggs to this cluster—adding luster to his cluster, so to speak. The male is territorial in the sense that he defends the eggs from other males who compete for the protective function. Although the behavior may be to guard eggs against predation, Parker's (1970a) explanation is more convincing: A male gains a high sexual selective advantage by reserving an egg mass so that he is able to mate with a female repeatedly, and perhaps additional females. He achieves a higher fertilization rate than do other males.

4. *Noncontact guarding phases.* This involves any phase of behavior in which a male remains close to a female but not in contact, and guards her from other males (Parker, 1970a). This behavior is commonly seen in insects. Several species of odonates show a guarding phase and in *Plathemis lydia,* males never begin guarding without first copulating with the female (Jacobs,

1955). A male may mate with another female if she enters his territory, and then he attempts to guard both, but in high male densities takeovers are common and copulation always precedes guarding by the second male. Jonathan Waage (1979) showed that males of the damselfly, *Calopteryx maculata,* actually remove sperm from the females' spermatheca with the aedeagus before inseminating the female, indicating that selection pressure for guarding must be strong in this and other species with the capability. Judging by the length of the aedeagus of other insects, such as fleas (cf. Rothschild and Traub, 1971), we might predict similar dual functions of sperm removal and insemination.

Noncontact guarding phases are also seen in bark beetles (Scolytidae), where the male guards the burrow entrance while the female constructs a brood gallery, and in wood-boring wasps *(Trypoxylon).* Various species show cooperative behavior between male and female, during which time mating probably occurs only between the pair (Richards, 1927). This occurs in some dung beetles (*Geotrupes* sp.), other scarabaeid beetles, and some crickets and flies.

5. *Avoidance of takeover.*   Takeovers may reduce the amount of sperm a male transmits, genitalia may be damaged during active rejection of other males and forced dislocation, time is wasted until the first male's eggs are laid, and sperm competition is increased. Therefore, in conditions of high male density, as in swarms, sexual selective advantage is gained by avoiding takeovers and the adaptations may take several forms:

a. Clasping organs of the male genitalia or other appendages may be modified to improve clasping. The prothoracic legs of many male insects are so adapted as in the Dytiscidae. Male sepsid flies have spined femora used in grasping females. The internal surfaces of the antennae of male fleas are modified for adhesion to the second abdominal sternite of the female (Rothschild and Hinton, 1968). Even cryptic coloration and spines may reduce takeovers (Parker, 1970a).

b. Specialized rejection responses occur when males are in the passive phase, guarding a female or a territory. For example, male grasshoppers when mounted may kick at other males or stridulate, and many male cerambycid beetles stridulate.

c. Emigration from areas of high intramale competition is common in swarms of Nematocera. Once a male and female meet, they drop to the ground to copulate. When male densities are high in the dung flies, *Scatophaga stercoraria,* males carry females away from dung and copulate in surrounding grass (Fig. 15.10) (Parker, 1971). Although this loses time, as copulation takes longer in these cooler conditions (Fig. 15.11), fitness is improved by the reduced probability of takeover and reduced disturbance (see also Borgia, 1980, 1982).

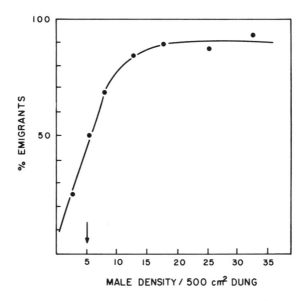

FIGURE 15.10   Percentage of dung fly pairs that emigrate to the surrounding grass after meeting on dung, in relation to the density of searching males present on the dung surface. Arrow indicates a possible threshold density. From Parker (1971). Reprinted from the *Journal of Animal Ecology* by permission of Blackwell Scientific Publications.

LIFE HISTORY GENDER DIFFERENCES

As Parker (1970a) says, there must be an evolutionary balance between the two conflicting sexually selected adaptations: (1) to mate with many females and to be successful in sperm competition, and (2) to prevent insemination by subsequent males until the ejaculate is fully utilized. Thus an enormous

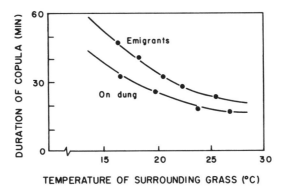

FIGURE 15.11   Duration of copulation of the dung flies, *Scatophaga stercoraria* (Scatophagidae) on dung compared to those that emigrate to the surrounding grass, in relation to the temperature of the surrounding grass. After Parker (1971). Reprinted from the *Journal of Animal Ecology* by permission of Blackwell Scientific Publications.

amount of social behavior is linked directly to avoidance of sperm competition or its reduction and the form of social behavior is influenced profoundly by the types of resources that are exploited (Borgia, 1979, 1980; Wade and Arnold, 1980). Even basic life history differences between males and females can be understood in relation to sexual selection. Some examples follow.

Emergence Times of Males and Females and Relative Size

In order to inseminate as many females as possible, it is adaptive for males to emerge earlier than females so that females may be discovered as soon as they emerge (Wiklund, 1995). In many species, males discover females before they emerge, aggregate around a female, and mate as soon as the genital pore can be reached by a male. This has been observed in the parasitoid wasps, *Megarhyssa* (Heatwole et al., 1964; Crankshaw and Matthews, 1981). Males mate with females in their pupal cases in the butterfly, *Heliconius charitonia* (Gilbert, 1975), and males of the thrips species, *Limothrips denticornis,* mate with many female prepupae (Lewis, 1973). A necessary prerequisite of earlier emergence of males is more rapid development or lower-temperature thresholds for development of the pupal stage. Darwin (1871) and Wallace (1867) therefore considered small size of male insects to be adaptive, "for the smaller males would be first matured, and thus would procreate a large number of offspring which would inherit the reduced size of their male parents, whilst the larger males from being matured later would leave fewer offspring" (Darwin, 1871, p. 628). The majority of insect species have males smaller than females.

Territoriality

Wherever resources are *moderately concentrated* and relatively patchy, a male can increase the quota of females he inseminates by defending for himself an area that either (1) many females pass through in order to oviposit, or (2) many females are attracted to because oviposition sites or food for adults are available. The males wait and inseminate females that enter the territory. Takeovers are avoided since additional males are slow to enter an established territory. In *highly concentrated* resources such as dung, territoriality is impossible because competition is too severe; other adaptations to avoid takeovers are necessary and some of these have already been described. Where resources are *evenly dispersed* or *diffuse,* a male is unable to gain an advantage over others, as female densities are highly unpredictable (R. R. Baker, 1972).

Davies (1978) studied an example where territorial males courted as many as 20 times more females than did nonterritorial males. Males of the speckled wood butterfly, *Pararge aegeria* (Satyridae), defended sunspots on the forest floor, and intraspecific male intruders were driven away. Only 60% of males had territories, while the others flew in the tree canopy searching for mates. Sunspots were worth defending because females were attracted into them (presumably because the larval food plants, grasses, grow where sun reaches the woodland floor), and many more females were encountered by territorial males per unit time than nonterritorial males.

Another example was studied by R. R. Baker (1972) using the butterfly *In-achis io*. He found that territories were located preferentially at sites through which females were likely to pass on their way to oviposition sites. For example, the corner of a field surrounded by a hedge or trees had many females directed there by these barriers (Fig. 15.12). A male defending a corner territory (1 in Fig. 15.12) had four females pass through, a male on the field edge (2) had three females traverse, but a territory away from the field edge may have only one (3) or zero females (4) pass through. The female quota of the corner territory was four times higher than some midfield territories. Sexual selection must play an important part in the evolution of this form of territoriality, and it seems that it should be regarded as a step beyond territoriality at the oviposition site. Strong competition at the oviposition site would select for males who leave the site and establish territories at approaches. These gain the virgin females, and if sperm competition is prevented by a mating plug or other adaptation, males at the oviposition site gain relatively few females that can be inseminated successfully. Heavy predation at oviposition

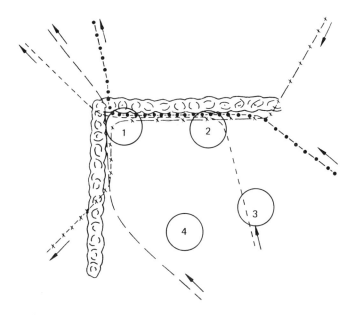

FIGURE 15.12 Location of male territories of *Inachis io* in a field through which females flew to reach oviposition sites. Reprinted with permission from R. R. Baker (1972) and Blackwell Scientific Publications Limited.

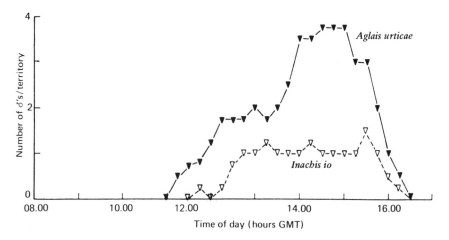

FIGURE 15.13   Relationship between time of day and number of males per territory for *Aglais urticae* and *Inachis io*. Reprinted with permission from R. R. Baker (1972) and Blackwell Scientific Publications Limited.

sites may also lead to territoriality and mating away from them, as in the lek behavior of Hawaiian Drosophilidae (Spieth, 1968, 1974).

Contrary to the selective pressure for mating away from the oviposition site, if sperm competition cannot easily be avoided by a mating plug or other device, the male will improve its quota of females if it defends females while ovipositing to ensure against takeovers. However, the density of males may be high and territories may be difficult to defend, as seems true in *Aglais urticae*, also studied by R. R. Baker (1972). Thus at optimum sites, as the day progresses, more and more males may occupy a given area, whereas away from oviposition sites territories may be defended successfully all day, as in *Inachis io* (Fig. 15.13). Again the type of territory that is maintained by males of a certain species must be in a state of equilibrium between selective pressures for gaining females before they reach the oviposition site or at the site.

An alternative territorial behavior is seen in perhaps the only territorial acridoid grasshopper, *Ligurotettix coquilletti*, which feeds preferentially on creosote bush, *Larrea divaricata*, in American hot deserts (Otte and Joern, 1975). Males and females are cryptic when on the grey stems of bushes during the day and feed on leaves during the night. Here males establish territories on larger bushes and may occupy such a territory for up to 39 days. The larger bushes are most likely to harbor females and the male's quota of females is increased by defending feeding sites for females, especially when densities of grasshoppers are low.

Maier and Waldbauer (1979a,b) have found a dual strategy in males of syrphid flies in the subfamily Milesiinae. In the mornings when nectar is abundant in flowers and it is cool, males defend territories in a patch of flowers and wait for females to feed in their territory. By 10 or 11 A.M., nectar supplies are depleted and hygrothermal stress is increasing rapidly. Males set up territories at oviposition sites in rot hollows in nearby trees which the females begin to visit.

Territoriality in insects is very common and takes many different forms (Table 15.3) if a territory is defined broadly as any defended area (see Hinde,

TABLE 15.3 Classification of Types of Territoriality in Insects Based on Resources Defended, and Some Examples. (For Further Examples, See Baker,1983; Fitzpatrick and Wellington, 1983; Thornhill and Alcock, 1983; Papaj, 1994).

Territory Type	Insect Order	References
Mating territory not obviously related to resource	Orthoptera	Alexander (1961), G. K. Morris (1971, 1972), Cade (1979)
	Lepidoptera	Baker (1972), Davies (1978)
Lek	Odonata	Campanella and Wolf (1974)
	Hymenoptera	Dodson et al. (1969), Kimsey (1980)
	Lepidoptera	Willis and Birch (1982)
	Diptera	Spieth (1968, 1974)
Mating territory associated with oviposition sites	Odonata	Jacobs (1955), Moore (1964), Pajunen (1966), Waage (1974), Campanella (1975)
	Lepidoptera	R. R. Baker (1972)
	Diptera	Varley (1947), Maier and Waldbauer (1979a,b)
	Hymenoptera	Alcock (1975)
Mating territory associated with feeding sites	Orthoptera	Otte and Joern (1975)
	Diptera	Maier and Waldbauer (1979a,b)
	Hymenoptera	Dodson and Frymire (1961), Pechuman (1967), Alcock (1979b)
Mating territory associated with adult emergence site	Hymenoptera	Wilson (1961), Lin (1963), King et al. (1969), Alcock (1979b)
Chemical defense of larval feeding site	Coleoptera	Oshima et al. (1973)
	Diptera	Pritchard (1969), Prokopy (1972)
	Hymenoptera	Salt (1937), Price (1970c, 1972), Vinson and Guillot (1972)
Physical defense of eggs and/or young	Homoptera	Wood (1976, 1978)
	Hemiptera	Kirkpatrick (1957), Odhiambo (1959)
	Coleoptera	Swaine (1918), Schedl (1958), Barr (1969)
	Hymenoptera	Wilson (1961), Dias (1975)
Defense of nest and foraging trails	Isoptera	Wilson (1971), Stuart (1972)
	Hymenoptera	Wilson (1971), Hölldobler (1976), Hölldobler and Wilson (1990)
Defense of food	Homoptera	Whitham (1979)
	Hemiptera	Sweet (1963)
	Coleoptera	Pukowski (1933)

1956, for defense, and discussions on definitions in Stokes, 1974, and Kaufmann, 1983). Many species establish territories for mating purposes, and these are frequently associated with feeding sites for adults. Other territorial behavior where food for the progeny is defended is very common. Territoriality independent of food and breeding sites may take the form of a lek. Spieth (1968, 1974) described this kind of behavior in some Hawaiian Drosophilidae where conspicuously patterned males gather together and each defends a small territory. This combined advertisement by males presumably attracts more females per male than is possible by individual displays. Lekking male moths of the species *Estigmene acrea* are also conspicuously colored, but an additional attraction is the male pheromone release from the large coremata extruded at the end of the abdomen (Willis and Birch, 1982). Other lekking insects include euglossine bees (Dodson et al., 1969; Kimsey, 1980), some dragonflies (Campanella and Wolf, 1974), and perhaps some hilltopping butterflies (Shields, 1967). Clearly, territoriality has a profound impact on the population dynamics of insect species, as it is a form of contest competition (see Chapter 21), and Alexander (1961) has recognized the close relationship between the evolution of territoriality and the evolution of social behavior.

POLYGYNY

Where species are territorial there is always the probability that territory quality varies greatly, or that few areas are suitable for establishment, or that few males have survived relative to survival of females. In each case the selective pressures may be sufficient so that the mating system becomes habitually polygynous—a male has two or more mates at the same time (see Emlen and Oring, 1977, for types of polygyny). Darwin (1871) found no evidence of polygyny of any kind in insects. However, as mentioned earlier in the chapter, some dragonflies are known to guard two females on occasion, having mated with each. Also, many bark beetle species are polygynous, with each female constructing a brood gallery, so they radiate out from the nuptial chamber constructed by the male (Fig. 15.14). The degree of polygyny varies enormously in this group, with one male having 60 females in some *Xyleborous* species, down to one male and two females in some *Ips,* and monogamy in *Scolytus* (Parker, 1970a). In the case of bark beetles, the sex ratio may be biased toward females by the heavier mortality of males, who initiate gallery construction. Heavy resin flow or predation may be more serious on males than on females. Janzen (1973c) discusses the evolution of polygyny in acacia ants.

Polygyny is rare in insects presumably because they are generally short-lived, and males are so rarely involved with brood rearing, so the opportunities for a male to associate with several mates at a time are strictly limited.

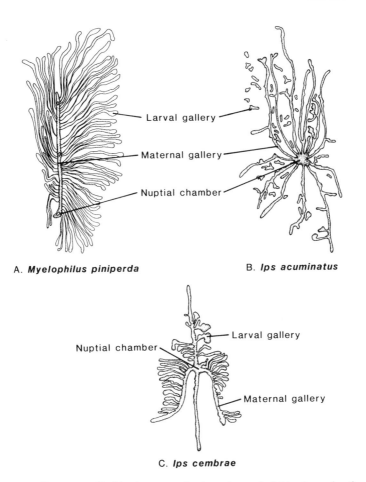

A. *Myelophilus piniperda*

B. *Ips acuminatus*

C. *Ips cembrae*

FIGURE 15.14   Gallery systems of bark beetles constructed in the cambium and adjoining tissues of conifers: (A) the common pine shoot beetle, *Myelophilus piniperda,* which is monogamous, with a nuptial chamber at the base and a single maternal gallery with lateral larval galleries; (B) pine bark beetle, *Ips acuminatus,* which is polygynous, with each female constructing a maternal gallery from the nuptial chamber; (C) *Ips cembrae,* a polygynous species that attacks larch trees. After illustrations in Novák et al. (1976).

Polygyny is also rare in birds (Lack, 1968; Orians, 1969). Lack (1968) estimated that only 2% of all birds are polygynous—only 1% of nidicolous species, but 8% of nidifugous species. Polygyny occurs where food is abundant, and therefore a female alone can find enough food for her young, or where young are precocious in finding food, as in the nidifugous species. Monogamy, however, is most common in birds, as the male plays an important role in gathering food.

By contrast, polygyny is common in mammals, where the female must take a major part in feeding the young (Orians, 1969). The evolution of the ultimate extreme in polygyny has been modeled by Bartholomew (1970) for the pinnipeds.

Once polygyny has evolved and is not attributable initially to sex-ratio differences due to mortality of males, competition between males must be even stronger than in monogamous matings. In birds, males evolve brilliant plumage or displays, which probably make them more vulnerable to predation, so the sex ratio is altered and polygyny is reinforced (Lack, 1968). In mammals there is strong selection for increased strength and size and sexual dimorphism can become extreme.

The relationships between the sexes and the searching for mates have ramifications involving many aspects of insect ecology, which have been touched on only briefly here. More detailed treatments and other aspects of this subject are provided by Krebs and Davies (1978, 1991), Matthews and Matthews (1979), Blum and Blum (1979), Lloyd (1980, 1982), Roff (1992), Stearns (1992), and Leather and Hardie (1995).

Ecological Genetics

The process of natural selection depends on the existence of genetic variation in populations. One of Darwin's nagging problems was to discover where this variation originated, since the then-current concept of heredity through pangenesis implied a blending process. Each generation apparently became less variable than the one previous to it. With the discovery of particulate inheritance through the pioneering work of Mendel, and the genetic process of mutation and recombination, the genetic sources of variation were identified, its maintenance was understood, and Darwin's dilemma was solved. Variation in natural populations must also be influenced by environmental factors, since these act as forces in natural selection. Thus ecological genetics is the study of environmental influences on the genetics of populations—genetic aspects of how populations are adapted to their environment.

Given Darwin's dilemma and the subsequent realization that variation is not lost during reproduction, a natural step was to ask:

1. How much genetic variation exists in populations?
2. How is it maintained?
3. Is this variation indeed adaptive?

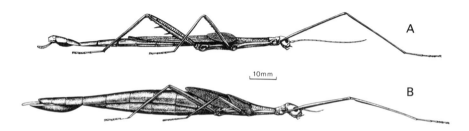

Large stick-insect, *Didymuria violescens* (Phasmatodea: Phasmatidae): (A) fully winged male; (B) flightless female. The species is a serious defoliator of *Eucalyptus* trees in the mountain forests of New South Wales and Victoria in Australia. From I. D. Naumann, *The insects of Australia: A textbook for students and research workers,* 2nd ed., p. 403. Copyright © 1991 by the Commonwealth Scientific and Industrial Research Organisation (Division of Entomology). Used by permission of the publisher, Cornell University Press. Drawing by F. Nanninga.

VARIATION IN POPULATIONS

Variation in populations was observed most easily in species that showed polymorphism in the phenotype externally, and much ecological genetics has concentrated on these multipatterned and multishaped species (see examples in Ford, 1964, 1975; Creed, 1971; Clarke, 1978). When differences in relative abundance of morphs were observed between populations, an attempt was made to interpret them in terms of selective factors acting in different ways on each interbreeding group. These qualitative differences in populations are useful in interpreting the population dynamics of the species and genetic strategies for survival in individuals and populations. Also, in Chapter 24 it will be seen that species that disperse extensively become subject to speciation, either by accumulating genetic differences as populations disperse as in the bees *Hoplitis* spp., or by crossing areas that later become barriers to gene flow as in the stoneflies, *Allocapnia* spp. For this to happen, there must be intraspecific differences between populations, and the study of population differences within a species is an important part of biogeography, particularly if the factors that cause and maintain these differences can be discovered.

POLYMORPHISMS

Polymorphisms manifested as external morph differences are relatively easy to study, but the number of species with two or more readily observed morphs is limited. However, polymorphisms in chromosome structure (**chromosomal polymorphism**) and in the migration rates of enzymes (**gene polymorphisms**) when subjected to an electric potential (electrophoresis) are also observable. The latter type of polymorphism can be studied in practically all organisms and results derived from use of this method are now providing a major thrust into the ecological genetics of organisms. Examples of each type of polymorphism will be given after genetic polymorphism has been defined. After the study of some cases of polymorphism, the basic evolutionary questions mentioned above can be addressed.

Balanced and Transient Polymorphism

Genetic polymorphism is the occurrence together in the same population of two or more discrete forms of a species in proportions greater than can be maintained by recurrent mutation alone (cf. Ford, 1940, 1964). This polymorphism may be stable or balanced, or it may be transient. **Balanced polymorphism** may be defined in general terms as genetic polymorphism maintained by contending selective forces, so that the frequency of each morph reflects the relative strength of these forces (cf. Ford, 1964). A more restrictive definition that describes the genetic mechanism for maintenance of polymorphism is given by Ford (1964) and Mayr (1963)—genetic polymorphism in which there is a balance of opposed advantage and disadvantage where the heterozygote is favored compared with both homozygotes. Therefore, in balanced polymorphism, as long as the opposing selective pressures remain, polymorphism is maintained. In contrast, **transient polymorphism** is a temporary state in a population in which a rare gene with an unopposed advantage spreads until its former normal allele is reduced to the status of a mutant (Ford, 1964, 1975).

POLYMORPHISM IN *PAPILIO DARDANUS*

By studying one of the most externally polymorphic insects, the swallowtail butterfly, *Papilio dardanus,* two of the basic questions in ecological genetics can be answered: How is polymorphism maintained, and is it adaptive? However, the answers are specific to this case. Also, possible pathways in the evolution of this polymorphism can be suggested. *Papilio dardanus,* the caterpillars of which feed on citrus plants, is restricted to the Ethiopian biogeographic realm (see Chapter 24). Within this range, eight races are recognizable (Fig. 16.1) and are distinguished by differences in the black markings on the wings of males and the morphology of the male genital armature (a **race** is defined as a class of individuals with common characteristics). The color pattern of males is always black and yellow and differs only in detail from race to race. They have tails on the hind wings characteristic of the genus *Papilio* (Fig. 16.2). Interbreeding occurs between races at the edges of their distributions indicated by shading in Fig. 16.1. In the races *meriones* on Madagascar and *humbloti* on the Comoro Islands, the females are malelike. In contrast, continental races have highly polymorphic females; for example, race *polytrophus* has 13 morphs alone. In fact, according to Sheppard (1962) there are 31 morphs, several of which are shared by four or five races. Most of the studies, particularly the genetics of morphs, have been executed by Clarke and Sheppard (1960a,b, 1963) and Sheppard (1961, 1962).

How has natural selection produced this large variety of morphs in a single species? Ecological factors and particularly sympatric species must be examined. *Papilio dardanus* adults fly during the day; they are palatable, and therefore, subject to heavy predation by birds in particular. As seen in Chapter 7, one antipredation strategy available to palatable prey is Batesian mimicry, and this is the strategy involved in *P. dardanus* females. Since several models exist sympatrically, an equal number of mimics could evolve. Throughout the geographic range of *P. dardanus,* different species of models have been exploited, or geographic variation of a broadly distributed model

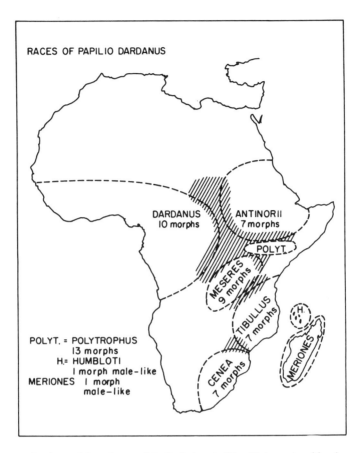

RACES OF PAPILIO DARDANUS

DARDANUS
10 morphs

ANTINORII
7 morphs

POLYT.

MESERES
9 morphs

TIBULLUS
7 morphs

H.

MERIONES

POLYT. = POLYTROPHUS
13 morphs
H.= HUMBLOTI
I morph male-like
MERIONES I morph
male-like

CENEA
7 morphs

FIGURE 16.1   Distribution of the eight races of *Papilio dardanus* in Africa with the number of female morphs in each race indicated. Shaded areas indicate zones of interbreeding between races. After Sheppard (1962).

has been mimicked, and thus many morphs have evolved. Each mimetic pattern is found only where the model is also present. There are a few exceptions where mimicry is poor, apparently because models became rare as a result of past unfavorable environmental conditions. The similarity between model and mimic is usually carried to a remarkable degree (e.g., Fig. 16.2). One distasteful model, *Amauris echeria*, is widely distributed on the higher plateau areas of Africa, and in Kenya it shows a local distribution as in Fig. 16.3. For each subspecies of the model there is a different morph that mimics it: race *polytrophus*, morph *cenea* mimics *A. echeria echeria*; race *polytrophus*, morph *ochracea* mimics *A. echeria septentrionalis*; and race *antenorii*, morph *cenea*like mimics *A. echeria steckeri*.

FIGURE 16.2 Three specimens of *Papilio dardanus* and a model *Amauris niavius:* (*a*) nonmimetic male from the mainland of Africa (race *dardanus*); (*b*) from Madagascar (race *meriones*) showing that males of different races differ only in detail (Nonmimetic females appear very similar to the males but have a dark patch about one-third up the costal margin of the forewing); (*c*) a model, *Amauris niavius niavius;* (*d*) a female *Papilio dardanus dardanus* (morph *hippocoon*), which mimics the *Amauris* sp. in coloration, pattern, and shape. Courtesy of James G. Sternburg.

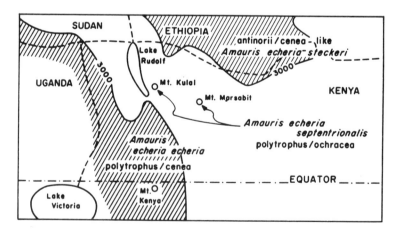

FIGURE 16.3 Map of the higher plateau of East Africa showing local distribution of the distasteful model *Amauris echeria* and the mimic *Papilio dardanus.* Contour is 3000 ft. By each subspecies of *Amauris* the mimetic form is designated as race/morph. Thus for each subspecies of *Amauris* there is a distinct morph of the mimetic species. Constructed from information in Sheppard (1962).

EVOLUTION OF POLYMORPHISM

Repeated examples like this provide almost certain proof that predators are the agents in natural selection, producing in many cases precise mimetic patterns of distasteful models. Protection of mimics from predation has been thoroughly documented for the North American nymphalid butterflies in the genus *Limenitis* (e.g., Brower, 1958a,b; Platt and Brower, 1968; Platt et al., 1971). But how do the selection pressures work, and how does the variation arise that lends itself to selection for mimicry? There is a big difference between the primitive pattern and shape, and that of the mimicking morph, and it is inconceivable that such perfect mimics could be produced by a single chance mutation. It is equally inconceivable that it could occur by the gradual change of the primitive pattern. There is no way in which intermediate forms could be selected, for because the predators would not confuse them with any models and the adaptive nature of the primitive pattern, whatever that might be, would also be forfeited. Sheppard's (1962) reasoning is followed here. Assume that there is a range of color patterns numbered from 0 to 5 on an arbitrary scale (Fig. 16.4). Then the relative fitness of each color pattern can be plotted on the ordinate. If a distasteful model is assumed to have a color pattern at 3, maximum fitness of a mimetic morph will be achieved by evolving an exact copy of the model, or acquisition of a color pattern 3. A mimetic morph with a pattern at 2.5 (Fig. 16.4a) will have less than maximal fitness but one that is greater than any other morph, with a pattern from 2.5 to 0 or 3.5 to 5, since in these cases the mimetic resemblance fades. Any pattern more like the model than the mimic at 2.5 will be selected for, since such an insect will be less often distinguished from the model by a predator. As the color pattern becomes less like the model, the selective advantage declines until it is completely unlike the model, and therefore no further change will produce a disadvantage relative to the mimic at 2.5.

If the pattern happens to become more cryptic as it diverges from that of the model at 3, the disadvantage of being a poor mimic or a nonmimic may be converted into an advantage, as predation will once again be avoided, say at 0.3 (Fig. 16.4b). Thus two morphs may exist that will be selected for simultaneously—one cryptic and one mimetic. Therefore, a highly cryptic morph at 0.3 is unlikely to become mimetic unless a single mutation moves the pattern from 0.3 to between 2.5 and 3.5, because all patterns between and beyond these values have a lower selective advantage, as they will be protected neither by crypsis nor mimicry. As seen in Chapters 7 and 8, if the model is not very abundant and becomes rarer than the mimic, the selective advantage for the mimic will be much reduced because the predators will be less able to learn that the model is distasteful (Fig. 16.4c). Conversely, if the model is very abundant, the mimic will gain a large selective advantage (Fig. 16.4d), and less exact fits to the model will still have a selective advantage so that the evolution of mimicry is more likely under these conditions. The same contrasting results will be seen where the model is only moderately unpalatable and where the model is highly unpalatable.

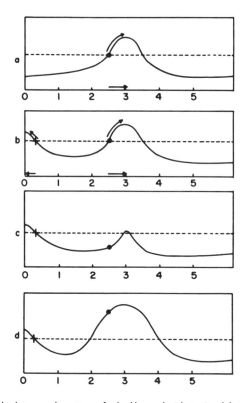

FIGURE 16.4   Relationship between color patterns of palatable morphs (abscissa) and their relative fitness (ordinate) when the model has a color pattern designated as 3. Color patterns are assumed to be similar on contiguous parts of the pattern range and more and more different as distance on the arbitrary scale increases. If a morph has a color pattern of 3, it is a perfect mimic. Color patterns at 2.5 and 2.0 and so on, and 3.5 and 4.0 and so on, show decreasing mimetic resemblances to the model. Therefore, as shown in (a), the morph with a pattern at 2.5 can increase fitness by approaching 3, but below 2.5 or above 3.5 fitness declines steadily as the resemblance fades. The following relationships are figured: (a) when a model has a color pattern at 3; (b) when a model has a color pattern at 3, but there is also a selective advantage of a cryptic pattern at 0; (c) when the model is at a relatively low population density, or is not very unpalatable; (d) when the model is at a high population density, or very unpalatable. Arrows indicate direction of selection. The dashed line indicates the relative fitness of the mimetic morph in (a) and (b), and the cryptic pattern in (b), (c), and (d). The solid circle represents the position of the mimetic morph and a cross the cryptic morph. From Sheppard (1962).

Suppose that the two patterns at 0.3 and 2.5 were controlled by a single pair of alleles. In Fig. 16.4d, where the model is very abundant, the allele producing the cryptic pattern at 0.3 would be at a disadvantage and would tend to be eliminated by natural selection. However, as the mimic becomes commoner, its advantage will decrease and may be converted to a disadvantage before the cryptic pattern is lost, in which case both morphs will be maintained in the population, in a state of balance—balanced or stable polymorphism. If the model is very abundant, the selective advantage to the mimetic morph will continue until all cryptic morphs are selected out of the population and the species becomes monomorphic and mimetic. In the case shown in Fig. 16.4c, the opposite trends would be seen. The advantage of the pattern at 0.3 might also be assumed to be due to another model with this pattern. Then the population would have two morphs, both mimetic, and the situation in *Papilio dardanus* can be envisaged where there are many morphs equivalent to the many models present within the range of a race.

To establish a new mimetic form, supposing that the species is monomorphic at 0.3, a new mutant must produce a color pattern of at least 2.5 in case (b), at least 3 in case (c), and at least 2 in case (d) (Fig. 16.4) to ensure that it is not immediately selected out of the population. That is, the chance of a mutant producing a new morph increases with increasing model population density and increasing unpalatability. In fact, for a new morph to be selected for, as opposed to the cryptic morph, it must have a selective advantage slightly higher than that of the cryptic morph. Once a mutant produces a morph that is selected for, selection will improve the fit of the mimic to the model by accumulation of small genetic changes that all together confer great similarity of the mimic to the model. Thus selection will gradually move the pattern in the direction of the arrows in examples (a) and (b) (Fig. 16.4).

This line of reasoning led Sheppard to make certain theoretical conclusions, which are actually supported by the genetic data that he and Clarke secured. Sheppard (1962) states that:

1. A model must be common, easily recognized, and well protected against predators.
2. The evolution of a mimetic form usually requires a large initial phenotypic change, because a small change is not likely to be advantageous. Consequently, most mimetic patterns are largely controlled by a single gene.
3. If the model is not very abundant compared with the mimetic species, or if it is not well protected, a stable polymorphism is likely to result.
4. The abundance of a mimetic form in a polymorphic species will depend on:
   (a) The abundance of the model.
   (b) Its degree of distastefulness.
   (c) The number of mimics of other species occupying the same model.
   (d) The strength of the selective force imposed by visually hunting predators (in the absence of predation, mimicry, or cryptic coloration can have no advantage).
5. The likelihood of mimicry evolving will depend not only on the availability of suitable models and the occurrence of suitable mutations, but also on how many other mimics of the available models are present. Thus when there are no other mimics, quite a poor copy of a model might be advantageous and become established. (For support of this contention, see Brower et al., 1971.) But when another species already copies the model in the area, the mimicry might have to be initially very

good to be advantageous, or it might never be advantageous if as many mimics were already present as could be "accommodated" by the model in terms of relative abundance of all mimics and the model.

6. Once the mimicry is established, the rest of the genetic constitution can change to produce a more perfect resemblance—supergenes are involved. A **supergene** is formed when mutually advantageous genes become closely associated on the same chromosome, by selection for this beneficial union, so that separation of the genes becomes so rare that they act together as if they were a single gene (see Ford, 1964).

7. If dominance were initially absent, it would tend to evolve, the new mimetic form being dominant. Matthews (1977) also discusses the evolution of mimicry.

INDUSTRIAL MELANISM AND TRANSIENT POLYMORPHISM

The action of visually hunting predators in the maintenance of polymorphism has been studied in detail in the peppered moth, *Biston betularia*. The typical morph of the moth (*B. betularia typica*) has black speckled markings on a white background and is very cryptic on light surfaces provided by lichens on tree trunks, on which they rest during the day (Fig. 16.5) (Kettlewell, 1959; Bishop and Cook, 1975). In Manchester, England, a heavily industrialized town, a black form of the moth was found in 1848 and was considered to be a rare specimen. However, within 40 years the black morph had become more common than the typical morph, resulting in much speculation on the causes for this change. This example was discussed briefly in Chapter 2.

FIGURE 16.5 Peppered moth, *Biston betularia* (Lepidoptera: Geometridae), showing the *typica* morph (left), the *carbonaria* morph (right), and the "inchworm" caterpillar below. From P. W. Price, *Biological evolution*. Copyright © 1996 by Saunders College Publishing. Reprinted by permission of the publisher.

Kettlewell (1955, 1956, 1961, 1973) argued that insectivorous birds were the agents of natural selection, which picked out the most conspicuous members of a peppered moth population. On lichen-covered tree trunks the typical morph was concealed and the black or melanic morph, named *carbonaria,* was prominent and selectively preyed upon by birds. Therefore, the black morph remained very rare. However, the selection pressure changed when the environment changed as a result of industrialization in the late eighteenth and early nineteenth centuries. Polluted air killed lichens on tree trunks and soot settled on all surfaces, resulting in a dramatic darkening of surfaces on which the moths rested. Now the *carbonaria* morph was more cryptic than the *typica* morph, and birds selected the light-colored moths more frequently than the black moths, and the *carbonaria* morph become more and more common in industrialized areas as a result. This was the evolutionary process in action, actually observed by humans, and subject to experimental investigation.

To support his arguments, Kettlewell designed experiments to test them. In one experiment equal numbers of *typica* and *carbonaria* morphs were placed on trees in Deanend Wood in Dorset in an unpolluted area in southern England. Observers then sat in a hide to watch for birds preying on the moths and to record the morph selected by each bird species. In this location where the *typica* morph was much more cryptic than the *carbonaria* morph, the latter was selected by each of five species of bird more frequently, with on average six times more *carbonaria* being taken than typical (Table 16.1). Natural selection was observed. Kettlewell also used mark–release–recapture experiments to test his hypothesis. Male moths were marked on the underside of the wing (concealed while at rest), released, and then recaptured using light traps and assembly traps with virgin females as bait. These experiments were performed in a polluted area near Birmingham and an unpolluted area in Dorset. In all experiments melanic moths survived better in the polluted area and typical moths survived better in unpolluted areas (Table 16.2), showing

TABLE 16.1   Predation of *Biston betularia* Morphs, Observed From a Hide, by Five Bird Species in Deanend Wood, an Unpolluted Woodland.

Bird Species	Number of Typica Taken	Number of Carbonaria Taken
Spotted flycatcher, *Muscicapa striata*	9	81
Nuthatch, *Sitta europaea*	11	40
Yellow hammer, *Emberiza citrinella*	0	20
Robin, *Erithacus rubecula*	2	12
Thrush, *Turdus ericetocum*	4	11
Total	26	164
Percent taken	13.68	86.32

SOURCE: Data from Kettlewell (1956).

TABLE 16.2 Results from Kettlewell's Experiments on Selective Predation by Birds on *typica* and *carbonaria* Morphs of *Biston betularia* Using the Mark–Release–Recapture Technique.

	Typica	*Carbonaria*	*Insularia*[a]	*Total*
Polluted woodland:				
Wild Birmingham population				
Number captured	53	486	20	559
Percent	9.48	86.94	3.58	
Number released	64	154	9	227
Number recaptured	16	82	2	100
Percent recaptured	25.00	53.25	22.22	
Unpolluted woodland:				
Wild Deanend wood population				
Number captured	359	34	21	414
Percent	86.71	8.21	5.07	
Number released	496	473	15	984
Number recaptured	62	30	4	96
Percent recaptured	12.50	6.34	26.67	

[a] *Insularia* is a dark morph intermediate between *typica* and *carbonaria*. The small sample sizes used in this morph make results less reliable than in the other two morphs.

SOURCE: Data from Kettlewell (1956).

that the high percentage of *carbonaria* moths in wild populations near Birmingham and the high percentage of *typica* moths in unpolluted areas were probably maintained by selective bird predation. Kettlewell elegantly supported his hypothesis with these experiments.

Another telling piece of evidence in support of the polymorphism being maintained by selective predation is that since 1960 stringent air quality control policies have resulted in reduced pollution, and the frequency of the melanic moths has declined in some urban areas (Fig. 16.6) (Bishop and Cook, 1975, 1980). This illustrates the dynamic state of the evolutionary process, in a changing environment (see also Clarke et al., 1985; Cook et al., 1986).

With further studies on *Biston betularia* the clear picture portrayed by Kettlewell has become complicated by the discovery of morph frequencies that cannot be explained by selective predation by birds alone. Melanism confers selective advantage in other ways. For example, whereas the *carbonaria* morph is most common in polluted areas in the north of England, the *insularia* morph is most common in industrialized southern Wales. The reasons for this difference are not clear (Bishop and Cook, 1980). In southern England, despite reduced pollution, the frequency of the *carbonaria* morph increased (Lees and Creed, 1975). The highest frequencies of melanic morphs in a population are at 90–95%, suggesting that there always remains some advantage to a polymorphic population over a monomorphic melanic population. This would not be predicted on the basis of Kettlewell's studies. Frequency-dependent selection and heterozygote advantage, including physiolog-

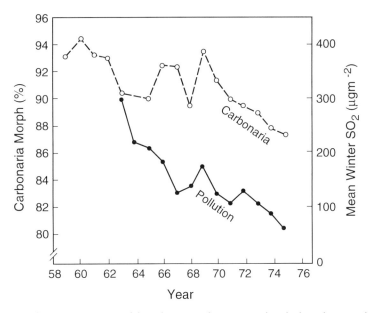

FIGURE 16.6 Change in representation of the *carbonaria* morph near Liverpool, England, in relation to reduced pollution, indicated by sulfur dioxide levels, from 1959 to 1975. Based on Bishop and Cook (1980). From P. W. Price, *Biological evolution*. Copyright © 1996 by Saunders College Publishing. Reprinted by permission of the publisher.

ical factors, may be involved in the maintenance of polymorphism (e.g., Lees et al., 1973; Steward, 1977; Bishop and Cook, 1980; Mikkola, 1984). Or, perhaps moths rest on birch trees with white bark and black crevices, in which both morphs are more or less cryptic (e.g., Grant and Howlett, 1988)?

In the foregoing examples of external polymorphism, there is good evidence to show that more than one morph can be maintained in a population by the selective pressure from visually hunting predators. An adaptive advantage for each morph exists under slightly different environmental conditions. By contrast, many internal polymorphisms may be maintained by influences other than predators. The study of chromosomal and gene polymorphisms has broadened the scope of ecological genetics tremendously and has led to important discoveries on one of the original basic questions: How much genetic variation exists in populations? These studies have also enabled insight into genetic strategies in response to differing environmental conditions.

CHROMOSOMAL POLYMORPHISM

Chromosomal polymorphisms have been found to be extremely common in some species of *Drosophila* and other flies (e.g., Wright, 1978). Crossing-over may be suppressed within an inversion and thus favorable combinations of genes can be maintained indefinitely. Therefore, in predictable environments inversion polymorphisms are likely to be common, and in unpredictable environments they should be rare, since genetic plasticity is most adaptive in harsh environments (see Parsons and McKenzie, 1972). For example, populations in the center of the range of *Drosophila robusta* have seven to nine inversions, whereas peripheral populations have only one to six inversions (Carson, 1958a,b, 1968). Inversion polymorphism in *Drosophila willistoni* is greatest at the center of its range and decreases rapidly toward the margins (Dobzhansky et al., 1950; Da Cunha and Dobzhansky, 1954; Da Cunha et al., 1959). It is also significant that cosmopolitan species such as *Drosophila melanogaster* have low levels of polymorphism or none (Parsons and McKenzie, 1972), when the genetic system for an opportunistic way of life is best kept in a form that ensures high variability in the population. However, not all species show these patterns, suggesting that other forms of genetic adaptation exist (e.g., Crumpacker and Williams, 1974).

The adaptive significance of many inversion morphs is not understood, but some show changes in frequency that are related to change in temperature and other factors. Dobzhansky (1948) found that there were consistent changes in the frequency of chromosomal morphs in *Drosophila pseudoobscura* (Fig. 16.7). The seasonal and altitudinal changes in frequency of morphs are produced by natural selection. The carriers of the different gene arrangements have different adaptive values and are subject to very high selective pressures (see Wright and Dobzhansky, 1946; Dobzhansky, 1947). For example, at 25°C, experiments were started with 50% of each chromosome type, Chiricahua (CH) and Standard (ST), and there was a highly significant increase in frequency of ST at the expense of CH. The relative selective values were estimated at 0.70 : 1.00 : 0.30 for ST/ST, ST/CH, and CH/CH, respectively, so that equilibrium was reached at 70% Standard and 30% Chiricahua, indicating the strong selective advantage of the Standard chromosome

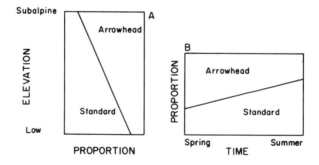

FIGURE 16.7 General trends in frequency of the chromosomal inversion morphs Standard and Arrowhead in *Drosophila pseudoobscura*: (A) with elevation in the Yosemite region of the Sierra Nevada, California; (B) with time from spring to summer at Aspen and Mather close to the western boundary of Yosemite National Park, during 1947. Presumably, the trend is reversed during the fall and winter.

type at this temperature. Dobzhansky noted that the morphs may be regarded as equivalent to the ecotypes of plants (Turesson, 1922), with the Standard chromosome morph conferring an advantage in warm situations and the Arrowhead morph conferring an advantage in cooler situations. Changes in these chromosome-type frequencies have now been studied for over 30 years (Anderson et al., 1975.) *Drosophila pseudoobscura* has several generations per year, so the three-decade record represents frequencies over perhaps 100 generations (Fig. 16.8). There are some remarkable features about these patterns. At these widely separated sites, very different frequencies of the inversion polymorphisms exist. Also, some morphs have remained surprisingly stable over 100 generations. Standard is usually the most common morph along the west coast of North America but becomes rare inland. Arrowhead is dominant in Utah and the adjacent states of Colorado, Arizona, and New Mexico (Mettler et al., 1988).

Clines in relative frequency of inversion morphs have also been observed by Brncic (1962, 1968), Mayhew et al. (1966), and Tonzetich and Ward (1973a,b). The last authors showed that one inversion morph conferred greater pupal survival in hot, dry conditions, while individuals with a different morph survived better in cooler, moister conditions. Nickerson and Druger (1973) showed that inversion polymorphisms also affect the fecundity, longevity, and competitive fitness in *Drosophila*. These polymorphisms permit a population to adapt to a variable environment by efficiently accommodating environmental variation (Parsons and McKenzie, 1972). This is possible and adaptive only if the generation time is relatively short compared to the rate of change in the environment. But by using *Drosophila* species with a short generation time, much progress has been made in understanding the adaptive nature of variability and the genetic systems that can influence the amount of this variability in response to environmental factors.

GENE POLYMORPHISM

The basic question of how much genetic variation exists in a population might never have been answered without the fundamentally different approach initiated by Lewontin and Hubby (1966). These authors demonstrated that differential migration rates of proteins during gel electrophoresis can be used to study enzyme or gene polymorphism at a single locus. Thus by sampling many randomly selected loci, an estimate of the total gene polymorphism in individuals and populations could be obtained. Since genes are the basic substance of variation, this new approach offered great promise and has indeed provided tremendous impetus to the field of ecological genetics. Even now the potential of the method has not been fully realized.

The early results were remarkable. Lewontin and Hubby (1966) found that an average population of *Drosophila pseudoobscura* was polymorphic for 30% of all loci and an average individual was heterozygous for 12% of

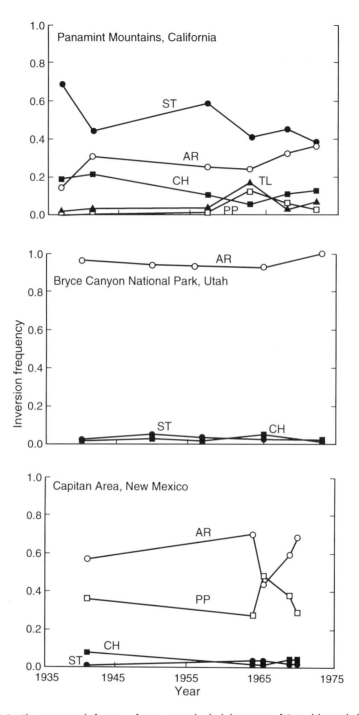

FIGURE 16.8 Thirty-year record of inversion frequencies on the third chromosome of *Drosophila pseudoobscura* in three localities: California, Utah, and New Mexico. Inversion types are as follows: ST, standard; AR, Arrowhead; CH, Chiricahua; TL, Tree Line; PP, Pikes Peak. Names usually denote the locality at which the inversion was first discovered. Data from Anderson et al. (1975). From P. Hedrick, *Genetics of populations*. Copyright 1983 by Jones and Bartlett Publishers, Boston. Reprinted with permission.

its loci. Even these high values that document the existence of extreme variation in populations were considered to be low estimates. Lewontin (1974)
postulates that only 25% of gene morphs may be identified by electrophoresis
(see Harris, 1970, for explanations). The high levels of enzyme polymorphism have been found repeatedly for many taxa: for example, *Drosophila*
spp. (Prakash, 1969; Prakash et al., 1969; Kojima et al., 1970; Richmond,
1972; Yang et al., 1972; Ayala et al., 1972a,b, 1974); the marine arthropod
Limulus (Selander et al., 1970); the mollusc *Tridacna* (Ayala et al., 1973);
birds (Ohno et al., 1969); mice (Petras et al., 1969; Berry and Murphy,
1970); and humans (Harris, 1971; Lewis, 1971; Harris and Hopkinson,
1972). However, some species show little or no enzyme polymorphism. The
reason for this and its significance are not understood for *Drosophila simulans* (Berger, 1970), but in northern elephant seals lack of polymorphism may
be due to fixation of alleles in small populations resulting from heavy mortality inflicted by sealers in the nineteenth century (Bonnell and Selander, 1974).
The amount of genetic variation in populations has been reviewed by
Selander (1976), Nevo (1978, 1983), Wright (1978), Hedrick (1983), and
Mettler et al. (1988).

ADAPTIVE ROLE OF GENE POLYMORPHISM

With the discovery of high genic polymorphism came a new dilemma. Could
such high levels of variation be maintained by natural selection or was much
of it adaptively neutral? (See King and Jukes, 1969; Johnson, 1973;
Lewontin, 1974; Kimura, 1979; Price, 1996a, for discussion.) Johnson (1973,
1974) concluded after reviewing the literature that this polymorphism is maintained by selection and the high levels of enzyme polymorphism may provide
"metabolic flexibility in a changeable environment." Bryant (1974) examined
eight sets of data (five on invertebrates, three on vertebrates) for geographic
variation in polymorphisms in relation to temporal variation in climate. He
found that approximately 70% of the variation could be accounted for by
variations in climate and concluded that shifts in morph frequency were therefore adaptive. It was interesting that Bryant found the frequency of morphs in
poikilotherms slightly better correlated with environmental factors than in
homeotherms, perhaps because the latter are buffered physiologically better
against climatic change.

The solution of one scientific problem usually leads to the creation of another. Although the debate on selective neutrality has not been resolved, there
is considerable evidence that many enzyme polymorphisms show frequency
shifts that are correlated with observed environmental changes. The inference
is that the shift is caused by different selective regimes. If so, what adaptive
functions do the enzymes perform? What is their biochemical role? Gillespie
and Kojima (1968) suggested that since substrates for enzyme action are
likely to differ over a range of habitats, enzyme polymorphism is adapted
to a variety of substrates. G. B. Johnson (1971, 1974) hypothesized that the

enzymes are associated with regulatory reactions in metabolism. Since environmental factors ultimately influence the cellular environment, polymorphism for enzymes enables cellular homeostasis despite these changes. If Gillespie and Kojima are correct, enzymes that work on external substrates, particularly those derived from food, should be more polymorphic than enzymes for substrates that are produced metabolically within the organisms. This was indeed the pattern found in 13 species of *Drosophila*. Following this line of reasoning, Powell (1975) anticipated that enzymes related to variable substrates such as food would show high levels of heterozygosity, while nonregulatory enzymes involved with structural and ribosomal patterns would have low levels. Also, regulatory enzymes may be influenced by variation in physical factors such as temperature, in which case heterozygosity would provide better homeostasis. Indeed, this pattern was found (Table 16.3), with *Drosophila* and other insects showing particularly high heterozygosity for variable substrate and regulatory enzymes.

There is a growing body of data supporting the hypothesis that genetic polymorphism increases as environmental heterogeneity increases. Hedrick et al. (1976) reached this conclusion in their review, as did Nevo (1978), who reviewed data on 243 species. As an example of this relationship, Steiner's (1977) work on Hawaiian *Drosophila* species is relevant. Steiner argued that species utilizing many host plants as food in the larval stage should exhibit greater enzyme polymorphism than species using one plant species because of the larger number of substrates utilized. Using data on at least 12 enzyme loci and 18 species, he did indeed find a positive relationship. He also found a positive relationship for genetic variability in populations of *Drosophila setosimentum* on an altitudinal gradient of 2500 feet, indicating that high variability is adaptive in harsher, more changeable environments. After studying

TABLE 16.3   Mean Heterozygosity in Animal Taxa for Enzymes with Different Functions.

| Organism | *Enzyme Category* | | |
	Variable Substrate	*Regulatory Enzymes*	*Nonregulatory Enzymes*
Drosophila	0.205	0.210	0.086
Other insects	0.289	0.281	0.094
Noninsect invertebrates	0.169	0.100	0.122
Fishes	0.063	0.110	0.066
Amphibians	0.118	0.227	0.062
Reptile	0.079	0.039	0.039
Birds	0.088	0.096	0.151
Mammals	0.048	0.056	0.032
Average	0.175	0.161	0.073

SOURCE: P. Hedrick, *Genetics of populations.* Copyright 1983 by Jones and Bartlett Publishers, Boston. Reprinted with permission. After Powell (1975).

38 species and 162 populations over 93 localities in Israel, Nevo (1983, p. 314) noted the general pattern of genic diversity in nature. "Genic diversity of allozymes varies nonrandomly among loci, populations, species and habitats and is at least partly correlated with and propelled by environmental heterogeneity in space and time as predicted by Darwinian evolutionary theory." He examined genic variation in populations in relation to 15 life history and ecological characteristics in plants, molluscs, insects, amphibians, one reptile, and mammals. As a broad generalization he found that high levels of genetic variation were associated with species with the following traits: large population size, widespread geographical range, high fecundity, generalized habitat use, aboveground existence (as opposed to fossorial species), and patchy or continuous population distribution.

Ultimately, all this genetic polymorphism should be related to the biochemical function of the enzymes concerned. This will provide a much more rigorous analysis of the adaptive nature of such polymorphism and contribute significantly to a resolution of the question of how much genetic variation is adaptive. This is why enzymes such as alcohol dehydrogenase (ADH) in *Drosophila melanogaster* have received so much attention, because larvae live in rotting fruit, where alcohols are an important environmental factor. The link between biochemical activity, environment, and the adaptive value of genetic polymorphism is clear. ADH activity can be selected for rapidly, with levels of activity almost doubling in three generations (Ward and Hebert, 1972), and ADH activity and survival of larvae in the presence of alcohol are correlated (McDonald and Avise, 1976). However, although the different activities of the two enzymes ADH^F (fast) and ADH^S (slow) are known in relation to activity, quantity, substrate specificity, heat stability, and so on, it is not yet clear how the polymorphism is maintained in natural populations (Hedrick et al., 1976).

Another enzyme, α-glycerophosphate dehydrogenase (α-GPDH), is important in metabolism of insect flight muscle, one enzyme morph functioning better at low temperatures and the other at higher temperatures in *Drosophila melanogaster* and the butterfly, *Colias meadii* (Miller et al., 1975; Johnson, 1976). Somero (1978) reviews the temperature adaptation of other enzymes.

Further evidence for the action of natural selection on gene frequencies comes from a different approach used by Ayala (1974). If genic polymorphisms are studied in several populations, then if they are under the influence of stabilizing selection, frequencies should be very similar, whereas if no such selection operates, frequencies should differ appreciably. Ayala presented data supporting the former case, providing several tables with frequencies of different alleles at a locus (one of which is given in Table 16.4). The frequencies are too similar for alleles to be changing in a random way. Neutralists may argue that not enough time has elapsed since divergence for frequencies to have changed much and that gene flow was sufficient to swamp random changes in frequency. Ayala countered these arguments by showing that at some loci the frequencies were very different between localities, indicating

TABLE 16.4   Allelic Variation at the EST-7 Locus in Eight Natural Populations of *Drosophila tropicalis,* in Central and South America. Numbers Indicate the Proportion of Individuals in a Population with a Given Allele.

	Alleles				
Locality	96	98	100	102	105
Catatumbo	0.04	0.08	0.48	0.32	0.04
Barinitas	0.02	0.14	0.64	0.18	0.02
Caripito	0.00	0.11	0.47	0.35	0.05
Tucupita	0.05	0.09	0.77	0.09	0.00
Santiago	0.00	0.15	0.43	0.35	0.05
Santo Domingo	0.01	0.11	0.46	0.30	0.09
Mayagüez	0.02	0.16	0.51	0.28	0.04
Barranquitas	0.00	0.06	0.34	0.41	0.16

SOURCE: Reprinted by permission from Francisco J. Ayala, "Biological evolution: Natural selection or random walk?" *Amer. Sci.,* 1974, **62**:692–701.

that there has been enough time for populations to diverge and that there is not enough dispersal of individuals to make gene flow an important factor. Natural selection must be maintaining frequencies in independently evolving populations through stabilizing selection.

There is no doubt that a large amount of genetic variability exists in natural populations, that at least some of it is maintained under the influence of differing selective regimes, and that it is adaptive. The full significance of such great variability has yet to be understood. As seen in Chapter 5, biochemistry and ecology must play mutually supportive roles in this important area. The building blocks of the community—the gene, the cell, the organism, and the population—can be integrated into a unified population biology by the study of ecological genetics. Indeed, it would now seem unreasonable to ignore ecological genetics in any study on population dynamics, competition, and biogeography, and many other aspects of ecology could be enlightened by this approach.

For further reading, excellent reviews are available on ecological genetics of *Drosophila* (Parsons and McKenzie, 1972) and enzyme polymorphism in general (LeCam et al., 1972; Johnson, 1973, 1974; Selander and Johnson, 1973; Hedrick et al., 1976; Selander, 1976; Nevo, 1978, 1983; Hedrick, 1983; Mettler et al., 1988). Many examples of studies in ecological genetics are given by Ford (1964, 1975), Creed (1971), Brussard (1978), Nei and Koehn (1983), and Real (1994). Wright (1968, 1969, 1977, 1978), Crow and Kimura (1970), Dobzhansky (1970), Kimura and Ohta (1971), Lewontin (1974), Spiess (1977), Roughgarden (1979), and Hedrick (1983, 1984) provide comprehensive treatments of population genetics. Mettler and Gregg (1969), Wilson and Bossert (1971), and Mettler et al. (1988) give good introductions to the subject.

Chapter 17

Population Dynamics: Conceptual Aspects

The subject of population dynamics is central to all of ecology. The field runs as the main theme through ecology because any individual with its own traits acts within the context of the population, any population functions in relation to other trophic levels and other populations of species in the community, and natural selection works on populations. In his Robert H. McArthur Award lecture, Murdoch (1994, p. 272) wrote: "Population regulation underlies most other ecological problems of interest, such as the dynamics of diseases, competition, and the structure and dynamics of communities. It is also integral to much that is of interest to evolutionary biologists since the major natural selective forces shaping life histories and behavior, for example, also affect population dynamics. Less obviously, perhaps, variants in the outcome of regulation (e.g., stability and population cycles) bear directly on the assumptions and range of application of evolutionary models. . . . Population regulation is still the central dynamical question in ecology."

If population dynamics is the centerpiece of ecology, we should be able to demonstrate that the field necessarily encompasses a large majority of ecological topics. That is, most of ecology is necessary for a comprehensive understanding of population dynamics, *and* population dynamics is a necessary concern in most aspects of ecology. This position is developed in three chapters,

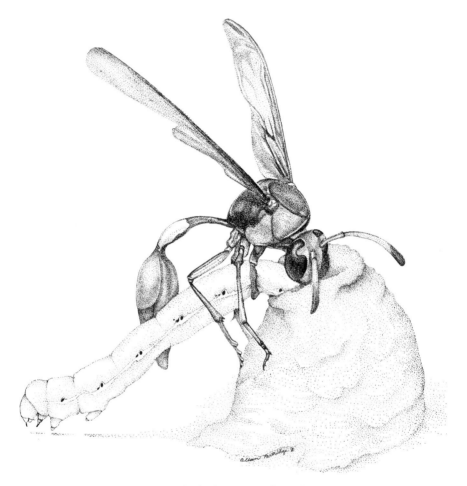

Potter wasp (*Eumenes*) provisioning its nest with a lepidopteran caterpillar on which its larvae will feed, illustrating that insects act as important mortality factors in the population dynamics of other insects. Drawing from a photograph by Anthony Bannister in Tweedie (1973). Drawing by Alison Partridge. Reprinted with permission.

beginning with this chapter on conceptual aspects of population dynamics, moving to modeling in Chapter 18, and to synthesis in Chapter 19, which makes the case that much in this book is relevant to understanding dynamics and regulation of populations.

If we accept population dynamics as a central theme, the field of ecology can be advanced dramatically by the development of theory. This is a tremendous challenge, even within the field of insect ecology, because dynamical patterns are so diverse, the number of factors that interact are no doubt multiple, and even the methodology is debated. Yet, surely, our motivation should be the development of pluralistic theory, with population dynamics as a theme, but with most of ecology contributing its essential components to that theme (cf. Cappuccino, 1995; Price and Hunter, 1995). "Although the study of population dynamics has come a long way since Malthus, explaining the stability and persistence of populations remains one of the most difficult challenges confronting twentieth-century ecologists" (Cappuccino, 1995, p. 3).

In this chapter we develop a historical perspective, to discuss some of the concepts developed over the decades. It will alert us to the many factors of potential importance and the kinds of debates relating to methodology for collection and interpretation of data. These themes will be picked up again in Chapters 18 and 19. The entire history of studies on insect population dynamics cannot be covered here and some earlier publications should be consulted for the excellent contributions by some of our pioneers in insect ecology (e.g., Bodenheimer, 1930, 1931, 1938, 1958; Allee et al., 1949; Andrewartha and Birch, 1954; Schwerdtfeger, 1968).

KEY FACTORS

Writing in 1957, Morris noted that in order to understand population dynamics and to test the major theories on natural control, it is necessary to undertake frequent population sampling and life table development. After a series of life tables has been developed, covering a wide range of conditions, it is likely that one or two **key factors** will be revealed that are mainly responsible for population changes. Then concentration on these factors, and disregard for relatively constant sources of mortality, may result in much less extensive sampling programs, and much more time spent on experiments and behavioral studies, in order to understand the mechanisms resulting in population change.

As far as I can determine, this is the first use of precisely the term *key factor,* and, of course, the basis for understanding the population dynamics of any organism lies in the identification of these key factors. The big debates in the literature of population dynamics have been on which factors contribute most to population fluctuations: density-dependent versus density-independent factors (see below for definitions); extrinsic versus intrinsic factors; quality versus quantity of food, and so on.

The definition of the term *key factor* has differed according to author, and this difference has led to important differences in analysis and some misunderstanding (see also Podoler and Rogers, 1975). Solomon (1949) defined a key factor as the main controlling factor affecting a population. He said (p. 25), "It is often possible to pick out one or two 'key' controlling factors, which are chiefly responsible for setting a limit to the density of a species in the areas where it is common." And a little later, "Any important part of the whole complex [of factors] which is outstandingly variable or unfavorable is likely to act as a 'key' controlling factor." Allee and Park (1939) had discussed limiting factors on populations but did not use the term *key factor*. This view of the key factor has been accepted and used by the majority of entomologists using the life table approach to analyzing population dynamics (e.g., Varley et al., 1973; Podoler and Rogers, 1975).

On the other hand, since his 1957 paper, Morris (e.g., in 1959) has used the term *key factor* for an influence that causes a degree of mortality that is closely related to changes in population density from generation to generation, and which therefore has predictive value. Morris (1959, p. 587) emphasized that "no attempt is made to establish cause and effect" (see also Morris, 1963a).

The distinction between these definitions is an important one and is emphasized by Varley et al. (1973) and Podoler and Rogers (1975). Solomon's definition invokes a cause-and-effect relationship between key factor and population change. Morris's definition implies no such relationship, and identifies a key factor as the best predictor of population change.

Both definitions are viable and useful, but for different purposes. For predictive modeling purposes Morris's approach is adequate, for the model can be successful in predicting population change without incorporating mechanisms resulting in this change. But for a real understanding of population dynamics it is important to identify key factors as Solomon defines them. This, however, necessitates much more than performing regression analysis on each mortality factor against insect population size. Once good correlations are established, extensive experimentation and behavioral studies are necessary to determine a cause-and-effect relationship and the mechanisms involved in the operation of the factor. Unfortunately, in so many studies, a simple correlation is accepted as evidence of a regulatory mechanism without adequate detailed testing—the use of **process studies** as Morris (1969) calls them: studies on the processes involved in population change.

The identification of key factors is central to the understanding of population dynamics. Many factors and combinations of factors can act in a regulatory role, and these are discussed in the following pages and summarized in Table 17.1. In Chapter 18 key factors will be discussed again in terms of modeling populations.

TABLE 17.1   Concepts on Population Regulation, a Historical Perspective.

A. Exogenous (extrinsic) population processes
  1. Density-dependent factors:
    (*a*) Predation—
        Hosts and parasitoids (Nicholson 1933, 1954a,b, 1957, 1958;
            Nicholson and Bailey, 1935).
        On microtine populations (Pitelka et al., 1955; Pearson, 1964).
        On plants (Harper 1968, 1977).
        On Kaibab deer herd (Allee et al., 1949).
        Herbivores in general (Hairston et al., 1960).
    (*b*) Food—
        Nicholson and Bailey model assumes that predator is food limited.
        Decomposers, producers, and predators in general—"The Étude." (See
            Hairston et al., 1960. See also Murdoch, 1966c; Ehrlich and Birch,
            1967; Slobodkin et al., 1967, for debate.)
        Birds (Lack, 1954).
        Predator populations (Elton, 1942).
        Ungulate populations (Caughley, 1970, n.b. comments on Kaibab deer).
        Nutrient recovery hypothesis in microtine cycles (Pitelka, 1959, 1964;
            Schultz, 1964; Batzli and Pitelka, 1970).
        Supplemental food (Fordham, 1971).
    (*c*) Combination of predation and food (Readshaw, 1965).
    (*d*) Space, that is, favorable habitats—usually pressure to leave favorable
            habitat has social interaction as proximate factor and so could be
            classified under B3.
        Muskrat populations (*Ondatra zibethica*) (Errington, 1946, 1967).
        Chaffinch (*Fringilla coelebs*) in Holland (Glas, 1960).
        Ground-dwelling Ichneumonidae (Price 1970b,c, 1971a).
        Emigration (e.g., Lidicker, 1962), but may be caused by food shortage
            or other factors.
  2. Density-independent factors:
    (*a*) Weather—
        Weather influence on *Thrips imaginis* (See Davidson and Andrewartha,
            1948a,b. See also Smith, 1961; Reddingius, 1971).
        Weather and the time available to reproduce (Andrewartha and Birch,
            1954).
        Release of the spruce budworm, *Choristoneura fumiferana* (Greenbank,
            1956), and lodgepole needle miner, *Recurvaria starki* (Stark, 1959).
    (*b*) Random changes in population due to random influences (Cole, 1951,
            1954a; Leslie, 1959).
  3. Intermediate between 1 and 2:
        Imperfect density dependence (Thompson, 1939; Milne, 1957a,b,
            1962).
        Urceolaria mitra and *Enchytraeus albidus* (Reynoldson, 1957).
        Red squirrel, *Tamiasciurus hudsonicus* (Kemp and Keith, 1970; cf.
            Lack, 1954, on spruce cone crop and crossbill populations).

B. Endogenous (intrinsic) population processes (all are probably density-dependent)
  1. Pathological effects in response to crowding (see Emlen, 1973, for
        discussion):

(Continued)

TABLE 17.1   *(Continued)*

    (*a*) Shock disease (e.g., Frank, 1957).
    (*b*) Adrenopituitary exhaustion (Christian, 1950, 1959; Christian and
        Davis, 1964).
  2. Processes with a genetic component:
    (*a*) Density-dependent increase in—
        (1) Proportion of congenitally less viable individuals (Chitty, 1957,
             1960).
        (2) Aggressive behavior (Chitty, 1957, 1967; Krebs, 1970).
    (*b*) Genetic breakdown of population during flush phase (Carson, 1968; see
        also Ford and Ford, 1930; Ford, 1957, 1964; Ayala, 1968).
    (*c*) Polymorphic behavior for dispersal and such in—
        (1) *Malacosoma pluviale* (Wellington, 1957a, 1960, 1964, 1977, 1979;
             Wellington et al., 1975).
        (2) *Microtus* spp. in Indiana (Myers and Krebs, 1971).
    (*d*) Changes in development times, dispersal rates, and such, in *Porthetria*
        dispar (Leonard, 1970a,b).
    (*e*) Genetic feedback (Pimentel, 1961b, 1968).
  3. Social interaction:
    (*a*) Evolution of social checks on population (Wynne-Edwards, 1962, 1964,
        1965; see also Hamilton, 1964a,b).
    (*b*) Intraspecific competition for space [see (*d*) under A1].
    (*c*) Aggressive behavior [see (*a*)(2) under B2; Calhoun, 1952].
  4. Dispersal (see Johnson, 1966, 1969).
    (*a*) The adaptive nature of dispersal (Johnson, 1966, 1969; Gadgil, 1971).
    (*b*) Evidence for density-regulated dispersal in insects [see (*d*) under A1]—
        (1) Broom insects (Dempster, 1968; Waloff, 1968a,b).
        (2) *Hyphantria cunea* (R. F. Morris, 1971b).
        (3) *Ascia monuste phileta* (Chermock, 1946; Nielsen, 1961; Johnson,
             1969).
        (4) *Phryganidia californica* (Harville, 1955).
        (5) Pierid butterflies (Shapiro, 1970).

EXOGENOUS AND ENDOGENOUS FACTORS

There are two basic concepts in population regulation: (1) that there are factors external to the population that influence population numbers (Table 17.1, Part A); and (2) that factors change within the population that affect numbers and produce regulation (Table 17.1, Part B). Within the first category there is the classic argument between proponents of regulation being caused by density-dependent factors: the Nicholson–Basiley school (Table 17.1, Section 1), and proponents that maintain that these factors are not important, or the Davidson, Andrewartha, and Birch school (Table 17.1, Section 2). However, Andrewartha and Birch (1954) found little use for the terms *density dependent* and *density independent*. They concluded that there is no component of the environment such that its influence is independent of population density, but there is no need to regard density-dependent factors as having any special importance. Since the concepts are much in use at present they are used here.

DENSITY DEPENDENCE AND INDEPENDENCE

A **density-dependent** factor is any factor in which its adverse effect increases, or its beneficial influence decreases, as a percentage of the population, as the population increases in density (Fig. 17.1). Conversely, any factor that shows no relationship in its influence on the population to the density of that population is a **density-independent** factor. These definitions have created some confusion in the literature because of their emphasis on *effect*. Almost any factor can have a density-dependent *effect*, such as climate or weather, as noted by Smith (1935) and Allee et al. (1949) long ago. Therefore, it is unrealistic to categorize factors unless their effects are first measured, and certainly no general statements are valid, such as food will act in a density-dependent way and weather in a density-independent manner.

A simplifying and more operational definition of the terms has been developed by Royama (1992) such that there can be no confusion about the terms. His emphasis is on the *state of existence of a factor:* Is the factor influenced by population density, or not? Is food supply influenced by density? Is weather or climate influenced by density? Royama's (1992, p. 21) definition is as follows: "If the state of existence of an ecological factor, identified in terms of its measure or parameter, is, in turn, influenced by the population density of the animal, the factor is said to be density-dependent. Otherwise, it is said to be density-independent." As population density increases, per capita food supply is likely to decline in a measurable way. Food is likely to be a density-dependent factor if competition prevails. But the same cannot be said for weather because population density does not have a measurable effect on it—

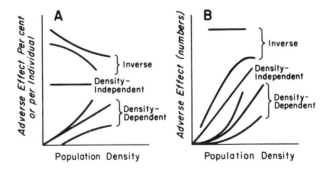

FIGURE 17.1 Types of response to changes in density (A) expressed as a percentage response to increasing populations and (B) expressed as the numbers of individuals affected with increasing populations. From Solomon (1969) and reprinted in 1973.

it is a density-independent factor. Note, however, that resources in a habitat may be density dependent or density independent, as Royama (1992, p. 22) explains: "For example, in a predator–prey interaction, the number of prey individuals at a given point in time—the measure of food supply for the predator at that point of time—is influenced by the predator density at the previous point in time. Thus, the prey population is a density-dependent factor for the predator with a time lag. On the other hand, annual production of seeds by a tree as a source of food may not be influenced by the abundance of the animals feeding on them. Then, the annual supply of seeds is a density-independent factor for the animals."

The importance of these terms and how they are employed become apparent when we consider what population regulation means.

POPULATION REGULATION

"Regulation is defined as the return of a population to equilibrium density. An operational definition of regulation is convergence to a single density by subpopulations which have been manipulated previously to different densities. The equilibrium density may be fixed or variable" (Murdoch, 1970, p. 497). A regulated population would fluctuate about some trend but would not "drift unboundedly away from it" (Royama, 1992, p. 14). Thus, in response to long-term trends such as global change or environmental deterioration, there may be a general decline in population size, even though the population is regulated (Fig. 17.2). Conversely, if regulation is around a stable equilibrium, the population will persist indefinitely. In general, regulation occurs around some reference point, whether it is changing or fixed (Fig. 17.2).

For regulation to occur, there must be a negative correlation between net reproductive rate and population density. Thus, in a general way, all regulation is density-dependent (Royama, 1992).

The three factors often considered as the most likely to operate in a density-dependent way, and to regulate populations are:

1. Predation or parasitism will increase.
2. Suitable food per capita will become more and more scarce, and increased mortality or decreased reproduction will result.
3. Space for living or breeding will also become more and more limiting.

ROLE OF NATURAL ENEMIES

Supporting evidence for **predators and parasitoids as regulating factors** come from various sources. The models of predation (Lotka–Volterra and Nicholson–Bailey; see Chapter 8) suggest that prey populations can be regu-

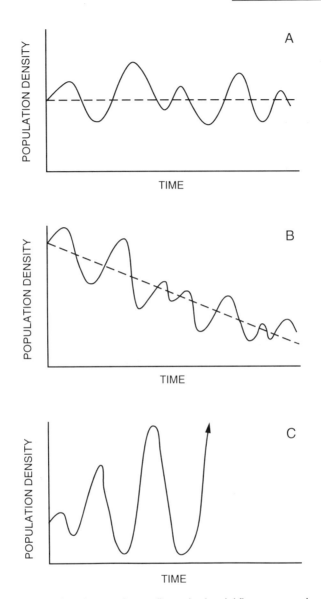

FIGURE 17.2 The concept of population regulation is illustrated as bounded fluctuations around a constant equilibrium density (A), bounded fluctuations in response to a long-term trend such as environmental deterioration (B), and an unbounded and unregulated population (C), as also seen in Fig. 8.2 and the Nicholson–Bailey equation.

lated completely by predators or parasitoids. Heavy mortality may be caused by parasitoids in insect populations (e.g., Baltensweiler, 1968; Waloff, 1968a; Price and Tripp, 1972; Faeth and Simberloff, 1981a,b; Washburn and Cornell, 1981), and Berryman (1996) has argued that parasitoids are probably the agents generating cycles in forest Lepidoptera.

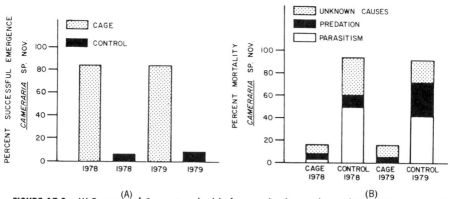

(A)   (B)

FIGURE 17.3   (A) Emergence of *Cameraria* moth adults from caged and uncaged (control) trees; (B) comparison of mortality factors on *Cameraria* larvae in leaf mines on caged and uncaged (control) trees. From S. H. Faeth and D. Simberloff, Population regulation of a leaf-mining insect, *Cameraria* sp. nov., at increased field densities, *Ecology*, 1981, **62**:620–624. Copyright © 1981 by the Ecological Society of America. Reprinted with permission.

Faeth and Simberloff (1981a) caged a small oak tree and increased density of a gracillariid leaf-mining moth, *Cameraria* sp. nov., in the cage relative to that on an uncaged control tree. Emergence of adults in the next generation was much higher in the cage than on the uncaged tree in two consecutive years (Fig. 17.3A), and there was thus no evidence of a density-dependent effect of intraspecific competition. However, predators and parasitoids accounted for about 60–70% mortality of the moth larvae in the uncaged tree, whereas they were largely excluded from the caged population (Fig. 17.3B).

In an infestation of the cynipid gall wasp, *Xanthoteras politum,* on rapidly growing oak shoots sprouting after a fire, parasitoids and inquilines were consistently the major mortality factor (Washburn and Cornell, 1981). Their impact on the host increased from causing 31% mortality to 74% mortality in three consecutive years and probably caused the local extinction of the gall wasp (Fig. 17.4).

The large literature on biological control also indicates the importance of enemies. Where insect species escape into new areas, in the absence of natural enemies, they become pests (Elton, 1958), but regulation can be reestablished by importing natural enemies (DeBach, 1964, 1974). A remarkable example of population regulation concerns the California red scale, *Aonidiella aurantii,* a pest of citrus crops, and its parasitoid, *Aphytis melinus* (Murdoch, 1994). Scale populations are as stable as any insect populations studied and fluctuate around an equilibrium density at something less than 1% of densities before the biological control agent, *Aphytis,* was introduced (Fig. 17.5). In the 1.5 years illustrated, the scale would have passed through four or five generations and the parasitoid 8–10 generations. The high population in the refuge occurred on the bark of branches and the trunk of trees, while the exterior population lives on the outer canopy. Even though the exterior population is less stable than the refuge population, it fluctuates within less than

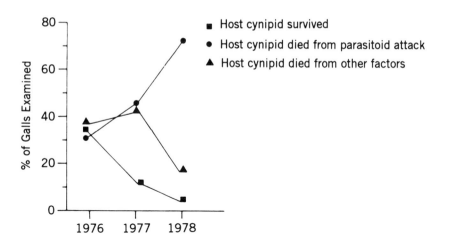

FIGURE 17.4 Trends in survival of the gall wasp. *Xanthoteras politum*, and morality factors, over 3 years. From J. O. Washburn and H. V. Cornell. Parasitoids, patches, and phenology: Their possible role in the local extinction of a cynipid gall wasp population, *Ecology*, 1981, **62**:1597–1607. Copyright © 1981 by the Ecological Society of America. Reprinted with permission.

2 orders of magnitude, and in the refuge fluctuations are well under one order of magnitude. The actual mechanism of such tight regulation is not understood but may involve size-dependent attack by *Aphytis* (cf. Fig. 20.4), needing a careful integration of studies on behavior and life history evolution with population dynamics.

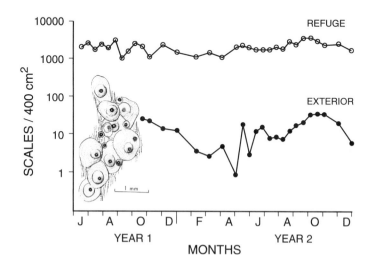

FIGURE 17.5 Densities of the California red scale, *Aonidiella aurantii* over 1.5 years, representing about 4–5 scale generations and about 8–10 parasitoid generations. Insert shows the scales. From W. W. Murdoch, Population regulation in theory and practice, *Ecology*, 1994, **75**:271–287. Copyright © by the Ecological Society of America. Reprinted with permission.

When insects are introduced for biological control of weeds, lack of success in a project can often be attributed to interference from natural enemies. In a review of 23 projects on the biological control of weeds, Goeden and Louda (1976) estimated that half had been rendered unsuccessful by the action of natural enemies. In addition, where insecticides have been used for insect control, increased mortality of enemies has frequently resulted in the release of pest species populations, and the emergence of "man-made pests" (Ripper, 1956; Moreton, 1969).

Hairston et al. (1960) argued for predation being the prime controlling influence on herbivores from a completely different standpoint. In their study, which is fondly named "the Étude," they argue as follows. The accumulation of fossil fuels is negligible, so all energy fixed by photosynthesis must flow through the biosphere. Therefore, all organisms taken together must be food limited. Any population that is not resource limited must be limited to a level below that set by its resources. The obvious lack of depletion of green plants by herbivores is an exception to the general picture. Therefore, producers are neither herbivore nor catastrophe limited; they must be limited by their own exhaustion of resources. Since we have seen that herbivores can deplete their food supply when numerous enough, that is, in the absence of predators, and since there usually is little evidence of this in general, herbivores must be controlled by predation (but see below). So predation has been repeatedly implicated as an important population-regulating agent.

Pathogens can also be exceedingly important agents in population regulation, especially at high insect population densities where transmission of the pathogen has a high probability (Anderson and May, 1980). Virus disease was the major factor in causing a population crash of the gray larch budmoth, *Zeiraphera griseana,* in the Engadin Valley in Switzerland (Auer, 1961). Accidental introduction of a virus on epidemic populations of the introduced European spruce sawfly, *Diprion hercyniae,* in eastern Canada, ended the epidemic in less than 10 years over a large geographic area (Clark et al., 1967). Virus epidemics on the western tent caterpillar, *Malacosoma pluviale,* also occur only at high caterpillar densities, and high mortality occurs after stressful conditions for the insects (Wellington, 1962). Here interaction between weather and enemies becomes very important (see also Myers, 1988, 1993).

ROLE OF FOOD

How important is **food?** In the predator–prey models where there is a mutual interaction, clearly food for the predator is important. Thus the Lotka–Volterra and Nicholson–Bailey models support the idea of food being a critical factor. Hairston et al. (1960) also cite food as the limiting factor for decomposers, producers, and predators in general, but papers debating the validity of this argument should be consulted (e.g., Murdoch, 1966c; Ehrlich and Birch, 1967; Slobodkin et al., 1967). For example, although the authors claim that only herbivores are predator limited, they may be limited by the amount of suitable food in tolerable microclimates (see Chapters 5 and 6).

In fact, the question of suitability of food, and how it varies in time and space, has become a central issue in understanding the population dynamics of insect herbivores. Therefore, considerable emphasis is placed on this subject in the following pages.

A cautionary note is provided by McNeill (1973), who studied the grass mirid, *Leptopterna dolabrata,* because he found two sets of interactive factors involving food quality. These accounted for the majority of population change. First, he found nitrogen to be limiting in the diet, which influenced fecundity, and there was competition for high-nitrogen feeding sites. Second, there was an interaction between density and weather that caused mortality of late-stage nymphs, which might also be related to competition for high-nitrogen feeding sites. We see nutrition, competition, and weather all playing important and interactive roles, and, as in many studies, it may be unrealistic to search for one or two factors only as explanations for population regulation.

One curious result that McNeill (1973) obtained was a negative intercept for the regression of loss of late nymphs on the log of numbers of instar III (Fig. 17.6). It appeared that this key factor did not operate at below a certain

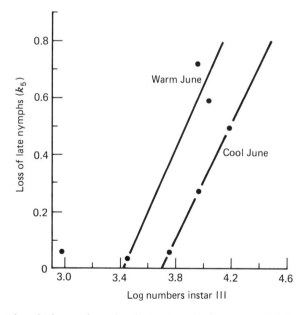

FIGURE 17.6   Relationship between the number of instar III nymphs of the grass mirid, *Leptopterna dolabrata,* and loss of late nymphs (k_5) on the food plant, *Holcus mollis.* Modified from S. McNeill, *Journal of Animal Ecology,* 1973. Published by Blackwell Scientific Publications Limited.

mirid density, and this threshold was lower during warm Junes than in cool Junes. McNeill suggested that because nitrogen levels were declining rapidly in the leaves of the food plant, *Holcus mollis,* during this time, there would be competition for feeding sites of high nutrient status. This would be particularly severe at high mirid densities, when many individuals would be forced to stay on leaves of low nutrient status. These individuals would be weakened and mortality would be greatest when the population came under climatic stress. This may work in two ways: (1) metabolic rate increases during warm weather and so increases nitrogen demand, thus aggravating the deficit; or (2) lower relative humidity may cause weakened individuals to die of dehydration after ecdysis.

As nitrogen levels in leaves declined, rapid growth and gonad development took place in the bug, making nitrogen requirements maximal. Seeds were becoming available at this time, and a switch in diet occurred, from leaves to seeds high in nitrogen. But again these high-nitrogen sites were limited, so when populations were high, a small proportion of females obtained adequate nitrogen and fecundities, on average, were greatly reduced. They ranged from a mean of 15.4 eggs laid per female when populations were high at 2.14 per square meter, to 67.0 eggs per female when populations were low at 1.20 per square meter.

Another result that should caution us against any simple categorization of mortality factors relates to the population dynamics of the winter moth, *Operophtera brumata.* Varley and Gradwell (Varley et al., 1973) clearly demonstrated that winter mortality was a major factor and much of this was due to asynchrony between egg hatching time and availability of food (see also Hunter, 1990, 1992a,b). The question "Why isn't the synchrony better between eclosion of larvae and bud burst?" could not be answered by regression analysis. A different tack was taken by Feeny (1968, 1969, 1970; Feeny and Bostock, 1968), who demonstrated convincingly that early eclosion enabled maximal exploitation of very early foliage which avoided foliage of low nutrient and water status later in the season (Figs. 5.1 and 5.3). Thus the immediate, or proximal, cause of mortality of first instar larvae was lack of food, but the ultimate and evolutionary cause of mortality was nutritionally poor food later in the season, which caused strong selection for eclosion at the earliest possible time (cf. Fig. 12.3). Evidently, the selective pressure has been so strong that the risks of precocious eclosion are negated. If we are to understand the processes involved in population dynamics, we must identify contemporary and evolutionary factors that influence population change (Mayr, 1961; Orians, 1962).

Animals in general contain much higher concentrations of nitrogen than plants (cf. Fig. 12.2). Therefore, herbivores are likely to be nitrogen limited, seeking to concentrate nitrogen as rapidly as possible (McNeill and Southwood, 1978; Mattson, 1980; Slansky and Scriber, 1985). For example, Slansky (1974) provides figures for percent nitrogen in the food plant and in the herbivore (Table 17.2), showing how much more concentrated nitrogen is

TABLE 17.2   Nitrogen Contents of Lepidopteran Insects and Their Food Plants.

Insect Species	Stage	Percent Nitrogen in Larva	Percent Nitrogen in Plant
Chilo suppressalis	Larva	6.7	1.4
Lymantria dispar	Instar III	11.0	5.6
	Last instar	8.2	2.5
Malacosoma neustria	Instar III	11.0	5.6
	Last instar	9.2	2.5
Phalera bucephala	Larva	5.3	2.5
Stilpnotia salicis	Instar III	11.3	10.1
	Last instar	11.0	10.1

SOURCE: Data from Slansky (1974).

in the herbivores. One exception is *Stilpnotia salicis*. Also, in every case except *Stilpnotia* the energy/nitrogen ratios in food plants were higher than in the lepidopterous larvae that fed on them (Table 17.3). This indicates that herbivorous insects frequently are likely to encounter nitrogen shortages sooner and more frequently than energy shortages, not to mention many other less studied nutrients in which differences between plants and herbivores may be even greater (cf. Clancy, 1992; Clancy and King, 1993; Clancy et al., 1988, 1993).

The relationship between aphid population growth and soluble nitrogen in the leaves is very clear (Dixon, 1970, 1973, 1985). Population growth is rapid when leaves are developing in the spring (Fig. 17.7). But as concentration of soluble nitrogen in the phloem sap falls, adults become progressively smaller, birth rate drops, and reproduction may stop on mature leaves. When leaves begin to senesce in the fall, active movement of soluble nitrogen out of the leaf provides a rich nitrogen supply for aphids and reproduction commences again. In some aphids, migration between trees and herbaceous plants

TABLE 17.3   Ratios of Energy to Nitrogen in Lepidopterous Larvae and Their Food Plants.

Insect Species	Energy/Nitrogen	
	Food	Larva
Agrotis orthogonia	0.82	0.65
Chilo suppressalis	3.05	0.81
Lymantria dispar (last instar)	1.71	0.66
Malacosoma neustria (last instar)	1.71	0.59
Phalera bucephala	1.71	1.02
Pieris rapae	1.41	0.64
Stilpnotia salicis (last instar)	0.42	0.49

SOURCE: Data from Slansky (1974).

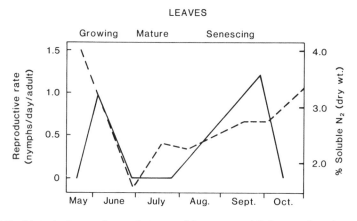

FIGURE 17.7 Relationship between the reproductive rate of the sycamore aphid, *Drepanosiphum platanoides* (solid line) and the percent of soluble nitrogen (dashed line) in the leaves of sycamore. Modified from Dixon (1970).

ensures continued reproduction through the summer, because during this season herbaceous plants grow rapidly and provide a high-nitrogen diet for the aphids.

Other studies that indicate nutrient shortages in herbivore populations include Greenbank (1963) and Kimmins (1971) on spruce budworm, Slansky (1974) and Slansky and Feeny (1977) on the imported cabbage butterfly, Morris (1967) on the fall webworm, and McClure (1980) on the elongate hemlock scale (see also McNeill and Southwood, 1978; Lawton and McNeill, 1979; Mattson, 1980; Scriber and Slansky, 1981; Slansky and Rodriguez, 1987). In careful studies on the western spruce budworm, *Choristoneura occidentalis,* on Douglas fir, *Pseudotsuga menziesii,* and other host plants, Clancy and colleagues have shown the great importance of micronutrients, especially the critical role of mineral/nitrogen ratios (Clancy, 1992; Clancy and King, 1993; Clancy et al., 1988, 1993). "These experimental results lead me to hypothesize that the critical role for host plant N is to determine the amount of food ingested, which in turn affects the amounts of other nutrients (and allelochemicals) consumed. Thus, the balance of N, minerals, carbohydrates, vitamins, fatty acids, and defensive compounds in foliage is likely to be the most important factor influencing herbivore survival and reproduction" (Clancy, 1992, p. 343). The challenge to really understanding budworm dynamics and host plant interactions is highlighted by the many different factors that appear to be involved, discovered by other researchers such as Cates and colleagues, including nitrogen and terpene interactions and carbohydrate and terpene interactions (e.g., Cates and Redak, 1988; Cates et al., 1983, 1987; Zou and Cates, 1994).

Greenbank (1963) summarized some of his research on the spruce budworm as follows, which was discussed briefly in Chapter 6:

1. Infestations of spruce budworm generally began in extensive areas of mature balsam fir during several successive years of dry and sunny weather. White (1974) interprets this as stressful conditions on host plants.

2. During development of high populations, increase is more rapid in mature stands than in immature stands, and the first severe defoliation is in mature stands.

3. Generally, mature balsam fir trees flower in alternate years, but in several years of dry and sunny weather staminate flower production occurs every year. Thus, there is an association of heavy and frequent flower production with an increase in spruce budworm populations.

4. Ovipositing females lay more eggs on the taller, more exposed trees in the stand, which also produce the most flowers. The distribution of eggs correlates coincidentally with the distribution of staminate flowers.

5. First instar larvae establish hibernacula preferentially in the cups formed by old staminate flower bracts, and survival is high in these sites.

6. In the spring, staminate flowers provide succulent food for the newly emerged larvae when the bursting of flower buds precedes bursting of vegetative buds.

7. Larval development is more rapid when the diet consists of staminate flowers. The advantage in development is greatest in sunny conditions when staminate flowers warm up more rapidly and retain heat longer than compact vegetative buds and needles.

8. No evidence suggests that a diet of pollen affects fecundity of females.

In relation to Greenbank's observations, studies by Kimmins (1971) on the amino acid composition of leaves of balsam fir and white spruce are interesting. He noted that amino acid levels in new foliage of flowering fir trees were highest. New foliage of nonflowering trees contained more amino acids than old foliage, and balsam fir contained more than white spruce. These findings correlate well with those of Durzan and Lopushanski (1968), who found that larvae fed on red or white spruce were smaller than those fed on the preferred host, balsam fir. Also, Miller (1957) and Blais (1952, 1953) showed that budworm larvae fed on old needles produce smaller pupae and adult females with lower fecundity.

It is also interesting to note that White (1969, 1974, 1976, 1978, 1993) regards physiological stress of the host plant as a key to understanding many insect herbivore outbreaks, as discussed in Chapter 6. When plants are placed under moisture stress, protein synthesis is decreased, with an increase in total nitrogen in aerial parts of the plant and changes in the relative amounts of certain amino acids. White (1969) found that outbreaks of psyllids, *Cardiaspina densitexta,* and others throughout Australia were strongly correlated with rapidly increased stress in host plants. Similar events seemed to be important in outbreaks of *Selidosema suavis* on *Pinus radiata* in New Zealand,

Neocleora herbuloti on exotic pines in South Africa, *Bupalus piniarius* on pine in the Netherlands, *Nepytia plantasmaria* on Douglas fir in California, and *Anacamptodes clovinaria* on mountain mahogany in Idaho (White, 1974). White (1976) extended his theory to plagues of locusts. In a similar way, heavy flowering of balsam fir occurs during extended periods of hot, dry weather; moisture stress may be a factor; and amino acid levels are high (Kimmins, 1971). Both weather, flowering, and nutrition are favorable to larval development and survival, and population growth in general.

These considerations also relate to the plant apparency hypothesis (Feeny, 1976; Rhoades and Cates, 1976) discussed in Chapters 5 and 6. Plants that are bound to be found by herbivores, for example, late successional trees in temperate-zone forests, have generalized defenses, including low nutrient status. Thus only during times of physiological imbalance or stress, when protein synthesis is reduced, does nitrogen become sufficiently abundant to allow rapid increase in the plant's herbivore populations (McNeill and Southwood, 1978). Also, it is the long-lived, bound-to-be-found species that will live through several seasons of stress, so population outbreaks are particularly likely. In annuals and biennials, nitrogen may be less limiting to start with, and the plants may succumb, or populations may be depleted during a series of stressful seasons, so that herbivore population growth may be counteracted.

POPULATION RELEASE

With these ideas in mind, it becomes clear that the concept of **population release** from low population levels should receive much more attention than in the past, when population suppression from potentially high levels was emphasized. The latter emphasis may have resulted from the life table approach, which emphasized mortality factors rather than survival factors (see Chapter 13). Figure 17.8 summarizes the concept of population release with particular reference to the spruce budworm, but the factors need be changed little to make them of general application. This picture bears similarities to the nutrient recovery hypothesis of Pitelka (1964) and Schultz (1964) for the lemming, *Lemmus trimucronatus* (Fig. 17.9), already discussed in Chapter 6.

FEEDING COMPENSATION

The scenario in Fig. 17.8 and the discussion above may have to be modified when more is known about relationships between nitrogen levels in plants, consumption rates, and assimilation efficiencies in herbivores, as discussed in Chapter 12. For example, Slansky (1974) and Slansky and Feeny (1977) have shown for the imported cabbage butterfly, *Pieris rapae*, that on low-nitrogen host plants consumption rates by larvae are increased, and as a result, suffi-

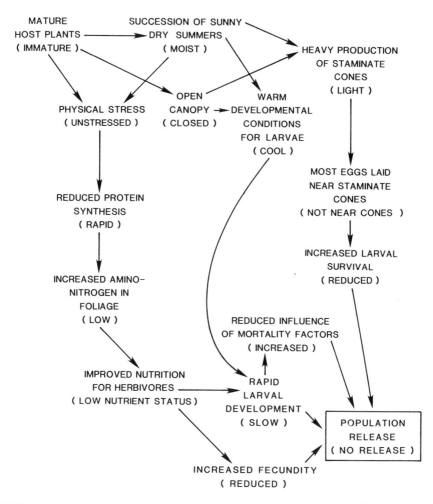

FIGURE 17.8 Concept of population release illustrated by the interacting factors relating to the spruce budworm, starting with the coupled conditions of mature trees with a series of sunny, dry summers. Conditions in parentheses relate to populations under normal conditions which do not result in population release.

cient nitrogen is assimilated but at a lower gross growth efficiency (the efficiency of conversion of ingested dry matter into tissue dry matter, *ECI*). Thus larvae can feed on a wide range of nitrogen levels, yet attain the same size at the same growth rate.

Two forms of compensation for low nitrogen levels were evident:

1. Larvae tended to increase the dry matter consumption rate, which compensated for reduced nitrogen content of food (Fig. 17.10). This compensation was not complete, however, so that the rate of nitrogen

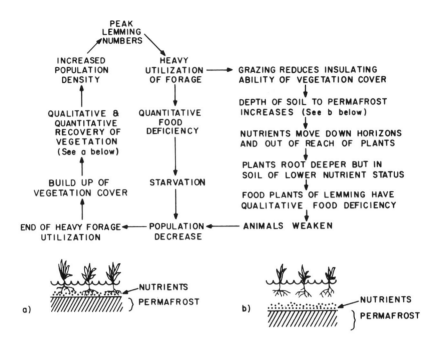

FIGURE 17.9   Summary of the nutrient recovery hypothesis of Pitelka (1964) and Schultz (1964) for the lemming, *Lemmus trimucronatus*, at Point Barrow, Alaska (derived in part from Krebs, 1964). Nutrients may be leached from the rooting medium as shown in (b), or they may be diluted by being dispersed through a larger volume of soil. As grazing is reduced, vegetation cover increases, the depth of permafrost decreases, nutrients are concentrated in the rooting zone (a), and a qualitative recovery of vegetation results.

intake was less on low-nitrogen plants, although the range in rates of consumption (1.2–2.8 mg N/day) was much lower than the range in nitrogen content of plants (1.5–5.0% dry weight) (Fig. 17.11).
2. Larvae on low-nitrogen plants had a higher efficiency of nitrogen utilization (Fig. 17.12).

The result of these two compensatory reactions was a more uniform nitrogen accumulation rate and growth rate, independent of plant nitrogen content. It must be remembered, however, that these experiments were performed using crucifer plants, which are hard to find and defended against generalist herbivores by toxic mustard oils. Whether such effects would be found in insects feeding on temperate tree leaves remains an open question. However, certain properties of host plants, particularly in those that are apparent, may make increased consumption rates hard to achieve for a small herbivore. Resin in coniferous trees, toughness of leaves, and chemical feeding deterrents may all play a role. For example, the three species of plant in the Cruciferae on which the nitrogen accumulation rate was lower than expected for *Pieris*

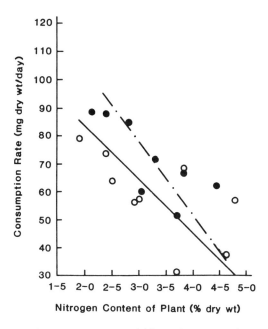

FIGURE 17.10   Relationship between nitrogen content of different plant species and varieties and rate of consumption of dry matter by *Pieris rapae* larvae. Open circles and solid line relate to one experiment and solid circles and dotted-dashed line to another. From F. Slansky and P. Feeny, Stabilization of the rate of nitrogen accumulation by larvae of the cabbage butterfly on wild and cultivated food plants, *Ecol. Monogr.*, 1977, **47**:209–228. Copyright © 1977 by the Ecological Society of America. Reprinted with permission.

rapae all contained cyanogenic chemicals or alkaloids that were not present in the other food plants (Slansky and Feeny, 1977).

Fertilizer, especially nitrogen, added to naturally growing plants may offer an experimental approach that would help to define the importance of food quality. But the results vary from one experiment to another. Fertilizer resulted in increased insect populations, for example, in the cone moth *Laspeyresia strobilella* in Norway (Bakke, 1969) and in the aphid, *Myzus persicae,* on tobacco in Maryland (Woolridge and Harrison, 1968). Other experiments show decline in insect populations with fertilizer application, for example, Oldiges (1958), Schwenke (1961), Rudnew (1963), Stark (1965), Haukioja and Niemelä (1976), and Shaw and Little (1972). A review of results from over 450 studies is provided by Waring and Cobb (1992).

Of course, fertilizer may have a multiplicity of effects on a plant. It may improve the defensive capabilities of the plant (Haukioja and Niemelä, 1976) or reduce stress and thus reduce free amino acids (White, 1974), or it may increase levels of available nitrogen in foliage and other plant parts (Woolridge and Harrison, 1968; Bakke, 1969; Slansky and Feeny, 1977). Also, a moderate dose of fertilizer may result in higher insect populations than a high dose

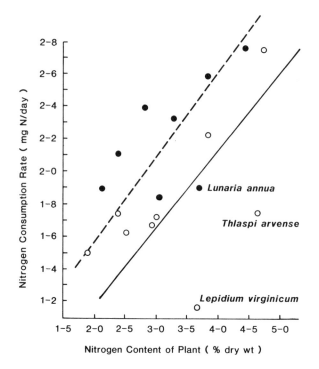

FIGURE 17.11   Relationship between nitrogen content of different plant species and varieties and rate of nitrogen consumption by the herbivore *Pieris rapae*. Open circles and solid line relate to one experiment and solid circles and dashed line to another. Larval growth was exceptional on the three named plants, probably because of unusual secondary chemicals in these species. These exceptional relationships are also seen in Figs. 17.10 and 17.12. From F. Slansky and P. Feeny, Stabilization of the rate of nitrogen accumulation by larvae of the cabbage butterfly on wild and cultivated food plants, *Ecol. Monogr.*, 1977, **47:**209–228. Copyright © 1977 by the Ecological Society of America. Reprinted with permission.

(Woolridge and Harrison, 1968). Thus, fertilizer experiments cannot help us to understand the role of nutrition in insect population dynamics unless we understand the detailed physiological consequences in the plant. This has usually not been studied in conjunction with insect population studies.

PREDATION AND FOOD

In some population studies a combination of **predation and food** has been found to play a major role. For example, Readshaw (1965) studied the stick insect *Didymuria violescens* (Phasmatidae) in New South Wales and Victoria in Australia. He presents a theory based on his data that the stick insect exists under two systems of control: one at low density when increase is limited

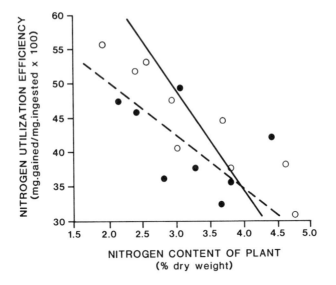

FIGURE 17.12   Relationship between nitrogen content of different plant species and varieties and the nitrogen utilization efficiency of *Pieris rapae* larvae. Open circles and solid lines relate to one experiment and solid circles and dashed lines relate to another. From Slansky (1974); reprinted with permission.

largely by egg parasitoids or birds or both; another at high density when increase is ultimately prevented by intraspecific competition for food. Birds are ineffective during peak years because of their limited total response to increase in numbers of the prey. (This supports the discussion in Chapter 8; cf. *ES* in Fig. 8.10B.) A similar situation seems to prevail in the population dynamics of the psyllid, *Cardiaspina albitextura,* on redgum, *Eucalyptus blakelyi* (Clark et al., 1967). Populations may remain low and stable for many years under the effective influence of natural enemies and then suddenly escape this influence and rise to where food becomes a limiting factor.

NUTRITION AND ENEMIES

Another combination of factors that is likely to be important is **nutrition and enemies.** The plant apparency hypothesis relates to this because late-successional plants, with low nutrition, digestibility reducers, and tough leaves force herbivores to feed longer and thus become more prone to attack by their enemies (see Chapter 5). Lawton and McNeill (1979) make this point, emphasizing that insect herbivores are trapped between "the devil" (natural enemies) and "the deep blue sea" (a sea of foliage which is usually nutritionally poor).

Rhoades (1979) has also advocated a modification of White's (1969, 1974, 1976) hypothesis on plant stress and herbivore population release, for stress also influences the defensive chemistry of the plant. In general, unapparent plants with cheap toxic chemical defenses increase defenses under stress, whereas apparent plants with heavy commitment to quantitative defenses decrease their concentrations under stress. Rhoades (1979) provides substantial evidence for this hypothesis, which may contribute importantly to the population release hypothesis. Haukioja (1980) provides a real example concerning defense in *Betula pubescens* trees and the population dynamics of the moth, *Oporinia antumnata,* in Fennoscandia.

ROLE OF SPACE

Any insect species whose individuals are territorial has evolved the trait in response to shortage of resources and competition for these resources. The resources in short supply are likely to be food or nesting, or oviposition sites, so defense of space around such supplies also indicates the importance of other factors. Territorial defense in insects is very common (see Chapter 15), indicating that the **role of space** in the population dynamics of insects must be considerable.

Errington (1946, 1967), studying muskrats, showed that early colonizers select the best sites that provide food and cover, and as the population increases young muskrats are forced through territorial aggression to establish territories beyond the limits of the best sites. In these suboptimal sites, heavy predation, shortage of food, and a harsher climate cause very high mortality (Fig. 17.13).

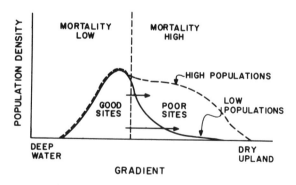

FIGURE 17.13   Summary of Errington's view of population regulation in muskrats and other organisms. Solid line indicates distribution of individuals during low populations. As populations increase more individuals move to harsher sites where mortality is high, so that the distribution indicated by the dashed line during high populations is only temporary.

An analogous situation is observed in insect populations. Price (1970b,c, 1971a) has observed ichneumonid parasitoids that search for sawfly cocoons buried in the forest litter. The females are able to detect the presence of others by an odor left on the litter, and they avoid this odor, as it seems to be repellent. Therefore, the maximum density of females that can exist in an area is not closely related to the host density. Rather, each female needs an amount of space so that contacts with the odors of others are not so frequent as to drive her from the area. Host density is ultimately limiting, as cocooned larvae are the food source; thus higher host densities do permit higher parasitoid densities, but the females become sensitive to crowding long before the highest host populations are reached. In favorable sites where hosts are abundant, populations of parasitoids will therefore increase rapidly, but many females will leave the site, due to their avoidance behavior. The result is that at high population levels many unfavorable sites are colonized by these parasitoids, whereas at low population levels only the most favorable sites are occupied (Fig. 17.14; cf. Fig. 17.13 on muskrat populations).

Aphids in the genus *Pemphigus* are strongly territorial and engage as first instar larvae in prolonged kicking and pushing battles for the best sites for gall initiation (Whitham, 1979). *Pemphigus betae* females on *Populus angustifolia* fight for the basal position on the leaf while they are only 0.6 mm long. But the consequences for the winners are, on average, much higher production of young per gall compared with galls formed more distally on the leaf, much lower mortality while establishing galls, and much higher ultimate weight of the stem mother, which initiates gall formation (Fig. 17.15 and Whitham, 1978). Space is a vital resource to these aphids, but this results in the evolutionary potential for the host plant to limit suitable space to the point where aphids are relatively rare compared to the number of leaves on a tree (Whitham, 1978, 1980, 1981; see also Chapter 15).

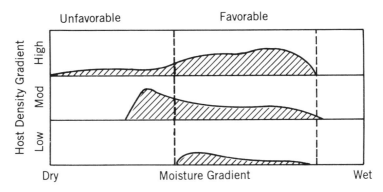

FIGURE 17.14   Distribution of the ichneumonid parasitoid, *Pleolophus basizonus*, in space defined by a moisture gradient and a host density gradient. As host density increases, parasitoid numbers increase and the dry marginal habitats become occupied only under this population pressure. The height of the distribution curves indicates the proportion of all species represented by *P. basizonus*, not actual abundance. After Price (1971a).

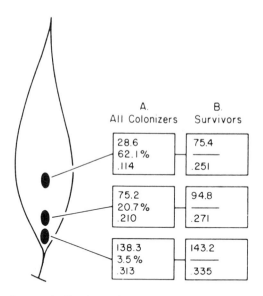

FIGURE 17.15   Effect of position of galls on leaves of *Populus angustifolia* on galling aphid fitness when three galls are formed on a leaf. The measures of fitness are boxed and represent mean number of aphids per gall (top), percent of stem mothers that died during colonization (middle, column A only), and mean dry weight of stem mothers (bottom). Column A provides estimates for all colonizing aphids; column B relates only to those that survive to produce galls. From T. G. Whitham, Habitat selection by *Pemphigus* aphids in response to resource limitation and competition, *Ecology*, 1978, **59**:1164–1176. Copyright © 1978 by the Ecological Society of America. Reprinted with permission.

ROLE OF WEATHER

The proponents of regulation by **density-independent factors,** such as mortality caused by weather or insecticide, have much evidence to draw on for support. Davidson, Andrewartha, and Birch have been the strongest proponents for population control by means of weather. Since weather is not influenced by population density, it is a density-independent factor. Andrewartha and Birch's (1954) book *The distribution and abundance of animals* is a treatise supporting their hypothesis that organisms are primarily limited by a shortage of time when weather conditions are so favorable that the population can increase. Therefore, the population never has enough time to reproduce up to the carrying capacity, competition for food does not result, and density-dependent factors are unimportant in population regulation. (Does this mean that most species are *r* selected?) This is in direct opposition to Nicholson's arguments.

The data that inspired Davidson, Andrewartha, and Birch were collected over 14 years, 1932–1946. The subject species was the apple blossom thrips, *Thrips imaginis,* and it was sampled on rose blossoms in the garden of the Waite Agricultural Research Institute, University of Adelaide, Australia.

Davidson and Andrewartha (1948a,b) found that the yearly abundance curve of thrips typically showed a rapid rise to peak abundance and then a rapid decline (Fig. 17.16). They concluded that competition plays little or no part in determining the maximum density reached in late November or early December (spring). Rapid reproduction was possible only for a limited period during spring and early summer. Insects increased rapidly but so did the flowers available to them, and the favorable period normally ended long before *Thrips* reached the carrying capacity of the flowers. In fact, they found that a combination of four weather factors explained 78% of the total variance in the annual maximum density attained by the thrips population. These four factors combined must have been far more important than any others. The most important factor was winter temperatures, which affected the earliness of appearance of blossoms, especially affecting the length of time for reproduction. The second most important factor was rainfall during September and October, which promoted development of host plants and reduced the mortality of pupal thrips in the soil. The third factor was winter temperatures of the preceding year. Presumably, they promoted survival of thrips just as factor 1. Finally, spring temperatures (during September and October) were

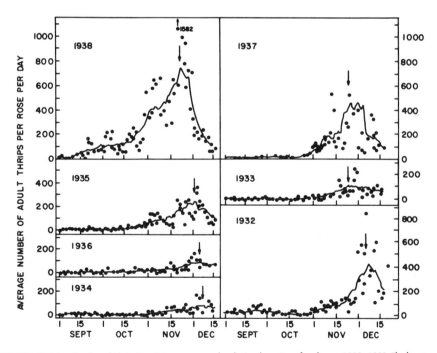

FIGURE 17.16   Number of *Thrips imaginis* per rose per day during the spring of each year 1932–1938. The line is a 15-point moving average, and the arrow indicates the maximum abundance in each year. From H. G. Andrewartha and L. C. Birch, *The distribution and abundance of animals*. Published by the University of Chicago Press. Copyright © 1954 by the University of Chicago Press. All rights reserved.

important. These increased the rate of development of thrips and host plant. But these temperatures were usually sufficient during this period, so it was the least influential of the four factors.

However, there are some difficulties in the interpretation that weather has a controlling influence (see also Smith, 1961; Reddingius, 1971). The carrying capacity of the environment is hard to determine. The roses did not actually serve as a breeding site for the thrips but acted as a trap, which indicated the density of the population in the area. We do not know how crowded the flowers were that supported breeding populations! Perhaps breeding sites were limiting the population. Also, the most important factor influencing thrips populations was winter temperature. This also influenced host-plant growth and flower production. Therefore, it was influencing the carrying capacity and population levels indirectly. Nicholson called such an influence a density-legislative factor. Thus weather can influence the food supply and thus density-dependent processes, as previous examples indicate. The third most important factor was the winter temperature of the preceding year. Since winter temperature is most closely correlated with maximum population, Davidson and Andrewartha really found that the population in the current year is dependent on the population the year before, probably a density-dependent effect! Finally, Adelaide is, practically speaking, on the edge of the Great Victoria Desert and *Thrips imaginis* is at the end of its range in Adelaide. Therefore, climatic stress is most likely to be seen in this population, although in the case of *Thrips* density-dependent factors cannot be ruled out.

Probably density-dependent factors are more important at the edge of the range of a species and density-dependent factors at the center (Fig. 17.17). Support for this statement was provided by Whittaker (1971), who studied the population dynamics of the cercopid, *Neophilaenus lineatus,* near the edge of its range and in more southerly sites in England. In the northern pop-

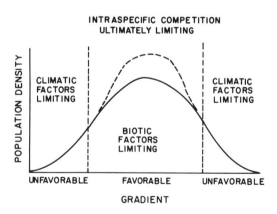

FIGURE 17.17 Hypothetical relationship between the major regulating factors on a species throughout the geographic range of the species.

ulations he found no evidence of density-dependent factors, whereas in the south the generation mortality was density-dependent, and a dipteran parasitoid acted in a density-dependent manner.

Another case where weather seems to exert a controlling influence is on the spruce budworm, *Choristoneura fumiferana,* in New Brunswick. Greenbank (1956, 1963) has suggested that a series of dry and sunny summers is beneficial to survival of the budworm and leads to a population release (see details above). The main reasons are that staminate flower production is higher, and these flowers act as a better food source for larvae, development is quicker, and old flowers provide safe overwintering sites. Stark (1959) concluded that climatic release resulted in outbreaks of the lodgepole needle miner in the Canadian Rocky Mountains and reviews the literature on the subject. Wellington (1954a,b) proposed a comprehensive theory of climatic release.

In between the density-dependent control camp and the density-independent control camp sit some people who wish to make peace and combine the two theories. Without doubt, both types of control are important. Milne (1957a,b) has been the main proponent of a joint theory. Reynoldson (1957) showed that weather can be critical to one species and competition to another in the very same sewage tank. LeRoux et al. (1963) and Watt (1963) concluded that the major factors influencing populations of five agricultural crop pests in Quebec were, for three species, migration of adults, parasitism, and bird predation, all likely to be density-dependent factors, and for the other two, frost and other weather conditions, which are density-independent factors. Among the many species of insect herbivores on broom, both density-dependent and density-independent factors were important (Waloff, 1968b). Intra- and interspecific competition, egg sterility at high densities, predation, parasitism, and dispersal were significant influences on some species, and in the same community other species were most affected by density-independent factors: plant aging that affected nutrition, difficulty in establishment of first instar larvae, and frosts.

Other examples show that weather controls the amount of food where food is limiting, so that these factors interact to influence the herbivore population, as suggested for *Thrips imaginis, Leptopterna dolabrata,* and all the studies discussed by White (1969, 1974, 1976, 1978), because weather conditions caused stress of host plants, and probably *Operophtera brumata,* since weather determines phenology of oak leaf production and its synchrony with larval eclosion.

Finally, in relation to density-independent processes, Cole's papers (1951, 1954a) should be consulted, as he suggests that population fluctuation (in particular population cycles) may be most appropriately described by assuming that there are so many influences on a population that the outcome is a random fluctuation. However, Moran (1954) takes exception to Cole's analysis, as only the mean distance between peak populations was considered. Moran concluded that lynx and snowshoe hare populations oscillate in a nonrandom manner. A symposium on cycles in animal populations (Hewitt, 1954) may be consulted for further details.

ENDOGENOUS POPULATION PROCESSES

Now the possibility must be investigated that **endogenous population processes** are important in population regulation. Density-dependent pathological effects have been assumed to be important. Here the pathological effects of stress on animals are involved. Pathogenic organisms such as bacteria or viruses are not included, as these would fit into the exogenous population processes, since the effects of a pathogen are likely to increase with density just as predation does. At high population densities animals are subjected to increased stresses in finding food and cover and escaping predators. They may be forced to fight with other animals more frequently to obtain sufficient food or a mate. The body adapts physiologically to these stresses in vertebrates under the stimulus of increased hormone secretion from the adrenal and pituitary glands. These glands can compensate for stress only within certain limits, and when the stresses become very great, Christian (1950, 1959; Christian and Davis, 1964) has postulated that death occurs through exhaustion of the adrenopituitary system, not through the action of food shortage, predators, or disease organisms. *Shock disease* is apparently also a manifestation of the same syndrome or perhaps a pathological condition that impairs the animal's ability to store glycogen. This disease has been observed in Minnesota in showshoe hare cycles (Green and Larson, 1938; Green and Evans, 1940) and in meadow vole populations in Germany (Frank, 1957). (For further discussion, see Emlen, 1973).

Chitty (1957, 1960), who studied the vole, *Microtus agrestis,* at Lake Vyrnwy in Wales, found that in a natural population of field voles that was declining, individuals were intrinsically less viable than their predecessors and that changes in external mortality factors, such as weather and predators, were insufficient to account for the increased probability of death. Chitty (1960) proposed the hypothesis that all species are capable of regulating their own population densities without destroying the renewable resources of their environment and without requiring enemies or bad weather to keep them from doing so. Under appropriate circumstances, indefinite increase in population density is prevented through a deterioration in the quality of the population. Chitty also pointed out that it is unsafe to assume that population density has no effect on the physiology of the individual or on the genetics of the population. Therefore, it is most improbable that the action of physical factors is independent of population density, as Andrewartha and Birch have stated. He therefore postulated that the effects of "independent" events, such as weather, become more severe as numbers rise and quality falls.

This hypothesis overcomes the difficulties in the density-dependent versus the density-independent control argument. The hypothesis is based on a set of known facts on population cycles:

1. Declines in population can take place in a favorable environment.
2. High population density is not sufficient to start an immediate decline.
3. A low population is not sufficient to halt a decline.
4. The vast majority of animals die from unknown causes, males more rapidly than females.
5. The death rate can be greatly reduced by placing animals in captivity.
6. The adult death rate is not abnormally high during the years of maximum abundance.

These facts are consistent with the idea that susceptibility to natural hazards increases among generations descended from animals affected by adverse environmental conditions. In 1960, Chitty wrote that the mechanism of this process had yet to be discovered.

Chitty (1967) stated that mechanisms for the self-regulation of animal numbers are thought to be a consequence of selection under conditions of mutual interference, in favor of genotypes that have a worse effect on their neighbors than vice versa. The most puzzling aspect of small mammal populations during their periodic declines in numbers is the severity of mortality at relatively low densities after peak abundance. Now he suggested that there was no evidence that the animals are less viable than normal, although a decrease in some components of fitness may be expected. Aggressive behavior is the most likely characteristic to improve survival under conditions where mutual interference is inevitable. Chitty pointed out that individuals that have not been selected under conditions of intraspecific strife are likely to be mutually tolerant, and populations composed of such individuals should reach abnormal abundance before such selection takes place. Such populations may occur when new areas are invaded, when an existing source of heavy mortality is removed, or when populations have been seriously reduced by natural or artificial catastrophes. At high populations aggressive individuals will be relatively more fit, and will be selected for, but as populations decline this trait will become maladaptive as energy is wasted. Thus populations may decline more than expected before aggressiveness is selected out of the population. Thus Chitty's theory depends on behavioral, physiological, and genetic changes in the population.

Carson (1968) has described the likely changes in genetic material in a population during a population increase, or flush, and the population crash. Small changes in the gene pool of a population may actually initiate a flush. Carson maintained a laboratory population of *Drosophila melanogaster* which was inbred and contained recessive mutant markers. About 160

Drosophila per generation could be maintained on a unit of food. He altered the genetic content of the population by adding a single male fly whose mother was from the population and whose father was from an unrelated wild-type laboratory strain. Without any change in food or space a population flush resulted and in nine generations the population had increased to three times its size (477 individuals). Even though food was unchanged, population growth was exponential. This change must have been due initially to higher Darwinian fitness of some of the genotypes following hybridization. After the peak the population fell abruptly and oscillated around a lower level.

In general, there seem to be four stages to the flush and crash according to Carson:

1. Initially, new genetic material is injected into the gene pool, or environmental influences change so that mortality in the population is reduced or fecundity is increased.
2. A flush in population results during the period of reduced natural selection and is accompanied by a breakdown in the genetic basis of fitness in the population.
3. More stringent environmental conditions imposed on the population cause a crash in numbers because of severely lowered fitness in the population.
4. As the crash reduces the population size toward the original level, it may carry the population to a size even smaller than at the inception of the cycle. This may be because of the lag between the appearance of stringent conditions and the multiplication by selection of individuals carrying genotypes with greater Darwinian fitness relative to these new conditions. This supports Chitty's point that low populations are not sufficient to halt a decline.

There may result considerable differences between the genotypes at stage 1 and the end of stage 4. The classic case of a natural population showing evolution in a similar way to that postulated by Carson was observed by H. D. and E. B. Ford and reported in 1930. They studied an isolated population of the marsh fritillary butterfly, *Melitaea (Euphydryas) aurinia,* in Cumberland, England, where changes in wing pattern were relatively easy to observe (Fig. 17.18). They studied the population for 19 years (Ford, 1964) and records of the population condition had been left by collectors during the previous 36 years. So preserved specimens covering a period of 55 years were available for study! Collectors of butterflies look for unusual forms. Therefore, the record was good for checking variability in the population. Also, the marsh fritillary has a particularly intricate pattern that reflects changes in genotype very sensitively. The general trends reported in population change in numbers and variability (Ford and Ford, 1930; Ford, 1957, 1964) are summarized in Fig. 17.19.

FIGURE 17.18 Marsh fritillary, a checkerspot butterfly, studied by Ford and Ford (1930). All specimens are males, with a dorsal view of a typical form (top left), a side view with the underside of the wing exposed, feeding on a devil's-bit scabious inflorescence (top right), an aberrant form with observable differences in markings from the typical form (bottom left), and a very bizarre male with few scales and almost no markings. From P. W. Price, *Biological evolution*. Copyright © 1996 by Saunders College Publishing. Reprinted by permission of the publisher.

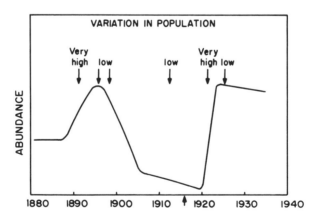

FIGURE 17.19 Diagrammatic representation of population size of the marsh fritillary butterfly studied by Ford and Ford (1930), based on accounts by these authors, and Ford (1957, 1964). Note that when selective pressure is released and populations are increasing rapidly, very high variation is seen in the population. In the period 1924–1935 the uniform type was recognizably distinct from that which prevailed during the period previous to 1920, showing that evolution had occurred in a span of 4 years. The arrow on the abscissa indicates the time at which the Fords began their own studies.

It is remarkable that in four to five generations, 1920–1924, a visible evolution in wing pattern took place, and this was probably accompanied by pleiotropic gene effects that shifted the adaptive core of the population in response to the new environmental conditions prevailing in 1924. This shift was the result of the incredible diversity of genotypes that were subject to the selection in 1923 and 1924 when the population flush ended and the population stabilized, so that mortality increased to about 99% (assuming that a female lays 200 fertile eggs on average). During the flush phase, as described by Ford (1964, pp. 14–15),

> an extraordinary outburst of variability took place. Hardly two specimens were alike, while marked departures from the normal form of the species in color pattern, size and shape were common. A considerable proportion of these were deformed in various ways; the amount of deformity being closely correlated with the degree of variation, so that the more extreme departures from normality were clumsy upon the wing or even unable to fly.

Little evolution occurred between 1896 and 1920, although the environment probably changed considerably during this time. A sudden flush permitted an increase in genetic diversity and rapid adaptation to the environment to take place so that a higher population could be maintained because of the greater fitness of individuals after stringent selection. The slow decline in abundance after 1924 may reflect a change in environment without a corresponding change in the butterfly gene pool. Tetley (1947) also found increased variability in an increasing population of the nymphalid butterfly, *Argynnis selene*. This type of evolution is harder to observe in organisms other than butterflies, but there is little doubt that in insects that undergo large and rapid population fluctuations there is an opportunity for populations to evolve very rapidly.

ROLE OF DISPERSAL

Polymorphic behavior for activity and **dispersal** may also be important in population dynamics. Wellington's (1957a,b, 1960, 1964, 1977, 1979) (Wellington et al., 1975) studies on the western tent caterpillar, *Malacosoma pluviale*, demonstrate how influential qualitative aspects of populations can be. Wellington studied populations of the caterpillars and adults near Victoria, British Columbia, on Vancouver Island. He found the population was polymorphic in behavior. Some individuals were inactive; females oviposited near their birthplace, produced mostly inactive offspring, exploited favorable habitats (Fig. 17.20A and B), and were killed by harsh climatic conditions (Fig. 17.20C). Other individuals were more active, females oviposited much farther from their place of origin, and always produced a higher proportion

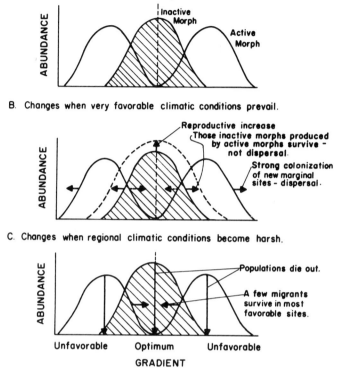

A. During moderately favorable regional climatic conditions.

B. Changes when very favorable climatic conditions prevail.

C. Changes when regional climatic conditions become harsh.

FIGURE 17.20 Diagrams of population trends in western tent caterpillar populations under three sets of conditions. Inactive morph distribution is shaded; active morph distribution is unshaded.

of vigorous individuals among their progeny. These active migrants tended to colonize more severe habitats (Fig. 17.20A and B), or replenish the vigor of other populations (Fig. 17.20B). During harsh climatic conditions individuals in marginal habitats were killed and those in favorable habitats survived (Fig. 17.20C) to recolonize all the vacant areas as conditions permitted.

The spatial dynamics of this species are extremely important. Wellington has emphasized the qualitative variables in a population and how they are necessary in understanding the change in population numbers in any one site. What has been emphasized in Fig. 17.20 is the spatial relationships of a species in an area and how important it is to sample an extensive area to obtain a clear picture of what is happening in the population. Populations do move from place to place, and expand and contract; they do not just go up and down. Thus setting up a single sample plot in an area can produce some rather bewildering results and, probably in some cases, spurious conclusions (see also R. F. Morris, 1971b; Price, 1971b; Price et al., 1990).

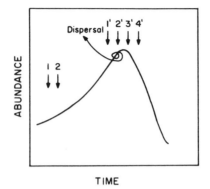

FIGURE 17.21 Diagrammatic representation of trends in the gypsy moth showing when dispersal occurs. Generations 1 and 2 feed in uncrowded conditions; generations 1' to 4' exist in a food shortage.

Leonard (1970a,b) has reported qualitative changes in populations of gypsy moth, *Porthetria dispar*. He has evidence that moth populations are numerically self-regulating because of a shift in the quality of individuals induced by changes in nutrition. This shift is rapid and does not depend on natural selection, but rather, on a switch in the behavior of the first instar larvae (Fig. 17.21). In generation 1, larval food is plentiful and the resulting females produce eggs with large food reserves: a maternal effect on the next generation. In generation 2, large first instar larvae hatch from eggs, settle rapidly and feed after eclosion, molt four times in males and five times in females, and produce females of high nutritional status. In contrast, in generation 1', larval food starts to be in short supply, and females result that produce smaller eggs with reduced food reserves. In generation 2', larvae are small and active and do not settle readily to feed. First instar developmental time is much longer with a longer prefeeding phase. The larvae are more sensitive to interference and readily drop and hang on a silken thread and thus became more readily dispersed by wind. Female adults are winged but cannot fly so that larvae on silk is the only possible dispersal phase. The same larval behavior is produced by crowded situations or starvation. At the end of their development these larvae usually have molted five times for males and six times for females. This was the most readily observed factor that led to detailed studies by Leonard which provided this understanding of the population dynamics. Thus, in generations 3' and 4', those larvae that do not disperse eat the remains of the food and a population crash results. But high dispersal has already occurred and new sites with ample food had been colonized through this high dispersal at high densities. Food plays a critical role in the fluctuation, but the important point is that there is a built-in mechanism that permits the population to respond to crowding or food shortage early enough so that many disperse before death by starvation becomes a critical factor. Thus the role of dispersal has a strong endogenous mechanism involved (see also Taylor, 1981a,b).

Dispersal is adaptive. It permits individuals to spread and escape crowding as the population increases, and it enables the colonization of new sites. New food sources can be exploited by dispersed individuals. It permits exchange of gene material between populations, and outcrossing produces a diverse array of genotypes in a parent's progeny. In the long run at least, this is usually advantageous, as a diverse gene pool promotes better population adjustment to the environment than does a restricted one, and thus promotes the welfare of individuals in that population. However, dispersal is not an unmitigated benefit to the individual. Since an individual insect that disperses from one food source frequently has a small chance of discovering a distant food source in tolerable environmental conditions, under most conditions it will be maladaptive to disperse more than very locally. It helps to consider the fitness of individuals that have a choice of dispersing. Those that are fittest leave many progeny or leave many of their genes in the gene pool. When food and other environmental factors are favorable, it is best to exploit this situation. Favorable conditions promote survival and the leaving of progeny. As conditions worsen, either through food shortage or other factors, and the chances of leaving viable progeny *in situ* decrease, dispersal becomes a profitable undertaking. Ultimately, when conditions are so severe that no progeny can survive, although dispersal is very costly, the probability of leaving viable progeny is increased by dispersal, and dispersing individuals will be, on average, infinitely fitter than the residents. Thus any adaptation that permits an individual to respond to worsening conditions before they become intolerable will be strongly selected for. Dispersal by the gypsy moth's first instar larvae is a good example.

Dempster (1968) gives dispersal figures for adults of the herbivore psyllid, *Arytaina spartii*, which feeds on broom, *Sarothamnus scoparius*. He also provides data on the mean fecundity of females that did not disperse (Table 17.4). Note the strongly density-dependent response in dispersal and

TABLE 17.4   Population Dynamics of the Psyllid *Arytaina spartii* on Broom at the Imperial College Field Station, Silwood Park, England.

Year	Number of Adults Per Bush	Number Dispersing, Caught in Suction Trap	Mean Fecundity of Remaining Females
1959	316	118	—
1960	3880	850	146.3
1961	7792	10,572	6.4
1962	223	131	267.5
1963	74	26	—

SOURCE: Dempster (1968).

reproduction. The 1961 population was twice that in 1960, but the number dispersing was about 13 times greater. The dispersing females did not have much to lose by leaving, since those that remained only produced an average of 6.4 eggs per female. Of these eggs, which amounted to a mean of 49,992 eggs per bush, only 223 adults for 0.4% survived for the next generation. In fact, those females that dispersed and discovered other broom bushes had a tremendous advantage over those that remained. There is an inflection in the dispersal curve with increasing numbers, suggesting that dispersal rate increased particularly rapidly near the peak abundance levels.

This example raises the interesting question of how selection is likely to modify dispersal time, dispersal rate, and amplitude of dispersal in a population. (See also Johnson, 1969; Dingle, 1972, 1978, 1996; Rainey, 1976; Rabb and Kennedy, 1979; Drake and Gatehouse, 1995, for extensive treatments of migration and dispersal.) This question has already been discussed in Chapter 14, especially in relation to water-striders, *Gerris,* and milkweed bugs, *Oncopeltus.* Gadgil (1971) presented a theory of dispersal based on a set of models he created. By considering such parameters as population density, carrying capacity, the fraction of the population dispersing, and any sensitivity to density in the fraction dispersing, he has been able to make some general statements about adaptive dispersal in relation to variability of the carrying capacity *(K)* in time and space.

Consider variability of K with time while keeping the spatial relationships of K constant. Such a situation would be a pattern of bodies of water in an area, each body representing a separate K, and each altering in K with time out of phase with other bodies of water. Some bodies of water would dry up sooner than others; rain and runoff water would replenish some and not others. So how are aquatic insects likely to adapt a dispersal pattern that is best equipped to exploit this variation in food resources with time? Note that carrying capacity is the ability of the environment to support a population, so while water availability is used in this example, it must be assumed that plenty of water carries with it plenty of food.

In any given pool or puddle, therefore, K may suddenly change from low to high after rainfall and the population would move from a crowded to an uncrowded situation. During the crowded situation it would seem to be adaptive to disperse, as the chances of finding a pool with a lower exploitation relative to the carrying capacity would be good. However, this newly discovered site in an uncrowded state may rapidly dry up so that the fitness of the individual could be lower than if it had stayed put. In this situation it is probably just as adaptive for an individual to leave an uncrowded situation as a crowded one, as the condition of the newly colonized site is so unpredictable. Leaving an uncrowded site and arriving in a crowded one may actually improve the fitness of an individual if the conditions in the two ponds is reversed quite rapidly. Therefore, the best strategy in this situation is for the population to maintain a relatively high magnitude of dispersal at all density levels. This strategy is seen in corixids studied by Brown (1951), where those in ephemeral situations had high dispersal rates and those in permanent bodies of water had low dispersal rates. This also seems to be the strategy of *Malacosoma pluviale*. Note that the western tent caterpillar is not only a forest-dwelling species, but colonizes isolated trees that may be dispersed patchily.

In contrast, it is possible to conceive of a condition where variability in K with time is synchronized throughout a fairly large area. Here the conditions of crowding are likely to be similar in each situation, so that frequent dispersal cannot be adaptive, as a dispersing insect is likely to settle in another crowded area. The best strategy is to keep dispersal to a minimum as long as food is available, but to have a very sensitive dispersal-triggering mechanism so that dispersal is initiated just before the population is doomed through starvation. A good dispersal mechanism is also needed. Such a situation is likely to occur where climate has an effect on a food source, and the feeding population, over a wide area. For example, cone crop production and seed production would vary over wide areas, as would male cone production in relation to spruce budworm, and even vegetative production on trees that are host to the gypsy moth.

Perhaps it is because of the relatively static spatial relationship of forest and shade trees and the synchronous nature of the fluctuation of insect populations that feed on these trees that the most clear-cut density-dependent dispersal has been observed in these herbivores, forest insects, and insects in similar situations. Some examples include the gypsy moth, *Porthetria dispar* (Leonard, 1970a,b), the California oak moth, *Phryganidia californica* (Harville, 1955), the fall webworm, *Hyphantria cunea* (R. F. Morris, 1971b), the broom psyllid, *Arytaina spartii* (Dempster, 1968; Waloff, 1968a,b), and perhaps the spruce budworm, *Choristoneura fumiferana* (Morris, 1963b).

GEOGRAPHIC SCALE POPULATION CHANGE

Morris (R. F. 1971b) provides interesting data on the fall webworm. He conducted very extensive road censuses of the conspicuous webs in areas of New Brunswick and Nova Scotia and plotted the number of nests per 100 road miles against the percentage of agricultural land in the census route (Fig. 17.22). Fall webworm feed on deciduous trees and survive best on open grown trees, that is, in agricultural sites. But in coastal areas, where densities are highest, forested areas sustain almost as high a population as the agricultural sites. This is not so in other regions. R. F. Morris (1971b, pp. 1530–1531) offers these possible explanations:

1. There may be an interaction between vegetation and climate. Populations reach higher levels in humid coastal areas with longer developmental seasons. Under these favorable conditions the webworm may be less sensitive to land use and vegetation pattern than in drier continental areas.

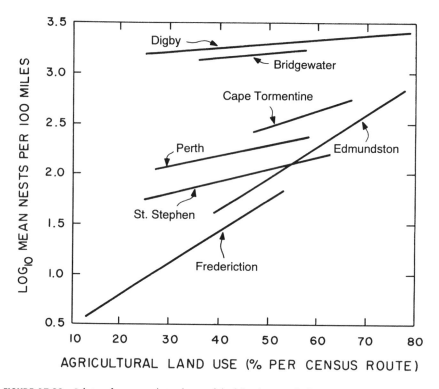

FIGURE 17.22 Relation of mean population density of the fall webworm to land use in seven census areas for the years 1957–1968. As the percent of agricultural land increased, so the population density increased. Note that in the moist coastal areas around Digby and Brigewater in Nova Scotia, populations were almost as high in more forested areas as in more agricultural areas, and these were the two areas that showed very high populations in general. These areas contrast, in particular, with the inland sites of Fredericton and Edmundston, New Brunswick. For location of sites, see R. F. Morris (1971a). From Morris R. F. (1971b).

2. As high populations develop on the favorable routes within an area, there may be a tendency for more nests to appear on unfavorable routes as a result of moth dispersal, so that the population density is evened out. Morris (p. 1531) stated:

> In 1961, density exceeded 100 nests per mile on some of the best coastal routes but remained below 10 nests per mile inland. Relatively high nest populations were observed in the Bridgewater area in that year along the sides of very narrow woods roads through pure stands of oak. In view of the poor exposure, as well as the unfavorable nutritional quality of oak, it is probable that the moths responsible for the nests came from agricultural land outside the oak stands.

Thus there exists evidence for density-dependent dispersal.

3. Other possibilities exist, such as changes in genetic quality associated with high density, but it is not yet known whether they influence the dispersal habits of the moths. Certainly, dispersal is playing an important part in the population dynamics of the fall webworm.

Morris has found in his studies that climate, land use, tree species, tree density, population density, and genetic change all influence the dynamics of the fall webworm, not to mention parasitoids, predators, and disease. (See Morris and Fulton, 1970a,b, and references therein; Morris, 1971a,b; see also Chapter 18.)

The importance of sampling insect populations over extensive geographic areas in order to understand the role of dispersal in population dynamics is illustrated by the studies of Wellington and Morris, discussed previously. Other extensive sampling studies provide new insights on population dynamics. The Rothamsted insect survey that covers the British Isles provides a picture of both seasonal change and annual change in insect species abundance and distribution (Taylor, 1979; Taylor and Taylor, 1979, and references therein). Within-season sampling shows very dramatic shifts in population density during each dispersal phase of such species as the grain aphid, *Sitobion avenae,* and the rose-grain aphid, *Metopolophium dirhodum* (Fig. 17.23). Such patterns are not repeated every year in many insects, which show great changes in density and location between years (Fig. 17.24). In this comparison of moth species, *Spilosoma luteum* has fairly predictable areas of high density in the south but very unpredictable distribution in the north. *Euxoa nigricans* has unpredictable distribution and density throughout its range. *Xanthorhoë fluctuata* seems to be quite stable in space and density, but pockets of low density shift around, probably caused by parasites. *Callimorpha jacobaeae* has quite predictable centers of distribution, although these move slightly from year to year. The challenge in understanding the mechanisms resulting in these patterns is considerable, and clearly, an understanding of dispersal patterns is essential.

As dispersal becomes more extensive and effective, the problem of understanding population dynamics increases. The most dramatic example involves the desert locust, *Schistocerca gregaria,* which requires integrated studies covering continents (e.g., Betts, 1976; Rainey, 1982; Showler, 1995, and references therein). Highly seasonal rain, unpredictable in terms of location and timing, stimulates development of vegetation that provides food for hatching locust nymphs. Once swarming behavior develops in local source areas with favorable rainfall and vegetation (see the areas in Arabia, northern Somalia, Ethiopia, Kenya, and Pakistan, Fig. 17.25), major atmospheric influences cause flying swarms to move downwind, as the arrows indicate, over thousands of miles. However, patterns may be emerging (Rainey, 1982), since the source areas in the 1954 swarm development may act repeatedly in this way, enabling a much more focused study on the environmental factors involved with the development of the gregarious phase and large swarms of adults.

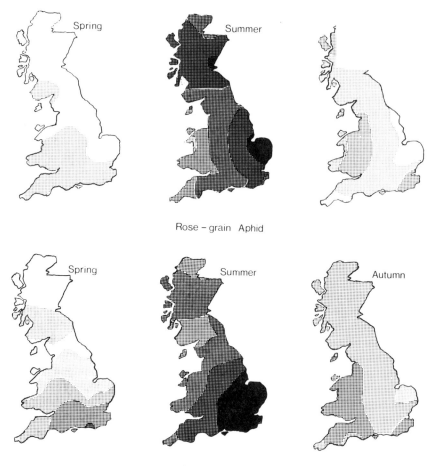

FIGURE 17.23 Change in population densities through the year of the rose-grain aphid, *Metopolophium dirhodum* (above), and the grain aphid, *Sitobion avenae* (below), based on mean densities from 1969 to 1974 in the British Isles. Density of shading corresponds to density of aphids. From Taylor (1979); with permission.

SYNOPTIC POPULATION MODEL

Is it possible to create a general theory for population dynamics? Statements on which regulating factor is likely to be most important in natural populations have been avoided, since each is likely to be significant for certain populations at certain times. Probably for each population there is a unique combination of factors regulating populations, if for no other reason because each population probably has a unique gene pool.

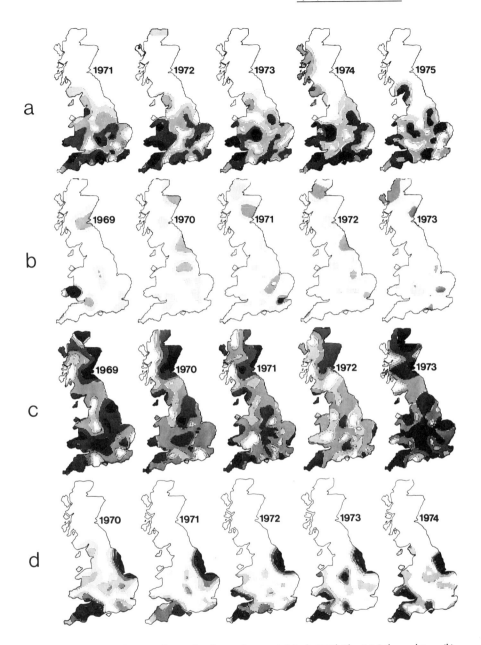

FIGURE 17.24 Pattern changes in density of moths over 5-year periods in the British Isles: (a) *Spilosoma luteum;* (b) *Euxoa nigricans;* (c) *Xanthorhoë fluctuata;* (d) *Callimorpha jacobaeae.* Shading density corresponds to insect density. From Taylor (1979); with permission.

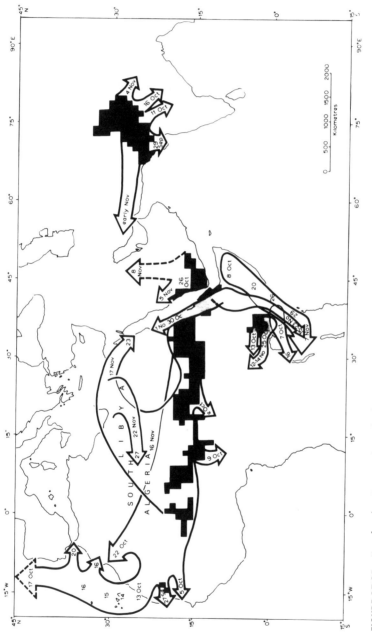

FIGURE 17.25 Map of southern Europe and central and northern Africa to India, showing the inferred movements (arrows and dates) during October and November of the desert locust, *Schistocerca gregaria*, and the areas of high egg and nymph densities (in black). From Rainey (1982); with permission.

However, one general model developed by Southwood (1977a) and Southwood and Comins (1976) integrates many of the factors discussed in this chapter, Chapter 14, and Chapter 8. This model is reconsidered in Chapter 19. The model synthesizes three aspects of the life history for each of a range of insect species in relation to the population density of that species:

1. Natality is low at low densities because mates are hard to find, increases to a peak, and then begins to decline because of intraspecific competition (Fig. 17.26).
2. The effect of enemies increases with herbivore density and then declines as populations escape from this regulating factor. This is because the overall response of predators to prey density is sigmoid, based on the functional response of individuals and the numerical response of the population (Fig. 17.26).
3. Intraspecific competition, emigration, and disease all increase with increased density, producing mortality in addition to the effects on natality mentioned in item 1 (Fig. 17.26).

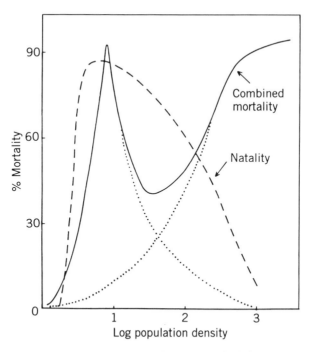

FIGURE 17.26 Generalized relationships between population density of a herbivore species, natality, mortality caused by enemies (with a peak at moderately low population density), and intraspecific competition (with a peak at high population density). From Southwood (1975); with permission.

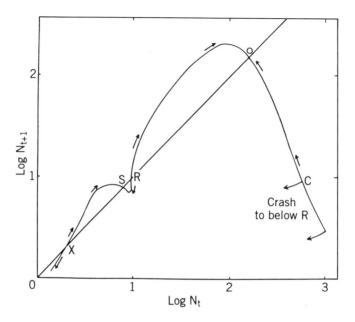

FIGURE 17.27   Generalized relationship between population size in generation t, (log N_t), and in generation $t + 1$, (log N_{t+1}). The line at 45° represents no change between the two generations, points above this line show population increase, and points below show decrease. X, extinction point; S, stable equilibrium point; R, population release point; O, point of oscillation around an equilibrium density; C, crash point. From Southwood (1975); with permission.

The net result of these interacting factors is that the rate of increase of the population changes dramatically with increasing density (Fig. 17.27). Populations below X go extinct. Populations above X increase rapidly until enemies become effective and a stable equilibrium is established at S, with the population at an endemic level. When the effect of enemies is disrupted or environmental conditions become particularly favorable for reproduction, the population escapes the stabilizing influence of enemies at R and increases to epidemic proportions until intraspecific competition, emigration, and disease stabilize the population at O. However, shortage of food and disease may lead to massive mortality and low natality, causing the population to crash to C and endemic levels.

The range of species superimposed on this scenario covers life history types from r-selected to K-selected species (see Chapter 14). The r-selected species colonize new resources very effectively and reproduce so rapidly that natural enemies frequently play an insignificant role in their population dynamics. In contrast, K-selected species are relatively poor colonizers, escape from enemies is less frequent, population growth is slow, and enemies play an important role in the dynamics of these species. The **synoptic model,** which integrates these factors, involves a three-dimensional plot of population density, population growth, and habitat stability, with the continuum for *r*-selected to *K*-selected species (Fig. 17.28). This model predicts that for strongly *r*-selected species in unstable habitats, natural enemies have practically no impact on population growth, and there is practically no biotic check on increase to epidemic levels. As species become more *K*-selected, the influence of natural enemies becomes more important in stabilizing populations and the "natural enemy ravine" deepens. Both the probability of epidemics and their intensity decline. The implications of this model for agricultural pests and biological control of these pests should be clear (Price, 1981b).

The extensive literature of insect population dynamics includes symposium volumes on population dynamics and regulation (den Boer and Gradwell, 1970; Watson, 1970; Duffey and Watt, 1971; Anderson et al., 1979; Barbosa and Schultz, 1987; Watt et al., 1990; Cappuccino and Price, 1995), books on biological control (DeBach, 1964, 1974; Huffaker, 1971; Van den Bosch and Messenger, 1973; Huffaker and Messenger, 1976; Ridgway and Vinson, 1977; Papavizas, 1981); pest management (Rabb and Guthrie, 1970; Geier et al., 1973; Price, Jones and Soloman, 1974; Metcalf and Luckmann, 1975, 1982, 1994; Knipling, 1979; Nordlund et al., 1981), insect migration and

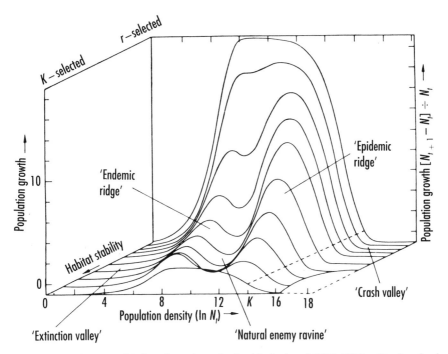

FIGURE 17.28   Synoptic model of population change developed by Southwood (1975, 1977) and Southwood and Comins (1976). Reprinted with permission from Southwood and Comins (1976) and Blackwell Scientific Publications Limited.

dispersal (Johnson, 1969; Rainey, 1976; Dingle, 1978b, 1996; Rabb and Kennedy, 1979; Drake and Gatehouse, 1995), and more general treatments of population dynamics (Richards, 1961; LeRoux et al., 1963; Solomon, 1964, 1969; Southwood, 1966, 1968; Clark et al., 1967; Van Emden, 1972a,b; Varley et al., 1973; Dempster, 1975; Pimentel, 1975; Begon et al., 1996a,b). Tamarin (1978) provides an overview of major concepts and debates on population regulation.

Chapter 18

Population Dynamics: Modeling

MODELS

A **model** can be defined as "an imitation and representation of the real world" (Ruesink, 1975, p. 358; see also Ruesink and Onstad, 1994). It should represent a simplified view of the system being modeled, yet capture the essence of the system such that the model has explanatory and predictive power.

As we have seen many times in this book, models play an important role in developing conceptual themes in ecology. They have powerful heuristic value, stimulating thought and development of modifications in order to simulate natural systems more effectively. The Lotka–Volterra model of predator–prey interactions (Chapter 8) still plays a very important role as a basis for more complex and realistic models today. The Nicholson–Bailey model is still in frequent use in modified form. Holling's (1959a) use of a human tapping for sandpaper disks as a model of a predator searching for prey resulted in a better understanding of fundamental aspects of predation and the disk equation describing the functional response of a predator to prey density (Chapter 8).

Models have more than heuristic value. They can help to identify critical data that need to be collected to complete an adequate analysis. They can be used to simulate experiments that might take too long or be too expensive to perform, or on too grand a scale for feasible experimentation in the field. They can help in predicting problems with pest species and in the identification of times in the life cycle that are vulnerable in terms of pest regulation. They can help to make the real, very complex world tangible, and in essence, relatively simple.

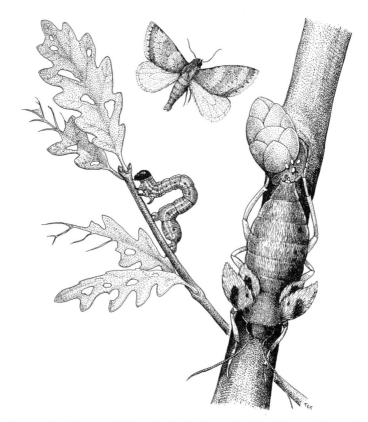

Winter moth, *Operophtera brumata*, with the wingless female, the winged male, and the caterpillar. Based on Novák et al. (1976). Drawing by Tad Theimer.

In the development and use of models it is clear that compromises are always necessary. A model of very general application, such as the Lotka–Volterra or Verhulst–Pearl models, will usually not fit many real systems well, although they serve as a basis from which modifications can make them relevant to such systems. A very general model is likely to make assumptions that do not apply to many real systems. On the other hand, a detailed, specific model on one system has only a narrow use, and its complexity is harder to comprehend and manipulate. In this chapter population models are discussed in terms of different approaches to modeling specific insect populations.

KEY FACTOR ANALYSIS

Morris was largely responsible for setting the analysis of insect population dynamics on a sound footing. With Miller (Morris and Miller, 1954) he modified life tables for use on natural insect populations, paid special attention to the development of accurate sampling techniques (Morris, 1955, 1960), and initiated the development of key factor analysis using life table data (Morris, 1959, 1963a,b; and see also Chapters 13 and 17).

Key factor analysis is explained by Morris (1963a) in a series of graphs starting with a typical fluctuating population in time (Fig. 18.1–18.8). This can be plotted to demonstrate how a population N_{n+1} is related to the preceding generation, N_n (Fig. 18.2). Regular oscillations would show on this plot as ellipses if successive generations are connected, and the direction of change would be clockwise. The relationship between N_{n+1} and N_n is curvilinear and the variance increases as N increases and is not independent of the mean population density. Therefore, population data are transformed to log N, providing a linear relationship between N_{n+1} and N_n (Fig. 18.3) and a variance independent of the mean (see Morris, 1955). As Morris (1959) points out, changes in animal populations tend to be geometric, not arithmetic (see Chapter 13), so it is more realistic to express N as a logarithm. In practice it is usual to employ the transformation $\log(N + 1)$ so that zero records can be handled easily ($\log_n 0 = 1$).

By correlation analysis the formula for the line, which provides the intercept, the slope ($b = 0.5$ in Fig. 18.3), and the correlation coefficient r, can be calculated from which the variance accounted for, r^2, by the line can be derived. In this case $r^2 = 0.3$, or 30% of the variation in log N_{n+1} is accounted for by log N_n alone. This is an important step in the analysis since it calculates the dependence of generation $n + 1$ on generation n, which are bound to be related. Once this dependence is estimated, further analysis can reveal how the addition of mortality factors improves the predictability of the model beyond that of the equation for the line in Fig. 18.3, which yielded $r^2 = 0.3$.

The slope of the line ($b = 0.5$ in Fig. 18.3) indicates to what extent the factor operates in a density-dependent way. If a factor is perfectly density dependent, b would equal 1; if perfectly density independent, b would equal 0.

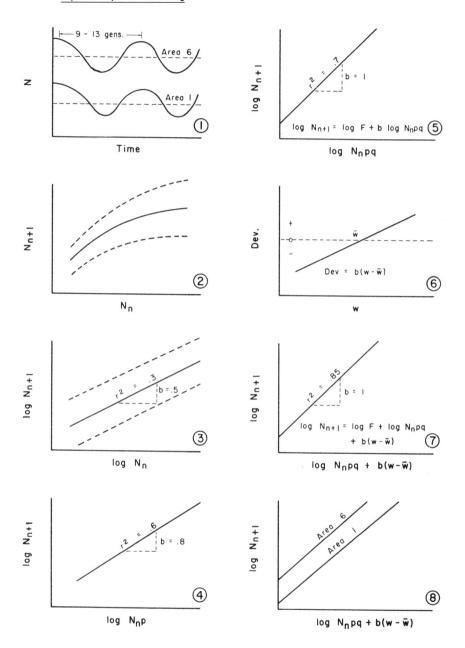

FIGURES 18.1–18.8   Steps in key factor analysis, according to Morris (1963a), discussed in the text. From Morris (1963a); with permission.

The next step is to introduce important factors one at a time in order to estimate the improvement of r^2. Parasitism may well be an important factor, as in the black-headed budworm, *Acleris variana* (Morris, 1959). If p is the proportion of N_n that survives parasitism, N_np will be the number of larvae that escape parasitism. This factor may improve the predictive value of the linear equation to $r^2 = 0.6$ with a slope of $b = 0.8$, so much more of the variance and density dependence in the system is explained (Fig. 18.4).

Predation might be tested next. Here, q is the proportion of N_np that survives predation, so that N_npq will survive both predation and parasitism, assuming that there is no interaction between predation and parasitism. Then the slope of the line may be increased to 1.0, in which case all density-dependent factors have been accounted for and r^2 may have increased to 0.7 (Fig. 18.5).

Thus an equation that predicts the size of N_{n+1} will be

$$\log N_{n+1} = \log F + b_d \log N_npq$$

where F, the intercept, is the effective rate of increase of the insect from generation to generation when the proportion of the population affected by a key factor is zero. The slope of the line b_d refers to the slope generated by density-dependent factors.

An accounting for 70% of the variation in N_{n+1} by N_npq is a good result, but density-independent factors may well explain some of the scatter around the regression line. Weather would be an obvious possibility, and some general synoptic index of weather might be tested. The positive and negative deviations from the line in Fig. 18.5 are plotted against an index of weather (w), and $\bar{w}$ is the weather index at zero deviation (Fig. 18.6). The equation for the deviation is

$$\text{dev.} = b_i(w - \bar{w})$$

where b_i is the slope of the regression for density-independent factors. So this equation can be added to the predictive equation to include the density-independent factor—weather. Thus the equation becomes

$$\log N_{n+1} = \log F + b_d N_npq + b_i(w - \bar{w})$$

Since $b_d = 1$, it can be omitted from the equation as in Fig. 18.7. Perhaps 85% of the variation in N_{n+1} is accounted for by the equation. Equations of this type have been generated for the spruce budworm (Morris, 1963b), the European spruce sawfly (Neilson and Morris, 1964), the black-headed budworm (Miller, 1966), and the jack-pine budworm (Clancy et al., 1980).

Morris (1963a) showed that the fall webworm data suggest that different areas with very different densities may show very similar predictive equations, with only F differing substantially (Fig. 18.8). Therefore, he stresses that more work should be concentrated on understanding F, which seems to be related in the webworm to climate and perhaps to vegetation. He also states that this kind of analysis is extremely helpful in determining which factors are worth more detailed study. Under no circumstances would Morris consider that this analysis is sufficient to determine the true nature of population regulation in insects. This broad approach must be used in conjunction with what Morris (1969) called "process studies," which attempt to explain in detail the mode of action of a mortality factor. These will be discussed when Morris's work on the fall webworm, *Hyphantria cunea*, is described later in the chapter.

Note that this key factor approach uses data on one stage of the life cycle in successive generations, and thus it contrasts sharply with the detailed life table approach, which studies all stages. Although it is less thorough than the life table approach, it may be preferable under a number of circumstances:

1. When a rare insect is studied so that it cannot be easily sampled at all stages of the life cycle.
2. When insect populations are low, as are most for various periods of time, so that a full census is impracticable.
3. When resources of manpower are restricted, the key factor approach provides for a well-organized scheme of sampling and analysis.
4. When life table studies have shown that a few factors are largely responsible, key factor analysis may provide all the data necessary for a continued accurate monitoring of the species. This might then allow a more diverse array of sites to be sampled so that greater breadth of understanding could be obtained, or a shift in emphasis to the essential process studies.

The paper by Morris (1959) on single-factor analysis stimulated the development of another analytical technique by Varley and Gradwell (1960, 1968, 1970; Varley et al., 1973), which in many respects is similar to Morris's method. Haldane (1949) had proposed that mortality factors could be expressed as the difference between logarithms of numbers per unit area before and after the action of a mortality factor. In this way the "killing power" of successive mortality factors could be compared and they would be additive. Thus total mortality, K, would equal the sum of all mortality factors or k-values:

$$K = k_1 + k_2 + k_3 = k_4 + \cdots + k_n$$

For each generation, each k-value and K can be plotted against time and the key factor can then be identified by inspection and by further graphical and regression analysis. Here it is usually assumed that there is a causal relationship between changes in k-values and K if regression analysis reveals a significant relationship, or if a k-value varies as K varies in a graphical analysis. Thus Solomon's (1949) definition of a key factor is accepted.

As an example, Varley and Gradwell (1960) had sampled winter moth (*Operophtera brumata*) populations from 1950 to 1959 and had estimated for each generation the following k-values:

k_1—*winter disappearance.* This factor includes all mortality between a sample taken in November–December, when the adult moths are sampled as they climb oak trees to lay eggs, and the next sample, which evaluates abundance of fully fed larvae in May which fall into traps on the ground. In addition, it is assumed that females each lay a constant number of eggs. Thus the k-value is an estimate of female mortality, variation in fecundity, mortality of eggs, and first to fourth instar larvae. It is an estimate of the mortality for the majority of the active life of the insect and all the time spent in the tree on which the larvae feed.

k_2—mortality due to *Cyzenis albicans*, a tachinid larval parasitoid.

k_3—other larval parasitoids.

k_4—microsporidian disease.

k_5—pupal predators.

k_6—*Cratichneumon culex*, an ichneumonid pupal parasitoid.

The results of their key factor analysis clearly identified k_1 as the key factor (Fig. 18.9). However, change in k_1 is usually greater than change in K, and k_5 may compensate partially for change in k_1. The graphical analysis shows that detailed studies on k_1 and k_5 would be well worthwhile.

Varley et al. (1973) list the ways in which population change and age-specific mortality can be expressed when generations do not overlap (Table 18.1). The value of logarithmic transformations can be appreciated, although large differences in abundance occur, as even the small values are not too low to be recorded readily on a graph. The k-values are calculated by estimating log N of each stage and substracting log N from each earlier stage. Thus, from Table 18.1,

$$k\text{-value (eggs)} = \log N_E - \log N_{L1} = 1.0$$

Density-dependent and density-independent relationships can be identified by plotting k-values against log N for each generation studied. When this is done for the winter moth data, k_1 appears to be density-independent and k_5 is density-dependent (Fig. 18.10).

Varley and Gradwell (1968) argue that errors in estimating k-values are likely to be similar to those in estimating N or K, and therefore they are not independent variables. Thus density-dependent factors can be identified only if log (survivors) is plotted against log N for each mortality factor and the slope of the regression is significantly different from 1.

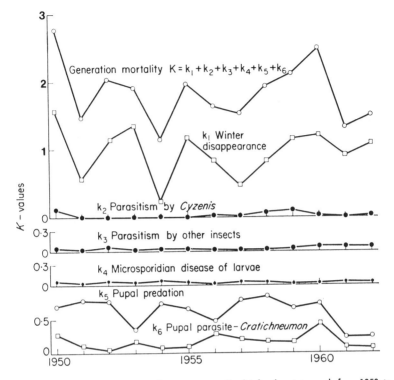

FIGURE 18.9   Generation mortality (*K*) and its components (*k₁–k₆*) for the winter moth from 1950 to 1962. Reprinted with permission from Varley et al. (1973) and Blackwell Scientific Publications Limited.

k-Factor Analysis

Podoler and Rogers (1975) have suggested another way for identification of key factors which appears to be more revealing than Varley and Gradwell's method. They suggest that *k*-values should be plotted on the abscissa and *K* on the ordinate. Then the *k*-value that provides the steepest slope is the key factor, where $b_{min} = 0$ and b_{max} will usually be close to 1. However, if a *k*-factor is much more variable than *K*, then *b* can be greater than 1. In this method more emphasis is placed on the slope than on the correlation. This seems a good approach, since in Varley and Gradwell's data and analysis k_1 does not correlate well with log *N* but it does with *K* (Fig. 18.11; cf. Fig. 18.10).

The validity of the methods developed by Morris and by Varley and Gradwell have been debated in many publications (e.g., Hassell and Huffaker, 1969; Morris and Royama, 1969; Maelzer, 1970; St. Amant, 1970; Luck, 1971; Royama, 1981a,b 1992, 1996; Murdoch, 1994), which should

TABLE 18.1 Ways of Expressing Population Change and Age-specific Mortality when Stages of the Insect Do Not Overlap.

Line	Eggs, N_E	Small Larvae, N_{L1}	Large Larvae, N_{L2}	Pupae, N_P	Adults, N_A	
1 Population	1000	100	50	20	10	
2 Number dying in interval	900	+50	+30	+10		Sum: 990 dead
3 Percent mortality	90	+5	+3	+1		Sum: 99% mortality
4 Successive percent mortality	90	50	60	50		
5 Successive percent survival	10	50	40	50		
6 Fraction surviving	0.1	×0.5	×0.4	×0.5		Product = 0.01 survival
7 Log population	3.0	2.0	1.7	1.3	1.0	
8 k value	1.0	+0.3	+0.4	+0.3		Sum: $K = 2.0$

SOURCE: Varley et al. (1973); with permission from Blackwell Scientific Publications Limited.

487

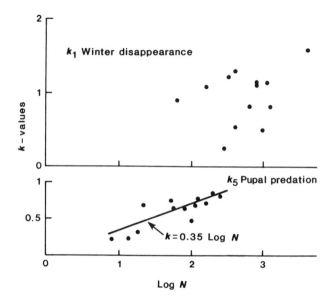

FIGURE 18.10 Relationships between log *N* and *k* values for *k₁* and *k₅*. Reprinted with permission from Varley et al. (1973) and Blackwell Scientific Publications Limited.

be consulted by the critical reader. Certainly, the power of regression analysis for understanding population regulation is severely limited (Eberhardt, 1970; Price, 1971b; Royama, 1992, 1996), and broader studies are essential (Wellington, 1957b, 1977; R. F. Morris, 1969, 1971a). As McNeill (1973, p. 502) stresses:

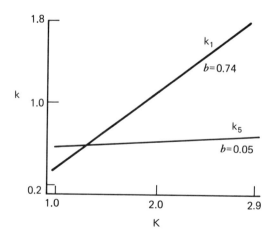

FIGURE 18.11 Relationships between *K* and *k* showing a steep slope (*b* = 0.74) for *k₁* and a shallow slope (*b* = 0.05) for *k₃* for the two mortality factors, winter disappearance and pupal predation on the winter moth in England. Reprinted with permission from Podoler and Rogers (1975) and Blackwell Scientific Publications Limited.

> Life-table analysis . . . does not tell anything about the underlying mechanisms and factors governing the control of density of the population studied: it simply provides us with a tool to indicate which mortalities are important, and whether they are density-dependent or not. Correlation with weather or other environmental variables provides us with further clues as to possible factors governing the mortalities in question; the proof that any of these possible factors do act must be obtained from independent observation and experimentation.

The methodology of key-factor analysis has been treated here because it has played a major role in the development of the study of population dynamics. However, there are good reasons to take the advice of those advocating alternative or additional methodologies. For example, Royama (1977, 1981a,b, 1984, 1992, 1996) has been a long-standing proponent for much more rigorous analytical techniques, and many authors have noted the requirement for experimental approaches (e.g., Morris, 1969; Murdoch, 1970; McNeill, 1973; Harrison and Cappuccino, 1995; and other papers in Cappuccino and Price, 1995). The sage advice of Royama (1996, pp. 92–93) should be taken to heart.

> The essence of the study of population dynamics is to analyze an observed process into major components and then to synthesize them to recreate the dynamics. In general, different factors play different roles in determining the process dynamics. For judging which components are major, the criteria are multiple and subtle beyond the simplistic idea of key factor analysis. The analysis of population processes, in which life table (or survivorship) data provide basic information, requires a much deeper insight than a simple index might provide. In this sense, there will be no simple alternative.
>
> We must look into a more comprehensive method, or a system of methods, to analyze each set of population process data at hand, rather than into a simple and easy-to-use tailor-made method; each set of data may be unique and may, accordingly, require a unique treatment. In other words, there is no easy road to the analysis of population processes.

With such an individualistic approach to analysis for each organism and population studied, it may become more challenging to compare the results from many studies in order to search for pattern. Therefore, with a sense of caution about the generalities that may be reached, let us examine how the methods described above may reveal patterns in nature.

Podoler and Rogers (1975) used their method of *k*-factor analysis on life table data for 17 species of insect and six species of vertebrate. With these 23 sets of data, perhaps we should be able to reach a conclusion on which theories on population regulation seem to be most important, and whether vertebrate and invertebrate populations differ substantially in their vulnerability to certain environmental factors. However, examination of the data indicates that every factor is likely to be important in one population or another (Table 18.2), and generalizations are difficult to make. Even populations of the same

TABLE 18.2 Important Factors in Regulation of Animal Populations, Based on Analyses by Podoler and Rogers (1975). Values in Tables are the Slopes (b) of the Correlation Between K and the Mortality Factor Listed (k's) with $b_{min} = 0$ and b_{max} Close to 1.

	Invertebrate Herbivores				Invertebrate Parasites				Vertebrate Herbivores, Total Four Species		Vertebrate Predators, Total Two Species	
	Natural Populations Total 12 Species		Laboratory Populations Total Three Species		Natural Populations Total One Species		Laboratory Populations Total 1 Species					
	Number of Species	Range in Slopes	Number of Species	Range in Slopes	Number of Species	Slope	Number of Species	Slope	Number of Species	Range in Slopes	Number of Species	Slope
A. Factor												
Parasitoids on eggs	3	−0.01, +0.19										
Parasitoids on other than eggs	8	−0.02, +0.74	3	+0.06, 0.63								
Predation	2	−0.66, +2.12										
Disease	1	+0.15	1	+0.05								
Parasitism + predation + disease	3	−0.03, +0.62			1	+0.33						
Food—starvation	1	+0.07	1?	+0.02								
Food—nutrition	2	+0.07, 0.29							2	+0.45, 0.63	1	+1.54
Competition	1	+0.29			1	+0.02						
Place to live	1	+0.02										
Fecundity	3	+0.07, 0.99					1	+0.75	1	+0.07		
Host-plant resistance	1	+0.02										
Emigration	1	+0.95										
Weather	4	+0.01, 0.27										
B. Factors not identified												
Egg mortality	4	+0.02, 0.36	1	+0.65							1	−0.26
Early mortality of young									2	+0.52, 0.85		
C. Adult mortality	3	−0.40, +0.28							1	+0.63		
Complex of factors not separated	8	−0.03, +0.74	2	+0.43					2	+0.21, 0.56	2	−0.24, +0.68

species in different locations may have major differences in regulating factors, as in the winter moth. In England the major regulating factor is winter mortality, mostly due to asynchrony between food availability and hatch of first instar larvae (Varley et al., 1973). This was also true for populations in Canada until introduced parasitoids became the key factor after 1958 (Embree, 1965, 1971).

We might be misled into concluding that parasitoids are the most important factor regulating herbivore populations since they were apparently influential either on eggs or other stages in 75% of the species studied. However, it is not yet clear if parasitoids cause host population changes or if they respond to them, like the predators of cycling microtine and lagomorph populations in the arctic. Some examples from the literature on biological control indicate that parasites can be important regulators, but relatively few studies have provided data before and after release of parasites. The studies by Embree (1965) come closest to this ideal. Life table studies on the winter moth in Nova Scotia were initiated in 1954 when the first releases of parasites were made. However, parasites did not become abundant until 1959, and it is since this time that the pest has declined in importance from a very serious defoliator to one of the less common defoliators of hardwoods. As the parasite became the key factor, the pest species was regulated (Fig. 18.12). However, we should be cautious about using this example to validate the concept of parasitoid regulation in all other cases. For example, Varley et al. (1973, p. 126) state that parasites of the winter moth in England "respond to the changes in their host's densities rather than cause them."

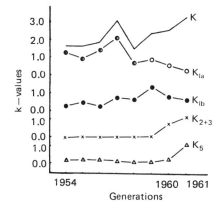

FIGURE 18.12 Generations of winter moth in Nova Scotia studied by Embree (1965) from 1954 to 1961, and the k-values for the following factors: k_{1a}, mortality of eggs and migrating larvae; k_{1b}, larval mortality; k_{2+3}, parasitism by *Cyzenis* and *Agrypon*; k_5 pupal mortality. Note that as K increased in 1959 this was the first year that parasites were abundant (although this is not registered on the figure), and in 1960 and 1961 parasites became important mortality factors, as did pupal mortality. Reprinted with permission from Podoler and Rogers (1975) and Blackwell Scientific Publications Limited.

But another case supports the conclusion that parasitoids can regulate host populations. *Olesicampe benefactor* was released in the early 1960s in Canada to attempt biological control of the larch sawfly, *Pristiphora erichsonii* (Ives, 1976). As parasitism by *O. benefactor* reached over 80%, the sawfly began to decline. Only mortality caused by *O. benefactor* had a density-dependent component, and in simulations, in the presence of *O. benefactor* and its hyperparasitoid, sawfly populations never reached outbreak proportions. In their absence, populations frequently reached epidemic levels (Ives, 1976).

Much of the debate on the validity of the analytical methods described above revolves around the ability of each to identify density-dependent, density-dependent, and delayed density-dependent factors. But the distinction between these factors may become unimportant because interactions between factors may be extremely important, as Andrewartha and Birch (1954) and Smith (1961) have emphasized. Examples relating to *Leptopterna dolabrata* and *Operophtera brumata* were provided in Chapter 17.

Several points are worthy of note in Table 18.2. A large proportion of studies have not separated out all components of mortality, so mortality is evaluated for a complex of factors. The difficulties in separating factors may be enormous, but nevertheless desirable, and experimental approaches may be essential. This separation is important if we are to isolate major factors working in a population. For example, fecundity seems to be frequently evaluated in a complex of factors, most of which cause mortality. But fecundity differences can give important clues to nutritional deficiencies or competition in previous stages. Moreover, fecundity may be easily estimated by dissection of a sample of females in many species, or by allowing caged females to oviposit, as discussed in Chapter 14.

Another set of factors that has been frequently combined is parasitism, predation, and disease. But particularly from a biological control or pest management point of view, it is important in the long run to know which factors are the more influential.

Laboratory studies using the life table approach, which potentially could contribute important insights into population regulation, have touched on very few of the factors acting on natural populations (Table 18.2). There seems to be a lot of scope for an experimental approach using animals which can also be studied under natural conditions.

Although nutrition as a factor in population regulation has not been studied for as long as many of the factors in Table 18.2, it is significant that by 1975 four studies in this table identified its importance: McNeill (1973) on the grass mirid, Metcalfe (1972) on the West Indian cane fly, Sinclair (1970) on wildebeest, and Sinclair (1973) on buffalo. All these species are herbivores, as we should expect, for there is a great difference between plant chemicals and those required by animals. Thus herbivores are likely to be nutrient limited, while predators and parasitoids feeding on animals will have the nutrient content of their food much more closely matched to their needs, so food limitation is more likely to be quantitative than qualitative, a pattern seen in Table 18.2. The role of nutrition in population regulation was discussed in Chapter 17.

The test of our understanding of population dynamics really lies in our ability to construct predictive models based on natural processes in the population, which are accurate enough to be useful. Three such models will be discussed rather briefly. The original literature needs to be studied carefully if a real understanding of the models is to be obtained.

WINTER MOTH MODEL

Varley and Gradwell (1968) and Varley et al. (1973) sampled a population of winter moth on or under five oak trees, *Quercus robur,* in Wytham Wood, Berkshire, England. The oaks stood among 20 large oaks in a pure stand. The flightless females were trapped as they climbed up the oak trunks in November and December to lay eggs in the tree canopy. One-fourth of the females were intercepted on each tree; some were dissected to estimate fecundity. The eggs hatch at about the time the oak buds open in spring, the larvae feed on the leaves, and are fully grown in May. They drop to the ground on silken threads, spin a cocoon, and pupate in the soil, where they remain until emergence as adults later in the year. The falling larvae were trapped in two 0.5-m^2 trays per tree and dissected to estimate the number parasitized by Diptera and Hymenoptera. Survival of parasites to the adult stage was estimated using two 0.5-m^2 emergence traps per tree.

Varley and Gradwell constructed life tables from these samples and analyzed them using their method of key factor analysis discussed earlier in the chapter. Winter mortality, k_1, was the major factor causing population change and was density-independent. Lacking process studies, particularly details on the influence of temperature on flushing oak buds and eclosion of winter moth eggs, this factor could not be modeled, so it was calculated empirically for each year and included in the model. Here

$$k_1 = \log N_E - \log N_{L1}$$

where N_E is the number of eggs and N_{L1} is the number of full-grown larvae.

Varley et al. (1973, p. 122) state, "The only way in which a population might be regulated is by some negative feed-back process; for instance, by a density dependent factor." The major density-dependent factors identified were:

1. Pupal predation, k_5, where

$$k_5 \simeq 0.35 \log N_{L1}$$

This formula was obtained from a regression of the k-value for each generation on log N_{L1}, where the intercept is zero (cf. Fig. 18.10).

2. Parasitism by insects other than *Cyzenis* and *Cratichneumon*, k_3, which showed a weak inverse density-dependent relationship:

$$k_3 \simeq 0.1 - 0.031 \log N_{L2}$$

where N_{L2} is the larvae surviving attack by *Cyzenis*. The formula is calculated as for k_5.

3 **and 4.** Delayed density-dependent factors of parasitism by *Cyzenis*, k_2, and *Cratichneumon*, k_6:

$$k_2 = \log N_{L1} - \log N_{L2}$$

and

$$k_6 = \log(C + N_{L2}) - \log N_A$$

where C is the number of *Cratichneumon* adults and N_A is the number of winter moth adults.

Thus the model for the year 1950 is complete. For estimates for the following years the same formulas can be used, but k_2 and k_6 must be calculated by the parasite submodels (see below), for they cannot be measured directly. From the 1950 data we can calculate the number of adult parasitoids that will emerge in 1951.

The parasite submodels are developed as follows. When the adult density of parasitoids is known, their area of discovery (a) can be calculated from the Nicholson–Bailey model using the formula

$$a = \frac{1}{P} \log_e \frac{N}{S}$$

where P is the number of searching adult parasites, N the number of hosts, S the number of hosts surviving parasitism, and a remains constant throughout the simulation. Then k_2 and k_6 in the next generation can be calculated by substituting in the formula

$$k = \frac{aP}{2.3}$$

where the factor 2.3 is used to convert the value of aP to the k-value expressed on the basis of $\log_{10}$.

TABLE 18.3 A Model of the Interaction Between Winter Moth and Its Parasites.

Cyzenis	Winter moth	Cratichneumon

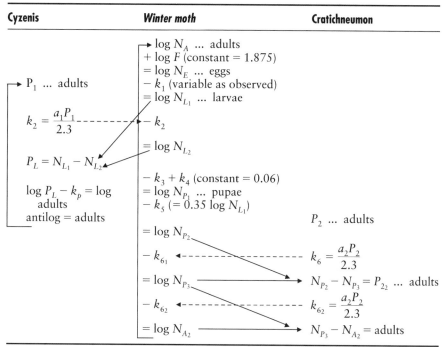

SOURCE: Varley et al. (1973); with permission from Blackwell Scientific Publications Limited.

The full model is given in Table 18.3. Each k-factor is taken in turn and used to calculate the density of the winter moth in the next developmental stage. Thus, given $\log N_A$, the density of adults, these will lay eggs for which a constant $\log F = 1.875$ is used, so that $\log N_E$, the density of eggs, can be calculated. From this figure, k_1 is subtracted, as observed each year to calculate $\log N_{L_1}$, the density of full-grown larvae. *Cyzenis* attack, k_2, is then subtracted, as calculated by the submodel on the left-hand side of Table 18.3. The process is continued until the number of adults in the next generation is predicted, and the cycle can be started again and repeated until the simulation is completed at year 1968.

Considering the assumptions of the model and the approximations used, the fit to actual data is quite good (Fig. 18.13). However, it must be remembered that the majority of mortality covered by the key factor k_1 was measured each year and was not modeled. So the model population could not deviate substantially from the observed data. Process studies on the key factor, and modeling of all stages included in k_1, would be a desirable next step.

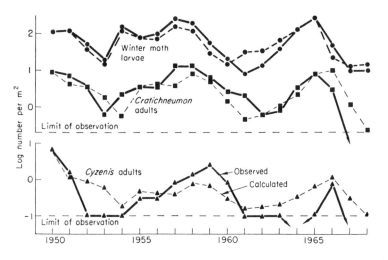

FIGURE 18.13   Observed changes in densities of winter moth and its parasites (solid lines) compared to the changes predicted from the winter moth model (dashed lines) given in Table 18.3. Reprinted with permission from Varley et al. (1973) and Blackwell Scientific Publications Limited.

Also, process studies and more detailed modeling of the parasitoids would be desirable so that the area of discovery calculation could be avoided (see Hassell, 1980). However, the process of modeling, using equations for each developmental stage in succession, is routinely used for other models, as we shall see in the next two examples.

FALL WEBWORM MODEL

The fall webworm is univoltine at the northern end of its range in Canada, and this range does not extend much beyond the northern New Brunswick border. It overwinters as a pupa, emerges as an adult in June and July, oviposits on many species of tree, and the larvae build a web or nest and feed colonially (Fig. 18.14).

Morris (1964, 1969) selected the fall webworm for study for several important reasons:

1. The species usually does not reach the limit of its food supply in Canada; it is regulated by less obvious factors. Morris (1964) noted that outbreak species, which deplete their food supply periodically, are rather exceptional herbivores, so apparent generalizations from numerous studies on such species should be weighed against findings on perhaps a more typical insect herbivore.

FIGURE 18.14 Aspects of the life cycle of the fall webworm, *Hyphantria cunea.* A female laying a clutch of eggs on a leaf (top left), a web constructed by colonial larvae (top right), a caterpillar walking to work in the morning (bottom right), and a spread adult male (bottom left). Based on Novák et al. 1976. Drawing by Tad Theimer.

2. The nests are large and conspicuous when built by fifth instar larvae, and with only one generation a year in New Brunswick and Nova Scotia, an annual census can be conducted rapidly to obtain an index of population size.

3. The rapid census also permits extensive censuses to be taken in diverse regions, so the eventual models have broad generality and an important comparative element between regions. For location of study areas, Morris (1971a) should be consulted.

4. Rapid population estimates enable most of the year to be devoted to process studies.

5. Colonies can be established as needed in the field or laboratory under different densities or physical conditions. Laboratory rearings enable an evaluation of any stress factor on any stage in development, both immediate and delayed.

As we should expect for a temperate organism close to the northern edge of its geographical range, Morris (1964) found physical factors to be particularly important in the population dynamics of the webworm. Temperature in late summer was identified as a key factor, particularly in increasing populations. When Morris (1964) plotted deviations from normal temperature in August and September in eastern and central Canada against time, and also plotted times of peak abundance of the webworm, the relationship became clear (Fig. 18.15). This appears to be another example of climatic release of a population discussed in Chapter 17. A series of warmer than normal developmental periods for the webworm resulted in an increase to peak densities; cooler periods tended to depress populations again.

Morris concentrated a considerable proportion of his efforts on understanding how temperature affects webworm populations. His findings show both direct and very subtle indirect effects, some of which are discussed in general terms here (as in Morris, 1969). The details of the modeling process are given in only one example. This emphasis on a physical factor will counterbalance the major concern for modeling biotic interactions in the winter moth model.

Temperature affects the webworm's development by providing sufficient heat, above the webworm's developmental threshold of 51°F, to permit development to the adult stage. Thus it is essential to keep account of the cumulative heat during the active season, which can be expressed as degree days (°D) above 51°F. For each day the maximum and minimum temperatures are recorded and the change in temperature is assumed to follow a sine curve between these extremes (Arnold, 1960). Then the area below the curve and

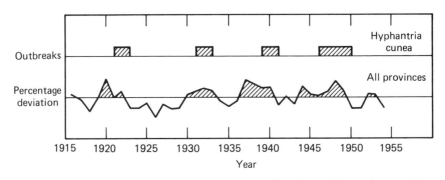

FIGURE 18.15   Relationship between *Hyphantria cunea* population peaks and positive deviations from mean August and September temperatures in eastern and central Canada. From Morris (1964); with permission.

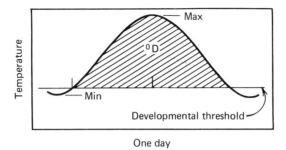

One day

FIGURE 18.16 Sine-curve method of calculating degree days from records of maximum and minimum temperature per day developed by Baskerville and Emin (1969).

above the developmental threshold of the insect under study is calculated, which provides an estimate of the total degree days for that day (Fig. 18.16). Totals per day are summed throughout the season. The method used by Baskerville and Emin (1969) is accurate and rapid, and they provided a table of degree days that is applicable to the fall webworm.

Morris (1969) identified 10 ways in which heat can play a role in the population dynamics of the webworm.

1. *The rate at which heat is accumulated in any year, t, has direct effects on survival in t and fecundity in t + 1.* The webworm's optimum temperature for development is about 80°F. If heat is accumulated rapidly, with temperatures close to this optimum, development is rapid, larval survival is high, pupae are large, and fecundity is high. All these factors are reduced as departures from the optimum increase (see Morris and Fulton, 1970a, for details).

2. *The amount of heat in t has direct effects on survival to the census period in t.* Pupae require a certain amount of heat in the spring to develop to the adult stage, and larvae require a certain amount of heat to reach the pupal stage before winter. Emergence times of adults and occurrence of fifth-instar larvae and pupae can therefore be plotted against the heat accumulated each year. In a warm year such as 1961, enough heat was available for even the latest adults to emerge, lay eggs, and for larvae to grow and pupate. Only a minute proportion did not have enough time to reach the pupal stage (Fig. 18.17). However, in a cold year, such as 1962, when there is a much lower total heat accumulation, progeny of late-emerging adults may not even reach the fifth instar before the end of the developmental season. Thus many nests would go unrecorded because they were too small (see Morris and Bennett, 1967; Morris and Fulton, 1970a, for details).

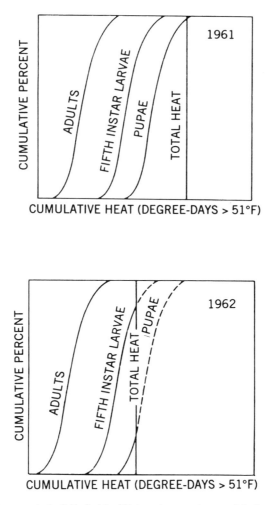

FIGURE 18.17   Occurrence in the field of adults, fifth-instar larvae, and pupae of *Hyphantria cunea* in relation to heat accumulation in the warm year 1961 and the cold year 1962. From Morris (1969).

3. *Heat in t has direct effects on survival after the census period in t and hence on the initial population density of t + 1.*   Because the census is taken on the fifth-instar larvae, and at the peak occurrence of this instar, much of the mortality shown in Fig. 18.17 will occur after the census. This mortality will have a direct influence on the number of adults that emerge in the following year, which will be large in 1962 but small in 1963. Therefore, survival of larvae and pupae is poor in 1962, for reasons given in item 2 above, but the population continues to decline in 1963 because of the small number of adults produced at the beginning of that year (see Morris and Bennett, 1967; Morris and Fulton, 1970a, for details).

These two direct effects of heat can be modeled easily provided that the heat requirements of each stage of the insect are measured accurately. The process is described briefly here (see Morris and Fulton, 1970a, for details). The developmental threshold temperature for the webworm pupae was estimated by rearing pupae at various temperatures to obtain developmental velocities (V) for a range of temperatures (Fig. 18.18). When these velocities are extrapolated to the ordinate, the threshold can be read off as 51°F. Several experiments led to the same conclusion.

Now it is necessary to calculate how much heat a pupa requires before it emerges as an adult—the thermal constant of the pupa, K_p, which, however, is applicable only to a certain population in a specific year since it is affected by natural selection, as we should expect:

$$K_p = D(T - t)$$

where D is the number of days required for development at a certain temperature T and t is the threshold temperature. Therefore, substitution in the formula with $D = 100/V$ yields

$$K_p = \frac{100(T - t)}{V}$$

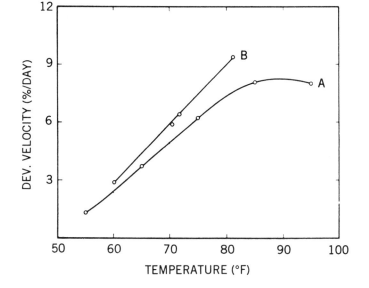

FIGURE 18.18   Pupal developmental velocity for *Hyphantria cunea*: (A) pupae after winter diapause; (B) pupae in the laboratory without diapause. Note that extrapolation of each line toward the origin (in A using 65°–75° region only) results in an estimate of 51°F for the developmental threshold. From Morris and Fulton (1970a); with permission.

The term $(T - t)/V$ is given by the slope of the line in Fig. 18.18, so the equation for K_p can be expressed as

$$K_p = \frac{100}{b}$$

For line A, $b = 0.262$. Therefore, K_p is about 382°D for this particular population in this particular year.

Similar processes can be used for calculating the thermal constants for all other stages, and for these the constants appear to be uniform no matter which population or year is involved. One correction factor is required, however, because in the field radiant heat may be important to development, but this is not measured in a Stephenson screen shelter, in which temperature is recorded in the field. The effect of direct sunlight will increase leaf temperatures, on which eggs are situated above ambient temperature, and nest temperatures, in which larvae feed will also be higher. This effect can be estimated by comparing laboratory and field rearings, and we see a gain of only about 45°D during instars I–V through radiant heat. Therefore, thermal constants estimated in the laboratory can be corrected for predicting field development (Table 18.4), and these can be tested against observed development (Table 18.5).

The agreement between observed and predicted heat requirements is very good considering the many sources of variation that Morris and Fulton (1970a) discuss. The effect of temperatures on development is modeled quite accurately, and the general form of the model is extremely simple:

$$\text{total heat requirement} = K_p + K_A + K_E + K_L + \cdots$$

TABLE 18.4   Laboratory Thermal Constants Corrected for Average Effects of Radiant Heat to Predict Field Development from Stephensen Screen Records Analyzed by the Sine Method.

Stage	Laboratory (°D)	Field (°D)	Field (°D) (Cumulative)
Pupa (K_p)	K_p	K_p − 65 (radiant heat)	K_p − 65
Adult (K_A)	40	40	K_p − 25
Egg (K_E)	260	225	K_p + 200
Larva (70% R.H.) (K_L)			
To instar V	410[a]	390[a]	K_p + 590
To instar VI	575[a]	530[a]	K_p + 730
To prepupa	800	(755)	K_p + 955
Prepupa	75	75	K_p + 1030
Total	K_p + 1175	K_p + 1030	

[a]Note that these figures are not used in the summation for total thermal heat required.
SOURCE: Morris and Fulton (1970a).

TABLE 18.5   Comparison Between Observed Heat Requirements in Field Census Areas and Predicted Heat Requirements [from Column "Field (°D) (Cumulative)" in Table 18.4].

Climate	Stage	Observed (°D)	Predicted (°D)	Difference (%)
Continental ($K_p = 350°D$)	To instar V	900	940	4
	To instar VI	1160	1080	7
	To pupa	1420	1380	3
Maritime ($K_p = 450°D$)	To instar V	980	1040	6
	To instar VI	1250	1180	6
	To pupa	1520	1480	3

SOURCE: Morris and Fulton (1970a).

The next step is to superimpose mortality or survival on this basic model (see Morris and Fulton, 1970a), but this will not be treated in detail here. We will return to a list of ways in which temperature affects the webworm. The following are indirect effects.

4. *Heat in t has indirect effects on survival in t, operating through food quality.* In cold years, development is prolonged and larvae must feed on older foliage. Survival is much reduced and fecundity of females is relatively low (Fig. 18.19). In the Maritime Provinces, larvae usually feed on mid- to late-summer foliage, but as in the winter moth, there must be strong selective pressure for early emergence (see Morris, 1967, for details).

5. *Heat in t has indirect effects on fecundity in t + 1, operating through food quality.* Females reared on early foliage produced about 600 eggs, those on midsummer foliage about 60% of this number, and those on late foliage produced only 20% (Fig. 18.19)—a dramatic reduction in a cool season (see Morris, 1967, for details).

6. *Heat in t has indirect effects on survival in t + 1, operating through a transmitted maternal influence on population quality.* When the progeny from females reared as in item 5 above are all reared under identical conditions on a deficient synthetic diet, a further effect due to maternal food quality is observed (Fig. 18.20). When the parent had fed on early foliage, some larvae survived and a few adults emerged to produce about 50% of the potential number of eggs. When parents had fed on midsummer foliage no viable adults were produced, and when fed on late foliage the second generation did not reach the fifth instar. Similar results were obtained when reared on foliage in the field. Maternal food quality had an important effect on the viability of eggs and larvae. Morris (1969, p. 21) observed that "when two equally cold years occur in succession, a realistic model should provide for lower survival in the second year because the resistance of the population to additional nutritional stress has been lowered" (see Morris, 1967, for details).

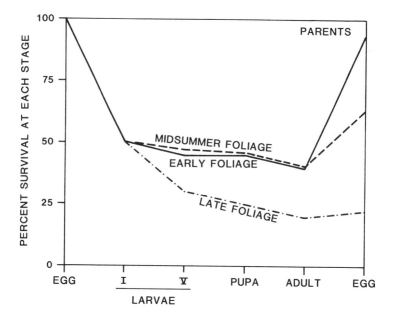

FIGURE 18.19   Effects of foliage age on *Hyphantria cunea* survival and fecundity. From Morris (1969); with permission.

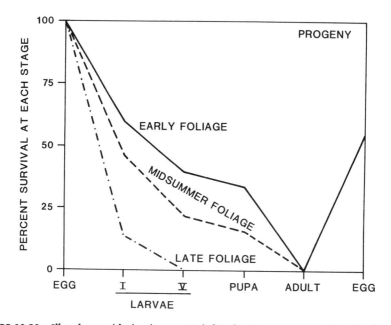

FIGURE 18.20   Effect of maternal food quality on survival of *Hyphantria cunea* progeny and fecundity of surviving adults. From Morris (1969); with permission.

7. *Heat in t has indirect effects on survival in t + 1, operating through the influence of natural selection on heat requirements.* Morris and Fulton (1970b) found that heat requirements for diapause termination in pupae of the webworm were under genetic control. An offspring–parent regression for $\log_e K_p$ gave a slope of 0.60, where perfect correspondence would yield a slope of 1.0. We saw in Fig. 18.17 that there is a high selection pressure in cold years against the progeny of adults that emerge late in the season or those that require the highest K_p. Therefore, after a cold year the mean K_p required by a population is reduced (Fig. 18.21). Morris (1969) pointed out that the rate of change in K_p from generation to generation was very rapid but should not be unexpected as moths live only about 8 days, so the probability is high that moths with similar K_p values mate. Mating is highly assortative.

Selection pressure will work in the opposite direction during long warm seasons. Genotypes that complete development too early in the season lose weight at about 10% per month at 65°F before cold weather sets in, and they are exposed for a longer period to predators and parasitoids.

8. *Heat in t − 1, t − 2, . . . effects population quality in t.* Since population quality changes because of maternal effects and natural selection, it is probably influenced by events in several generations into the past. For example, area A is coastal with a long developmental season, and area B is inland, where the season is shorter with greater deviations from the mean as in 1956 and 1958 (Fig. 18.22). Populations in both areas increased during a series of warm years to a peak of about 100 nests per mile in area A and 10 nests per mile in area B. The cold summer in 1962 was unusual in the coastal region and the population was reduced by 99%. But in area B the population had

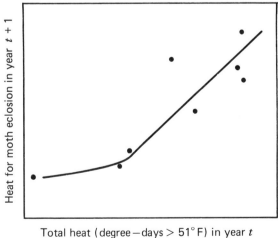

Total heat (degree−days > 51°F) in year *t*

(y-axis: Heat for moth eclosion in year t + 1)

FIGURE 18.21   Relationship between mean heat required for eclosion from pupae in *Hyphantria cunea,* comparing heat required in year *t* and year *t* + 1. From Morris (1969); with permission.

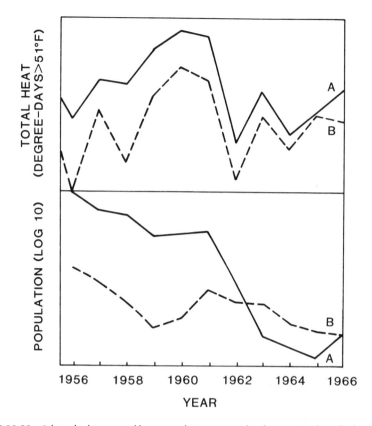

FIGURE 18.22 Relationship between total heat accumulation in a coastal study area (A) and an inland area (B) and the population densities of *Hyphantria cunea* in these areas, for the period 1956–1966. From Morris (1969); with permission.

undergone a similar stress in 1956 and 1958 and only a slight reduction in the population resulted. Heat requirements for eclosion were much higher in A than in B before the cold summer, but similar afterward. The population in area A evolved very rapidly. The longer the population goes without stress, the less resistant it is to stress, so that after a crash, in years of webworm scarcity, as in 1963–1965, more nests can be found in the harsh interior of New Brunswick than in the usually favorable maritime locations.

9. *Heat may affect the influence of parasitoids.* As webworm populations declined in the early 1960s, the parasite *Campoplex validus* increased in both areas A and B simultaneously, although densities of hosts differed dramatically. Morris (1969) considered the possibility that since *Campoplex* has a lower K_p than the webworm, synchronization of the parasite might have improved due to weather changes, and this would be independent of host density (see Morris, 1976b,c, for details).

10. *Heat may affect the amount of encapsulation of Campoplex eggs.* The internal eggs of *Campoplex* were more frequently encapsulated in the years after 1960 than before. Morris (1969) found that large host larvae encapsulated eggs more readily than small ones. Thus if the weather changes such that *Campoplex* is better synchronized with its host, more eggs may be laid but in larger hosts, so encapsulation is more common (see Morris, 1976a, for details).

It is clear that the process studies used by Morris have enabled a very detailed comprehension of many ways in which heat influences the population dynamics of the webworm. They also show how and when heat should be measured for modeling purposes and the following estimates of heat are pertinent:

1. Total heat in year t and also heat in year $t + 1$ up to the census period.
2. Heat in past years as it affects population quality in two ways, as described previously.
3. Seasonal distribution of heat.
4. Accumulation of heat above different thresholds as it affects synchronization of the webworm and food quality, and the webworm and its parasitoids.

Morris (1969, p. 27) concludes with the statement:

The model should represent a higher level of biological meaning than could be achieved through regression analysis based on field data alone. My hope is that it will be good enough for reliable simulation studies, with the object of learning whether or not density dependence represents an essential aspect of the webworm's system of regulation and what would happen if sequences of warm years extended beyond their normal expectancy. As a result of the effects of natural selection on heat requirements, webworm populations that are increasing during a series of warm years become progressively less able to take advantage of these favorable conditions [i.e., they need more and more heat for the same amount of development]. It will be instructive to learn how much the genetic parameters in the model, by themselves, contribute to population stability. Finally, it can be shown that population density is related to land use, vegetation types [see R. F. Morris, 1971b] and other variables. . . . It will, therefore, be worthwhile to employ simulation and minimization techniques to see whether cultural manipulation of the environment can be used feasibly to reduce webworm damage.

Morris was getting close to answering many basic questions on the theory of population dynamics in relation to the fall webworm: the role of density dependence, genetic quality of the population and intrinsic regulation of population size, and the importance of understanding phenology and host-plant quality.

CEREAL LEAF BEETLE MODEL

The cereal leaf beetle, *Oulema melanopus,* was accidentally introduced from Europe into North America and is now a major pest on small grain crops, mainly wheat, oats, and barley in the central United States (Helgesen and Haynes, 1972; Tummala et al., 1975). Adults overwinter in waste places in litter, emerge in April, and feed on grasses and grains. They mate and oviposit from mid-April to June. Larvae drop to the ground and enter the soil to pupate. New adults emerge and feed in late June and then move to overwintering sites.

Ruesink (1975) and Ruesink and Onstad (1994) discuss the steps in construction of a complete descriptive model and refer specifically to the cereal leaf beetle. Model development consists of four basic steps:

Step 1. The main object is to understand and to predict changes in population size with time, so the system is restricted to the life stages of the leaf beetle. All other factors act on these stages of the leaf beetle (e.g., weather, natural enemies, and host-plant condition), but they are not modeled directly—only their total effect on the survival of each stage.

Step 2. The systems graph of the cereal leaf beetle should contain one component per life stage of the insect as in Fig. 18.23, each connected to the next by inputs and outputs. The only exception to this is where third-instar larvae are parasitized by *Tetrastichus julis,* a major parasite, which is also modeled.

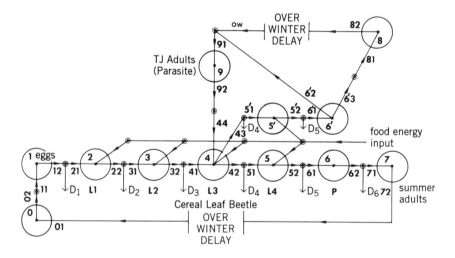

FIGURE 18.23 Systems graph for the cereal leaf beetle developed by Tummala et al. (1975). Components of the life history from eggs, larvae (L1, L2, L3, L4), pupae (P), and adults are arranged horizontally and the parasitoid model for *Tetrastichus julis* forms the upper section of the systems graph. Reprinted with permission from Tummala et al. (1975) and the Entomological Society of America.

Step 3. Here the biological data are gathered and modeled so that they can be incorporated into the general scheme given by the systems graph. For the cereal leaf beetle a dynamic-deterministic-discrete model (Tummala et al., 1975) was employed using difference equations (see Ruesink, 1975, and Ruesink and Onstad, 1994, for other types of model). The discrete form of the model enables the simple handling of time lags common in development of several life stages, and the discrete steps are made up of physiological time units, as in Morris's models, rather than days or months. The developmental threshold used for the beetle was 48°F and each step in the difference equation was equal to 60°D. Given this step size, three such steps were needed to complete egg development (state variables ψ_{11}, ψ_{12}, and ψ_{13}; see Table 18.6), one each for the five larval components (state variables ψ_{21}, ψ_{31}, ψ_{42}, ψ_{51}, and ψ'_{51} for parasitized fourth-instar larvae), seven steps for the pupa (state variables ψ_{61}–ψ_{67}), and two steps for the new adults and old adults (state variables ψ_{71} and ψ_0, respectively). Equations are then developed to interrelate

TABLE 18.6 State Variables Used in the Cereal Leaf Beetle Model.

Name	Description
ψ_0	Spring adult cereal leaf beetle density
$\left.\begin{array}{l}\psi_{11}\\\psi_{12}\\\psi_{13}\end{array}\right\}$	Cereal leaf beetle egg density
ψ_{21}	First-instar cereal leaf beetle density
ψ_{31}	Second-instar cereal leaf beetle density
ψ_{41}	Third-instar cereal leaf beetle density
ψ_{51}	Unparasitized fourth-instar cereal leaf beetle density
$\left.\begin{array}{l}\psi_{61}\\\cdot\\\cdot\\\cdot\\\cdot\\\cdot\\\psi_{67}\end{array}\right\}$	Unparasitized cereal leaf beetle pupa density
ψ_{71}	Summer adult cereal leaf beetle density
ψ_{81}	Density of diapausing *Tetrastichus julis*
ψ_{91}	Adult *Tetrastichus julis* density
ψ'_{51}	Parasitized fourth-instar cereal leaf beetle density
$\left.\begin{array}{l}\psi'_{61}\\\cdot\\\cdot\\\psi'_{67}\end{array}\right\}$	Parasitized cereal leaf beetle pupa density
E'_{ij}	Parasites per pest individual in stage *ij*

SOURCE: Tummala et al. (1975); with permission from the Entomological Society of America.

inputs, outputs, and the state of each component for each state variable. In their simplest form equations are as follows:

$$\psi_{13}(n + 1) = \psi_{12}(n)$$

if no mortality occurs between state variables ψ_{12} and ψ_{13}. Or if mortality occurs,

$$\psi_{21}(n + 1) = (1 - k_1)\psi_{13}(n)$$

where $1 - k$ is the age-specific survival.

Step 4. From these equations for subcomponents the systems model equations can be developed by equating the output of one with the input of the next (Table 18.7). For example, the third 60°D step in the egg stage can be modeled as

$$\psi_{13}(n + 1) = \psi_{12}(n)$$

TABLE 18.7   Systems Model for the Cereal Leaf Beetle.

$$\psi_0(n + 1) = a\psi_0(n)$$
$$\psi_{11}(n + 1) = b\psi_0(n)$$
$$\psi_{12}(n + 1) = \psi_{11}(n)$$
$$\psi_{13}(n + 1) = \psi_{12}(n)$$
$$\psi_{21}(n + 1) = (1 - k_1)\psi_{13}(n)$$
$$\psi_{31}(n + 1) = (1 - k_2)\psi_{21}(n)$$
$$\psi_{41}(n + 1) = (1 - k_3)\psi_{31}(n)$$
$$\psi_{51}(n + 1) = (1 - k_4)\,[1 - f_2(\psi_{41}(n),\psi_{91}(n))]\psi_{41}(n)$$
$$\psi_{61}(n + 1) = (1 - k_5)\psi_{51}(n)$$
$$\psi_{62}(n + 1) = \psi_{61}(n)$$
$$\psi_{63}(n + 1) = \psi_{62}(n)$$
$$\psi_{64}(n + 1) = \psi_{63}(n)$$
$$\psi_{65}(n + 1) = \psi_{64}(n)$$
$$\psi_{66}(n + 1) = \psi_{65}(n)$$
$$\psi_{67}(n + 1) = \psi_{66}(n)$$
$$\psi_{71}(n + 1) = c^*\psi_{71}(n) + (1 - k_6)\psi_{67}(n)$$
$$\psi_{81}(n + 1) = \psi_{81}(n) + \psi'_{64}(n)E'_{64}(n)[1 - D(n)]$$
$$\psi_{91}(n + 1) = d\psi_{91}(n) + \psi'_{64}(n)E'_{64}(n)D(n)$$
$$\psi'_{51}(n + 1) = (1 - k_4)f_2(\psi_{41}(n),\psi_{91}(n))\psi_{41}(n)$$

$$E'_{51}(n + 1) = \frac{f_1(\psi_{41}(n),\psi_{91}(n))}{f_2(\psi_{41}(n),\psi_{91}(n))\psi_{41}(n)}$$

$$\psi'_{61}(n + 1) = (1 - k_5)\psi'_{51}(n)$$
$$\psi'_{62}(n + 1) = \psi'_{61}(n)$$
$$\psi'_{63}(n + 1) = \psi'_{62}(n)$$
$$\psi'_{64}(n + 1) = \psi'_{63}(n)$$
$$E'_{61}(n + 1) = (1 - r_p)E'_{51}(n)$$
$$E'_{62}(n + 1) = E'_{61}(n)$$
$$E'_{63}(n + 1) = E'_{62}(n)$$
$$E'_{64}(n + 1) = E'_{63}(n)$$

SOURCE: Tummala et al. (1975); with permission from the Entomological Society of America.

when apparently there is no measurable mortality. In the next 60°D step, the density of first-instar larvae can be calculated as

$$\psi_{21}(n + 1) = (1 - k_1)\psi_{13}(n)$$

when k_1 is the mortality of leaf beetle eggs. Once k_1 has been estimated by field observations, as explained later, $\psi_{21}(n + 1)$ can be calculated. In exactly the same way, the densities of second, third, and pupal instars can be calculated using the equations

$$\psi_{31}(n + 1) = (1 - k_2)\psi_{21}(n)$$

$$\psi_{41}(n + 1) = (1 - k_3)\psi_{31}(n)$$

$$\psi_{61}(n + 1) = (1 - k_5)\psi_{51}(n)$$

Only ψ_{51}, which estimates the density of unparasitized fourth-instar larvae, is more complex and is given by the equation

$$\psi_{51}(n + 1) = (1 - k_4)[1 - f_2(\psi_{41}(n), \psi_{91}(n))]\psi_{41}(n)$$

where k_4 is the mortality of third-instar leaf beetle larvae and the expression $f_2(\psi_{41}(n), \psi_{91}(n))$ gives the proportion of the ψ_{41} population that is parasitized. Details of the calculation are provided by Tummala et al. (1975).

Thus the systems model is gradually built up and all that is needed now are values for the mortalities k_1 to k_5 for the part of the model already discussed (Table 18.8). These values can be estimated from life table data and regression analysis as Helgesen and Haynes (1972) did for the cereal leaf beetle. They found that egg mortality ranged from 0 to 31% when field-collected eggs were incubated in the laboratory, and the variability was apparently not related to differences in host plant, locality, or density. Since handling and laboratory conditions probably increased mortality, the best estimate for k_1 was considered to be 10% mortality, or 0.10, which was used as a constant in the systems model (see Table 18.8).

The other values for k were estimated from field samples by regressions of percent mortality of each larval stage on the density of the population estimated as eggs per square foot (Fig. 18.24). Two hosts—wheat and oats—produced different results in some cases, so results had to be treated separately. For example, in first-instar larvae, mortality was constant at 35% in wheat but it increased with density in oats, so k_2 (wheat) = 0.35 and k_2 (oats) is variable (Table 18.8) but can be estimated from the regression once the input density of first-instar larvae is known. Mortality varies very little in the second instar on both hosts and k_3 can be set as a constant for both crops ($k_3 = 0.30$). Mortality in instar III does not vary much and k_4 again can be given a constant value of 0.45. In the fourth instar, mortality is highly

TABLE 18.8   Standard Values for the Parameters of the Cereal Leaf Beetle (CLB) Model (TJ = *Tetrastichus julis*).

Name	Description	Value
a	Spring adult survival	0.70
b	CLB eggs/ CLB female/60DD	22.00
c	Summer adult survival	1.00
d	TJ adult survival	0.60
e_1	Max eggs/TJ adult/60DD	20.00
e_2	Max TJ eggs/CLB larva/60DD	5.00
e_3	TJ searching constant	100.00
r_p	TJ mortality inside CLB	0.00
t_{sy}	Time when TJ first shows	6.00
t_e	Time when CLB leaves oats	16.00
t_d	Time when TJ diapause starts	13.00
k_1	Mortality of CLB eggs	0.10
k_2	Mortality of CLB L_1	Variable
k_3	Mortality of CLB L_2	0.30
k_4	Mortality of CLB L_3	0.45
k_5	Mortality of CLB L_4	Variable
k_6	Mortality of CLB pupae	0.40
k_7	Mortality of overwintering CLB	0.50
k_8	Mortality of overwintering TJ	0.50
c_p	Exponent in parasitism equation	0.75

SOURCE: Tummala et al. (1975); with permission from the Entomological Society of America.

dependent on density and k_5 is therefore variable and must be estimated from the regression line once the input density of unparasitized fourth-instar larvae is known. In this way the parameters of the model, as in Table 18.8, are estimated. Thus with an input density of ψ_0, which is determined from field sampling, the size of the next generation can be calculated. The time of each stage in the field can be predicted from the data on accumulated °D above the developmental threshold (°D > 48°F).

Tummala et al. (1975) tested the validity of this model against what was known from the field as follows:

1. Each female should lay about 60 eggs per season.
2. If no parasites are present, the leaf beetle population should increase about six times per year at very low densities and plateau at about 1100 eggs/ft² to the total seasonal input.
3. If the leaf beetle density is high, the *Tetrastichus julis* population should increase about 30 times per year at low *T. julis* population levels.

Using the standard values for model parameters given in Table 18.8, the predictions were for 62 eggs per female per season, an increase in population of 7.5 times, with a plateau of 1250 eggs/ft², and for *T. julis* a population in-

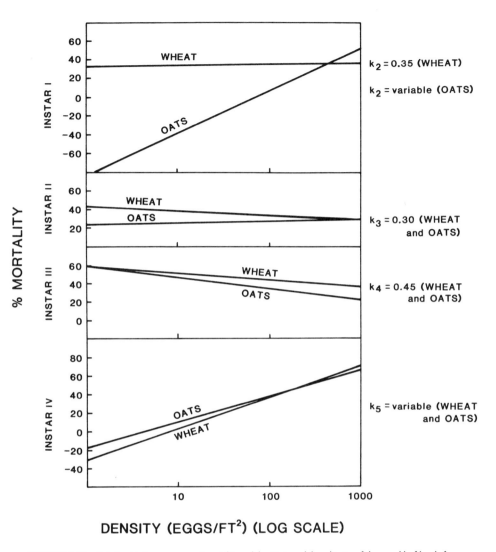

FIGURE 18.24 Relationship between percent mortality and density in each larval instar of the cereal leaf beetle from which standard values for the parameters in the model are derived (cf. Table 18.8). Reprinted with permission from Helgesen and Haynes (1972) and from the *Canadian Entomologist.*

crease of 25.9 times. All results were realistic and supported the validity of the model.

As Ruesink (1975) and Ruesink and Onstad (1994) emphasized, a more generally applicable model, which will fit many more specific locations, can be obtained by submodeling the parameters that are given as constants in Table 18.8, as Morris did. This more versatile model requires much more study and understanding, but it may be used in a great variety of geographic and climatic conditions and in various other environmental circumstances.

However, even with the constant parameters, the model can be very valuable. With the computerized system, which receives information about insect population densities and age structure, the host crop, and the abiotic environment, insect populations can be forecasted and recommendations regarding control can be made (Haynes et al., 1973). Haynes et al. (1973) should be consulted for illustrations of the flow of information in a research program to develop a pest management system, the basic components of that system, and the way it is applied to a real operational pest management system.

In addition to its use in a forecasting system, the model can predict the results of many different tactics that might be tried to manage the pest. It can be estimated how much mortality at a certain stage would be needed when toxic chemicals were applied to reduce the impact of the pest on the crop by 50 or 75 or 90%, for example. Or a simulation can be run to see if it is worth manipulating conditions to improve parasitoid survival so that parasitism is increased by 10 or 20%. Thus endless simulation "experiments" can be run on a computer using the model, and the best set of control devices can be integrated into the pest management system.

It seems now that the legitimate goal of all studies on population dynamics, be they for basic or applied reasons, should be the development of models based on a sound understanding of the systems involved, which therefore have predictive value. Recent developments in this field are considered in Anderson et al. (1979), Conway (1981), Metcalf and Luckmann (1982, 1994), Lauenroth et al. (1983), and Goodenough and McKinion (1992).

Chapter 19

Population Dynamics: Synthesis

Can we find some broad themes by which to obtain a unifying view of insect population dynamics? Is there a way to penetrate the vast literature with its diverse hypotheses and examples? Will it ever be possible to develop theory that is generally accepted? Some would argue that we already have enough theory on population dynamics and that we understand the principles involved. Therefore, this chapter presents some tentative steps, following a simple path, which may actually be much more complicated, or perhaps even imaginary.

First, we may gain some insights from considering successes in understanding nature and the basis of that success. The conclusion reached will be that an evolutionary perspective provides various themes that unify the very diverse ecological studies in the literature and provides a broad comparative basis from which to approach population dynamics.

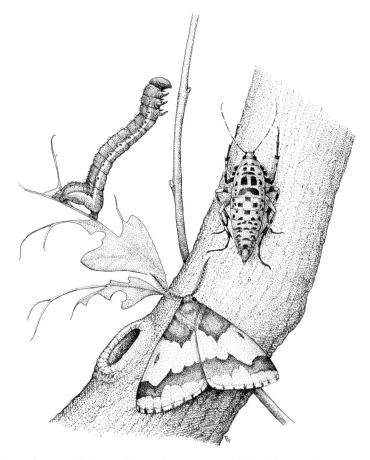

Mottled umber moth, *Erranis defoliaria,* a widely spread species in Europe, which includes leaves of many deciduous trees and shrubs in its diet. Based on Novák et al. (1976). Drawing by Tad Theimer.

DARWIN'S EXAMPLE

At the end of *The origin of species* Darwin (1859, pp. 489–490) wrote:

> It is interesting to contemplate an entangled bank, clothed with many plants of many kinds, with birds singing on the bushes, with various insects flitting about, and with worms crawling through the damp earth, and to reflect that these elaborately constructed forms, so different from each other, and dependent on each other in so complex a manner, have all been produced by laws acting around us. These laws, taken in the largest sense, being Growth with Reproduction; Inheritance which is almost implied by reproduction; Variability from the indirect and direct action of the external conditions of life, and from use and disuse; a Ratio of Increase so high as to lead to a Struggle for Life, and as a consequence to Natural Selection; entailing Divergence of Character and the Extinction of less-improved forms. Thus, from the war of nature, from famine and death, the most exalted object which we are capable of conceiving, namely, the production of higher animals, directly follows. There is grandeur in this view of life, with its several powers, having been originally breathed into a few forms or into one; and that, whilst this planet has gone cycling on according to the fixed law of gravity, from so simple a beginning endless forms most beautiful and most wonderful have been, and are being evolved.

Darwin, no doubt, wrote this passage after taking a walk from Down House along the lanes with raised banks along them, covered with rich vegetation and inhabited by many animals. As he contemplated that entangled bank he must have realized that five simple laws could explain the richness and diversity of life on earth, but also the unity in design. Through the bewildering array of plants and animals, and the extensive literature on how they came to be, Darwin had cut a simple path, revealing a simple theory of evolution. Only five laws accounted for the unity and diversity of life.

EVOLUTIONARY BASIS FOR POPULATION DYNAMICS

Thus an evolutionary and phylogenetic approach to ecology may well provide the kinds of pattern, linkage, and comparative power needed when complex issues are debated. Examples of comparisons among related species were discussed for life history traits in Chapter 14. As a result of a strong evolutionary perspective, the literature on the adaptive nature of life histories has developed rapidly even though general theory has yet to emerge (cf. Stearns, 1992; Roff, 1992). In Chapter 9, cladistic analysis provided a more rigorous test of hypotheses on radiation of parasites than the ad hoc approach, pulling examples from here and there. The general successes in evolutionary ecology and behavioral ecology provide cases in which there are clear predictions and clear tests, as we saw in Chapter 15, and in Krebs and Davies (1991). The theory of sex allocation or sex ratio theory is well developed: when and why sex ratios change as resources change, the competitive environment shifts, and from one species compared to another (Charnov, 1982). Charnov emphasized the need for *selection thinking,* which of course originated with Darwin. "We ask, when, or under what environmental, social, or life-history conditions natural selection favors one or the other form of sexuality. We test our theory [= hypothesis] by arranging experiments, geographic or taxonomic comparisons to see if selection acts as we think it does or if the sexuality is matched to the appropriate environmental condition. Thus we seek to understand sexuality in terms of the ultimate causes (the *why* questions) rather than the proximate mechanisms (the physiological *how* questions). Seeking answers to the way in which nature is structured, in terms of *why* questions, I shall term *selection thinking*" (Charnov, 1982, p. 3).

Can we apply selection thinking to insect population dynamics?

FLIGHTLESS LEPIDOPTERA

A feel for this approach may be obtained by considering the evolution of flightlessness in insects, discussed broadly by Roff (1990, 1994) and treated more narrowly by Barbosa et al. (1989) and Hunter (1995a,b). Barbosa et al. considered the life history traits of forest-inhabiting flightless Lepidoptera. They noted considerable convergence in traits in three families: the Geometridae, Lymantriidae, and Psychidae. Selection thinking would force us to question why such convergence should exist. The traits included with flightless females were ballooning on a silken thread by dispersing larvae, polyphagy, univoltinism, and overwintering eggs or larvae. They then argued that ecological factors that probably allowed or selected for such convergence were habitat stability, resource persistence, convergence in tree chemical defense, and phenological variability in bud break. As a consequence of the evolved traits and ecology, these kinds of species can become eruptive and defoliate forests, as we have discussed for winter moth, gypsy moth, fall cankerworm, and other species. Hunter (1995) pursued this argument with a phylogenetic analysis. She discovered seven independent origins of reduced wings in temperate forest Macrolepidoptera, and these tended to occur only in lineages with overwintering eggs and spring-feeding larvae. In the Lymantriidae wing reduction originated once, in a genus *Orgyia,* the tussock moths. The gypsy moth was not included because, although females do not fly, they have well-formed wings. In the Geometridae wing reduction originated six times, and five of these involved winter-active adults with the majority of species in the Operophterini (with five macropterous and four micropterous species) and the Bistonini (with four macropterous and 16 micropterous species). Winter-active adults includes those species whose entire adult life occurs in the cold season, when leaves are not available on deciduous trees (October to April in north temperate latitudes). The Bistonini includes the genera *Biston, Lycia, Phigalia, Paleacrita,* and *Erannis* (McGuffin, 1997) (Table 19.1). Naturally, if females are wingless and active in winter, they will probably lose the ability to feed, and lack of adult feeding in Lepidoptera has been shown to be correlated with eruptive population dynamics (Miller, 1996; Tammaru and Haukioja, 1997). Thus considerable progress has been made in understanding the origins and consequences of wing reduction in Lepidoptera while providing one interesting avenue into the core of the population dynamics literature.

But selection thinking should push us into asking about the ultimate causes that may explain why species evolve with larval feeding early in the spring. One answer was developed by Feeny (1970) as discussed in Chapter 5 in relation to the oak tree, *Quercus robur,* and the winter moth, *Operophtera brumata.* An additional consideration was emphasized by Niemalä and Haukioja (1982). Some forest trees flush all leaves rapidly in the spring (Fig. 19.1). *Quercus robur* and *Prunus padus* cease annual shoot growth in June in Finland, when shoot growth is initiated only in May. Other species grow throughout the summer up to September, such that new leaves are available throughout most of the season for insect herbivores: for example, *Populus tremula, Alnus,* and *Betula* species. Surely, much of the selective advantage to early feeding, and oviposition in the autumn or winter, must be related to tree species phenology, especially when most leaves develop in a few weeks in spring. Are there more outbreak species on rapidly flushing host species such as oak because synchrony with this flush is so critical? Hunter (1992a) has even shown that for *Quercus robur* trees within a wood, those that flush leaves rapidly have higher densities of larvae and suffer greater leaf damage than those that flush slowly. Are conifers such as pines, firs, and spruces susceptible to outbreak species for similar reasons? Do we see less extreme dynamics of leaf-feeding species with declining latitude because seasons are prolonged? How far can we progress with this bottom-up, resource-supply approach to insect population dynamics?

Probably there is, in the development of these ideas on flightless Lepidoptera, the foundation for an empirically and factually based theory, narrowly focused on the syndrome of evolved traits coupled with the ecological consequences in population dynamics. According to Hunter (1995a), such a theory may be relevant to about 32 species of macrolepidopterans. Not many species, but a synthesis nonetheless, and an important contribution to a pluralistic theory on insect population dynamics. The working hypothesis could take the form of a flow diagram from environmental conditions to evolved traits in lepidoptera (Fig. 19.2). This is an "ecological theater and evolutionary play" approach. Clearly, the many origins of reduced wings indicates lack of major phylogenetic groups involved with the syndrome.

TABLE 19.1 Examples of Macropterous and Micropterous Moth Species in the Geometridae.

Subfamily and Tribe; Genus and Species	Common Name	Winged or Micropterous	Total Fecundity	Winter-Active Adults	Position of Eggs	Over-Wintering Stage	Number of Plant Genera Utilized	Population Dynamics
Ennominae Bistonini								
Biston betularia	Peppered moth	Winged	Up to 670	No	Bark crevices	Pupa	18	Severe localized outbreaks
Lycia ursaria		Winged	150–200	No	Under bark	Pupa-egg	13	
Lycia rachelae		Microp.	150–200	Yes	Under bark	Pupa-egg	11	
Phigalia titea		Microp.	80–140	Yes	Trunks, branches, twigs	Pupa-egg	8	Defoliation of forest
Phigalia strigataria		Microp.	80–140	Yes	Trunks, branches, twigs	Pupa-egg	1	
Phigalia plumigeraria		Microp.	80–140	Yes	Trunks, branches, twigs	Pupa-egg	8	Local defoliation
Paleacrita vernata	Spring cankerworm	Wingless	250	Yes	Bark crevices	Larva, pupa-egg[a]	4+	Frequent pest outbreaks
Paleacrita merriccata		Wingless	250	Yes	Bark crevices	Pupa-egg	?	
Erannis tiliaria	Linden looper	Microp.		Yes	Bark, twigs	Egg	15	Severe defoliation
Laurentiinae Operophterini								
Operophtera brumata	Winter moth	Microp.		Yes	Bark	Egg	8+	Severe defoliation
Operophtera bruceata	Bruce spanworm	Microp.		Yes	Bark	Egg	9+	Serious outbreaks
Oenochrominae								
Alsophila pometaria	Fall cankerworm	Wingless	100	Yes	Twigs, branches	Egg	10+	Extensive outbreaks

[a] Larvae overwinter in cells in soil, pupation is in late winter, and eggs are laid before leaves emerge.
SOURCE: Data from McGuffin (1977), W. L. Baker (1972), Arnett (1993).

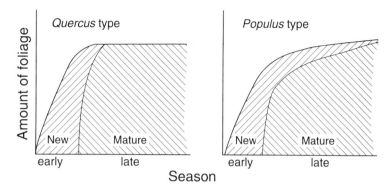

FIGURE 19.1 Two types of foliage production recognized by Niemalä and Haukioja (1982) showing the *Quercus* type, in which all leaves flush early in the season, and the *Populus* type, in which new leaves are produced for much of the growing season. From Niemalä and Haukioja (1982); with permission from the Royal Entomological Society, London.

GALL-FORMING SAWFLIES

Another tactic in the development of empirically and factually based theory is to take a single taxon such as a genus, and to explore as fully as possible in a phylogenetically united group the linkage between evolved traits and ecological consequences. This we have attempted for the gall-forming genus *Euura* (Price et al., 1995b) (Fig. 19.3).

First, Craig et al. (1989) established the tight ovipositional preference and larval performance linkage discussed in Chapter 15. The ultimate mechanism causing females to oviposit on long shoots was the much higher survival of their progeny in this class of shoots.

Second, we wished to know what the proximate mechanism might be, by which a female was able to select long shoots. A phenolic glucoside, tremulacin, was found to be the oviposition stimulant (Roininen et al., 1997), and highest concentrations were known to occur in the longest shoots (Price et al., 1989).

Third, we monitored populations for over a decade and found them to be highly predictable and stable relative to outbreak species (Price et al., 1995b). This stability was driven by resource supply, with strong population limitation imposed by the availability of young, vigorous long shoots in a willow population. Thus we found a mechanism that imposed a low carrying capacity, K, on the population, a bottom-up influence that determined the latent condition of the population.

Fourth, we examined attack and success of natural enemies, the top-down influences on herbivore populations, especially parasitoids. We found the parasitoids to be very passive and opportunistic in the system, attacking where they could but generally with very limited access because the gall provided protection by its size and toughness (Price and Clancy, 1986a; Craig et al., 1990).

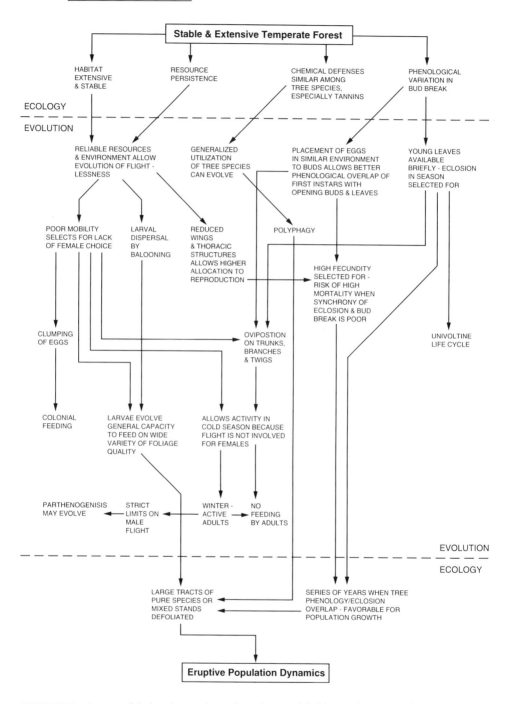

FIGURE 19.2   Summary of the hypothesis on the population dynamics of flightless Lepidoptera in north temperate forests.

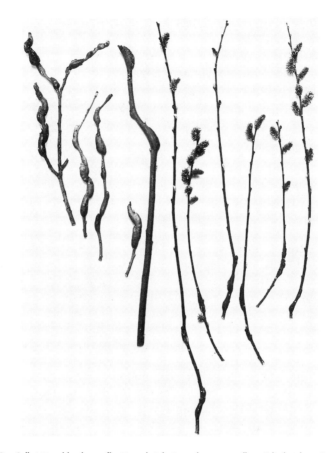

FIGURE 19.3 Galls initiated by the sawfly, *Euura lasiolepis*, on the arroyo willow, *Salix lasiolepis*. Five stems on the left are from one willow clone and six stems on the right are from an adjacent genotype. The contrast in gall size and suppression of shoot growth and inflorescence production is evident. From Price (1992a). From *Biology of insect-induced galls*, edited by J. D. Shorthouse and O. Rohfritsch. Copyright © 1992 by J. D. Shorthouse and O. Rohfristch. Used by permission of Oxford University Press, Inc.

The next step was to examine the extent to which these clear patterns were relevant to other species of *Euura*. We studied several species in Finland: the bud galler, *Euura mucronata* on *Salix cinerea*; and the stem gallers, *Euura amerinae* on *Salix pentandra* and *Euura atra* on *Salix alba*. In the United States we examined the bud galler close to *Euura mucronata* on *Salix scouleriana*, a new species of midrib galler on *Salix exigua*, a stem galler *Euura exiguae* on *Salix exigua*, and another, *Euura s-nodus*, on *Salix interior*. Finally, on Hokkaido, Japan, we studied a bud galler similar to *Euura mucronata* on *Salix sachalensis*. In all cases, the results were closely similar to those on

Euura lasiolepsis: females attacked long shoots preferentially, and larvae survived best in long shoots, whether the galls were formed on the stems themselves, buds, or midribs.

We then extended the comparative approach to closely related genera of gall-forming sawflies: the leaf-edge gallers in the genus *Phyllocolpa* and the leaf-lamina gallers, *Pontania*. All the species of *Phyllocolpa* we have studied showed patterns consistent with those in *Euura*. In *Pontania* some species show the same strong pattern, and others show a weaker pattern or no such pattern.

Certainly, for the genus *Euura* and probably for *Phyllocolpa,* we have a strong basis for developing theory on their population dynamics (Price et al., 1995b). The working hypothesis follows the argument developed earlier on the flow of influences from phylogenetic constraints to adaptive syndromes, as discussed in Chapter 14, and the degree of preference-performance linkage treated in Chapter 15. Those species with high ovipositional preference and larval performance linkage are likely to have a phylogenetic constraint: the female oviposits where the larva or nymph spends most of its life feeding, as in the case of gall-forming insects in general. Therefore, the adaptive response includes the evolution of female behaviors that maximize larval performance by preferences for high-quality resources such as vigorous plant parts (the plant vigor hypothesis, discussed in Chapter 6). Here larvae perform better than on plant modules of low vigor. These adaptations and others constitute the adaptive syndrome of the species (Fig. 19.4) (cf. Price et al., 1990). Because of these evolved characters encompassed in the phylogenetic constraints and adaptive syndromes, there are ecological consequences which may be called **emergent properties** because the ecology depends on the evolutionary history of the species. These emergent properties include the population dynamics of the species. If females, such as those of *Euura lasiolepsis,* are highly selective for vigorous plant modules in which to oviposit, it is likely that such resources are in short supply, and populations are constrained by the bottom-up effect of adequate food availability. Thus species of this kind are likely to be uncommon or rare, as defined by resource supply (Fig. 19.4). This scenario contrasts with those species in which females make no ovipositional choices relevant to the quality of food for progeny, such as flightless Lepidoptera, already discussed in this chapter. Females cannot evaluate food quality for larvae because they oviposit before larval food is available to assess (Fig. 19.4).

We have called this flow of influences from phylogenetic constraints to adaptive syndromes to emergent properties the *phylogenetic constraints hypothesis* in general (Price, 1994a) and applied it specifically to *Euura* species with high preference-performance linkage. There are over 100 species in the genera of *Euura* and *Phyllocolpa* worldwide to which this hypothesis probably applies.

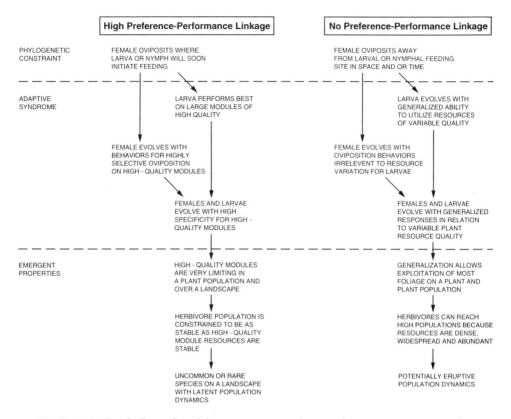

FIGURE 19.4 Flow of influences from phylogenetic constraints to adaptive syndromes, to emergent properties for species with a high ovipositional preference and larval performance linkage and for species with no preference-performance linkage. From Price (1994a); with permission from the Society of Population Ecology, Tokyo.

BARK BEETLES

A completely different set of life history traits, host-plant characteristics, and environmental conditions prevail when bark beetles in the family Scolytidae are considered. Most bark beetle species bore in the cambial layer of angiosperm and coniferous trees (cf. Fig. 15.14) and attack as trees begin to die and lose their natural defenses. Most species are nonaggressive, in the sense that healthy trees are not attacked and killed. In a small minority of cases, however, some species become epidemic and attack and kill many coniferous trees that would otherwise survive. These are the aggressive species. High mortality in large and valuable trees has created a strong incentive and the financial support for extensive studies, resulting in well-developed hypotheses and a sound foundation for theory on the eruptive population dynamics of bark beetles.

Female or male beetles bore through the bark of a tree to the cambial layer, forming a nuptial chamber in which mating occurs and from which maternal galleries are constructed along which eggs are laid (Fig. 15.14). In the aggressive species, adults inoculate the cambium with a pathogenic fungus that kills living host tissue (Whitney, 1982). The fungi are parasitic on the tree and mutualistic with the bark beetle, frequently involving blue-stain fungi in the genus *Ceratocystis* and beetles in the genus *Dendroctonus*, although many other associations are recorded (Whitney, 1982). The fungal associate is transported from tree to tree by successive successful generations of beetles while the fungus kills host tissue, suppresses host defenses, and allows the beetle to breed successfully. When large numbers of beetles attack a tree, it is killed, many progeny in the cambium of the dying tree are produced, and such trees are likely to act as focal points from which an epidemic results (Fig. 19.5).

The cause of the interaction between *Dendroctonus* bark beetles and the host tree was summarized by Raffa and Berryman (1980) (Fig. 19.6). Resin

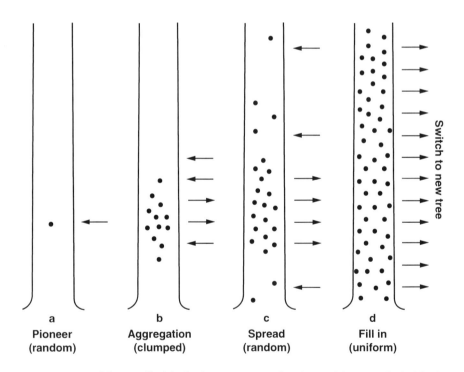

FIGURE 19.5   Spatial dynamics of bark beetle colonization starting with random attack by one or a few bark beetles (a), aggregation if initial attack is not suppressed by the tree (b), and general utilization of the bark (c and d). Arrows indicate attraction to a tree based on pheromones at low densities and repulsion as density of attack increases. From A. A. Berryman in J. B. Mitton and K. B. Sturgeon (eds.)., *Bark beetles in North American conifers: A system for the study of evolutionary biology.* Copyright © 1982. By permission of the University of Texas Press.

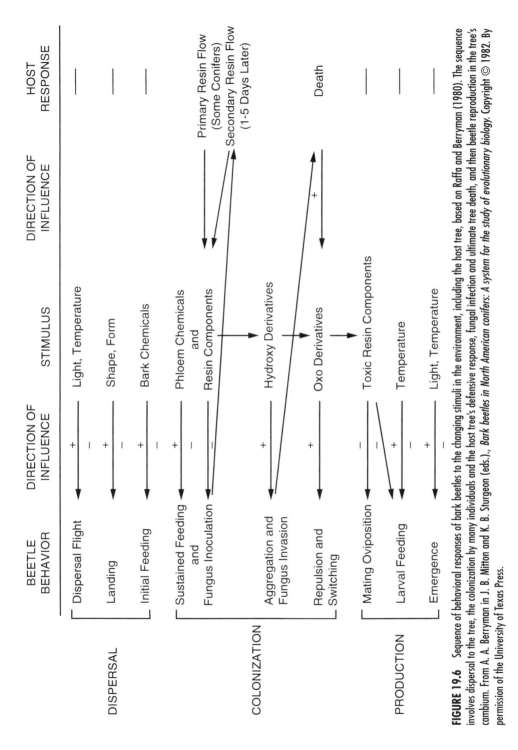

FIGURE 19.6 Sequence of behavioral responses of bark beetles to the changing stimuli in the environment, including the host tree, based on Raffa and Berryman (1980). The sequence involves dispersal to the tree, the colonization by many individuals and the host tree's defensive response, fungal infection and ultimate tree death, and then beetle reproduction in the tree's cambium. From A. A. Berryman in J. B. Mitton and K. B. Sturgeon (eds.)., *Bark beetles in North American conifers: A system for the study of evolutionary biology.* Copyright © 1982. By permission of the University of Texas Press.

flow is the major defense of conifers, which "pitches out" the beetles, but the inoculated fungus kills the tree and suppresses the flow.

There are two important points in the development of theory on these kinds of species. "Three death following bark beetle attack is unique" (Whitney, 1982, p. 195). "Understanding bark beetle population dynamics rests on a sound knowledge of the interaction between beetle populations and the defense systems of their hosts. Other factors, such as parasitism [parasitoids and insect pathogens], predation, and competing species may modify this behavior . . . but will not change the general qualitative properties of the system" (Berryman, 1982, p. 313). Therefore, bark beetle population dynamics requires a special approach relating to the special features of the system. Theory on the population dynamics of insects requires a pluralistic theory because so many very different life histories and environmental conditions exist in nature.

Berryman's (1982) model captures the essential interplay between variable host tree resistance to attack and beetle population size (Fig. 19.7). Pine trees

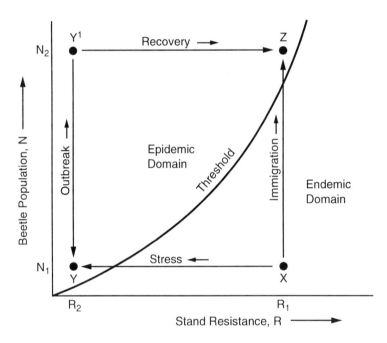

FIGURE 19.7   Berryman's (1982) threshold model of bark beetle population dynamics. Stands of trees are usually highly resistant to attack (R_1), so bark beetle populations are low (X). Under various forms of stress, resistance may decline to R_2, crossing a threshold at which beetles can breed successfully in most trees, and an outbreak results (Y). Other details are in the text. From A. A. Berryman in J. B. Mitton and K. B. Sturgeon (eds.)., *Bark beetles in North American conifers: A system for the study of evolutionary biology.* Copyright © 1982. By permission of the University of Texas Press.

are usually highly resistant to bark beetle attack when growing vigorously, say at R_1, so that beetle populations are very low, existing on the rare trees weakened by a lightning strike, a pathogenic fungus, or other isolated events. The host plant–beetle interaction may well maintain populations at an endemic state over many years (X in Fig. 19.7). Then resistance in a tree population may decline to R_2 under the influence of general stresses such as droughts and water stress, insect defoliation causing nutrient stress, or extensive fires. In weakened trees reproduction is very successful, the carrying capacity of the forest is increased dramatically, and bark beetle mortality during flight declines because of proximity of suitable trees, resulting in an outbreak to N_2. At these populations even healthy trees may be killed by mass attack, such that the outbreak may become epidemic and spread over a much larger area than would be possible without the pathogenic fungal mutualist (cf. Fig. 19.5).

As stress factors in a forest ameliorate, resistance of living trees increases toward R_1, creating a situation such as depicted at Z, where an epidemic exists in resistant trees because of mass attack. However, a small decline in population density may render mass attacks less effective, the threshold for maintenance of an epidemic would be passed, and the beetle population would decrease to point X and N_1. The threshold for a population moving from an endemic to an epidemic state obviously is low when tree resistance is low, and increases as resistance increases, as indicated in Fig. 19.7.

Another set of interactions may result in an epidemic state while tree resistance remains generally high, with populations moving directly from X to Z in Fig. 19.7. Population increase may occur locally because of local damage to trees, resulting in an epicenter of high densities, which then disperse into healthy forest areas. With mass immigration, even resistant trees at R_1 can be attacked successfully by aggressive species.

The thresholds between endemic and epidemic populations vary according to how aggressive a bark beetle species is (Fig. 19.8). Highly aggressive beetle species such as the mountain pine beetle, *Dendroctonus ponderosae*, which attacks lodgepole pine, *Pinus contorta*, and other pines in the western United States, can reach epidemic levels when tree resistance is high, moving from X to Z in Fig. 19.7. Several species of the blue-stain fungus genus *Ceratocystis*, the most aggressive pathogens carried by bark beetles, are associated with this aggressive bark beetle (Whitney, 1982). Less aggressive species such as the fir engraver, *Scolytus ventralis,* which attacks grand fir, *Abies grandis,* require considerable lowering of resistance in host trees before an epidemic is reached (Fig. 19.8). This beetle species is not associated with a *Ceratocystis* symbiont, although other fungal species cooccur with the beetle. Then the nonaggressive bark beetles attack only trees or tree parts with highly reduced resistance, and generally attack as secondary factors after trees are severely weakened or killed by pathogenic fungi, defoliation, heavy winds, snowstorms, or fire. The nonaggressive species are in the majority in the family Scolytidae and clearly represent the ancestral condition of the group.

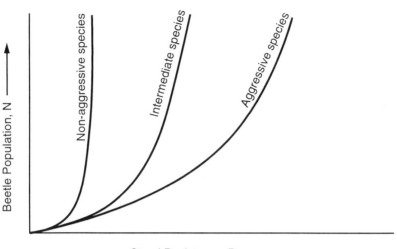

FIGURE 19.8 Thresholds applicable to Berryman's model (Fig. 19.7) relating to highly aggressive bark beetle species, intermediate species, and nonaggressive species. From A. A. Berryman in J. B. Mitton and K. B. Sturgeon (eds.)., *Bark beetles in North American conifers: A system for the study of evolutionary biology.* Copyright © 1982. By permission of the University of Texas Press.

Berryman's model (1982) and the knowledge represented therein represent a robust hypothesis at least, probably well advanced toward factually based theory. Not all the mechanistic details are understood, such as the mechanisms involved with changes in host resistance, and these should be explained in a fully developed theory.

Applying to bark beetles the approach using phylogenetic constraints and their consequences, there is a most interesting escape from constraints enabled by the acquisition of a mutualistic fungus, especially a member of the genus *Ceratocystis*. Evolution of the scolytids involved a lifestyle that penetrated to the "heart" of the tree host—the cambial layer between wood and bark. Usually, this nutritious layer is well defended, but when defenses are down, bark beetles could exploit the zone. The evolution of the life history of scolytids had created a phytogenetic constraint involving the exploitation of a resource, the normally highly defended cambial layer of the host tree (Fig. 19.9). In healthy forests, trees with poor defenses would be rare, the resources would be limiting, the carrying capacity of the environment would be low, and bark beetle species would be patchily distributed and relatively rare with latent population dynamics. The chance acquisition of a mutualistic fungus by a few bark beetle species, which was pathogenic and frequently lethal in tree hosts, resulted in a massive increase in the potential carrying capacity of the forest and an eruptive population dynamics (Fig. 19.9). The number of bark beetles in the family Scolytidae is about 480 species in North America,

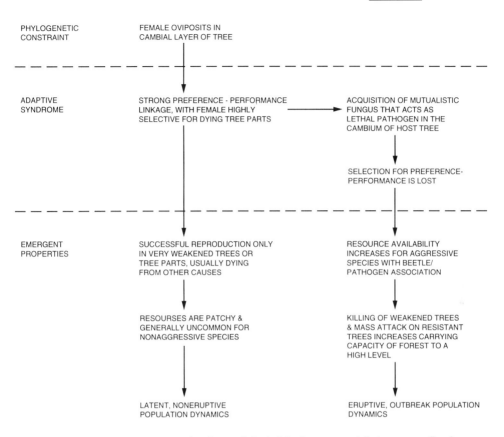

FIGURE 19.9 Phylogenetic constraints hypothesis applied to bark beetles, starting with the lineage originally utilizing dead or dying trees and tree parts. Escape from this constraint resulted from the acquisition of a mutualistic fungus lethal to trees, which greatly increased the carrying capacity of the forest and the potential for damaging outbreaks.

with 460 species nonaggressive and about 20 that are aggressive. Five percent of species are aggressive in the genera *Dendroctonus, Scolytus,* and *Ips.*

GRASSHOPPERS

Now we move to a different habitat, usually open grasslands, and a completely different kind of herbivore, which feeds on herbs and grasses. The grasshoppers are remarkable for many reasons, but one feature is striking: we know a great deal about the population dynamics of members of the family Acrididae, the short-horned grasshoppers, but almost nothing about dynamics in the Tettigoniidae, the long-horned grasshoppers and katydids. Both groups feed on grasses, herbs, shrubs, and other woody plants, they may be found in the same habitats, and most members of each family overwinter in the egg stage.

Yet in an authoritative treatment of grasshoppers (Chapman and Joern, 1990) and in a chapter on population dynamics (Joern and Gaines, 1990), tettigoniid species were rarely discussed. We can ask why there is this disparity in concern, and the obvious answer is that acridids are serious pests of rangelands and agricultural landscapes, and tettigoniids are not. One group exhibits eruptive population dynamics in many of its species, and the other group apparently has sparse and more latent population dynamics. Selection thinking should prompt the question on why such a dichotomy should exist.

Using the phylogenetic constraints hypothesis, we should focus on the evolution of morphological, life history, and behavioral traits relating to the way females lay eggs. Acridid females have a short robust ovipositor but a highly extensible abdomen, such that egg pods can be buried in the ground (Fig. 19.10). Warm, dry locations with reduced plant cover are commonly used. Contact between the ovipositing female and the plants on which the nymphs will feed, several weeks to many months later in some cases, is absent—there is no preference-performance linkage. We should anticipate the possibility of a set of interactions as in Fig. 19.4, resulting in the potential for eruptive population dynamics. Very few species oviposit in plants. In the Tettigoniidae a greater range of oviposition sites is used, from the ground, to the base of plants, to living plant parts such as leaves, leaf sheaths, and stems. Ovipositors used for egg-laying in plants are well developed (Fig. 19.10); eggs are laid singly or in small batches, usually in moist surroundings. In many species eggs absorb water from the substrate during development (Scholtz and Holm,

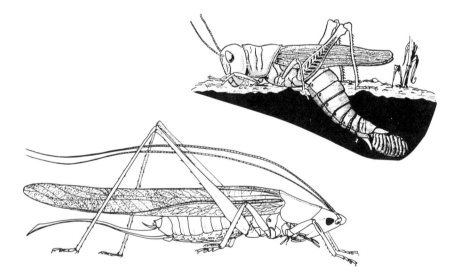

FIGURE 19.10 Contrasting designs of grasshoppers. A short-horned grasshopper, the brown locust *Locustana pardalina* (Orthoptera: Acrididae: Acridinae), ovipositing in the soil (right) is a highly eruptive plague locust in southern Africa. A long-horned grasshopper, *Ruspolia differens* (Orthoptera: Tettigoniidae), also from Africa, with its long antennae and long ovipositor used for inserting eggs singly into plant tissues. From Scholtz and Holm (1985).

1985). Where species are associated with particular host-plant species, as in the katydids or bush katydids (subfamily Pheneropterinae), they are usually green and cryptic, nocturnal, and oviposit in plant tissue, for example by inserting a flat ovipositor into the edge of a leaf. In these katydids especially we may expect a preference-performance linkage and a flow of influences, resulting in latent population dynamics (Fig. 19.4).

The large-scale eruptive pattern of the desert locust, *Schistocerca gregaria*, was discussed briefly in Chapter 17. However, many other highly eruptive acridids occur around the world, and it is worth looking for common themes in these kinds of species (Table 19.2). Geographic distribution of some species in Africa, Asia, and India is illustrated in Fig. 19.11. They become abundant when food is plentiful, and most show a phase change from a solitary phase to a gregarious phase. Gregarious feeding can result in serious crop damage and mass flights result in widely dispersed damage and oviposition. Females lay egg clutches protected in a coating or by a plug, several centimeters in the ground, eggs hatch a few weeks later, nymphs wriggle to the surface and start feeding, usually on nearby grasses or herbs (Uvarov, 1966, 1977; Skaife et al., 1979). Unusually heavy rains, for example, along the Red Sea coasts of Eritrea, Saudi Arabia, Sudan, and Yemen in 1992 in relation to the desert locust, *Schistocerca gregaria*, result in a large increase in the carrying capacity of forage and populations increase (e.g., Showler, 1995; see also Nailand and Hanrahan, 1993). Increasing populations may continue for two or three generations and result in a phase change, from solitary to gregarious, several months after the rains. An outbreak results and swarms of adults migrate with prevailing winds (Farrow, 1993; Drake and Gatehouse, 1995). Green plants are necessary for continued reproduction, at least in the African migratory locust (Price and Brown, 1990). Even in uniform crops such as maize or wheat, females tend to aggregate in clearings where they can bask in the sun and oviposit (R. E. Price, 1991), resulting in the clumped emergence of nymphs.

TABLE 19.2   Highly Eruptive Grasshopper Species from Around the World

Species	Common Name	Location
Anacridium melanorhodon	Tree locust	Africa
Austriocetes cruciata	Plague grasshopper	Australia
Chortoicetes terminifera	Australian plague locust	Australia
Dociostaurus maroccanus	Moroccan locust	Western Turkey
Gastrimargus musicus	Yellow-winged locust	Australia
Locusta migratoria	African migratory locust	Africa, Asia, Australia
Locustana pardelina	Brown locust	Southern Africa
Melanoplus sanguinipes	Migratory grasshopper	North America
Melanoplus spretus	Rocky Mountain grasshopper	North America
Nomadacris guttulosa	Spur-throated locust	Australia
Nomadacris septemfasciata	Red locust	Tropical Africa
Nomadacris succincta	Bombay locust	India
Schristocerca gregaria	Desert locust	North Africa, Asia

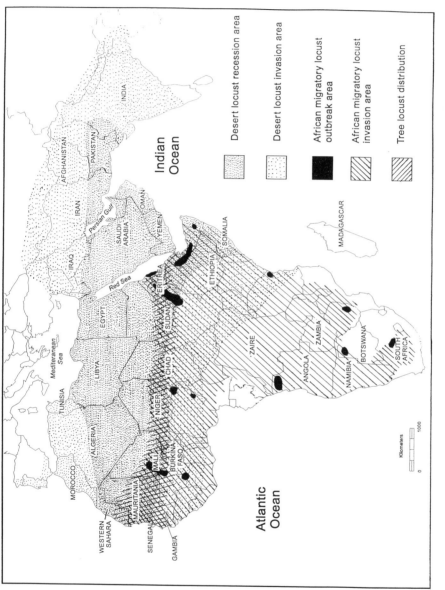

FIGURE 19.11 Geographic distributions of the desert locust, *Schistocerca gregaria*, the African migratory locust, *Locusta migratoria*, and the tree locust, *Anacridium melanorhodon*. Note the relatively localized outbreak areas of the African migratory locust but the very extensive areas invaded by swarms of flying adults. From Showler (1995); reprinted by permission of the Entomological Society of America.

Desert locust recession area

Desert locust invasion area

African migratory locust outbreak area

African migratory locust invasion area

Tree locust distribution

All the features of a group of species with no preference-performance linkage are illustrated by the locusts. However, many acridid grasshoppers remain uncommon, but the reasons are obscure. Natural enemies are a possible top-down influence on uncommon species, but we know too little about their impact (Joern and Gaines, 1990). Dempster (1963) concluded that natural enemies of grasshoppers seldom have more than a small dampening effect at high densities (cf. Berryman, 1982, on bark beetles). Belovsky and Joern (1995) studied prairie grasshoppers in different sites but with similar methods using long-term field observations and experiments. An interplay of factors, based on plant resource supply, yielded a conclusion as to why population regulation differed in the two localities.

"Our studies indicate that one of the key mechanisms that changes over time and between sites for herbivore populations is the influence of plant quality and quantity. Plant quality and quantity impact density-independent survival and reproduction and determine the strength of food competition. For example, given two populations of the same herbivore with identical functions relating predatory mortality to prey density and abiotic-induced mortality, the population with better food resources will more likely be food-limited, whereas a population with poorer food resources will more likely be predator-limited" (Belovsky and Joern, 1995, p. 382).

COMMON FEATURES AMONG EXAMPLES

As examples of the use of an evolutionary approach to the understanding of population dynamics, we have examined about as diverse a range in phytophagous insect taxa as can be represented in only four cases: flightless Lepidoptera, gall-forming sawflies, bark beetles, and grasshoppers. Yet a phylogenetic basis for latent or eruptive population dynamics can be traced in each case. Similar cases could be developed for many other kinds of insect species (Table 19.3). There is an evolutionary basis for understanding and predicting the *potential* of a taxon to show latent or outbreak population dynamics, but this does not mean that the species *actually* demonstrate their predicted dynamical patterns. Ecological factors may suppress the potential to outbreak through the action of natural enemies, for example. But those species known to show outbreak population dynamics seem to be aggregated in the taxa in which a preference-performance linkage is absent, or has been lost, as in the eruptive bark beetles.

The common features we have seen in the four examples provided may prove to be a relatively robust set of generalizations on which to ground the development of theory.

1. In each case there is a very strong bottom-up influence on population dynamics from the food plants. Understandably, food supply is a critical ingredient in population dynamics theory.

TABLE 19.3 Examples of Taxonomic Groups Well Represented by Species with Eruptive Population Dynamics, or Serious Pest Species, Some Examples of Representative Species, and an Apparently Key Evolved Characteristic that Results in the Lack of a Linkage Between Ovipositional Preference and Larval Performance.

Taxonomic Group	Examples	Key Characteristics
Phasmida: Walking sticks	*Didymuria violescens* *Ctenomorphodes tessulatus* *Podacanthus wilkinsoni*	Eggs dropped to ground, some moved by ants, nymphs search at random for food (Readshaw, 1965)

Orthoptera:
  Acrididae
  Grasshoppers/
  locusts
	Locusta migratoria
	Locustana pardelina
	Schistocerca gregaria
		Eggs laid in soil weeks to months before nymphs hatch and feed (Chapman and Joern, 1990)

Homoptera:
  Cercopidae
  Froghoppers/
  spittlebugs
	Deois flavopicta
	Zulia entreriana
	Aeneolamia selecta
		Eggs laid in soil, not on plants; nymphs hatch and find the nearest food plant (tropical) (Fontes et al., 1995)

Homoptera:
  Coccoidea
  Scale insects
	Lepidosaphes ulmi
	Quadraspidiotus perniciosus
	Saissetia oleae
		First-instar "crawlers" disperse by walking or passively in the wind; females wingless and sedentary

Coleoptera:
  Chrysomelidae
  Leaf beetles
  Galeracinae
	Diabrotica undecimpunctata
	Diabrotica barberi
	Trirhabda virgata
		Eggs laid in soil and not on plant, so ovipositing female cannot evaluate plant quality (Borror et al., 1989)

Coleoptera:
  Alticinae
  Flea beetles
	Phyllotreta striolata
	Epitrix hirtipennis
	Disonycha pluriligata
		Eggs laid in soil and not on plant, so ovipositing female cannot evaluate plant quality (Borror et al., 1989)

Coleoptera:
  Scolytidae
  Bark beetles
	Dendroctonus pseudotsugae
	Ips pini
	Scolytus multistriatus
		Mutualistic pathogenic fungus enables females to attack and reproduce in most trees (Fig. 19.9)

Lepidoptera:
  Tortricidae
  Tortricid moths
Lepidoptera:
  Olethreutid
  moths
	Choristoneura fumiferana
	
	Sciaphila duplex
	Zeiraphera improbana
		Eggs laid in late summer, larvae begin feeding in early spring (Furniss and Carolin, 1977)

Lepidoptera:
  Pyralidae
  Snout moths
	Eldana saccharina
	Diatraea saccharalis
	Elasmopalpus lignoselus
		Oviposition on dead foliage or bracts; feeding in stem (Conlong, 1990)

Lepidoptera:
  Geometridae
  Measuring
  worms
	Operophtera brumata
	Alsophila pometaria
	Erannis tiliaria
		Eggs laid on bark by flightless females (Table 19.1)

Lepidoptera:
  Lymantriidae
  Tussock moths
	Lymantria dispar
	Orgyia pseudotsugata
	Leucoma salicis
		Eggs laid on bark, some females flightless (Furniss and Carolin, 1977)

Lepidoptera:
  Saturniidae
  Emperor moths
	Coloradia pandora
	Gonimbrasia belina
	Imbrasia cyntherea
		Unspecific oviposition on leaves or bark (Scholtz and Holm, 1985)

Lepidoptera:
  Noctuidae
  Cutworms/
  armyworms
	Euxoa auxiliaris
	Spodoptera exempta
	Pseudoletia unipuncta
		Large larvae move extensively among plants (Drake and Gatehouse, 1995)

2. Top-down influences from carnivores in general were seen to be inconsequential, inadequately studied, or variable depending on plant resource quality and quantity.
3. For eruptive species there was no preference-performance linkage, although the circumstances in each case were completely different.
4. An understanding of life history traits and behavior, coupled with details on host plant and herbivore interactions, provides the basis for understanding and predicting general patterns in population dynamics.

BROADENING THE PERSPECTIVE AND THE COMPARATIVE APPROACH

With these general themes in mind we can extend the arguments in various ways in an effort to broaden our working hypotheses, with the goal of providing the basis for more general theory. We should examine other taxonomic groups, a broad range of geographic and latitudinal locations, and the variation in preference-performance linkage.

Taxonomic Groups with Eruptive Species Well Represented

A sampling of taxa with outbreaking species well represented is provided in Table 19.3. In each case the key characteristics of the group that prevent the development of a strong preference-performance linkage are given. Many orders and families are represented, but there always appears to be a lack of close association between female ovipositional preferences and how larvae or nymphs perform.

Broadening the scope of these comparisons will be helpful. Testing the phylogenetic bases for divergence among lineages into life cycles that allow or promote eruptive dynamics, versus those that result in latent dynamics, would be most valuable.

Temperate and Tropical Comparisons

The same kinds of patterns are seen in temperate and tropical latitudes and in dry and wet climates (Table 19.3). Tropical and temperate latitudes contain highly eruptive grasshoppers. Pyralids, noctuids, and cercopids are serious pests in the tropics, as they are in the temperate. Outbreak species occur in wet and dry climates.

In a similar way, what little comparative work has been completed shows that those species with high preference-performance linkage in the tropics, such as membracids, cicadellids, and gall-formers, show patterns of plant utilization remarkably similar to those in temperate regions (e.g., Price et al., 1995b). Where preference-performance associations are strong, insect species respond positively to vigorous plant growth at all latitudes (cf. Prada et al., 1995; Vieira et al., 1996; and temperate evidence in Chapter 6 in the section "Plant Vigor Hypothesis"). Clearly, further testing of these patterns and mechanisms is needed, especially in tropical and desert environments, from which so few comparative data are available.

Gradient of Preference–Performance Linkage

So far, we have considered the two extremes on a gradient from high to no preference-performance linkage. Many species fall between these extremes or do not fit the patterns well. Much more research is needed on these kinds of species. For example, most leaf miner species have females that oviposit just where the larvae will begin to feed, so we should anticipate a tight preference-performance linkage and latent population dynamics. Yet many leaf-miner species are serious pests and highly eruptive, such as in several genera of Lepidoptera and Hymenoptera. What are the mechanisms involved? Some gall-forming species in the gallflies oviposit on leaves and larvae crawl to the sites where gall formation is initiated (Gagné, 1989). Does this small reduction in preference-performance linkage reduce the selectivity of females and increase the potential for more eruptive population dynamics? Are there ultimate factors that account for so many uncommon species that show no eruptive population behavior? Many questions remain unanswered.

Perhaps the answers lie more in the proximate factors that influence populations than in the ultimate factors. Proximate factors would include those contemporary ecological processes working on populations, such as natural enemies, weather, and food supply, which may vary significantly from place to place or in time from one season to the next. Experimental approaches are important in unraveling and sorting all the possible factors involved.

EXPERIMENTAL APPROACHES: PROXIMATE FACTORS

One of the strong limitations on understanding insect population dynamics has been the shortage of experiments. The problem has been recognized for a long time, but a sound experimental approach has yet to take a central place in this field. Forty years ago Nicholson (1957, p. 326) emphasized the need for experiments and provided a methodological approach. "The population is subdivided into replicate subpopulations. In one group of subpopulations the density is reduced, in another increased, and in the third left unaltered. The densities are then followed through time, and a return to the original, unaltered density by the experimental subpopulations is evidence for regulation. Such convergence also demonstrates density dependence. If the populations do *not* converge, it is said that the population is not regulated over the experimental time interval and here, therefore, is an empirical criterion for distinguishing regulated and nonregulated populations. Some populations may be regulated infrequently and for brief periods, and we might miss such regulation during the experiment."

Murdoch (1970) also promoted the use of such experiments, calling them **convergence experiments**, while Harrison and Cappuccino (1995) called them **density-perturbation experiments**. An example of a density-perturbation experiment is provided by Cappuccino (1992), who won the Mercer Award from the Ecological Society of America in 1993 for this paper. The methods were as follows. All galls were removed from 20 discrete subpopulations of the stem-galling fly, *Eurosta solidaginis* (Diptera: Tephritidae), on goldenrod, *Solidago altissima* (Asteraceae). Insects within galls were reared to adults, so that emergence was synchronized with that in the field. Ten goldenrod patches were assigned to a density-augmentation treatment, in which twice the number of flies were released in 1989 compared to the number that had emerged in the 1988 generation. Ten other patches were assigned to a density-reduction treatment in which half the number of flies were released in 1989 compared to the 1988 generation. Parasitoids (*Eurytoma gigantea*) and stem-boring enemies (*Mordellistena unicolor*) were reared from galls and released in the plots from which galls had been collected. Therefore, populations of carnivores were not manipulated in terms of densities. Galls formed by released female flies in 1989 were collected in May 1990 after one generation, and insects therein were reared. Galls were assigned to one of five categories: survived, if an adult gall-maker emerged; early larval death, if nothing emerged and dissection of galls revealed no evidence of enemies; killed by *Eurytoma*, if a *Eurytoma* adult emerged; killed by *Mordellistena*, if one of these black beetle adults emerged; and killed by avian predators, chickadees and woodpeckers, if a pecked hole were evident in the gall. *Eurosta* mortality was expressed in terms of the killing power, or k value, for total mortality and for the individual mortality factors (cf. Varley et al., 1973; and Chapter 18).

Cappuccino found that the change in gall densities achieved by the perturbations ($\log_{10}N_{1989} - \log_{10}N_{1988}$) ranged from -1.407 to 0.105, meaning that most subpopulations had decreased in density, and only two showed density increases. When change in mortality from the 1988 to the 1989 generation was plotted against change in population size, a pattern of temporal density dependence was observed (Fig. 19.12). The more the population had been decreased, the less mortality was observed. By far the strongest mortality factor was early larval death, ranging from 31.8 to 82.4% of individuals in a population with a mean mortality of 62.8% of individuals. Plotting ($\log_{10}N_{1989} - \log_{10}N_{1988}$) against ($k_{1989} - k_{1988}$) for each of the four mortality factors, it became clear that only larval death was regulating the population (Fig. 19.13). (Note that $\log_{10}1 = 0$ or no change in population; $\log_{10}0.5 = -0.30$ is a 50% reduction in population; $\log_{10}0.1 = -1.0$ is a 90% reduction; $\log_{10}0.05 = -1.30$ is a 95% reduction; and $\log_{10}0.01 = -2.00$ is a 99% reduction.) From these results it appears to be likely that *Eurosta* densities are limited by the number of adequate resources for gall

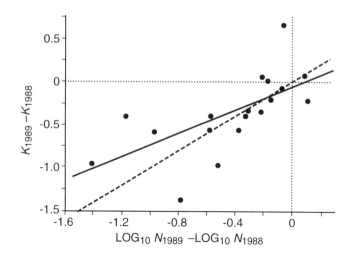

FIGURE 19.12   General results from Cappuccino's (1992) perturbation experiment, in which change in *Eurosta* mortality from 1988 to 1989 ($k_{1989} - k_{1988}$) is plotted in relation to density ($\log_{10}N_{1989} - \log_{10}N_{1988}$). Each solid circle represents one subpopulation. The solid line is the linear regression, with $r_2 = 0.399$ and $p < 0.0049$. The dashed line represents complete compensation in mortality at a given density, or perfect regulation. Dotted lines indicate no density change and no change in mortality. From N. Cappuccino, The nature of population stability in *Eurosta solidaginus*, a nonoutbreaking herbivore of goldenrod, *Ecology*, 1992, **73:**1792–1801. Copyright © 1992 by the Ecological Society of America. Reprinted with permission.

initiation and larval survival. The populations are regulated, in effect, by the carrying capacity of the host-plant population (Cappuccino, 1992). Bottom-up regulation of herbivore insect density is evidently strong in this case.

Another density-perturbation experiment was conducted by Harrison (1994) using a species known to reach outbreak proportions, the western tussock moth, *Orgyia vetusta*. The host plant was bush lupine, *Lupinus arboreus*, at Bodega Bay, California. On one lupine stand populations of caterpillars were very high and defoliated bushes, but at nearby sites populations were low. Females are flightless, eggs are laid on stems at the pupation site, and overwintering is in the egg stage. Therefore, an interesting comparison can be made between the results for a species with presumably high preference-performance linkage, the gall-formers studied by Cappuccino, and the outbreak species, with no preference-performance linkage studied by Harrison.

Harrison conducted three experiments: one on density, one on dispersal, and another on heterogeneity. In the density experiment she selected 60 uninfested lupine bushes and assigned them randomly to the caged or uncaged

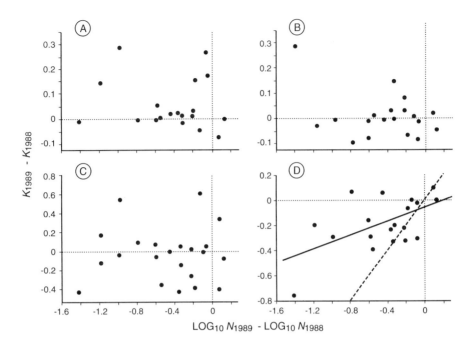

FIGURE 19.13   Specific mortality factors in Cappuccino's (1992) perturbation experiment plotted as in Fig. 19.12: (A) predation by chickadees and woodpeckers; (B) predation by *Mordellistena;* (C) parasitism by *Eurytoma;* (D) early larval death. Only in (D) is a significant linear regression observed, with $r^2 = 0.283$ and $p < 0.0191$, indicating density regulation through early larval death. Symbols are as in Fig. 19.12. From N. Cappuccino, The nature of population stability in *Eurosta solidaginus,* a nonoutbreaking herbivore of goldenrod, *Ecology,* 1992, **73:**1792–1801. Copyright © 1992 by the Ecological Society of America. Reprinted with permission.

treatment. Six density treatments were used, each replicated five times within the caged and uncaged treatments. Egg masses were collected with about 1900 eggs per mass, and realistic densities were used to set up the experiment with treatments of about 25, 50, 100, 200, 400, and 800 larvae per bush. Bushes were censused twice per week for the number of larvae and pupae present and pupae were weighed. The number of ants, *Formica lasiodes,* was censused also. The ant is an abundant predator of early instar tussock moth larvae.

The results from this experiment included larval survival higher in caged plants than in uncaged plants, probably because of ant predation in the uncaged treatment. In the caged treatments there was density-dependent mortality, presumably due to food shortage—quantitative limitation of resources. Strong bottom-up forces were important. Clearly, being absolutely sure of the cause of loss of individuals is difficult, even in well-designed experiments on low bushes in which all larvae can be observed. However, the experiments provided a strong probability that interpretations were correct.

In the dispersal experiment, Harrison studied emigration at high densities by releasing 30,000 eggs in the middle of two replicate circles. They were constructed with circular paths around the plots and radial paths so that larvae could be censused, but without limiting their dispersal over upper lupine branches that touched over the paths. Larvae were censused every week, and median dispersal distance was calculated. She found that early instar larvae hardly moved from the central location of release, and ultimately, the median distance moved by the older larvae was only 2 m, even though the central bushes were defoliated. The results showed the very slow spread of an eruptive population.

The heterogeneity experiment tested if areas of lupine without tussock moths were suitable as food. Larvae were bagged on bushes in three different areas and reared until pupation. Pupae were collected, weighed, and sexed. The results showed that all locations provided suitable forage for larvae, indicating that limitations on the magnitude of the eruptive area were not resulting from bottom-up effects or the abiotic environment.

Harrison's conclusions from these experiments were as follows: (1) population size is limited by the quantity of plant food per unit of area, not by qualitative factors, a result to be expected for an outbreak species based on the discussions in this chapter; (2) on a larger scale, the population is limited by its incapacity to move extensively—"the 'stickiness' of the moth population" (Harrison, 1994, p. 32); and (3) natural enemies did not prevent the population from reaching the carrying capacity of the foliage available. Possibly, ants restrict areas where outbreaks occur, but they do not prevent outbreaks in general.

Comparing the results from these density-perturbation experiments by Cappuccino and Harrison is revealing. Both studies used systems that were readily censused and manipulated, providing much more reliable interpretations of data than is possible in life table studies in unmanipulated conditions or on species in large trees. Both studies show strong bottom-up effects on herbivore populations, which were the most important regulating factors in each case. Cappuccino studied an endophytic noneruptive insect that was regulated by variation in plant quality affecting survival of early larval instars. The carrying capacity of the plant population was very much below the number of stems apparently available for oviposition. Harrison found for the exophytic species of tussock moth that populations were affected most by the absolute density or biomass of food for late-instar larvae. Larvae die or experience reduced fecundity resulting from shortage of food. These results are consistent with the phylogenetic constraints hypothesis illustrated in Fig. 19.4. They are also consistent with general results from many experimental studies on population regulation.

GENERALIZATIONS FROM
DENSITY-PERTURBATION EXPERIMENTS

The kinds of density-perturbation experiments advocated by Nicholson (1957), Murdoch (1970), and Harrison and Cappuccino (1995) have been reported only about 60 times in major journals since 1970. Covering herbivorous insects, other invertebrates, fish, amphibians, birds, and mammals, the studies may seem too broad to provide clear generalizations. Nevertheless, Harrison and Cappuccino (1995) found clear general patterns (Table 19.4).

First we should note the poor representation of studies on herbivorous insects (nine in all, in 25 years), and the great opportunity available to contribute importantly to this field. Second, more than one kind of regulating force was examined in many of the studies, so a total of 84 attempts to detect regulation in a population are included in Table 19.4. Sources for the table are provided in the original publication. A notable general result is that 78.6% of the 84 cases showed direct density dependence. Another is that bottom-up regulation was much more common than top-down density dependence—89.4% versus 38.5% of cases. Among studies on herbivorous insects bottom-up regulation was found in 66.7% of cases, but only 33.3% of studies showed top-down regulation from carnivores.

Lateral forces did not include those that directly involved species at other trophic levels, such as exploitation competition for food, which would enter into the bottom-up category. Thus interference competition, territoriality, cannibalism, and dispersal were grouped as lateral forces. Density-dependent dispersal before resources became limiting were identified by Ohgushi and Sawada (1985a) as important in regulation of *Henospilachna nipponica* populations, as discussed in Chapter 13. Lateral forces in regulation were generally more commonly observed than top-down effects (79.2% versus 38.5%).

TABLE 19.4   Number of Density-Perturbation Studies (*N*) that Examined Regulatory Forces Involving Bottom-up, Top-down, and Lateral Effects, and the Percentage of Studies that Documented Direct Density Dependence. Overall Percentages are Shown in Italic at the Bottom of Each Type of Regulation.

	Bottom-up		Top-down		Lateral		Total	
	N	*%*	*N*	*%*	*N*	*%*	*N*	*%*
Invertebrates	25	88.0	10	30.0	9	77.7	44	72.7
Herbivorous insects	9	66.7	6	33.3	5	80.0	20	60.0
Fish, amphibians, reptiles	18	94.4	1	100.0	9	77.8	28	89.3
Birds, mammals	4	75.0	2	50.0	6	83.3	12	75.0
Total	47	*89.4*	13	*38.5*	24	*79.2*	84	*78.6*

SOURCE: Harrison and Cappuccino (1995), with permission.

Harrison and Cappuccino (1995, p. 135) ended their analysis with the following statement. "Several striking conclusions emerge from this survey of experimental studies. First, the results appear to refute the idea that density dependence is rare in natural populations. Second, they indicate that resources are a much more common regulating force than natural enemies. Third, the fact that these conclusions hold for herbivorous insects as well as other taxa would seem to overthrow both of the traditional opposing views: that populations of herbivorous insects are wholly unregulated, and that they are usually regulated by enemies!"

Cautionary notes are offered by the authors on how strong our conclusions can be when based on a relatively small number of studies, and when such studies are not a random sample of systems to study, nor a random sample of all studies undertaken, for many no doubt remain unpublished because of negative data. Harrison and Cappuccino (1995, pp. 136–137) then enter into an important discussion on guidelines for experimental studies of population regulation. "How long do experiments need to run?" "How large should experimental units be and should they be caged?" "What treatment levels should be used?" "How far can experiments get us?" (p. 141), and other considerations.

These experimental studies, and their review by Harrison and Cappuccino, contribute importantly to a synthesis on population dynamics. As the authors state, valuable additions to our understanding could be made with more experimental studies on species with eruptive and latent population dynamics. More effort toward finding general patterns in bottom-up and top-down regulation in populations of various kinds of species in a diverse array of environments would be rewarding.

The question of length of experiments is a critical one because population change may be cyclical, and many effects, such as disease, may be sporadic even though important. Should we perform experiments for one generation, multiple generations, or through a population cycle lasting perhaps 10–15 years?

TIMING OF OUTBREAKS AND PERIODICITY: PROXIMATE FACTORS

Repeated patterns of population change that produce population cycles of a certain periodicity offer beguiling potential for pinning down the mechanisms involved. Hence the search for periodicity in populations and the study of factors potentially involved has been a long-standing theme in insect population dynamics. Some potential factors causing cycles have already been discussed, including the nutrient recovery hypothesis and the induced defense hypothesis in chapter 6; and parasitoids, diseases, and Wellington's studies on behavioral polymorphism in the western tent caterpillar in Chapter 17.

Myers (1988) provides an excellent treatment of population cycles in forest Lepidoptera, noting at least 18 species in North America and Europe that cycle with average periodicities of 8–11 years (Table 19.5). These are all outbreak species, with common characteristics identified by Myers (1988), including frequently high fecundity, eggs in clusters, and passive dispersal by

TABLE 19.5   Species of Eruptive Forest Lepidoptera that Show Periodicity in Outbreaks.

Species	Periodicity[a]			Number of Sites[b]
	X	N	SE	
Autumnal moth (*Oporinia autumnata*) ⎫	8.8	12	0.5	M
Winter moth (*Operophtera brumata*) ⎬	9			M
Northern winter moth (*Operophtera fagata*) ⎭				
Douglas-fir tussock moth (*Orgyia pseudotsugata*)	10			M
Forest tent caterpillar (*Malacosoma disstria*)	8–12			M
Western tent caterpillar (*Malacosoma californicum pluviale*)	8.2	6	0.7	1
Eastern spruce budworm (*Choristoneura fumiferana*)	~35			M
Western spruce budworm (*Choristoneura occidentalis*)	28	6	4.0	2
	12.8	6	3.0	M
Larch budmoth (*Zeiraphera diniana*)	8.5	15	0.3	1
Pine looper (*Bupalus piniarius*)	7.9	7	1.0	1
Lymantria fumida	7.0	5	0.3	1
Gypsy moth (*Lymantria dispar*)	~8			
Black-headed budworm (*Acleris variana*)	8.0	1		1
Winter moth (*Operophtera brumata*)	8.0	1		1
Western hemlock looper (*Lambdina fiscella lugubrosa*)	8.8	5	0.3	M
False hemlock looper (*Nepytia freemani*)	11.3	4	1.9	M
Fall webworm (*Hyphantria cunea*)	8.0	4	0.7	2
Saddled prominent (*Heterocampa guttivitta*)	11.2	6	1.2	1
Kotochalia junodi	6–8			
Eastern hemlock looper[c] (*Lambdina fiscellaria*)	12–16			
Green striped forest looper[c] (*Melanolophia imitata*)	8–10			
Jack pine budworm[c] (*Choristoneura pinus*)	6–8			
Pine butterfly[c] (*Neophasia menapia*)	10–16			
Saddle backed looper[c] (*Ectropis crepuscularia*)	8–12			
Western black-headed budworm[c] (*Acleris gloverana*)	8–16			
Western false hemlock looper[c] (*Nepytia canosaria*)	10–16			

[a] X, mean, range, or approximate periodicity; N, number of outbreaks counted in the determination of the mean periodicity; SE, standard error.

[b] M, more than 3.

[c] Additional species are from McNamee (1987), listed in Myers (1988).

SOURCE: Myers (1988), with permission. Based on compilations by Myers (1988) and McNamee (1987). Source references are provided by Myers (1988).

larvae, although no general pattern is evident (Table 19.6). Cycles are frequently synchronous in different geographic areas and even among different species of Lepidoptera, suggesting a broad influence such as weather. There is no evidence that tropical insects show regular cycles.

What are the forcing factors or causative mechanisms involved in population cycles? Myers (1988) evaluated each of the major hypotheses in turn: (1) the **variation in insect quality hypothesis,** including genetic variation and behavior as proposed by Chitty and Wellington (see Chapter 17); (2) the **climatic release hypothesis** (Chapters 6 and 17); (3) the **food quality deterioration hypothesis,** involving induced defenses in plants, including increased defenses and reduced nutrient quality (Chapter 6); (4) the **food quality improvement hypothesis** or plant stress hypothesis (Chapter 6); and (5) the **disease susceptibility hypothesis,** invoking the interplay between increased pathogen transmission as insect density increases, reduced resistance as food becomes limiting, and selection for more resistant genotypes, with the added possibility of sublethal pathogen effects on insect vigor and fecundity. The feature in common among these hypotheses is that regulatory mechanisms are likely to work through delayed density dependence, with gradual influence long enough to generate cycles of 10 years or so. Several hypotheses suggest cumulative influences passing from one insect generation to another, either through natural selection and heredity, or through maternal effects on subse-

TABLE 19.6   Genera of Lepidoptera Representing Species with Eruptive and Cyclic Population Dynamics, Their Characteristics, and the Kind of Host Utilized, Deciduous or Coniferous.

Genus	Family	Fecundity[a]	Egg Distribution	Larval Dispersal
Deciduous Hosts				
Heterocampa	Notodontidae	200 (500)	30–300	Gregarious
Hyphantria	Arctiidae	400 (1000)	Masses	Gregarious
Lymantria	Lymantriidae	300 (800)	Masses	Ballooning
Malacosoma	Lasiocampidae	150–225 (250)	Masses	Gregarious
Oporinia (= Epirrita)	Geometridae	120 (250)	Spread	Ballooning
Operophtera	Geometridae	150	Spread	Ballooning
Symmerista	Notodontidae	300	50	Gregarious
Coniferous Hosts				
Acleris	Tortricidae	53 (83)	Singly	Ballooning
Bupalus	Geometridae	220	2–25	Wander
Choristoneura	Tortricidae	175	5–50	Ballooning
Lambdina	Geometridae	59 (122)	2–3	Wander
Melanolophia	Geometridae	80	Singly	Solitary
Orgyia	Lymantriidae	110–150 (275)	Masses	Ballooning
Pieris	Pieridae	?	5–20	Local
Zeiraphera	Tortricidae	20–180 (350)	1–7	Wander

[a] Maximum fecundity is given in parentheses.
SOURCE: Myers (1988), with permission.

quent generations, as suggested by Chitty, Wellington, Leonard (Chapter 17), and Morris (Chapter 18).

The mechanisms involved with maternal effects and delayed influences on population density have been examined in detail by Rossiter (1992, 1994, 1995). The **maternal effects hypothesis** on insect herbivore population fluctuations involves the following scenario. Herbivore density or environmental factors are likely to have an effect on population quality. Female quality (in

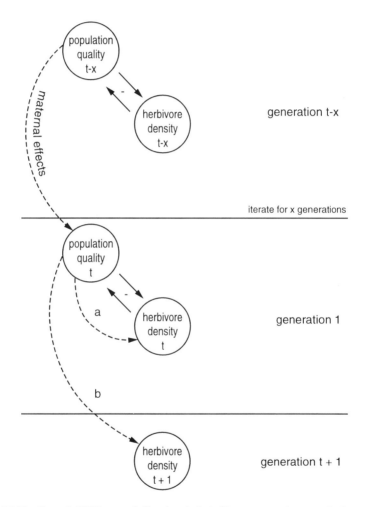

FIGURE 19.14   Rossiter's (1994) maternal effects hypothesis, in this case representing a negative impact of herbivore density in generation $t - x$ on population quality. Maternal effects then affect progeny in the next generation (t) through a change in survival of progeny (a) and in the subsequent generation ($t + 1$) through fecundity of progeny. Note the clearly illustrated delayed density dependence in this model. From M. C. Rossiter, Maternal effects hypothesis of herbivore outbreak, *BioScience*, 1994, **44:**752–763. Copyright © 1994 by the American Institute of Biological Sciences. Reprinted with permission.

generation $t - x$) may improve as density decreases and deteriorate as densities increase. These changes may affect subsequent generations through maternal effects on the population quality of the next generation *(t)* and even the subsequent generation $(t + 1)$ through changes in fecundity of the female's progeny (Fig. 19.14). Thus effects in generation $t - x$ have delayed effects for at least two generations (t and $t + 1$).

Maternal effects have been documented in many insect species (Mousseau and Dingle, 1991) and in many orders. Some examples of effects were summarized by Rossiter (1995) and appear in Table 19.7. In relation to eruptive herbivore population dynamics, Rossiter (1994) identified 12 species in which one or more characteristics of the hypothesis had been demonstrated: changes in population quality across generations, maternal effects themselves, the presence of delayed density dependence, or a density-related shift in food quality (Table 19.8). All the genera and most of the species have been mentioned in this book in relation to population dynamics, including nine insect species and three mammal species. The maternal effects hypothesis is in need of more experimental study, for there are many intriguing facets of potential importance (Rossiter, 1995). However, the hypothesis is well developed and remains as an important component in the development of synthesis in insect population dynamics.

Myers (1988) reached conclusions on each of the five hypotheses examined, recognizing problems with each and the need for more detailed analyti-

TABLE 19.7   Maternal Effects in Insects Documented in the Literature, Arranged According to the Phylogenetics of Insects Represented.

Order	Number of Genera	Offspring Traits Affected	Examples
Collembola	1	Development, weight, fecundity	Springtails
Orthoptera	6	Diapause, color, behavior, lipid reserves, development time	Crickets, locusts
Psocoptera	3	Diapause	Psocids
Hemiptera	6	Diapause, development time, wing polymorphism	True bugs
Homoptera	5	Diapause	Hoppers
		Wing polymorphism, sex ratio	Aphids
Coleoptera	4	Diapause, dispersal, body size, growth rate	Beetles
Lepidoptera	5	Diapause, growth, survival, toxin resistance	Moths
Diptera	12	Diapause, toxin resistance, survival	Mosquitos, midges
		Body size, development time	Flies
Hymenoptera	8	Diapause	Parasites

SOURCE: Rossiter (1995) with permission. Based on information in Mousseau and Dingle (1991).

TABLE 19.8   Rossiter's (1994) List of Species of Insects and Mammals for which Evidence Supports the Maternal Effects Hypothesis.[a]

Common Name	Taxonomic Name	Shift in Population Quality Across Generations	Maternal Effects	Delayed Density Dependence	Density-Related Shift in Food Quality
Gypsy moth	*Lymantria dispar*	X	X	X	X
Autumnal moth	*Epirrata autumnata*	X	X	X	X
Douglas fir tussock moth	*Orgyia pseudotsugata*	X	?	X	X
Spruce budworm	*Choristoneura fumiferana*	?	X	X	No
Western tent caterpillar	*Malacosoma californicum pluviale*	X	X	X	?
Larch budmoth	*Zeiraphera diniana*	X	P	X	X
Mountain pine beetle	*Dendroctonus ponderosae*	X	?	X	?
Fall webworm	*Hyphantria cunea*	X	X	X	X
Locust	*Locusta migratoria*	X	X	?	P
Vole	*Microtus pennsylvanicus*	X	X	?	P
Snowshoe hare	*Lepus americanus*	X	?	X	X
Lemming	*Lemmus trimucronatus*	X	?	?	X

[a] X, supporting evidence is available; ?, no data available; P, data suggest but do not confirm support; No, evidence indicates lack of density-related change in food quality.

SOURCE: Data from Rossiter, with permission of the author and the American Institute of Biological Sciences. Copyright © 1994 by the American Institute of Biological Sciences. Sources are provided in Rossiter (1992, 1994).

cal research. She favored the disease susceptibility hypothesis, although "I may simply be emphasizing the importance of disease because it is the least known factor in studies of insect cycles" (Myers, 1988, p. 231; see also Myers, 1993).

In the synthesis of hypotheses on insect population dynamics, we need to assemble a wide variety of approaches, including those in the foregoing chapters relating to plant–herbivore interactions, predators, parasites and mutualists, demography, life histories, and behavior, and those in subsequent chapters, especially on competition. The development of theory on population dynamics depends on a large effort to test hypotheses more carefully and rigorously (cf. Myers and Rothman, 1995). These hypotheses include those on the evolutionary basis for population dynamics and the ecological circumstances that affect timing, periodicity, and magnitude of population change. The challenge is great, but the opportunities are exciting and the rate of progress is accelerating.

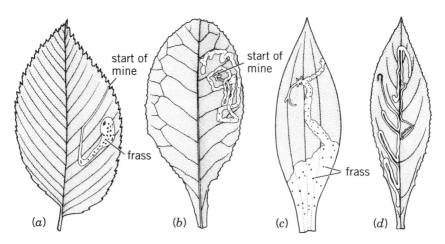

Leaf mines of four species of agromyzid fly showing (a) the linear-blotch mine of *Agromyza aristata*, (b) the linear mine of *Chromatomyia primulae*, (c) the linear-blotch mine of *C. gentianella*, and (d) the linear mine of *Phytomyza senecionis*. From Gullan and Cranston (1994); with permission. Drawing by Karina Hansen McInnes.

IV

Communities and Distributions

Species adapt to exploit a certain set of necessary resources (Chapter 20). Those with closely similar requirements may compete for them (Chapter 21). Therefore, the resources available in each community in turn influence the number of species in the community and their relative abundance (Chapter 22). The number and range of resources tend to increase with time and on latitudinal gradients from the poles to the equator, resulting in the patterns seen in diversity of communities (Chapter 23). All factors, both contemporary and historical, that influence the presence and absence of a species at a particular site contribute to the determination of the geographic range and paleological record for the species (Chapter 24).

Chapter 20

The Niche Concept and Division of Resources

To the practiced eye and to those who are familiar with organisms in the field, even very similar species can be differentiated. The differences between species have really been the focus of taxonomists and ecologists alike. The obvious corollary to differences in shape and size is that organisms differ in what they do. The larger the differences between species, or the more the differences in what animals do in a given area, the greater the number of species that can live together in the same area. Darwin (1859) called this the *principle of divergence.*

DEFINITIONS OF THE NICHE

Just as the obvious differences between species demanded that they be given different names, so there was a necessity for an ecological classification of the roles of these organisms in the environment, and Grinnell in 1971 was one of the first to use the word **niche** to describe the place of an organism in the environment. Grinnell pointed out that the California thrasher was dependent

Do predators reduce prey numbers enough, such that insect herbivore species do not compete for plant resources? Coccinellid beetles prey on aphids. The aposematic oleander aphid, *Aphis nerii*, is toxic to vertebrate predators, having sequestered toxic compounds from its host plant, but the lady beetle, *Cheilomenes lunata*, preying on the aphid is also toxic and aposematic, having synthesized its own toxins. After Burton and Burton (1975) and a photograph by Anthony Bannister. From Gullan and Cranston (1994); with permission. Drawing by Karina Hansen McInnes.

on a certain set of physical factors and biotic factors in the environment and that these factors together defined the niche of the species. Elton (1927), however, placed emphasis on the function of an organism in relation to other organisms—an animal's "place in the biotic environment, its relation to food and enemies." He was not so concerned with the physiological adaptation of a species to a particular set of environmental conditions. The difficulties with this classical concept of the niche are that often species have more than one role in the environment, for example, larvae and adults of many holometabolous insects, or species that show sexual dimorphism in shape or behavior. To be useful the niche should be defined in terms of the species rather than subdivision within a species. Also, the species' niche may be modified by the presence and absence of other species, so that it is necessary to differentiate between the **actual niche** and the **potential niche** of a species (i.e., the niche occupied when in the presence of competing species and the niche occupied in their absence). Equivalent terms used in the literature are **realized niche** and **fundamental niche,** respectively. The role of a species probably changes little when other species are present, but its success and abundance may be greatly modified by other species. The role of a species also presents problems when we wish to quantify it. How does one measure the role of a species?

The more modern niche concept, developed independently by Pitelka (1941), Macfadyen (1957), and Hutchinson (1957), attempts to circumvent these problems by making the niche a quality of the environment rather than a quality of a species. Pitelka (1941) wrote, "The niche of a particular species is obviously difficult to define and entails an extensive study of environmental relations. However, one refers to the niche most conveniently in terms of the place within the environment characteristically frequented by the species." Macfadyen's (1957) concept of the niche was "that set of ecological conditions under which a species can exploit a source of energy effectively enough to be able to reproduce and colonize further such sets of conditions."

n-DIMENSIONAL HYPERVOLUME

Hutchinson has formalized the definition so that it becomes clear how to determine the niche requirements of a species. If the temperature requirements of a species are considered, it is possible to determine the tolerance range of a species in relation to this resource, or on this dimension of the niche—a one-dimensional niche (see Fig. 20.1). If the species eats seed, it can eat a certain range of seed sizes within those available in the habitat, and combined with temperature, we have a two-dimensional niche—an area. Assume that this species lays its eggs in plant stems. A third dimension might be the width of the stem of the host plant required by an ovipositing insect, and a three-dimensional niche is described—a volume. By adding more essential resources to this niche volume, the niche can be defined as an *n*-dimensional

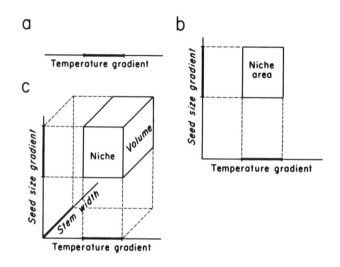

FIGURE 20.1   Diagrammatic representation of the niche, with the range over which a species can utilize each resource marked with a heavier line: (a) niche defined by one dimension; (b) niche defined by two dimensions; (c) niche defined by three dimensions. The niche volume represents the space in which a species can survive and reproduce indefinitely.

hypervolume in the environment, the perimeter of which circumscribes the space in which a species can reproduce indefinitely.

NICHE EXPLOITATION PATTERN

Conceptualizing the niche as a quality of the environment also helps in the realization that a species is constantly adapting to exploiting that environment, and modifications of the **niche exploitation pattern,** a phrase first used by Root in 1967, are required as the environment changes. Root (1967) studied this pattern in the blue-gray gnatcatcher, *Polioptila caerulea,* in California to see how the species copes with a constantly changing environment. Interestingly, the gnatcatcher seemed to be adapted primarily to those features of the environment that changed least through the season, and these adaptations differed considerably from those in the other coexisting species. Root found that the blue-gray gnatcatcher differed most from other coexisting insectivorous birds in its foot length and chord of wing. Associated with its smaller size was its utilization of smaller foraging perches, meaning that gnatcatchers forage in the periphery of the canopy in woody plants, irrespective of species or vegetation type. Its smaller wings enable high maneuverability for catching most prey encountered in the foraging zone defined by perch size. At the beginning of the breeding season insects were most abundant in the chaparral vegetation, and territories were established in bushes. Later in the season the insects became more abundant in nearby oak woods. As the insect abundance moved, so the foraging of the gnatcatchers moved, and thus they essentially remained in a niche with sufficient food by making this shift. The niche was therefore not defined in terms of plant species but rather by the perch size and abundance of insects, and any plant that sustained a good insect population and provided adequate small perches was included in the niche exploitation pattern.

Of necessity, though, this pattern was pliable, especially to cope with the extra stress of feeding young in the nest. When individuals were feeding to maintain themselves, a large majority of time was spent feeding on insects on foliage, the mode of feeding to which they are clearly adapted. However, when they needed extra food for young, food-catching routines that were energetically more expensive were seen more and more frequently, for example, catching insects in the air and gleaning on herbs (Table 20.1). In fact, the species became much more of a generalist in its feeding behavior to obtain sufficient food. The niche exploitation pattern was very pliable, to meet the changing environmental demands imposed by the presence of young and changes in food supply. It is interesting to wonder just how this pliability is related to the mental capacity of an organism and how pliability, intelligence, and longevity interact. The longer a species lives, presumably the more pliable it must be, and perhaps more intelligent, and perhaps insects, by living briefly, have circumvented the need for a plastic behavioral system. Nevertheless, we should recall, as examples, the lability of exploitation patterns in pollinating insects (Chapter 11), the complexities of site choice by *Pemphigus* aphids (Chapter 15), and the variable responses of caterpillars to plant nitrogen content (Chapter 17). However, whatever the species' mental capacity is, the niche exploitation pattern is likely to change with changing density, as seen in Chapter 17.

TABLE 20.1   Feeding Pattern of the Blue-Gray Gnatcatcher in March, and in Late June and July, Expressed as the Percent of Each Type of Feeding in the Total Feeding Time. In March, Foliage was only Partially Developed and Much Time was Spent Feeding on Twigs. In June and July During Self-Maintenance, Most Time was Spent Feeding on Insects on Foliage, but while Feeding Young a Shift was Made to More Feeding in the Air and on Herbs.

Feeding Location	March	June and July	
		Self-Maintenance (%)	Feeding Young (%)
Foliage	14.5	78.5	57.2
Air	21.0	11.9	26.1
Twigs	64.5	7.7	6.9
Herbs	0.0	1.9	9.8

SOURCE: Root (1967).

NICHE, HABITAT, AND ECOTOPE

Two important points emerge from Root's study. The niche exploitation pattern not only describes the limits on the species breeding ability that would be described by the Hutchinsonian niche concept, but also describes the distribution of abundance of a species within these limits. A population response to a range of qualities in any resource is superimposed on the Hutchinsonian niche, which greatly increases the value and precision of niche description. This population response is incorporated into the definition of the niche by Whittaker et al. (1973). The second point is that a species distribution may be defined by parameters of a very different scale. On the one hand, perch size determines distribution within a community, but food availability determines distribution between habitats. Whittaker et al. reserve the term *niche* for within-community responses; they recognize the habitat as another scale to which species respond and that these combined responses constitute the "ecotope" of a species.

Whittaker et al. (1973) reason as follows:

1. The term *niche* should be used only for a species resource utilization within a community.
2. The term **biotope** should be used for the environment in which a community exists.
3. The term **habitat** applies to the abiotic environment in which a species exists. (However, as we have just seen in this chapter, habitat also involves biotic characteristics, and particularly for animals, these will probably have to be included in the definition of habitat.)
4. Thus a species existence and distribution is defined by factors on two variables, its niche and its habitat, and where the total resources utilized by a species are considered, the term **ecotope** is suggested. "We suggest that henceforth it represent the species relation to the full range of environmental and biotic variables affecting it."

These terms allow a distinction to be made between rather different determinants of species distribution, and by further definition, Hutchinson's definition of the niche can be improved on, since population size in a series of resource units is taken into account; something that is done when using Levin's niche breadth formula discussed later in this chapter but which was not incorporated explicitly into the definition of the niche. Thus Whitaker et al. suggest the following:

1. The variables of the physical and chemical environment that form spatial gradients in a landscape or area define as axes a **habitat hyperspace.** The part of this hyperspace a given species occupies is its **habitat hypervolume.** The species' population response to habitat variables within this hypervolume describes its **habitat.**

2. The variables by which species in a given community are adaptively related define as axes a **niche hyperspace.** The part of this hyperspace in which a species exists is its **niche hypervolume.** The species' population response within its niche hypervolume describes its **niche.**
3. The variables of habitats and niches may be combined to define as axes an **ecotope hyperspace.** The part of this hyperspace to which a given species is adapted is its **ecotope hypervolume,** and population responses in this hypervolume define the **ecotope** of the species.

Examples of the application of these concepts may be seen in Price (1971a) and McClure and Price (1976). In his niche analysis of a parasitoid community exploiting sawfly cocoons on the ground, Price (1971a) recognized the need to distinguish between *distribution within plots,* that is, niche differences, such as distributions on a litter moisture gradient, and seasonal abundance; and *distribution between plots,* or habitat differences, such as responses to host density, host species, and vegetation type. McClure and Price (1976) studied the leafhoppers coexisting on American sycamore in the genus *Erythroneura.* The niche factors considered were temporal utilization, distribution within the canopy, occurrence on leaves of various sizes and ages, and location of feeding sites on the leaf. The habitat factors were occurrence on trees along a moisture gradient, latitudinal distribution of leafhoppers in relation to that of sycamore, and distribution within geographic localities. Thus intracommunity and intercommunity differences between species were distinguished.

Kulesza (1975) was uncomfortable with the distinction between niche and habitat made by Whittaker et al. (1973), largely because they distinguished between niche differences among species evolving primarily in response to other species in the community, while habitat differences evolved mainly in response to environmental gradients external to the community. Kulesza argued that species' interactions frequently extend beyond intracommunity interactions such that one species is displaced into another habitat by a superior competitor, as seen in arthropods such as whirligig beetles (Istock, 1967) and crayfish (Bovbjerg, 1970). Conversely, abiotic environmental factors such as temperature or humidity may well influence microhabitats within communities. Whittaker et al. (1975) agreed that strict delimitation of evolutionary influences is not possible or useful. However, it remains imperative in such studies to define terms accurately and the dimensions over which the studies are made.

NICHE BREADTH

Once the species population response to a gradient of conditions is known, the niche of a species can be calculated using simple formulas. Resources that the organism needs to survive and reproduce must be selected and its

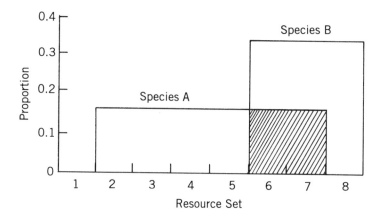

FIGURE 20.2 Hypothetical distributions of species A with a broad niche and of species B with a narrower niche, on a resource set subdivided into eight resource units. The species have the same proportional similarity, which measures the shaded zone, but species A overlaps B more than B overlaps A.

distribution on each resource sampled in turn. Each resource can be considered as a composite of many smaller resources so that this set makes up one dimension of the niche. A simple case would be to consider the available temperature range of a species as the **resource set** from − 10 to + 40°C, and this set can be divided up into units of 10°C each; the abundance of the species can be sampled in each temperature class (cf. Fig. 20.2). The **niche breadth** of the species on this resource set can be quantified by Levins's (1968) formula:

$$B = \frac{1}{\sum_{i=1}^{s} p_i^2(S)}$$

where p_i is the proportion of a species found in the ith unit of a resource set, S the total units on each resource set, $B_{max} = 1.0$, and $B_{min} = 1/S$. B can be scaled to vary between 1 and 0 (e.g., Colwell and Futuyma, 1971; Fager, 1972), so that it can be used to compare species distributions on resource sets with different numbers of units. This formula, or slight modifications of it, have been used frequently in niche analysis studies (e.g., Levins, 1968; Pianka, 1969; Price, 1971a).

PROPORTIONAL SIMILARITY

The **proportional similarity** between species distributions on a resource set can be quantified simply by the formula

$$\text{PS} = \sum_{i=1}^{n} p_{mi} \quad \text{or} \quad \text{PS} = 1 - \frac{1}{2} \sum_{j=1}^{n} | p_{ij} - p_{hj} |$$

where p_{mi} is the proportion of the less abundant species of a pair in the ith unit of a resource set with n units, and p_{ij} and p_{hj} are the proportions of species i and h, respectively, in resource unit j. $\text{PS}_{max} = 1$ and $\text{PS}_{min} = 0$.

NICHE OVERLAP

Finally, the **niche overlap** of one species with another can be quantified by a formula from Levins (1968):

$$\alpha_{ij} = \sum_{h=1}^{n} p_{ih} p_{jh}(B_i)$$

where α_{ij} is the niche overlap of species i over species j, p_{ih} and p_{jh} the proportion of each species in the hth unit of a resource set, and B_i the niche breadth of species i. These formulas adequately quantify the differences between species A and B in Fig. 20.2. Species A has a larger niche breadth than species B, species A and B have the same proportional similarity, and A will have a greater niche overlap on B than B will on A.

These formulas, or similar formulas, have been used commonly in niche analysis studies of communities or guilds. Although each formula has conceptual difficulties, they are still extremely useful in quantifying the niche of species and interactions between species. By studying several dimensions of the niche in this way, a picture may be obtained of the **fundamental niche** (=potential niche) of a species, that is, the niche that would be occupied in the absence of competing species, how this niche is modified by coexisting species, the **realized niche** (=actual niche), and how resources are partitioned between guild members.

COMPETITIVE DISPLACEMENT OR EXCLUSION

Two species with very similar environmental needs are unlikely to coexist for long, because sooner or later, competition for food will lead to the local extinction or displacement of the species that has a slightly inferior competitive ability. This has been expressed in varying ways, probably starting with Grinnell in 1904, although Darwin obviously understood the principle. A sampling of attitudes on **competitive displacement,** or **competitive exclusion,** is presented below (see also Brown, 1958b; Cole, 1960). Grinnell (1904) stated: "Every animal tends to increase at a geometric ratio, and is checked only by limit of food supply. It is only by adaptations to different sorts of food, or modes of food getting that more than one species can occupy the same locality." Later Grinnell (1971) wrote: "It is, of course, axiomatic that no two species regularly established in a single fauna have precisely the same niche relationships." Gause (1934) found that competition for common food in protozoa led to a complete displacement of one species by another. Hutchinson and Deevey (1949) stated it another way: "Two species with the same niche requirements cannot form steady-state populations in the same region." MacArthur (1958) wrote: "To permit coexistence it seems necessary that each species, when very abundant, should inhibit its own further increase more than it inhibits the others." DeBach's (1966) review of the subject should also be consulted.

An example of competitive exclusion follows, studied by DeBach and Sundby (1963), Podoler (1981), and Luck and Podoler (1985). The California red scale, *Aonidiella aurantii,* is a common pest of citrus trees in southern California around Santa Barbara, Los Angeles, Long Beach, and San Diego. Around 1900, purely by chance, a parasitoid *Aphytis chrysomphali* was introduced from the Mediterranean region. This species spread and became an effective parasitoid on the scale, but it was certainly more effective in the milder climates of the coastal areas (Fig. 20.3). In 1948 a new species of *Aphytis, lingnanensis,* was obtained from southern China and was reared and colonized in all areas where *Aphytis chrysomphali* occurred. A. *lingnanensis* became more abundant than A. *chrysomphali* and by 1958, A. *chrysomphali* was almost completely displaced from the entire area. By 1961, A. *chrysomphali* could be found in only three areas, where it represented 32, 6, and 14% of the total *Aphytis* population.

In 1956 and 1957, another species of *Aphytis, melinus,* was imported from India and West Pakistan, areas with relatively large annual climatic fluctuations resembling those of the interior citrus areas of southern California. This was done in the hope that biological control would be improved in the interior, where A. *lingnanensis* was not really effective, although it had displaced A. *chrysomphali. Aphythis melinus* was liberated in all the citrus areas in southern California and it was soon clear that this species was displacing A. *lingnanensis* in the interior. At the original release sites, competitive displacement of A. *lingnanensis* by A. *melinus* occurred within about 1 year, or eight or nine generations. By 1961, A. *melinus* accounted for 94–99% of the total *Aphytis* population in large areas in the interior.

DeBach and Sundby (1963) studied the direct competitive interactions of these species, and their paper makes interesting reading. But this example should be remembered when competition is discussed (Chapter 21). Lack of competition permitted broad exploitation, in this case by the first parasitoid species, A. *chrysomphali,* and competition forced the species back into a more specialized exploitation pattern. This is a clear-cut example of species with very similar niches being unable to coexist, resulting in competitive exclusion of one by the other. DeBach and Sundby (1963) make the claim that

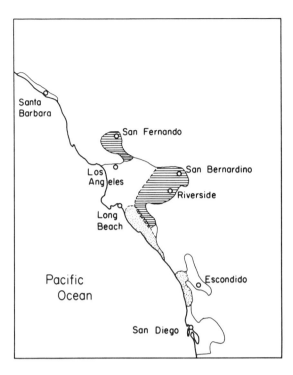

FIGURE 20.3 Citrus-growing areas of southern California and the distribution of *Aphytis* species from 1948 to 1961. In 1948, *A. chrysomphali* occurred in all citrus areas. By 1959 this species had been largely displaced in all areas by *A. lingnanensis*, but populations remained in the dotted zones. In 1961 *A. chrysomphali* was very rare in dotted zones, and *A. melinus* had displaced *A. lingnanensis* from many areas in the interior (hatched). After DeBach and Sundby (1963).

food was never a limiting factor for these parasitoids, yet the competitive exclusion principle obviously implies competition for a common resource in limited supply. Perhaps all hosts were not available to the parasitoids for various reasons, or as Huffaker and Laing (1972) point out in an interesting discussion on DeBach and Sundby's studies, interference competition may have been important. Pheromone interactions and response of parasitoids to increasing density may indicate that space was limiting rather than hosts, as suggested for parasitoids of sawfly cocoons (Price, 1970c, 1981a). Huffaker and Laing conclude that competition must be involved for displacement to occur. Bess and Haramoto (1958) describe a similar case of competitive exclusion in parasitoids of the oriental fruit fly.

Further studies have thrown light on the mechanisms involved with competitive displacement of *Aphytis lingnanensis* by *A. melinus*. Podoler (1981) studied functional, numerical, and combined responses of these species to host density under three variable temperature regimes: 18–24°C, 24–29°C, and 29–35°C. The functional response of *A. lingnanensis* was stronger than that of *A. melinus* at all temperatures, although *A. melinus* showed greater improvement in its functional response at the highest temperature. Conversely, the numerical response of *A. melinus* was stronger than that of *A. lingnanensis* over most of the host density range at the highest temperature. Using the effect of temperature on sex ratio, known from studies by Kfir and Luck (1979), Podoler then calculated the combined responses of the species to host density. A dramatic effect of temperature on sex ratio was that the proportion of females in *A. lingnanensis* is reduced from 60.6% to 17.8% in the lowest and highest temperatures used, whereas for *A. melinus* the reduction was smaller, from 66.1% to 50.6%. The net result of functional, numerical, and sex-ratio responses at different temperatures was that *A. lingnanensis* showed a stronger response to host density at 18–24°C, and *A. melinus* responded more effectively in the hotter environment at 29–35°C. These results correspond to those observed in the field, where *A. melinus* displaced *A. lingnanensis* from the hottest citrus-growing areas first.

Another important consideration in the interaction between *A. melinus* and *A. lingnanensis* is the size of host that is utilized (Luck et al., 1982; Luck and Podoler, 1985). Each host scale naturally grows from small to large during its lifetime, and *A. melinus* is able to mature females on smaller scales than is *A. lingnanensis*. In addition, if third-instar scales are considered alone, many more small scales exist available to *A. melinus* than large ones available to both *A. melinus* and *A. lingnanensis*, especially on the poorer-quality substrates, wood and leaves, on which the hosts feed (Fig. 20.4). Thus *A. melinus* can attack scale insects before they become available to *A. lingnanensis*, and a smaller size class of scales, so that it preempts resources and has a refuge from competition that *A. lingnanensis* lacks. Luck and Podoler (1985) discuss several other important factors involved with the competitive displacement of *A. lingnanensis* by *A. melinus*, and these studies have contributed significantly to an understanding of the mechanisms involved.

Another example of competitive displacement concerns the interaction of the mediterranean fruit fly, *Ceratitis capitata*, and the oriental fruit fly, *Dacus dorsalis*, in the Hawaiian Islands (Keiser et al., 1974). The "med fly" was widespread in the islands early in the twentieth century in the absence of *D. dorsalis*. The latter species was accidentally introduced into Hawaii in the 1940s and has become common and generally distributed since, while *C. capitata* exists in a restricted range at higher elevations and in refugia from competition in fruits such as coffee. In competition experiments using guava fruits, *D. dorsalis* produced many more progeny than *C. capitata*, in a ratio of about 99:1, no matter which species oviposited first in the fruit (Keiser et al., 1974). Even when species oviposited at opposite ends of guava fruits, *D. dorsalis* produced more progeny than *C. capitata*. The species now seem to coexist by *C. capitata* existing in more temperate zones on the islands and in fruits infrequently attacked by *D. dorsalis*.

There are at least two cases, however, where it appears (or used to appear) that species can exploit the same niche and coexist. Ross (1957) found that

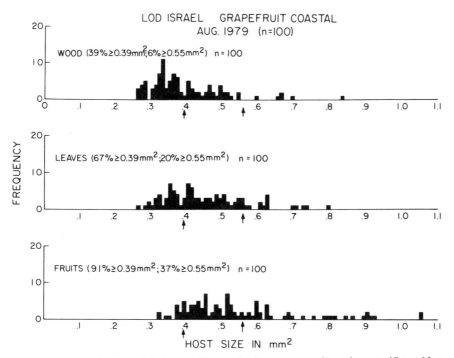

FIGURE 20.4   Size distribution of third-instar California red scale present on wood (top), leaves (middle), and fruits (bottom) on grapefruit trees at Lod, Israel. Arrows beneath the abscissas on each graph indicate the smallest host on which *Aphytis melinus* can produce a female progeny (left arrow) and on which *A. lingnanensis* can produce a female progeny (right arrow). Numbers in parentheses indicate the percentage of scales above the minimum size for *A. melinus* first and *A. lingnanensis* second. From Luck and Podoler (1985); with permission.

six species in the leafhopper genus *Erythroneura* lived together on sycamore leaves in Illinois (*Platanus occidentalis*, or plane tree). Savage (1958) argued that Ross had shown that the leaf hoppers occupied the same habitat but had not demonstrated that the same niche was exploited. In response, Ross (1958) supported his claim, and many authors have used Ross's example in support of their own arguments (e.g., Hairston et al., 1960; Hutchinson, 1965; Ayala, 1970; Edington and Edington, 1972). McClure's studies (1974, 1975) (McClure and Price, 1975, 1976) now indicate that geographical distributions of the species may differ sufficiently to enable coexistence and that niche differences are not significant but habitat differences are, using the distinction made by Whittaker et al. (1973). The species studied are illustrated in Fig. 20.5. Their distributions along a latitudinal gradient, from 43 to 35°N in Illinois, Kentucky, and Tennessee, show that each has a peak of abundance at a unique latitude (Fig. 20.6). Each species is presumably adapted to slightly different climatic conditions, and in these favorable sites they outcompete other species.

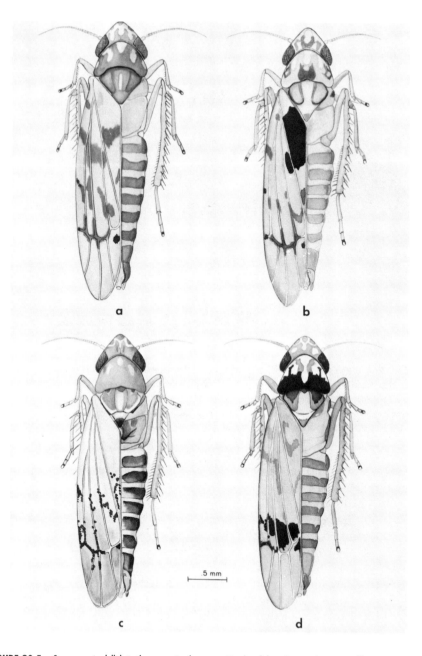

FIGURE 20.5   Sycamore cicadellids in the genus *Erythroneura*. Mature adult males are shown: (a) sibling species — *E. arta, E. ingrata, E. lawsoni, E. torella,* and *E. usitata,* which can be distinguished only on the basis of male genitalia; (b) *E. bella;* (c) *E. hymettana;* (d) *E. morgani.* Drawings by Alice Prickett. Reprinted with permission from McClure (1975) and the Entomological Society of America.

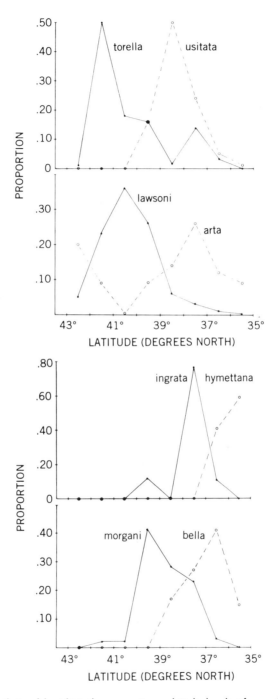

FIGURE 20.6 Distribution of the eight *Erythroneura* species on a latitudinal gradient from northern Illinois (42.5°N) to the Smoky Mountains (35.5°N), showing the proportion of individuals collected per species found in each of the eight sample zones. From M. S. McClure and P. W. Price, Ecotope characteristics of coexisting *Erythroneura* leafhoppers (Homoptera: Cicadellidae) on sycamore. *Ecology,* 1976, **57**:928–940. Copyright © 1976 by the Ecological Society of America. Reprinted with permission.

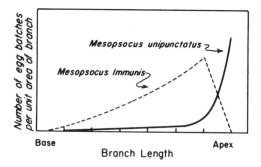

FIGURE 20.7   Abundance of egg batches of *Mesopsocus immunis* and *M. unipunctatus* along the length of a larch branch. After Broadhead and Wapshere (1966).

In northern England, Broadhead (1958) found that two psocid species in the genus *Mesopsocus* coexisted in larch plantations. The species were similar in size, body form, phenology, and food preferences. They occurred on the same twigs and often at high density in almost equal numbers. There was no apparent difference between the ecologies of these species. By 1966, however, Broadhead and Wapshere had found that there was a significant difference between the species, and their paper illustrates the detailed work that enabled them to find this difference. When they checked the density of egg batches along a branch of a larch tree they found that the species distributions differed (Fig. 20.7). In fact, the species were ovipositing in rather different sets of microhabitats. *Mesopsocus immunis* laid most eggs in axils of dwarf side shoots, and *M. unipunctatus* laid most eggs in girdle scars and leaf scars (Fig. 20.8). It appears that the population of each species of *Mesopsocus* was lim-

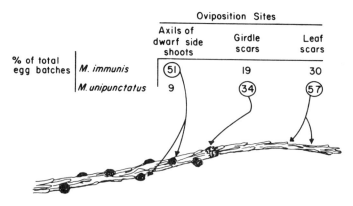

FIGURE 20.8   Difference in oviposition sites between *Mesopsocus immunis* and *M. unipunctatus* on a larch branch. The partial refuges from competition are indicated by the circled numbers and arrows indicating the sites. Leaves are not shown. Data from Broadhead and Wapshere (1966).

ited by the number of oviposition sites that it could utilize on a larch twig. When populations became too high, density-dependent emigration took place, caused by intraspecific competition, and there was no evidence of interspecific competition. Thus these species have attained a stable coexistence by diverging in their oviposition sites, which are in shorter supply that food.

NICHE PARTITIONING

If species have very similar niche exploitation, what modifications or adaptations enable them to coexist? These differences can clearly be very subtle, as in the case of the psocids on larch, and actually there are many evolutionary answers that result in coexistence. These include character displacement, habitat selection, microhabitat differences, temporal differences, and diet. Examples of such **niche partitioning** follow.

Character Displacement

Brown and Wilson (1956) suggested that where two species overlap in their distributions, the differences between them are accentuated in the zone of overlap and are weakened or lost entirely in the parts of the range outside this zone. Morphological, behavioral, and ecological or physiological characters may diverge in this fashion, and Brown and Wilson have proposed the term **character displacement** for divergence in the overlap zone. They site several examples, including the rock nuthatches *Sitta neumayer* and *Sitta tephronota* in Asia (Vaurie, 1951), Darwin's finches, *Geospiza fuliginosa*, and *G. fortis* on the Galápagos Islands (Lack, 1947b), and the ants *Lasius flavus* and *L. nearcticus* in North America (Wilson, 1955).

The ants *L. flavus* and *L. nearcticus* occur most commonly in wooded areas, overlapping in their distributions in the eastern states, but only *L. flavus* occurs in the west. Where the species are sympatric, they differ in at least seven characters: antennae length, ommatidium number, head shape, degree of worker polymorphism, relative lengths of palpal segments, cephalic pubescence, and queen size. In western North America, where *L. nearcticus* is absent, *L. flavus* converges in all characters (Brown and Wilson, 1956). For example, queen-size variation in the two species, as measured by head-length and head-width relationships, is very similar in allopatric colonies but distinct in sympatric colonies (Wilson, 1955) (Fig. 20.9). Brown and Wilson (1956) suggest that this character displacement is related to competition and ecological displacement between the species. They have similar food requirements, but *L. flavus* occurs in open, dry forest with little leaf litter, and *L. nearcticus* colonizes moist, dense forest with thick leaf litter. Unfortunately, habitat preferences for *L. flavus* are not well known in the west and are worthy of study.

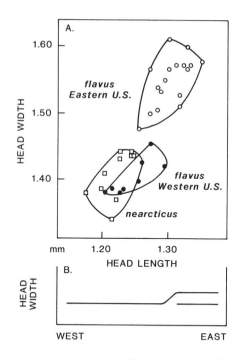

FIGURE 20.9   (A) Relationships between head length and head width in queens of *Lasius nearcticus* in the eastern United States and *L. flavus* when allopatric in the west and sympatric in the east; (B) same relationship with head width of each species plotted on geographical location from west to east. Modified from Wilson (1955).

Other examples of character displacement among the insects suggested by Brown and Wilson (1956) include ants in the genus *Rhytidoponera* in Australia, slave-making *Formica* ants of the Holarctic realm, and perhaps scarabaeid beetles in the genus *Eucanthus* in North America. These cases are in need of closer study now that there is growing discontent with the data presented in support of character displacement and morphological differences between coexisting species.

Null Hypotheses

Several authors have noted that hypotheses alternative to that of evolutionary divergence of morphological characters among coexisting species should be addressed, and these alternatives are usually more parsimonious. P. R. Grant (1972, 1975) has suggested that in the rock nuthatches each species shows clinal variation, and the clines do not change in character in the zone of sympatry—they maintain the same slope as in the allopatric regions. These kinds of doubts about the validity of character displacement are reviewed by Strong et al (1979), who argue that differences between species in a community should not be analyzed with the view that competition is important, but rather, that the **null hypothesis** should be tested—that communities are assembled at random from a source of species available to colonize the particular habitats under study. For the three communities they analyzed in detail, the null hypothesis could not be rejected. Simberloff and Boecklen (1981) also tested the null hypothesis for many communities, including insect communities of parasitic wasps, tiger beetles, aculeate wasps, and bees, and communities of wandering, crab, and orb-weaving spiders. They found that the null hypothesis could not be rejected in the majority of cases, but reasoned that this does not prove the absence of competition but rather, that the cited cases claiming competition's role in producing divergent morphologies do not stand a rigorous statistical test. Whether tests for the null hypothesis are appropriate will be debated into the future (e.g., Grant and Abbott, 1980; Grant, 1984; Simberloff, 1984). The important point is that claims for character displacement and morphological divergence must be substantiated by statistical tests of the null hypothesis; alternative and falsifiable hypotheses should be tested in addition to that on the influence of competition; and ideally, experimental evidence should be provided in support of the conclusions reached (Simberloff and Boecklen, 1981). Similar caution should be used in interpreting data on niche differences, as in the following examples. If we measure enough niche variables, we are bound to find differences between species, but it is very hard to understand the reasons for these differences, and competition should not be invoked as a cause without compelling evidence. This subject will be revisited in Chapter 21.

Habitat

Differences between species in **habitat selection** and utilization can be seen in insects and spiders. The species of *Erythroneura* discussed earlier in this chapter are distributed along a latitudinal gradient, with each species apparently adapted to one point of latitude better than the other species.

Price (1971a) has shown that three species of ichneumonid are able to survive together because of different responses to habitats defined by moisture content in the litter and host density (Fig. 20.10). A similar response was found by Turnbull (1966), in which the lycosid spiders *Pardosa milvina* and *P. sexatilus* showed niche segregation on a moisture gradient.

Microhabitat

Microhabitat differences may also be sufficient to permit coexistence, as seen in *Mesopsocus* species, discussed earlier. O'Neill (1967) found that seven species of diplopod coexist in decaying material in maple–oak forests in central Illinois. However, each was found to be dominant in only one of seven microhabitats: heartwood at center of logs, superficial wood of logs, outer

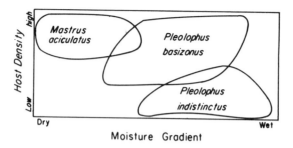

FIGURE 20.10 Zones of dominance occupied by three species of ichneumonid in a space defined by moisture and host density gradients. After Price (1971a).

surface of logs beneath bark, under log but on log surface, under log but on ground surface, within leaves of litter, and beneath litter on ground surface. Two parasitic mites on the damselfly, *Cercion hieroglyphicum*, colonize different parts of the host's body, one on the thorax and another on the abdomen (Mitchell, 1968). Two congeneric flea beetles on crucifers, *Phyllotreta cruciferae* and *P. striolata*, occupy different surfaces of leaves of *Brassica oleracea* (Tahvanainen, 1972) (Fig. 20.11). The preference for shady microhabitats in open sites shown by *P. striolata* correlates nicely with the greater abundance of this species on shaded wild crucifers in meadows and woods. Fitzgerald (1973) found three species of bark-mining moth in the genus *Marmara* (Gracillariidae) on green ash, subdividing the bark resources into three. Other species attacked the main stem at the base of the tree and the root collar. The upper parts of the tree were divided by one species feeding in the periderm and the other feeding in the cortex of the bark. Bark beetles also commonly differ in the part of the tree they most commonly exploit (Fig. 20.12). The sibling species, *Drosophila heedi* and *D. silvarentis*, both use as

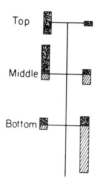

FIGURE 20.11 Percentage distribution of *Phyllotreta cruciferae* (left) and *P. striolata* (right) between top leaves, upper surface (solid bars), and lower surface (hatched bars) of middle and bottom leaves of *Brassica oleracea*. Note the strong dominance of *P. cruciferae* in sunny locations and of *P. striolata* in shaded locations. From Tahvanainen (1972).

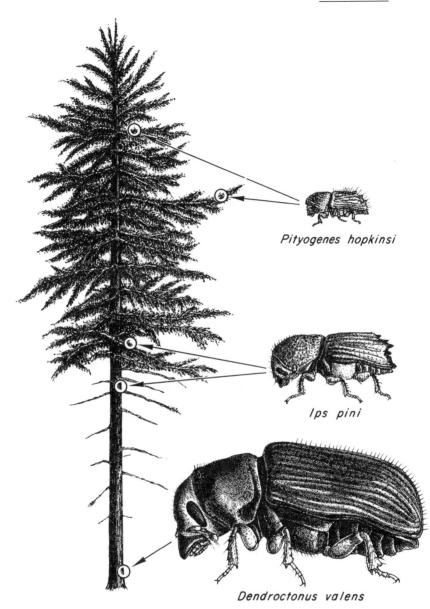

Pityogenes hopkinsi

Ips pini

Dendroctonus valens

FIGURE 20.12   Relationship between size of bark beetles and the area and thickness of bark they attack, on white pine, *Pinus strobus*, in the northeastern United States and eastern Canada. The relative sizes of the beetles are shown. The red terpentine beetle, *Dendroctonus valens*, is restricted to the base of trees. The pine engraver, *Ips pini*, attacks the upper trunk and large branches and is usually much more abundant than *D. valens*. *Pityogenes hopkinsi* enters thin bark on smaller branches, which provide a very large area for attack, and this species is considered to be one of the most abundant bark beetles of the northeast. Drawing by Alice Prickett.

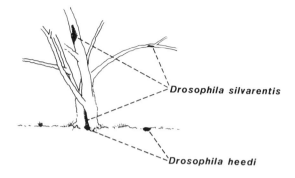

FIGURE 20.13 Niche differences between *Drosophila heedi* and *D. silvarentis* both using sap oozing from the tree, *Myoporum sandwicense*, in Hawaii. From K. Y. Kaneshiro et al., Niche separation in a pair of homosequential *Drosophila* species from the island of Hawaii, *Am. Nat.*, 1973, **107:**766–774. Published by the University of Chicago Press. Copyright © 1973 by the University of Chicago. All rights reserved.

oviposition sites sap oozing from wounds of the tree, *Myoporum sandwicense,* in Hawaii (Kaneshiro et al., 1973). However, *D. heedi* breeds only in flux puddles on the ground, whereas *D. silvarentis* oviposits predominantly on sap fluxes on the tree trunk and branches (Fig. 20.13).

Time of Day or Season

Another way of subdividing resources is by utilizing these resources at **different times.** Linsley et al. (1963b) found that some bees do this in the Great Basin when collecting nectar and pollen from the evening primrose, *Oenothera clavaeformis* (Fig. 11.5). The figure shows that *Andrena rozeni* forages only when the flower is new and *Andrena chylismiae* forages only when the flower is old, and thus they segregate time completely; but *Andrena raveni* forages at both times and thus competes directly with both species. Time may be divided between species on a seasonal basis as well as on a daily basis. Clench (1967) found that coexisting hesperiine butterflies tend to fly at different periods of the year, each species being associated in time with the flowering of a different plant species, so that competition for nectar was effectively eliminated. Istock (1973) collected data suggesting that population peaks of two dominant waterboatman species (Corixidae) in ponds of northern Michigan are segregated in time. Thus coexistence is permitted at least in part by one species breeding early in the season and one breeding late in the season (Fig. 20.14).

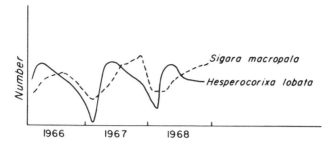

FIGURE 20.14   Population size of two corixid species over 3 years showing tendency for population increase of each species to occur at a different time of the year. After Istock (1973).

Food

Dietary differences between coexisting species may be observed frequently. Two species of nectar-feeding mites, *Rhinoseius colwelli* and *R. richardsoni*, which fight to the death when they meet, exploit mutually exclusive flowering plant species in Costa Rica (Colwell, 1973). Differences in the size and form of the chelicerae of soil-dwelling mites result in use of different particle sizes of food and other factors associated with their diet (Wallwork, 1958). Aquatic insects show a similar partitioning of resources. MacKay and Kalff (1973) have provided one example observed in caddis fly larvae (Trichoptera) in Quebec, Canada. Ninety percent of *Pycnopsyche gentilis* larvae occurred in fallen leaves and 10% in detritus. *Pycnopsyche luculenta* occurred in detritus (50%) or in leaves close to detritus (50%) and fed less on leaves if twigs were available. Thus competition is almost completely eliminated.

Fellows and Heed (1972) found extreme host specificity in larvae of *Drosophila* species inhabiting fermenting cactus stems in the Sonoran Desert. For each cactus species there seemed to be one resident species of *Drosophila* that frequently represented over 90% of the specimens reared from a sample. This specificity is accomplished by well-developed host-plant discrimination by feeding and presumably ovipositing adults, and competitive dominance of the resident species' larvae over larvae of other species. Thus competitive exclusion of invading species may still operate in these situations and selection for increased specificity of the invading species may be under way. However, there are clearly other strategies available to *Drosophila* species on rotting cacti, although narrow niche exploitation seems to be the predominant one.

In spiders the foraging strategy may be different between coexisting species, and this results in different diets. Such a case is described by Kiritani et al. (1972) in rice paddies in Japan. *Lycosa pseudoannulata* is a hunting spider, *Oedothorax insecticeps* and *Enoplognatha japonica* are web spinners that hunt occasionally, and *Tetragnatha* sp. is a typical web-spinning spider. In the period 1968–1970, each species utilized a different percentage of each food type available to it (Table 20.2). Calculating the proportional similarities between all pairs of species shows that diets were from 48 to 77% similar and that there is not a clear-cut segregation of foods between species, although the observed differences are probably an important factor in coexistence. These sorts of results would cause the observer to look more closely at dietary constituents, or to start looking at other dimensions of the niche to find other differences. For example, Kiritani et al. found that *Oedothorax* was most active around noon and thus would catch insects active at this time, whereas the other species were more crepuscular in activity. Also, webs may be placed in different microhabitats, so that although the same food is utilized, different segments of it are exploited. For example, Enders (1974) has shown that the orb-web spiders, *Argiope aurantia* and *A. trifasciata* place their webs at different heights.

The relationships between the ecological niches of organisms and the conceptual basis for niche theory are surveyed by Whittaker and Levin (1975). To what extent niche differences between species are selected for by competition, or other causes, or are the result of more fortuitous processes has been debated (e.g., Connell, 1980; Lawton and Strong, 1981; Schoener, 1982; Price et al., 1984, Strong et al., 1984b; Denno et al., 1995), and the controversy will be discussed again in Chapter 21.

TABLE 20.2   Percent Composition of Various Foods in the Diet of Four Species of Spider. The Proportional Similarity (*PS*) in Diets of Species Pairs are Given.

		Spider Genus			
Food Type		Lycosa (%)	Oedo-thorax (%)	Enoplog-natha (%)	Tetrag-natha (%)
Green rice	Adult	26	4	12	16
leafhopper	Nymph	27	35	28	10
Brown planthopper	Adult	10	5	12	12
	Nymph	15	19	12	6
Spiders		9	16	3	2
Miscellaneous arthropods		13	21	33	54
		100	100	100	100

$$PS = 0.73 \quad PS = 0.75 \quad PS = 0.75$$
$$PS = 0.77$$
$$PS = 0.57$$
$$PS = 0.48$$

SOURCE: Kiritani et al. (1972).

VACANT NICHES AND SPECIES PACKING

Part of the answer to the question of whether niche partitioning results from the coevolution of competitors (e.g., Connell, 1980) or from individualistic species colonization of resources, independent from competition, must lie in the extent to which species are "packed" into communities. Are species tightly packed, like sardines in a can, or is much niche space actually unutilized, like balloons in the air? These analogies are crude because all sardines or all balloons belong to the same "species," but if we accept each sardine as a species, or each balloon, the images are suitable enough.

We therefore enter into the somewhat controversial area involving the identification of vacancy of niches on resource gradients or on arrays of discrete resources. What is a **vacant niche**, after all? Yet the question of whether communities are saturated with species or whether they are unsaturated is central to community ecology, as discussed in Chapters 22 and 23. If the species defines the niche, as Elton (1927) proposed, there can be no vacant niches. However, the Hutchinsonian perspective of the niche, as a property of the environment, provides us with an avenue for exploring species packing in a novel way.

Hutchinson (1957, p. 422) himself contemplated the empty niche in relation to "the problem of the saturation of the biotop." He used the example of corixid waterboatmen, for he knew of 15–20 species in Europe in the subfamily Corixinae, but of only three species in the Nilghiri Hills in southwestern India. "The question raised by cases like this is whether the three Nilghiri Corixinae fill all the available niches which in Europe might support perhaps 15 or 20 species, or whether there are really empty niches. Intuitively one would suppose both alternatives might be partly true, but there is no information on which to form a real judgment. The rapid spread of introduced species often gives evidence of empty niches, but such rapid spread in many instances has taken place in disturbed areas. The problem clearly needs far more systematic study than it has been given." Hutchinson clearly accepted the concept of the empty niche, but was not aware of a strong analytical approach to identifying such niches, although he did emphasize the need for more formal study.

Mayr (1963, p. 87) also saw the empty niche as an unresolved problem. "Why are so many apparently empty niches not filled?" he asked. "There are 27 or 28 species of woodpeckers in Borneo and Sumatra and none in similar forests in the New Guinea region. The important niche filled by woodpeckers in the Oriental region seems to be almost entirely vacant east of Weber's line. The opening of the Suez Canal permitted more than a dozen Red Sea fishes to invade the Mediterranean, apparently without displacing any native species. The invaders must have found partly unoccupied niches. Invading species only rarely displace a native species from its niche. In most cases they occupy at least in part what appears to have been a previously unfilled niche. . . . Whenever a local biota can be enriched this provides to me *de facto* that vacant niches have existed."

Mayr (1963) discusses again the vacant niche in relation to speciation: "There is evidence that there are vacant niches or partly vacant niches even in well-balanced faunas. The spectacular success of faunal transfers illustrates this point . . ." (pp. 573–574). "Abundant empty niches are available in diversified islands when the first colonists arrive and this invites rapid adaptive radiation. . . . We see again and again that an incipient species can complete the process of speciation only if it can find a previously unoccupied niche" (p. 574).

More recently, Strong et al. (1984a, p. 93) said: "Good evidence of 'empty niches' can be found in phytophage communities." Thus, while there are doubts about the existence of empty niches expressed by some (e.g., Colwell and Fuentes, 1975; Herbold and Moyle, 1986; Colwell, 1988), many authors have used the concept to address critical questions in ecology and evolution, as the quotations suggest.

Habitat Template

The view of the habitat as a template on which species live and evolve, developed by Southwood (1977b) in his presidential address to the British Ecological Society, is a crucial perspective with which to enter investigations on the vacant niche. If we can identify with clarity the essential habitat and the essential resources for each species, we can discern if there are empty spaces among them. Then we can ask if such spaces are sufficiently large to permit another species to live there, if that species colonized the habitat, if species are tightly packed enough to compete with each other, or if there is sufficient niche space among them that competition is unlikely.

The most well-defined habitat templates are those for parasites (Price, 1990). The host species provides the habitat and the resources for whichever stage of the parasite is utilizing the host. Related host species provided replicated habitats for parasites in space and time, with small host-species differences, providing a strongly comparative analysis of the extent to which communities of parasites are saturated or unsaturated on each host. For example, herbaceous plants all provide the same set of basic resources which are available to be used by insect herbivores: leaves, stems, roots, flowers, and fruits. A wild carrot plant provides the same resources as a wild parsley plant. Therefore, we can compare directly niche occupation of herbivores on these plants because they offer very similar templates available for colonization.

Niche Space in Parasite Communities

The first formal analysis, which revealed much open space in communities, was made by Lawton and Price (1979), using the principle of the plant as the habitat template, with each species of host plant within a family offering a similar set of resources to colonizing herbivore species. Additional principles adopted were as follows:

1. Such replicated habitats should contain the same number of ecological niches for specialized herbivores such as leaf-mining flies (Agromyzidae) used in the analysis (cf. Table 9.1—77% of agromyzid species attack only one or two host-plant species).
2. Therefore, species of host plants with similar geographic ranges should support approximately equally rich leaf-mining insect faunas.
3. We can estimate what a saturated community may be like in two ways:
   (a) By taking the plant species with the richest agromyzid fauna and the largest geographical range, we can argue that this is a conservative estimate of the richness that could be achieved, with enough evolutionary time, on all plant species with an equivalent geographic range.
   (b) Perhaps a more accurate estimate is to take each plant part, find the largest number of species supported by that part on one host-plant species and then add up the richest assemblages on each plant part. After all, we know that several species can coexist on one plant part empirically, so presumably, more than one ecological niche exists. This total number of herbivore species would provide a higher estimate of what a saturated community may contain.
4. Thus we develop expectations based only on what species of herbivores actually do in nature—only on current components of species richness. We do not extrapolate beyond what empirical studies clearly reveal.
5. Once we focus on a particular herbivore taxon, we can consider special cases of plant species that may reduce the probability of herbivore colonization and check for these effects. For example, most agromyzid flies pupariate in the soil, so plants in aquatic habitats are hard to colonize. Most agromyzids mine leaves, so finely dissected leaves may be more difficult to utilize. These effects can be tested for.
6. Other effects can be measured, such as the number of species of the most likely competitors on each host plant, such as leaf-mining Lepidoptera, or the number of parasitoid species associated with each plant species, to estimate other factors producing "resistance" associated with plant species habitats to colonization by herbivores (see Lawton and Price, 1979, for details).

With these general approaches the published records of agromyzid flies on the British Umbelliferae (Apiaceae) were analyzed. The central result was that there was a species–area relationship, but a weak one, accounting for only 32% of the variance of agromyzid species related to plant geographic area (Fig. 20.15). We should note in the figure the following points:

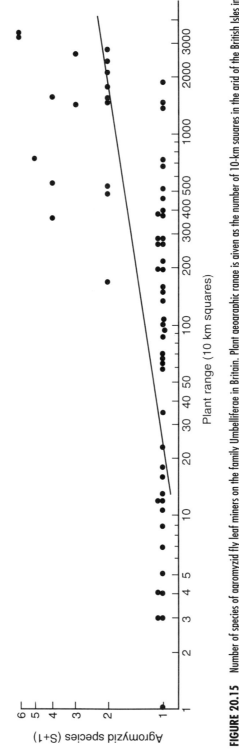

FIGURE 20.15 Number of species of agromyzid fly leaf miners on the family Umbelliferae in Britain. Plant geographic range is given as the number of 10-km squares in the grid of the British Isles in which the plant species is recorded. From Lawton and Price (1979).

1. Plants with less than 160 10-km squares of geographic range did not support any agromyzid species. Whether vacant niches are available on these plant species is hard to determine unless there are species in Europe that could be introduced.

2. Many plant species over 160 10-km squares in distribution have no agromyzid species—17 species in all—even though in this geographic range many plant species do support agromyzids. In fact, in this geographic range there are as many plant species without agromyzids as those with one or more species (17 vs. 17 species). This suggests that there are probably more empty niches than occupied niches in these assemblages.

3. Many common plant species support only one agromyzid species, while others of similar geographic range support four or five times as many. Again, this reveals much empty niche space.

4. Even the richest assemblages on *Heracleum* and *Angelica* are undersaturated relative to the number of species on herbaceous plants outside the Umbelliferae—*Ranunculus, Senecio,* and *Taraxacum*—each with eight species. Eight species per plant is one estimate of a saturated community (3a above), and all umbellifers fall below this richness.

5. Another estimate of what a saturated community may consist of is derived from the addition of plant parts with the highest species richness per plant (3b above). Across all herbaceous plant species in the British Isles we can add these components: 7 species of leaf blade miners on one plant species, 4 leaf midrib miners, 3 stem borers, and 3 stem miners, for a total of 17 agromyzid species. Thus 17 species of agromyzid may constitute a reasonable estimate of a saturated community, a richness two times higher than that seen in any plant species in the British Isles.

6. Tests of species richness of competitors in the leaf-mining Lepidoptera and richness of parasitoids did not help to explain this apparent undersaturation of communities.

Lawton and Price (1979) concluded that there is much available niche space for new species of agromyzids to colonize. Assemblages on the majority of plants are undersaturated, and perhaps all are. We can reasonably estimate what the high and low ends of species richness might be in saturated communities. Vacant niches can be shown to exist and can be quantified in an objective manner.

Even more convincing are the studies that evaluate whole insect assemblages on one plant species that grows over a wide geographic range. Lawton and colleagues (1982, 1984b), Lawton et al. (1993), and Martins et al. (1995) have sampled insect herbivores on bracken fern (*Pteridium aquilinum*) in South Africa, New Guinea, the United States, Australia, England, and Brazil (Fig. 20.16). Each assemblage is plotted on a guild matrix, with each guild of chewers, suckers, miners, and gallers, associated with each resource,

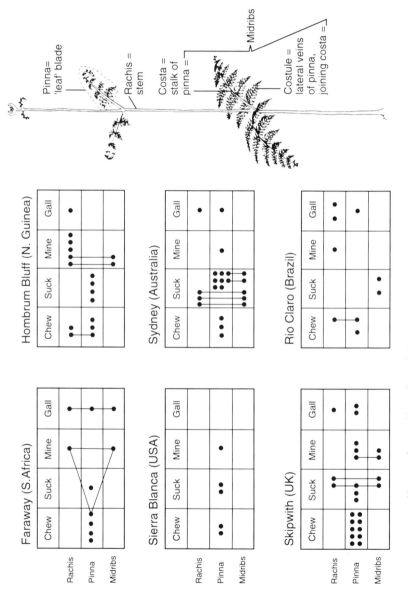

FIGURE 20.16 Guild matrices for assemblages of insect herbivores (plus a mite species in South Africa) on bracken fern in six geographical regions. The number of species in each guild is indicated by solid circles. Lines join a species that exploits more than one part of the plant. From J. H. Lawton et al. in R. E. Ricklefs and D. Schluter (eds.), *Species diversity in ecological communities*. Published by the University of Chicago Press. Copyright © 1993 by the University of Chicago. All rights reserved. A bracken fern frond is illustrated on the right with plant parts noted. From Strong et al. (1984a).

rachis, pinna, and midribs. Within each guild the number of species present is marked with a closed circle. Lines join circles if a species utilizes more than one resource or if it uses two modes of exploitation, such as chewing and mining. No two assemblages are remotely alike. As bracken fern has spread around the world, a distinct herbivore fauna has developed from independent sources in each locality.

Note that the richness of species in each assemblage on bracken fern differs from a high of 21 species at Shipwith Common (U.K.), to 17 species near Sydney, Australia, to 14 species at Hombrum Bluff, New Guinea, down to only 5 species on the isolated peak of Sierra Blanca, New Mexico. The richness of the subtropical assemblage at Rio Claro, Brazil, is much lower than that in northern England at Shipwith Common! And look at all the empty niches!

At Shipwith, 10 pinna chewers coexist, but only two in Rio Claro. Near Sydney eight pinna suckers coexist, but none at Rio Claro. If we take the richest guild from all locations and add them all to obtain a maximum possible assemblage size (e.g., 10 pinna chewers at Shipwith, + 8 pinna suckers near Sydney, + 5 rachis miners at Hornbrum Bluff, etc.), the total assemblage would be 36 coexisting species! Even the richest assemblage of 21 species has many empty niches, and most have over 50% of niches empty, according to this approach. "Local communities do not become saturated with species" (Lawton et al., 1993, p. 180).

Another case of a widely distributed plant concerns the soybean *(Glycine max)*, which has been grown agriculturally in many parts of the world. New insect assemblages have formed at each geographic locality. Turnipseed and Kogan (1976) summarized the niche occupation of insect herbivores in three regions, North America, Central and South America, and the Orient, where the plant originally grew. Strong et al. (1984a) created a matrix of localities and soybean resources, with the number of herbivore species in the top left of each box and the number of vacant niches in the bottom right of each box (Fig. 20.17).

For every plant part listed in Fig. 20.17, and in every region, vacant niches were available, even in the place of origin of the plant species and even though very large areas were sampled. Of course, in any one local community there would be many more unoccupied sites and much larger deficiencies below levels expected in saturated communities.

In whichever way we look at insect herbivore assemblages, using plant species as the habitat template, the evidence is that many vacant niches exist and assemblage size is well below community saturation. Whether these patterns result from slow speciation rates of herbivores, or high extinction rates, we do not know. The surprising fact seems to be that despite the great diversity of herbivorous insects, the capacity of plant assemblages to support them is much greater than is now utilized. This raises the interesting question of how important competition is likely to be in structuring communities, or

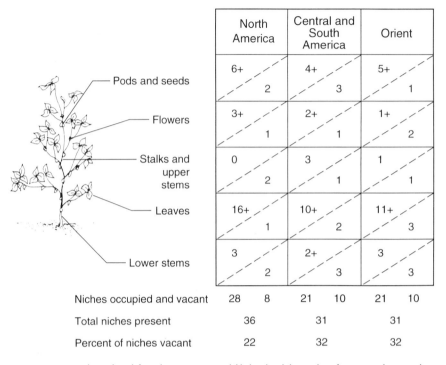

	North America		Central and South America		Orient	
Pods and seeds	6+	2	4+	3	5+	1
Flowers	3+	1	2+	1	1+	2
Stalks and upper stems	0	2	3	1	1	1
Leaves	16+	1	10+	2	11+	3
Lower stems	3	2	2+	3	3	3
Niches occupied and vacant	28	8	21	10	21	10
Total niches present	36		31		31	
Percent of niches vacant	22		32		32	

FIGURE 20.17   Soybean plant (left) with its resources available listed and the number of species exploiting each resource (top left in each box) in North America, Central and South America, and the Orient. A + indicates a minimum estimate. In the lower right of each box is the number of vacant niches estimated. Below the figure the total number of niches occupied and vacant is given, the total number of niches present, and the percent of niches that are vacant. From Strong et al. (1984a).

whether insect species are assembled on resources without much interaction among them. These topics are investigated further in Chapters 21 and 22.

DETECTING SATURATION OF ASSEMBLAGES

The foregoing discussion raises an interesting question of how we detect whether an assemblage or a community is saturated or unsaturated. What is a saturated community? Is it simply full of all the species available to colonize the habitat template, even though many vacant niches remain available? Or is it a community that is tightly packed with every niche occupied? Of course, we are now deeply into community ecology, discussed in Chapter 22, but it is the ecological niche, its identity, and the number available in the habitat template that is basic.

Terborgh and Faaborg (1980) developed a simple technique to test for saturation in assemblages of organisms. Cornell (1985, 1993) applied their approach to the analysis of assemblages of cynipid gall wasps in California, and this specific example will be used to explain the test and the result. Cynipid species were censused locally on seven oak species, with local samples taken over the geographic range of each oak species. Thus estimates of local species richness were obtained—α diversity—and the total number of species over a geographic area was estimated—γ diversity (Whittaker, 1972). Cynipids are highly specific to individual host-plant species and their galls are relatively conspicuous and easy to census. Thus each host-plant species acted as an island available for colonization by these gall wasps.

The expectation for a saturated community would be that as γ diversity increases on oak species with larger and larger geographic ranges, and richness increases, at some point the α diversity stops increasing and remains at an asymptote (Fig. 20.18). The α diversity departs from the maximum possible number of species defined by the γ diversity because all niche space is utilized at the asymptote and additional species cannot colonize the local habitat template. If the assemblages are unsaturated, we would expect richness to increase in a linear manner, because vacant niches remained in the habitat template of each oak species (Fig. 20.18).

The results Cornell (1985) obtained clearly indicated unsaturated assemblages of cynipids on oaks (Table 20.3, Fig. 20.18). Richness increased in a linear manner with increasing oak species geographical range, and no asymptote or saturation level of species per local habitat was observed. Presumably, many vacant niches remain in each assemblage of cynipids on oaks, but using this approach, we cannot tell how many.

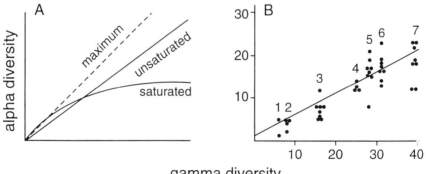

FIGURE 20.18 Hypothetical patterns (A) and the actual pattern (B) for saturated and unsaturated communities when α diversity is plotted against γ diversity. The maximum richness indicates that α and γ diversities are equal. In (B), each point represents a sampled site and each species of oak is given in Table 20.3. The linear regression accounts for 73% of variance in α diversity accounted for by γ diversity. Slightly modified from H. V. Cornell, Species assemblages of cynipid gall wasps are not saturated, *Am. Nat.*, 1985, **126**:565–569. Published by the University of Chicago Press. Copyright © 1985 by the University of Chicago. All rights reserved.

TABLE 20.3   Data from Cornell's (1985) Study on Seven Oak Species and the Cynipid Gall Wasp Assemblages on these Oaks. The Code Numbers for Each Oak Species Appear on Fig. 20.18.

Code	Oak Species (Quercus)	Geographic Range (mi²)	Number of Sites Sampled	Cynipid Alpha Diversity (mean ± SE)	Gamma Diversity
1	Q. dunnii	3,700	2	3 ± 2	6
2	Q. tomentella	500	4	4 ± 0.71	7
3	Q. agrifolia	15,500	8	7.38 ± 0.80	16
4	Q. durata	2,500	4	12.75 ± 0.48	25
5	Q. lobata	33,600	8	16 ± 1.35	29
6	Q. douglasii	21,600	8	17 ± 1.10	30
7	Q. chrysolepis	26,600	8	18 ± 1.43	39

NOTE: Range, number of sites sampled, mean cynipid α diversity, and γ diversity for the seven California oak species censused. Gamma and α diversity are given as species richness.
SOURCE: Cornell (1985).

Clearly, the identification of the frequency of vacant niches in assemblages, and unsaturated communities, bears directly on the subjects of the next two chapters: competition in communities (Chapter 21) and community organization (Chapter 22).

Intraspecific and Interspecific Competition

LIMITING FACTORS

If the probability is accepted that many populations increase until they reach the carrying capacity of the environment, the inevitable result is that the resource in shortest supply becomes the **limiting factor.** Liebig in 1840 expressed this in what is now called *Liebig's law of the minimum* and expressed in modern terms (Odum, 1971), "Under 'steady-state' conditions the essential material available in amounts most closely approaching the critical minimum needed, will tend to be the limiting one. This law is less applicable under 'transient-state' conditions when the amounts, and hence the effects, of many constituents are rapidly changing." Actually, excesses may also become limiting, such as excessive moisture for terrestrial organisms or excessive lead or copper in the soil for plants, and it is the tolerance of organisms to a range of concentrations of a resource that is critical. In 1913, Shelford expressed this in what is now called *Shelford's law of tolerance,* which states: "Absence or failure of an organism can be controlled by the qualitative or quantitative de-

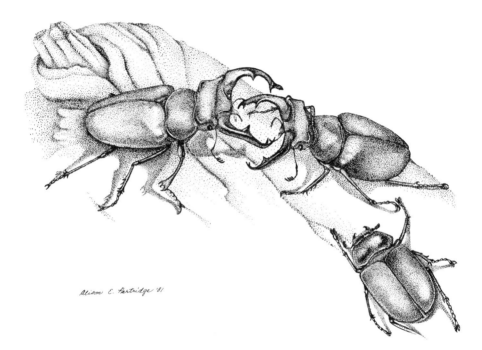

Competition for resources other than mates is discussed in this chapter, but the two male stag beetles (*Lucanus cervus*) fighting for a mate (lower right) serve as a reminder that sexual selection is a potent force in nature, as discussed in Chapter 15. Drawing by Alison Partridge.

ficiency or excess with respect to any one of several factors which may approach the limits of tolerance for that organism" (Odum, 1971).

These two laws actually relate to individual organisms and the survival of an individual in a given set of conditions, independent of others in the same niche. They relate to physiological limits to existence. Clearly, limiting factors that operate because they are in excess will not be influenced by the number of organisms present. But limiting factors, such as food and shelter, that are minimal will become more and more limiting for survival the more individuals there are making an active demand on this resource. Of course, physiological limits such as temperature cannot be influenced directly by competition. The resource that is most limiting will be competed for by organisms in a given habitat, and this competition will be most severe between organisms that have the most similar demands—individuals within the same species that are involved in intraspecific competition. The difference between the exponential growth curve and the logistic curve may be due solely to intraspecific competition for the resource that determines the carrying capacity (the exceptions are in microbial cultures, where the growth rate falls off because of intolerance to accumulating metabolites, such as alcohol or other toxins).

PHENOTYPIC DIVERSITY

Density-dependent factors often involve intraspecific competition for food as seen in Chapter 17, so that any adaptation that will reduce this competition will be strongly selected for. Individuals that are most different from the majority of the population will suffer least from competition because their requirements will be slightly different and not sought after by so many individuals. Therefore, intraspecific competition will select for a diverse array of qualities within a population, resulting in minimized competition. Phenotypic divergence between individuals will be held at a maximum leading to a broad-based exploitation pattern on the part of the population (Fig. 21.1a). An obvious evolutionary solution that reduces intraspecific competition by broadening the exploitation pattern is for polymorphisms to develop in the population. An example is the sexual dimorphism, both morphological and behavioral, seen in many birds and insects (Fig. 21.1b). Here polymorphism results in two or more adaptive modes in a population. However, much sexual dimorphism in insects is probably related more to sexual selection (see Chapter 15) involving competition for mates rather than competition for other limiting resources.

DEVELOPMENTAL POLYMORPHISM

Perhaps the ultimate strategy in avoiding intraspecific competition has been achieved in the developmental polymorphism in holometabolous insects (those insects that have a radically different morphology and ecology in the

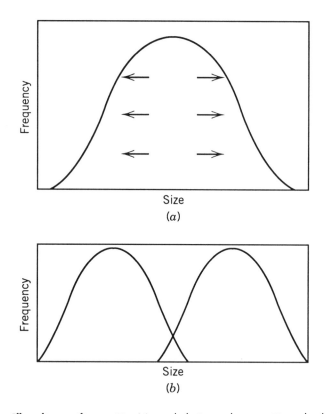

FIGURE 21.1   Effects of intraspecific competition: (a) natural selection to reduce competition tends to broaden the frequency distribution of any character involved in the competitive interaction; (b) dimorphism in characters that results in a broader exploitation pattern in the species.

larval and adult stages). Since there are about 750,000 species of named insects and 85% of these are holometabolous, there are many examples to choose from. A larval flea feeds in the debris of the host nest and the adult feeds on the host itself (for details, see Chapter 9). A larval fly may mine the mesophyll tissue of a leaf and the adult feeds on sugary solutions such as honeydew or nectar. In hemimetabolous insects (those in which the morphology and the ecology of the nymphs are very similar to those of the adult), the immature stages are very vulnerable. They usually feed in exposed situations just as the adults, yet they do not have wings that permit them rapid escape. They are also constrained from evolving into forms that would permit better concealment, because they must remain very similar to the adult. This is perhaps the reason for the commonness of hopping in hemimetabolous species, as this is the best escape method for exposed wingless nymphs and for the overwhelming diversity of defensive secretions produced by members of these orders.

As soon as the larval morphology and ecology can diverge from that of the adult, because of holometabolous development, an obvious adaptive strategy is to diverge as far as possible away from the adult, into concealed, equable, lucrative subniches. Intraspecific competition is completely subdued between parent and offspring, and automatically higher population levels can be maintained on a given resource. Also, the adult must adapt to provisioning its progeny with food, so that the larva adapts into a very efficient eating machine—an energy transformer. This is Darwin's (1859) **principle of divergence** operating only within species, so that more individuals per species can coexist per unit area because of this divergence. The success of the aculeate Hymenoptera really hinges on this provisioning ability, which varies from laying an egg on or in a host in the parasitic Hymenoptera and the digger wasps, right up to the highly developed provisioning in the social insects, wasps, bees, and ants, where food is brought into a central nursery. The larva is a sedentary metabolic factory and energy store, while the adult is a highly mobile and sensitive transport system, both maintaining the present factory and setting up new ones. Thus both forms can be exquisitely adapted to their individual functions and achieve a tremendous efficiency in the utilization of resources (See Wilson, 1971, 1975; and Hölldobler and Wilson, 1990, for accounts of division of labor in social insects.)

CONTEST COMPETITION

Basically, there are two types of intraspecific competition. Nicholson (1954b) made a clear distinction between them and fully realized their population consequences. He gave the name **contest competition** to those situations where the winner in the competition "obtains as much of the governing requisite as it needs for survival and reproduction" and the loser "relinquishes the requisite to its successful competitors." Nicholson used the example of plants competing for sunlit space, where, as the plants grow, fewer and fewer survive and the surviving plants get all the sunlight they need. But now that more is known about plant competition, this is not really a clear-cut contest. A much better example is that of territoriality of any sort, where as long as the defender is successful, it has an exclusive resource on which to feed. Territories of birds, at least, are usually adjusted in size according to food abundance, so that each territory owner has enough food for maintenance and reproduction. Where breeding sites are limited, as with cliff-nesting birds, a young bird must contest the right to maintain that breeding site either by direct aggression or by intimidation displays. But there is no doubt that contest is involved. In fact, in any breeding population the most favorable sites are often limited, so that contest ability becomes an important part of the social structure of the population. (See Chapter 15 for a discussion of territoriality in insects.)

In nonterritorial animals, such as the gallinaceous birds and the large ungulates, a social hierarchy or peck order is present, and this also is a manifestation of contest competition, where the dominant individuals get all they need and the subordinates take what is left, if there is any. In *Polistes* wasps a dominance hierarchy of a truncated form appears to exist among the foundress females on a single nest, where the queen gains her state by dominating other reproductives, and, should she be killed or die, a subordinate female may gain this dominance (Evans and Eberhard, 1970).

The population result of contest competition is that the population increases in an area until the carrying capacity is reached and because each organism has enough to live on the population can maintain this level indefinitely (Fig. 21.2a). The term **interference competition** is more commonly used now than is *contest competition* and is applied to both intraspecific and interspecific competition. It is defined by R. S. Miller (1967) as "any activity which either directly or indirectly limits a competitor's access to a necessary resource or requirement." Interference competition is discussed in experiments with *Drosophila* species, for example, in which although food is abundant, one species is outcompeted by "conditioning" of the medium, presumably by waste products released into the medium (see Budnik and Brncic, 1974; Gilpin, 1974). Pheromone trails of parasitoids would be another example, as would many other examples of allelopathy. Strictly speaking, however, we should retain the term *contest competition* because interference does not ensure that the winner gains a sufficient supply of food—it merely increases its chance of getting sufficient food. Thus contest competition is a refined type of interference competition that does not cover the range of possibilities in interference competition.

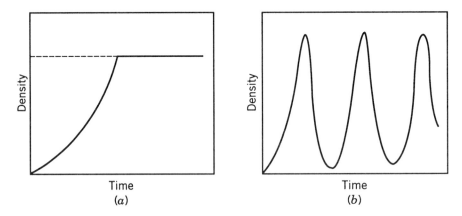

FIGURE 21.2   Population trends in (a) contest competition and (b) scramble competition.

SCRAMBLE COMPETITION

The other sort of intraspecific competition that Nicholson (1954b) distinguished is **scramble competition.** Here all members of a population have equal access to the limited resource and a free-for-all results. Nicholson pointed out that the characteristic of scramble competition is that success is commonly incomplete, so that some, and at times all, of the requisite secured by the competing animals takes no part in sustaining the population. The energy available is dissipated by individuals that obtain insufficient energy for survival. Nicholson studied the effect of scramble competition on laboratory populations of the sheep blowfly, *Lucilia cuprina,* because these were a serious threat to the sheep industry in Australia. When few larvae fed on 1 g of homogenized bullock's brain, production of adults was high, and the highest number of adults (16.5) was produced when 30 larvae were placed on the bullock's brain (Fig. 21.3). Above this density, production of adults fell off rapidly until none were produced if 200 or more larvae were placed on the medium. At around the 180 larvae level, very few adults were produced; these would lay few eggs on the 1 g of medium, and the progeny would be successful and produce a large brood of adults again. If this interaction is viewed in time, with the amount of food kept constant in time, a series of peak populations will result from this scramble type of competition, in contrast to the population pattern resulting from contest competition (Fig. 21.2b). Thus the best way to control sheep blowfly on sheep carcasses is to let nature take its course and allow females to lay enough eggs

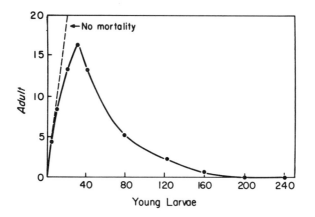

FIGURE 21.3   Effect of competition for food among larvae of *Lucilia cuprina* upon the number of adults produced. Numbers are expressed per 1 g of homogenized bullock's brain. After Nicholson (1954b).

to overpopulate the carcass so that few or none survive. Then there are fewer adults to lay eggs on sores on living sheep, where the larvae do the real damage. But this scrambling for food can be adaptive, and in some circumstances an advantage over territorial partitioning of available resources. This is so when resources are locally superabundant relative to the requirements of a breeding pair, just as a carcass is to breeding blowflies under natural conditions.

The term more commonly used now for scramble competition is **exploitation competition** or competition for a resource once access to it has been achieved and maintained. In the case of competing *Drosophila* populations both exploitation competition and interference competition seem to be involved (Budnik and Brncic, 1974; Gilpin, 1974). When species compete, the same forms of competition are apparent. Gill (1974) has argued that the results of pure exploitation competition between species can be predicted from knowledge of the population parameters r and K of the species in monoculture. However when interference competition is evident, a third parameter, α, which is an estimate of the amount by which the species impairs the reproductive rate of a competitor, must be involved. Thus natural selection may work to promote fitness in any of three ways—through selection for increased r, increased K, or increased α.

INTERSPECIFIC COMPETITION

The study of interspecific competition has a longer history than that of intraspecific competition. Gause's (1934) classic studies on interspecific competition between *Paramecium caudatum* and *Paramecium aurelia* are worth examination. Both species feed on the same food; in Gause's experiments this was the pathogenic bacterium *Bacillus pyocyaneus*. Since *P. aurelia* was much smaller than *P. caudatum*, to make a direct comparison between the species, Gause expressed the populations as "volumes." The volume was actually the number of *P. caudatum* unaltered, and the number of *P. aurelia* multiplied by a factor of 0.39. He placed 20 individuals of a species in a 10-mL tube with a constant bacterial food source, and monitored the population growth. He also placed 20 of each species in a single microcosm to see how the populations interacted. It is clear from Fig. 21.4 that *P. aurelia* had a much stronger population-depressing effect on *P. caudatum* than *P. caudatum* had on *P. aurelia*. Thus *P. caudatum* was pushed to extinction.

COMPETITION EQUATIONS

This difference in the power of interaction had already been incorporated into the population growth formula by Lotka (1924), Volterra (1926), and Gause (1934) on theoretical grounds. The equations do not permit any steady

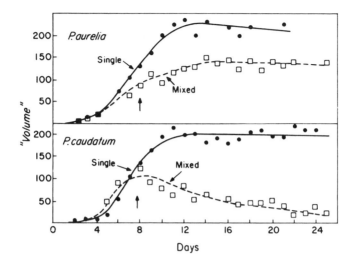

FIGURE 21.4 Growth of single and mixed populations of *Paramecium aurelia* and *P. caudatum* in the competition experiments conducted by Gause. "Volume" refers to the actual numbers of *P. caudatum* and the numbers of *P. aurelia* multiplied by a factor of 0.39. After Gause (1934).

state to develop except in the remotely likely case where the population-depressing effects of one species on the other is the same for both species, or more realistically, where a density-dependent intraspecific competition halts growth before the competitor is eliminated. The rate of growth of the first species in a mixed population, dN_1/dt, is equal to the potential increase of the population (r_1N_1) multiplied by the degree to which the carrying capacity is already filled, or the number of still-vacant places $[K_1 - (N_1 + N_2)/K_1]$. But perhaps N_2 consumes more food than N_1 so that N_2 will actually reduce the growth of N_1 more strongly than mere numbers can indicate. A factor must be introduced to represent this interaction term, and the equations for population growth of two competing species become

$$\frac{dN_1}{dt} = r_1N_1\frac{K_1 - (N_1 + \alpha N_2)}{K_1}$$

$$\frac{dN_2}{dt} = r_2N_2\frac{K_2 - (N_2 + \beta N_1)}{K_2}$$

where α and β are coefficients of competition and are proportional to the population-depressing effect of one species on the other. Gause and Witt (1935) point out that if an individual of species N_2 eats twice as much food as does an individual of species N_1 per unit time, it will restrict the growth of species 1 twice more than species 1 will depress its own growth. Then α will

be 2. But the reverse action of the first species on the second will be twice fee-
bler than the action of the second species on itself ($\beta = \frac{1}{2}$). If $\alpha = 2$, the carrying capacity (i.e., number of individuals that can survive on a given amount
of food) of K_2 must be less than K_1 because species 2 needs more food per individual than does species 1. Therefore, the species with the greater carrying
capacity, species 1, will survive, while species 2 will go to extinction, independent of the initial concentrations of the species. Thus if populations of species
1 are plotted against populations of species 2, the outcome can be predicted if
K, α, and β are known (Fig. 21.5a). When these conditions hold, as populations increase to beyond the carrying capacity, K_2, species 2 will always be
pushed toward extinction. Of course, the opposite set of conditions leads to
extinction of species 1, and where α and β are large, either species can win,
depending on the initial combination of species (Fig. 21.5b). A stable equilibrium between the two species can exist if population growth is reduced by
density-dependent intraspecific competition that halts growth before the competitor is eliminated (i.e., results of competition are frequency dependent)
(Fig. 21.5c). In general, species 1 wins when $\alpha < K_1/K_2$ and $\beta > K_2/K_1$, and
species 2 wins when $\alpha > K_1/K_2$ and $\beta < K_2/K_1$.

Philip (1955) provides a good example of the essential difference between
successful and unsuccessful competitors. Here although species 2 has a higher
fecundity than species 1, it is sufficient to offset the only slightly greater tolerance of species 1 to low levels of food supply. Philip's example looks very
similar to the actual population changes in Gause's *Paramecium caudatum–
P. aurelia* experiments. Since the larger *P. caudatum* is very likely to suffer
from food shortage earlier than *P. aurelia,* the Lotka, Volterra, and Gause
competition model may well describe this interaction quite accurately. As
might be expected, Andrewartha and Birch (1953, 1954), whose views were
discussed in Chapter 17, have criticized the Lotka–Volterra equations, but the
reader should also consult Philip (1955), who has defended them.

COMPETITION BETWEEN *TRIBOLIUM* SPECIES

Even in the *Paramecium caudatum–P. aurelia* situation, where *P. aurelia* always wins, there are complex interactions that are not directly related to
competition for food. Gause's (1934) book *The Struggle for Existence* should
be consulted for details. Some of the complicating factors in interspecific
competition can be seen by looking at Park's work on the flour beetles, *Tribolium confusum* and *T. castaneum.* Chapman had only introduced *Tribolium* as a laboratory animal in 1928 (Chapman, 1928) and Park at the
University of Chicago was already publishing papers on *Tribolium* in 1932.

Park (1948) showed that in *Tribolium* there is rarely a clear-cut winner
in competition. Given a certain set of conditions, if many replicates are
run, a certain probability in the outcome can be expressed, but for any single

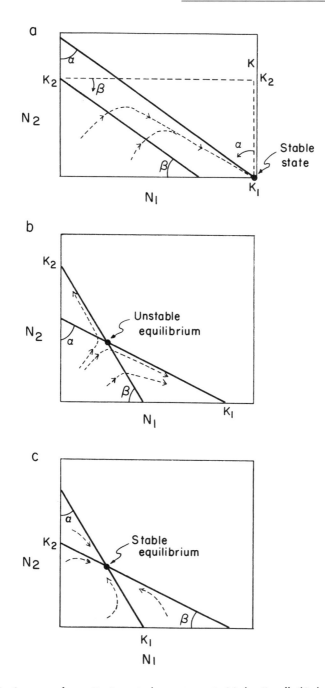

FIGURE 21.5 Some types of competitive interaction between two species (a) when $K_1 > K_2$, (b) where α and β are both large, and (c) where density-dependent intraspecific competition prevents growth of each species beyond the equilibrium point, so that stability is reached.

replicate, predictability of the result is low. Park kept all populations in dark incubators at a constant temperature of 29.5°C and at a relative humidity of 60–75%. As long as the flour medium in which beetles lived was changed frequently, populations of individual species would last indefinitely at a fairly stable density, *T. confusum* being more dense at equilibrium than *T. castaneum* (Fig. 21.6a).

When both species were reared together, both species never survived. Out of 74 trials, *T. confusum* became extinct eight times and *T. castaneum* 66 times. It appears that the species with the higher carrying capacity is most likely to win in competition. But this victory is achieved because of the interaction of another organism. The extinction curve of *T. castaneum* typically

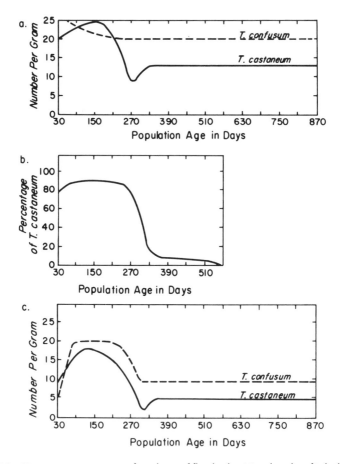

FIGURE 21.6 Diagrammatic representation of populations of flour beetles: (a) total number of individuals per gram of medium in individual cultures, with *T. confusum* maintaining a higher population than *T. castaneum*; (b) extinction curve of *T. castaneum* in the presence of *T. confusum*; (c) total number of adults per gram of medium in individual cultures in the presence of *Adelina tribolii*. After Park (1948).

shows a dramatic drop in numbers around the 300th day (Fig. 21.6b). Even in single culture, *T. castaneum* suffers a large mortality around the 300th day (Fig. 21.6a). This is actually caused by a pathogenic coccidian parasite, *Adelina tribolii*. In single culture *T. castaneum* recovers, but in the presence of *T. confusum*, *T. castaneum* is unable to recover and goes extinct soon after the 300th day. On those occasions when *Adelina* does not cause a severe drop in numbers, *T. castaneum* causes *T. confusum* to go extinct, that is, in 8 times out of 74. Indeed, in *Adelina*-free cultures, *T. castaneum* won in competition 12 out of 18 times.

This is not a trivial finding. Here, a third party, an endemic pathogen in this case, alters the competitive balance between two organisms because of its differential impact. There are probably abundant cases where a similar situation exists, but it has not been demonstrated in many. There is a great need for the study of such interactions. For example, how much influence do herbivores have on the competitive balances in plant communities? Really, the whole process of succession of plants and animals, at least after the first stage of colonization, depends on an independent organism creating a situation that favors one species over another. Also, when a plant evolves to reduce herbivore pressure, how does this alter the competitive balance between two herbivores exploiting this plant? We would do well to consider the third-party effect on competition more often and in much more detail than we have to date (see Price, 1980, and Price et al., 1986, for further discussion).

When we realize that *Adelina* is an endemic disease and that *T. confusum* wins 89% of the competitive encounters when this is present, it seems that *T. castaneum* has a poor chance of surviving as a species if this is a general phenomenon. Perhaps *T. castaneum* is a fugitive species, constantly colonizing food that has not been discovered by *T. confusum*. Before accepting this view we should, as Park did, examine their relative competitive prowess under different conditions to see if the balance is changed. Perhaps in the 1948 experiments conditions were optimal for *T. confusum* and marginal for *T. castaneum*. Park (1954a,b) set up a factorial design experiment of *Adelina*-free cultures with three temperatures, 24, 29, and 34°C, and two humidities, 30 and 70% relative humidity, with replicates in the six treatments. He obtained the results given in Table 21.1. There are indeed situations in which *T. castaneum* wins quite regularly. These are in moist conditions at moderate and high temperatures, and thus *T. castaneum* does have a refuge from the superior competitive ability of *T. confusum* found in the 1948 experiments. Their niches defined by two resources differ (Fig. 21.7). *Tribolium confusum* is dominant in one part of the available resources and *T. castaneum* in another. If conditions change in time, for example, from temperate and moist to cold and dry, and if this change is reversed frequently enough so that there is not enough time for one species to be driven to extinction, the two species may coexist indefinitely. As Hutchinson (1953) pointed out, this may be the case for many pairs of species, particularly for fairly short generation species, such as many insects.

TABLE 21.1 Conditions and Results of Experiments Conducted by Park (1954a,b) on Competition between *Tribolium confusum* and *T. castaneum*. The Prediction Column Indicates Those Results that could be Predicted from a Comparison of the Carrying Capacities of the Two Species in Single Species Cultures—See Text to Follow.

Treatment	Conditions	Temperature (°C)	Relative humidity (%)	Result: Percent of Replicates Won	Prediction
I	Hot and moist	34	70	*T. castaneum* wins 100	X
II	Hot and dry	34	30	*T. confusum* wins 90	✓
III	Temperate and moist	29	70	*T. castaneum* wins 86	✓
IV	Temperate and dry	29	30	*T. confusum* wins 87	✓
V	Cold and moist	24	70	*T. confusum* wins 69	X
VI	Cold and dry	24	30	*T. confusum* wins 100	✓

It would certainly be gratifying if we could predict the outcome of competition between these species by looking at individual characteristics. Understanding of the process of competition would be advanced if characteristics that resulted in competitive superiority could be identified. Some population characteristics of the species grown in pure culture, with the same factorial design as before, provide some clues (Fig. 21.8). If the hypothesis is adopted that the species that has the higher population at carrying capacity is most likely to win in competition, we see that *T. confusum* should always win at 30% R.H. but only at 70% R.H. when at 34°C. In fact, the predicted result comes close to the actual result, although all six results cannot be predicted. But being able to predict two-thirds of the results is encouraging. That still

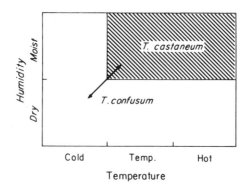

FIGURE 21.7 Ecological niches to *Tribolium confusum* and *T. castaneum* defined by two gradients: temperature and humidity. Note that as these physical parameters fluctuate first one species may be favored and then another, as indicated by the arrows.

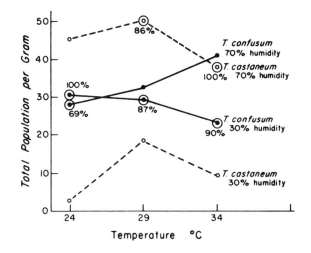

FIGURE 21.8 Analysis to test the hypothesis that the species which can exist at the higher carrying capacity in pure culture will win in competition. Equilibrium numbers in total population per gram for each species grown in pure culture are plotted for each humidity on a temperature gradient. For each treatment the outcome of competition in other experiments is shown by a circle around the winning species with the percentage of replicates that it won. Four of the six results can be predicted correctly. Data from Park (1954b).

leaves two perplexing exceptions, and this really illustrates how elusive the quality of competitive ability is. Many questions remain:

1. What is the physiological explanation for changes in the carrying capacity of the medium at different temperatures and humidities?
2. Why is the carrying capacity of equivalent habitats different for species as similar as *T. confusum* and *T. castaneum*?
3. What is this intangible element that prevents us from a perfect prediction in competition?
4. What makes a species dominant?
5. Why are animals such as murine rodents, the old world rats and mice, so successful?

Frank, a student of Thomas Park's, used an aquatic system for studying competition so that interaction was more easily observed and availability of food was more easily monitored. Thus F. Frank's papers (1952, 1957) make interesting reading and should be consulted for more information on laboratory studies on competition. More recently it has been possible to explain the outcome of competition between two species on the basis of monoculture characteristics (Goudriaan and de Wit, 1973), an encouraging result. Also, Gill (1974) claims that where this prediction is not possible, as in the *Tribolium* studies, interference competition was probably involved.

Certainly, several other attributes of *Tribolium* species need to be considered in relation to their competitive interaction. Cannibalism within and between species is an important factor influencing the population dynamics of single-species and two-species cultures (Mertz, 1972; Fox, 1975). Sinha (1968) has shown that each *Tribolium* species survives best on a different range of fungi growing in stored products. Although some fungi supported both species through the life cycle, others would support only *T. castaneum* and a different set would support only *T. confusum*. The beetle's feeding niches differed, with each species having a refuge from competition. Another important factor is the dispersal ability of the species. *Tribolium castaneum* is a strongly dispersing species and *T. confusum* is considered to have low dispersing capacity (Dingle, 1984). Whereas *T. castaneum* was the less successful competitor in several situations studied by Park, its better dispersal ability would ensure colonization of new resource patches before *T. confusum* established itself in them.

Although Park demonstrated the intensity of competition between *Tribolium* species in simple arenas, we should be circumspect about extending this view of competition to natural systems. In nature, resources would be much more complex, with many niche dimensions represented in any one resource patch, so the probability of species occupying very different niche hypervolumes is high. Given that these beetles differ in susceptibility to pathogens, temperature and humidity preferences, the fungi they survive on well, and dispersal ability, we are left to wonder if they would ever really meet and interact in a natural system. Much of the biology of *Tribolium* species has been reviewed by Sokoloff (1972, 1974, 1977).

COMPETITION IN SEED WEEVILS

A similar history of change in our perception of competition may be seen in experiments using the bruchid seed weevils, *Callosobruchus chinensis* and *C. maculatus*. Utida (1952) found that *C. maculatus* was more successful than *C. chinensis* in competition (see also Utida, 1953). But by 1957 Yoshida was able to show that *C. maculatus* usually went extinct in competition experiments with *C. chinensis*, using methods very similar to Utida's. Fujii (1965) speculates on the reasons for this change of result, but the reasons are still not clear. However, Fujii undertook a series of experiments that have refined our view of competition between these species. First, he studied competition under well-aerated and poorly aerated conditions (Fujii, 1965), finding that in the former condition *C. chinensis* always won, but under the latter condition *C. maculatus* always won. Later, Fujii (1967) found that at high humidity *C. chinensis* won, while at lower humidity *C. maculatus* won in competition. Further studies (Fujii, 1969, 1970, 1975) have refined the understanding of competition between these species. Utida (1972) also showed that flight polymorphism in *C. maculatus* is under the influence of density-dependent effects. At low densities, a dark morph that cannot fly predominates but at relatively high densities, high temperature, or dry food, a lighter morph capable of flight becomes more common. Again, we see that as more ecological niche factors are studied, the less likely it is that intense competition would occur if these species were sympatric in nature.

The long-term, intensive, and elegant studies by Reynoldson and his associates on natural populations of planarians have contributed substantially to a growing recognition that while species can compete in nature, there are many factors that reduce the severity of competition. (See Price, 1975c, for a review, and Reynoldson et al., 1981, for more recent references). Some kinds of interaction, in addition to those discussed already, are considered in the following paragraphs.

FREQUENCY-DEPENDENT COMPETITION

In 1958, MacArthur stated that to permit coexistence of species, it is necessary that each species inhibits its own increase in population more than it inhibits that of the other species, as seen in one of the competition models derived from the Gause, Lotka, and Volterra equations. In other words, competition between species is frequency dependent. When frequencies of the potentially dominant species are low, it is a good competitor; when frequencies are high, it is a poor competitor. This frequency dependence has been observed in animals and plants.

Ayala (1971) set up competition experiments between larvae of *Drosophila pseudoobscura* and *D. willistoni,* starting each experiment with different ratios of each species. He compared the input ratios with the output ratios, or the results of the competitive reaction. Without frequency dependence we should expect a one-to-one relationship between the input and output ratios, but Ayala found that there was a significant and inverse relationship between the initial frequency of a species and its competitive fitness (Fig. 21.9). Stable equilibrium would result at the intersection of the regression lines and the predicted line where fitness is 1 for each species. This equilibrium is quite different for different strains of *D. willistoni,* just as different strains of *Tribolium* have very different competitive abilities (e.g., Park et al., 1964).

De Wit (1960, 1961) had conducted similar experiments on plants much earlier and in one case found a similar result (de Wit, 1961). He tested competition between ryegrass, *Lolium perenne,* and clover, *Trifolium repens,* and plotted the input ratios of length of stolon of clover and the number of tillers of ryegrass at the start and end of the growing season (Fig. 21.10). Presumably, as clover becomes more abundant in the mixture than 50 times the ryegrass, it becomes less fit relative to ryegrass. Marshall and Jain (1969) found similar results at certain densities of competing species of wild oats (*Avena*). Intraspecific competition during mate selection also has been found to be frequency dependent in species of *Drosophila* and *Mormoniella* (e.g., Grant et al., 1974, and references therein).

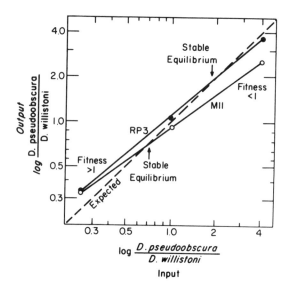

FIGURE 21.9 Frequency-dependent competition in *Drosophila* species. Line RP3 represents the outcome of competition (output) at various initial ratios (input) between strain 211 of *D. pseudoobscura* and strain RP3 of *D. willistoni*. Line M11 is for the same strain of the former species and strain M11 of *D. willistoni*. The expected line indicates a slope of 1 with an input/output ratio of 1, or a fitness of 1. After Ayala (1971).

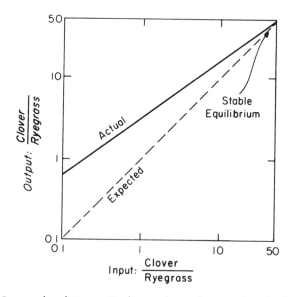

FIGURE 21.10 Frequency-dependent competition between clover and ryegrass. Note that clover competes much more successfully at low input ratios than at high, and a stable equilibrium is reached when clover is 50 times more abundant than ryegrass. After de Wit (1961).

NONEQUILIBRIUM TYPES OF COEXISTENCE

These examples show that competing species can maintain a stable equilibrium and therefore they can coexist indefinitely. But it is also possible for competing species to coexist without developing a stable equilibrium. Hutchinson (1953) described two situations where this may occur. First, where catastrophes frequently create new empty habitats for colonization, two very similar species may arrive at the same time, and one will tend to outcompete the other. Before competition goes to completion a new habitat becomes available; it is colonized by both species and a catastrophe destroys the former habitat and this process may continue. It will appear that the two species coexist. Hutchinson points out, however, that probably "the tendency of the weaker species to disappear by competition must be balanced by a tendency for it to spread a little more easily than the stronger; it must in fact be a fugitive species." This argument is supported by studies on *Tribolium*. In a similar vein, Skellam (1951) concluded from theoretical considerations that competitors with different dispersal rates could continue to coexist if the species with the faster dispersal rate had a lower reproductive rate.

A second nonequilibrium type of coexistence of very similar species may be seen where several generations occur in a year, and thus each generation may be exposed to rather different environmental conditions. One species may be favored in one generation and the other in another, and competition will never run to completion. This situation can be visualized for *Tribolium castaneum* and *T. confusum* and *Callosobruchus* species during temperature, humidity, or carbon dioxide fluctuations as described earlier in the chapter. Thus coexistence may result in these species, whereas Hutchinson points out that species with long generation times, such as a year or more, should demonstrate the competitive exclusion principle, as should very short-lived species such as bacteria, where competition will be so rapid as to be completed before the environment changes.

An interesting case of coexistence has been described by Istock (1967) concerning two whirligig beetles, *Dineutes nigrior* and *D. horni*. The former species has a more northerly distribution; it occurs more commonly on ponds than on lakes and appears to be adapted to boreal conditions in the absence of competition. *Dineutes horni* is more temperate; it coexists with other species throughout its range and therefore must be a successful competitor, and it is most commonly found in lakes. During high spring waters, adults of *D. horni* swam into ponds along channels from the lakes and oviposited. Their larvae were more successful than were the resident *D. nigrior* larvae, and competitive displacement was evident. However, before *D. nigrior* was completely displaced, the larvae of both species pupated, and the adults emerged and dispersed by flight in the late summer. This flight erased the advance in competition made by *D. horni* during larval growth in the pond, since a late summer census showed *D. nigrior* to be more abundant. Presumably, in the region as a whole more *D. nigrior* than *D. horni* were produced.

Thus competitive displacement was only transient because the competitive edge was lost by the adults of this holometabolous insect. Might this not be a common phenomenon in such organisms?

RESOURCE CONDITIONS

Another important influence on the probability of competition being effective in communities is the way in which resources occur in nature. Patterns of resource availability differ dramatically (Fig. 21.11). The broad categories of rapidly increasing, pulsing, steadily renewed, constant, and rapidly decreasing cover a wide range of resource types available for insects. This theme is developed in more detail by Price (1984b).

An example of rapidly increasing resources is foliage for herbivores in temperate deciduous trees and in annual and perennial plants. Leaves are absent at the start of the warm season, followed by a flush of foliage, subsequent production of more leaves, and finally, a rapid decline in foliage in the fall. This pattern has been described for stinging neetle (Davis, 1973), soybean (Price, 1976), and bracken fern (Lawton, 1978). Where several generations of a herbivore species occur each year, increase in population may be rapid enough to result in overexploitation of leaves, as in *Erythroneura* leafhoppers, with four generations per year (McClure, 1974; McClure and Price, 1975), discussed in Chapter 20. But where species have an annual life cycle it is impossible to develop a numerical response to increasing resources and it is most unlikely that competition will be important, since numbers at the beginning of the season probably will be small due to heavy winter mortality. In his studies on the community of insects on bracken, Lawton (1984a,b) found no evidence of interspecific competition in field observations and experiments.

Pulsing resources exist briefly and the probability of colonization of any one resource patch is low even for one species, so the chances for competition becoming important are low. For example, *Depressaria* moths oviposit on unopened umbels of wild parsnip. Any one umbel is available for only 2 or 3 days and so can easily pass through this available period without being discovered by a moth. In some dense patches of wild parsnip, all umbels together were available for only 13 days in a season because plants developed rapidly and synchronously from overwintered rosette to senescent plant. The net result was that even when populations were extremely high, only 52% of plants were attacked and on these plants less than 50% of the umbels were attacked, leaving 75% of resources unutilized (Thompson and Price, 1977). Of course, such ephemeral resources are only relevant to specialized organisms that cannot switch feeding among ephemeral resources. But as we have seen in Chapter 9, many insects have little or no choice in the plants they feed upon. The view of Andrewartha and Birch (1954) that time is usually more limiting than other resources is certainly relevant to insects exploiting pulsing resources, and in many cases to those using rapidly increasing resources.

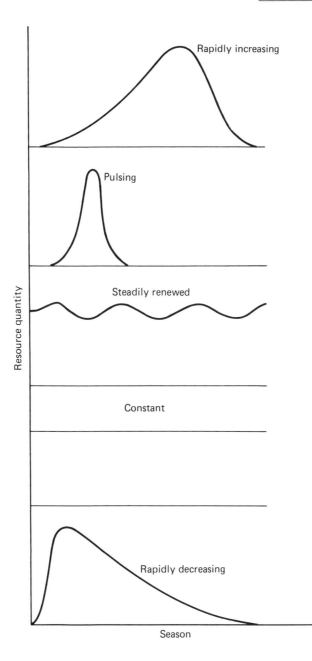

FIGURE 21.11 Types of resources that will have different effects on the probability of competition occurring for them. Reproduced with permission from Price (1984b).

Steadily renewed resources include, for example, resources for animal parasites such as ectoparasitic insects on birds and mammals, and leaves for herbivores in tropical rain forest. If colonization is effective and reproduction rapid, we may expect competition to become important in communities based on such resources. Marshall (1981) reviews the many cases of competitive exclusion in ectoparasitic insects and cases where realized niche occupation declined in the presence of another ectoparasite. On the other hand, many resources may be difficult to colonize, resulting in typically low population numbers and few individuals colonized. Marshall (1981) states that under natural conditions on healthy hosts the populations of ectoparasites are low and there is frequently no obvious sign of competition. He cites several studies indicating the absence of interspecific competition (Beer and Cook, 1958; Marshall, 1977; Overal, 1980). Many hosts may have only one parasite species exploiting a particular resource. For example, in West Malaysia 67% of bat species have less than two ectoparasite species (Marshall, 1981), and even those species with two or more may have low probabilities of any two species becoming associated on a single host. In relation to leaves utilized by insects in tropical rain forest, competition does not seem to be important in critical tests made by Strong (1981, 1982, 1984a). For hispine chrysomelid beetles that utilize the rolled leaves of *Heliconia* plants in Costa Rica, resources remain superabundant. This seems to be because young rolled leaves develop rapidly and open up before their resources can be exploited even to the extent of 10% of what is available. Any one leaf is an ephemeral resource, although a patch of *Heliconia* plants produces new rolled leaves persistently through the year.

Constant resources are those that are uninfluenced by utilization or seasonal change. Tree holes for hole nesting or breeding insects are an example where the holes are not made by the insects; they are not destroyed by the insects, and numbers per unit area are likely to remain constant through time. Where holes are scarce, competition for them is likely to be strong. Syrphid flies in the subfamily Milesiinae utilize rot holes in trees as breeding sites in which the aquatic rat-tailed maggots develop. After feeding at flowers in the morning, males of many species, such as *Mallotta posticata* and *Spilomyia decora,* aggressively defend such tree rots against males of the same and other species and mate with females visiting to oviposit in the rot hole (Maier and Waldbauer, 1979a,b). The first male to arrive at a tree rot is normally able to drive others away, but there is no doubt that tree rots are in short supply and competition for them is fierce.

Rapidly decreasing resources are produced during a short period of time and then decline through the rest of the season. Seeds for granivores in desert environments fit this category where a pulse of seeds is released after the rainy season and subsequently declines as seeds are eaten, become buried, and finally germinate in the next season. All granivores in the desert seem to compete for this declining resource involving rodents, birds, and ants (and probably lygaeid bugs, although studies have not been extensive, but see Sweet, 1963, 1964). This literature has been reviewed by Brown et al. (1979). Many specialized insectivores are also faced with rapidly declining numbers as an insect population changes from an available to an unavailable stage, such as from larva to pupa for larval parasitoids or from cocoon to adult for cocoon parasitoids. Members of the cocoon parasitoid guild were forced into intense competition as cocoons was depleted by parasitization and emergence of adults until 100% of available cocoons were utilized and the parasitoids could persist only by hyperparasitizing others (Price and Tripp, 1972, and Fig. 21.12).

Even the very simplistic view of resources undertaken here, ignoring the many other factors involved in population regulation discussed in Chapters 17 to 19, should caution against any generalizations that competition will usually be an important influence in community organization. The challenge is to identify patterns in when it will be and will not be a significant factor. The even greater challenge is to identify if interspecific competition has been an important force in the past influencing the evolutionary divergence of species into different ecological niches.

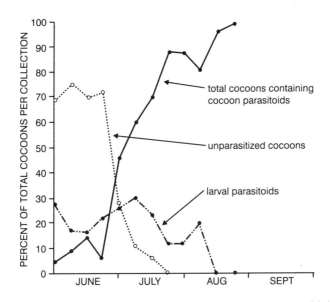

FIGURE 21.12 Change through spring and summer of unparasitized cocoons, showing their rapid decline in late June and early July, and parasitized cocoons, showing that almost all resources were utilized by mid-August, forcing strong competition between species. After Price and Tripp (1972).

HOW IMPORTANT IS COMPETITION IN NATURE?

Problems with identifying the presence of character displacement were discussed in Chapter 20 and by Strong et al. (1979) and Simberloff and Boecklen (1981). Connell (1980) argues cogently that there is little strong evidence that competitors have shaped each other's ecological niches, suggesting that both theory and empirical evidence indicate that only in low-diversity communities will competition be an important evolutionary force. Connell has challenged ecologists to test for competition rigorously outlining the methods whereby the influence of competition can be unequivocally assessed. Lawton and Strong (1981) note the similarity in community patterns between vertebrates and folivorous insects but emphasize the "general infrequency of interspecific competition and lack of competitive organization in folivorous insect communities" (p. 321). They suggest that other factor will usually be more important in defining presence and abundance of species in folivorous communities—autecological responses to weather, phenology, patchiness, host-plant quality, and the action of predators and parasitoids.

Clearly, the role of interpsecific competition as an important influence in communities has been seriously challenged in the literature, much of the challenge coming from those studying invertebrates, including insects. Insect ecologists can play a major role in resolving the debate on the importance of competition because of the relative ease with which experimental studies can be undertaken, at a time when experimentation is critical to the further development of this field.

Experimental analysis of competition in communities has become more common, particularly in the study of insect communities. The studies by Seifert and Seifert (1976, 1979), Strong (1981, 1982), Wise (1981), and Lawton (1984a) illustrate the potency of the experimental approach to interspecific competition and the methods that should be emulated many times in the future. Both Connell (1983) and Schoener (1983, 1985) undertook large surveys of good experiments capable of detecting competition among plants, invertebrates, and vertebrates in aquatic and terrestrial systems. "Competition was found in 90% of the studies and 76% of their species, indicating its pervasive importance in ecological systems" (Schoener, 1983, p. 276). "Competition was found in most of the studies, in somewhat more than half of the species, and in almost two-fifths of the experiments" (Connell, 1983, p. 682). These authors reached somewhat different conclusions because they used different methods of analysis (Schoener, 1985), but the clear overall conclusion is that competition is very common and important in the structuring of communities. So why have some authors, such as Lawton and Strong (1981), argued that interspecific competition is unlikely to be important in some kinds of communities?

A brief historical overview will place this question in perspective. Early experiments by Gause, Park, Utida, and others discussed in this chapter were very influential in generating interest in competition and creating a view of competition as a central organizing force in communities. However, the experiments were conducted in laboratories with little or no concern about possible niche differences in nature between species forced to compete in essentially one ecological niche. In the 1960s and 1970s many field studies were undertaken, although they tended to be observational rather than experimental, and emphasis was on resource partitioning (cf. Denno et al., 1995). The implication was that if niche segregation between species were found, it resulted from competition in the past. These studies reinforced the view that competition was pervasive in nature and a strong component of community organization.

In the latter part of the 1970s and early 1980s, experimental studies of insect communities started to show the lack of competition as an organizing force, with those of Seifert and Seifert (1976, 1979), Wise (1981), Strong (1981, 1982, 1984a) and Lawton (1984a) being especially influential. Connell (1980) also noted that good evidence that competition had shaped communities in the past was largely lacking. There was only "the ghost of competition past" (Connell, 1980, p. 131), not tangible evidence.

However, with the reviews on competition by Connell (1983) and Schoener (1983, 1985) the subject of the importance of contemporary competition in natural communities remained controversial. Now, a broader picture and a sense of pattern in nature is emerging for insect communities, especially relating to phytophagous insects, based on an extensive review by Denno et al. (1995) and the authors' personal experience with experimental studies of competition. First, let us examine some studies showing lack of detectable competition in communities, and then end the chapter with the views expressed by Denno et al. (1995) and some of the relevant evidence.

Heliconia Bract Communities

Communities of insects develop in the bracts of *Heliconia* species commonly found in wet Neotropical forests. Each bract contains flowers, and rainwater accumulates, forming an aquatic microcosm with some typical insect inhabitants (Fig. 21.13). Common denizens of these phytotelmata (Maguire, 1971), or plant-held waters, are members of the genera *Cephaloleia* (hispine chrysomelid beetles); *Gillisius* (hydrophilid beetles); *Quichuana* (rat-tailed maggots—syrphid Diptera); another syrphid genus, *Copestylum*; a stratiomyiid, *Merosargus*; and a few other genera and families. Extensive studies on such communities were conducted by the Seiferts over many sites in Central and South America (Seifert and Seifert, 1976, 1979; Seifert, 1984).

The most telling evidence on how these kinds of communities were organized came from experiments in which communities were established in new, uncolonized bracts. Using the communities illustrated in Fig. 21.13 and one

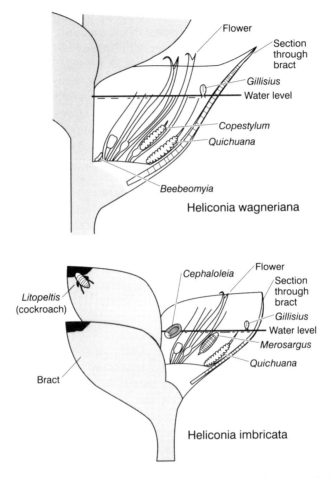

FIGURE 21.13   Side views of stylized bracts and insect communities in two species of *Heliconia* studied by Seifert and Seifert (1976): *Heliconia wagneriana* (above) and *Heliconia imbricata* (below). From R. P. Seifert and F. H. Seifert, A community matrix analysis of *Heliconia* insect communities, *Am. Nat.,* 1976, **110**:461–483. Published by the University of Chicago Press. Copyright © 1976 by the University of Chicago. All rights reserved.

on another *Heliconia* species, different combinations of insect species were placed in bracts at normal or higher than normal densities and allowed to interact, usually for 6 or more days. The bracts were then opened and the number of insects per genus recorded. Of the 36 interspecies interactions tested in the three communities, 6 proved to be competitive, 6 were facilitative, and 24 showed no interaction; and the facilitation was stronger than competition. When hispine chrysomelids occurred in the communities, feeding on the bracts, they increased detritus in the water, which was fed on by fly larvae, and open wounds on bracts provided good feeding sites for hydrophilid beetle larvae. Stratiomyid fly larvae were found only in bracts with hispine beetles present. Seifert (1984) noted that densities of insects in bracts are usually low and species came from divergent phylogenetic lineages. Species do not compete frequently because lineages diverged in resource utilization in the distant past, not because competition selected for niche segregation once the species co-occurred frequently.

Tenebrionid Communities

Even when species in the same genus coexist, competition may still be hard to detect. Wise (1981) conducted elegant removal experiments to detect the existence of competitive release after a dominant member of the community was removed. The community consisted of five species of tenebrionid beetle, four in the genus *Eleodes* and one in the genus *Asidopsis*. *Eleodes obscurus* adults were the dominants in terms of biomass, and these were removed from caged arenas for 3 years, with a paired arena left unmanipulated as a control. Six 10 by 10 m plots were used in all with three treatment plots and three control plots. All species have larvae that live in the ground, and these were not manipulated. Results showed that in the removals none of the other species increased in density, and adult size of the remaining species did not increase.

Hispine Chrysomelids in Leaf Rolls

Another community of closely related beetle species was studied by Strong (1981, 1982, 1984a). Up to eight species of hispine chrysomelids may be found at one site, living in the rolled leaves of *Heliconia* plants. Five species may be found in a single leaf roll. Strong (e.g., 1981, 1982, 1984a) studied interference competition among adults and exploitation competition among larvae. The adults and larvae are long-lived, eat the same food, and use the same rolled-leaf habitat. Therefore, we should expect strong interference competition. However, Strong found that adults are not aggressive intraspecifically or interspecifically. When experimental communities were established that minimized leaf and habitat heterogeneity, there was no segregation of species observed, and there was actually more aggregation of species observed than segregation. Densities of hispine beetles are usually very low relative to the apparent carrying capacity of the food and habitat resources. Any one leaf may be rolled and habitable for only a few days, so repeated colonization of new leaf rolls is essential. Therefore, resources for larvae were apparently never limiting. All herbivory on leaves averaged only 1.53% of leaf area. The term *harmonious coexistence* used by Strong (1982) was suitable for these hispine chrysomelid assemblages.

Herbivores on Bracken Fern

Lawton's research on bracken fern herbivores, discussed in Chapter 20 in relation to vacant niches, was also instrumental in developing the basis for the
arguments in Lawton and Strong (1981) and Strong et al. (1984a). "We conclude that interspecific competition is generally too feeble and sporadic to be
the most important or consistent mechanism structuring communities of phytophagous insects" (Strong et al., 1984a, p. 157).

Lawton (1984a,b) used two approaches to the study of interspecific competition among bracken fern herbivores. One was a test for density compensation in natural communities and the other was a direct test of interspecific
competition. If interspecific competition is important in communities, we
should expect densities of individuals per species to be higher when few
species are present than when many species are present in a community. Density compensation should be observed. In two tests of density compensation,
Lawton (1984b) found quite the reverse. As numbers of species in an assemblage declined, the densities per species declined. Within the assemblage at
Skipwith common, Lawton examined pairwise species densities on individual
bracken fronds. Negative correlations between species densities would be expected if competition were playing a role in community organization. In six
pairwise species comparisons, on different dates and years, making 17 comparisons in all, 14 showed no significant trend, 2 showed positive trends in
densities, and only 1 showed a negative trend. As with hispine chrysomelids
in *Heliconia* leaf rolls, insect herbivores on bracken occur at low densities per
frond usually, with ample resources for those present. Such very "open" assemblages are particularly tangible when the resource and the organisms utilizing that resource are so clearly displayed in space and time, as are the insect herbivores on plants.

Overview

Now that we are well into the 1990s let us leave the last words in this chapter to Denno et al. (1995, pp. 297–298), who have taken a broad look at
competition among phytophagous insect species. They evaluated evidence on
193 pairwise species interactions, which represented all the major feeding
guilds: chewers, suckers, free feeders, borers, miners, gallers, and so on. "Interspecific competition occurred in 76% of interactions, was often asymmet-
ric, and was frequent in most guilds (sap feeders, wood and stem borers, seed
and fruit feeders) except free-living mandibulate folivores. Phytophagous insects were more likely to compete if they were closely related, introduced, sessile, aggregative, fed on discrete resources, and fed on forbs or grasses. Interference competition was most frequent between mandibulate herbivores
living in concealed niches. Host plants mediated competitive interactions

TABLE 21.2 Abbreviated Table, Based on Denno et al. (1995), on the Frequency of Interspecific Interactions in Phytophagous Insect Communities. Only Well-Represented Interactions are Included.

Guild (Number of Study Systems)	Interaction				Symmetry		Consequences of Interaction			Mechanism		
	Comp.	Facilit.	None	Frequency	Sym.	Asym.	Excl.	Loc. Dis.	Fit. Red.	Explt.	Interf.	Both
Haustellate mouthparts												
Free living (26)	36	2	3	22	16	14	7	19	33	16	5	15
Total (haustellate) (29)	42	3	3	27	16	20	7	24	34	17	10	15
Mandibulate mouthparts												
Free living (21)	30	2	21	4	0	27	1	8	22	26	3	1
Subtotal 1 (external feeders) (21)	30	2	21	4	0	27	1	8	22	26	3	1
Stem borer (3)	13	0	1	7	0	13	0	13	0	0	13	0
Wood borer (8)	14	1	0	13	3	8	0	10	5	0	4	10
Seed/pod/cone/bract feeder (17)	21	0	0	5	1	18	1	13	18	5	8	8
Subtotal 2 (internal feeders) (31)	51	1	6	25	4	42	1	37	25	5	26	20
Subtotal 3 (mandibulate guild interactions) (13)	16	2	0	7	0	11	2	11	8	7	8	1
Total (mandibulate) (65)	97	5	27	36	4	80	4	56	55	38	37	22
Total (haustellate × mandibulate) (10)	8	3	5	1	0	7	1	6	6	2	5	1
Grand total (104)	147	11	35	64	20	107	12	86	95	57	52	38

SOURCE: Based on Denno et al. (1995). With permission from the *Annual Review of Entomology*, vol. 40, © 1995, by Annual Reviews Inc.

more frequently than natural enemies, physical factors, and intraspecific competition. Sufficient experimental evidence exists to reinstate interspecific competition as a viable hypothesis warranting serious consideration in future investigations of the structure of phytophagous insects communities."

The review included the studies discussed above by the Seiferts, Wise, Strong, and Lawton. Denno had concentrated his studies on sucking insects, as indicated in Chapter 14, and McClure had worked on cicadellids (See Chapter 20), scale insects, and adelgids, all sucking insects. Hence we see perspectives on competition developed based on different kinds of herbivores. However, there is some agreement on the following: (1) free-feeding mandibulate herbivores are less likely to compete than members of other guilds (cf. Strong on hispine chrysomelids), and (2) closely related species are most likely to compete (cf. the very different lineages represented in communities in *Heliconia* bracts and on bracken fern). The summary table developed by Denno et al. (1995) is given in very abbreviated form in Table 21.2. Patterns are becoming clearer on the role of competition, at least in phytophagous insect communities.

Some details on Table 21.2 are needed. Haustellate, or sucking insects, included the Homoptera, Hemiptera and Thysanoptera. Mandibulate, or chewing insects, included all the other major orders, including larval Diptera with mouthhooks or buccopharyngeal armatures. Also reviewed were interactions between mandibulate guilds, such as between free-living and leaf-mining guilds, and interactions between haustellate and mandibulate guilds. Evidence for and against competition was divided into the kind of interaction (competition, facilitation, or none). The frequency of the interaction is given only as the number of cases of frequent interaction in Table 21.2. Symmetry of the interaction was divided into more or less equal negative interactions between species (symmetrical) or very uneven impact (asymmetrical). The consequences of the interaction were divided into competitive exclusion, local displacement, and fitness reduction or population change. The mechanism involved in competition was identified as exploitation or interference competition or both. Only examples are provided in Table 21.2, but all the totals and subtotals for different kinds of interactions are included. The grand total of 104 studies includes a total of 193 pairwise species interactions, with 147 showing competition, 11 indicating facilitation, and 35 with no interaction.

Chapter

22

Community Development, Structure, and Organization

COMMUNITIES AND ASSEMBLAGES

In Chapter 2 the community was defined as a set of coexisting interacting populations. If the populations of different species in a given area are truly interdependent, this is the unit of classification that deserves careful study. Here concern is focused on the presence, and absence, of species in a given area and not on total distribution of species. Study is concentrated on a woodlot, a field, a rotting log, a tree hole, or slime flux, or practically any unit that contains several species. Thus the community is perhaps the largest unit in an ecological classification in which relationships can be studied in fair detail. Once study of the biome is undertaken, a systems analysis approach is usually necessary and the guild may be the smallest unit of study that can realistically be treated. Thus community ecology offers the last, but biggest, chance of looking at the interrelations of organisms with each other, and their physical setting in detail.

It is essential first to establish the reality of the community. MacArthur and MacArthur (1961) discovered that a given structural diversity of a biotope usually supports a certain diversity of bird species. That is the number of

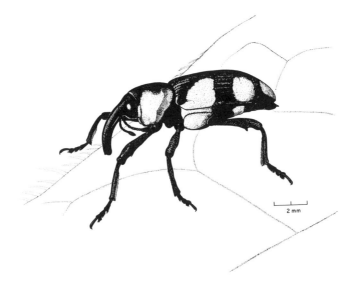

One of the weevils, *Ambates cretifer* (Curculionidae: Baradinae), which is an important herbivore on *Piper* species at La Selva, Costa Rica. From R. J. Marquis in P. W. Price et al. (eds.), *Plant–animal interactions.* Copyright © 1991 John Wiley & Sons, Inc. Reprinted by permission of John Wiley & Sons, Inc. Drawing by Alice Prickett.

species, and their relative abundance is dictated by the biotope. If this diversity of birds is so predictable from biotope characteristics, it suggests that each biotope must be filled up with bird species and no more could colonize the site. If the biotopes were not filled, many points below the regression line between bird species diversity and foliage height diversity should have been observed. But the diversity index, H' (see the definition in Chapter 23), has two important components, the number of species and the evenness of abundance (equitability) of these species, so that it is difficult using this method to see what is actually limiting. Number of species per biotope is a simpler value to look at first.

In Chapters 20 and 21 the terms *community* and *assemblage* have been used almost interchangeably. The term **assemblage** simply means a group of species in the same locality or on the same resource, without the implication that individual species interact with each other. Therefore, when discussing groups of insect species utilizing a plant species, for example, the term *assemblage* would be applicable if nothing is known about the interactions, or if a study reveals that there is no interaction, as is the case apparently in the hispine chrysomelids in *Heliconia* leaves and the phytophagous insect assemblage on bracken fern. However, the term **community** would be more appropriate for some of the communities in the phytotelmata of *Heliconia* studied by the Seiferts because facilitation and competition were demonstrated, even though in a minority of cases. Based on Chapter 21, there is clearly a debate on the extent to which groups of insects warrant the term *assemblage* or the term *community*. In this chapter we examine the many hypotheses developed that attempt to account for organization in communities or its apparent lack in assemblages.

DEFAUNATION OF ISLANDS

A study on arthropod communities supports the view that biotopes have a certain limit for species. Wilson and Simberloff set out to test some of the theories of island biogeography proposed by MacArthur and Wilson in 1963 and 1967. In 1966, Wilson and Simberloff carefully found all the species of arthropods (insects and spiders) that existed on each of several islands of red mangrove (*Rhizophora mangle*), off the west coast of Florida, in Florida Bay. In late 1966 and early 1967, they fumigated each island by putting tents over the island and pumping in methyl bromide. All arthropods were killed except a very few individual wood borers (Wilson and Simberloff, 1969). Since then they have monitored the recolonization of the sites (Simberloff and Wilson, 1969, 1970; Simberloff, 1978), and some of the colonization curves are given in Figure 22.1. The shape of the colonization curve appears to be quite typical of all but the islands farthest from the mainland. Colonization is rapid at first; the number of species overshoots the equilibrium number before defaunation and then drops rather rapidly to an equilibrium state. The more

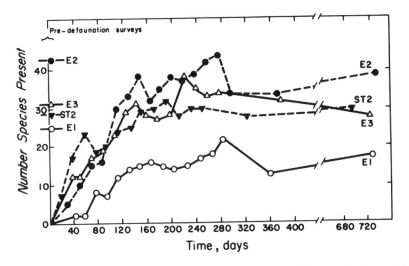

FIGURE 22.1 Colonization curves for four small mangrove islands in the lower Florida Keys after faunas had been exterminated by fumigation with methyl bromide. Numbers of species present before defaunation are given for each island on the ordinate. The near island, E2, had the most species; the farthest island, E1, the fewest; and the intermediate islands, ST2 and E3, intermediate numbers. From Simberloff and Wilson (1970).

distant the island is from the species pool of potential colonizers, the slower is the process. The islands seem to have a definite limit for a certain number of species. This limit will impose a sorting action on the arrivals; only those that can exploit a discrete set of resources can survive, and those that cannot will be departures from the community, through extinction. Thus there must be a self-organizing quality in the community, which therefore justifies its study as a natural unit in the hierarchy of ecological assemblages of organisms.

STAGES IN ACCUMULATION OF SPECIES

Wilson (1969) postulates four stages in the organization of a community: noninteractive, interactive, assortative, and evolutionary (Fig. 22.2). Noninteractive conditions exist when the population densities and numbers of species are too low to cause competition for resources and there is insufficient biomass for predators and parasites to exploit effectively. The interactive equilibrium is due to competition, the extinction of some species, and the establishment of predators and parasitoids. Many species will go extinct through this interaction, but others will colonize the area, and during this assortative phase those species that can coexist and utilize the resources most efficiently will remain. The assortative phase sorts species for efficient coexistence, and this efficiency in the community permits more species to pack in.

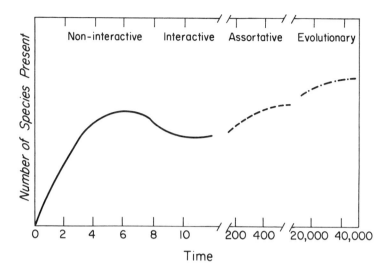

FIGURE 22.2   Hypothetical sequence of equilibrium number of species in a community with time. The time scale represents relative periods of time that it would take to reach the various equilibrial states. From Wilson (1969).

Finally, coevolution of species makes their total exploitation more efficient still and there will be a further increase in coexisting species. This evolutionary aspect is shown, perhaps, by Southwood's (1961) data on the number of insect species on different tree species in Europe. For example, in Britain, the longer and more abundant a tree species has been in the islands, the more species exist on that tree (Fig. 22.3).

Studies by Heatwole and Levins support the findings of Wilson and Simberloff, the theory of island biogeography, and the contention that communities are organized in a predictable way (e.g., see Heatwole and Levins, 1972a,b, 1973; Levins and Heatwole, 1973). They found that equilibrium numbers of species were reached rapidly on small islands of the Puerto Rican Bank and that species turnover rates were high. In spite of the high immigration and extinction rates, trophic structure of the community remained remarkably stable, suggesting that only species that could interact effectively with other existing species were able to establish on these islands. The interactive phase in colonization is important in sorting species. However, Simberloff (1976) claimed that such structuring was not evident in the mangrove island study.

This approach to community study need not apply only to oceanic islands. Janzen (1968, 1973a) has emphasized that the host plants of insects can be regarded as islands in space and time. Southwood's data support this view and Opler's (1974) study, "Oaks as evolutionary islands for leaf-mining insects," is particularly relevant. A plant crop is clearly an island and many natural areas form islands on continents. Vuilleumier (1970) has published a paper, "Insular biogeography in continental regions," dealing with the paramo

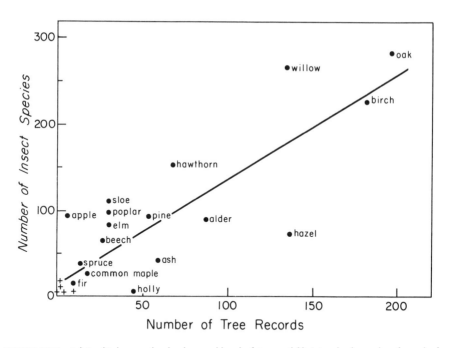

FIGURE 22.3 Relationship between the abundance and length of time available (given by the number of records of Quaternary remains) for colonization by insects and the number of herbivorous insect species present on trees in Britain. "+" indicates numbers of insects on introduced tree species. From Southwood (1961); reprinted by permission of Blackwell Scientific Publications.

islands in Ecuador, Colombia, and Venezuela in relation to bird communities. Brown (1971) also regarded mountaintops as islands in relation to mammal communities. Although not alluding to island biogeography specifically, Mani (1968) provides much information on insects at high altitudes, of interest to the continental island biogeographer. Culver et al. (1973) and Culver (1982) have studied the islands that are available to cave-dwelling organisms. Thus "island biogeography" is just as important to mainland situations, since many communities represent effective islands. (For an application of island biogeography to agricultural crops and pest management, see Price and Waldbauer, 1975, 1994).

DETERMINANTS OF EQUILIBRIUM NUMBER

For any given community five major factors can be identified that determine the number of species it contains. First there is the historical factor: the time it has been available for colonization. Then two external factors are important: the number of potential colonizers, that is, the size of the species pool from which colonization can occur, and the distance from the source of colonizers. Finally, there are two internal factors that relate to the structural diversity of the biotope: the size of the biotope, which influences the structural diversity of the biotope, and interaction between species leading to extinction of some and survival of others, which depends in part on the structural diversity.

The factors within the community that regulate the number of species present are discussed here and external factors are discussed in Chapter 24.

QUALITY AND QUANTITY OF RESOURCES

There is a fundamental need for species to exploit an unshared resource. Therefore, the numbers of species depends on the diversity of resources within the biotope. Since this diversity within biotopes of similar structure is likely to be similar, the number of species they can support should be similar, just as MacArthur and MacArthur (1961) have shown for bird species and Patrick and Strawbridge (1963) for diatoms in waters of similar quality. In Chapter 20 it was seen that the range of quality in a resource was important in determining the number of species that could exist on that type of resource. It might be argued, however, that the larger the quantity of the resource, the more species it can sustain. This is not supported by Rosenzweig's (1971) models, where increasing a resource (enrichment) decreased the stability of a two-species system and therefore increased the chance of one of the species going extinct. Similar results were obtained by Riebesell (1974) when he considered enrichment in relation to competition models. Also, Patrick and Strawbridge (1963) argue that dominant species are often characteristic of faunas and floras that have an excess of nutrients, or light, or water. For example, temperate forests have one or two dominant species and here the sites have reserve nutrients, whereas in the tropical forests many species are supported in lateritic soils of very low nutrient status. Woodwell's (1974) studies suggest that where nutrients are in short supply niche segregation of species is necessarily greater. Terborgh (1973) argues that the high numbers of species seen in deserts of North America is the result of the diversity of ways in which plants have evolved to partition the water, which is in short supply. Stephenson (1973) has shown that increased fertilization of old-field communities reduced species richness and evenness, and Yoda et al. (1963) found that less fertile soils supported a larger number of small plants than did fertile soils. Pollution frequently has a similar effect since large amounts of one substance are deposited. For example, Harman (1972) found that pollutants reduced the number of benthic substrate types and therefore reduced the number of freshwater molluscs present. Rosenzweig and Abramsky (1993) provide a detailed analysis of the relationships between productivity and the number of species supported. They found much support for highest species richness at intermediate levels of productivity.

Thus in many cases, large quantities of a resource favor one or two species and not others, so that dominance develops and species diversity is reduced. This was seen among parasitoids, where as host density increased, *Pleolophus basizonus* became dominant and several other species of cocoon parasitoid became extinct, or reduced in numbers. The diversity changed as seen in Fig. 22.4. There is close similarity between this trend and the colonization curve for an island community. Therefore, within the community the quantity and quality ranges of resources have an organizing influence, and these will be influenced in turn by the size of the habitat, which is ultimately connected with the structural diversity of the habitat. The quality range of resources influences the number of species supported, while the quantity of the resource affects the population size of each species (Fig. 22.5). Therefore, the internal factors influencing the number of species that can exist in a community can be redefined more accurately: (1) the number of resources available, (2) the quality range (or size range) in each resource, and (3) the quantity of a resource in each quality category.

The number of resources in each community and the abundance of each must therefore become the central concerns for research on community organization in ecological time. We see that the most incisive studies of communities emphasize the detailed consideration of resources available to community members. The detail required demands that either very simple communities are studied, such as in the effluent of a hot spring, or that the community is broken down into subunits—the component community, or guild, defined in Chapter 2.

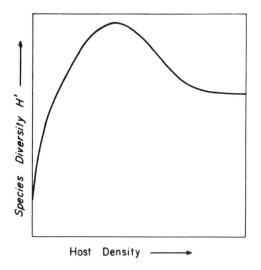

Host Density ⟶

FIGURE 22.4 Trend in species diversity in the parasitoid fauna as the food resources (host cocoons) become more abundant. The reason for the decline in diversity is the increased dominance of one species, *Pleolophus basizonus*. (See Chapter 23 for the definition of H'.) After Price (1970b).

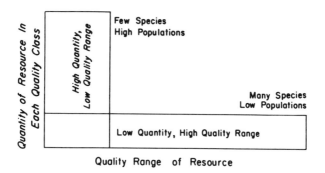

FIGURE 22.5   Relationships between abundance of resources and range of resources in a community, and their effects on numbers of species present and population sizes. Quality ranges may include size ranges of seeds for granivores, nectar depths in flowers for pollinators, range of bark thicknesses available to bark beetles, and such. The two extremes in resource distribution have equal amounts of energy available.

COMPONENT COMMUNITIES

As Root (1973) points out, member species of a component community are adapted to similar conditions, they may interact frequently and thus tend to influence the ecology and evolution of each other. They are all part of "a relatively distinct set of interactions" (p. 96). Once a component community is identified, the resources available can be measured relatively easily and accurately, especially in insects feeding on living plants. The degree to which communities are subdivided into guilds depends on the detail required. Herbivores in a community might be divided into a stem-boring, a leaf-feeding, and a root-feeding guild. But the more precise use of the term *guild* requires that the leaf-feeding guild be broken down into guilds of species that eat leaves in a similar manner. Thus Root (1973) divided the foliage feeders on collards (*Brassica oleracea* var. *acephala*) into three distinct guilds (Fig. 2.1):

1. The pit-feeding guild, members of which chew small pits into a leaf surface, composed mainly of small chrysomelid and curculionid beetles.
2. The strip-feeding guild, members of which chew relatively large holes in the leaf or strip pieces off the leaf edge, composed mainly of lepidopterous larvae and slugs.
3. The sap-feeding guild, members of which have piercing–sucking mouthparts with which they suck plant juices, composed principally of aphids but also including cercopids, cicadellids, and mirids.

Just as members of a component community have closer ecological links with each other than with those of other such communities, so members of the same guild are intimately associated. We have here a hierarchy of functional ecological relationships. Here, organization within component communities will be discussed first, and then whole communities. Some whole communities are composed of one predominant plant species and the organisms associated, as in hot spring effluents discussed later in this chapter (e.g., Brock, 1967; Wiegert, 1973; Collins, 1975a). Not all whole communities are compound communities.

One of the most thoroughly studied component communities is that based on the plant family Cruciferae. Pimentel (1961b) studied the differences between arthropods in pure stands of *Brassica oleracea* and in diverse communities of old-field vegetation with *B. oleracea* plants situated 9 ft from the nearest conspecific plant. He found more species of arthropod associated with *B. oleracea* in the pure stand than in the diverse community, and, although many more predators and parasite species were present in the pure stand, only here did herbivore populations reach outbreak levels. Flea beetles, lepidopteran larvae, and aphids all reached high populations in the pure stand, showing that members of each guild, as Root (1973) defined them, were capable of becoming important species in the component community.

ENEMY IMPACT HYPOTHESIS

Pimentel concluded that outbreaks were prevented in the diverse community for three main reasons. First, although there were fewer species on *B. oleracea* in the diverse community, the total flora and fauna in this community was much more diverse than in the pure stand. The diversity of herbivore species available as food for predators and parasites always provided alternative supplies when species fluctuated in abundance, providing greater stability of food available to the primary carnivore trophic level. Second, more species of parasite and predator provide a greater probability that prey species will be maintained at or near equilibrium density (as defined by Holling, 1965; see Chapter 8), and therefore species will not reach the threshold density for population escape. Third, predators in the diverse community, particularly spiders, were more polyphagous than predators in pure stands, which were mainly specialists on aphids—coccinellids and syrphids. Thus Pimentel supported the long-held view among ecologists that diversity leads to stability, and this through more favorable conditions in diverse communities for enemies of the herbivores. He supported the **enemy impact hypothesis** on community organization. Much support for this hypothesis has been generated since (e.g., Lawton and Strong, 1981; Strong, 1984b; Strong et al., 1984a; Russell, 1989).

RESOURCE CONCENTRATION HYPOTHESIS

Root (1973) studied similar *B. oleracea* plots in a similar location, found similar results, but reached dissimilar conclusions. Root noted that natural enemies could not account for the differences between pure stands and diverse communities. On the *B. oleracea* plants themselves predators and parasitoids were more diverse in the pure stands, and their densities were higher. In addition, predators fed mostly on sap feeders and young strip-feeders, which constituted a small proportion of the herbivore biomass, whereas adult flea beetles, which were the dominant herbivores, were rarely eaten by predators. The enemy impact hypothesis was discounted by Root in this example. Therefore, the differences between pure stands and plants in diverse communities must be due to other factors.

Root (1973) noted that there were several studies, including Pimentel's and his own, indicating that the concentration or dispersion of food resources had a direct influence on insect populations. In his **resource concentration hypothesis** he proposed that herbivores, particularly specialists, (1) are more likely to find hosts that are concentrated; (2) once arrived, are likely to remain there; and (3) reproductive success is likely to be greater. Some species will find all their needs within the plot and populations will undergo rapid and extensive increase. Other species may find adequate oviposition sites but inadequate adult food, and members of these may tend to emigrate from the patch. Therefore, herbivore biomass will become invested in a few specialist species and the equitability component of species diversity will be low where resources are concentrated (Fig. 22.6). (See the beginning of Chapter 23 for a discussion on measuring diversity.) As populations of the favored species increase, their dominance will be heightened by interspecific competition in a community with resources of high quantity and low-quality range (Fig. 22.5).

Low equitability in the pure stand therefore results from a strong numerical response to resources in some species and a weak one in others. It therefore becomes important to understand the response of individuals in each species to each condition of resources found in communities in order to understand differences in dominance, or the presence of a herbivore under certain conditions and its absence under others.

The flea beetles were important herbivores both in pure stands and in diverse stands with *Phyllotreta cruciferae* usually dominant, and *Phyllotreta striolata* usually ranking third to fifth in biomass (milligrams per 100 g of collard foliage). The strip-feeding *Pieris rapae* larvae frequently ranked second in biomass. These three species will be used as examples to investigate specific qualities of species which influence their numerical response to resources. Specifically, three questions can be posed:

1. Why was *Phyllotreta cruciferae* more dominant than *P. striolata*?
2. Why was *Phyllotreta cruciferae* more dominant than *Pieris rapae*?
3. Why did other crucifer specialists among flea beetles, such as *Phyllotreta bipustulata* and *Psylloides napi*, fail to colonize collards when they were present in the vicinity?

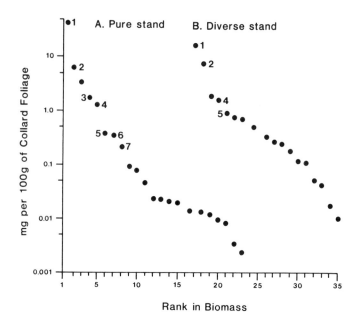

FIGURE 22.6   Relationship between biomass and rank in biomass of herbivores on collards—a dominance–diversity curve. Note that (1) the dominant species, *Phyllotreta cruciferae,* is about four times more abundant in the pure stand (A) than in the diverse stand (B); (2) dominance of *P. cruciferae* over *Pieris rapae* is much greater in (A) (by about seven times) than in (B) (by about two times); and (3) the linear part of the relationship, indicating strong dominance, is much more strongly developed in (A) than in (B). Numbers against dots indicate positions of some important species: 1, *Phyllotreta cruciferae;* 2, *Pieris rapae;* 3, *Myzus persicae;* 4, *Phyllotreta striolata;* 5, *Plutella maculipennis;* 6, *Rhopalosiphum maidis;* 7, *Rhopalosiphum pseudo-brassicae.* Modified from Root (1973).

The answers have been provided by further work by Root and his students Tahvanainen and Cromartie, who have shown that several factors are important:

1. Habitat preferences related to host-plant distribution (Tahvanainen, 1972; Hicks and Tahvanainen, 1974)
2. Microhabitat preferences (Tahvanainen, 1972; Hicks and Tahvanainen, 1974)
3. Phenology of host plant in relation to phenology of the herbivore (Tahvanainen, 1972)
4. Stand size of host plant in relation to the distinctive behavior of each herbivore species (Cromartie, 1975a)
5. Vegetational background in which host plants exist (Tahvanainen and Root, 1972; Cromartie, 1975a)

INDIVIDUALISTIC RESPONSE HYPOTHESIS

These factors, which will be treated in order, illustrate the individuality of each species and the complexity of factors that must be understood if community organization is to be described adequately. These studies tend to support the **individualistic response hypothesis** of community development, in which the community is an assemblage of species whose individuals react to resources in an idiosynchratic manner, largely independent of the other species present.

Tahvanainen (1972) studied the distribution of flea beetles on three crucifer plants: toothwort (*Dentaria diphylla*), a perennial herb that occurs in moist woodlands; yellow rocket (*Barbarea vulgaris*), a common biennial weed in open disturbed habitats; and collard (*Brassica oleracea* var. *acephala*), a biennial cultivated for its leaves and grown in experimental gardens kept free of weeds. The distribution of the four flea beetle species mentioned above differed considerably (Table 22.1). *Phyllotreta bipustulata* was specific to *Dentaria diphylla,* and since it will attack other species under suitable microhabitat conditions, microclimate appears to determine the presence or absence of this beetle. Individuals remain on the underside of lower leaves of the host plant, are slow in movement, and are relatively sedentary. They live in deep shade and move little from the perennial food plants which exist over long time spans in the eastern deciduous climax vegetation community.

Psylloides napi were the dominant flea beetle on *Barbarea vulgaris*, and phenological barriers and behavior may contribute to its low incidence on the other hosts. Young *Barbarea* rosettes were colonized soon after they were available in late summer, when the leaves of *Dentaria* have died back.

TABLE 22.1 Percentage of Each Species of Flea Beetle in the Flea Beetle Fauna on Three Species of Cruciferae Based on Flea Beetle Biomass. Percentages for Two Species are not Given, so Totals do not Reach 100%.

	Host Plant		
Flea Beetles	**Dentaria diphylla (One Plot)**	**Barbarea vulgaris (Mean of Five Plots)**	**Brassica oleracea (Mean of Four Plots)**
Phyllotreta bipustulata	91.2	0	0
Psylloides napi	0.1	84.6	≤0.1
Phyllotreta striolata	6.3	10.3	13.1
Phyllotreta cruciferae	1.4	0.2	86.7
Total	99.0	95.1	99.8

SOURCE: From data in Tahvanainen (1972).

Brassica plants were planted in the spring, thereby missing major colonization in the fall. Also, *Psylloides napi* is a relatively sedentary species that stays under the basal leaves of the host plant except after emergence of the new generation in late summer. Beetle phenology in relation to host-plant phenology is clearly important in determining the abundance of herbivores, although other factors, such as suitability of host as a breeding site, may also play a role.

Phyllotreta striolata was the second most abundant beetle on all host plants, whereas *Phyllotreta cruciferae* was less abundant on *Dentaria* and *Barbarea* but strongly dominant on *Brassica*. Tahvanainen's (1972) studies indicate that *P. striolata* selects shaded sites whereas *P. cruciferae* selects exposed sites (Fig. 20.11), and therefore the latter species becomes the dominant species in the open-grown stands of collards on which insolation is high.

Cromartie (1975a) investigated the effect of stand size on the response by herbivores. He provided potted collard plants in stands of a single plant (30 plots), 10 plants (five plots), and 100 plants (two plots) for colonization by herbivores. Samples from 30 plants for each treatment showed that *Pieris rapae* became less frequent as plot size increased, whereas *Phyllotreta cruciferae* and *Phyllotreta striolata* became more frequent, with the former beetle more frequent in the largest plots and the latter species more frequent than the other in intermediate-size plots (Table 22.2).

Cromartie (1975a,b) noted that *Pieris rapae* failed to colonize hosts in woodland or woodland edge habitats, but in the open it was the most successful colonizer in general, being common at all host-plant plot sizes and colonizing plants within a complex background of associated plant species where *Phyllotreta cruciferae*, *Phyllotreta striolata*, and alate aphids were poor colonists. Cromartie (1975a) suggested that the strong flying *Pieris rapae* females could reach isolated plants easily and that mutual interference between females may occur. With females relatively evenly spaced over an area, fewer females per plant occurred in large plots than small ones, so frequency of attack would be lower in these plots.

TABLE 22.2   Frequency of Occurance of Species of Herbivore in Collard Plots of Different Sizes on August 6, 1973. Figures Represent the Proportion of Plants Out of the 30 Sampled per Plot Type which had the Herbivore Species Present.

	Number of Plants per Plot		
Herbivore Species	*1*	*10*	*100*
Pieris rapae	1.00	0.97	0.73
Phyllotreta cruciferae	0.03	0.17	0.97
Phyllotreta striolata	0.30	0.67	0.73

SOURCE: From data in Cromartie (1975a).

ASSOCIATIONAL RESISTANCE

Tahvanainen and Root (1972) examined the role of plants associated with collards in the population ecology of *Phyllotreta cruciferae,* by use of field experiments. In 1969 they planted two collards per clay pot and 25 pots per plot for a monoculture, and two collards per pot plus a tomato seedling, with 25 pots for a diverse stand. Tomato is not a host for *Phyllotreta cruciferae.* Even with such a slight change in vegetational diversity, colonization was slower and eventual population sizes were lower in the diverse stand (Table 22.3).

Another experiment in 1970 showed even more dramatic differences between monocultures and diverse stands when two noncrucifer plants were planted in the diverse stands. Tomato and tobacco plants were placed alternately within rows of collards and plots with this arrangement were interspersed with monocultures. *Phyllotreta cruciferae* colonized the monoculture much more rapidly and populations increased to more than three times the size of those in diverse stands (Table 22.4).

In choice experiments in the laboratory *Phyllotreta* adults showed a strong preference for feeding on collard leaves in the absence of tomato leaves and leaves of ragweed (*Ambrosia artemisiifolia*), the latter being a common weed in the old fields used in Pimentel's and Root's studies, which also occurred around the plots used by Tahvanainen and Root (Table 22.5). There was also an indication that beetles were repelled by the odors of tomato and ragweed, and Tahvanainen and Root (1972) concluded that the nonhost plants interfered with finding host plants and feeding on them, and that in diverse natural communities there existed an **associational resistance** to herbivore populations which reduces the probability of herbivore population eruptions compared to monocultures. Associational resistance is thus an important component in the population dynamics of herbivores, and so it

TABLE 22.3   Colonization of Collards by *Phyllotreta cruciferae* Represented by the Number of Individuals per Plot (Mean of about 25 Replicates) when Collards are Grown Alone and with Tomatoes.

| | Sampling Dates | | | | | | | | | | |
| | June | | | | | July | | | | | |
Treatment	16	18	23	26	28	4	9	13	17	23	30
Collards alone	0.12	0.40	3.88	6.92	5.72	7.40	4.84	9.32	16.56	23.12	46.00
Collards and tomatoes	0.00	0.08	1.04	3.96	2.24	3.80	0.36	2.32	12.04	10.04	25.45

SOURCE: Tahvanainen and Root (1972); with permission.

TABLE 22.4   Populations (Number per 10 g Dry Weight of Collard) of *Phyllotreta cruciferae* in Collard Monocultures and Diverse Stands with Collards, Tomato, and Tobacco (Mean of Three Plots in Each Treatment).

| Treatment | Sampling Date | | | | |
	June 26th	July 13th	July 24th	August 6th	August 18th
Monoculture	5.5	10.5	34.1	36.3	22.0
Diverse stand	0.6	4.6	9.6	12.4	6.6

SOURCE: From data in Tahvanainen and Root (1972).

plays a significant role in community organization, especially in those communities composed largely of relatively small, specialized organisms sensitive to chemical and small physical differences in the environment.

Atsatt and O'Dowd (1976) recognized several roles that associated plants may play in influencing the functional or numerical responses of herbivores to their host plants. Just as Tahvanainen and Root (1972) had done, they emphasized the role of *repellent plants* in reducing herbivore pressure. For example, the grasses *Agrostis* and *Festuca* are grazed heavily in the absence of *Ranunculus bulbosus,* which contains a lactone, ranunculin, which is an irritant of skin and mucous membranes, and grazing pressure of cattle is reduced with increasing density of buttercups (Fig. 22.7). Associated plants may also act as *attractants* to herbivores or as *decoys.* Attractants will dilute the impact of herbivores on one plant species without reducing herbivore fitness; but decoys contain the appropriate attractants for specialist herbivores, so eggs are laid, but toxic chemicals may be lethal to feeding larvae. For example, in the presence of the normal host, *Solanum tuberosum,* the Colorado

TABLE 22.5   Numbers of Feeding Holes Made by 15 *Phyllotreta cruciferae* in Collard Leaves in 12 Hours in Choice Experiments Between Collard Leaves Alone and Collard Leaves Plus Associated Plants or Plant Odors. Figures are the Means of 10 Trials per Comparison.

Choice Experiment	Mean Number of Feeding Holes	Mean Ratio
Collard alone	72.6	5.5:1
Collard and tomato leaf	13.2	
Collard alone	88.9	4.3:1
Collard and ragweed leaf	20.5	
Collard alone	68.9	4.1:1
Collard and tomato leaf odor	16.7	
Collard alone	87.5	6.8:1
Collard and ragweed leaf odor	12.8	

SOURCE: From data in Tahvanainen and Root (1972).

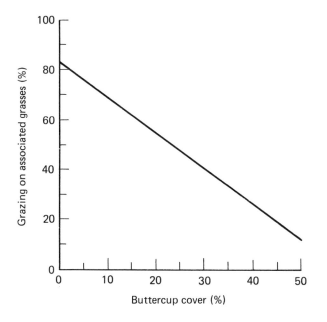

FIGURE 22.7 Selective grazing by cattle on the grasses *Festuca* and *Agrostis* in association with increasing cover by the buttercup, *Ranunculus bulbosus*. Modified from Atsatt and O'Dowd (1976).

potato beetle, *Leptinotarsa decemlineata,* laid more eggs on *Solanum nigrum* than on the normal host, but larvae survived poorly on this toxic decoy species (Hsiao and Fraenkel, 1968).

Another class of associated plants recognized by Atsatt and O'Dowd is those species that promote the welfare of parasites and predators of herbivores—*insectary plants*. They either provide nectar, shelter, or alternative hosts. For example, wild blackberries (*Rubus ursinus*) support the leafhopper, *Dikrella cruentata,* the egg stage of which is heavily parasitized by *Anagrus epos*. When grape leaves become available for oviposition by the grape leaf hopper, *Erythroneura elegantula,* large numbers of *Anagrus* transfer to grape and attack these eggs, thus significantly reducing herbivore pressure on grapes (Doutt and Nakata, 1965, 1973). Associational resistance, then, may take various forms and include associated plants that act as repellents, decoys, or insectaries for the herbivore's enemies. Presence or absence of these plants will have profound effects on the organization of herbivores in component communities.

Tahvanainen and Root (1972) emphasized the importance of associational resistance in regulating the *environmental capacity* to support herbivores, for although *Phyllotreta cruciferae* was subjected to similar levels of predation and parasitism in simple and diverse communities, the populations were much higher in the former. Thus although the populations were exposed to similar mortality factors (*limitation*), the communities provided dissimilar environmental capacities (*determination*). Factors that limit populations and those that determine the upper limit of density are both important. The authors state that "under the same relative intensities of substractive factors (limitation), completely different densities may be maintained in two localities or at different times in the same locality if the environmental capacities (levels of determination) are different" (p. 339). Morris's (1963a) studies on the fall webworm illustrate this important point, so frequently ignored in population studies of herbivores. He felt that climate and vegetation were important in determining the differences in population level between sites and concludes that "the study of differences in mean density in different areas and in different stand or crop environments deserves more attention than it has received, since it is the basis of any attempt to reduce insect damage through cultural methods" (p. 19). The understanding of population dynamics of individual species is vital to an understanding of community organization. The **individualistic response hypothesis** on community development requires more study, although it is a well-weathered concept developed by Gleason (e.g., 1917, 1926; McIntosh, 1985, 1995).

RESOURCE QUALITY

Not only may associational resistance differ from one location to another but plant quality within a species may be highly variable, with varying influences on the herbivore community. For example, allocation of plant biomass to various parts may vary enormously within a species according to the environment in which the plants grow. Under stress through root competition, *Senecio vulgaris* shows reduced reproductive effort (Harper and Ogden, 1970). *Veronica peregrina* responds to competitive stress in a dramatic manner (Linhart, 1974). Without competition, plants may reach 40 cm high and produce several hundred capsules, whereas under high competitive stress plants may be only 2–3 cm tall and produce only one capsule each. These differences in allocation and stature of plants will undoubtedly have an impact on the herbivore response and rates of colonization and utilization. Rathcke (1976b) found that stem-boring herbivores were more aggregated than their host plants because stem borers preferred large-diameter stems (Table 22.6), although large stems were less abundant than small stems. Thus resource concentration as a human would see it is very different from that experienced by a female insect ovipositing into stems.

Plant quality may also differ chemically as discussed in Chapters 5 and 6. Jones (1962) indicated that bird's food trefoil, *Lotus corniculatus*, is polymorphic for the presence or absence of a cyanogenic glucoside, and that acyanogenic plants grow and clone more rapidly and produce more seeds. The higher the level of glucoside content in the morph, the less they were eaten by herbivores such as slugs, snails, caterpillars, and a vole. Cates (1975) also found that wild

TABLE 22.6   Percent Total Exploitation of Stems by Stem-Boring Insects in each Stem Diameter Class in a Tall Grass Prairie Plant Community in Illinois.

Plant Species	Stem Diameter (mm)				
	1	2	3	4	5
Ambrosia trifida	20	100	100	100	100
Coreopsis tripteris	7	43	76	89	43
Helianthus rigidus	3	5	26	62	—
Solidago altissima	0	2	2	27	—
Monarda fistulosa	0	10	28	100	—
Total number of stems	289	575	399	210	8

SOURCE: Rathcke (1976b); with permission from *American Midland Naturalist*/University of Notre Dame.

ginger, *Asarum caudatum,* was polymorphic for growth rate, seed production, and palatability to slugs. In both cases, each morph would affect functional and numerical responses of herbivores in different ways, and thus populations with different morph ratios would support herbivore communities differing in dominance–diversity relationships. The unpalatable morphs would reduce both responses by herbivores, but also flowering phenology was delayed, thus changing synchrony of flower and seed feeders.

Plant variability and polymorphism may take many different forms, which affect herbivores in diverse ways. Resource availability must be examined, as it changes in time and space, and a fine line must be drawn between resources that at first sight appear to be one of a kind but which actually are very different to the insects concerned.

Root and his co-workers knowingly selected collard plants for their studies since they change little in form during the growing season and no reproduction occurs in the first year, when observations were made. The collards effectively represented a single resource, leaves, since no stem borers were observed, and roots were not examined. Even the component community was simplified. What happens, then, in a more complex component community when plants flower and fruit, and therefore offer several resources for exploitation by herbivores? This question was examined in detail by Thompson (1978, and unpublished; Thompson and Price, 1977).

RESOURCE DIVERSITY HYPOTHESIS

Thompson (1978, and unpublished) found that the wild parsnip, *Pastinaca sativa,* provided resources for herbivores in a different way when it grew at low densities (no conspecific individuals within 3 m) than when it grew at high densities (30–40 flowering individuals per square meter). The effective resources proved to be young leaves, mature leaves, and dead leaves (stems were hardly exploited), and the umbels provided flowers, nectar and pollen, immature (green) seeds, and mature (brown) seeds—seven resources in all. Low-density plants allocated relatively more above ground biomass initially to leaves, and then to umbels, and less to stems than high-density plants, and the resources were available to herbivores for a longer period of time at low density. Thus the low-density plants provided resources that were more available than the high-density plants, and herbivore densities were higher on the low-density plants as a result. Thus Root's resource concentration hypothesis was supported to the extent that herbivore densities were highest where resources were most available; but Thompson's results did not support the corollary that specialized herbivores reach higher densities in pure or dense stands compared to low-density stands. This was because at high densities, competition among plants increased allocation to stems and reduced allocation to the resources the available herbivores utilized, and the time over which these resources were available was reduced.

For each resource, values of 0–3 were given for availability of that resource, meaning that for 0, less than 25% of plants provided that resource on a particular sampling date; for 1, 25–49% of plants; for 2, 50–74% of plants, and for 3, 75–100% of plants. Thus for a total resource availability index, the value for each resource was summed—with seven resources the maximum total possible was 21. The number of herbivore species collected per gram of plant on a sampling date correlated closely with the resource availability index (Fig. 22.8) and accounted for an impressive 79% of the variation in species number. These data support the **resource diversity hypothesis.**

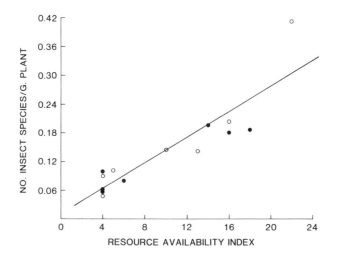

FIGURE 22.8   Relationship between number of insect species per gram of wild parsnip plant and the resource availability index. Solid circles, in high-density stands; open circles, in low-density stands. Figure by J. N. Thompson; reprinted with permission.

Of course, collards and wild parsnip offered very different kinds of resources, and the herbivore faunas were distinctive, as outlined here:

1. Collard leaves were available to herbivores throughout the growing season, from early June to September in Root's plots, and each leaf was persistent for a long time. By contrast, any one resource on a wild parsnip was available for a relatively short period of time. Young and mature leaves were available only in June. Flowers were available for about 6 weeks in low-density plants but for only 4 weeks on high-density plants.
2. Collard leaves were therefore available to several generations of some herbivores, so the full consequences of resource concentration were realized—that is, high initial colonization and high reproductive success of colonists. On wild parsnip only one generation of any herbivore was supported, so only colonization was in response to resource concentration, with no time for dominance to develop to the extent seen on collards (cf. comments on resource conditions in Chapter 21).
3. Species with the highest reproductive rates, such as flea beetles and aphids on collards, were absent from the wild parsnip fauna, so even if resources had been available longer, dominance would not have become so strongly developed as in collards.

The studies on collards and parsnip illustrate the many factors that influence community organization of specialist herbivores. They emphasize the need for a detailed understanding of how resources are displayed, the individualistic responses of herbivores to these resources, the life cycle of the herbivores relative to the duration of resources, and the plant habitat and associated plants. Part of community ecology must involve an understanding of the kinds of species that are assembled, not only their dominance and diversity relationships.

HYPOTHESES INVOKING SPACE OR TIME

On a larger scale, both in space and time, component communities have been studied comparatively in order to understand why those on some plant species are speciose, whereas those on others are relatively depauperate. Even among trees, when size differences between species are not very pronounced, such differences exist. Southwood (1960a,b, 1961) suggested the hypothesis that the number of insect species associated with a tree species is positively related to the cumulative abundance of that tree in recent geological history because abundant species offer greater opportunities for colonization by herbivores than rare species. For British trees and their insect fauna, Southwood (1961) found that abundant and widely distributed trees during the Quaternary period had the largest insect faunas, a relationship that accounted for

72% of the variance (Fig. 22.3). A comparative analysis of trees and their faunas in Cyprus, Russia, and Sweden supported the hypothesis. Southwood (1960a,b) argued that just as greater exposure to insecticide yielded more rapid resistance among insects, so greater exposure to a tree species would increase the chances of its resistance to insects being overcome, so that a new host race becomes established. Thus the evolutionary process is involved, although Southwood (1960b) recognized that these evolutionary steps could occur rapidly, as has been discussed by Bush (1975a,b).

Additional evidence in support of Southwood's hypothesis came from the Hawaiian islands, where, again, the relative abundance of trees before large-scale disturbance by humans correlated with the number of insects associated with them (Southwood, 1960a). Three important points make these data particularly valuable:

1. A large part of the flora and fauna of Hawaii is endemic, so the plant–insect relationships are likely to be unique to Hawaii and must have evolved in situ. Colonization by herbivores from the mainland must be ruled out as a major factor in ecological time.
2. Many herbivore species were known to be specific to a tree species or a genus of trees; an evolutionary event must have enabled this colonization.
3. Both the numbers of species-specific herbivores and more generalized species showed the trend of increased herbivore species on tree species of greater abundance (Fig. 22.9).

These results strongly support Southwood's hypothesis, although two different processes may be involved. One process includes an evolutionary step onto a new host with reproductive isolation of the population on the new host, so creating a species-specific herbivore. The other process involves a nonevolutionary extension of host range by a nonspecific herbivore. The Hawaiian data indicate that both processes are accelerated when hosts are abundant.

Southwood's hypothesis contains two additive components: the time a tree species has existed in geological history and its abundance during this time. However, the need for differential lengths of geological time in the accumulation of herbivore species in a component community has been questioned. Opler (1974) had evidence that the oak species in California were of similar age, yet the number of herbivore species on each differed markedly from one to another when the leaf-mining Lepidoptera were considered as a guild. These leaf-miners are usually host specific, and Opler noted that successful colonization of a new host leads to the evolution of a new herbivore species. But the area an oak species covered, being another way of estimating abundance, correlated very well with the number of leaf-miner species ($r^2 = 0.85$), and log area versus log species gave an even better fit ($r^2 = 0.90$). The latter relationship is in agreement with the predictions of island biogeography

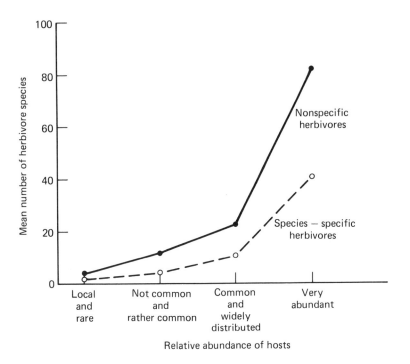

FIGURE 22.9 Relationship between number of herbivore species associated with a host tree species and the relative abundance of the host tree species in the Hawaiian islands. Both species-specific and nonspecific insects follow the same trend. Based on data provided in Southwood (1960a).

(MacArthur and Wilson, 1963, 1967), which account for the number of species on an island by the island's size without regard to geological age of the island. Thus Opler regarded oaks as evolutionary islands available for colonization by leaf-mining Lepidoptera. The larger these islands were, the greater was the probability of an evolutionary step by a herbivore onto the island.

Strong (1974a,b) also argued that Southwood's analysis of the British tree fauna did not distinguish between host-plant age in the region (i.e., the time for colonization) and host-plant distribution—a composite index of abundance was used. Also, no account was taken of very recent host-plant distribution. Yet since Southwood's analysis, the theory of island biogeography was published, which demonstrated the importance of current area in influencing the number of resident species. Therefore, Strong undertook an analysis of the relationship between current tree species area and the number of herbivore species per host tree species in Britain. He found a significant relationship ($r^2 = 0.61$, Fig. 22.10). Unfortunately, however, present range correlated with the number of fossil records of tree species ($r^2 = 0.27$), so a clear

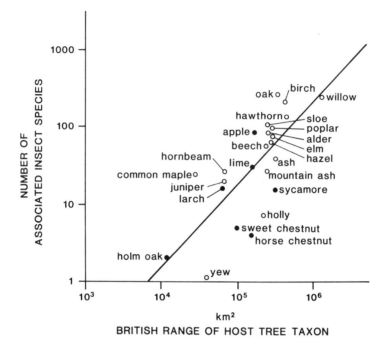

FIGURE 22.10 Species–area relationship of the insects on British trees. From Strong (1974a); with permission.

distinction between the two effects cannot be made. Strong found that with multiple regression of first range, and then number of fossil records, upon insect species richness of each tree, the first regression was highly significant but the improvement in the correlation when the second factor was added was not significant. When the reverse processes were undertaken a highly significant relationship with number of fossil records was seen, but a significant proportion of the residual variance was accounted for by the present host range. Thus the most parsimonious explanation for different numbers of species in component communities on trees in Britain appeared to be the present area of the tree as the most important single factor.

Birks (1980) used new estimates of residence times of trees in Britain and supported the contention that time is an important factor in species enrichment. We therefore have two hypotheses that need to be distinguished: the **time hypothesis,** involving geological time as a necessary factor in species richness, and the **island** or **patch-size hypothesis,** which predicts that species richness is accounted for adequately by the area over which the resource is distributed.

The lessons to be learned from detailed community studies, discussed earlier—that each species responds in a unique way to available resources—should be applied to the data analyzed by Southwood, Opler, and Strong. The faunas on British trees are made up of two fundamentally different types of species: the host-specific herbivores and the more generalized species. The former species have colonized through an evolutionary event, whereas the latter have colonized without one. A reanalysis of Southwood's data on British trees using Strong's method, but distinguishing between specialists and generalists, would be valuable. We should expect specialist species numbers to correlate better with tree abundance in time (but not necessarily fossil abundance as used by Southwood and Strong, because of problems with the fossil record; see Strong, 1974a), and generalists to correlate better with present-day abundance.

It is therefore interesting to examine the further evidence that Strong mustered in support of his thesis that component communities on plants reach an equilibrium for species within ecological time. In addition to the insects on British trees and leaf miners on oaks in California, Strong et al. (1977) cite as supportive examples parasitic fungi on British trees (Strong and Levin, 1975), insects on cacao (Strong, 1974c), insects on sugarcane (Strong et al., 1977), and mites on rodents in North America (Dritschilo et al., 1975). Mites on rodents are very generalized in their host attack pattern, fungi on trees tend to occur on many hosts, and insects on exotic crop plants such as sugarcane and cacao are most unlikely to be specialists. Therefore, we should expect rapid colonization of hosts by these generalists, in ecological time, for no speciation events are involved.

However, for host-specific herbivores, an evolutionary equilibrium is involved rather than an equilibrium through the simple relationship between colonization and extinction in ecological time as Strong argues. Opler (1974) emphasized the stability of this evolutionary equilibrium for the oak leaf-miner guilds, where colonization and extinction rates are low. For example, Opler pointed out that Santa Cruz Island has been separated from the California mainland for several hundred thousand years (it is about 20 miles from the mainland). The leaf-miner guild on *Quercus agrifolia* on the mainland consists of 14 species and on the island consists of 12 species. From Opler's species-area relationship, 12 species at equilibrium would normally be found on an island well over 100 square miles in area, yet Santa Cruz Island is less than 10 squares miles. Clearly, the component community of leaf miners on *Quercus agrifolia* on the island has not equilibrated in ecological time, since for an island that size only two or three species should be expected at equilibrium. Opler calculated that the relaxation time for this island fauna to reach its expected species-area relationship is approximately 1 million years. But the possibility that repeated extinctions are counterbalanced by repeated colonizations from the mainland in ecological time cannot be discounted completely in this case.

HABITAT HETEROGENEITY HYPOTHESIS

As the distribution of plants increases over a wider area there is accompanying this the probability that the number of habitats the plant species occurs in will also increase. Herbivores will probably be adapted to different habitats, and therefore there will be a positive correlation between the heterogeneity of habitats a plant species occupies and the number of herbivore species recorded on that plant—the **habitat heterogeneity hypothesis.** For example, Fowler and Lawton (1982) showed that the area a plant species occupied had no effect on the leaf-mining fly (Agromyzidae) species number when the effect of habitat number occupied by a plant was removed.

With the present state of knowledge it seems that the differences in opinion on the mechanisms resulting in number of species in component communities could be resolved, but the data available at present are inadequate for the task. But, in summary, the following points need to be kept in mind:

1. Where host-specific herbivores are present in the community, evolutionary time must be involved with community development.
2. Tree abundance increases the probability of colonization by specialists involving new species formation, so the most abundant trees have the largest component communities. Abundance at the present time and in geological time for trees in Britain are correlated, and both correlate with size of the insect fauna.
3. For nonhost-specific herbivores and other such parasites, abundance of hosts at the present time is clearly the significant factor influencing colonization and extinction rates of species, and therefore the equilibrium number of species in the component community. But the area covered by a host also correlates well with the number of habitats it occurs in, so the island or patch-size hypothesis and the habitat heterogeneity hypothesis should be distinguished if possible.
4. With the possibility for rapid speciation through host shifting, evolutionary time and ecological time may be very similar.

As with other debates in ecology, each proponent of a view is probably correct under certain conditions, but these may be more limiting than originally suspected.

WHOLE COMMUNITIES

Studies on whole, but simple, communities permit a detailed understanding of species interactions and their effects on community organization, just as with studies on component communities. One such thoroughly studied whole community is that in thermal spring effluents in Yellowstone National Park, investigated by Collins et al. (1976). An integrative paper on the system, based on a census of an entire spring community, is the basis for this discussion.

The water source was at 43°C and the effluent ran over a convex bank, cooling with distance from the source to a little below 25°C at the perimeters of the flow. Primary production was by a mucilaginous blue-green algal-bacterial mat, with the alga growing best at above 40°C. The major herbivore was a brine fly, *Paracoenia turbida* (Diptera: Ephydridae), larvae of which grow best at 25–35°C and require temperatures below 40°C. Other herbivores were common only at temperatures below 30°C. The major carnivores were two water mites, *Partnuniella thermalis* (Acarina: Protosiidae) and *Thermacarus nevadensis* (Acarina: Hydrodromidae), and a long-legged fly, *Tachytrechus angustipennis* (Diptera: Dolichopodidae).

The eggs of *Partnuniella* hatch into parasite larvae which jump onto brine fly adults as they walk on the algal mat. Later stages feed on eggs of brine flies and other insects. The other mite feeds on eggs, while the dolichopodid fly adult runs on the algal mat surface and feeds on eggs and young larvae. The entire community food web is illustrated in Fig. 22.11.

The simplicity of the system and the detailed and long-term nature of the studies permitted the identification of the major organizing influences in this community. As the algal mat grew, it thickened, reducing light penetration and nutrient diffusion to lower layers, thus inhibiting primary production. Also, a thick mat reduced water flow and water temperature until ultimately, algal dams were formed, creating cool patches free of flowing water. Then cool patches of algal mat provided oviposition sites for brine flies which saturated them with eggs, most eggs being inserted into the mat and thus unavailable to egg predators. Heavy feeding on algal mat by brine fly larvae resulted in the eventual breakdown of the algal dams, and renewed flow of hot water killed larvae and pupae. Brine flies persisted by rapid colonization of newly

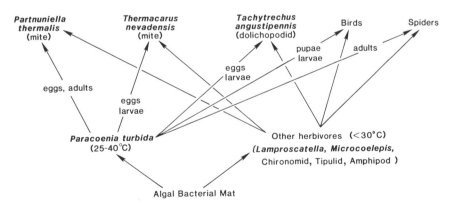

FIGURE 22.11   Food web in a high-productivity alkaline thermal spring in Yellowstone National Park, studied by Collins et al. (1976).

available patches of cool algal mat. The algal mat persisted because much of it was in water too hot for flies, or because it was underwater and unavailable to flies. The highly clumped and rapidly changing distribution of flies made exploitation by the parasitic mite difficult; the grooming behavior of flies removed more than half the mites; and most eggs were buried in the algal mat, unavailable to mites.

Collins et al. (1976) concluded that (1) although high standing crops of algae for fly larvae and adult flies and eggs for mites persisted in the system, (2) most existed in refuges from exploitation by consumers such that (3) consumers were resource limited, and (4) had little influence on the lower trophic level. Consequently, algae and flies reached high densities with strong density-dependent feedback on population growth. They say: "The fact that refuges apparently define the entire biomass structure of the spring ecosystem emphasizes the importance of understanding the kinds of mechanisms limiting the impact of consumers and defining refuges for prey." They recognized four general mechanisms:

1. Evolutionary specialization by the consumer may limit the range of conditions it can exploit below the range that is available. For example, the brine fly can oviposit in exposed algal mats but is unable to exploit algae under a moving-water film. Also, the narrow range of temperatures over which larvae of the brine fly develop best precludes effective exploitation of cooler and hotter algal mats and may be an evolutionary response to shortage of resources at higher temperatures and excessive competition from other herbivores at low temperatures.

2. Although primary production is high relative to herbivore populations, much of it may be of low-nutrient status or otherwise qualitatively inferior. Collins (1975a) noted, again in Yellowstone National Park, that another brine fly, *Ephydra thermophila*, oviposited principally in areas of high-density interstitial algae—*Euglena mutabilis* and *Chlamydomonas* spp.—in which larvae would grow. However, the major primary producer, forming the algal mat, was a filamentous dark-purple *Zygonium* sp., which apparently provided abundant food. Collins suspected that *Zygonium* differs substantially from the interstitial algae in its relative abundance of nutrients needed by the flies. If this be so, fitness is apparently greater when a narrow adaptive peak is maintained on interstitial algae than when a more generalized strategy which would include both algal types is adopted.

Mechanisms 1 and 2 above differ in that mechanism 1 specifies a reduced availability of suitable food through specialization in response to other environmental factors, while mechanism 2 specifies reduced suitability of food through specialization to maximize efficiency of food utilization.

3. Evolved defenses of the prey species may limit consumer impact (as discussed in Chapters 5 and 7). The parasitic mites in the hot spring community

are limited by host availability because of the brine fly's grooming behavior and insertion of eggs into the algal mat.

4. Consumers are limited by the uncertainty of resource distribution in space and time. For example, female mites could not predict the future locations of hosts on which their larvae could feed, and this greatly limited larval success. Improvements in larval searching may be restricted because, as Collins (1975b) points out, a larger larva that may search more effectively, and over a longer period, would also be removed more easily during host grooming. Thus "randomness in host distribution may be evolutionarily insoluble" (Collins et al., 1976) and therefore of great significance as a regulatory factor.

These four mechanisms may be further generalized into the statement that "the biomass structure of many complex terrestrial and aquatic systems may largely reflect the sizes of refuges from consumers. Likewise, consumers in general may be resource limited much more frequently than one would conclude from a casual evaluation of resource abundance" (Collins et al., 1976). The common theme in this chapter is again emphasized—that a detailed knowledge of resources in relation to the consumers present is required before community organization can be understood.

Although the consumers in the hot spring community seem to play a more passive role than we often expect of herbivores and carnivores, the authors emphasize their important regulatory role. The brine fly invades thick mats of low productivity, breaks them down with a resulting highly productive mat, and, on average, a greater productivity per square meter than in the absence of the fly. In a similar way, the predators and parasites on the brine fly reduce the density-dependent loss of immatures through scramble competition, and so increase the ratio of biomass of adult flies produced per gram of algal mat consumed. The efficiency of conversion of algae is increased. The importance of consumers as regulators, rather than energy movers, was also noted in Chapter 12.

The component community studies emphasized the details of individualistic responses of herbivores where discrete resources were rather easily defined, while studies on the hot spring community stressed the subtle nature of resource availability as an organizing influence in the community. Much of the literature discussed in this chapter was developed in the 1970s, for they were important formative years in insect community ecology, but we may well ask what emerged in the 1980s. A brief consideration of some studies follows.

In the 1980s there was a strong impetus for experimental approaches to community ecology. Agriculturally based habitat templates using several plant species seemed to be particularly tractable for studying insect herbivore communities because plant community structure could be composed, plant densities regulated, and polycultural practices in many farming systems were attractive alternatives to the agricultural monocultures dominating much of Europe, North America, Australia, and equivalent regions.

For example, in the arena of experiments on insects in polycultural systems, much research energy was reflected in the reviews of Risch et al. (1983), Altieri and Letourneau (1984), Sheehan (1986), and Andow (1988, 1991). Consideration of a few studies, eight in all, will illustrate how the various hypotheses discussed in this chapter relate to agricultural and other systems, and the different approaches adopted. In a summary of these eight studies, we may be surprised to see the general picture that emerges.

Bach (1988a,b) studied the response of chrysomelid beetles to patch size of squash (*Cucurbita maxima*) plots and the effect of associated plants. Each species of beetle responded to patch size in a different way. The western corn rootworm, *Diabrotica virgifera*, increased in density with patch size. The southern corn rootworm, *Diabrotica undecimpunctata*, showed no density change with patch size, and the striped cucumber beetle, *Acalymma vittatum*, declined in density with patch size. For the striped cucumber beetle, associated plants around plot edges had their strongest effect in small patches, in which beetle densities were highest. Associated plants seemed to promote growth of squash, and beetles responded to the larger plants apparently more than to a pure effect of associated plants.

Bach's studies illustrate the subtle nature of insect herbivore responses to vegetation quality. They support a patch-size effect, but individualistic responses to patch size by beetle species. One species showed a positive response to resource concentration, and associated resistance was not demonstrated, although there was an indirect influence through change in host-plant size. Kareiva (1983) reviews studies on the response of herbivores to size of host-plant patches, and all possible patterns are recorded. With increasing host patch size there may be increases, decreases, or no change in herbivore number per plant.

Letourneau (1986) was interested in comparing monocultures of squash with traditional polycultures used in tropical Mexico which mix squash, corn (= maize), and a legume such as cowpea. The pyralid moth, *Diaphania hyalinata*, aphids, *Diabrotica* species, and *Acalymma* species were generally higher in the monoculture, indicating an effect of associated resistance and resource concentration. But the squash bug, *Anasa tristis*, was more abundant in the polyculture, perhaps attracted by the more shaded structurally complex habitat. Again, responses were individualistic. Stanton (1983) reviewed studies on effects of plant diversity on herbivores and showed that all possible responses had been found.

Letourneau (1987) also took a careful look at the effects of monocultures and polycultures on enemy impact on herbivores in Mexico. In the polycultures of maize, squash, and cowpea, Malaise traps collected more parasitoids than in monocultures, and actual parasitoid attack on the pyralid moth, *Diaphania hyalinata*, was higher than in monocultures. This study is one of the few to measure actual impact by enemies in different cropping systems.

In Turrialba, Costa Rica, Risch (1981) studied traditional agricultural systems using corn, squash, and beans (*Phaseolus vulgaris*) in polycultures and monocultures. Responses of six species of chrysomelid beetles were examined, in the genera *Acalymma, Ceratoma, Diabrotica,* and *Paranapiacaba.* In the polyculture beetle numbers were lower than in monocultures, supporting the resource concentration hypothesis and associational resistance, resulting from different movement patterns of adult beetles. Differences in parasitism and predation were not apparent between monocultures and polycultures, so the enemy impact hypothesis was not supported. Studies on beetle movement showed higher rates of emigration from polycultures, avoidance of host plants shaded by corn, and a higher frequency of movement in polycultures, all factors reducing tenure time in a polycultural patch.

In a review on effects of agricultural crop diversity on pest species, Risch et al. (1983) covered 150 studies and 198 herbivore species. Fifty-three percent of species were less abundant in more diverse plant species systems, 18% were more abundant, 9% showed no difference, and 20% showed varying responses. Therefore, polycultural systems appear to be favorable for reducing pest numbers most of the time and Risch's (1981) system may be representative of many.

Response of two lepidopteran species to collard patch size was studied by Maguire (1983). The imported cabbage worm, *Pieris rapae,* was more numerous per plant in small patches than in large patches, as Cromartie (1975a) had found. But the diamond-back moth, *Plutella xylostella,* was more abundant in the large patches. Patch size had an inconsistent effect and the herbivores responded individualistically.

Rosebay willowherb or fireweed (*Chamerion angustifolium*) is commonly found in England and Scotland, and its insect community was studied by MacGarvin (1982) in relation to size of natural patches. There was a weak positive patch size effect in the number of phytophagous insect species present. On a per plant basis the number of species and the number of individuals increased with patch size, suggesting a resource concentration effect and a reduced extinction rate with larger patches. However, some insect species showed positive slopes with number per plant and patch size; others, negative slope; and some, no significant pattern at all, suggesting individualistic responses to patch size.

Phytochemical effects on community structure have not been studied very frequently, but Jones and Lawton (1991) found that plant chemistry had very little explanatory power regarding insect species richness per host-plant species in the umbellifers of Britain. However, Berenbaum (1981) showed that the major chemical defenses in the group, furanocoumarins, play a role in defining the nature of the community relative to its composition of generalists and specialists. Among umbellifers, some, including wild carrot, *Daucus carrota,* have no furanocoumarins; some have only linear furanocoumarins, such as the water parsnip, *Sium suave;* and some have both linear and the more complex angular furanocoumarins, such as angelica, *Angelica atropurpurea.* One species, poison hemlock, *Conium maculatum,* has linear furanocoumarins and piperidine alkaloids, making it chemically very distinct from other members of the family. With this escalation of defenses there is a change from a preponderance of generalist herbivores to a preponderance of specialist herbivores. The percentage of all insect individuals collected represented by specialists increased from under 25% on wild carrot to over 90% on poison hemlock. Clearly, when phytochemistry differs considerably between members of a plant family, as in the Umbelliferae (= Apiaceae), effects on herbivores may be strong even though numbers of species show little or no pattern, as found by both Berenbaum (1981) and Jones and Lawton (1991).

Finally, the eighth study has been selected to illustrate that plant genotype can have an effect on the proportion of species present and density of individuals per species (Fritz and Price, 1988). Willow cuttings from several genotypes were grown in pots and exposed to galling sawflies: leaf gallers, leaf folders, petiole gallers, and stem gallers. Genotypes of plants had significant effects on number of individual insects per species per plant and their presence or absence. In 1986 plant genotype accounted for the following percentages of variation in density of sawflies per species: leaf galler, 9%; petiole galler, 28%; leaf folder, 48%; stem galler, 31%. In addition, species densities per plant were positively correlated for most pairwise comparisons of species. This was because some genotypes grew more vigorously than others and those were more heavily attacked. So species tended to aggregate on plants because of similar responses to resource variation. They did not segregate across resources, as if competition had driven divergence of resource utilization. These results supported field observations and experiments reported by Fritz et al. (1986, 1987) showing positive association of species and plant genotype as an important determinant of community structure.

Can anything general be said about such a wide range in the eight kinds of studies undertaken and discussed here? Several important points need to be recognized.

1. Most studies indicated individualistic responses of species to resource variation (Table 22.7). This suggests that assemblages of insects are involved rather than communities. Competition, for example, was never invoked as an important determinant of species distributions or abundances.

2. There is evidence that all hypotheses considered can play a role in assemblage structure. However, we cannot be sure that a hypothesis will predict the correct effect. For example, associational resistance, or enemy impact, may or may not be important.

3. With the emphasis on studying individualistic responses of herbivores to resources comes a better understanding of the kinds of species in assemblages. Some are colonizers of small patches, some of large patches. Specialists and generalists separate out depending on the phytochemistry of hosts. Some species aggregate because of common responses to resource quality.

TABLE 22.7 Summary of Eight Studies on Communities Discussed in the Text.[a]

Publication	Individualistic Response	Community or Assemblage	Resource Concentration	Associational Resistance	Enemy Impact	Patch Size	Resource Diversity	Competition
Bach (1988a,b)	×	A	×	—		×		(−)
Letourneau (1986)	×	A	×	×				(−)
Letourneau (1987)	×	A			×		×	(−)
Risch (1981)	×	A	×	×	—			(−)
Maguire (1983)	×	A				×		(−)
MacGarvin (1982)	×	A	×			×		(−)
Berenbaum (1981)	×	A						(−)
Fritz and Price (1988)	×	A						(−)

[a] ×, hypothesis was tested and supported; —, hypothesis was tested and rejected; (−), factor not invoked as important; A, species probably acting as an assemblage rather than a community.

But clearly, there is still much room for progress. With so much individual-istic response evident, we need to understand why individuals respond to re-sources in a certain way and what the consequent population dynamics are after arrival. The need for understanding behavior of individuals is critical. Kareiva (1983, 1986, 1989) has emphasized the need for studies on insect foraging behavior and the development of theory on herbivore searching. In most plant-based assemblages of insects, repeated colonization and the bal-ance between rates of immigration and emigration are of central concern. For agricultural systems, temperate vegetation, dry tropical vegetation, and oth-ers, the seasonality of plant growth forces recurring bouts of colonization by insects. How insects make decisions and their behavior during colonization need much more attention.

DOMINANCE–DIVERSITY RELATIONSHIPS

Because of the complexity of compound communities, some shortcuts in their study may be needed. The detailed information gathered on component com-munities and very simple whole communities discussed above cannot usually be obtained on the many species present in compound communities of in-sects. Some of these shortcuts include the use of species-area curves and dom-inance–diversity relationships as predicted by the logarithmic series, the log-normal distribution, the random niche boundary hypothesis, and the niche preemption hypothesis (Whittaker, 1970b, 1972). The dominance-diversity relationships now have very limited use (see May, 1975; Hengeveld et al., 1979), particularly if the total number of species present is small, as measures of diversity (such as H' discussed in Chapter 23) tend not to distinguish be-tween the random niche boundary hypothesis, the niche preemption hypothe-sis, and the log-normal series (Fig. 22.12). With large samples the log-normal series is statistically inevitable and thus provides no insights into community organization. Even for the biologically meaningful distributions, interpreta-tion of the processes that result in these distributions, from the distributions themselves, is "a tactic of weak inference" (Whittaker, 1972). Some form of index of resource availability must be obtained as well as the dominance-diversity relationships of consumers

SYNOPTIC INDICES OF RESOURCES

Another shortcut has been to use a simple synoptic index of resources in a community and to observe how this correlates with the organisms present. The indices that have proved to be most useful are plant species diversity and foliage height diversity. But for rather specialized organisms such as insects

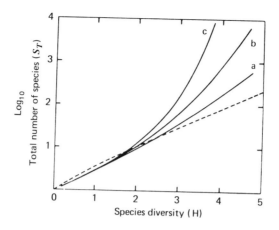

FIGURE 22.12 Comparison of the random niche boundary distribution (broken-stick model) (dashed line) and log-normal distributions for various dominance–diversity relationships (lines a, b, and c) showing that at low diversities ($H = 0 - 2.5$), with relatively small numbers of species present, the two models are essentially identical. From May (1975); with permission.

these synoptic indices may be inadequate because of the many subtle responses of organisms to resources and the subtle nature of resource availability, as discussed above.

Murdoch et al. (1972) set out to test the applicability of the synoptic indicators of resources, plant species diversity, and foliage height diversity to an understanding of insect communities. They studied three old fields abandoned in 1920 in Michigan. Field 1 was dry and dominated by the grass *Poa pratensis* and had a low plant species diversity compared to fields 2 and 3. Field 2 was a dry upland site with little grass (but one of the major species was *Poa compressa*), dominated by forbs such as *Lespedeza* spp. and *Rumex acetosella*. Field 3 was a damp meadow with relatively high production of forbs (e.g., *Hieracium* spp. and *Daucus carota*) and grasses (particularly *Poa pratensis*). The Homoptera were studied in the fields as the dominant group of insect herbivores by sweep net samples. The plant community was described by estimating plant cover per species in 1-m² quadrats for plant species diversity and using a point-quadrat method for foliage height diversity and plant species diversity. At each point a long needle was passed at 45° to the vertical through the vegetation and each contact with a plant recorded as to species and height in three height classes: 0–6 in., 6–18 in., and above 18 in.

In general, the correlations between foliage height diversity and plant species diversity, and these indices with insect species diversity, were weak when within-field comparisons were made. When correlations were significant they accounted for less than 50% of the variance, and when tested over

2 years such correlations were not significant in both. The methods using synoptic indices were not sufficient for predicting the detailed distribution of insect species. However, the gross differences between fields had profound influences on the insect community so that between-field comparisons yielded significant correlations: plant species diversity with foliage height diversity (e.g., $r^2 = 0.66$), plant species diversity with insect species diversity (e.g., $r^2 = 0.52 - 0.72$), and foliage height diversity with insect species diversity (e.g., $r^2 = 0.71$).

Based on an earlier discussion in this chapter and other considerations, we can see why the correlations within fields cannot be expected to be strong:

1. Associational resistance is not measured by the synoptic indices, foliage height diversity, and plant species diversity (see Tahvanainen and Root, 1972; Atsatt and O'Dowd, 1976).
2. Stand-size differences of species between sample sites are inadequately represented in synoptic indices, although stand sizes have marked influences on the species diversity of herbivores (Tahvanainen and Root, 1972; Root, 1973; Cromartie, 1975a).
3. Different species of host in the same plant family may have very different forms and height, yet support the same herbivores, or very similar spectra of herbivores (cf. flea beetles on *Dentaria diphylla*, a very small low plant, with the larger *Barbarea vulgaris* and the even larger *Brassica oleracea* in Tables 1 and 2 in Tahvanainen, 1972).
4. The addition of one plant species in a community, for example, a crucifer, while changing plant species diversity little, may add a large number of specific herbivore species and significantly increase insect species diversity.
5. Conversely, the addition of many species of plants, so changing plant species diversity significantly, may add no new herbivore species. For example, no cicadellids, usually the most abundant herbivores in old-field communities, are specialists on annual grasses. Therefore, annual grasses play a very different role from perennial grasses in insect community organization, at least in relation to cicadellids.
6. Density of foliage may increase in all strata and thus change resource concentration significantly, without influencing foliage height diversity.
7. Although Singer (1971) demonstrated the importance of plant structure in determining suitability of hosts, the relevant differences between potential hosts are too small to be detected by a measure of foliage height diversity. For example, oviposition by the butterfly, *Euphydryas editha*, may ultimately be determined by the availability of leaves near the ground. Thus some plants, such as *Plantago erecta*, make good oviposition plants, whereas others, such as *Orthocarpus densiflorus*, make poor oviposition plants because of differences in structure (Fig. 22.13), although both are suitable larval food plants (Singer, 1971).

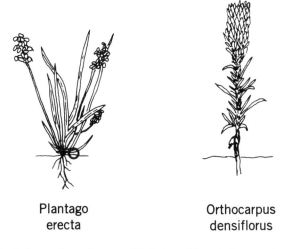

Plantago
erecta

Orthocarpus
densiflorus

FIGURE 22.13   Two food plants of butterfly larvae of *Euphydryas editha* in the form experienced by ovipositing females: *Plantago erecta* showing leaves close to the ground, the undersides of which are utilized as oviposition sites; *Orthocarpus densiflorus* lacking leaves close to the ground and suitable oviposition sites. From M. C. Singer, *Evolution*, 1971, **25**:383–389.

Many other factors could be listed as detractants from the plant species diversity or foliage-height diversity relationships with insect species diversity. In terms of habitat suitability for a single herbivore, Singer (1972; see also Chapter 15) has provided an understanding of the complexity of resources available to adults and larvae of *Euphydryas editha*. Now that we understand some of the details of individualistic species responses to resources, it should be clear that broad general approaches to community organization will probably seldom aid in understanding, and that a concentration on resources and response by organisms to these resources is a necessary approach (see also Kareiva, 1983).

To summarize this chapter, seven hypotheses have been defined, all of which may be important at some time or place in insect community organization either singly or in combination.

In the order that they have been mentioned, these are:

1. Enemy impact hypothesis
2. Resource concentration hypothesis
3. Individualistic response hypothesis
4. Resource diversity hypothesis
5. Time hypothesis
6. Island or patch-size hypothesis
7. Habitat heterogeneity hypothesis

All but hypothesis 3 are discussed by Price (1983, 1984a,b), with more references in support of each hypothesis provided and with some dissenting views expressed. Other hypotheses are also discussed. It is noteworthy that so many of the hypotheses involve resources as the basis for understanding community organization and that competition is not invoked as a major organizing influence (c.f. Lawton and Strong, 1981; Strong et al., 1984b).

The challenge in future work will be to determine for insect communities which hypotheses are valid and when. Studies designed to test several hypotheses simultaneously will be essential, and these will yield a very fruitful area of research in the coming years.

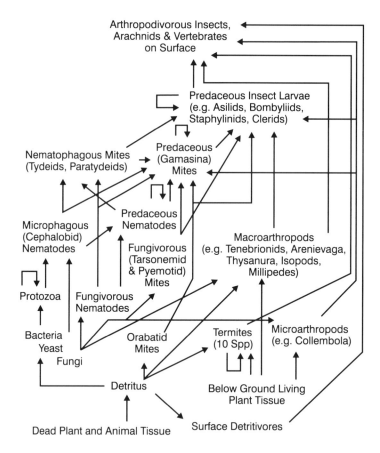

FIGURE 22.14   Simplified food web of trophic interactions within sandy soils in the Coachella Valley, California. Arrows turning back on a taxon, such as with termites and protozoa, indicate that cannibalism is present, as in the following figures. Note that this is actually only a component, or subweb, of the full trophic system, also involving living plants and animals above ground. The single taxon "termites" is represented by 10 species, each with its own complex trophic linkages, and "tenebrionids" are represented by 14 species. From Polis (1991b); with permission.

Community ecology is a much debated topic at present, and several publications convey this situation, for example, Cody and Diamond (1975), Price et al. (1984), Strong et al. (1984a,b), Diamond and Case (1986), Kikkawa and Anderson (1986), Lawton and MacGarvin (1986) Esch et al. (1990), Polis (1991a), and Richlep and Schluter (1993).

On a sobering note to complete this chapter, let us contemplate the complexity of interactions that enter into the presence and absence and relative abundance of organisms in any usual terrestrial community. We should also reflect on how much we have learned and advanced beyond early developments in this field of community ecology.

We have tended to simplify nature in this book, as one needs to do, in order to capture some basic processes and the essence of what matters in nature. However, toward the end of these considerations, it is healthy to

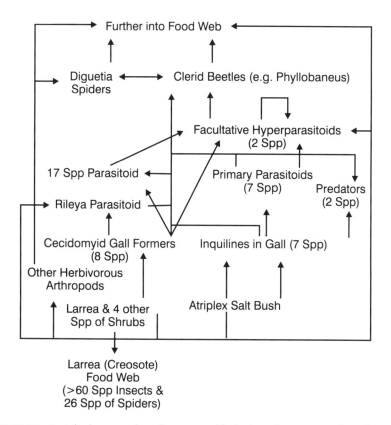

FIGURE 22.15 Reminder that apparently small components of food webs can be very speciose, here taking a single guild of herbivores, the gallers, on a single plant species, the saltbush, *Atriplex canescens*, in the Coachella Valley. In fact, 67 species interact with cecidomyiid galls (Hawkins and Goeden, 1984), and many of the carnivores also enter into subwebs based on creosote bush, *Larrea divaricata* (cf. Schultz et al., 1977). From Polis (1991b); with permission.

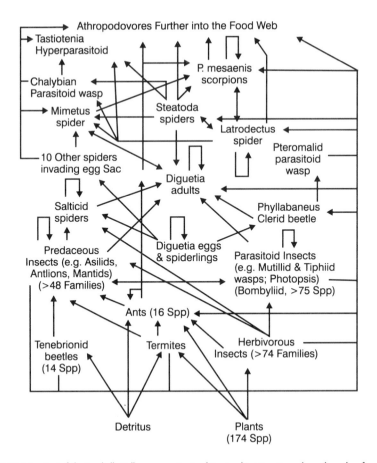

FIGURE 22.16 Part of the Coachella Valley community, involving trophic interactions above the soil surface and focusing on predaceous arthropods, especially the spiders, *Diguetia mohavea* and *Latrodectus hesperus*. From Polis (1991b); with permission.

reflect on how simplification distorts our view of nature. We have discussed bottom-up versus top-down influences in food webs, and three-trophic-level interactions, as if each trophic level is discrete. In this chapter we have discussed communities of insects, usually existing on one trophic level as herbivores on plants. Where do reality and abstraction actually intersect?

Reality shows us that whole communities are commonly extremely complex and that food web diagrams seldom, if ever, capture the full network of interacting factors. Many species of animals, as consumers, act as omnivores, feeding on several trophic levels, and their densities are influenced by resources "across the trophic spectrum, the herbivore and detrital channels, other habitats . . . " (Polis and Strong, 1996).

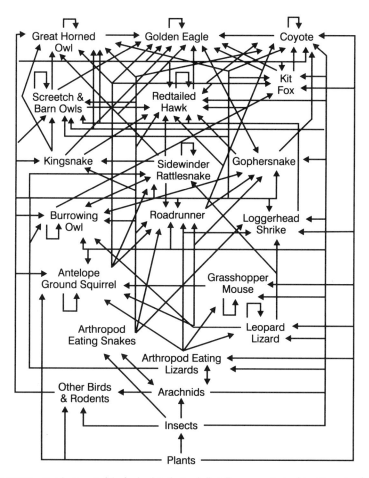

FIGURE 22.17   Vertebrate part of the food web in the Coachella Valley community involving 96 species, developed by Polis. Lines with an arrow at both ends indicate looping of interactions through mutual predation. Note the absence of a top predator. From Polis (1991b); with permission.

Polis (1991b) provides particularly telling analyses of food webs in the Coachella Valley in Riverside County, California. Starting from the ground up, Polis provides a simplified food web, in reality a subweb in the trophic system, of interactions in sandy soils (Fig. 22.14). Then he takes one example of a plant species, the saltbush, *Atriplex canescens,* and only one guild of herbivores, the gall-forming insects, to illustrate the complexity of linkages involved (Fig. 22.15). Aboveground, there are also complexities involving the subweb of predaceous arthropods (Fig. 22.16), and 96 vertebrate species (Fig. 22.17). If this kind of realism is general, how should we approach the analysis of communities in the future? Will we find general patterns and general hypotheses, and will factually based theory eventually emerge?

Oksanen (1991, p. 294), never known to be pusillanimous in ecological circles, has also warned us that progress in understanding communities is more apparent than real:

> Much of the apparent progress in community ecology amounts to little more than re-inventing the wheel, albeit with technical improvements. Many central ideas in the field were stated by A. K. Cajander at the turn of the century. Thereafter, community ecology has moved back and forth between competition-centered and individualistic views of community structure. The chief problems have probably been inappropriate methodology (lack of rigorous formulation and/or critical experimental testing of theories) and a tendency to work only on subcommunities that are delimited by taxonomic criteria. Although these problems are beginning to be remedied, a new one has emerged: ignorance of history and a tendency to re-invent old ideas under new, flashy names. This problem is potentially as dangerous as the old ones and must be tackled by improving our first-hand contact with the classics in our field.

Chapter 23

Diversity and Stability

Diversity is synonymous with variety and is generally used in ecological circles to relate to the variety, or number, of species in a community, on an environmental gradient, or some other logical unit of the environment. When we consider a variety of habitat types or resources we tend to use the term *heterogeneity,* but there is certainly no rule on wording in ecology relating to variety.

The simplest measure of diversity, and the most commonly employed, is to use the number of species present. We saw several examples in Chapter 22 when community structure was discussed. However, abundance of each species is an interesting additional aspect of the community and we may wish to show the abundance or biomass of each species as another component of diversity. Such relationships can be plotted as dominance–diversity curves, as we saw for the insect community on collards (Fig. 22.6). If the abundance of

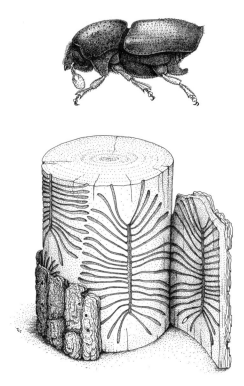

Are there vacant niches available for insect species arriving in a new environment? The smaller European elm bark beetle, *Scolytus multistriatus,* found its niche on American elm, *Ulmus americana,* and with the aid of the tree pathogen that it transported, *Ceratocystis ulmi,* its activities resulted in the death of most elm trees in the United States. Infection of the tree results from adult feeding on twigs, during which inoculation of the fungus occurs. Fungal infection travels from the crown, eventually killing the tree. The adult is above, and the maternal and larval tunnels in new wood and bark are below. Based on Novák et al. (1976). Drawing by Tad Theimer.

each species is plotted against its rank in abundance, such diagrams may be called rank-abundance plots (e.g., Begon et al., 1996a; Stiling, 1996). Rank-abundance plots provide a full display of the number of species present and the abundance of each species. We can see how many common and how many rare species are present in each community. However, the comparison of many communities may become cumbersome, such that simple indices of diversity become popular, although one number as a diversity index for a community clearly hides much information.

SHANNON–WEAVER DIVERSITY INDEX

Diversity may be defined by use of a formula. Many indices of diversity have been proposed (see Pielou, 1969, 1975; Poole, 1974; Begon et al., 1996a; Stiling, 1996), but the one most commonly used in the past is the **Shannon–Weaver diversity index** H', where

$$H' = -\Sigma \, p_i \log_e p_i$$

and p_i is the proportion of the ith species in the total sample. If two species are unevenly abundant, their diversity is lower than when they are equally abundant (Table 23.1). Also, adding species increases the diversity, so that a community with three species is likely to have a diversity higher than a community with only two species, even though the latter may be evenly distributed. Thus number of species (species richness) in the community and their evenness in abundance (or equitability) are the two parameters that define H'. As species are added, diversity increases, and as species become evenly distributed in abundance, diversity increases. In a diverse situation, species cannot be very dominant and in a low diversity community one or two species will be much more abundant than others. This index has commonly been used because it is easily calculated, it includes the two components of diversity that are intuitively important, and these components can readily be separated since the number of species can be counted, and their evenness, J', can be estimated from the formula

$$J' = \frac{H'}{H'_{max}} = \frac{H'}{\log_e S}$$

TABLE 23.1   Examples of How Diversity Measured by H' Changes with the Evenness of Distribution of Species and the Number of Species Present (or Species Richness).

	Species 1	Species 2	Species 3	H'
Two species	90	10	—	0.33
Two species	50	50	—	0.69
Three species	70	15	15	0.82
Three species	33.3	33.3	33.3	1.10

where S is the number of species present. H'_{max} may also be calculated by assuming that all p_i's are equal in the calculation of H' (see Lloyd and Ghelardi, 1964; Lloyd et al., 1968). In general, the popularity of this kind of index has faded, partly because species number and abundance per species are so important that they each need explicit treatment, and partly because the search for pattern often proceeds in the absence of abundance data. Some examples of how data are analyzed and presented may be found in Strong et al. (1984a,b), Gaston (1988), Gaston and Lawton (1988), Lawton (1990, 1991), Ricklefs and Schluter (1993), Brown (1995), and Eggleton et al. (1996).

DIVERSITY IN PLANT SUCCESSION

How does diversity change from one community to another? Whittaker (1969) gives a good summary of the way in which plant species diversity changes in response to many different gradients in time and space. As far as plant succession is concerned, he says that diversity increases during succession but may in many cases decrease from some stage before the climax into the climax itself. He gives 10 references to this trend (see also Loucks, 1970). In a plant succession from bare ground caused by fire to a mature oak forest on Long Island, New York, Whittaker and Woodwell (1968, 1969) found the trends shown in Fig. 23.1. That is, the more extreme early stages of succession have fewer plants adapted to such harsh environmental conditions; species diversity, production, and biomass are low, but these rise to a plateau (cf., Fig. 2.10). Species diversity reaches a maximum at about 50 years in the early forest stage, when some of the grasses and forbs of earlier successional

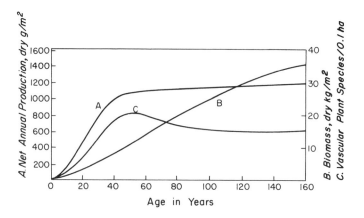

FIGURE 23.1   Trends in (A) production, (B) biomass, and (C) plant species diversity (richness) in forest succession after fire on Long Island, New York. Data from Whittaker and Woodwell (1968, 1969). Reprinted from R. H. Whittaker, *Communities and ecosystems*, by permission of Macmillan Publishing Co., Inc. Copyright © 1970 by Robert H. Whittaker.

stages are still present. These eventually fail in the community and the plant diversity declines slightly. Of course, many factors change on successional gradients in terms of physiological relationships among plants (Bazzaz, 1979), the chemical constituents of plants relative to herbivores, and many other traits, with some effects passing up the trophic system to carnivores (Table 23.2). In response, arthropod species change (Price, 1984c), species richness increases, and the range of population dynamics of many residents probably increases also (P. W. Price, 1991a,b, 1994b).

DIVERSITY ON LATITUDINAL GRADIENTS

A similar trend to that of the first 50 years in the forest studied by Whittaker and Woodwell can be seen in species diversity on a latitudinal gradient from the high harsh latitudes to low latitudes with an equable climate (Fig. 23.2). This trend is shown by insects, molluscs, vertebrates, plants, and many other taxa (e.g., Table 23.3; Stevens, 1989), although some freshwater inverte- brates, phytoplankton, and some parasitoids seem to be exceptions. These sorts of trends have long fascinated biogeographers and ecologists. Why is there such a diversity of species in the tropics? What are the critical factors that promote, or permit, this diversity? How does natural selection affect community organization? Connell and Orias (1964), Pianka (1966), Uetz (1964), Pielou (1975), Rohde (1978a), and Thiery (1982) have summarized some of the hypotheses that have been proposed to explain this diversity gra- dient, and here Pianka's explanations are followed. Wilson (1974), Brown and Gibson (1983), Rohde (1992), Rosenzweig (1995), and Begon et al. (1996) also consider some of the hypotheses.

HYPOTHESES ON SPECIES RICHNESS

1. Time

Proponents of this hypothesis assume that all communities tend to diversify with time, and therefore the older communities have more species than do newer ones. For example, temperate regions have much younger communities than do tropical regions because of recent glaciations. Thus temperate faunas and floras are impoverished because (1) species that could live in temperate regions have not migrated back from unglaciated areas, and (2) species have not had time to evolve to enable exploitation of temperate areas. South- wood's (1961) data on insects on trees in Britain support this hypothesis (see Chapter 22 and Fig. 22.3). The longer the time available, the more species adapt to the resource. Also, Wilson's (1969) hypothesis on the four stages in community development—the noninteractive, interactive, assortative, and evolutionary—subscribes to the same view. Simpson (1964) has argued

TABLE 23.2 Examples of Patterns in Ecological Succession Relevant to Insect Herbivore Population Dynamics in Moist North Temperate Vegetation Dominated by Forest.[a]

Characteristic	Early Succession	Late Succession	Source
		Plants (P)	
1. Dominant species	Annual herbs	Large trees	Bazzaz (1968, 1975)
2. Individual longevity	Short, 1–5 yr	Long, 50–500 yr	Odum (1959)
3. Size	Small	Large	Odum (1959)
4. Physiological rates	Generally high	Relatively low	Bazzaz (1979)
5. Apparency	Low	High	Feeny (1976)
6. Chemical defense	Toxins	Digestibility reducers	Feeny (1976), Rhoades and Cates (1976)
7. Spatial Distribution	Patchy	Extensive tracts	Loucks (1970), Pickett and Thompson (1978), Shugart and West (1981)
8. Biomass per unit area	Low	High	Odum (1969), Whittaker (1970)
9. Biomass of leaves per plant species	Low	High	Ovington (1962); Whittaker and Woodwell (1968, 1969)
10. Number of species per unit area	Low but increasing rapidly	High	Bazzaz (1975), Whittaker (1970)
11. Structural diversity	Low	High	Southwood et al. (1979), Lawton (1983)
		Herbivores (H)	
1. Insect species richness	Low	High	Lawton (1983), Southwood et al. (1979), Leather (1986)
2. Host plant species specificity	High	Low	Futuyma (1976)
3. Palatability of plant food	High	Low	Reader and Southwood (1981)
4. Flight capacity	High	Can be very low or absent in females	Barbosa et al. (1989), Roff (1990)

5. Evolution of egg load per plant	Can be high	Can be absent	Thompson (1983), Price et al. (1990)
6. Linkage of ovipositional preference and larval performance	?	Can be low or absent	Price et al. (1990)
7. Potential population size	Low	High	Price (1992)
8. Potential population persistence	Low	High	Price (1992)
9. Local abundance	Low?	High?	Gaston (1988), Gaston and Lawton (1988)
10. Regional distribution	Low?	High?	Gaston (1988), Gaston and Lawton (1988)
11. Population variation	Low?	High?	Gaston (1988), Gaston and Lawton (1988)
12. Population dynamics	Usually latent?	Many species epidemic	Price et al. (1990)
13. Adult size	Small	Large	Niemalä el al. (1981)
14. Feeding pattern	On vigorously growing plant parts, with toxins, if specialists (cf. H2)?	Toxin avoiders, if generalists, feed on oldest leaves (cf. H2)?	Cates (1980, 1981)
		Carnivores (C)	
1. Species richness of vertebrate predators	Low	High	Odum (1950), Johnston and Odum (1956)
2. Population size of vertebrate predators	Low	High	Johnston and Odum (1956), Odum (1959)
3. Species richness of parasitoids	Low	High	Hawkins (1988), Hawkins et al. (1990)
4. Mortality caused by parasitoids	Low?	High?	Price and Pschorn-Walcher (1988), Gross and Price (1988), Hawkins and Gross (1992)
5. Probability of host limitation by parasitoids	Low	Relatively high	Hawkins and Gross (1992)
6. Host specificity in parasitoids	High	Low	Hawkins et al. (1990), Sheehan (1991)

[a] Only endpoints are listed; the original literature needs to be consulted for details and trends in middle succession. Speculative generalizations are followed by a question mark. Most generalities about herbivores apply to the Lepidoptera.
SOURCE: P. W. Price (1991a). Reprinted by permission of the Entomological Society of America.

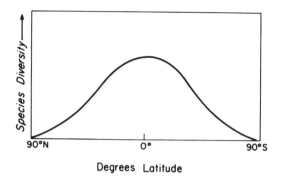

FIGURE 23.2   Trend in species diversity from the north pole (90°N), across the equator (0°), to the south pole (90°S).

against this hypothesis on the grounds that warm temperate regions have had a long undisturbed history, from the Eocene to the present, yet contain fewer species than the tropics. Also, if the time hypothesis were correct, the steepest gradient in species diversity should occur in recently glaciated temperate zones. In North American mammals at least, the gradient is not steep, so that the theory is not supported by data on mammals. More data are needed, but Wilson (1974) should be consulted for a reanalysis. Based on a group of marine parasites, Rohde (1978a,b) supports this hypothesis, although he concluded in 1992 that it was the greater *effective* evolutionary time, or speed of evolution, in the tropics that resulted in higher diversity.

2. Spatial heterogeneity

There might be a general increase in environmental complexity on the latitudinal gradient toward the tropics. The more heterogeneous and complex the physical environment becomes, the more complex and diverse will be the plant and animal communities supported by that environment. Certainly, to-

TABLE 23.3   Latitudinal Trends in Numbers of Species (Richness) in Insect Taxa Showing an Increase with Decreasing Latitude.

Beetles[a]		Ants[b]		Dragonflies[c]	
Labrador	169 species	Alaska	7 sp	Nearctic	59 species
Massachusetts	2000 species	Iowa	73 sp	Neotropical	135 species
Florida	4000 species	Trinidad	134 sp		

[a] From Clarke (1954).
[b] From Fischer (1960).
[c] From Williams (1964).

tal diversity due to topographic relief and, consequently, the number of habitats increase toward the tropics. Temperature gradients and moisture gradients (ranges) are much longer in the tropics than in temperate regions. This will be seen in Chapter 24 when the Holdridge (1947, 1967) (Holdridge et al., 1971) system of classifying world plant formations is considered (cf. Fig. 24.17). Unfortunately, this diversity of habitats does not explain why there should be so much diversity within habitats in the tropics. Vegetative heterogeneity presumably depends on microhabitat diversity, which has not been studied, although Janzen's (1967c) paper refers to this and invokes this as the causative agent in restricting plant dispersal in the tropics, thus increasing diversity. But Janzen's paper is better dealt with under climatic stability. Once plant species diversity increases, the animal species diversity follows, according to MacArthur and MacArthur's (1961) study, as the structural diversity of the community will increase. So some emphasis should be placed on plant species diversity.

3. Competition hypothesis

Dobzhansky (1950) and Williams (1964) have proposed that natural selection in temperate zones is controlled mainly by the physical environment, whereas in the tropics biological competition is more important in evolution. Therefore, in the tropics there is a much greater restriction of food types and habitat requirements, and thus more species can coexist in a given habitat. Tropical species will be more finely adapted and have narrower niche exploitation patterns than those in temperate regions. In temperate zones, mortality is often catastrophic and indiscriminate, that is, density independent, such as drought and cold, and r-selection for increased fecundity and development rate will be important, rather than selection for competitive ability or K-selection seen in tropical species (see Chapter 14). What little empirical evidence exists indicates no greater specificity in tropical than in temperate insects (Beaver, 1979; Price, 1980, 1991c) and gill flukes (Rohde, 1978c, 1979, 1981).

4. Predation hypothesis

Paine (1966) has claimed that there are more predators and parasites in the tropics and that these suppress prey populations sufficiently to reduce competition between them. Reduced competition then permits coexistence of additional prey species, which therefore permit additional predators in the system. According to this hypothesis, competition among prey organisms is less intense in the tropics than in temperate areas, and this predicts the opposite to the competition hypothesis. However, there is evidence to support the predation hypothesis. Grice and Hart (1962) have shown that the proportion of predatory species in the marine zooplankton increases along a latitudinal diversity gradient. Janzen (1970, 1973b) and Harper (1969) also present information suggesting that herbivores, particularly plant predators, are necessary agents in maintaining or permitting high plant species diversity, and Elton (1973) argued that long-term predation pressure has resulted in the very low densities of insect herbivores in tropical rain forests. Menge and Sutherland (1976) reasoned that the competition and predation hypotheses were complementary, because competition seems to be more important at higher trophic levels, while predation plays a stronger role at lower trophic levels.

5. Climatic stability

Klopfer (1959), Klopfer and MacArthur (1960), and others have suggested that regions with stable climates allow the evolution of finer specializations and adaptations than do regions with more changeable climate, because resources remain more constant in the stable conditions. Just as in the competition hypothesis, smaller niches are predicted and therefore closer species packing. Although Janzen (1967c) says that his paper, "Why mountain passes are higher in the tropics," is not an attempt to explain tropical species diversity, his paper clearly relates to the climatic stability hypothesis. Janzen suggests that greater sensitivity to change is promoted by less frequent contact with change. Therefore, in a stable environment small changes in conditions will be as effective as large changes in temperate regions. Since there are more small changes in a habitat than large ones, this will result in more species being able to colonize that habitat and coexist in it. Sanders (1968, 1969) found more species in deep-sea environments, where conditions are relatively stable, than in shallow waters, where physical factors are much more variable. Randolph (1973) showed that snails tended to be generalists in variable environments and more specialized in stable environments. The fact that tropical environments are more predictable than are temperate ones is shown in Costa Rica, for example. Although the climate is much more uniform than in temperate regions, eight seasons in the year are recognized, as opposed to four in temperate latitudes. The more precise fitting of population activities to more predictable environmental conditions, the less will be the ability to tolerate different conditions outside the usual habitat. Therefore, the smaller will be the change required to form a barrier to colonization, which leads to narrow niche requirements and many species per unit area. Stevens (1989) showed that there is a general trend of reduced latitudinal range in distribution toward the equator, calling it **Rapoport's rule,** with low-latitude species typically having narrower environmental tolerances and higher endemicity than those of high-latitude species.

6. Productivity hypothesis

This states that greater production results in greater diversity; that is, a broader base to the energy pyramid permits more species in that pyramid. However, in Chapter 22 it was seen that increase in any resource that is normally limiting leads to decreased stability and species diversity. If increased production is obtained by increasing the number of species coexisting, this clearly will increase species diversity in the first trophic level and all subsequent trophic levels. But this does not resolve the problem of why more species coexist in lower latitudes. However, Connell and Orias (1964) followed an argument that is conceptually satisfying, which correlates high productivity with increased speciation. They considered whole ecosystems that are mature and that are not limited by absolute physical space. (See also the hypothesis discussed next, that area is a major factor in biotic diversity.) A simplified version of their argument is given in Fig. 23.3, where the comparison is between a stable environment (the climatic stability hypothesis is involved) and an unstable environment.

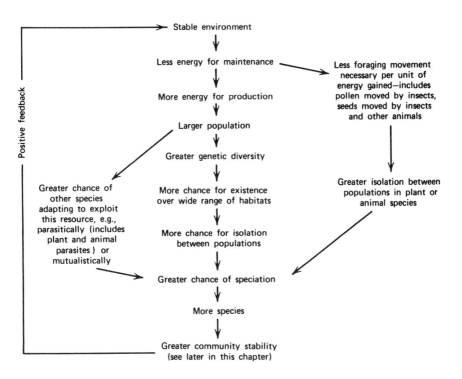

FIGURE 23.3 Simplified representation of the hypothesis by Connell and Orias (1964) that greater production in stable environments leads to the evolution of more species.

Three additional hypotheses have been advanced that were not dealt with by Connell and Orias (1964) or Pianka (1966).

7. Area available as primary source of species diversity

Terborgh (1973) would agree with Connell and Orias (1964) that existence of a species over a large area increases the chances of isolation between populations with consequent speciation. Terborgh therefore argues that the largest areas of climatic similarity will thus have the greatest species diversity. He notes that there is symmetry of climates on a pole-to-pole gradient, but only in the equatorial climatic zone do we see the symmetrically opposite climates adjacent. Therefore, in area these contiguous zones must be larger than any other. In addition, the temperature gradient between the equator and the poles is nonlinear because cloudiness at the equator reduces mean temperature, and cloud-free areas beyond this zone increase mean temperature relative to insolation (Fig. 23.4). Then by combining the areas with similar climate, the amount of habitat available for species to spread into can be estimated. The difference between equatorial and temperate zones is striking (Fig. 23.5), with a correspondingly greater chance for speciation to proceed in the former. This difference can also be appreciated by looking at the Holdridge system of biome classification discussed in Chapter 24, if one remembers that log scales are used in the plot. Terborgh brings together enough

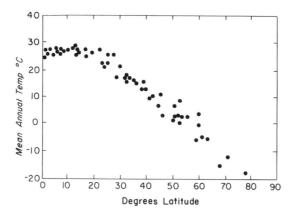

FIGURE 23.4   Mean annual temperatures (°C) of low elevation, mesic, continental locations on a latitudinal gradient. Note the wide band of similar temperatures between 20°N and 20°S of the equator. From J. Terborgh, On the notion of favorableness in plant ecology, *Am. Nat.*, 1973, **107**:481–501. Published by the University of Chicago Press. Copyright © 1973 by the University of Chicago. All rights reserved.

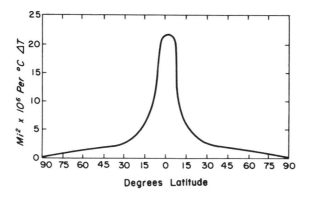

FIGURE 23.5 Millions of square miles of earth's surface between 1°C isotherms of mean annual temperature as a function of latitude. Note the much greater area of surface around the equator. Considerations of landmasses only would change the pattern little. From J. Terborgh, On the notion of favorableness in plant ecology, *Am. Nat.,* 1973, **107**:481–501. Published by the University of Chicago Press. Copyright © 1973 by the University of Chicago. All rights reserved.

supporting evidence to make this hypothesis worthy of detailed consideration. Terborgh's hypothesis is very similar to Darlington's (1959) contention that most dominant animals have usually evolved in the largest areas and that species tend to diffuse out from these equatorial zones, thus creating a species diversity gradient (see Chapter 24 and Fig. 24.28).

8. Resource limitation

In Chapter 22 we saw that Patrick and Strawbridge (1963) suggested that the high diversity of plants in tropical forests may be due to the inability of species to develop dominance in soils of very low nutrient status. Low-nutrient status is ensured by high temperatures and high rainfall, with consequent rapid recycling or leaching of nutrients. Thus the theory contrasts with the competition theory, since here competition is thought of as being suppressed by lack of sufficient nutrient resources. However, Woodwell's (1974) studies suggest that because of nutrient limitation, niche specialization is increased, and more species can coexist as a result. The resource limitation and competition hypotheses become closely linked. "The most diverse tropical forests, grasslands and shrublands occur in relatively unproductive habitats" (Tilman and Pacala, 1993, p. 24), but the underlying mechanisms are not understood. Similar patterns are given for animals by Rosenzweig and Abramsky (1993), with no compelling hypothesis on why species richness should decline from the peak with increasing productivity.

9. Animal pollinators

In humid parts of the world, particularly the tropics, wind pollination is ineffective and most plants are pollinated by animals: insects, birds, and bats. Even some grasses which are typically wind pollinated throughout most of the world, are probably pollinated by insects in the tropics (Bogdan, 1962; Soderstrom and Calderon, 1971, and references therein). Particularly with bee pollination the probability of reproductive isolation between plant populations is greatly increased, with a resultant increase in speciation rates (see Grant, 1949; Dressler, 1968; Dodson et al., 1969, and Chapter 11). Increased speciation rates will produce greater species diversity in tropical regions, where the proportion of animal-pollinated plants is the highest.

No doubt other hypotheses will be proposed and combination off hypotheses may provide convincing arguments in support of the development of latitudinal gradients as Rosenzweig and Abramsky (1993) have claimed. But much more field work must be done before we can determine which hypotheses or combinations of these are the most realistic in the development of theory. There are many difficulties to be overcome, and the main one is to find the same or similar habitats over a considerable part of the latitudinal gradient so that direct comparisons can be made between the habitats in low and high latitudes. Here the mangrove islands that Wilson and Simberloff have been studying may provide a neat solution to the problem. Mangrove islands extend from about 27°N on the west coast of Florida down to about 23°S on the Brazilian coast. Islands of the same size provide habitats that are more or less identical except for the climatic component. If there is no difference in animal species diversity with latitude, diversity must depend on plant species diversity, and the search for causative agents could be narrowed. Therefore, we can hope for some interesting new data on this problem of latitudinal gradients in species diversity which will relate to the time hypothesis, the competition hypothesis, the predation hypothesis, climatic stability, and possibly the productivity hypothesis. Recent results indicate that animal species diversity does not seem to change with latitude on mangrove islands (Simberloff, personal communication), which suggests that the causes of plant species diversity alone must be understood. Given this possibility and Baker's (1970) request for synthesis of hypotheses after he had considered those that relate to plant species diversity, a partial synthesis has been attempted (Fig. 23.6) drawing from the references already cited and from Smith (1973). Connell and Orias' hypothesis should be integrated with the figure but was omitted to avoid congestion. Most interactions have already been discussed in this chapter.

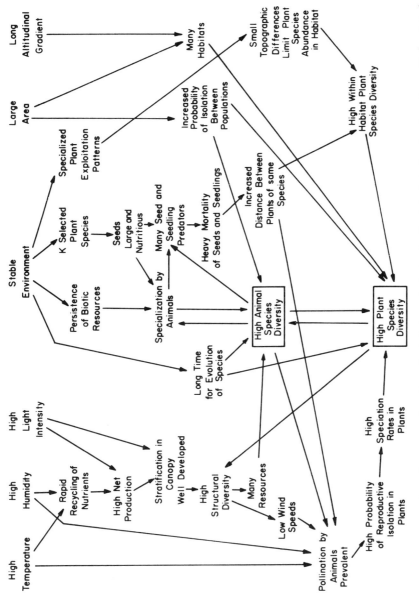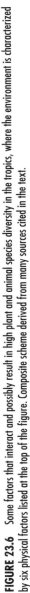

FIGURE 23.6 Some factors that interact and possibly result in high plant and animal species diversity in the tropics, where the environment is characterized by six physical factors listed at the top of the figure. Composite scheme derived from many sources cited in the text.

ANOMALOUS LATITUDINAL TRENDS

There are at least two groups of insects that illustrate the unusual trend of greater species diversity in temperate regions: the aphids and ichneumonids. Eastop (1972) noted that aphids are less abundant in the tropics than in temperate regions and species are more generalized in the tropics. He estimated that 80% of aphid species occur in the north temperate region (Eastop, 1978). Psyllids show a similar trend although less strongly developed, with about 60% of species in temperate regions (Eastop, 1978). The reasons for this trend in species diversity are not clear. Several authors have noted a decline in species diversity of ichneumonid parasitic wasps with decreasing latitude (Owen and Chanter, 1970; Townes, 1972; Owen and Owen, 1974; Janzen and Pond, 1975; Heinrich, 1977; Janzen, 1981, Gauld, 1986; Gauld et al., 1992; Gauld and Gaston, 1994). Several hypotheses have been suggested to account for this trend.

1. Ichneumonids are adapted to cool, moist conditions and have radiated extensively in such areas, in cool temperate climates in northern latitudes or at high elevations in the tropics (Heinrich, 1977).
2. Ichneumonid species tend to be microhabitat specific rather than host specific. Thus although more hosts are available in the tropics, diversity of ichneumonids does not increase because there is no greater diversity of microhabitats into which species can radiate (Owen and Owen, 1974).
3. Levels of predation, particularly due to ants, are much higher in the tropics. Parasitized hosts frequently become sluggish and have prolonged development times, making them particularly susceptible to predation, and making parasitism as a way of life more precarious in tropical than in temperate communities (Rathcke and Price, 1976).
4. Tropical woody plants tend to be more toxic than temperate species, and the allelochemicals involved may pass up the food web and affect parasitoids negatively (cf. Chapter 5). The *nasty host hypothesis* states that "tropical insect herbivores are less available to parasitoids than are extra-tropical hosts because their tissues are, on average, more chemically defended than are the tissues of extra-tropical hosts" (Gauld and Gaston, 1994, p. 284; Gauld et al., 1992).
5. As hosts become richer in species, abundance per host will decline, making it increasingly difficult to discover them. Thus on a latitudinal gradient of increasing host numbers toward the tropics, parasitoid diversity will increase with host diversity and then decline, for faced with rare hosts, selection for less specificity in host exploitation will be strong (Janzen, 1981).

Janzen's hypothesis may also account for the trends seen in aphids (see also Dixon et al., 1987) and psyllids mentioned above and in bark and ambrosia beetles described by Beaver (1979), where tropical species are less specific than temperate species. Dissenting views on parasitoid species richness in the tropics have been expressed by Hespenheide (1979) and Morrison et al. (1979), but clearly, better field research is needed to test among hypotheses and to check the validity of the earlier observations. Hawkins (1994, p. 45) summarized a massive volume of information on complete parasitoid communities around the world, concluding that "overall, exophytic hosts generally support relatively depauperate parasitoid assemblages towards the tropics, whereas better concealed hosts support similarly sized complexes everywhere or slightly richer assemblages in the tropics."

LOCAL ASSEMBLAGES IN THE TROPICS

Most of the hypotheses on diversity in latitudinal gradients were developed by researchers other than insect ecologists, except for the competition and pollination hypotheses. Most of the hypotheses on community organization have not involved insect communities very much, except for the important contributions on competition or its absence, discussed in Chapter 21. But if local insect assemblages are as open and uninteractive as the studies by the Seiferts (Seifert and Seifert, 1976, 1979; Seifert, 1984) and Strong (1981, 1982, 1984a) indicate, we must wonder about the extent to which insects can reinforce or undermine current hypotheses on communities in the tropics. Hawkins's (1994) extensive analysis of local species richness of parasitoids on host insects, at the appropriate scale for understanding community richness and organization, illustrates one of the kinds of data needed to make significant contributions on insect community ecology in tropical and temperate zones. He reinforced the intuitively unexpected results discussed above, on a much broader taxonomic scale: parasitoid communities in the tropics are either depauperate, no richer, or only slightly richer than their temperate counterparts.

Some kinds of insect herbivore assemblages also appear to be depauperate in the wet tropics. Local samples of gall-forming insects from many parts of the world have revealed a remarkable peak in local species richness centered on warm temperate latitudes and in sclerophyllous vegetation types (Fig. 23.7). A hypothesis to account for this pattern has been proposed by Fernandes and Price (1991) which links low-nutrient-status soils to sclerophyllous low-nutrient-status plants. Gall-formers cause concentrations of nutrients, high phenolics that concentrate on the outside of the galls act as effective defenses, and natural enemy impact by parasitoids and fungi is reduced in dry sites.

Other studies, such as those by Elton (1973, 1975) and Price et al. (1995c), have shown high species richness in tropical insect herbivore assemblages but very low abundance per species (Fig. 23.8). "The numbers in the counts were low, less than about 1–2 animals per m^3 of habitat. The diversity of species was very great, so that the population density of most species was

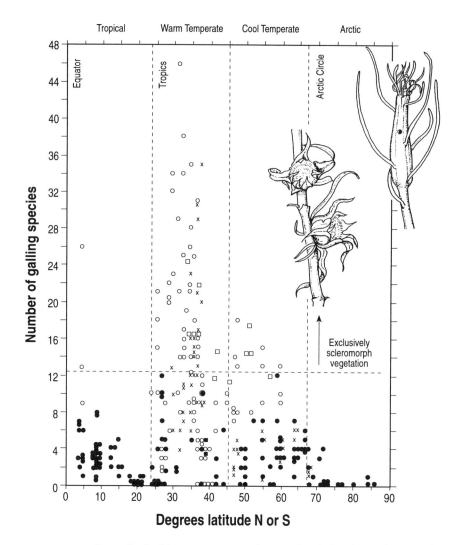

FIGURE 23.7   Distribution of local gall-forming insect species richness on a latitudinal gradient north or south of the equator, with all samples corrected for altitude and plotted as if at sea level. Note the strong peak of richness in warm temperate latitudes in scleromorph vegetation (open circles). Solid circles, samples in mesic sites on nonscleromorph vegetation; ×, samples in relatively mesic sites on scleromorph vegetation. Samples from fynbos vegetation in South Africa were collected as an independent test of the pattern (12 open squares), and from riparian woodland (two solid squares). The three open circles from the tropics are from campina vegetation along the Rio Negro River, Amazonia, a scleromorph vegetation on poor white sands. The inserted galls were formed by cecidomyiid gallflies on rabbit brush, *Chrysothamnus nauseosus*. From Price et al. (1996), based on Price et al. (1977).

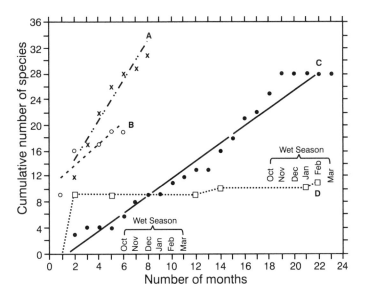

FIGURE 23.8 Cumulative number of lepidopteran species recorded on three *Erythroxylum* species when 40–45 plants were carefully inspected in cerrado vegetation near Brasilia, Brazil. Line A and ×'s, number of morphospecies found from June to December 1992; line B and open circles, number of morphospecies for the period May to October 1993; line C and solid circles, number of species reared to adults for the period May 1991 to March 1993. For comparison, data from the temperate are shown in line D and open squares for lepidopteran morphospecies on three leguminous host plants in Upper Sonoran vegetation in Arizona. The wet season in the dry tropical area is indicated. Both vegetation types represent dry savannah with highly seasonal precipitation. From Price et al. (1995c).

extremely low" (Elton, 1973, p. 101). This pattern is also seen in canopy-dwelling beetles in the wet tropics around Manaus, Amazonia, and at the Tambopata Reserved Zone, Peru (Erwin, 1988). In fact, any insect ecologist entering tropical rain forest for the first time is impressed by how hard it is to find insect herbivores at ground level and by their local rarity. This results in the few detailed studies on herbivore assemblages on related plants in the wet tropics, such as those completed by Marquis (1991; Marquis and Braker, 1994) on the fauna on *Piper* (Piperaceae) at the La Selva Biological Station in the Atlantic Lowland forest of Costa Rica. Other studies at La Selva are discussed by Hespenheide (1994) and by others in the volume on La Selva edited by McDade et al. (1994). For example, Marquis (1991) censused 200 marked plants at night every month for 18 months. Geometrid moth species were considered to be "common" if individuals were found five or more times during this period on a particular host-plant species, and rare if found less than five times or in less formal observations. Of the 52 combinations of plant species and abundances of geometrids on each species, 58% of herbivores were rare per plant species. The weevils (Curculionidae), less host-plant specific than the geometrids, had 43% of species rare on a per plant species basis in 110 combinations, even though some species were grouped because field inspection could not separate them.

The emerging picture of local assemblages of insects in the tropics, be they herbivores or parasitoids, or beetles with unknown ecology, and other studies, such as by Wolda (e.g., 1983, 1996; Wolda et al., 1996), pose problems on how they are organized. Let us ask some questions central to hypotheses on species richness and abundance in the tropics. Are species more specialized in the tropics? Is there more competition in the tropics? Are there fewer vacant niches in the tropics? Is there more biotic interaction in the tropics? (Cf. P. W. Price, 1991c.) These questions will be addressed in turn.

Are species more specialized in the tropics?

We tend to view rich tropical communities as full of specialized species, tightly packed, with narrow niches. This concept is encapsulated in the **niche compression hypothesis,** by MacArthur and Wilson (1967), which states that as more species become packed into a community, the occupied habitat per species shrinks. For insect herbivores the prediction would be that species become more specialized in their utilization of plant taxa as insect species richness increases. Where is the evidence that insect herbivores are more specialized in the tropics?

Comparison of tropical and temperate butterfly and bruchid weevil host-plant specificity yielded almost identical patterns (P. W. Price, 1991c). Gall-forming insects are universally specific, usually to one host-plant species. Aphids, bark and ambrosia beetles, and membracids seem to be less specialized in the tropics than in temperate latitudes (Eastop, 1972; Beaver, 1979; Wood, 1984). Preliminary analyses on specificity of herbivores at La Selva by Marquis and Braker (1994) indicate higher specificity of butterflies and acridid grasshoppers in the wet tropical site, but based on Wood's research, lower specificity of tree hoppers (Membracidae) (e.g., Wood and Olmstead, 1984).

Broad generalizations are obviously premature, except to repeat the point made in Chapter 9 that parasitic ways of life tend to select for specificity, and many taxa show high specificity in temperate regions. To date there seems to be little or no compelling evidence for insect communities that higher specificity generally exists at lower latitudes.

Is there more competition in the tropics?

Too few good experimental studies of competition have been conducted in the tropics, and a virtual absence of direct comparative studies leaves us unable to answer this question (Fig. 23.9). However, competition is so commonly demonstrated in temperate latitudes that it is hard to imagine that strong latitudinal trends will ever be revealed. In Schoener's (1983) review,

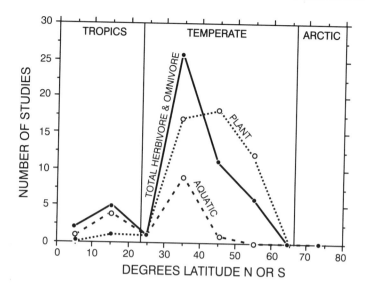

FIGURE 23.9   Distribution of 100 studies on competition in relation to latitude. The total number of studies in each 10° of latitude are given for terrestrial plants (49 studies), terrestrial herbivores (23 studies) plus terrestrial omnivores (12 studies) plus aquatic herbivores (16 studies) (51 studies total). Aquatic herbivores are also shown independently from other groups. Data from Shoener (1983). From P. W. Price in P. W. Price et al., *Plant–animal interactions*. Copyright © 1991 John Wiley & Sons, Inc. Reprinted by permission of John Wiley & Sons, Inc.

only eight studies were in tropical latitudes, while 92 were in the temperate zones. Even so, he found that competition was evident in 90% of the studies and 76% of species, indicating how commonly observed competition is in the higher latitudes.

Are there fewer vacant niches in the tropics?

Are tropical insect communities saturated with few or no vacant niches and no space for new species to colonize? The evidence suggests that many empty niches exist, as we discussed in Chapter 20, for herbivore assemblages on bracken fern and soybean in the tropics. Also, many insect species have been successfully introduced into tropical latitudes for the purposes of biological control, with little or no recorded impact on resident species at the same trophic level (Simberloff, 1981). Establishment of insects for biological control appears to be just as successful in tropical and temperate latitudes (De Bach, 1964b), indicating no difference in "community resistance" to colonizing species. Recall also the evidence for many vacant niches among large mammal herbivores in the tropics, discussed in Chapter 4, created by Pleistocene extinctions in the New World.

Again, generalizations would be premature. However, equivalent generalizations about community structure have been made many times (e.g., Dobzhansky, 1950; MacArthur and Wilson, 1967) and have entered into a practical dogma on latitudinal gradients in the organization of community structure. That insect assemblages frequently seem not to conform with such generalizations should become a subject of serious investigation.

Is there more biotic interaction in the tropics?

With so much competition evident in temperate latitudes and more or less equivalent rates of success in biological control in tropical and temperate latitudes, it appears that strong biotic interactions are ubiquitous. We must add to these considerations the evidence discussed above that insect herbivore populations are generally low in the tropics relative to temperate regions. Coley and Aide (1991) did show significantly higher herbivory in tropical versus temperate broad-leaved forests, but 10.9% compared to 7.5% leaf area damaged per year shows that the differences are small. No significant trends in successes in biological control attempts on latitudinal gradients were observed in the 158 cases classified by De Bach (1964b) (Fig. 23.10).

In sum, there are more questions than answers in terms of how insect assemblages or communities "work," or are organized, in the tropics. They do not conform to dogma, suggesting that there is no validity to generalizations on community structure in the tropics, and certainly that much effort should be focused on the comparative study of the mechanisms resulting in tropical and temperate insect assemblages.

DEPENDENCE OF STABILITY ON DIVERSITY

Why have the measurement and causation of diversity occupied a large part of ecological thinking? Primarily, it is because of the long-held belief that diversity leads to stability, and thus it is related to one of the central themes of ecology: homeostasis or balance in the system. May (1973a) considers that stability can be identified where there is a "tendency for population perturbations to damp out," thus "returning the system to some persistent configuration." Holling (1973) defines stability in the same way.

There is some evidence that diversity leads to stability (but other studies discussed later suggest no such direct link). Pimentel (1961a) tested the concept experimentally by planting collards *(Brassica oleracea)*, some in pure stands and some in the middle of an old-field community, so that the collards represented one species in about 300 species present in the mixed community. He sampled the arthropods each week for 15 weeks, so that the species abun-

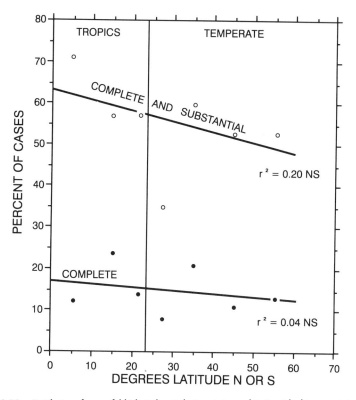

FIGURE 23.10 Distribution of successful biological control attempts to regulate insect herbivore pests in relation to latitude, based on 158 studies categorized by DeBach (1964b). He classed the cases into complete, substantial, or partial control, and here data are presented for the complete and the complete plus substantial control. In neither case is there a significant latitudinal trend. Based on data from DeBach (1964b). From P. W. Price in P. W. Price et al., *Plant–animal interactions.* Copyright © 1991 John Wiley & Sons, Inc. Reproduced by permission of John Wiley & Sons, Inc.

dance and trophic position in the community were known. Pimentel found that outbreaks of aphids, flea beetles, and Lepidoptera occurred in the single-species planting, but they were not observed in the mixed-species planting. The outbreaks were due to much higher numbers of pests and much longer persistence of populations. Although the actual causes of this difference were impossible to determine (but see Root, 1973, for a detailed analysis and hypotheses on similar systems and also see Chapter 22), it appears that the difficulty of host-plant finding and the increased efficiency of predators and parasitoids in the mixed species planting all contributed to the maintenance of population stability. Pimentel concluded that species diversity and complexity of association among species were essential to the stability of the community (see also Chapter 22).

ROLE OF PREDATORS

Paine (1966) suggested that animal diversity is related to the number of predators in the system and their efficiency in preventing single species from monopolizing some important, limiting requisite. In the marine rocky intertidal zone that he studied at Mukkaw Bay, Washington State, space was the main limiting factor, as observed in Connell's (1961) study of competition between *Balanus* and *Chthamalus*. Here the top predator was a starfish, *Pisaster,* which fed on chitons, limpets, bivalves, barnacles, and *Mitella* (a goose-necked barnacle). When *Pisaster* was present, 15 species actually coexisted in the rocky intertidal zone. Paine set aside a treatment area that he kept free of *Pisaster* and observed the change in the community. The barnacle *Balanus glandula* rapidly became a dominant species, but within a year *Balanus* was being crowded out by the mussel, *Mytilus,* and the goose-necked barnacle, *Mitella.* The space became occupied by the most efficient exploiters and the other species were outcompeted, as one should predict for competition for a single large resource. After 3 years the community had been reduced in species from 15 to 8 just because of the removal of a single predator species, *Pisaster,* and the simplification process had not run its course. Several of the species that became extinct were not in the *Pisaster* food chain at all, but were crowded out by species released from predation, so that *Pisaster* had an extensive influence on the intertidal community.

KEYSTONE SPECIES

A species with this sort of control on community structure Paine later called a **Keystone species** (1969a,b). A Keystone species can be identified where (1) a primary consumer is capable of monopolizing a basic resource and outcompetes other species, and (2) this primary consumer is itself preferentially consumed by the keystone species. A top predator can play an important role in maintaining community diversity and stability by preventing overt competition in the trophic level below by always feeding on the most abundant species. However, where interactions between members of a trophic level are not intense, as among protozoa in pitcher-plant communities, predators (mosquito larvae) tend to decrease diversity by their activity (Addicott, 1974). Murdoch and Oaten (1975) review the stabilizing factors involved in predator–prey population dynamics.

DEPENDENCE OF DIVERSITY ON STABILITY

In other cases stability clearly leads to diversity rather than the reverse. The larger a patch of resources persists, the more species are likely to colonize that patch (see Chapter 22). In agricultural systems, annual crops provide an

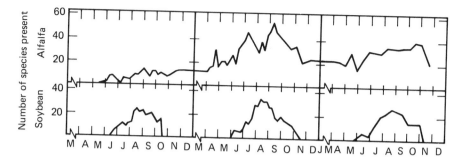

FIGURE 23.11   Comparison of number of foliage-dwelling spiders in a perennial alfalfa crop (above) planted in 1978, and an annual soybean crop (below) planted each year 1978–1980. Modified from Culin and Yeargan (1983).

unstable resource for arthropods, whereas perennial crops provide longer persistence of resources and greater microclimate stability. We should therefore expect lower diversity and stability in arthropods in annual crops compared to perennial crops. One study that supports this prediction compared foliage-dwelling spiders in annual soybean and perennial alfalfa fields from planting time through 3 years (Culin and Yeargan, 1983). In alfalfa, numbers of species accumulated gradually and stabilized around 30 species, whereas in soybean each year it showed a rapid increase in species after planting, a decline towards the end of the season, and the number of species present reached above 30 only briefly (Fig. 23.11).

DISTURBANCE AND PERTURBATION

Many studies, however, indicate that there are seldom simple relationships between diversity and stability. Many authors have provided evidence that disturbance is an important factor in maintaining diversity of species because disturbances open up new resources for colonists and prevent competitively dominant species from monopolizing resources (e.g., Loucks, 1970; Dayton, 1971; Grime, 1977, 1979; Grubb, 1977; Benzing, 1978a,b; Connell, 1978, 1979). Connell (1978) distinguishes between hypotheses on production and maintenance of diversity that invoke equilibrium and nonequilibrium conditions. Also, an effect causing stability at one trophic level may lead to less stability at another trophic level. For example, Hurd et al. (1971) found that perturbation of a community in the form of addition of inorganic fertilizer caused greater effects in the simpler plant community than in the more complex one. However, it appears that at the herbivore and carnivore trophic levels, resistance to perturbation was greatest in the simpler community. (This interpretation has been challenged by Harger, 1972, but see the defense by Hurd et al., 1972.) Hairston et al. (1968) found that increase in the diversity of food items increased the stability of species in the next trophic level, but three species at this level were less stable than two.

Thus there seems to be no simple and predictable link between diversity and stability. In his review of an extensive literature on this subject, Goodman (1975) concluded that empirical evidence has not supported the diversity-stability hypothesis. Thiery (1982) also found environmental instability to be a "troublesome concept," with no obvious guidelines for addressing its role in communities and not enough critical studies to evaluate the many hypotheses relating to this topic. May (1973a) even generalized by saying that increased complexity in communities leads to instability.

Clearly, more detailed studies are needed, which will eventually lead to a more complex but realistic theory on the ways in which diversity and stability interact. The feeding linkages in a food web must be understood and described so that comparisons between studies take into account similarities and differences in the role of each organism present. For example, Pimm and Lawton (1978) and Pimm (1979a) found in their models that species feeding at several trophic levels had a destabilizing effect on food webs, and the more trophic levels present, the less stable was the food web (Pimm and Lawton, 1977). Pimm (1979b, 1980) also distinguished between effects on stability caused by small population perturbations and return times to equilibrium, and large perturbations that could lead to extinction of a species in a food web. In the latter case, Pimm (1979b) found that generally increased food-web complexity does not result in more stability when a species goes extinct. This conclusion was qualified later (Pimm, 1980) by the finding that when plants or herbivores in a food web go extinct a more complex food web is more stable, but when carnivores go extinct a less complex food web is more stable. (See also Pimm, 1982, for an overview of this subject.)

DEFINITION OF STABILITY

Part of the problem involved with identifying broad patterns in diversity and stability is also associated with the definition of stability, which differs widely among authors (Orians, 1974; Harrison, 1979). In this chapter the concept of stability has related to the stability of both populations and communities. A stable population is one showing low variance in numbers through time, and an unstable population will have high variance (May, 1973a). When a population is disrupted it tends to return to its former equilibrium size. A stable community has both stable populations and a number of species present that remains roughly constant through time. But environmental stability is more difficult to define and has at least two components. **Constancy** relates to lack of change in the environment, whereas the term **contingency** applies to periodic changes in conditions that are therefore predictable (Colwell, 1974; see also Stearns, 1981). Thiery (1982) then reserves the term *stability* as a relative term that includes both constancy and contingency of environmental factors. It then becomes clear that many insects may respond to great environmental inconstancy by going into a resting stage—diapausing or aestivating—thereby making the environment more constant during the times in which they are active. Holling (1973) has pointed out the need for distinguishing between **resilience** and stability. He emphasized the commonness of unstable populations in forest insects such as the spruce budworm, but nevertheless how well these species and other components in these systems persist. Extreme climatic fluctuations do not cause extinction of species, and in this sense the systems are resilient. The concept of resilience contrasts with Murdoch's (1970) concept of **population inertia,** which is "the tendency for a population to resist changes away from its current density" (p. 497). Thus stable populations may result from one of two major influences that need to be distinguished: tight regulation or high inertia. Many of these terms are discussed in Levin (1975) and Westman (1978).

The further development of understanding on the relationships between diversity and stability will come from well-designed experiments in natural environments. Development of hypotheses has wavered from detecting direct positive links between diversity and stability (MacArthur, 1955), to direct negative effects (May, 1973a), to some positive and some negative effects (Pimm, 1980). In the hypothetical world much is possible. In the real world, insect ecologists can play an important role in developing rigorous tests using simple and diverse arrays of plants, insect herbivores, and their enemies to study the dynamical properties of food webs.

Paleoecology, Biogeography, and Biodiversity

GEOGRAPHIC INFLUENCES ON DISTRIBUTION

In this book it has been taken for granted that organisms are adapted to a certain set of conditions that provide the minimal physiological needs: sufficient moisture, heat, oxygen, food, and such. The requirements of organisms have an important influence on where they live and how they are distributed. Distribution of organisms in relation to each other in the community and in relation to the structural features of the community has been discussed. It therefore seems to be a natural step to examine how the larger features of the environment affect the distribution of organisms. These larger features are those recognized as significant by a geographer: mountain ranges, river valleys, rift valleys, oceans, seas, islands, continents, and the spatial relationships

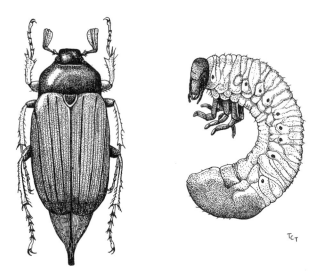

Scarab beetle, the common cockchafer or maybug, *Melolontha melolontha,* adult male (left) and larva (right), found in Europe. The larvae feed on roots and have been called rookworms in England because rooks apparently favor the grubs as food. The species is a member of a taxon with very extensive adaptive radiation. Based on Novák et al. (1976). Drawing by Tad Theimer.

among these geographical features, their formation, and their erosion. All these factors have a strong influence on the distribution of organisms. The most extensive influences on distribution are treated first and then more local influences.

AUSTRAL DISJUNCT DISTRIBUTIONS

In 1839–1843, Hooker was an assistant surgeon on an antarctic voyage and he was able to study the flora and collect plants in Tasmania, New South Wales, New Zealand, Tierra del Fuego and its archipelago (Feugia), the Falklands Islands, and Kerguelen Island between South Africa and Australia, due south of India. Thus he had a good look at, and first-hand experience with, the circumantarctic flora. Hooker was struck by the similarity of the plants in these southern lands separated by vast stretches of ocean, where he found many genera peculiar to the south in common, and even species in common. Hooker felt that these disjunct distribution patterns could not be accounted for by long-distance dispersal, and he believed that the circumantarctic flora was the split-off remnant of a once much larger austral vegetation that evolved and flourished under warmer conditions and greater proximity between the southern land masses. These ideas he discussed in the volumes *The botany of the antarctic voyage of H.M. discovery ships Erebus and Terror in the years 1839–1843*, published between 1847 and 1860.

BIOGEOGRAPHY OF CHIRONOMID MIDGES

Much more recently similar distributions have been seen in insects, for example, in Brundin's (1960, 1967) studies of chironomid midges conducted between 1953 and 1963. These are small, nonbiting, primitive flies (Diptera), the larvae of which are aquatic. Some have a red blood pigment and are known as bloodworms. In the southern hemisphere the larvae occur in cool mountain streams. The world distribution of two tribes is shown in Fig. 24.1. In the Tribe Podonomini many genera occur in the southern temperate regions and only one genus and one species occur in the north. This clearly indicates that the center of evolution for the tribe was in the south. Not only is the tribe represented on three widely separated landmasses—South America, Australia, and New Zealand—but also each has species of the same genera (Table 24.1), indicating very close phylogenetic lines, so close that it is inconceivable that the evolution in the tribe occurred in anything but a situation that permitted considerable movement of genetic material between these landmasses.

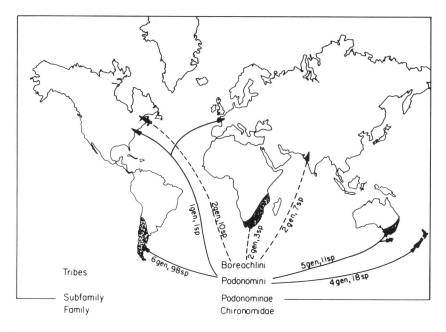

FIGURE 24.1   Distribution of chironomid midges of the subfamily Podonominae, tribes Boreochlini and Podonomini. Note that the Boreochlini are distributed mainly in Africa, Eurasia, and North America, whereas the Podonomini are most abundant in South America, Australia, and New Zealand. After Brundin (1960, 1967).

POSSIBLE EXPLANATIONS FOR DISJUNCT DISTRIBUTION

The question is: How is it possible for such closely related species to occur on such widely disjunct areas of land? There are at least four possible ways in which closely related species can appear in widely separated parts of the world. All four possibilities have been proved to be important for one group of organisms or another in explaining and understanding their biogeography, but only the fourth provides an adequate explanation for the distribution of the Podonomini:

TABLE 24.1   Examples of the Close Relationships in Genera of the Tribe Podonomini (Chironomidae).

	Number of Species Per Genus			
Genus	South America	South Africa	Australia	New Zealand
Parochlus	30	0	4	10
Podonomus	38	0	2	3
Podochlus	22	0	2	4

Transport by humans

The first possibility is remote and can almost certainly be discounted; that is, they were transported by humans either purposely or accidentally. Although humans have had a pronounced effect on the distribution of carabid beetles by transporting them in ship's ballast (see Lindroth, 1957, 1963, 1971) and the distribution of pests by transporting the economic crops (e.g., see Elton, 1958), there is no way in which the present distribution can logically be explained by invoking human activities. If people had been responsible, we could predict many more species present in North America and Eurasia and it would be hard to explain why none of the Tribe Podonomini had arrived in South Africa.

Migration

The second possibility is more plausible but still most unlikely. That is, that there was a migration of populations across the tropics, from South America perhaps, over the Bering Land Bridge and into eastern Eurasia, and down into Australia and New Zealand, and the present distribution is a relic of the much more extensive, distribution. There are two strong reasons for rejecting this argument: (1) If such a migration had occurred, many more species would have remained in the north temperate, where suitable conditions for chironomids exist just as they do in the south; and (2) movement of populations from south to north and around the Pacific Ocean would provide many opportunities for geographical barriers to reduce gene flow between contiguous populations which would lead to speciation. Many such events would result in species that arrived in Australia and New Zealand which would be very distantly related to those in South America, whereas they are actually closely related species.

Aerial plankton

The third possibility is that these small, highly mobile flying insects drifted as aerial plankton from one continent to another, or in a series of hops from one oceanic island to another, on the prevailing westerly winds in these temperate latitudes. Johnson (1969) provides much information that shows the importance and commonness of this form of movement. Again at least two reasons prevent acceptance of this suggestion for the chironomids studied by Brundin: (1), we would expect South America and Africa to have more closely related species, just as Australia and South America have congeneric species, and (2) colonization by drift over such a vast area would be a rare phenomenon.

Only one or two specimens are likely to arrive on a new continent, and thus they would represent a minute sample of the gene pool of the parent population. This minute sample could not possibly represent the full genetic diversity of the population it left. This principle, that the founders of a new colony or population contain only a small fraction of the total genetic variation of the parental population or species, is known as the *founder principle* (see Mayr, 1963). Under a new set of environmental conditions there would probably be a rapid divergence of genetic stock between the new and parent population, rapid speciation in a previously uncolonized area, and resultant large differences between species, rather than the small differences actually seen.

Continental drift

The fourth possibility is the most reasonable explanation for the chironomids. It is thought that the evolution of the subfamily occurred on a single landmass, which later split up to produce the disjunct austral distribution seen at present. This concept of continental drift has been discussed since 1620, when Bacon suggested that North and South America had been joined to Europe and Africa at one time. Only since about 1965 has sufficient information accumulated to make evidence incontrovertible that the earth's crust is formed of numerous plates that are moving relative to each other, thus causing the continents to move. The new theory of plate tectonics goes a long way to explaining the present positions of the continents, the formation of many of the world's mountain ranges, and practically all major earthquakes. Just what is causing movement of the plates is not yet understood. Hurley's (1968) and Hammond's (1971a,b) papers describe many of the important aspects of plate-tectonic theory and show how it is one of those great theories that brings a large and diverse set of observations into an organized concept.

PLATE TECTONICS

Hurley (1968) discusses the evidence that leads to the conclusion that the present continents were once assembled into two great landmasses, Laurasia in the north and Gondwanaland in the south (together called Pangaea) (Fig. 24.2). These landmasses have been split and driven apart by slow upwelling from the earth's mantle, which has resulted in the present upwellings being midway between continents in the middle of the oceans. This is clearly shown in the National Geographic Society map of the Atlantic Ocean floor (Fig. 24.3), where a midoceanic ridge is cut longitudinally by a rift valley. Rocks on the ocean floor become progressively younger as the rift valley is approached, indicating that this is the site of the upwelling that causes the plates to move. As the American plate has moved westward, part of the Nazca plate has been forced down into the earth's mantle again, and the Peru–Chile trench is the site of subduction. Thus congestion of the earth's crust at this junction of plates has resulted in formation of the Andes. The complex structure of the Alps in Europe is thought to be the result of repeated crunching and shearing of the African plate against the Eurasian plate. The Indo–Australian plate that crunched against Asia produced the Himalayas. Volcanic and earthquake activity is concentrated at plate margins, which are clearly the most unstable parts of the earth's crust (e.g., Dietz and Holden, 1970; Matthews, 1973; Wilson et al., 1974).

Thus the plate-tectonic theory unites under one concept those elements of the world's geography that are critically important in influencing the present distribution of plants and animals: (1) creation of the continents and oceans, (2) formation of mountain ranges, and (3) production of volcanic islands, all providing geographic isolation that prevents gene flow between populations thus promoting speciation.

To return to the chironomids studied by Brundin, all that remains to be known is how Gondwanaland split up before one can explain the present distribution. It appears that the first split in Gondwanaland occurred along what is now the Mid-Indian Ocean Ridge in the Permain about 280 million years ago. In the Triassic the now Mid-Atlantic Ridge split North America from Europe and Africa, a process that was completed in the Cretaceous, 135 million years ago, when South America split from Africa. Seafloor spreading has separated these continents more and more and is continuing to do so at about 2 cm/year.

Evidence suggests then that Africa was widely separated from the rest of Gondwanaland before South America, Antarctica, and Australia drifted far apart (cf., Fig. 24.2). This explains why South America and Australia have more closely related chironomids than does South Africa. This pattern is also evident in the insect taxa Ephemeroptera (Edmunds, 1972), Plecoptera (Illies, 1965, 1969), Curculionidae (Kuschel, 1962, 1969), Mycetophilidae (Monroe, 1974), Formicidae (Brown, 1973), and many others (Gressitt, 1965, 1970; Mackerras, 1970). It is also interesting to note, in the light of the breakup of Gondwanaland, that some primitive insects show affinities between India and Australia (e.g., ledrine leafhoppers; Evans, 1959) and India and South America (e.g., termites; Sen-Sarma, 1974). By reference to continental drift, many other plant and animal distributions have been explained (e.g., Axelrod, 1972, and references therein. Mackerras (1970) states that most orders of insects, from the mayflies to the beetles, have closely related species in South America and Australia, but the early groups to evolve are usually representative of this pattern. The distributions of more recently evolved taxa are more easily explained by invoking dispersal over land or sea, sometimes under the influence of wind and ocean currents. Therefore, the biogeographic patterns of these taxa differ in many features from the more primitive groups.

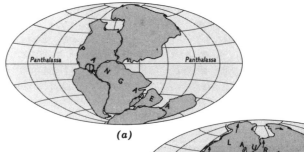

(a)

(b)

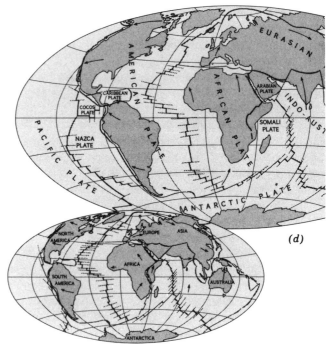

(d)

(e)

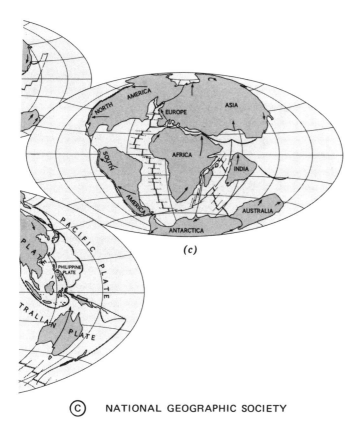

(c)

FIGURE 24.2 Movement of the earth's landmasses due to tectonic forces, top to bottom: (*a*) 200 million years ago Pangaea existed as the single landmass on earth surrounded by a single ocean Panthalassa; (*b*) 135 million years ago Laurasia and Gondwanaland have divided up Pangaea, and Gondwanaland is splitting up, with India moving toward Laurasia; lines indicate edges of plates where upwelling or downward movement of the earth's crust occurs; (*c*) 65 million years ago South America and Africa have split to form the Atlantic Ocean; (*d*) the present, with plates delineated; (*e*) 50 million years in the future Australia will still be moving northward, the Atlantic and Indian oceans will continue to widen, and much of California will have become an island. Arrows indicate direction of land movement. From Matthews (1973).

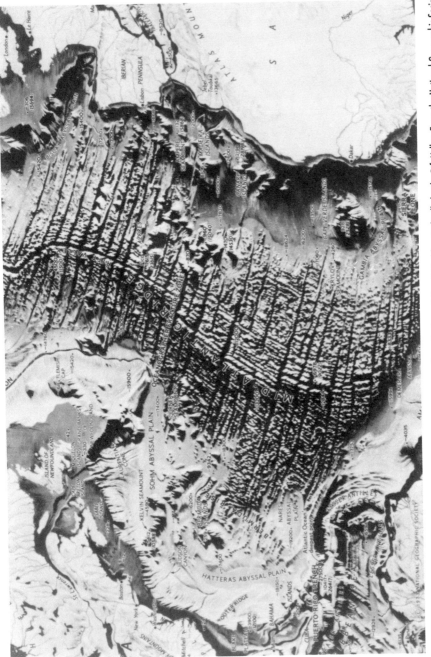

FIGURE 24.3 Relief map of a portion of the Atlantic Ocean floor showing the Mid-Atlantic Ridge divided longitudinally by the Rift Valley. From the National Geographic Society, June 1968.

BIOGEOGRAPHIC REALMS

Once the continents have drifted apart, it is clear that they are effectively isolated, so that continued evolution on these continents will be more or less independent of what is going on elsewhere, and endemic faunas and floras will result that are characteristic of that continent. Wallace, in his book *The geographical distribution of animals,* printed in 1876, recognized six basically dissimilar distributions or groupings of the existing animals, which are the six biogeographic realms still recognized today with very little alteration (Fig. 24.4). However, the Nearctic and Palearctic regions are more similar than the others and may be considered as one Holarctic Realm. The boundaries fall more or less around the continents, although the changes in fauna reflect response to barriers other than those most obvious to us. For example, the Sahara desert has proved to be as much of a barrier as the Mediterranean Sea between the Holarctic fauna and the Ethiopian fauna. The Sierra Madre have provided the barrier between the Nearctic and Neotropical, not the isthmus of Panama. The Himalayas and adjoining mountain ranges cut off the Oriental Realm from the Palearctic, not the Indian Ocean and the South China Sea.

It took Wallace 1110 pages to describe the faunas of each realm, so that it is hard to characterize briefly the essential components of each, but a grossly abbreviated list of vertebrates (e.g., see Darlington, 1957; Kendeigh, 1961) and insects (Wallace, 1876) is provided:

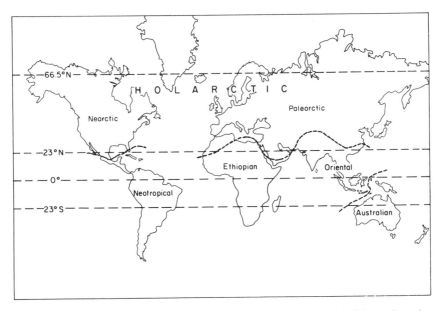

FIGURE 24.4   Biogeographic realms of the world delimited by major barriers to dispersal of plants and animals.

1. *Australian Realm.*   Absence of placental mammals and a preponderance of marsupials, parrots, honeyeaters, birds of paradise, emu, cassowary, lyrebird, and kiwi. *Xenica, Heteronympha* (Satyridae), *Ogyris, Utica* (Lycaenidae), *Ceratognathus* (Lucanidae), *Stigmodera,* and *Ethon* (Buprestidae).
2. *Neotropical Realm.*   Sloths, armadillos, anteaters, tapirs, rhea, toucans, and hummingbirds. Families Brassolidae, Heliconidae, Eurygonidae (Lepidoptera), *Oxychila* (Cicindellidae), and *Conognatha* (Buprestidae).
3. *Ethiopian Realm.*   Giraffes, antelopes, zebras, elephant, hippopotamus, rhinoceros, chimpanzee, gorilla, and the cats and dogs—lions, leopards, jackals, foxes, and so on. *Amauris* (Danaidae), *Acraea* (Acraeidae), *Anthia* (Carabidae), and *Tithoes* (Cerambycidae).
4. *Oriental Realm.*   Tree shrews, orangutan, and gibbon. *Erites* (Satyridae), *Amnosia* (Nymphalidae), *Odontolabris* (Lucanidae).
5. *Holarctic Realm.*   Nearctic vultures, turkeys, mockingbirds, vireos, wood warblers, rattlesnakes, suckers, and catfish. Differences in insects largely at the specific level. Palearctic corn crake, great bustard, genus *Fringilla,* and some fish genera in the Percidae, Salmonidae, and Cyprinidae.

Most differences between Nearctic and Palearctic regions are at the specific and generic levels.

COMPLEMENTARITY OF TAXA

One feature of distribution that Darlington (1957) pointed out is the complementarity of taxa. He applies this term to "dominant, ecologically equivalent, presumably competing groups of animals that occupy complementary areas." The obvious example is the world distribution of mammals, where the placental mammals dominate all biogeographic realms except the Australian, and here the marsupials are dominant. Another case can be seen in the Order Rodentia, which contains among many others, two very large families: Cricetidae (five subfamilies) and the Muridae (six subfamilies). Each family contains a large subfamily, Cricetinae, the New World mice and rats in North and South America, and Murinae, the Old World mice and rats in Africa and temperate and tropical Eurasia. That is, each subfamily has colonized a complementary area, and by being ecologically similar they tend to compete when members of one taxon invade the area of the other.

There is an example among the insects where the ant genus *Crematogaster* is dominant in the Old World while the genus *Iridomyrmex* and closely related genera are dominant in Australia and in the south of South America (Brown, 1973, and Fig. 24.5). The latter group will be referred to as "*Iridomyrmex.*" Notice again that as in the chironomid midges that Brundin

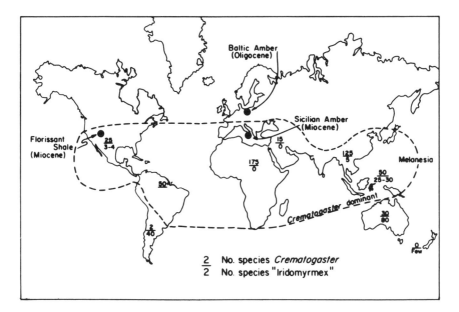

FIGURE 24.5 Complementary distributions of two groups of ants, the genus *Crematogaster* and the taxon "*Iridomyrmex*." Numbers of species are given above a line for the former and below a line for the latter. Sites of fossil records are indicated by solid circles. From Brown (1973) and personal communication.

studied, there is a closer relationship between taxa in South America and Australia than with either of these taxa and that in Africa. Although *Crematogaster* and "*Iridomyrmex*" are not closely related, Brown (1973) states that "they have entered a very similar adaptive zone," *Crematogaster* has a weak sting that is spoon shaped. It contains the exit of a gland that secretes a repellent viscous poison. The gaster can be brought overhead and pointed forward when defense is necessary. "*Iridomyrmex*" has no sting but a gland at the tip of the abdomen secretes terpenoids which when exposed to air become gummy, so that enemies can be immobilized. The gaster can also be lifted over the head and aimed forward. These two taxa are ecologically equivalent in many ways (see Brown, 1973, for details)—colonies are large; they attend Homoptera on stems and foliage; and they form long dense columns from nest to food.

In Baltic Amber (Oligocene), and Florissant Shale in Colorado, one-third of all fossil ants found were "*Iridomyrmex*." No *Crematogaster* were discovered in these deposits or in the Sicilian Amber. It looks as if *Crematogaster* has evolved later than "*Iridomyrmex*" and is actually replacing "*Iridomyrmex*." Thus their complementary distributions reflect a dynamic process and one often involving competitive displacement. These are the sorts of interactions that are going on between realms. There is a continuing movement of animals and plants.

ADAPTIVE RADIATION OF THE SCARABAEOIDEA

Another intriguing case involving complementarity of a different kind concerns the scarabaeoid beetles: an enormously rich taxon with diverse habitat and food utilization patterns in larvae and adults (Table 24.2). This superfamily includes stag beetles (Lucanidae), passalids, dung beetles, white grubs, some carrion beetles, the sacred scarab of ancient Egypt, hercules beetles, and the heaviest beetle of all, the goliath beetle (*Goliathus*) (Fig. 24.6). The group has been radiating for about 200 million years apparently, since the Triassic and the existence of the unified continents of Pangea.

Taking a paleoecological view of the new opportunities for adaptive radiation in the group through its long history provides an unusual perspective. Perhaps the dinosaurs provided new ecological zones for utilization in the form of dung, carrion, skin, and bone. Certainly, the emergence of the large mammalian herbivores provided these resources, and the massive radiation of flowering plants increased opportunities for feeding on plant detritus and on living roots, shoots, leaves, and flowers. Tracking the course of the radiation of the scarabaeoid beetles involves the history of almost half the record of life in terrestrial environments (see Price, 1996a for more details on the history of life). The story has been developed and summarized by Scholtz and Chown (1995).

The primitive condition for the group appears to be larvae feeding in humus in the ground and adults feeding on fungi. This link with organic matter and associated organisms, coupled with the necessary burrowing of larvae and adults, with hypogeal development of larvae, set the ecology and the course for the entire radiation. One phylogenetic path expanded after the Triassic–Jurassic extinctions at about 208 million years before the present (the Mesozoic clade in Table 24.2). Breakthroughs into new adaptive zones occurred onto substrata such as wood, roots, mycorrhiza, dung, keratin, carrion, and predation on other carrion insects, and inquilinism in ant and termite nests (Fig. 24.7). Substrates and feeding ecology are all associated with decaying organic matter and associated organisms, plus living plant parts such as roots in close proximity to detritus.

A remarkable, virtual recapitulation of this phylogeny occurred in another adaptive radiation after the Cretaceous–Tertiary extinctions at 65 million years ago (the Cenozoic clade in Table 24.2). Almost all the adaptive zones utilized in the older group were exploited in this clade (Fig. 24.7). Here we have a complementarity of taxa, with two clades radiating in a more-or-less symmetrical way onto new resources as they became available. Even though the lineages separated millions of years apart, their basic phylogenetic heritage and the new ecological opportunities resulted in complementary adaptive radiations.

TABLE 24.2   Taxonomic Units in the Scarabaeoidea, the Feeding Ecology of Each, and the Number of Genera and Species Per Taxon. The Mesozoic Clade Radiated after the Triassic–Jurassic Extinctions and the Cenozoic Clade Radiated after the Cretaceous–Tertiary Extinctions.

	Feeding Ecology			
Taxon	Larvae	Adults	Genera	Species
Mesozoic Clade				
Glaresidae	?	?	1	50
Passalidae	Wood/humus	Wood	40	500
Diphyllostomatidae	?	Humus	1	3
Lucanidae	Wood	Flowers or do not feed	100	750
Glaphyridae	Roots/humus	Humus/flowers	10	30
Trogidae	Keratin	Keratin	3	300
Pleocomidae	Roots/mycorrhiza	Do not feed	1	33
Bolboceratidae	Fungi	Fungi	40	350
Geotrupinae	Fungi/humus/dung	Fungi/dung	25	150
Taurocerastinae	Humus/dung	Dung	2	2
Lethrinae	?	?	1	80
Hybosoridae	Humus/dung	Carrion and carrion insects	28	275
Ceratocanthidae	Fungi/humus (inquilines)	Fungi (inquilines)	25	150
Ochodaeidae	?	?	8	80
Cenozoic Clade				
Aegialiinae	Humus/fungi	Humus	4	50
Aphodiinae	Humus/roots/dung	Humus/fungi/ dung/carrion	100	2500
Aulonocneminae	Wood	Wood	4	50
Scarabaeinae	Humus/fungi/ dung/carrion	Humus/fungi/ dung/carrion	200	4500
Orphninae	Roots	Spores	10	100
Melolonthinae	Humus/roots/dung	Leaves/flowers or do not feed	500	10000
Dynastinae	Humus/roots	Roots	225	1400
Rutelinae	Humus/roots/wood	Leaves	200	4100
Osmoderminae	Wood	Plant saps and oozes	2	10
Cetoniinae	Humus	Plant saps and oozes, flowers/ humus	400	3000
Cremastocheilinae	Humus/ants	Predaceous on ants	50	400
Valginae	Humus/termites	Predaceous on termites	30	200

SOURCE: Scholtz and Chown (1995), by permission of the Muzeum i Instytut Zoologii, Warsaw, Poland.

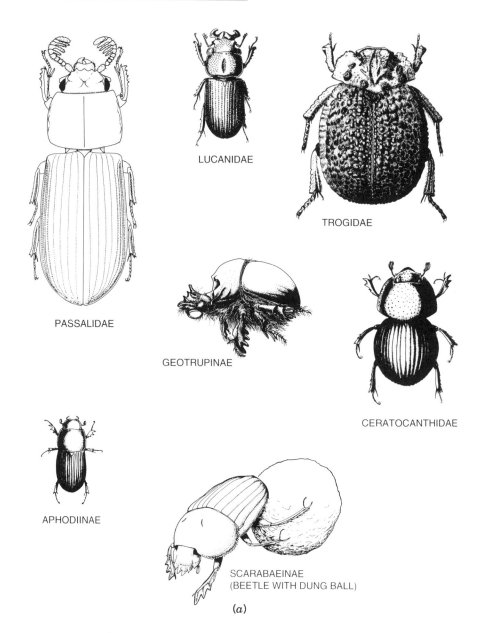

LUCANIDAE

TROGIDAE

PASSALIDAE

GEOTRUPINAE

CERATOCANTHIDAE

APHODIINAE

SCARABAEINAE
(BEETLE WITH DUNG BALL)

(a)

FIGURE 24.6 Examples of species from families and subfamilies in the superfamily Scarabaeoidea, with some habits of dung beetles represented. All examples are from Scholtz and Holm (1985). Illustrations are not to scale.

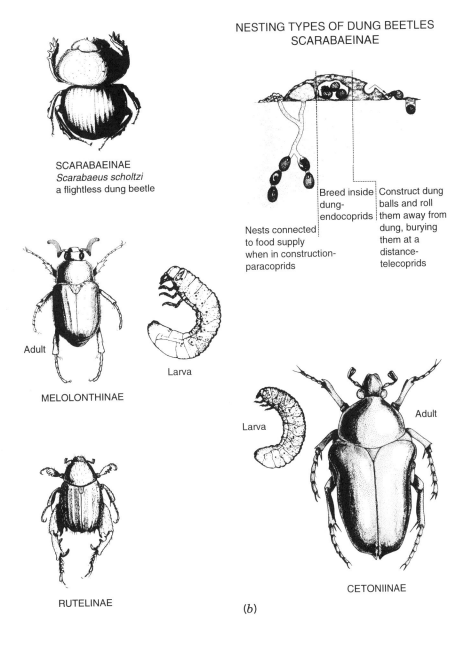

SCARABAEINAE
Scarabaeus scholtzi
a flightless dung beetle

NESTING TYPES OF DUNG BEETLES
SCARABAEINAE

Nests connected to food supply when in construction- paracoprids

Breed inside dung- endocoprids

Construct dung balls and roll them away from dung, burying them at a distance- telecoprids

Adult

Larva

MELOLONTHINAE

Larva

Adult

CETONIINAE

RUTELINAE

(b)

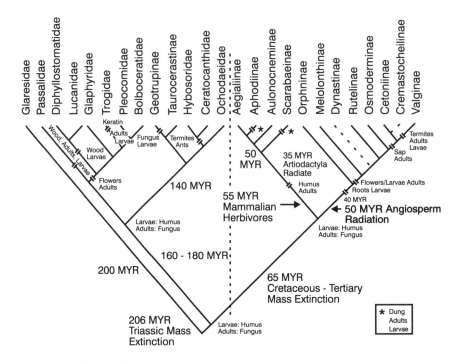

FIGURE 24.7 Phylogeny of the Scarabaeoidea. The Mesozoic clade is on the left and the Cenozoic clade is on the right. The vertical dashed line separates the clades. Major changes in the larval and adult diet are indicated as changes in ecological character states by double lines on the clades. Note that major new adaptive zones developed with the radiation of mammalian herbivores (55–35 million years ago) and the rapid expansion of angiosperms 50 million years ago. Note that the phylogeny begins with hypogean, humus-feeding larvae, and fungus-feeding adults. From Scholtz and Chown (1995); by permission of the Muzeum i Instytut Zoologii, Warsaw, Poland.

The complementary clades pose some interesting questions on why the later radiation should yield almost 10 times more current species than the earlier diversification (Table 24.2). Was the Mesozoic clade as diverse, but a flourishing scarabaeoid fauna went extinct with the dinosaurs? Or was the early clade so entrenched in its evolutionary pathways that opportunism had a very low probability as the angiosperms radiated? Note that a large part of the difference in species richness results from the subfamily Melolonthinae (Table 24.2), whose members, as adults, feed on flowers and leaves. These May or June beetles and chafers have become serious economic pests in pastures, lawns, and agricultural crops.

CONTINENTAL PATTERNS

Within each biogeographic realm more local influences are acting on species distribution—mountain ranges, deserts, prairies, and so on. For example, the small winter stoneflies of the genus *Allocapnia* are all restricted to the eastern United States and as far west as Missouri. Since the majority of the immature stages occupy cold rapid streams, it seems that the genus has been unable to spread across the rain shadow of the Rockies. This dry zone acts as a barrier to the dispersal of winter stoneflies (Ross and Ricker, 1971). The distribution of subspecies of the bee, *Hoplitis producta,* indicates that the desert of the Great Basin of Nevada (cool desert with sage brush) has acted as a barrier to dispersal (Fig. 24.8). The subspecies *gracilis* occupies California and Oregon; *subgracilis* occupies Washington; and *interior* occupies the Rocky mountain range down into Arizona and New Mexico (Michener, 1947). From Utah, California has been recolonized by individuals that do not interbreed with *gracilis,* and in fact three species are living in close proximity in California— *gracilis, bernardina,* and *panamintana.* Thus subspecies have encircled the Great Basin and in so doing have overlapped, and the overlapped populations have been sufficiently different; they have accumulated sufficient isolating mechanisms, so that they do not interbreed. That is, a new biological species has been created and the phenomenon is known as *circular overlap* (see Mayr, 1963, for other examples).

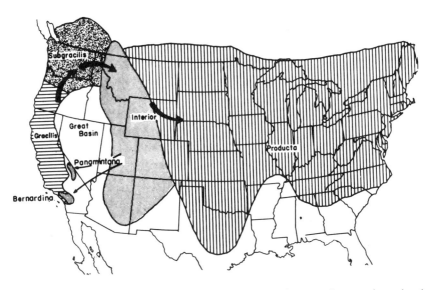

FIGURE 24.8   Distribution of subspecies of the megachilid bee, *Hoplitis producta,* in North America showing how the east has been colonized by dispersal around the Great Basin. This eventually led to circular overlap when two subspecies, *panamintana* and *bernardina,* derived from *interior* became established in California. From Michener (1947).

PLEISTOCENE GLACIATIONS AND
THE DISTRIBUTION OF WINTER STONEFLIES

Present geographical features cannot provide explanations for all distributions. For example, past glaciations have had a profound influence on speciation of birds (e.g., Mengel, 1964), amphibians (e.g., Smith, 1957), and insects, and present distributions reflect the strong influence that glaciation has had.

This is shown by the winter stoneflies of the genus *Allocapnia,* so called because all species emerge as adults during the winter or early spring (see Ross, 1965, 1967; Ross and Yamamoto, 1967; Ross and Ricker, 1971, for more details). The center of origin of the genus was the Appalachian Mountain system, which provided an abundance of habitats in the form of cold, rapid, spring-fed streams that the majority of the species require. A curious feature of the distribution is that although the Appalachians reach into Maine, many *Allocapnia* species are not distributed beyond Pennsylvania or New York State, and Ross has pointed out that in the eastern states this limit of distribution is marked by the southern boundary of the most extensive glaciations during the Pleistocene. For example, Nantucket Island, Martha's Vineyard, and Long Island are all composed at least partly of terminal moraines. Thus some distributions of *Allocapnia* species appear as in Fig. 24.9, indicating very little migration since the ice age. Other species have migrated north as the glaciers have receded, so that several species now exist in Canada.

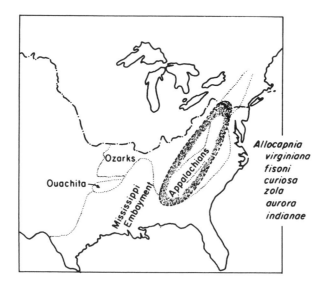

FIGURE 24.9 Approximate distributions (shaded outline) of *Allocapnia* species, which are restricted to the Appalachian Mountains. The southernmost extent of the Illinoian and Wisconsin glaciations is shown running north of the Ozark Plateau and the Ouachita Mountains up to Long Island. Martha's Vineyard and Nantucket Island are shown to the right of Long Island. The Mississippi Embayment was underwater until the coastal plain was uplifted in the Tertiary. After Ross and Ricker (1971).

However, before the ice age, in the late Pliocene, there was considerable tectonic uplifting in central North America. Subsequent erosion produced a series of spring-fed streams that permitted dispersal of *Allocapnia* species around the Mississippi Embayment and into the Ozark Plateau and Ouachita Mountains. Thus *Allocapnia rickeri* has a present distribution as in Fig. 24.10. Subsequent glaciation, particularly the Illinoian glaciation, which reached almost to the southern tip of Illinois, eroded away this central corridor for dispersal, leaving only the very narrow belt of uplands in southern Illinois. As the glaciers receded they left what is seen today, flat land with sluggish, muddy rivers, unsuitable for colonization by *Allocapnia*. Thus the species such as *rickeri* that had spread into the Ozarks were more or less cut off into two isolated sets of populations, and speciation occurred in some cases. As a result, sister species occur such as *Allocapnia recta* in the Appalachians and *Allocapnia mohri* in the Ozarks and Ouachita Mountains (Fig. 24.11).

Occasional dispersal from west to east has produced further speciation and has increased the diversity of species in the Appalachians. Also, as the glaciations receded, some populations migrated north and others became isolated in the south in the high mountains where cold streams persisted, such as in the Great Smoky Mountains, despite the general warming trends. Thus some closely related species pairs occupy southern and northern parts of the Appalachian chain.

Allocapnia rickeri

FIGURE 24.10   Approximate distribution of *Allocapnia rickeri*, which occupies the Appalachians, Ozarks, and Ouachita Mountains. After Ross and Ricker (1971).

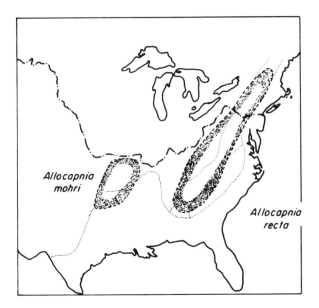

FIGURE 24.11   Approximate distributions of the closely related species *Allocapnia mohri* in the west and *A. recta* in the east, indicating how the narrow waist of mountains in southern Illinois, through which dispersal would be rare, has acted as a barrier to gene flow with consequent speciation of the isolated populations. After Ross and Ricker (1971).

Ross's studies clearly show how geographic features such as mountains, lowlands, and rivers, and past geological events, particularly mountain formation and glaciation, and juxtaposition of all these features, influence speciation and species distribution. Note that whereas mountains are normally thought of as barriers, in the case of the stoneflies it is the mountains that provided the habitats and corridors for dispersal. Thus it is important to understand the detailed ecological requirements of a group of species before trying to interpret reasons for their present distribution. Chown (1990) makes a particularly compelling case concerning insect species on islands in the southern Indian Ocean: that past events, including island age and glaciation history, play a large part in the determination of species richness on each island.

CLIMATE

Some species of insect show truly remarkable constancy in characters over half a million years, despite dramatic shifts in climate. Coope (1978) states that the fossil record shows no evidence of any morphological evolution during the last half million years at least. How, then, have species persisted in such changeable conditions? Coope argues that "insect species altered their geographical ranges on an enormous scale as the climatic zones shifted"

(p. 185). For example, an *Aphodius* beetle that lived in Britain during the last glaciation now occurs in Tibet (Coope, 1973) and a *Tachinus* species that co-occurred with *Aphodius* in Britain is now found in Siberia and North America (Ullrich and Coope, 1974) (see also Ashworth, 1977; Coope, 1977). It now seems that many species moved together with climatic changes without showing any morphological change. Even fossil remains from deposits 5.5 million years old show that species are very close to modern species (Matthews, 1976a,b), prompting Coope (1978) to assert that stability in morphology is the norm rather than the exception.

Elias (1994) provides evidence of beetles found in Britain during glacial or interglacial times in the late Quaternary—less than 500,000 years ago. Some are cold-adapted species and can now be found in arctic and alpine localities (Fig. 24.12) in Europe but not in England. Other species were thermophilous in Britain but are now found in Spain, or Sicily, or southern Europe and North Africa (Fig. 24.13). In North America a Southern Refugium south of

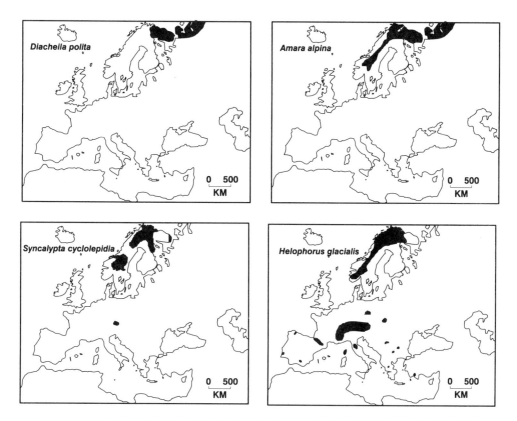

FIGURE 24.12   Current distributions of four beetle species that are cold-adapted which have been found in British glacial deposits. From S. A. Elias, *Quaternary insects and their environments,* Smithsonian Institution Press, Washington, D.C., p. 93 (data from G. R. Coope, 1986), by permission of the publisher. Copyright 1994.

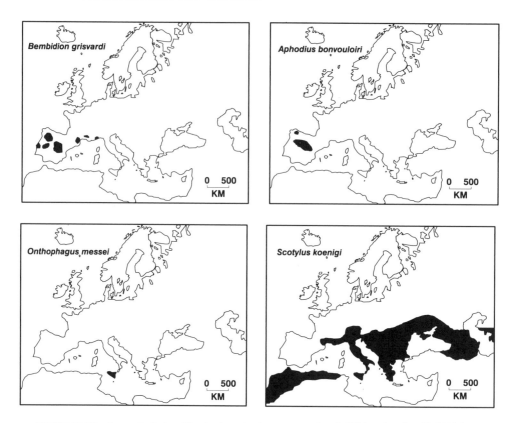

FIGURE 24.13 Current distribution of four beetle species that are warm-adapted which have been found in British interstadial and interglacial deposits. From S. A. Elias, *Quaternary insects and their environments*, Smithsonian Institution Press, Washington, D.C., p. 95 [data from C. H. Lindroth (ed.), 1960], by permission of the publisher. Copyright 1994.

the glaciation enabled some cold-adapted beetles to persist, which repeatedly migrated northward as the ice receded. Present distributions of some cold-adapted species are in Alaska and northern Canada, with one species, *Amara alpina*, also represented in the Rocky Mountains (Fig. 24.14).

Pleistocene glaciations have also had some more subtle effects on insect distributions, apparently. Niemelä and Mattson (1996) noted the strongly skewed intercontinental movement of forest insects, with a preponderance of species colonizing from Europe into North America. Counting the number of phytophagous forest insect species from Europe, now in North America, 266 species was an estimate (Table 24.3), compared to 28 species in Europe derived from the North American continent—practically a 10-fold bias in colonization. One might expect, as a null hypothesis, a more-or-less equal flow of species across the Atlantic given the long-term commercial links and shipping among the countries on either side.

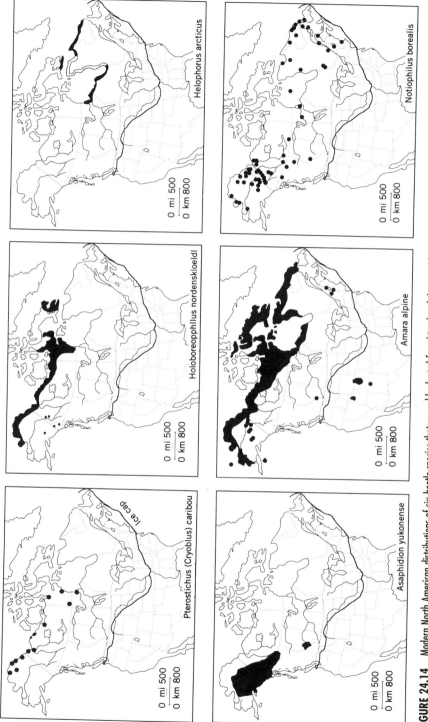

FIGURE 24.14 Modern North American distributions of six beetle species that are cold-adapted found in glacial deposits. These beetles probably survived glaciation in the Southern Refugium, south of the Laurentide Ice Cap, the limits of which are indicated in the top left figure. Slightly modified from S. A. Elias, *Quaternary insects and their environments*, Smithsonian Institution Press, Washington, D.C., p. 100 (data for *P. caribou* from G. E. Ball, 1966; map of *H. nordenskioeldi* after A. V. Morgan et al., 1984; map of *H. arcticus* after A. V. Morgan, 1989; map of *A. Yukonense* after A. V. Morgan et al., 1984; data for *A. alpina* from C. H. Lindroth, 1968; data for *N. borealis* from C. H. Lindroth, 1961), by permission of the publisher. Copyright 1994.

711

TABLE 24.3   Examples of Insects on Woody Plants Introduced to North America from Europe in the Four Orders Accounting for Most of the Cases.

Order	Number of Species from Europe	Examples
Homoptera	76	Adelgidae, 5 spp.; Aphididae, 14 spp.; Cicadellidae, 34 spp.; Eriococcidae, 3 spp.
Coleoptera	57	Chrysomelidae, 9 spp.; Curculionidae, 24 spp.; Scolytidae, 9 spp.
Lepidoptera	73	Geometridae, 7 spp.; Lymantriidae, 4 spp.; Tortricidae, 24 spp.; Yponomeutidae, 7 spp.
Hymenoptera	34	Diprionidae, 5 spp.; Siricidae, 2 spp.; Tenthredinidae, 23 spp.

SOURCE: Mattson et al. (1994).

Niemelä and Mattson (1996) erected several alternative hypotheses that might singly or in concert explain this "trade imbalance." Based on what we know in general about plant and herbivore interactions (cf. Chapters 5 and 6) the authors noted that the probability of being a successful colonizer will depend on a series of factors: (1) size of geographic ranges in areas of all potential hosts (i.e., those in the same genus or family), (2) the mean density or abundance of each host in this area, (3) the similarity of potential hosts to the original host plants in terms of phytochemistry, morphology, and ecology, and (4) the synchronization of host-plant phenology and insect phenology. Major elements in the argument follow:

1. **Higher rates of host-plant extinction during the Pliocene and Pleistocene in Europe left a richer flora of potential host plants in North America.** Species richness of trees in North America and Europe showed considerable disparities (Table 24.4). Clearly, North America has a richer plant fauna of potential hosts for colonists by about twofold or more. The reason for this difference may well involve compression of floristic zones during glaciation events to the south of Europe when long ranges of mountains running across the path of southern migration limited and fragmented floral elements, causing extinctions. The Cantabrian Mountains and Pyrenees blocked migration into Spain; and the Alps, Carpathians, and Caucasus ranges blocked southerly movement into southern Europe and the Middle East. In North America, mountain ranges run north and south in parallel with migrating plants and animals during glaciations, without constricting southerly movements and without the resultant extinctions.

2. **Plant extinctions during the Pleistocene may have selected for European insects with broader capacity to utilize host genera than those in North America.** A more benign environment in North America with fewer extinctions of plant species may have left the component communities of insect her-

TABLE 24.4   Numbers of Tree Genera and Species of Gymnosperms and Angiosperms in North America and Europe and the Species Per Genus of Each.

Taxon	North America		Europe	
	Genera	Species	Genera	Species
Gymnosperms	16	97	8	30
Species per genus		6.06		3.75
Angiosperms	143	503	78	256
Species per genus		3.52		3.28

SOURCE: Based on information in Niemelä and Mattson (1996).

bivores on these plants more or less intact and specific to their original hosts after the Pleistocene. However, in Europe, with extinctions of genera and compaction of vegetational types, host shifting by insect herbivores may have been common, resulting in broader host-utilization patterns and perhaps the retention of characters relevant to exploitation of host genera in North America, now extinct in Europe. Hence there would exist a disparate propensity for successful colonization by insect herbivores moving from east to west across the Atlantic.

3. **North American forest provide a larger, less fragmented island than those in Europe relative to colonization of phytophagous insects.** Deforestation in Europe has been much more extensive than in North America, with serious fragmentation especially along the coast, where insects may colonize initially. Also, the forests are composed principally of pines and spruces. In North America larger tracts of forest remain composed of pines, spruces, and many hardwood species, with increased likelihood for an insect to find a suitable host plant.

4. **Pleistocene and modern conditions in Europe selected for superior competitors (or colonizers?)** Insect herbivores in Europe have been exposed to more serious impact from disturbance during glaciations and modern times through repeated compaction of floras, northward migrations, and modern fragmentation. This has selected apparently for effective colonizing ability and perhaps a competitive edge against resident species in North America. Associated with these traits, polyploidy and parthenogenesis seem to be unusually common among European insect taxa. Polyploid insects appear to have strong colonizing ability, and parthenogenesis allows one individual colonizer to establish a population in a new and remote location.

5. **Synchrony of phenologies between plant and insect herbivores has a higher probability when insects colonize from higher to lower latitudes.** With temperate deciduous forest in Europe situated between about 43 and 60°N and in North America between 30 and 48°N, seasonal photoperiodic changes relevant to insect diapause and plant phenology are much greater in Europe than in North America. Day lengths in the north at 60°, for example, range

from 6 to 19 h, but at 30°N only between 10 and 14 h. Therefore, an insect colonizing from south to north would experience long day lengths in summer, too long to trigger diapause in time before the rapid onset of winter. It would be unlikely to survive. In the spring, the extraordinarily rapid flush of foliage in the north with rapidly increasing day lengths demands stringent physiological timing in herbivores. However, an insect moving from north to south would be likely to experience shorter day lengths that stimulate diapause in adequate time before the onset of winter.

Overall, Niemelä and Mattson (1996) have erected a sweeping and fascinating set of hypotheses that may account for the "trade imbalance" in phytophagous insects or trees. These hypotheses encompass the paleoecological and biogeographical aspects of insect ecology in a heuristic synthesis. Certainly, much research is needed to test the hypotheses, and all those erected by the authors have not been dealt with here. However, these ideas stimulate our imagination, draw us back in time through the millennia, and create a broader perspective in time and space. It is "the sensation of the mystical," after all, that is our source for good science in the future.

The realization that the centers of distribution of species shift so dramatically with changing climate should foster concern in ecological studies for an understanding of species geography. Two species in the same community may show very different geographic distributions, and they will be sampled in different places relative to the center of geographic distribution. Thus their relative abundance may reflect their different distributions and the fact that species in various parts of their range come under varying influences on population regulation (see Chapter 17). Haeck and Hengeveld (1979) and Hengeveld and Haeck (1982) show that for any location species within a taxon have geographical distributions that are marginal, submarginal, subcentral, and central relative to that location (Fig. 24.15). As the sampling location becomes more central in the geographic distribution of a species, the species tend to become more common and widespread, and species richness within a taxon increases (Hengeveld and Haeck, 1982). Thus any comparative ecological studies on species in the same location should consider their geographic distributions.

LIFE ZONES

Climate has a continuing influence on distribution of animals and plants. Merriam, who published his ideas in 1894 and 1898, made a long-accepted description of animal and plant distributions by defining what he called *life zones*. On an expedition to the San Francisco Peak area in Arizona he was impressed by the sharp zonation of plants and animals up the mountains, apparently under the sole influence of temperature. He formulated two laws about the control of temperature on the distribution of living things:

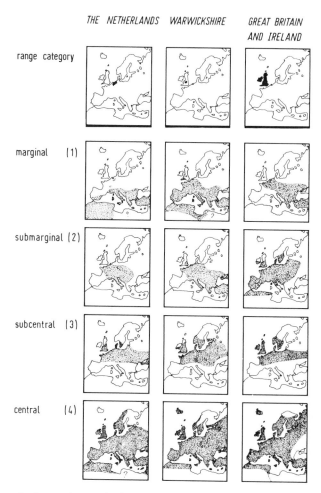

FIGURE 24.15 Classification of species distributions in relation to the sampling areas of the Netherlands, Warwickshire, and Great Britain and Ireland indicated in black in the range category maps. Illustrated geographical distributions of species are marginal, submarginal, subcentral, or central in relation to these sampling areas. Reprinted with permission from Hengeveld and Haeck (1982) and Blackwell Scientific Publications Limited.

1. The northward distribution of terrestrial plants and animals is governed by the sum of the positive temperatures (above 6°C) for the entire season of growth and reproduction.
2. The southward distribution is governed by the mean temperature of a brief period during the hottest part of the year.

Using these criteria he divided the continent into areas, or life zones, with similar temperature characteristics so that the belts up a mountain appeared as in Fig. 24.16. Tilting of the zones results because in the northern

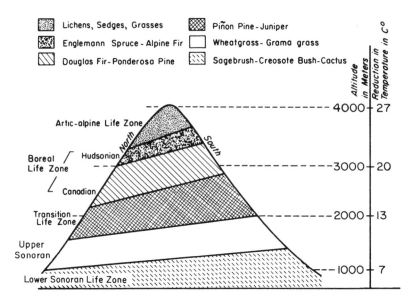

FIGURE 24.16   Life zones recognized by Merriam in the Rocky Mountains. South slopes in the northern hemisphere receive more insolation and thus ecotones are at slightly higher altitudes than on north slopes. Reprinted from A. S. Boughey, *Ecology of populations*, with permission of Macmillan Publishing Co., Inc. Copyright © 1968 by Arthur S. Boughey.

hemisphere, the south slope is warmer than the north slope. Merriam's system worked well in the mountainous west but could not be used effectively in the east. Pitelka's (1941) approach to animal distribution by defining major differences in the structural characters of the vegetation, the biomes, has improved on Merriam's life zone system.

BIOMES

Holdridge (1947, 1967; Holdridge et al., 1971) has gone one step further than Pitelka by producing a system that shows how biomes can be delimited, particularly plant formations, by using three climatic parameters: temperature, precipitation, and evaporation (Fig. 24.17). Mean temperature above 0°C declines from the equator to the polar regions, so temperatures can be plotted as parallel lines representing isotherms. The values are plotted on a logarithmic scale. Precipitation is calculated as the average total annual rainfall in millimeters and plotted at about 60° to the temperature axis, again on a log scale. The evaporation lines represent the balance between temperature and rainfall, and the values represent the number of times the actual rainfall could be evaporated in 1 year at sea-level atmospheric pressure, plotted on a

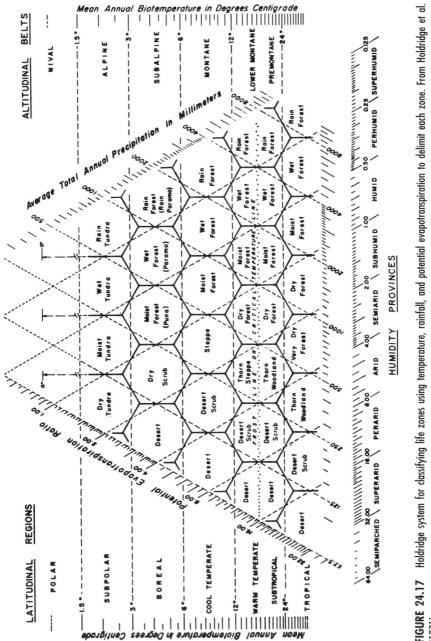

FIGURE 24.17 Holdridge system for classifying life zones using temperature, rainfall, and potential evapotranspiration to delimit each zone. From Holdridge et al. (1971).

log scale. Holdridge also gives the altitude belts that result from the equivalent effect to latitude, and he points out that only in the tropics will all altitudinal belts be encountered. This relates back to the tropical species diversity problem and the environmental diversity in the tropics, caused by the longer physical gradients present (Chapter 23). By measuring two parameters one can predict what the vegetation structure will be, and from the vegetation structure one can predict what sorts of animals will be present.

THEORETICAL BIOGEOGRAPHY OF ISLANDS

So far, oceanic islands have almost been ignored in this chapter, yet they are known to be important in speciation, providing good examples of adaptive radiation (e.g., Price, 1996a). Also, the theoretical aspects of island biogeography are much more advanced than those for continents. The theory of island biogeography was advanced by MacArthur and Wilson in 1963 and 1967. In Chapter 22 it was stated that Wilson and Simberloff (e.g., Simberloff, 1969, 1978; Simberloff and Wilson, 1969, 1970; Wilson, 1969; Wilson and Simberloff, 1969) tested this theory with defaunation studies on the mangrove islands off the Florida Keys.

Many people have observed a striking relationship between area of an island and the number of species of a given taxon it can support, for example, the number of species of land and freshwater birds of the Sunda Islands, the Philippines, and New Guinea (Fig. 24.18). These species–area curves were

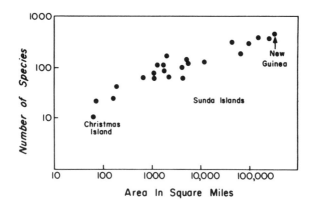

FIGURE 24.18   Relationship between area of island in square miles (log scale) and the number of land and freshwater bird species in the Sunda Island group, the Philippine Islands, and New Guinea. The islands are grouped close together and close to the Asian continent. From R. H. MacArthur and E. O. Wilson, *The theory of island biogeography.* Copyright © 1967 by Princeton University Press. Reprinted by permission of Princeton University Press.

discussed in relation to community organization (Chapter 22). These islands and archipelagoes are grouped close to one another and to the Asian continent, and the areas of the islands accounted for at least 90% of the variance about the mean. As a general trend, Hamilton et al. (1964) and Hamilton and Rubinoff (1967) have found that in most cases, island area accounts for 80–90% of the variation and elevation for another 2–15%. Actually, both factors can be interpreted functionally as contributing to the habitat diversity of an island. By calculating the habitat diversity for each island and relating it to bird species, Watson (1964) explained practically all variation in the Aegean Islands between Greece and Turkey in the Mediterranean.

But other groups of islands support fewer species than the number predicted from size alone. If area and species richness are plotted for the Moluccas, Melanesia, Micronesia, and Polynesia, the slope of the species–area curve is much steeper than for the Sunda Islands (Fig. 24.19). These islands are spread out over the Pacific Ocean, and clearly some are isolated, so that a good chance exists that all species that could have colonized these islands have not done so; that is, a distance factor is involved. In fact, the islands close to New Guinea (less than 500 miles), which has the richest bird fauna, can be grouped, and all are close to the saturation number of species. Islands greater than 2000 miles from New Guinea are those that deviate furthest from the expected species saturation line.

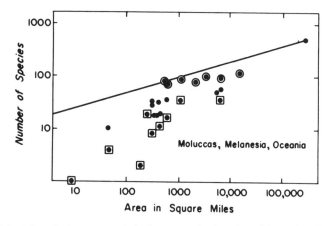

FIGURE 24.19   Relationship between area of island in square miles (log scale) and the number of land and freshwater bird species in the Moluccas, Melanesia, Micronesia, and Polynesia. Circled islands are less than 500 miles from New Guinea and conform to the species–area curve observed for the Sunda Islands. Squared islands are over 2000 miles from New Guinea and the relationship diverges significantly from that expected from the species–area curve. From R. H. MacArthur and E. O. Wilson, *The theory of island biogeography.* Copyright © 1967 by Princeton University Press. Reprinted by permission of Princeton University Press.

Species equilibrium, area, and distance

Therefore, MacArthur and Wilson (1963, 1967) proposed the theory of island biogeography to relate three components: (1) species numbers per island, (2) size of island, and (3) distance of island from source of colonists. MacArthur and Wilson pointed out that there are two components to the number of species on an island, regardless of its distance from the mainland or its area. For each island there will be an immigration rate of animal and plant species arriving on the island per unit time. Not all species that immigrate will survive, so there must also be an extinction rate. These opposing factors will act to produce an equilibrium number or species on any given island (Fig. 24.20). The immigration rate of species will decline as a function of species number because as more species become established, fewer immigrants will belong to new species. The line will be steeper, at first, as the most rapidly dispersing species will become established first, whereas the late, slowly dispersing arrivals will prolong the time of immigration and cause the rate to decline only slowly as point P is reached. P is the total possible number that can reach the island because that is all the species the mainland supports. The extinction curve has an exponential shape and a positive slope because as more species arrive there is a greater chance of some going extinct; competition between species will increase and population sizes may be reduced, both accelerating the rate of extinction.

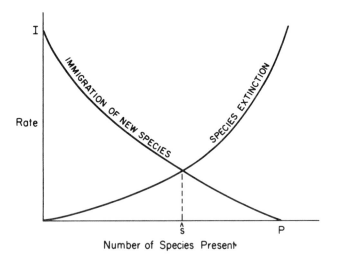

FIGURE 24.20   Relationship between the immigration and extinction rates on an island, which leads to an equilibrium number of species, $\hat{s}$, on the island. I is the initial immigration rate and P is the total number of species in the source of colonizers. From R. H. MacArthur and E. O. Wilson, *The theory of island biogeography*. Copyright © 1967 by Princeton University Press. Reprinted by permission of Princeton University Press.

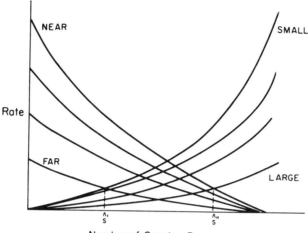

Number of Species Present

FIGURE 24.21 Effects of size of island and distance from source of colonizers on the equilibrium number of species. Near, large islands have many more species, $\hat{s}''$, than far, small islands, $\hat{s}'$. From R. H. MacArthur and E. O. Wilson, *The theory of island biogeography.* Copyright © 1967 by Princeton University Press. Reprinted by permission of Princeton University Press.

The distance of an island from the source of colonizing species and its size should have considerable impact on the relationship between immigration and extinction. The nearer the island is to the source of the species, the more colonists will arrive per unit time and the closer $\hat{s}$, the equilibrium number of species, approaches P, the number of species in the source area (Fig. 24.21). The larger the island is, the higher the carrying capacity will be for each species. More diverse habitats will permit species to exist in suitable enclaves from competition, and the net result will be reduced rates of extinction with increased island size.

This theory has supplied an impressive impetus to ecological research and is rapidly being extended to ecological island situations on continents, as described in Chapter 22 (Table 24.5). The theory also provides a new method of looking at ecological islands such as crops and the colonization of crops by insects. Do crop conditions permit equilibrium to develop, or is the colonization curve truncated by harvesting the crop? The theory will help in understanding the community organization of insects on crops and thus the most ecologically sound methods of managing insect pests (see Price and Waldbauer, 1975; Price, 1976).

BIODIVERSITY AND CONSERVATION

An important development using the theory of island biogeography has been in studies of biodiversity and its conservation (e.g., Wilson, 1988,

TABLE 24.5 Studies of Species-Area Relationships Showing a Large Range in Percentage of the Variance Accounted for ($r^2 \times 100$). Note that Studies with Strong Regressions are Consistent with the Theory of Island Biogeography although They have not Demonstrated that the Mechanisms Leading to this Relationship are the Same as Invoked in the Theory (see Connor and McCoy, 1979; Strong, 1979; Lawton and Strong, 1981).

Parasite Taxon	Host Island	$r^2 \times 100$	Source
Leaf miners	Oaks	90	Opler (1974)
Mites	*Peromyscus* spp.	86	Dritschilo et al. (1975)
Insects	Woody shrubs	85	Lawton and Schröder (1977)
Anthropods	*Astragalus sericoleucus*[a]	71–85	Tepedino and Stanton (1976)
Cynipids	Oaks (California)	72	Cornell and Washburn (1979)
Insects	Perennial herbs	71	Lawton and Schröder (1977)
Insects	Trees	61	Strong (1974a,b)
Insects	Sugarcane[a]	61	Strong et al. (1977)
Insects	Annual plants	59	Lawton and Schröder (1977)
Macrolepidoptera	Trees and shrubs	57	Neuvonen and Niemelä (1981)
Insects	Monocotyledons	51	Lawton and Schröder (1977)
Insects	Cacao[a]	47	Strong (1974c)
Mites	*Microtus* spp.	46	Dritschilo et al. (1975)
Cynipids	Oaks (Atlantic)	41	Cornell and Washburn (1979)
Mites	Cricetid rodents	37	Dritschilo et al. (1975)
Parasitoids	Umbelliferae	35	Lawton and Price (1979)
Agromyzids	Umbelliferae	32	Lawton and Price (1979)
Fungi	Trees	28	Strong and Levin (1975)
Microlepidoptera	Umbelliferae	24	Lawton and Price (1979)
Arthropods	*Phlox bryoides*[a]	10–16	Tepedino and Stanton (1976)
Leafhoppers	Trees	0.04	Claridge and Wilson (1976, 1978)

[a] Studies on area differences within a host species. All others compare area differences among host species.

1992). Wilson (1992) makes a compelling case for the importance of conserving biodiversity and the many highly threatened floras and faunas around the world. Yet there remains inexorable attrition in areas of high diversity and sensitivity, involving habitat destruction, degradation, and fragmentation.

As areas of natural vegetation decline, as in the case of tropical rain forest, as a major example, we can employ the theory of island biogeography to predict the associated decline in biodiversity. Using species–area relationships as in Fig. 24.18, we can reverse the area scale and plot the predicted decline in species number as island size declines through degradation of habitat (Fig. 24.22). The figure relates to reptiles and amphibians in the West Indies, but it provides a general picture of how biodiversity is lost in any taxon as area declines. We will see later that the rate of loss of insects in the tropics may well be similar to the rate illustrated in Fig. 24.22. As a general statement, or rule of thumb, there is a 50% loss of species with a 90% reduction in island area (Wilson, 1992).

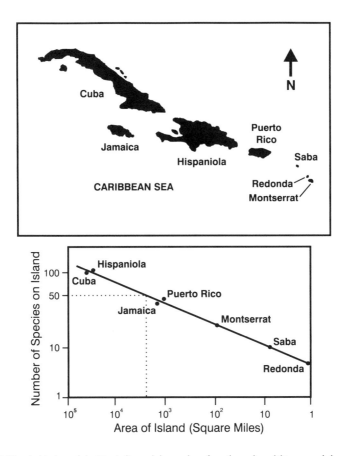

FIGURE 24.22 Archipelago of the West Indies and the number of reptiles and amphibians recorded on each island, relative to island area (log scale). Note that Cuba and Hispaniola have about 100 species recorded and have an area of about 40,000 square miles. When the number of species drops to 50 per island, the area has declined about 90%, to 4000 square miles, as indicated by dashed lines. This illustrates the rule of thumb that a reduction of 90% of area results in a loss of 50% of the species, approximately. Illustration by George Ward. Reprinted by permission of the publisher from *The diversity of life* by E. O. Wilson. Cambridge, Mass.: Harvard University Press, © 1992 by E. O. Wilson.

Using this generalization, various authors have estimated rates of extinction as tropical forest area is depleted. Following Wilson's (1992) argument, the predictions are as follows. There may be 10 million species living in tropical rain forest. By 1979 the rain forest was reduced to about 56% of its prehistoric area and cutting was proceeding at about 1% per year. By 1980 the rate of cutting had almost doubled to about 2% per year. This means that about 0.5% of the species are going extinct every year—or about 50,000 species. A conservative estimate would be 27,000 species per year, 74 species each day, and three species each hour! At present rates of destruction, 90% of the tropical rain forest could be destroyed in 45 years, say by the year 2040, when it is possible that 50% of tropical species will have gone extinct.

The equation for the species–area relationship, as in Figs. 24.18 and 24.22, is

$$S = CA^z$$

where S is the number of species present, C is a constant, A is the area of an island, be it a habitat island or an oceanic island, and z varies with the group of organisms under study. Species with a relatively high value of z, about 0.35, are poor dispersers, such as land snails and orchids, resulting in higher levels of endemicity. Therefore, as area is lost, the slope of species loss is relatively high. A relatively low value of z is observed for species with high dispersal, such as birds, where a z value may be 0.15. Where are insects likely to place within this range, especially those in tropical rain forest?

Erwin (1988) made some preliminary estimates of the endemicity of canopy-dwelling beetles in South America in the Amazonian Basin. In the Tambopata Reserved Zone in Peru, in one upland forest type and in two plots only 50 m apart, the overlap of shared canopy beetle species was only 8.7% of species. The sample included seven families of beetles and 126 species. This small similarity of beetle assemblages so close together suggests that beetles have specific habitat requirements and are distributed in very local patches of the canopy. Other samples from Manaus, Brazil, add support to the evidence that endemicity is high in these canopy assemblages. In four forest types 70 km or more apart, a total sample of 61 beetle families and 1080 species showed that only 1% of species were found in all four forest types.

Thus for tropical canopy-dwelling beetles at least, and probably for many insect herbivores, we should anticipate relatively high z values of around 0.30, the value used for the rule of thumb mentioned above. Several studies have shown the high richness of insect herbivores in various vegetation types but the very high number of rare species. Erwin (1988), for example, collected 3099 individual beetles at Tambopata and 1093 species, less than an average of three individuals per species. In the much drier savannah vegetation, called cerrado, in Brazil, increased sampling yielded more and more rare species with no indication that the assemblage had been sampled completely even after 2 years (Price et al., 1995c). Sampling weevils in tropical Panama, Wolda et al. (1996) noted the high degree of endemism and the high proportion of rare species. Wolda (1996) also noted impressive endemism in tropical cockroaches and pselaphid beetles (see also Wolda, 1983).

RELIABILITY OF SPECIES–AREA ESTIMATES

If we are to use the theory of island biogeography in an informed manner to predict extinction rates and to plan for areas of conservations, we should examine the confidence we can invest in the simple patterns of species and area.

Are the patterns reliable enough to use in planning the size of nature reserves or in predicting decline of species richness as reserves become more isolated from extensive tracts of similar flora and fauna?

First, it is important to note that as in Table 24.5, there is no guarantee that the species–area relationship will account for 50% of the variance in any particular case. The percentage of the variance accounted for ranged from 90 to 0.04% in the examples provided. In a much larger sample of cases given in Connor and McCoy (1979), the mean R^2 value in 100 studies was just less than 0.50 (Fig. 24.23) (Boecklen and Gotelli, 1984). For each taxon studied there was a wide range in the percent of variation in species number accounted for by area (Fig. 24.24). Clearly, the species–area relationship is not generally a strong predictor of species richness and should be used with caution in well-studied cases where reliability is high.

Boecklen and Gotelli (1984) recognized that determinants of species richness are commonly too complex to be predicted by area alone. One well-studied example serves to illustrate this point. The extensive studies by Liebherr (e.g., 1988, 1992, 1994) on the carabid beetle genus, *Platynus* (Fig. 24.25), have revealed that two species exist on Cuba and 14 on Hispaniola. As we have seen in Fig. 24.22, these islands support similar numbers of reptiles and amphibians, but island area fails to account for the large discrepancy in beetle species numbers. With Cuba's area of over 44,000 square miles and less than 30,000 square miles on Hispaniola, we would predict a smaller fauna of beetles on Hispaniola at about 66% of that found in Cuba, yet Hispaniola has seven times more species!

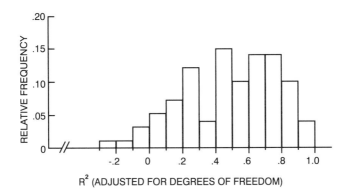

FIGURE 24.23 Frequency histogram of R^2 values for 100 log area–log species models in Connor and McCoy (1979) developed by Boecklen and Gotelli (1984). The R^2 value provides an estimate of the proportion of variation in species number accounted for by area in a linear regression. The mean R^2 value is 0.49. Reprinted from *Biological conservation*, Vol. 29, W. J. Boecklen and N. J. Gotelli, Island biogeographic theory and conservation practice: Species-area or specious-area relationships? pp. 63–80, copyright 1984, with kind permission from Elsevier Science Ltd, The Boulevard, Langford Lane, Kidlington OX5 1GB, UK.

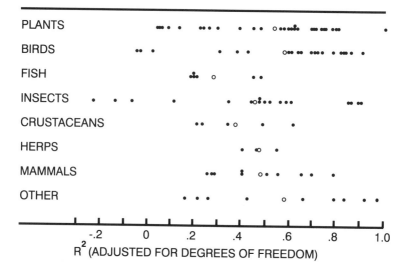

FIGURE 24.24 Distribution of R^2 values from Fig. 24.23 arranged according to taxon. Open circles are means for each taxon. Note the wide spread of values in the insects. Reprinted from *Biological conservation*, Vol. 29, W. J. Boecklen and N. J. Gotelli, Island biogeographic theory and conservation practice: Species-area or specious-area relationships? pp. 63–80, copyright 1984, with kind permission from Elsevier Science Ltd, The Boulevard, Langford Lane, Kidlington OX5 1GB, UK.

Liebherr's (1992) explanation for the differential richness of *Platynus* beetles invokes the concept of the taxon cycle developed by E. O. Wilson (1961). The concept develops predictions on phylogenetic development in a lineage after colonization of tropical oceanic islands. Generalized habitat use is likely for a successful colonizer, adapted to lowland and variable habitats. Dispersal, especially in dissected and mountainous terrain, will result in the evolution of new species adapted to upland mesic forest and ultimately to specialized endemics in the higher rain forests. Now, much of Cuba lies within a few hundred feet of sea level, whereas Hispaniola is a mountainous island with the Cordillera Central rising to over 10,000 ft. As a result, Cuba can support two lowland-dwelling generalized *Platynus* species, while Hispaniola has 14 species in five areas of endemism, all species wingless with narrow upland to mountain distributions. Habitat diversity and complex topography clearly play a major role in supporting species richness on an island.

PREDICTIONS FOR NATURE RESERVES

A common goal in conservation is to predict the area needed as a reserve to support and protect all or a large majority of extant species in an area, be the

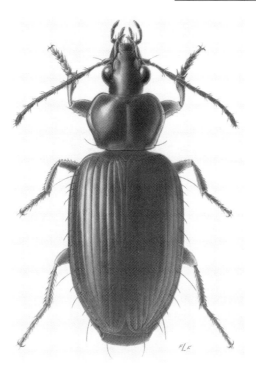

FIGURE 24.25 Carabid beetle, *Platynus marginissimus,* described by Liebherr (1992), distributed in tropical Mexico, Guatemala, and El Salvadore. Drawing by Frances L. Fawcett.

species mammals or birds or butterflies. The desire for this estimate indicates that areas large enough are not available, and extrapolation from smaller areas and species numbers are necessary to make the prediction. Take, for example, the study by Shreeve and Mason (1980) on 22 woodland areas in eastern England which supported a total of 26 butterfly species. Any one wood supported from 1 to 22, but none contained all 26. How much woodland would be needed to support them all, and what confidence can we have in the estimate?

The area required to support all species is called the **point estimate** (Boecklen and Gotelli, 1984) and is derived by extrapolating the species–area regression beyond the data points to the number of species to be preserved (Fig. 24.26). Then we would like to know the confidence we can have in any one area supporting a certain number of species, for which **prediction intervals** are calculated. These are similar to confidence intervals which provide the **average** number of species a given area will support, but for conservation we need to know a **particular** species number for a given area. Thus prediction intervals are wider than confidence intervals, but more realistic if we really want to conserve, say, 26 species of butterfly in one area.

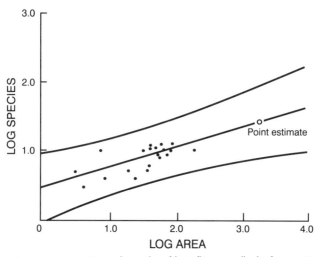

FIGURE 24.26   Species–area regression on the number of butterflies in woodlands of eastern England studied by Shreeve and Mason (1980). The regression line is extrapolated to the point estimate (open circle), and the 95% prediction intervals are also estimated. Reprinted from *Biological conservation,* Vol. 29, W. J. Boecklin and N. J. Gotelli, Island biogeographic theory and conservation practice: Species–area or specious-area relationships? pp. 63–80, copyright 1984, with kind permission from Elsevier Science Ltd, The Boulevard, Langford Lane, Kidlington OX5 1GB, UK.

When such estimates are made, we begin to see some of the problems involved (Fig. 24.26). The point estimate is more than one order of magnitude beyond the empirical data, and we have no idea if the linear species–area model is valid at this scale. In addition, the 95% prediction intervals for areas maintaining 26 butterfly species, either side of the point estimate, span over two orders of magnitude (Boecklen and Gotelli, 1984). If a planner wanted to be sure of conserving 26 butterfly species with 95% confidence, the safest area to design for would be two orders of magnitude larger than the point estimate!

Similar concerns with predictions in conservation relate to the loss of species as nature reserves become isolated from contiguous areas of a similar kind. In eastern England, some of the woodlands were more isolated than others in relation to butterfly habitat (Shreeve and Mason, 1980), so a distance effect on the equilibrium number maintained, as in Fig. 24.21, is likely to be involved. Thus as nature reserves become more isolated because of habitat destruction around their perimeters, increasing isolation and effective area reduction will result in the eventual loss of species. The equilibrium number of species will relax back to a lower equilibrium number ($\hat{s}$), or, in other words, there will be a collapse down to that lower $\hat{s}$ value. Models that predict the rate of loss of species as refuges become more isolated have been called **relaxation models** and **faunal collapse models,** and are summarized by Boecklen and Simberloff (1986).

We can imagine the decline in the number of species with isolation as in Fig. 24.19, with refuges and reserves becoming more and more distant from a "mainland," or major source of colonists. The species–area regression for remote islands is lower than that for close islands (Fig. 24.27). How many species will be lost from a reserve as it is more and more insularized, and what will be the rate of loss?

To put it bluntly, we do not know. For any taxon of concern, or group of taxa, the relevant data are usually unavailable on distance effects on species equilibria in mainland habitat islands. Rates of extinction are unknown relative to faunal collapse. Whichever extinction rate one wishes to use, the 95% prediction intervals for the percent loss of species per unit of time cover almost all possibilities, from 0 to 100% for most of the time scales employed (Boecklen and Simberloff, 1986).

It is now apparent that informed conservation planning and a predictive ecology must be based on much more detailed information on species requirements, habitat heterogeneity, and a unified concept and practice of ecosystem and biome function. Beyond this realization we should accept that equilibrium does not necessarily prevail and there is continuing migration of plants and animals, for example in response to post-Pleistocene events, as discussed in this chapter. If we are to predict the future, as any theoretical ecologist should wish to do, we need to be collecting much more relevant data now. At what scale should we concentrate our efforts? As a heuristic conclusion to this chapter, Darlington's (1959) broad perspective on biogeographic patterns should freshen our perspective on the dynamism of the natural world in terms of changing ecology and evolutionary patterns.

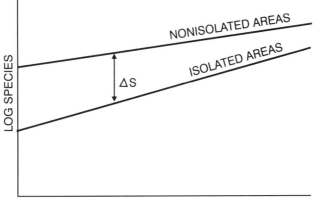

FIGURE 24.27 Conceptual view of faunal collapse (ΔS) as an area such as a nature reserve becomes isolated from contiguous habitat. Compare this figure with data on near and distant islands in Fig. 24.19. From W. J. Boecklen and D. Simberloff in D. K. Elliot (ed.), *Dynamics of extinction.* Copyright © 1986 John Wiley & Sons, Inc. Reprinted by permission of John Wiley & Sons, Inc.

EVOLUTION OF DOMINANT GROUPS

This hypothesis relates to continental biogeography in relation to competition, particularly the development of dominance in competitive interactions. It was Darlington (1959) who asked the question: Where do dominant animals evolve? Do they evolve in large or small areas, warm or cold, stable or less predictable climates? To be dominant an animal must possess adaptations that are general in nature and that will promote its survival in every environment it invades. Brown (1958a) contrasted this general adaptation of dominant species, which might increase an animal's efficiency in food utilization or rate of development independent of the local environment, with what he called special adaptations that would promote the welfare of an organism in relation to a particular set of environmental conditions. To occupy a large area, to be competitively superior, to be an effective colonizer, a species must have accumulated a large number of general adaptations. Darlington wanted to know under which conditions factors are likely to be favorable for the accumulation of such a set of general adaptations.

Darlington examined the evidence for the origin of several dominant or potent groups of animals. For example, all clues point to the evolution of *Homo sapiens* in the Old World tropics. In the elephants, mastodonts, and so on, the earliest fossils were found in Egypt, and Simpson (1940) has suggested that they radiated from North Africa. The family Bovidae, which includes cattle, antelopes, sheep, and goats, is the most recent family of hoofed animals to evolve, and now they are the most numerous and diverse in Africa. The Old World rats and mice, the murid rodents, are dominant practically wherever they have dispersed, and these are most diverse in the Old World tropics. In fact, the evidence shows that many dominant groups have arisen in the Old World tropics and have spread northward and southward. There has been considerably less movement in the opposite directions (Fig. 24.28). The most dominant groups seem to evolve in the largest areas with the most favorable climate. Here species will not have to specialize in tolerating physiological hardships, and they are more likely to evolve toward a strong competitive ability because so many other related species coexist. Darlington suggested further that the tropical climate might accelerate reproduction and mutation. A large area will contain many populations, and the larger the area, the more numerous and diverse will be the populations of any species. The gene pool will be very diverse, making the likelihood of combinations of genes that produce dominance much greater. Thus the theory also relates to the discussion of diversity and stability in Chapter 23. However, larger areas also contain more species and, independent of qualitative differences between species in biogeographic realms, there will be a greater probability of species moving more from species-rich areas than from areas lower in species diversity. This prediction from the theory of island biogeography appears to be a sufficient explanation for the initial exchange of land mam-

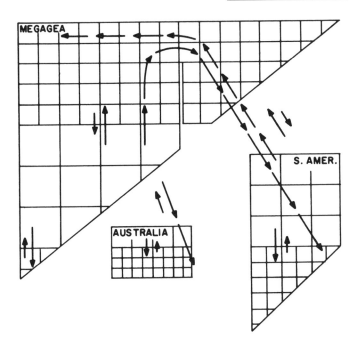

FIGURE 24.28   Diagram illustrating Darlington's concept that large land area in a favorable climate will lead to a large number of populations (represented by the large squares) per unit area and consequent evolution of new generally adapted species. These then disperse as indicated by the long arrows, with very little dispersal in the opposite direction, as indicated by the short arrows. Megagea is the landmass composed of Eurasia, Africa, and North America. From Darlington (1959).

mals between North and South America after the Panamanian land bridge appeared about 3 million years ago (Marshall et al., 1982).

There is much more to be said that is relevant to an understanding of biogeography, paleoecology, and biodiversity, and other works should be consulted. Textbooks on the subject include Darlington (1957), Udvardy (1969), Cox et al. (1976), Pielou, (1979) and Brown and Gibson (1983). Darlington's (1965) book *Biogeography of the southern end of the world* provides a discussion of an area that is of great importance, includes much more information on insects than his former book, and will introduce the reader to his major contributions on the biogeography of carabid beetles. Strongly ecological treatments may be found in MacArthur's (1972) *Geographical Ecology* and the section on geographical ecology in Kendeigh (1974). Symposia on regional biogeographies include South America (Fittkau et al., 1968, 1969), India (Mani, 1974), Australia (Keast et al., 1959), and Antarctica (Van Mieghem et al., 1965). Much of *Macroecology* by Brown (1995) relates to biogeography.

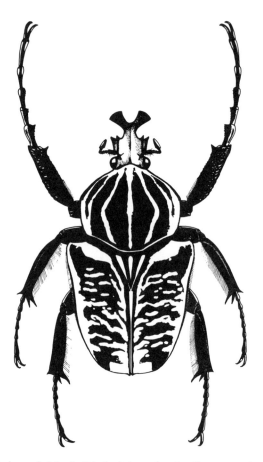

Male of the large goliath beetle, *Goliathus kirkianus,* from East Africa. Drawing by Steve Price.

References

Abrahamson, W. G. (ed.). 1989. *Plant–animal interactions.* McGraw-Hill, New York.

Addicott, J. F. 1974. Predation and prey community structure: An experimental study of the effect of mosquito larvae on the protozoan communities of pitcher plants. *Ecology* **55**:475–492.

Addicott, J. F. 1978. Competition for mutualists: Aphids and ants. *Can. J. Zool.* **56**:2093–2096.

Addicott, J. F. 1979. A multispecies aphid–ant association: Density dependence and species-specific effects. *Can. J. Zool.* **57**:558–569.

Addicott, J. F. 1981. Stability properties of 2-species models of mutualism: Simulation studies. *Oecologia* **49**:42–49.

Addicott, J. F. 1986. Variation in the costs and benefits of mutualism: The interaction between yuccas and yucca moths. *Oecologia* **70**:486–494.

Addicott, J. F., J. Bronstein, and F. Kjellberg. 1990. Evolution of mutualistic life-cycles: yucca moths and fig wasps. In F. Gilbert (ed.). *Genetics, evolution and coevolution of insect life cycles.* Springer-Verlag, New York, pp. 143–161.

Akre, R. D., L. D. Hansen, and R. S. Zack, 1991. Insect jewelry. *Am. Entomol.* **37**:91–95.

Alcock, J. 1973a. The foraging behavior of seed-caching beetle larvae (Coleoptera: Carabidae). *Bull. Ecol. Soc. Am.* **54**(1):45–46.

Alcock, J. 1973b. Cues used in searching for food by red-winged blackbirds *(Agelaius phoeniceus). Behaviour* **46**:174–188.

Alcock, J. 1975. Territorial behaviour by males of *Philanthus multimaculatus* (Hymenoptera: Sphecidae) with a review of territoriality in male sphecids. *Anim. Behav.* **23**:889–895.

Alcock, J. 1979a. *Animal behavior: An evolutionary approach.* 2nd ed. Sinauer, Sunderland, Mass.

Alcock, J. 1979b. The evolution of intraspecific diversity in male reproductive strategies in some bees and wasps. In M. S. Blum and N. A. Blum (eds.). *Sexual selection and reproductive competition in insects.* Academic Press, San Diego, Calif., pp. 381–402.

Alexander, R. D. 1961. Aggressiveness, territoriality, and sexual behavior in field crickets (Orthoptera: Gryllidae). *Behaviour* **17**:130–223.

Alexander, R. D., and W. L. Brown, Jr. 1963. Mating behavior and the origin of insect wings. *Occas. Pap, Mus. Zool. Univ. Mich.* **628**:1–19.

Alexander, R. M. 1971. *Size and shape.* Edward Arnold, London.

Alexopoulos, C. J. 1952. *Introductory mycology.* Wiley, New York.

Allee, W. C., A. E. Emerson, O. Park, T. Park, and K. P. Schmidt. 1949. *Principles of animal ecology.* W. B. Saunders, Philadelphia, Pa.

Allee, W. C., and T. Park. 1939. Concerning ecological principles. *Science* **89**:166–169.

Altieri, M. A., and D. K. Letourneau. 1984. Vegetation diversity and insect pest outbreaks. *CRC Crit. Rev. Plant Sci.* **2**(2):131–169.

Amman, G. D. 1977. The role of the mountain pine beetle in lodgepole pine ecosystems: Impact on succession. In W. J. Mattson (ed.). *The role of arthropods in forest ecosystems.* Springer-Verlag, New York, pp. 3–18.

Anderson, N. G. 1970. Evolutionary significance of virus infection. *Nature* **227**:1346–1347.

Anderson, R. M., and R. M. May. 1980. Infectious diseases and population cycles of forest insects. *Science* **210**:658–661.

Anderson, R. M., B. D. Turner, and L. R. Taylor (eds.). 1979. *Population dynamics.* Symp. Br. Ecol. Soc. 20, Blackwell Scientific Publications, Oxford.

Anderson, W., T. Dobzhansky, O. Pavlovsky, J. Powell, and D. Yardley. 1975. Genetics of natural populations. XLII. Three decades of genetic change in *Drosophila pseudoobscura. Evolution* **29**:24–36.

Andow, D. A. 1988. Management of weeds for insect manipulation in agroecosystems. In M. A. Altieri and S. M. Lieberman (eds.). *Weed management in agroecosystems: Ecological approaches.* CRC Press, Boca Raton, Fla., pp. 265–301.

Andow, D. A. 1991. Vegetational diversity and arthropod population response. *Annu. Rev. Entomol.* **36**:561–586.

Andrewartha, H. G., and L. C. Birch. 1953. The Lotka–Volterra theory of interspecific competition. *Aust. J. Zool.* **1**:174–177.

Andrewartha, H. G., and L. C. Birch. 1954. *The distribution and abundance of animals.* Univ. Chicago Press, Chicago.

Andrews, J. H. 1991. *Comparative ecology of microorganisms and macroorganisms.* Springer-Verlag, New York.

Andrzejewska, L. 1967. Estimation of the effects of feeding of the sucking insect *Cicadella viridis* L. (Homoptera-Auchenorrhyncha) on plants. In K. Petrusewicz (ed.). *Secondary productivity of terrestrial ecosystems.* Institute of Ecology, Polish Academy of Sciences, Warsaw, pp. 791–805.

Antonovics, J., and D. A. Levin. 1980. The ecological and genetic consequences of density-dependent regulation in plants. *Annu. Rev. Ecol. Syst.* **11**:411–452.

Applebaum, S. W. 1964. Physiological aspects of host specificity in the Bruchidae. I. General considerations of developmental compatibility. *J. Insect Physiol.* **10**:783–788.

Armbruster, W. S. 1983. *Dalechampia scandens* (Ortiguilla, Bejuco de Pan). In D. H. Janzen (ed.). *Costa Rican natural history.* Univ. Chicago Press, Chicago, pp. 230–233.

Armbruster, W. S., and G. L. Webster. 1979. Pollination of two species of *Dalechampia* (Euphorbiaceae) in Mexico by euglossine bees. *Biotropica* **11**:278–283.

Arndt, W. 1940. Der prozentuelle Anteil der Parasiten auf und in Tieren im Rahmen des aus Deutschland bisher bekannten Tierartenbestandes. *Z. Parasitenkd.* **11**:684–689.

Arnett, R. H. 1993. *American insects: A handbook of the insects of America north of Mexico.* Sandhill Crane Press, Gainesville, Fla.

Arnold, C. Y. 1960. Maximum–minimum temperatures as a basis for computing heat units. *Proc. Am. Soc. Hort. Sci.* **76**:682–692.

Ashworth, A. C. 1977. A late Wisconsin coleopterous assemblage from southern Ontario and its environmental significance. *Can. J. Earth Sci.* **14**:625–634.

Askew, R. R. 1961. On the biology of the inhabitants of oak galls of Cynipidae (Hymenoptera) in Britain. *Trans. Soc. Br. Entomol.* **14**:237–268.

Askew, R. R. 1971. *Parasitic insects.* American Elsevier, New York.

Atsatt, P. R., and D. J. O'Dowd. 1976. Plant defense guilds. *Science,* **193**:24–29.

Auer, C. 1961. Ergebnisse zwölfjähriger quantitativer Untersuchungen der Populationsbewegung des Grauen Lärchenwicklers *Zeiraphera griseana)* (Hb.) im Oberengadin, 1949–1958. *Mitt. Schweiz. Anst. Forstl. VersWes.* **37**:175–263.

Axelrod, D. I. 1972. Ocean-floor spreading in relation to ecosystematic problems. In R. T. Allen and F. C. James (eds.). *A symposium on ecosystematics.* Univ. Ark. Mus. Occ. Paper 4, Univ. Arkansas Press, Little Rock, Ark., pp. 15–68.

Ayala, F. J. 1968. Genotype, environment and population numbers. *Science* **162**:1453–1459.

Ayala, F. J. 1970. Competition, coexistence and evolution. In M. K. Hecht and W. C. Steere (eds.). *Essays in evolution and genetics.* Appleton-Century-Crofts, New York, pp. 121–158.

Ayala, F. J. 1971. Competition between species: Frequency dependence. *Science* **171**:820–824.

Ayala, F. J. 1974. Biological evolution: Natural selection or random walk? *Am. Sci.* **62**:692–701.

Ayala, F. J., J. R. Powell, and M. L. Tracey. 1972a. Enzyme variability in the *Drosophila willistoni* group. V. Genic variation in natural populations of *Drosophila equinoxalis*. *Genet. Res.* **20**:19–42.

Ayala, F. J., D. Hedgecock, G. S. Zumwalt, and J. W. Valentine. 1973. Genetic variation in *Tridacna maxima*, an ecological analog of some unsuccessful evolutionary lineages. *Evolution* **27**:177–191.

Ayala, F. J., J. R. Powell, M. L. Tracey, C. A. Mourão, and S. Pérez-Salas. 1972b. Enzyme variability in the *Drosophila willistoni* group. IV. Genic variation in natural populations of *Drosophila willistoni*. *Genetics* **70**:113–139.

Ayala, F. J., M. L. Tracey, L. G. Barr, and J. G. Ehrenfeld. 1974. Genetic and reproductive differentiation of the subspecies, *Drosophila equinoxalis caribbensis*. *Evolution* **28**:24–41.

Bach, C. E. 1988a. Effects of host plant patch size on herbivore density: Patterns. *Ecology* **69**:1090–1102.

Bach, C. E. 1988b. Effects of host plant patch size on herbivore density: Underlying mechanisms. *Ecology* **69**:1103–1117.

Baker, H. G. 1961. *Ficus* and *Blastophaga*. *Evolution* **15**:378–379.

Baker, H. G. 1970. Evolution in the tropics. *Biotropica* **2**:101–111.

Baker, H. G., and I. Baker. 1973a. Amino-acids in nectar and their evolutionary significance. *Nature* **241**:543–545.

Baker, H. G., and I. Baker. 1973b. Some anthecological aspects of the evolution of nectar-producing flowers, particularly amino acid production in nectar. In V. H. Heywood (ed.). *Taxonomy and ecology*. Academic Press, San Diego, Calif., pp. 243–264.

Baker, H. G., and I. Baker. 1975. Studies on nectar-constitution and pollinator-plant coevolution. In L. E. Gilbert and P. H. Raven (eds.). *Coevolution of animals and plants*. Univ. Texas Press, Austin, Texas, pp. 100–140.

Baker, H. G., and I. Baker. 1983a. A brief historical review of the chemistry of floral nectar. In B. Bentley and T. Elias (eds.). *The biology of nectaries*. Columbia Univ. Press, New York, pp. 126–152.

Baker, H. G., and I. Baker. 1983b. Floral nectar sugar constituents in relation to pollinator type. In C. E. Jones and R. J. Little (eds.). *Handbook of experimental pollination biology*. Van Nostrand Reinhold, New York, pp. 117–141.

Baker, R. R. 1972. Territorial behaviour of the nymphalid butterflies, *Aglais urticae* (L.) and *Inachis io* (L.). *J. Anim. Ecol.* **41**:453–469.

Baker, R. R. 1983. Insect territoriality. *Annu. Rev. Entomol.* **28**:65–89.

Baker, W. L. 1972. Eastern forest insects. *USDA For. Serv. Misc. Publ. 1175*. U.S. Government Printing Office, Washington, D.C.

Bakke, A. 1969. The effect of forest fertilization on the larval weight and larval density of *Laspeyresia strobilella* (L.) (Lepidoptera; Tortricidae) in cones of Norway spruce. *Z. Angew. Entomol.* **63**:451–453.

Bakker, K., I. H. Bosch, R. van Crevel, N. C. Michielsen, G. Drost, E. A. J. Wanders, V. Westhoff, and M. van Wijngaarden (eds.). 1974. *Meijendel: Duin-water-leven.* van Hoeve, The Hague, The Netherlands.

Baldwin, I. T., and J. C. Schultz. 1983. Rapid changes in tree leaf chemistry induced by damage: Evidence for communication between plants. *Science* **221**:277–279.

Baltensweiler, W. 1968. The cyclic population dynamics of the grey larch tortrix. *Zeiraphera griseana* Hübner (= *Semasia diniana* Guenée) (Lepidoptera: Tortricidae). In T. R. E. Southwood (ed.). *Insect abundance.* Symp. R. Entomol. Soc. London 4, Royal Entomological Society of London, London, pp. 88–97.

Banks, C. J. 1962. Effects of the ant *Lasius niger* (L.) on insects preying on small populations of *Aphis fabae* Scop. on bean plants. *Ann. Appl. Biol.* **50**:669–679.

Barbehenn, K. R. 1969. Host–parasite relationships and species diversity in mammals: An hypothesis. *Biotropica* **1**:29–35.

Barbosa, P., and J. C. Schultz (eds.). 1987. *Insect outbreaks.* Academic Press, San Diego, Calif.

Barbosa, P., V. Krischik, and D. Lance. 1989. Life-history traits of forest-inhabiting flightless Lepidoptera. *Am. Midl. Nat.* **122**:262–274.

Barr, B. A. 1969. Sound production in Scolytidae (Coleoptera) with emphasis on the genus *Ips. Can. Entomol.* **101**:636–672.

Bartholomew, G. A. 1970. A model for the evolution of pinniped polygyny. *Evolution* **24**:546–559.

Bartlett, B. R. 1961. The influence of ants upon parasites, predators, and scale insects. *Ann. Entomol. Soc. Am.* **54**:543–551.

Baskerville, G. L., and P. Emin. 1969. Rapid estimation of heat accumulation from maximum and minimum temperatures. *Ecology* **50**:514–522.

Bateman, A. J. 1948. Intra-sexual selection in *Drosophila. Heredity* **2**:349–368.

Bates, H. W. 1862. Contributions to an insect fauna of the Amazon Valley. *Trans. Linn. Soc. London* **23**:495–566.

Batra, L. R. (ed.). 1979. *Insect–fungus symbiosis: Nutrition, mutualism, and commensalism.* Allanheld, Osmun, Totowa, N.J.

Batra, L. R., and S. W. T. Batra. 1979. Termite–fungus mutualism. In L. R. Batra (ed.). *Insect-fungus symbiosis: Nutrition, mutualism, and commensalism.* Allanheld, Osmun, Totowa, N.J., pp. 117–163.

Batra, S. W. T., and L. R. Batra. 1967. The fungus gardens of insects. *Sci. Am.* **217**(5):112–120.

Batzli, G. O. 1974. Production, assimilation and accumulation of organic matter in ecosystems. *J. Theor. Biol.* **45**:205–217.

Batzli, G. O. 1983. Responses of arctic rodent populations to nutritional factors. *Oikos* **40**:396–406.

Batzli, G. O., and F. A. Pitelka. 1970. Influence of meadow mouse populations on California grassland. *Ecology* **51**:1027–1039.

Bawa, K. S., and M. Hadley (eds.). 1990. *Reproductive ecology of tropical forest plants.* UNESCO, Paris.

Bazzaz, F. A. 1968. Succession on abandoned fields in the Shawnee Hills, southern Illinois. *Ecology* 49:924–936.

Bazzaz, F. A. 1975. Plant species diversity in old-field successional ecosystems in southern Illinois. *Ecology* 56:485–488.

Bazzaz, F. A. 1979. The physiological ecology of plant succession. *Annu. Rev. Ecol. Syst.* 10:351–371.

Bazzaz, F. A. 1990. Plant–plant interactions in successional environments. In J. B. Grace and G. D. Tilman (eds.). *Perspectives on plant competition.* Academic Press, San Diego, Calif., pp. 239–263.

Beattie, A. J. 1985. *The evolutionary ecology of ant–plant mutualisms.* Cambridge Univ. Press, Cambridge.

Beaver, R. A. 1966. The development and expression of population tables for the bark beetle *Scolytus scolytus* (F.). *J. Anim. Ecol.* 35:27–41.

Beaver, R. A. 1979. Host specificity of temperate and tropical animals. *Nature* 281:139–141.

Beer, J. R., and E. F. Cook. 1958. The louse populations on some deer mice from western Oregon. *Pan-Pacif. Entomol.* 34:155–158.

Begon, M., J. L. Harper, and C. R. Townsend. 1996a. *Ecology: Individuals, populations and communities.* 3rd ed. Blackwell Scientific, Oxford.

Begon, M., M. Mortimer, and D. J. Thompson. 1996b. *Population ecology: A unified study of animals and plants.* 3rd ed. Blackwell Scientific, Oxford.

Belovsky, G. E., and A. Joern. 1995. The dominance of different regulating factors for rangeland grasshoppers. In N. Cappuccino and P. W. Price (eds.). *Population dynamics: New approaches and synthesis.* Academic Press, San Diego, Calif., pp. 359–386.

Belt, T. 1874. *The naturalist in Nicaragua.* Murray, London.

Belt, T. 1888. *A naturalist in Nicaragua.* 2nd ed. Bumpus, London.

Bentley, B. L. 1977. Extrafloral nectaries and protection by pugnacious bodyguards. *Annu. Rev. Ecol. Syst.* 8:407–427.

Bentley, B. L., and T. Elias (eds.). 1983. *The biology of nectaries.* Columbia Univ. Press, New York.

Bentley, S., and J. B. Whittaker. 1979. Effects of grazing by a chrysomelid beetle, *Gastrophysa viridula,* on competition between *Rumex obtusifolius* and *Rumex crispus. J. Ecol.* 67:79–90.

Bentley, S., J. B. Whittaker, and A. J. C. Malloch. 1980. Field experiments on the effects of grazing by a chrysomelid beetle *(Gastrophysa viridula)* on seed production and quality in *Rumex obtusifolius* and *Rumex crispus. J. Ecol.* 68:671–674.

Benzing, D. H. 1978a. The life history profile of *Tillandsia circinnata* (Bromeliaceae) and the rarity of extreme epiphytism among the angiosperms. *Selbyana* 2:325–337.

Benzing, D. H. 1978b. Germination and early establishment of *Tillandsia circinnata* Schlecht. (Bromeliaceae) on some of its hosts and other supports in southern Florida. *Selbyana* **5**:95–106.

Bequaert, J. 1921. On the dispersal by flies of the spores of certain mosses of the family Splachnaceae. *Bryologist* **24**:1–4.

Berenbaum, M. 1981. Patterns of furanocoumarin distribution and insect herbivory in the Umbelliferae: Plant chemistry and community structure. *Ecology* **62**:1254–1266.

Berenbaum, M. R. 1995. *Bugs in the system: Insects and their impact on human affairs.* Addison-Wesley, Reading, Mass.

Berenbaum, M. R., and A. R. Znagerl. 1992. Genetics of secondary metabolism and herbivore resistance in plants. In G. A. Rosenthal and M. R. Berenbaum (eds.). *Herbivores: Their interactions with secondary plant metabolites.* Vol. 2. 2nd ed. Academic Press, San Diego, Calif., pp. 415–438.

Berger, E. M. 1970. A comparison of gene–enzyme variation between *Drosophila melanogaster* and *D. simulans. Genetics* **66**:677–683.

Bernays, E. A. 1978. Tannins: An alternative viewpoint. *Entomol. Exp. Appl.* **24**:44–53.

Bernays, E. A. 1981. Plant tannins and insect herbivores: An appraisal. *Ecol. Entomol.* **6**:353–360.

Bernays, E. A., and R. F. Chapman. 1994. *Host-plant selection by phytophagous insects.* Chapman & Hall, New York.

Berry, R. J., and H. M. Murphy. 1970. The biochemical genetics of an island population of the house mouse. *Proc. R. Soc. London Ser. B* **176**:87–103.

Berryman, A. A. 1973. Population dynamics of the fir engraver, *Scolytus ventralis* (Coleoptera: Scolytidae). I. Analysis of population behavior and survival from 1964 to 1971. *Can. Entomol.* **105**:1465–1488.

Berryman, A. A. 1981. *Population systems: A general introduction.* Plenum Press, New York.

Berryman, A. A. 1982. Population dynamics of bark beetles. In J. B. Mitton and K. B. Sturgeon (eds.). *Bark beetles in North American conifers: A system for the study of evolutionary biology.* Univ. Texas Press, Austin, Texas, pp. 264–314.

Berryman, A. A. 1992. The origins and evolution of predator–prey theory. *Ecology* **73**:1530–1535.

Berryman, A. A. 1996. What causes population cycles of forest Lepidoptera: *Trends Ecol. Evol.* **11**:28–32.

Berryman, A. A., and R. W. Stark. 1985. Assessing the risk of forest insect outbreaks. *Z. Angew. Entomol.* **99**:199–208.

Bess, H. A., and F. H. Haramoto. 1958. Biological control of the oriental fruit fly in Hawaii. *Proc. Int. Congr. Entomol. 10 Montreal 1956* **4**: 835–840.

Betts, E. 1976. Forecasting infestations of tropical migrant pests: The desert locust and the African armyworm. *Symp. R. Entomol. Soc. London* **7**: 113–134.

Bidder, G. P. 1931. The biological importance of brownian movement (with notes on sponges and Protista).*Proc. Linn. Soc. London* **143**:82–96.

Biere, J. M., and G. W. Uetz. 1981. Web orientation in the spider *Micrathena gracilis* (Araneae: Araneidae). *Ecology* **62**:336–344.

Birch, L. C. 1948. The intrinsic rate of natural increase of an insect population. *J. Anim. Ecol.* **17**:15–26.

Birks, H. J. B. 1980. British trees and insects: A test of the time hypothesis over the last 13,000 years. *Am. Nat.* **115**:600–605.

Bishop, J. A., and L. M. Cook, 1975. Moths, melanism, and clean air. *Sci. Am.* **232**(1):90–99.

Bishop, J. A., and L. M. Cook. 1979. A century of industrial melanism. *Antenna* **3**:125–128.

Bishop, J. A., and L. M. Cook. 1980. Industrial melanism and the urban environment. *Adv. Ecol. Res.* **11**:373–404.

Black, L. M. 1979. Vector cell monolayers and plant viruses. *Adv. Virus Res.* **25**:191–271.

Blackman, M. W., and H. H. Stage. 1924. On the succession of insects living in the bark and wood of dying, dead and decaying hickory. *New York State Col. For. Syracuse Univ. Tech. Publ.* **17**:1–269.

Blais, J. R. 1952. The relationship of the spruce budworm *(Choristoneura fumiferana)* (Clem.) to the flowering condition of balsam fir *(Abies balsamea* (L.) Mill.). *Can. J. Zool.* **30**:1–29.

Blais, J. R. 1953. Effects of the destruction of the current year's foliage of balsam fir on the fecundity and habits of flight of the spruce budworm. *Can. Entomol.* **85**:446–448.

Blest, A. D. 1957a. The function of eyespot patterns in the Lepidoptera. *Behaviour* **11**:209–256.

Blest, A. D. 1957b. The evolution of protective displays in the Saturnioidea and Sphingidae (Lepidoptera). *Behaviour* **11**:257–309.

Blest, A. D., 1963. Longevity, palatability and natural selection in five species of New World saturniid moth. *Nature* **197**:1183–1186.

Blum, M. S. 1977. Behavioral responses of Hymenoptera to pheromones, and allomones and kairomones. In H. H. Shorey and J. J. McKelvey (eds.). *Chemical control of insect behavior: Theory and application*. Wiley, New York, pp. 149–167.

Blum, M. S. 1981. *Chemical defenses of arthropods*. Academic Press, San Diego, Calif.

Blum, M. S., and N. A. Blum (eds.). 1979. *Sexual selection and reproductive competition in insects*. Academic Press, San Diego, Calif.

Bodenheimer, F. S. 1930. Über die Grundlagen einer allgemeinen Epidemiologie der Insektenkalamitäten. *Z. Angew. Entomol.* **16**:433–450.

Bodenheimer, F. S. 1931. Der Massenwechsel in der Tierwelt. *Arch. Zool. Ital.* **16**:98–111.

Bodenheimer, F. S. 1938. *Problems of animal ecology.* Clarendon Press, Oxford.

Bodenheimer, F. S. 1958. *Animal ecology today.* Dr. W. Junk, The Hague, The Netherlands.

Boecklen, W. J., and N. J. Gotelli. 1984. Island biogeographic theory and conservation practice: Species-area or specious-area relationships? *Biol. Conserv.* **29**:63–80.

Boecklen, W. J., and D. Simberloff. 1986. Area-based extinction models in conservation. In D. K. Elliott (ed.). *Dynamics of extinction.* Wiley, New York, pp. 247–276.

Bogdan, A. V. 1962. Grass pollination by bees in Kenya. *Proc. Linn. Soc. London* **173**:57–60.

Boggs, C. L. 1995. Male nuptial gifts: Phenotypic consequences and evolutionary implications. In S. R. Leather and J. Hardie (eds.). *Insect reproduction.* CRC Press, Boca Raton, Fla., pp. 215–242.

Bond, A. B. 1980. Optimal foraging in a uniform habitat: The search mechanism of the green lacewing. *Anim. Behav.* **28**:10–19.

Bonnell, M. L., and R. K. Selander. 1974. Elephant seals: Genetic variation and near extinction. *Science* **184**:908–909.

Booth, W. C., G. G. Colomb, and J. M. Williams. 1995. *The craft of research.* Univ. Chicago Press, Chicago.

Borgia, G. 1979. Sexual selection and the evolution of mating systems. In M. S. Blum and N. A. Blum (eds.). *Sexual selection and reproductive competition in insects.* Academic Press, San Diego, Calif., pp. 19–80.

Borgia, G. 1980. Sexual competition in *Scatophaga stercoraria:* Size- and density-related changes in male ability to capture females. *Behaviour* **75**:185–206.

Borgia, G. 1982. Experimental changes in resource structure and male density: Size-related differences in mating success among male *Scatophaga stercoraria. Evolution* **36**:307–315.

Borror, D. J., C. A. Triplehorn, and N. F. Johnson. 1989. *An introduction to the study of insects.* 6th ed. W. B. Saunders, Philadelphia, Pa.

Botkin, D. B., and R. S. Miller, 1974. Mortality rates and survival of birds. *Am. Nat.* **108**:181–192.

Boucher, D. H. (ed.). 1985. *The biology of mutualism.* Oxford Univ. Press, New York.

Bovbjerg, R. V. 1970. Ecological isolation and competitive exclusion in two crayfish *(Orconectes virilis* and *Orconectes immunis). Ecology* **51**:225–236.

Bowers, M. D. 1992. The evolution of unpalatability and the cost of chemical defense in insects. In B. D. Roitberg and M. B. Isman (eds.). *Insect chemical ecology: An evolutionary approach.* Chapman & Hall, New York.

Bowers, M. D. 1993. Aposematic caterpillars: Lifestyles of the warningly colored and unpalatable. In N. E. Stamp and T. M. Casey (eds.). *Caterpillars: Ecology and evolutionary constraints on foraging.* Chapman & Hall, New York, pp. 331–371.

Bowers, W. S. 1991. Insect hormones and antihormones in plants. In G. A. Rosenthal and M. R. Berenbaum (eds.). *Herbivores: Their interactions with secondary plant metabolites.* Vol. 1. 2nd ed. Academic Press, San Diego, Calif., pp. 431–456.

Brattsten, L. B. 1979. Biochemical defense mechanisms in herbivores against plant allelochemicals. In G. A. Rosenthal and D. H. Janzen (eds.). *Herbivores: Their interaction with secondary plant metabolites.* Academic Press, San Diego, Calif., pp. 199–270.

Brattsten, L. B. 1992. Metabolic defenses against plant allelochemicals. In G. A. Rosenthal and M. R. Berenbaum (eds.). *Herbivores: Their interactions with secondary plant metabolites.* Vol. 2. 2nd ed. Academic Press, San Diego, Calif., pp. 175–242.

Brehm, B. G., and D. Krell. 1975. Flavonoid localization in epidermal papillae of flower petals: a specialized adaptation for ultraviolet absorption. *Science* 190:1221–1223.

Breznak, J. A. 1975. Symbiotic relationships between termites and their intestinal microbiota. *Symp. Soc. Exp. Biol.* 29:559–580.

Brncic, D. 1962. Chromosomal structure of populations of *Drosophila flavopilosa* studied in larvae collected in their natural breeding sites. *Chromosoma* 13:183–195.

Brncic, D. 1968. The effects of temperature on chromosomal polymorphism of *Drosophila flavopilosa* larvae. *Genetics* 59:427–432.

Broadhead, E. 1958. The psocid fauna of larch trees in northern England: An ecological study of mixed species populations exploiting a common resource. *J. Anim. Ecol.* 27:217–263.

Broadhead, E., and A. J. Wapshere. 1966. *Mesopsocus* populations on larch in England—the distribution and dynamics of two closely related co-existing species of Psocoptera sharing the same food source. *Ecol. Monogr.* 36:327–388.

Brock, T. D. 1967. Relationship between standing crop and primary productivity along a hot spring thermal gradient. *Ecology* 48:566–571.

Brodsky, A. K. 1994. *The evolution of insect flight.* Oxford Univ. Press, Oxford.

Bronstein, J. L. 1987. Maintenance of species-specificity in a Neotropical fig-pollinator wasp mutualism. *Oikos* 48:39–46.

Bronstein, J. L. 1988a. Limits to fruit production in a monoecious fig: Consequences of an obligate mutualism. *Ecology* 69:207–214.

Bronstein, J. L. 1988b. Mutualism, antagonism, and the fig-pollinator interaction. *Ecology* 69:1298–1302.

Bronstein, J. L., and K. Hoffman. 1987. Spatial and temporal variation in frugivory at a Neotropical fig, *Ficus pertusa. Oikos* 49:261–268.

Bronstein, J. L., and D. McKey (eds.). 1989. The comparative biology of figs. *Experientia* **45**:601–679.

Brooks, D. R., and D. A. McLennan. 1993a. *Parascript: Parasites and the language of evolution.* Smithsonian Institution Press, Washington, D.C.

Brooks, D. R., and D. A. McLennan. 1993b. Comparative study of adaptive radiations with an example using parasitic flatworms (Platyhelminthes: Cercomeria). *Am. Nat.* **142**:755–778.

Brower, J. V. Z. 1958a. Experimental studies of mimicry in some North American butterflies. I. The monarch, *Danaus plexippus,* and viceroy, *Limenitis archippus archippus. Evolution* **12**:32–47.

Brower, J. V. Z. 1958b. Experimental studies of mimicry in some North American butterflies. III. *Danaus gilippus berenice* and *Limenitis archippus floridensis. Evolution* **12**:273–285.

Brower, L. P. 1969. Ecological chemistry. *Sci. Am.* **220**(2):22–29.

Brower, L. P., and J. V. Z. Brower. 1964. Birds, butterflies, and plant poisons: A study in ecological chemistry. *Zoologica* **49**:137–159.

Brower, L. P., J. Alcock, and J. V. Z. Brower, 1971. Avian feeding behaviour and the selective advantage of incipient mimicry. In R. Creed (ed.). *Ecological genetics and evolution.* Appleton-Century-Crofts, New York, pp. 261–274.

Brower, L. P., P. B. McEvoy, K. L. Williamson, and M. A. Flannery. 1972. Variation in cardiac glycoside content of monarch butterflies from natural populations in eastern North America. *Science* **177**:426–429.

Brower, L. P., W. N. Ryerson, L. L. Coppinger, and S. C. Glazier. 1968. Ecological chemistry and the palatability spectrum. *Science* **161**: 1349–1351.

Brown, A. W. A. 1978. *Ecology of pesticides.* Wiley, New York.

Brown, E. S. 1951. The relation between migration-rate and type of habitat in aquatic insects, with special reference to certain species of Corixidae. *Proc. Zool. Soc. London* **121**:539–545.

Brown, J. H. 1971. Mammals on mountaintops: Nonequilibrium insular biogeography. *Am. Nat.* **105**:467–478.

Brown, J. H. 1995. *Macroecology.* Univ. Chicago Press, Chicago.

Brown, J. H., and A. C. Gibson. 1983. *Biogeography.* C.V. Mosby, St. Louis, Mo.

Brown, J. H., O. J. Reichman, and D. W. Davidson. 1979. Granivory in desert ecosystems. *Annu. Rev. Ecol. Syst.* **10**:201–227.

Brown, J. M., O. Pellmyr, J. N. Thompson, and R. G. Harrison. 1994. Mitochondrial DNA phylogeny of the Prodoxidae (Lepidoptera: Incurvarioidea) indicates rapid ecological diversification of yucca moths. *Ann. Entomol. Soc. Am.* **87**:795–802.

Brown, W. L., Jr. 1958a. General adaptation and evolution. *Syst. Zool.* **7**:157–168.

Brown, W. L., Jr. 1958b. Some zoological concepts applied to problems in evolution of the hominid lineage. *Am. Sci.* **46**:151–158.

Brown, W. L., Jr. 1960. Ants, acacias and browsing mammals. *Ecology* **41**:587–592.

Brown, W. L., Jr. 1973. A comparison of the Hylean and Congo–West African rain forest ant faunas. In B. J. Meggers, E. S. Ayensu, and W. D. Duckworth (eds.). *Tropical forest ecosystems in Africa and South America: A comparative review.* Smithsonian Institution Press, Washington, D. C., pp. 161–185.

Brown, W. L., Jr., and E. O. Wilson. 1956. Character displacement. *Syst. Zool.* **5**:49–64.

Brues, C. T. 1924. The specificity of food plants in the evolution of phytophagous insects. *Am. Nat.* **58**:127–144.

Brundin, L. 1960. *Transantarctic relationships and their significance, as evidenced by chironomid midges.* Almquist & Wiksell, Stockholm.

Brundin, L. 1967. Insects and the problem of austral disjunctive distribution. *Annu. Rev. Entomol.* **12**:149–168.

Brussard, P. F. (ed.). 1978. *Ecological genetics: The interface.* Springer-Verlag, New York.

Bryant, E. H. 1971. Life history consequences of natural selection: Cole's result. *Am. Nat.* **105**:75–76.

Bryant, E. H. 1974. On the adaptive significance of enzyme polymorphisms in relation to environmental variability. *Am. Nat.* **108**:1–19.

Bryant, J. P. 1987. Feltleaf willow–snowshoe hare interactions: Plant carbon/nutrient balance and floodplain succession. *Ecology* **68**:1319–1327.

Bryant, J. P., S. Chapin, and D. R. Klein. 1983. Carbon/nutrient balance in boreal plants in relation to vertebrate herbivory. *Oikos* **40**:357–368.

Bryant, J. P., J. Tahvanainen, M. Sulkinoja, R. Julkunen-Tiitto, P. Reichardt, and T. Green. 1989. Biogeographic evidence for the evolution of chemical defense against mammal browsing by boreal birch and willow. *Am. Nat.* **136**:20–34.

Bryant, J. P., K. Danell, F. Provenza, P. B. Reichardt, T. A. Clausen, and R. A. Werner. 1991. Effects of mammal browsing on the chemistry of deciduous woody plants. In D. W. Tallamy and M. J. Raupp (eds.). *Phytochemical induction by herbivores.* Wiley, New York, pp. 135–154.

Buchmann, S. L., and G. P. Nabhan. 1996. *The forgotten pollinators.* Island Press, Covelo, Calif.

Buchner, P. 1965. *Endosymbiosis of animals with plant microorganisms.* Wiley, New York.

Buckley, R. C. 1982. *Ant–plant interactions in Australia.* Dr. W. Junk, The Hague, The Netherlands.

Budnik, M., and D. Brncic. 1974. Preadult competition between *Drosophila pavani* and *Drosophila melanogaster, Drosophila simulans,* and *Drosophila willistoni. Ecology* **55**:657–661.

Burbott, A. J., and W. D. Loomis. 1969. Evidence for metabolic turnover of monoterpenes in peppermint. *Plant Physiol.* **44**:173–179.

Burdon, J. J. 1978. Mechanisms of disease control in heterogeneous plant populations: An ecologist's view. In P. R. Scott and A. Bainbridge (eds.). *Plant disease epidemiology.* Blackwell Scientific, Oxford, pp. 193–200.

Burdon, J. J. 1987. *Diseases and plant population biology.* Cambridge Univ. Press, Cambridge.

Burdon, J. J., and G. A. Chilvers. 1974. Fungal and insect parasites contributing to niche differentiation in mixed species stands of eucalypt saplings. *Aust. J. Bot.* **22**:103–114.

Burkholder, P. R. 1952. Cooperation and conflict among primitive organisms. *Am. Sci.* **40**:601–631.

Burnett, T. 1958. A model of host–parasite interaction. *Proc. 10th Int. Congr. Entomol.,* Vol. 2. pp. 679–686.

Burns, D. P. 1973. The foraging and tending behavior of *Dolichoderus taschenbergi* (Hymenoptera: Formicidae). *Can. Entomol.* **105**:97–104.

Burton, M., and R. Burton. 1975. *Encyclopedia of insects and arachnids.* Octopus Books, London.

Bush, G. L. 1975a. Modes of animal speciation. *Annu. Rev. Ecol. Syst.* **6**:339–364.

Bush, G. L. 1975b. Sympatric speciation in phytophagous parasitic insects. In P. W. Price (ed.). *Evolutionary strategies of parasitic insects and mites.* Plenum Press, New York, pp. 187–206.

Busvine, J. R. 1976. *Insects, hygiene and history.* Athlone, London.

Butler, C. G. 1943. Work on bee repellents: Management of colonies for pollination. *Ann. Appl. Biol.* **30**:195–196.

Butler, C. G. 1945. The influence of various physical and biological factors of the environment in honeybee activity: An examination of the relationship between activity and nectar concentration and abundance. *J. Exp. Biol.* **21**:5–12.

Butler, C. G., E. P. Jeffree, and H. Kalmus. 1943. The behaviour of a population of honeybees on an artificial and on a natural crop. *J. Exp. Biol.* **20**:65–73.

Cade, W. 1979. The evolution of alternative male reproductive strategies in field crickets. In M. S. Blum and N. A. Blum (eds.). *Sexual selection and reproductive competition in insects.* Academic Press, San Diego, Calif., pp. 343–379.

Cain, A. J., and P. M. Sheppard. 1954a. Natural selection in *Cepaea. Genetics* **39**:89–116.

Cain, A. J., and P. M. Sheppard. 1954b. The theory of adaptive polymorphism. *Am. Nat.* **88**:321–326.

Calhoun, J. B. 1952. The social aspects of population dynamics. *J. Mammal.* **33**:139–159.

Callan, E. M. 1964. Ecology of sand dunes with special reference to the insect communities. *Monogr. Biol.* **14**:174–185.

Calow, P. 1978. *Life cycles: An evolutionary approach to the physiology of reproduction, development and ageing.* Chapman & Hall, London.

Campanella, P. J. 1975. The evolution of mating systems in temperate zone dragonflies (Odonata: Anisoptera). II. *Libellula luctuosa* (Burmeister). *Behaviour* 54:278–209.

Campanella, P. J., and L. L. Wolf. 1974. Temporal leks as a mating system in a temperate zone dragonfly (Odonata: Anisoptera). I. *Plathemis lydia* (Drury). *Behaviour* 51:49–87.

Campbell, B. C., and S. S. Duffey, 1979. Tomatine and parasitic wasps: Potential incompatibility of plant antibiosis with biological control. *Science* 205:700–702.

Caouette, M. R., and P. W. Price. 1989. Growth of Arizona rose and attack and establishment of gall wasps, *Diplolepis fusiformans* (Ashmead) and *D. spinosa* (Ashmead) (Hymenoptera: Cynipidae). *Environ. Entomol.* 18:822–828.

Cappuccino, N. 1992. The nature of population stability in *Eurosta solidaginis*, a nonoutbreaking herbivore of goldenrod. *Ecology* 73:1792–1801.

Cappuccino, N. 1995. Novel approaches to the study of population dynamics. In N. Cappuccino and P. W. Price (eds.). *Population dynamics: New approaches and synthesis.;* Academic Press, San Diego, Calif., 3–16.

Cappuccino, N., and P. W. Price (eds.). 1995. *Population dynamics: New approaches and synthesis.* Academic Press, San Diego, Calif.

Carlisle, A., A. H. F. Brown, and E. J. White. 1966. Litter fall, leaf production and the effects of defoliation by *Tortrix viridana* in a sessile oak *(Quercus petraea)* woodland. *J. Ecol.* 54:65–85.

Carlisle, D. B., and P. E. Ellis. 1968. Bracken and locust ecdysones: Their effect on molting in the desert locust. *Science* 159:1472–1474.

Carpenter, S. R. (ed.). 1988. *Complex interactions in lake communities.* Springer-Verlag, New York.

Carpenter, S. R., and J. F. Kitchell. 1988. Consumer control of lake productivity. *BioScience* 38:764–769.

Carson, H. L. 1958a. The population genetics of *Drosophila robusta. Adv. Genet.* 9:1–40.

Carson, H. L. 1958b. Response to selection under different conditions of recombination in *Drosophila. Cold Spring Harbor Symp. Quant. Biol.* 23:291–306.

Carson, H. L. 1968. The population flush and its genetics consequences. In R. C. Lewontin (ed.). *Population biology and evolution.* Syracuse Univ. Press, Syracuse, N.Y., pp. 123–137.

Carter, M. C., and A. F. G. Dixon. 1982. Habitat quality and the foraging behaviour of coccinellid larvae. *J. Anim. Ecol.* 51:865–878.

Carter, W. 1973. *Insects in relation to plant disease.* 2nd ed. Wiley, New York.

Cates, R. G. 1975. The interface between slugs and wild ginger: Some evolutionary aspects. *Ecology* 56:391–400.

Cates, R. G. 1980. Feeding patterns of monophagous, oligophagous and polyphagous insect herbivores: The effect of resource abundance and plant chemistry. *Oecologia* **46**:22–31.

Cates, R. G. 1981. Host plant predictability and the feeding patterns of monophagous, oligophagous, and polyphagous insect herbivores. *Oecologia* **48**:319–326.

Cates, R. G., and G. H. Orians. 1975. Successional status and the palatability of plants to generalized herbivores. *Ecology* **56**:410–418.

Cates, R. G., and R. A. Redak. 1988. Variation in the terpene chemistry of Douglas-fir and its relationship to western spruce budworm success. In K. Spencer (ed.). *Chemical mediation of coevolution.* Academic Press, San Diego, Calif., pp. 317–344.

Cates, R. G., C. B. Henderson, and R. A. Redak. 1987. Responses of the western spruce budworm to varying levels of nitrogen and terpenes. *Oecologia* **73**:312–316.

Cates, R. G., R. A. Redak, and C. B. Henderson. 1983. Patterns in defensive natural product chemistry: Douglas fir and western spruce budworm interactions. In P. A. Hedin (ed.). *Plant resistance to insects.* American Chemical Society, Washington, D.C., pp. 3–19.

Caughley, G. 1970. Eruption of ungulate populations, with emphasis on Himalayan thar in New Zealand. *Ecology* **51**:53–72.

Center, T. D., and C. D. Johnson. 1974. Coevolution of some seed beetles (Coleoptera: Bruchidae) and their hosts. *Ecology* **55**:1096–1103.

Chandler, A. E. F. 1967. Oviposition responses of aphidophagous Syrphidae (Diptera). *Nature (London)* **213**:736.

Chandler, A. E. F. 1968a. Some host-plant factors affecting oviposition by aphidophagous Syrphidae (Diptera). *Ann. Appl. Biol.* **61**:415–423.

Chandler, A. E. F. 1968b. The relationship between aphid infestations and oviposition by aphidophagous Syrphidae (Diptera). *Ann. Appl. Biol.* **61**:425–434.

Chandler, A. E. F. 1968c. Some factors influencing the occurrence and site of oviposition by aphidophagous Syrphidae (Diptera). *Ann. Appl. Biol.* **61**:435–446.

Chaplin, S. B., and S. J. Chaplin. 1981. Comparative growth energetics of a migratory and nonmigratory insect: The milkweed bugs. *J. Anim. Ecol.* **50**:407–420.

Chapman, R. F. 1969. *The insects: Structure and function.* American Elsevier, New York.

Chapman, R. F., and E. A. Bernays (eds.). 1978. Insect and host plant. *Entomol. Exp. Appl.* **24**:1–566.

Chapman, R. F., and A. Joern (eds.). 1990. *Biology of grasshoppers.* Wiley, New York.

Chapman, R. N. 1928. The quantitative analysis of environmental factors. *Ecology* **9**:111–122.

Charnov, E. L. 1976a. Optimal foraging: Attack strategy of a mantid. *Am. Nat.* **110**:141–151.

Charnov, E. L. 1976b. Optimal foraging: The marginal value theorem. *Theor. Popul. Biol.* **9**:129–136.

Charnov, E. L. 1982. *The theory of sex allocation.* Princeton Univ. Press, Princeton, N.J.

Charnov, E. L., and W. M. Schaffer. 1973. Life-history consequences of natural selection: Cole's result revisited. *Am. Nat.* **107**:791–793.

Charnov, E. L., R. L. Los-den Hartogh, W. J. Jones, and J. van den Assem. 1981. Sex ratio evolution in a variable environment. *Nature* **289**:27–33.

Chermock. R. L. 1946. Migration of *Ascia monuste phileta* (Lepidoptera, Pieridae). *Entomol. News* **57**:144–146.

Cherry, R. H. 1993. Insects in the mythology of Native Americans. *Am. Entomol.* **39**:16–21.

Chevin, H. 1966. Végétation et peuplement entomologique des terrains sablonneux de la côte ouest du Cotentin. *Mem. Soc. Nat. Sci. Nat. Math. Cherbourg* **52**:8–137.

Chew, F. S. 1977. Coevolution of pierid butterflies and their cruciferous foodplants. II. The distribution of eggs on potential foodplants. *Evolution* **31**:568–579.

Chew, R. M. 1974. Consumers as regulators of ecosystems: An alternative to energetics. *Ohio J. Sci.* **74**:359–370.

Chitty, D. 1957. Self-regulation of numbers through changes in viability. *Cold Spring Harbor Symp. Quant. Biol.* **22**:277–280.

Chitty, D. 1960. Population processes in the vole and their relevance to general theory. *Can. J. Zool.* **38**:99–113.

Chitty, D. 1967. The natural selection of self-regulatory behaviour in animal populations. *Proc. Ecol. Soc. Aust.* **2**:51–78.

Chown, S. 1990. Possible effects of Quaternary climatic change on the composition of insect communities of the South Indian Ocean Province Islands. *S. Afr. J. Sci.* **86**:386–391.

Christian, J. J. 1950. The adreno-pituitary system and population cycles in mammals. *J. Mammal.* **31**:247–259.

Christian, J. J. 1959. The roles of endocrine and behavioral factors in the growth of mammalian populations. In A. Gorbman (ed.). *Comparative endocrinology.* Wiley, New York, pp. 71–97.

Christian, J. J., and D. E. Davis. 1964. Endocrines, behavior and population. *Science* **146**:1550–1560.

Clancy, K. M. 1992. Response of western spruce budworm (Lipidoptera: Tortricidae) to increased nitrogen in artificial diets. *Environ. Entomol.* **21**:331–344.

Clancy, K. M., and R. M. King. 1993. Defining the western spruce budworm's nutritional niche with response surface methodology. *Ecology* **74**:442–454.

Clancy, K. M., R. L. Giese, and D. M. Benjamin. 1980. Predicting jack-pine budworm infestations in northwestern Wisconsin. *Environ. Entomol.* 9:743–751.

Clancy, K. M., J. K. Itami, and D. P. Huebner. 1993. Douglas-fir nutrients and terpenes: potential resistance factors to western spruce budworm defoliation. *For. Sci.* 39:78–94.

Clancy, K. M., M. R. Wagner, and R. W. Tinus. 1988. Variation in host foliage nutrient concentrations in relation to western spruce budworm herbivory. *Can. J. For. Res.* 18:530–539.

Claridge, M. F., and M. R. Wilson. 1976. Diversity and distribution patterns of some mesophyll-feeding leafhoppers of temperate woodland canopy. *Ecol. Entomol.* 1:231–250.

Claridge, M. F., and M. R. Wilson. 1978. British insects and trees: A study in island biogeography or insect/plant coevolution? *Am. Nat.* 112:451–456.

Clark, L. R., P. W. Geier, R. D. Hughes, and R. F. Morris. 1967. *The ecology of insect populations in theory and practice.* Methuen, London.

Clarke, B. 1960. Divergent effects of natural selection on two closely related polymorphic snails. *Heredity* 14:423–443.

Clarke, B. 1962. Balanced polymorphism and the diversity of sympatric species. In D. Nichols (ed.). *Taxonomy and geography.* Syst. Assoc. Publ. 4, Systematics Association, London, pp. 47–70.

Clarke, B. 1978. Some contributions of snails to the development of ecological genetics. In P. F. Brussard (ed.). *Ecological genetics: The interface.* Springer-Verlag, New York, pp. 159–170.

Clarke, C. A., and P. M. Sheppard. 1960a. The evolution of mimicry in the butterfly *Papilio dardanus. Heredity* 14:163–173.

Clarke, C. A., and P. M. Sheppard. 1960b. Super-genes and mimicry. *Heredity* 14:175–185.

Clarke, C. A., and P. M. Sheppard. 1963. Interactions between major genes and polygenes in the determination of the mimetic patterns of *Papilio dardanus. Evolution* 17:404–413.

Clarke, C. A., G. S. Mani, and G. Wynne. 1985. Evolution in reverse: Clean air and the peppered moth. *Biol. J. Linn. Soc.* 26:189–199.

Clarke, G. L. 1954. *Elements of ecology.* Wiley, New York.

Clench, H. K. 1967. Temporal dissociation and population regulation in certain hesperiine butterflies. *Ecology* 48:1000–1006.

Cleveland, L. R. 1924. The physiological and symbiotic relationships between the intestinal protozoa of termites and their host, with special reference to *Reticulitermes flavipes* Kollar. *Biol. Bull. Mar. Biol. Lab. Woods Hole* 46:178–227.

Cleveland, L. R. 1925. The ability of termites to live perhaps indefinitely on a diet of pure cellulose. *Biol. Bull. Mar. Biol. Lab. Woods Hole* 48:289–293.

Cleveland, L. R., S. R. Hall, E. P. Sanders, and J. Collier. 1934. The wood-feeding roach *Cryptocercus*, its protozoa, and the symbiosis between protozoa and roach. *Mem. Am. Acad. Arts Sci.* 17:184–342.

Coats, J. R., R. L. Metcalf. P.-Y. Lu, D. D. Brown, J. F. Williams, and L. G. Hansen. 1976. Model ecosystem evaluation of the environmental impacts of the veterinary drugs phenothiazine, sulfamethazine, clopidol and diethylestibestrol. *Environ. Health Perspect.* **18**:167–179.

Coats, S. A. 1976. Life cycle and behavior of *Muscidifurax zaraptor* (Hymenoptera: Pteromalidae). *Ann. Entomol. Soc. Am.* **69**:772–780.

Cody, M. L. 1966. A general theory of clutch size. *Evolution* **20**:174–184.

Cody, M. L. 1971. Ecological aspects of reproduction. In D. S. Farner and J. R. King (eds.). *Avian biology.* Vol. 1. Academic Press, San Diego, Calif., pp. 461–512.

Cody, M. L., and J. M. Diamond (eds.). 1975. *Ecology and evolution of communities.* Belknap Press, imprint of Harvard Univ. Press, Cambridge, Mass.

Cole, J., G. Lovett, and S. Findlay (eds.). 1991. *Comparative analyses of ecosystems: Patterns, mechanisms, and theories.* Springer-Verlag, New York.

Cole, J. J., N. F. Caraco, G. W. Kling, and T. K. Kratz. 1994. Carbon dioxide supersaturation in the surface waters of lakes. *Science* **265**:1568–1570.

Cole, L. C. 1951. Population cycles and random oscillations. *J. Wildl. Manage.* **15**:233–252.

Cole, L. C. 1954a. Some features of random population cycles. *J. Wildl. Manage.* **18**:2–24.

Cole, L. C. 1954b. The population consequences of life history phenomena. *Q. Rev. Biol.* **29**:103–137.

Cole, L. C. 1960. Competitive exclusion. *Science* **132**:348–349.

Coleman, E. 1933. Pollination of the orchid genus *Prasophyllum*. *Victorian Nat.* **49**:214–221.

Coley, P. D. 1983. Herbivory and defensive characteristics of tree species in a lowland tropical forest. *Ecol. Monogr.* **53**:209–233.

Coley, P. D., and T. M. Aide. 1991. Comparison of herbivory and plant defenses in temperate and tropical broad-leaved forests. In P. W. Price, T. M. Lewinsohn, G. W. Fernandes, and W. W. Benson (eds.). *Plant–animal interactions: Evolutionary ecology in tropical and temperate regions.* Wiley, New York, pp. 25–49.

Coley, P. D., J. P. Bryant, and F. S. Chapin. 1985. Resource availability and plant antiherbivore defense. *Science* **230**:895–899.

Collins, N. C. 1975a. Population biology of a brine fly (Diptera: Ephydridae) in the presence of abundant algal food. *Ecology* **56**:1139–1148.

Collins, N. C. 1975b. Tactics of host exploitation by a thermophilic water mite. *Misc. Publ. Entomol. Soc. Am.* **9**:250–254.

Collins, N. C., R. Mitchell, and R. G. Wiegert. 1976. Functional analysis of a thermal spring ecosystem, with an evaluation of the role of consumers. *Ecology* **57**:1221–1232.

Collins, N. M., and J. A. Thomas (eds.). 1991. *The conservation of insects and their habitats.* Academic Press, London.

Colwell, R. K. 1973. Competition and coexistence in a simple tropical community. *Am. Nat.* **107**:737–760.

Colwell, R. K. 1974. Predictability, constancy, and contingency of periodic phenomena. *Ecology* **55**:1148–1153.

Colwell, R. K. 1984. What's new? Community ecology discovers biology. In P. W. Price, T. M. Lewinsohn, G. W. Fernandes, and W. W. Benson (eds.). *Plant-animal interactions: evolutionary ecology in tropical and temperate regions.* Wiley, New York, pp. 387–396.

Colwell, R. K. 1988. Ecology and biotechnology: Expectations and outliers. In J. Fiksel and V. T. Covello (eds.). *Safety assurance for environmental introductions of genetically-engineered organisms.* Springer-Verlag, Berlin, pp. 163–180.

Colwell, R. K., and E. R. Fuentes. 1975. Experimental studies of the niche. *Annu. Rev. Ecol. Syst.* **6**:281–310.

Colwell, R. K., and D. J. Futuyma. 1971. On the measurement of niche breadth and overlap. *Ecology* **52**:567–576.

Conlong, D. E. 1990. A study of pest–parasitoid relationships in natural habitats: An aid towards the biological control of *Eldana saccharina* (Lepidoptera: Pyralidae) in sugar cane. *Proc. S. Afr. Sugar Tech. Assoc.,* June:111–115.

Connell, J. H. 1961. The influence of interspecific competition and other factors on the distribution of the barnacle *Chthamalus stellatus. Ecology* **42**:710–723.

Connell, J. H. 1978. Diversity in tropical rain forests and coral reefs. *Science* **199**:1302–1310.

Connell, J. H. 1979. Tropical rain forests and coral reefs as open nonequilibrium systems. *Symp. Br. Ecol. Soc.* **20**:141–163.

Connell, J. H. 1980. Diversity and the coevolution of competitors, or the ghost of competition past. *Oikos* **35**:131–138.

Connell, J. H. 1983. On the prevalence and relative importance of interspecific competition: evidence from field experiments. *Am. Nat.* **122**:661–696.

Connell, J. H., and E. Orias. 1964. The ecological regulation of species diversity. *Am. Nat.* **98**:399–414.

Connor, E. F., and E. D. McCoy. 1979. The statistics and biology of the species–area relationship. *Am. Nat.* **113**:791–833.

Conway, G. 1981. Man versus pests. In R. M. May (ed.). *Theoretical ecology: Principles and applications.* 2nd ed. Sinauer, Sunderland, Mass., pp. 356–386.

Cook, L. M., G. S. Mani, and M. E. Varley. 1986. Postindustrial melanism in the peppered moth. *Science* **231**:611–613.

Cook, R. M., and B. J. Cockrell, 1978. Predator ingestion rate and its bearing on feeding time and the theory of optimal diets. *J. Anim. Ecol.* **47:** 529–547.

Cook, R. M., and S. F. Hubbard. 1977. Adaptive searching strategies in insect parasites. *J. Anim. Ecol.* **46:**115–126.

Coombs, C. W., and G. E. Woodroffe. 1973. Evaluation of some of the factors involved in ecological succession in an insect population breeding in stored wheat. *J. Anim. Ecol.* **42:**305–322.

Coope, G. R. 1973. Tibetan species of dung beetle from late Pleistocene deposits in England. *Nature* **245:**335–336.

Coope, G. R. 1977. Fossil coleopteran assemblages as sensitive indicators of climatic changes during the Devensian (last) cold stage. *Philos. Trans. R. Soc. London B* **280:**313–340.

Coope, G. R. 1978. Constancy of insect species versus inconstancy of Quarternary environments. In L. A. Mound and N. Waloff (eds.). *Diversity of insect faunas.* Symp. R. Entomol. Soc. London 9, pp. 176–187.

Coppinger, R. P. 1970. The effect of experience and novelty on avian feeding behavior with reference to the evolution of warning coloration in butterflies. II. Reaction of naive birds to novel insects. *Am. Nat.* **104:**323–335.

Cornell, H. 1974. Parasitism and distributional gaps between allopatric species. *Am. Nat.* **108:**880–883.

Cornell, H. V. 1983. The secondary chemistry and complex morphology of galls formed by the Cynipinae (Hymenoptera): Why and how? *Am. Midl. Nat.* **110:**225–234.

Cornell, H. V. 1985. Species assemblages of cynipid gall wasps are not saturated. *Am. Nat.* **126:**565–569.

Cornell, H. V. 1993. Unsaturated patterns in species assemblages: The role of regional processes in setting local species richness. In R. E. Ricklefs and D. Schluter (eds.). *Species diversity in ecological communities: Historical and geographical perspectives.* Univ. Chicago Press, Chicago, pp. 243–252.

Cornell, H. V., and B. A. Hawkins. 1995. Survival patterns and mortality sources of herbivorous insects: some demographic trends. *Am. Nat.* **145:**563–593.

Cornell, H. V., and J. O. Washburn. 1979. Evolution of the richness–area correlation for cynipid gall wasps on oak trees: A comparison of two geographic areas. *Evolution* **33:**257–274.

Cott, H. B. 1940. *Adaptive coloration in animals.* Oxford Univ. Press, New York.

Courtney, S. P. 1981. Coevolution of pierid butterflies and their cruciferous foodplants. III. *Anthocharis cardamines* (L.) survival, development and oviposition on different host plants. *Oecologia* **51:**91–96.

Courtney, S. P. 1982a. Coevolution of pierid butterflies and their cruciferous foodplants. IV. Hostplant apparency and *Anthocharis cardamines* oviposition. *Oecologia* **52:**258–265.

Courtney, S. P. 1982b. Coevolution of pierid butterflies and their cruciferous foodplants. V. Habitat selection, community structure and speciation. *Oecologia* **54**:101–107.

Courtney, S. P., and T. T. Kibota. 1990. Mother doesn't know best: Selection of hosts by ovipositing insects. In E. A. Bernays (ed.). *Insect–plant interactions*. Vol. 2. CRC Press, Boca Raton, Fla., pp. 161–188.

Covich, A. P. 1976. Analyzing shapes of foraging areas: Some ecological and economic theories. *Annu. Rev. Ecol. Syst.* **7**:235–257.

Cowie, R. J. 1977. Optimal foraging in great tits *(Parus major)*. *Nature* **268**:137–139.

Cowles, H. C. 1899. The ecological relations of the vegetation on the sand dunes of Lake Michigan. *Bot. Gaz.* **27**:95–117, 167–202, 281–308, 361–391.

Cox, C. B., I. N. Healey, and P. D. Moore. 1976. *Biogeography: An ecological and evolutionary approach*. 2nd ed. Wiley, New York.

Craig, G. B. 1967. Mosquitoes: Female monogamy induced by male accessory gland substance. *Science* **156**:1499–1501.

Craig, R. B., D. L. De Angelis, and K. R. Dixon. 1979. Long- and short-term dynamic optimization models with application to the feeding strategy of the loggerhead shrike. *Am. Nat.* **113**:31–51.

Craig, T. P., and S. Mopper. 1993. Sex ratio variation in sawflies. In M. R. Wagner and K. F. Raffa (eds.). *Sawfly life history adaptations to woody plants*. Academic Press, San Diego, Calif., pp. 61–92.

Craig, T. P., J. K. Itami, and P. W. Price. 1989. A strong relationship between oviposition preference and larval performance in a shoot-galling sawfly. *Ecology* **70**:1691–1699.

Craig, T. P., J. K. Itami, and P. W. Price. 1990. The window of vulnerability of a shoot-galling sawfly to attack by a parasitoid. Ecology **71**:1471–1482.

Craig, T. P., P. W. Price, and J. K. Itami. 1986. Resource regulation by a stem-galling sawfly on the arroyo willow. *Ecology* **67**:419–425.

Craig, T. P., P. W. Price, and J. K. Itami. 1992. Facultative sex ratio shifts in response to host plant quality in an herbivorous insect. *Oecologia* **92**:153–161.

Crankshaw, O. S., and R. W. Matthews. 1981. Sexual behavior among parasitic *Megarhyssa* wasps (Hymenoptera: Ichneumonidae). *Behav. Ecol. Sociobiol.* **9**:1–7.

Creed, E. R. (ed.). 1971. *Ecological genetics and evolution*. Blackwell Scientific, Oxford.

Crepet, W. L. 1979. Insect pollination: A paleontological perspective. *BioScience* **29**:102–108.

Cromartie, W. J. 1975a. The effect of stand size and vegetational background on the colonization of cruciferous plants by herbivorous insects. *J. Appl. Ecol.* **12**:517–533.

Cromartie, W. J. 1975b. Influence of habitat on colonization of collard plants by *Pieris rapae*. *Environ. Entomol.* **4**:783–784.

Crossley, D. A., Jr., and M. P. Hoglund. 1962. A litter-bag method for the study of microarthropods inhabiting leaf litter. *Ecology* **43**:571–573.

Croteau, R., A. J. Burbott, and W. D. Loomis. 1972. Biosynthesis of mono- and sesquiterpenes in peppermint from glucose-^{14}C and $^{14}CO_2$. *Phytochemistry* **11**:2459–2467.

Crow, J. F., and M. Kimura. 1970. *An introduction to population genetics theory*. Harper & Row, New York.

Croze, H. 1970. *Searching image in carrion crows*. Parey, Berlin.

Cruden, R. W. 1972a. Pollinators in high-elevation ecosystems: Relative effectiveness of birds and bees. *Science* **176**:1439–1440.

Cruden, R. W. 1972b. Pollination biology of *Nemophila menziesii* (Hydrophyllaceae) with comments on the evolution of oligolectic bees. *Evolution* **26**:373–389.

Cruden, R. W., S. M. Hermann, and S. Peterson. 1983. Patterns of nectar production and plant-pollinator coevolution. In B. Bentley and T. Elias (eds.). *The biology of nectaries*. Columbia Univ. Press, New York, pp. 80–125.

Cruden, R. W., S. Kinsman, R. E. Stockhouse, and Y. B. Linhart. 1976. Pollination, fecundity, and the distribution of multi-flowered plants. *Biotropica* **8**:204–210.

Crumpacker, D. W., and J. S. Williams. 1974. Rigid and flexible chromosomal polymorphisms in neighboring populations of *Drosophila pseudoobscura*. *Evolution* **28**:57–66.

Culin, J. D., and K. V. Yeargan. 1983. A comparative study of spider communities in alfalfa and soybean ecosystems: Foliage-dwelling spiders. *Ann. Entomol. Soc. Am.* **76**:825–831.

Culver, D. C. 1982. *Cave life: Evolution and ecology*. Harvard Univ. Press, Cambridge, Mass.

Culver, D., J. R. Holsinger, and R. Baroody. 1973. Toward a predictive cave biogeography: The Greenbrier Valley as a case study. *Evolution* **27**:689–695.

Cummins, K. W. 1973. Trophic relations of aquatic insects. *Annu. Rev. Entomol.* **18**:183–206.

Cummins, K. W. 1974. Structure and function of stream ecosystems. *BioScience* **24**:631–641.

Currey, J. D. 1970. *Animal skeletons*. Edward Arnold, London.

Cushing, C. E., K. W. Cummins and G. W. Minshall (eds.). 1995. *River and stream ecosystems*. Elsevier, New York.

Cushman, J. H., and J. F. Addicott. 1991. Conditional interactions in ant–plant–herbivore mutualisms. In C. R. Huxley and D. F. Cutler (eds.). *Ant–plant interactions*. Oxford Univ. Press, Oxford, pp. 92–103.

Cushman, J. H., and J. F. Addicott. 1989. Intra- and interspecific competition for mutualists: Ants as a limited and limiting resource for aphids. *Oecologia* **79**:315–321.

Cushman, J. H., and T. G. Whitham. 1989. Conditional mutualism in a membracid–ant association: Temporal, age-specific, and density-dependent effects. *Ecology* **70**:1040–1047.

Cyr, H., and M. L. Pace. 1993. Magnitude and patterns of herbivory in aquatic and terrestrial ecosystems. *Nature* **361**:148–150.

DaCosta, C. P., and C. M. Jones. 1971. Cucumber beetle resistance and mite susceptibility controlled by the bitter gene in *Cucumis sativus* L. *Science* **172**:1145–1146.

Da Cunha, A. B., and T. Dobzhansky. 1954. A further study of chromosomal polymorphism in *Drosophila willistoni* in its relation to the environment. *Evolution* **8**:119–134.

Da Cunha, A. B., T. Dobzhansky, O. Pavlosky, and B. Spassky. 1959. Genetics of natural populations. XXVIII. Supplementary data on the chromosomal polymorphism in *Drosophila willistoni* in its relation to the environment. *Evolution* **13**:389–404.

Dahlsten, D. L. 1967. Preliminary life tables for pine sawflies in the *Neodiprion fulviceps* complex (Hymenoptera: Diprionidae). *Ecology* **48**:275–289.

Dalton, S. 1975. *Borne on the wind.* Chatto & Windus, London.

Danell, K., and K. Huss-Danell. 1985. Feeding by insects and hares on birches earlier affected by moose browsing. *Oikos* **44**:75–81.

Danell, K., K. Huss-Danell, and R. Bergström. 1985. Interactions between browsing moose and two species of birch in Sweden. *Ecology* **66:** 1867–1878.

Darlington, P. J., Jr. 1957. *Zoogeography: The geographical distribution of animals.* Wiley, New York.

Darlington, P. J., Jr. 1959. Area, climate, and evolution. *Evolution* **13:** 488–510.

Darlington, P. J., Jr. 1965. *Biogeography of the southern end of the world.* Harvard Univ. Press. Cambridge, Mass.

Darwin, C. 1859. *On the origin of species by means of natural selection, or the preservation of favoured races in the struggle for life.* Murray, London.

Darwin, C. 1860. *The voyage of the Beagle.* Natural History Library, New York.

Darwin, C. 1871. *The descent of man and selection in relation to sex.* 2 vols. Appleton, New York.

Darwin, F. (ed.). 1892. *Charles Darwin: His life told in an autobiographical chapter and in a selected series of his published letters.* Appleton, New York.

Davidson, D. W. 1978. Experimental tests of optimal diet in two social insects. *Behav. Ecol. Sociobiol.* **4**:35–41.

Davidson, D. W., and B. L. Fisher. 1991. Symbiosis of ants with *Cecropia* as a function of light regime. In C. R. Huxley and D. F. Cutler (eds.). *Ant–plant interactions.* Oxford Univ. Press, Oxford, pp. 298–309.

Davidson, D. W., and S. R. Morton. 1981. Competition for dispersal in ant-dispersed plants. *Science* **213**:1259–1261.

Davidson, D. W., R. B. Foster, R. R. Snelling, and P. W. Lozada. 1991. Variable composition of some tropical ant–plant symbioses. In P. W. Price, T. M. Lewinsohn, G. W. Fernandes, and W. W. Benson (eds.). *Plant–animal interactions: Evolutionary ecology in tropical and temperate regions.* Wiley, New York, pp. 145–162.

Davidson, J., and H. G. Andrewartha. 1948a. Annual trends in a natural population of *Thrips imaginis* (Thysanoptera). *J. Anim. Ecol.* **17:** 193–199.

Davidson, J., and H. G. Andrewartha. 1948b. The influence of rainfall, evaporation and atmospheric temperature on fluctuations in the size of a natural population of *Thrips imaginis* (Thysanoptera). *J. Anim. Ecol.* **17:** 200–222.

Davies, A. J. S., J. G. Hall, G. A. T. Targett, and M. Murray. 1980. The biological significance of the immune response with special reference to parasites and cancer. *J. Parasitol.* **66:**705–721.

Davies, N. B. 1977. Prey selection and social behaviour in wagtails (Aves: Motacillidae). *J. Anim. Ecol.* **46:**37–57.

Davies, N. B. 1978. Territorial defence in the speckled wood butterfly *(Pararge aegeria):* The resident always wins. *Anim. Behav.* **26:**138–147.

Davis, B. N. K. 1973. The Hemiptera and Coleoptera of stinging nettle (*Urtica dioica* L.) in East Anglia. *J. Appl. Ecol.* **10:**213–237.

Dawkins, M. 1971a. Perceptual changes in chicks: Another look at the "search image" concept. *Anim. Behav.* **19:**566–574.

Dawkins, M. 1971b. Shifts of "attention" in chicks during feeding. *Anim. Behav.* **19:**575–582.

Day, P. R., 1974. *Genetics of host–parasite interaction.* W. H. Freeman, San Francisco.

Dayton, P. K. 1971. Competition, disturbance and community organization: The provision and subsequent utilization of space in a rocky intertidal community. *Ecol. Monogr.* **41:**351–389.

DeBach, P. (ed.). 1964a. *Biological control of insect pests and weeds.* Reinhold, New York.

DeBach, P. 1964b. Successes, trends, and future possibilities. In P. DeBach (ed.). *Biological control of insects pests and weeds.* Reinhold, New York, pp. 673–713.

DeBach, P. 1966. The competitive displacement and coexistence principles. *Annu. Rev. Entomol.* **11:**183–212.

DeBach, P. 1974. *Biological control by natural enemies.* Cambridge Univ. Press, New York.

DeBach, P., and H. S. Smith. 1941. Are population oscillations inherent in the host–parasite relation? *Ecology* **22:**363–369.

DeBach, P., and R. A. Sundby. 1963. Competitive displacement between ecological homologues. *Hilgardia* **34**:105–166.

Deevey, E. S., Jr. 1947. Life tables for natural populations of animals. *Q. Rev. Biol.* **22**:283–314.

DeJong, G. D. 1994. Insect cartoons: When do they appear in newspapers and magazines? *Am. Entomol.* **40**:149–151.

Delwiche, C. C. 1970. The nitrogen cycle. *Sci. Am.* **223**(3):136–146.

Dempster, J. P. 1960. A quantitative study of the predators on the eggs and larve of the broom beetle, *Phytodecta olivacea* Forster, using the precipitin test. *J. Anim. Ecol.* **29**:149–167.

Dempster, J. P. 1963. The population dynamics of grasshoppers and locusts. *Biol. Rev. Cambridge Philos. Soc.* **38**:490–529.

Dempster, J. P. 1967. The control of *Pieris rapae* with DDT. I. The natural mortality of the young stages of *Pieris. J. Appl. Ecol.* **4**:485–500.

Dempster, J. P. 1968. Intra-specific competition and dispersal: As exemplified by a psyllid and its anthocorid predator. In T. R. E. Southwood (ed.). *Insect abundance.* Symp. R. Entomol. Soc. London 4, pp. 8–17.

Dempster, J. P. 1971. The population ecology of the cinnabar moth, *Tyria jacobaeae* L. (Lepidoptera: Arctiidae). *Oecologia* **7**:26–67.

Dempster, J. P. 1975. *Animal population ecology.* Academic Press, London.

den Boer, P. J., and G. R. Gradwell (eds.). 1970. *Dynamics of populations.* Proc. Adv. Study Inst. Dynam. Numbers Popul. Osterbeek, The Netherlands. Centre for Agricultural Publications and Documentation (Pudoc), Wageningen, The Netherlands.

Denno, R. F. 1994. Life history variation in planthoppers. In R. F. Denno and T. J. Perfect (eds.). *Planthoppers: Their ecology and management.* Chapman & Hall, New York.

Denno, R. F., and T. J. Perfect (eds.). 1994. *Planthoppers: Their ecology and management.* Chapman & Hall, New York.

Denno, R. F., and M. A. Peterson. 1995. Density-dependent dispersal and its consequences for population dynamics. In N. Cappuccino and P. W. Price (eds.). *Population dynamics: New approaches and synthesis.* Academic Press, San Diego, Calif., pp. 113–130.

Denno, R. F., M. S. McClure, and J. R. Ott. 1995. Interspecific interactions in phytophagous insects: Competition reexamined and resurrected. *Annu. Rev. Entomol.* **40**:297–331.

Denno, R. F., G. K. Roderick, K. L. Olmstead, and H. G. Döbel. 1991. Density-related migration in planthoppers (Homoptera: Delphacidae): The role of habitat persistence. *Am. Nat.* **138**:1513–1541.

de Ruiter, L. 1952. Some experiments on the camouflage of stick caterpillars. *Behaviour* **4**:222–232.

de Ruiter, L. 1956. Countershading in caterpillars. *Arch. Neerl. Zool.* **11**:285–342.

Dethier, V. G. 1954. Evolution of feeding preferences in phytophagous insects. *Evolution* **8**:33–54.

Dethier, V. G. 1976. *The hungry fly: A physiological study of the behavior associated with feeding.* Harvard Univ. Press, Cambridge, Mass.

Detling, J. K., and M. I. Dyer. 1981. Evidence for potential plant growth regulators in grasshoppers. *Ecology* **62**:485–488.

de Wilde, J., and L. M. Schoonhoven (eds.). 1969. *Insect and host plant.* North-Holland, Amsterdam, pp. 471–810. Reprinted from *Ent. Exp. Appl.* **12**:471–810.

de Wit, C. T. 1960. On competition. *Versl. Landbouwk. Onderzoek.* **66**:1-82.

de Wit, C. T. 1961. Space relationships within populations of one or more species. In F. L. Milthorpe (ed.). *Mechanisms in biological competition.* Symp. Soc. Exp. Biol. 15, pp. 314–329.

Diamond, J., and T. J. Case (eds.). 1986. *Community ecology.* Harper & Row, New York.

Dias, B. F. 1975. Comportamento presocial de sinfitas do Brazil Central. I. *Themos olfersii* (Klug) (Hymenoptera: Argidae). *Stud. Entomol.* **18**: 401–432.

Dicke, M. 1988. Infochemicals in tritrophic interactions. Ph.D. thesis, Agricultural Univ. Wageningen, The Netherlands.

Dicke, M., T. A. van Beek, M. A. Posthumus, N. Den Dom, H. van Bokhoven, and Æ. de Groot. 1990a. Isolation and identification of volatile kairomone that affects acarine predator–prey interactions: Involvement of host plant in its production. *J. Chem. Ecol.* **16**:381–396.

Dicke, M., M. W. Sabelis, J. Takabayashi, J. Bruin, and M. A. Posthumus. 1990b. Plant strategies of manipulating predator–prey interactions through allelochemicals: Prospects for application in pest control. *J. Chem. Ecol.* **16**:3091–3118.

Dicke, M., J. C. van Lenteren, A. K. Minks, and L. M. Schoonhoven (eds.). 1990c. Semiochemicals and pest control: Prospects for new applications. *J. Chem. Ecol.* **16**:3017–3212.

Dicke, M., K. J. van der Maas, J. Takabayashi, and L. E. M. Vet. 1990d. Learning affects response to volatile allelochemicals by predatory mites. *Proc. Exp. Appl. Entomol.* **1**:31–36.

Dickenson, C. H., and G. J. F. Pugh. 1974. *Biology of plant litter decomposition.* Vols. 1 and 2. Academic Press, San Diego, Calif.

Dietz, R. S., and J. C. Holden. 1970. Reconstruction of Pangea: Breakup and dispersion of continents, Permian to present. *J. Geophys. Res.* **75**:4939–4956.

Dingle, H. 1972. Migration strategies of insects. *Science* **175**:1327–1335.

Dingle, H. 1978a. Migration and diapause in tropical, temperate, and island milkweed bugs. In H. Dingle (ed.). *Evolution of insect migration and diapause.* Springer-Verlag, New York, pp. 254–276.

Dingle, H. 1978b. *Evolution of insect migration and diapause.* Springer-Verlag, New York.

Dingle, H. 1981. Geographic variation and behavioral flexibility in milkweed bug life histories. In R. F. Denno and H. Dingle (eds.). *Insect life history patterns: Habitat and geographic variation.* Srpinger-Verlag, New York, pp. 57–73.

Dingle, H. 1984. Behavior, genes and life histories: Complex adaptations in uncertain environments. In P. W. Price, C. N. Slobodchikoff, and W. S. Gaud (eds.). *A new ecology: Novel approaches to interactive systems.* Wiley, New York.

Dingle, H. 1996. *Migration: The biology of life on the move.* Oxford Univ. Press, New York.

Dixon, A. F. G. 1970. Quality and availability of food for a sycamore aphid population. In A. Watson (ed.). *Animal populations in relation to their food resources.* Br. Ecol. Soc. Symp. 10, Blackwell Scientific Publications, Oxford, pp. 271–286.

Dixon, A. F. G. 1973. *Biology of aphids.* Edward Arnold, London.

Dixon, A. F. G. 1985. *Aphid ecology.* Blackie, Glasgow, Scotland.

Dixon, A. F. G., P. Kindlmann, J. Leps, and J. Holman. 1987. Why there are so few species of aphids, especially in the tropics. *Am. Nat.* **129**:580–592.

Dobzhansky, T. 1947. Genetics of natural populations. XIV. A response of certain gene arrangements in the third chromosome of *Drosophila pseudoobscura* to natural selection. *Genetics* **32**:142–160.

Dobzhansky, T. 1948. Genetics of natural populations. XVI. Altitudinal and seasonal changes produced by natural selection in certain populations of *Drosophila pseudoobscura* and *Drosophila persimilis. Genetics* **33**:158–176.

Dobzhansky, T. 1950. Evolution in the tropics. *Am. Sci.* **38**:209–221.

Dobzhansky, T. 1970. *Genetics of the evolutionary process.* Columbia Univ. Press, New York.

Dobzhansky, T., H. Burla, and A. B. da Cunha. 1950. A comparative study of chromosomal polymorphism in sibling species of the *willistoni* group of *Drosophila. Am. Nat.* **84**:229–246.

Dodson, C. H. 1975. Coevolution of orchids and bees. In L. E. Gilbert and P. H. Raven (eds.). *Coevolution of animals and plants.* Univ. Texas Press, Austin, Texas, pp. 91–99.

Dodson, C. H., and G. P. Frymire. 1961. Natural pollination of orchids. *Bull. Mo. Bot. Gard.* **49**:133–152.

Dodson, C. H., R. L. Dressler, H. G. Hills, R. M. Adams, and N. H. Williams. 1969. Biologically active compounds in orchid fragrances. *Science* **164**:1243–1249.

Donaldson, J. S. 1992. Adaptation for oviposition into concealed cycad ovules in the cycad weevils *Antliarhinus zamiae* and *A. signc̄   ̄tera:* Curculionidae). *Biol. J. Linn. Soc.* **47**:23–35.

Doutt, R. L., and J. Nakata. 1965. Overwintering refu⸠ (Hymenoptera: Mymaridae). *J. Econ. Entomol.* **58**:5⸠

Doutt, R. L., and J. Nakata. 1973. The *Rubus* l⸠

parasitoid: An endemic biotic system useful in grape-pest management. *Environ. Entomol.* 2:381–386.

Drake, J. A., H. A. Mooney, F. di Castri, R. H. Groves, F. J. Kruger, M. Rejmánek, and M. Williamson (eds.). 1989. *Biological invasions: A global perspective.* Wiley, New York.

Drake, V. A., and A. G. Gatehouse (eds.). 1995. *Insect migration: Tracking resources through space and time.* Cambridge Univ. Press, Cambridge.

Dressler, R. L. 1968. Pollination in euglossine bees. *Evolution* 22:202–210.

Dressler, R. L. 1982. Biology of the orchid bees (Euglossini). *Annu. Rev. Ecol. Syst.* 13:373–394.

Dritschilo, W., H. Cornell, D. Nafus, and B. O'Connor. 1975. Insular biogeography: Of mice and mites. *Science* 190:467–469.

Drooz, A. T. 1971. The elm spanworm (Lepidoptera: Geometridae): Natural diets and their effect on the F_2 generation. *Ann. Entomol. Soc. Am.* 64:331–333.

Duffey, E., and A. S. Watt (eds.). 1971. *The scientific management of animal and plant communities for conservation.* Br. Ecol. Symp. 11. Blackwell Scientific, Oxford.

Duffey, S. S. 1970. Cardiac glycosides and distastefulness: Some observations on the palatability spectrum of butterflies. *Science* 169:78–79.

Duffey, S. S., and G. G. E. Scudder. 1972. Cardiac glycosides in North American Asclepiadaceae, a basis for unpalatability in brightly coloured Hemiptera and Coleoptera. *J. Insect Physiol.* 18:63–78.

Dunlap-Pianka, H., C. L. Boggs, and L. E. Gilbert. 1977. Ovarian dynamics in helicoiine butterflies: Programmed senescence versus eternal youth. *Science* 197:487–490.

Durzan, D. J., and S. M. Lopushanski. 1968. Free and bound amino acids of spruce budworm larvae feeding on balsam fir and red and white spruce. *J. Insect Physiol.* 14:1485–1497.

Dwyer, G. 1995. Simple models and complex interactions. In N. Cappuccino and P. W. Price (eds.). *Population dynamics: new approaches and synthesis.* Academic Press, San Diego, Calif., pp. 209–227.

Dyer, M. I., and U. G. Bokhari. 1976. Plant–animal interactions: Studies of the effects of grasshopper grazing on blue grama grass. *Ecology* 57:762–772.

Eastop, V. F. 1972. Deductions from the present day host plants of aphids and related insects. *Symp. R. Entomol. Soc. London* 6:157–178.

Eastop, V. F. 1978. Diversity of the Sternorrhyncha within major climatic zones. *Symp. R. Entomol. Soc. London* 9:71–88.

Eberhardt, L. L. 1970. Correlation, regression, and density dependence. *Ecology* 51:306–310.

Eckhardt, R. D. 1979. The adaptive syndromes of two guilds of insectivorous birds in the Colorado Rocky Mountains. *Ecol. Monogr.* 49:129–149.

Edington, J. M., and M. A. Edington. 1972. Spatial patterns and habitat partition in the breeding birds of an upland wood. *J. Anim. Ecol.* 41:331–357.

Edmunds, G. F., Jr. 1972. Biogeography and evolution of Ephemeroptera. *Annu. Rev. Entomol.* **17**:21–42.

Edmunds, G. F., and D. N. Alstad. 1978. Coevolution in insect herbivores and conifers. *Science* **199**:941–945.

Edmunds, G. F., and D. N. Alstad. 1981. Responses of black pineleaf scales to host plant variability. In R. F. Denno and H. Dingle (eds.). *Insect life history patterns: Habitat and geographic variation.* Springer-Verlag, New York, pp. 29–38.

Edney, E. B. 1977. *Water balance in land arthropods.* Springer-Verlag, New York.

Edson, K. M., S. B. Vinson, D. B. Stoltz, and M. D. Summers. 1981. Virus in a parasitoid wasp: Suppression of the cellular immune response in the parasitoid's host. *Science* **211**:582–583.

Edwards, C. A., and G. W. Heath. 1963. The role of soil animals in breakdown of leaf material. In J. Doeksen and J. van der Drift (eds.). *Soil organisms.* North-Holland, Amsterdam, pp. 76–84.

Edwards, C. A., and B. R. Stinner (eds.). 1988. *Biological interactions in soil.* Elsevier, Amsterdam.

Edwards, C. A., D. E. Reichle, and D. A. Crossley, Jr. 1970. The role of soil invertebrates in turnover of organic matter and nutrients. In D. E. Reichle (ed.). *Analysis of temperate forest ecosystems.* Springer-Verlag, Berlin, pp. 147–172.

Edwards, P. J., and S. D. Wratten. 1987. Ecological significance of wound induced changes in plant chemistry. In V. Labeyrie, G. Fabres, and D. Lachaise (eds.). *Insects–plants.* Dr. W. Junk, Dordrecht, The Netherlands, pp. 213–218.

Edwards, P. J., S. D. Wratten, and H. Cox. 1985. Wound-induced changes in the acceptability of tomato to larvae of *Spodoptera littoralis:* A laboratory bioassay. *Ecol. Entomol.* **10**:155–158.

Edwards, P. J., S. D. Wratten, and S. Greenwood. 1986. Palatability of British trees to insects: Constitutive and induced defenses. *Oecologia* **69**:316–319.

Eggleton, P., D. E. Bignell, W. A. Sands. N. A., Mawdsley, J. H. Lawton, T. G. Wood, and N. C. Bignell. 1996. The diversity, abundance and biomass of termites under differing levels of disturbance in the Mbalmayo Forest Reserve, southern Cameroon. *Philos. Trans. R. Soc. London Ser. B* **351**: 51–68.

Ehrlich, P. R., and L. C. Birch. 1967. The "balance of nature" and "population control." *Am. Nat.* **101**:97–107.

Ehrlich, P. R., and L. E. Gilbert. 1973. Population structure and dynamics of the tropical butterfly *Heliconius ethilla. Biotropica* **5**:69–82.

Ehrlich, P. R., and P. H. Raven. 1964. Butterflies and plants: A study in coevolution. *Evolution* **18**:586–608.

Ehrlich, P. R., and P. H. Raven. 1967. Butterflies and plants. *Sci. Am.* **216**(6):104–113.

Eisner, T. 1970. Chemical defense against predation in arthropods. In E. Sondheimer and J. B. Simeone (eds.). *Chemical ecology.* Academic Press, San Diego, Calif., pp. 157–217.

Eisner, T., L. B. Hendry, D. B. Peakall, and J. Meinwald. 1971. 2,5-Dichlorophenol (from ingested herbicide?) in defensive secretion of grasshopper. *Science* 171:277–278.

Eisner, T., J. S. Johnessee, J. Carrel, L. B. Hendry, and J. Meinwald. 1974. Defensive use by an insect of a plant resin. *Science* 184:996–999.

Eisner, T., F. C. Kafatos, and E. G. Linsley. 1962. Lycid predation by mimetic adult Cerambycidae (Coleoptera). *Evolution* 16:316–324.

Eisner, T., R. E. Silberglied, D. Aneshansley, J. E. Carrel, and H. C. Howland. 1969. Ultraviolet video-viewing: The television camera as an insect eye. *Science* 166:1172–1174.

Elias, S. A. 1994. *Quaternary insects and their environments.* Smithsonian Institution Press, Washington, D.C.

Ellington, C. P. 1985. Power and efficiency in insect flight muscle. *J. Exp. Biol.* 115:293–304.

Ellington, C. P. 1991. Limitations on animal flight performance. *J. Exp. Biol.* 160:71–91.

Ellis, W. M., R. J. Keymer, and D. A. Jones. 1976. On the polymorphism of cyanogenesis in *Lotus corniculatus* L. VI. Ecological studies in the Netherlands. *Heredity* 36:245–251.

Ellis, W. M., R. J. Keymer, and D. A. Jones, 1977a. The defensive function of cyanogenesis in natural populations. *Experientia* 33:309–311.

Ellis, W. M., R. J. Keymer, and D. A. Jones. 1977b. On the polymorphism of cyanogenesis in *Lotus corniculatus* L. VIII. Ecological studies in Anglesey. *Heredity* 39:45–65.

Elmes, G. W. 1991. Ant colonies and environmental disturbance. *Symp. Zool. Soc. London* 63:1–13.

Elton, C. 1927. *Animal ecology.* Macmillan, New York.

Elton, C. 1942. *Voles, mice and lemmings; problems in population dynamics.* Oxford Univ. Press, Oxford.

Elton, C. S. 1958. *The ecology of invasions by animals and plants.* Methuen, London.

Elton, C. E. 1973. The structure of invertebrate populations inside neotropical rain forest. *J. Anim. Ecol.* 42:55–104.

Elton, C. E. 1975. Conservation and the low population density of invertebrates inside neotropical rain forest. *Biol. Conserv.* 7:3–15.

Embree, D. G. 1965. The population dynamics of the winter moth in Nova Scotia, 1954–1962. *Mem. Entomol. Soc. Can.* 46:1–57.

Embree, D. G. 1971. *Operophtera brumata* (L.), wintermoth (Lepidoptera: Geometridae). In *Biological control programmes against insects and weeds in Canada,* 1959–1968. Commonw. Inst. Biol. Control Tech. Bull. 4, pp. 167–175.

Emlen, J. M. 1966. The role of time and energy in food preference. *Am. Nat.* **100**:611–617.

Emlen, J. M. 1973. *Ecology: An evolutionary approach.* Addison-Wesley, Reading, Mass.

Emlen, S. T., and L. W. Oring. 1977. Ecology, sexual selection, and the evolution of mating systems. *Science* **197**:215–223.

Enders, F. 1974. Vertical stratification in orb-web spiders (Araneidae, Araneae) and a consideration of other methods of coexistence. *Ecology* **55**:317–328.

Engelmann, M. D. 1966. Energetics, terrestrial field studies, and animal productivity. *Adv. Ecol. Res.* **3**:73–115.

Erickson, J. M., and P. Feeny. 1974. Sinigrin: A chemical barrier to the black swallowtail butterfly. *Papilio polyxenes. Ecology* **55**:103–111.

Erlanson, C. O. 1930. The attraction of carrion flies to *Tetraplodon* by an odoriferous secretion of the hypophysis. *Bryologist* **33**:13–14.

Errington, P. L. 1946. Predation and vertebrate populations. *Q. Rev. Biol.* **21**:144–177, 221–245.

Errington, P. L. 1967. *Of predation and life.* Iowa State Univ. Press, Ames, Iowa.

Erwin, T. L. 1982. Tropical forests: Their richness in Coleoptera and other arthropod species. *Coleop. Bull.* **36**:74–75.

Erwin, T. L. 1988. The tropical forest canopy: The heart of biotic diversity. In E. O. Wilson (ed.). *Biodiversity.* National Academy Press, Washington, D.C., pp. 123–129.

Esch, G. W., A. O. Bush, and J. M. Aho (eds.). 1990. *Parasite communities: Patterns and processes.* Chapman & Hall, London.

Evans, D. L. 1990. Phenology as a defense: A time to die, a time to live. In D. L. Evans and J. O. Schmidt (eds.). *Insect defenses: Adaptive mechanisms and strategies of prey and predators.* State Univ. New York Press, Albany, N.Y., pp. 191–202.

Evans, D. L., and J. O. Schmidt (eds.). 1990. *Insect defenses: adaptive mechanisms and strategies of prey and predators.* State Univ. New York Press, Albany, N.Y.

Evans, F. C. 1956. Ecosystem as the basic unit in ecology. *Science* **123**:1127–1128.

Evans, H. E., and M. J. W. Eberhard. 1970. *The wasps.* Univ. Michigan Press, Ann Arbor, Mich.

Evans, J. W. 1959. The zoogeography of some Australian insects. In A. Keast, R. L. Crocker, and C. S. Christian (eds.). *Biogeography and ecology in Australia.* Monogr. Biol. 8, Dr. W. Junk, The Hague, The Netherlands, pp. 150–163.

Evoy, W. H., and B. P. Jones. 1971. Motor patterns of male euglossine bees evoked by floral fragrances. *Anim. Behav.* **19**:583–588.

Ewald, P. W. 1983. Host–parasite relations, vectors and the evolution of disease severity. *Annu. Rev. Ecol. Syst.* **14**:465–485.

Ewald, P. W. 1994. *Evolution of infectious disease.* Oxford Univ. Press, Oxford.

Faegri, K. 1978. Trends in research in pollination ecology. In A. J. Richards (ed.). *The pollination of flowers by insects.* Academic Press, London, pp. 5–12.

Faegri, K., and L. van der Pijl. 1971. *The principles of pollination ecology.* 2nd ed., rev. Pergamon Press, New York.

Faegri, K., and L. van der Pijl. 1979. *The principles of pollination ecology.* 3rd ed., rev. Pergamon Press, Oxford.

Faeth, S. H., and D. Simberloff. 1981a. Population regulation of a leaf-mining insect, *Cameraria* sp. nov., at increased field densities. *Ecology* **62**:620–624.

Faeth, S. H., and D. Simberloff. 1981b. Experimental isolation of oak host plants: Effects on mortality, survivorship, and abundances of leaf-mining insects. *Ecology* **62**:625–635.

Fager, E. W. 1972. Diversity: A sampling study. *Am. Nat.* **106**:293–310.

Farmer, E. E., and C. A. Ryan. 1990. Interplant communication: Airborne methyl jasmonate induces synthesis of proteinase inhibitors in plant leaves. *Proc. Natl. Acad. Sci. USA* **87**:7713–7716.

Farmer, E. E., and C. A. Ryan. 1992. Octadecanoid precursors of jasmonic acid activate the synthesis of wound-inducible proteinase inhibitors. *Plant Cell* **4**:129–134.

Farrell, B. D., and C. Mitter. 1993. Phylogenetic determinants of insect/plant community diversity. In R. E. Ricklefs and D. Schluter (eds.). *Species diversity in ecological communities.* Univ. Chicago Press, Chicago.

Farrow, R. A. 1993. Flight and migration in acridids. In R. F. Chapman and A. Joern (eds.). *Biology of grasshoppers.* Wiley, New York, pp. 227–314.

Feener, D. H. 1981. Competition between ant species: Outcome controlled by parasitic flies. *Science* **214**:815–817.

Feeny, P. P. 1968. Effect of oak leaf tannins on larval growth of the winter moth *Operophtera brumata. J. Insect Physiol.* **14**:805–817.

Feeny, P. P. 1969. Inhibitory effect of oak leaf tannins on the hydrolysis of proteins by trypsin. *Phytochemistry* **8**:2119–2126.

Feeny, P. P. 1970. Seasonal changes in oak leaf tannins and nutrients as a cause of spring feeding by winter moth caterpillars. *Ecology* **51**:565–581.

Feeny, P. 1975. Biochemical coevolution between plants and their insect herbivores. In L. E. Gilbert and P. H. Raven (eds.). *Coevolution of animals and plants.* Univ. Texas Press, Austin, Texas, pp. 3–19.

Feeny, P. 1976. Plant apparency and chemical defense. In J. W. Wallace and R. L. Mansell (eds.). *Biochemical interaction between plants and insects.* Plenum Press, New York, pp. 1–40.

Feeny, P. 1992. The evolution of chemical ecology: Contributions from the study of herbivorous insects. In G. A. Rosenthal and M. R. Berenbaum (eds.). *Herbivores: Their interactions with secondary plant metabolites.* Vol. 2. 2nd ed. Academic Press, San Diego, Calif., pp. 1–44.

Feeny, P. P., and H. Bostock. 1968. Seasonal changes in the tannin content of oak leaves. *Phytochemistry* 7:871–880.

Feeny, P. P., K. L. Paauwe, and N. J. Demong. 1970. Flea beetles and mustard oils: Host plant specificity of *Phyllotreta cruciferae* and *P. striolata* adults (Coleoptera: Chrysomelidae). *Ann. Entomol. Soc. Am.* 63: 832–841.

Feinsinger, P. 1983. Coevolution and pollination. In D. J. Futuyma and M. Slatkin (eds.). *Coevolution.* Sinauer, Sunderland, Mass., pp. 282–310.

Feir, D., and J.-S. Suen. 1971. Cardenolides in the milkweed plant and feeding by the milkweed bug. *Ann. Entomol. Soc. Am.* 64:1173–1174.

Fellows, D. P., and W. B. Heed. 1972. Factors affecting host plant selection in desert-adapted cactiphilic *Drosophila. Ecology* 53:850–858.

Fenner, F., and F. N. Ratcliffe. 1965. *Myxomatosis.* Cambridge Univ. Press, London.

Fernandes, G. W. 1990. Hypersensitivity: A neglected plant resistance mechanism against insect herbivores. *Environ. Entomol.* 19:1173–1182.

Fernandes, G. W., and P. W. Price. 1991. Comparison of tropical and temperate galling species richness: The roles of environmental harshness and plant nutrient status. In P. W. Price, T. M. Lewinsohn, G. W., Fernandes, and W. W. Benson (eds.). *Plant–animal interactions: Evolutionary ecology in tropical and temperate regions.* Wiley, New York, pp. 91–115.

Fischer, A. G. 1960. Latitudinal variations in organic diversity. *Evolution* 14:64–81.

Fittkau, E. J., J. Illies, H. Klinge, G. H. Schwabe, and H. Sioli (eds.). 1968, 1969. Biogeography and ecology in South America. Vols. 1 and 2. *Monogr. Biol.* 18–19, 1–946.

Fitzgerald, T. D. 1973. Coexistence of three species of bark-mining *Marmara* (Lepidoptera: Gracilariidae) on green ash and descriptions of new species. *Ann. Entomol. Soc. Am.* 66:457–464.

Fitzgerald, T. D. 1993. Sociality in caterpillars. In N. E. Stamp and T. M. Casey (eds.). *Caterpillars: Ecology and evolutionary constraints on foraging.* Chapman & Hall, New York, pp. 372–403.

Fitzpatrick, S. M., and W. G. Wellington. 1983. Insect territoriality. *Can. J. Zool.* 61:471–486.

Flor, H. H. 1956. The complementary genic systems in flax and flax rust. *Adv. Genet.* 8:29–54.

Flor, H. H. 1971. Current status of the gene-for-gene concept. *Annu. Rev. Phytopathol.* 9:272–296.

Fontes, E. G., C. S. S. Pires, and E. R. Sujii. 1995. Mixed risk-spreading strategies and the population dynamics of a Brazilian pasture pest, *Deois flavopicta* (Homoptera: Cercopidae). *J. Econ. Entomol.* 88:1256–1262.

Forbes, S. A. 1887. The lake as a microcosm. *Bull. Ill. Nat. Hist. Surv.* 15:537–550.

Ford, E. B. 1940. Polymorphism and taxonomy. In J. Huxley (ed.). *The new systematics.* Clarendon Press, Oxford, pp. 493–513.

Ford, E. B. 1957. *Butterflies.* 3rd ed. Collins, London.

Ford, E. B. 1964. *Ecological genetics.* 2nd ed. Methuen, London.

Ford, E. B. 1975. *Ecological genetics.* 4th ed. Chapman & Hall, London.

Ford, H. D., and E. B. Ford. 1930. Fluctuation in numbers and its influence on variation in *Melitaea aurinia,* Rott. (Lepidoptera). *Trans. R. Entomol. Soc. London* **78**:345–351.

Fordham, R. A. 1971. Field populations of deermice with supplemental food. *Ecology* **52**:138–146.

Formanowicz, D. R. 1982. Foraging tactics of larvae of *Dytiscus verticalis* (Coleoptera: Dytiscidae): The assessment of prey density. *J. Anim. Ecol.* **51**:757–767.

Foster, M. S. 1969. Synchronized life cycles in the orange-crowned warbler and its mallophagan parasites. *Ecology* **50**:315–323.

Foster, M. S. 1974. A model to explain molt-breeding overlap and clutch size in some tropical birds. *Evolution* **28**:182–190.

Fowler, S. V., and J. H. Lawton. 1982. The effects of host-plant distribution and local abundance on the species richness of agromyzid flies attacking British umbellifers. *Ecol. Entomol.* **7**:257–265.

Fowler, S. V., and J. H. Lawton. 1985. Rapidly induced defenses and talking trees: The Devil's advocate position. *Am. Nat.* **126**:181–195.

Fox, J. F., and J. P. Bryant. 1984. Instability of the snowshoe hare and woody plant interaction. *Oecologia* **63**:128–135.

Fox, L. R. 1975. Cannibalism in natural populations. *Annu. Rev. Ecol. Syst.* **6**:87–106.

Fox, L. R., and B. J. Macauley. 1977. Insect grazing on *Eucalyptus* in response to variation in leaf tannins and nitrogen. *Oecologia* **29**:145–162.

Fraenkel, G. 1959. The raison d'être of secondary plant substances. *Science* **129**:1466–1470.

Fraenkel, G. 1969. Evaluation of our thoughts on secondary plant substances. *Entomol. Exp. Appl.* **12**:473–486.

Frank, F. 1957. The causality of microtine cycles in Germany. *J. Wildl. Manage.* **21**:113–121.

Frank, J. H. 1967. The insect predators of the pupal stage of the winter moth, *Operophtera brumata* (L.) (Lepidoptera: Hydriomenidae). *J. Anim. Ecol.* **36**:375–389.

Frank, P. W. 1957. Coactions in laboratory populations of two species of *Daphnia. Ecology* **38**:510–519.

Frank, P. W. 1952. A laboratory study of intraspecies and interspecies competition in *Daphnia pulicaria* (Forbes) and *Simocephalus vetulus* O. F. Müller. *Physiol. Zool.* **25**:178–204.

Frank, P. W. 1968. Life histories and community stability. *Ecology* **49**:355–357.

Frankie, G. W., H. G. Baker, and P. A. Opler. 1974. Comparative phenological studies of trees in tropical wet and dry forests in the lowlands of Costa Rica. *J. Ecol.* **62**:881–913.

Free, J. B. 1968. Dandelion as a competitor of fruit trees for bee visits. *J. Appl. Ecol.* **5**:169–178.

Free, J. B. 1970. *Insect pollination of crops.* Academic Press, San Diego, Calif.

Freeland, W. J., and D. H. Janzen. 1974. Strategies in herbivory by mammals: The role of plant secondary compounds. *Am. Nat.* **108**:269–289.

Fretwell, S. D. 1969. The adjustment of birth rate to mortality in birds. *Ibis* **111**:624–627.

Fretwell, S. D. 1972. *Populations in a seasonal environment.* Princeton Univ. Press, Princeton, N.J.

Fretwell, S. D., and H. L. Lucas, Jr. 1970. On territorial behavior and other factors influencing habitat distribution in birds. I. Theoretical development. *Acta Biotheoret.* **19**:16–36.

Friend, J., and D. R. Threlfall. 1976. *Biochemical aspects of plant–parasite relationships.* Academic Press, London.

Frings, H., E. Goldberg, and J. C. Arentzen. 1948. Antibacterial action of the blood of the large milkweed bug. *Science* **108**:689–690.

Fritz, R. S., and P. W. Price. 1988. Genetic variation among plants and insect community structure: Willows and sawflies. *Ecology* **69**:845–856.

Fritz, R. S., and E. L. Simms (eds.). 1992. *Plant resistance to herbivores and pathogens: Ecology, evolution, and genetics.* Univ. Chicago Press, Chicago.

Fritz, R. S., C. F. Sacchi, and P. W. Price. 1986. Competition versus host plant phenotype in species composition: Willow sawflies. *Ecology* **67**:1608–1618.

Fritz, R. S., W. S. Gaud, C. F. Sacchi, and P. W. Price. 1987. Patterns of intra- and interspecific association of gall-forming sawflies in relation to shoot size on their willow host plant. *Oecologia* **73**:159–169.

Frohlich, M. W. 1976. Appearance of vegetation in ultraviolet light: Absorbing flowers, reflecting backgrounds. *Science* **194**:839–841.

Fujii, K. 1965. Studies on interspecies competition between the azuki bean weevil and the southern cowpea weevil. I. The reversal in competition result. *Res. Popul. Ecol.* **7**:43–51.

Fujii, K. 1967. Studies on interspecies competition between the azuki bean weevil, *Callosobruchus chinensis,* and the southern cowpea weevil, *C. maculatus.* II. Competition under different environmental conditions. *Res. Popul. Ecol.* **9**:192–200.

Fujii, K. 1969. Studies on the interspecies competition between the azuki bean weevil and the southern cowpea weevil. IV. Competition between strains. *Res. Popul. Ecol.* **11**:84–91.

Fujii, K. 1970. Studies on the interspecies competition between the azuki bean weevil, *Callosobruchus chinensis,* and the southern cowpea weevil, *C. maculatus.* V. The role of adult behavior in competition. *Res. Popul. Ecol.* **12**:233–242.

Fujii, K. 1975. A general simulation model for laboratory insect populations. I. From cohort of eggs to adult emergences. *Res. Popul. Ecol.* **17**:85–133.

Fullard, J. H. 1990. The sensory ecology of moths and bats: global lessons in staying alive. In D. L. Evans and J. O. Schmidt (eds.). *Insect defenses: Adaptive mechanisms and strategies of prey and predators.* State Univ. New York Press, Albany, N.Y., pp. 203–226.

Furniss, R. L., and V. M. Carolin. 1977. Western forest insects. *USDA For. Serv. Misc. Publ. 1339.* U.S. Government Printing Office, Washington, D.C.

Futuyma, D. J. 1973. Community structure and stability in constant environments. *Am. Nat.* **107**:443–446.

Futuyma, D. J. 1976. Food plant specialization and environmental predictability in Lepidoptera. *Am. Nat.* **110**:285–292.

Futuyma, D. J., and F. Gould. 1979. Associations of plants and insects in a deciduous forest. *Ecol. Monogr.* **49**:33–50.

Futuyma, D. J., and M. C. Keese. 1992. Evolution and coevolution of plants and phytophagous arthropods. In G. A. Rosenthal and M. R. Berenbaum (eds.). *Herbivores: Their interactions with secondary plant metabolites.* Vol. 2. 2nd ed. Academic Press, San Diego, Calif., pp. 439–475.

Futuyma, D. J., and M. Slatkin (eds.). 1983. *Coevolution.* Sinauer, Sunderland, Mass.

Futuyma, D. J., and S. S. Wasserman. 1980. Resource concentration and herbivory in oak forests. *Science* **210**:920–922.

Gadgil, M. 1971. Dispersal: Population consequences and evolution. *Ecology* **52**:253–261.

Gadgil, M., and W. H. Bossert. 1970. Life historical consequences of natural selection. *Am. Nat.* **104**:1–24.

Gagné, R. J. 1989. *The plant-feeding gall midges of North America.* Cornell Univ. Press, Ithaca, N.Y.

Galil, J., and D. Eisikowitch. 1968. On the pollination ecology of *Ficus sycomorus* in East Africa. *Ecology* **49**:259–269.

Galil, J., and D. Eisikowitch. 1969. Further studies on the pollination ecology of *Ficus sycomorus* L. (Hymenoptera, Chalcidoidea, Agaonidae). *Tijdschr. Entomol.* **112**:1–13.

Galilei, G. 1638. *Discorsi e dimonstrazioni matematiche, intorno à due nuove scienze attenenti alla mecanica ed i movimenti locale.* Elsevirii, Leyden. Translated by H. Crew and A. de Salvio. 1914. *Dialogues concerning two new sciences.* Macmillan, New York.

Gange, A. C., and V. K. Brown (eds.). 1996. *Multitrophic interactions in terrestrial systems.* Blackwell Science, Oxford.

Garrett, L. 1994. *The coming plague: Newly emerging diseases in a world out of balance.* Penguin, New York.

Gaston, K. J. 1988. Patterns in the local and regional dynamics of moth populations. *Oikos* **53**:49–57.

Gaston, K. J. 1991. The magnitude of global insect species richness. *Conserv. Biol.* **5**:283–296.

Gaston, K. J. 1992. Regional numbers of insect and plant species. *Funct. Ecol.* **6**:243–247.

Gaston, K. J. 1994. *Rarity.* Chapman & Hall, London.

Gaston, K. J., and J. H. Lawton. 1988. Patterns in the distribution and abundance of insect populations. *Nature* **331**:709–712.

Gaston, K. J., T. R. New, and M. J. Samways (eds.). 1993. *Perspectives on insect conservation.* Intercept, Andover, Hants, England.

Gauld, I. D. 1986. Latitudinal gradients in ichenumonid species richness in Australia. *Ecol. Entomol.* **11**:155–161.

Gauld, I. D., and K. J. Gaston. 1994. The taste of enemy-free space: Parasitoids and nasty hosts. In B. A. Hawkins and W. Sheehan (eds.). *Parasitoid community ecology.* Oxford Univ. Press, Oxford, pp. 279–299.

Gauld, I. D., K. J. Gaston, and D. H. Janzen. 1992. Plant allelochemicals, tritrophic interactions and the anomalous diversity of tropical parasitoids: the "nasty" host hypothesis. *Oikos* **65**:353–357.

Gause, G. F. 1934. *The struggle for existence.* Williams & Wilkins, Baltimore. Reprinted 1964. Hafner, New York.

Gause, G. F., and A. A. Witt. 1935. Behavior of mixed populations and the problem of natural selection. *Am. Nat.* **69**:596–609.

Geier, P. W., L. R. Clark, D. J. Anderson, and H. A. Nix (eds.). 1973. Insects: Studies in population management. *Mem. Ecol. Soc. Aust.* **1**:1–295.

Georghiou, G. P. 1972. The evolution of resistance to pesticides. *Annu. Rev. Ecol. Syst.* **3**:133–168.

Georghiou, G. P., and C. E. Taylor. 1977. Pesticide resistance as an evolutionary phenomenon. *Proc. 15th Int. Congr. Entomol., Washington, D.C.,* pp. 759–785.

Gershenzon, J. 1984. Changes in the levels of plant secondary metabolites under water and nutrient stress. In B. N. Timmermann, C. Steelink, and F. A. Loewus (eds.). *Phytochemical adaptations to stress.* Plenum Press, New York, pp. 273–320.

Gilbert, L. E. 1972. Pollen feeding and reproductive biology of *Heliconius* butterflies. *Proc. Natl. Acad. Sci. USA* **69**:1403–1407.

Gilbert, L. E. 1975. Ecological consequences of a coevolved mutualism between butterflies and plants. In L. E. Gilbert and P. H. Raven (eds.). *Coevolution of animals and plants.* Univ. Texas Press, Austin, Texas, pp. 210–240.

Gilbert, L. E. 1977. The role of insect–plant coevolution in the organization of ecosystems. In V. Labyrie (ed.). *Comportement des insectes et milieu trophique.* C.N.R.S., Paris, pp. 399–413.

Gilbert, L. E. 1991. Biodiversity of a Central American *Heliconius* community: Pattern, process, and problems. In P. W. Price, T. M. Lewinsohn, G. W. Fernandes, and W. W. Benson (eds.). *Plant–animal interactions: Evolutionary ecology in tropical and temperate regions.* Wiley, New York, pp. 403–427.

Gilbert, L. E., and P. H. Raven (eds.). 1975. *Coevolution of animals and plants.* Univ. Texas Press, Austin, Texas.

Gill, D. E. 1974. Intrinsic rate of increase, saturation density, and competitive ability. II. The evolution of competitive ability. *Am. Nat.* **108**:103–116.

Giller, P. S. 1980. The control of handling time and its effects on the foraging strategy of the heteropteran predator *Notonecta. J. Anim. Ecol.* **49**:699–712.

Giller, P. S., and S. McNeill. 1981. Predation strategies, resource partitioning and habitat selection in *Notonecta* (Hemiptera/Heteroptera). *J. Anim. Ecol.* **50**:789–808.

Gillespie, J. H., and K. Kojima. 1968. The degree of polymorphisms in enzymes involved in energy production compared to that in nonspecific enzymes in two *Drosophila ananassae* populations. *Proc. Natl. Acad. Sci. USA* **61**:582–585.

Gillett, J. B. 1962. Pest presure, an underestimated factor in evolution. In D. Nichols (ed.). *Taxonomy and geography.* Syst. Assoc. Publ. 4, Systematics Association, London, pp. 37–46.

Gilpin, M. E. 1974. Intraspecific competition between *Drosophila* larvae in serial transfer systems. *Ecology* **55**:1154–1159.

Gilpin, M., and I. Hanski (eds.). 1991. *Metapopulation dynamics: Empirical and theoretical investigations.* Academic Press, London.

Gittelman, S. H. 1978. Optimum diet and body size in backswimmers. (Heteroptera: Notonectidae, Pleidae). *Ann. Entomol. Soc. Am.* **71**:737–747.

Glas, P. 1960. Factors governing density in the chaffinch *(Fringilla coelebs)* in different types of wood. *Arch. Neerl. Zool.* **13**:466–472.

Gleason, H. A. 1917. The structure and development of the plant association. *Bull. Torrey Bot. Club* **44**:463–481.

Gleason, H. A. 1926. The individualistic concept of the plant association. *Am. Midl. Nat.* **21**:92–110.

Godfray, H. C. J. 1994. *Parasitoids: Behavioral and evolutionary ecology.* Princeton Univ. Press, Princeton, N.J.

Godfray, H. C. J., and M. P. Hassell. 1991. Encapsulation and host–parasitoid population biology. In C. A. Toft, A. Aeschlimann, and L. Bolis (eds.). *Parasite–host associations: Coexistence or conflict?* Oxford Univ. Press, Oxford.

Goeden, R. D., and S. M. Louda. 1976. Biotic interference with insects imported for weed control. *Annu. Rev. Entomol.* **21**:325–342.

Goh, B. S. 1979. Stability in models of mutualism. *Am. Nat.* **113**:261–275.

Golley, F. B. 1960. Energy dynamics of a food chain of an old-field community. *Ecol. Monogr.* **30**:187–206.

Golley, F. B. 1972. Energy flux in ecosystems. In J. A. Wiens (ed.). *Ecosystem structure and function.* Oregon State Univ. Press, Corvallis, Oreg., pp. 69–88.

Golley, F. B. (ed.). 1977. *Ecological succession.* Dowden, Hutchinson & Ross, Stroudsburg, Pa.

Golley, F. B., and J. B. Gentry. 1964. Bioenergetics of the southern harvester ant, *Pogonomyrmex badius*. *Ecology* **45**:217–225.

Goodenough, J. L., and J. M. McKinion (eds.). 1992. *Basics of insect modeling*. American Society of Agricultural Engineers, St. Joseph, Mich.

Goodman, D. 1974. Natural selection and a cost ceiling on reproductive effort. *Am. Nat.* **108**:247–268.

Goodman, D. 1975. The theory of diversity–stability relationships in ecology. *Q. Rev. Biol.* **50**:237–266.

Goudriaan, J., and C. T. de Wit. 1973. A re-interpretation of Gause's population experiments by means of simulation. *J. Anim. Ecol.* **42**:521–530.

Gould, F. 1991. The evolutionary potential of crop pests. *Am. Sci.* **79**:496–507.

Gould, F. 1994. Potential and problems with high-dose strategies for pesticidal engineered crops. *Biocontrol Sci. and Technol.* **4**:451–461.

Gould, F., G. G. Kennedy and M. T. Johnson. 1991. Effects of natural enemies on the rate of herbivore adaptation to resistant host plants. *Entomol. Exp. Appl.* **58**:1–14.

Gould, S. J. 1974. Size and shape. *Nat. Hist.* **83**(1):20–26.

Gould, S. J. 1977. *Ontogeny and phylogeny*. Belknap Press, imprint of Harvard Univ. Press, Cambridge, Mass.

Gould, S. J. 1980. *The panda's thumb*. W. W. Norton, New York.

Gould, S. J., and R. C. Lewontin. 1979. The spandrels of San Marco and the Panglossian paradigm: A critique of the adaptationist programme. *Proc. R. Soc. London Ser. B* **205**:581–598.

Grant, B., and R. J. Howlett. 1988. Background selection by the peppered mth (*Biston betularia* Linn.): Individual differences. *Biol. J. Linn. Soc.* **33**:217–232.

Grant, B., G. A. Snyder, and S. F. Glessner. 1974. Frequency-dependent mate selection in *Mormoniella vitripennis*. *Evolution* **28**:259–264.

Grant, P. R. 1972. Convergent and divergent character displacement. *Biol. J. Linn. Soc.* **4**:39–68.

Grant, P. R. 1975. The classical case of character displacement. *Evol. Biol.* **8**:237–337.

Grant, P. R. 1984. Interspecific competition inferred from patterns of guild structure. In D. R. Strong, D. Simberloff, L. G. Abele and A. B. Thistle (eds.). *Ecological communities: Conceptual issues and the evidence*. Princeton Univ. Press. Princeton, N.J.

Grant, P. R., and I. Abbott. 1980. Interspecific competition, island biogeography and null hypotheses. *Evolution* **34**:332–341.

Grant, V. 1949. Pollination systems as isolating mechanisms in angiosperms. *Evolution* **3**:82–97.

Grant, V. 1950. The flower constancy of bees. *Bot. Rev.* **16**:379–398.

Grant, V. 1975. *Genetics of flowering plants*. Columbia Univ. Press, New York.

Gray, J., and A. J. Boucot. 1994. Early Silurian nonmarine animal remains and the nature of the early continental ecosystem. *Acta Palaentol. Polon.* **38**:303–328.

Gray, J., and W. Shear. 1992. Early life on land. *Am. Sci.* **80**:444–456.

Greany, P. D., J. H. Tumlinson, D. L. Chambers, and G. M. Boush. 1977. Chemically mediated host finding by *Biosteres (Opius) longicaudatus*, a parasitoid of tephritid fruitfly larvae. *J. Chem. Ecol.* **3**:189–195.

Green, R. G., and C. A. Evans. 1940. Studies on a population cycle of snowshoe hares on the Lake Alexander area. *J. Wildl. Manage.* **4**:220–238, 267–278, 347–358.

Green, R. G., and C. L. Larson. 1938. A description of shock disease in the snowshoe hare. *Am. J. Hyg.* **28**:190–212.

Green, T. R., and C. A. Ryan. 1972. Wound-induced proteinase inhibitor in plant leaves: A possible defense mechanism against insects. *Science* **175**: 776–777.

Greenbank, D. O. 1956. The role of climate and dispersal in the initiation of outbreaks of the spruce budworm in New Brunswick. I. The role of climate. *Can. J. Zool.* **34**:453–476.

Greenbank, D. O. 1963. The development of the outbreak, pp. 19–23; Climate and the spruce budworm, pp. 174–180; Staminate flowers and the spruce budworm, pp. 202–218; Host species and the spruce budworm, pp. 219–223. In R. F. Morris (ed.). *The dynamics of epidemic spruce budworm populations*. Mem. Entomol. Soc. Can. 31, Entomological Society of Canada, Ottawa, Ontario.

Greene, E. 1989. A diet-induced developmental polymorphism in a caterpillar. *Science* **243**:643–646.

Gressitt, J. L. 1965. Biogeography and ecology of land arthropods of Antarctica. In J. van Mieghem, P. van Oye, and J. Schell. (eds.) *Biogeography and ecology in Antarctica*. Monogr. Biol. 15, Dr. W. Junk, The Hague, The Netherlands, pp. 431–490.

Gressitt, J. L. 1970. Subantarctic entomology and biogeography. *Pac. Insects Monogr.* **23**:295–374.

Grice, G. D., and A. D. Hart. 1962. The abundance, seasonal occurrence and distribution of the epizooplankton between New York and Bermuda. *Ecol. Monogr.* **32**:287–309.

Griffiths, D. 1980. The feeding biology of ant-lion larvae: Prey capture, handling and utilization. *J. Anim. Ecol.* **49**:99–125.

Griffiths, D. 1981. Sub-optimal foraging in the ant-lion *Macroleon quinquemaculatus*. *J. Anim. Ecol.* **50**:697–702.

Grime, J. P. 1977. Evidence for the existence of three primary strategies in plants and its relevance to ecological and evolutionary theory. *Am. Nat.* **111**:1169–1194.

Grime, J. P. 1979. *Plant strategies and vegetation processes*. Wiley, London.

Grinnell, J. 1904. The origin and distribution of the chestnut-backed chickadee. *Auk* **21**:364–382.

Grinnell, J. 1917. The niche-relationships of the California thrasher. *Auk* **34**:427–433.

Gross, P. 1993. Insect behavioral and morphological defenses against parasitoids. *Annu. Rev. Entomol.* **38**:251–273.

Gross, P., and P. W. Price. 1988. Plant influences on parasitism of two leafminers: A test of enemy-free space. *Ecology* **69**:1506–1516.

Grubb, P. J. 1977. The maintenance of species-richness in plant communities: The importance of the regeneration niche. *Biol. Rev.* **52**:107–145.

Guilford, T. 1990. The evolution of aposematism. In D. L. Evans and J. O. Schmidt (eds.). *Insect defenses: Adaptive mechanisms and strategies of prey and predators.* State Univ. New York Press, Albany, N.Y., pp. 23–61.

Gullan, P. J., and P. S. Cranston. 1994. *The insects: An outline of entomology.* Chapman & Hall, London.

Haddow, A. J., and R. C. M. Thomson. 1937. Sheep myiasis in south-west Scotland, with special reference to the species involved. *Parasitology* **29**:96–116.

Hadley, M. 1972. Perspectives in productivity studies, with special reference to some social insect populations. *Ekol. Pol.* **20**:173–184.

Haeck, J., and R. Hengeveld. 1979. Biogeografie en Oecologie: Over verschillen in de mate van voorkomen binnen het soortsareaal. *Vakbl. Biol.* **59**(3):26–31.

Hagerman, A. E., and L. G. Butler. 1991. Tannins and lignins. In G. A. Rosenthal and M. R. Berenbaum (eds.). *Herbivores: Their interactions with secondary plant metabolites.* Vol. 1. 2nd ed. *The chemical participants.* Academic Press, San Diego, Calif., pp. 355–388.

Hairston, N. G., F. E. Smith, and L. B. Slobodkin. 1960. Community structure, population control, and competition. *Am. Nat.* **94**:421–425.

Hairston, N. G., J. D. Allan, R. K. Colwell, D. J. Futuyma, J. Howell, M. D. Lubin, J. Mathias, and J. H. Vandermeer. 1968. The relationship between species diversity and stability: An experimental approach with protozoa and bacteria. *Ecology* **49**:1091–1101.

Hairston, N. G., D. W. Tinkle, and H. M. Wilbur. 1970. Natural selection and the parameters of population growth. *J. Wildl. Manage.* **34**:681–690.

Haldane, J. B. S. 1949. Disease and evolution. In Symposium sui fattori ecologici e genetici della speciazone négli animali. *Ric. Sci.* **19** (Suppl.), pp. 3–11.

Halford, F. M. 1897. *Dry-fly entomology.* Vinton, London.

Hamilton, T. H., R. H. Barth, Jr., and I. Rubinoff. 1964. The environmental control of insular variation in bird species abundance. *Proc. Natl. Acad. Sci. USA* **52**:132–140.

Hamilton, T. H., and I. Rubinoff. 1967. On predicting insular variation in endemism and sympatry for the Darwin finches in the Galápagos archipelago. *Am. Nat.* **101**:161–171.

Hamilton, W. D. 1964a. The genetical evolution of social behavior. I. *J. Theor. Biol.* **7**:1–16.

Hamilton, W. D. 1964b. The genetical evolution of social behavior. II. *J. Theor. Biol.* **7**:17–52.

Hammond, A. L. 1971a. Plate tectonics: The geophysics of the earth's surface. *Science* **173**:40–41.

Hammond, A. L. 1971b. Plate tectonics II: Mountain building and continental geology. *Science* **173**:133–134.

Hammond, P. M. 1992. Species inventory. In B. Groombridge (ed.). *Global biodiversity: Status of the earth's living resources.* Chapman & Hall, London, pp. 17–39.

Harborne, J. B. (ed.). 1972. *Phytochemical ecology.* Academic Press, San Diego, Calif.

Harborne, J. B. 1977. *Introduction to ecological biochemistry.* Academic Press, London.

Harborne, J. B. (ed.). 1978. *Biochemical aspects of plant and animal coevolution.* Academic Press, San Diego, Calif.

Harcourt, D. G. 1963. Major mortality factors in the population dynamics of the diamondback moth, *Plutella maculipennis* (Curt) (Lepidoptera: Plutellidae). *Mem. Entomol. Soc. Can.* **32**:55–66.

Harcourt, D. G. 1966. Major factors in survival of the immature stages of *Pieris rapae* (L.). *Can. Entomol.* **98**:653–662.

Harcourt, D. G. 1969. The development and use of life tables in the study of natural insect populations. *Annu. Rev. Entomol.* **14**:175–196.

Harger, R. 1972. Relative consumer species diversity with respect to producer diversity and net productivity. *Science* **176**:544–545.

Harley, J. L. 1969. *The biology of Mycorrhiza.* 2nd ed. Hill, London.

Harman, W. N. 1972. Benthic substrates: Their effect on fresh-water Mollusca. *Ecology* **53**:271–277.

Harper, J. L. 1968. The regulation of numbers and mass in plant populations. In R. C. Lewontin (ed.). *Population biology and evolution.* Syracuse Univ. Press, Syracuse, N.Y., pp. 139–158.

Harper, J. L. 1969. The role of predation in vegetational diversity. In *Diversity and stability in ecological systems.* Brookhaven Symp. Biol. 22, Brookhaven National Laboratory, Upton, N.Y., pp. 48–61.

Harper, J. L. 1977. *Population biology of plants.* Academic Press, London.

Harper, J. L., and J. Ogden. 1970. The reproductive strategy of higher plants. I. The concept of strategy with special reference to *Senecio vulgaris* L. *J. Ecol.* **58**:681–698.

Harper, J. L., P. H. Lovell, and K. G. Moore. 1970. The shapes and sizes of seeds. *Annu. Rev. Ecol. Syst.* **1**:327–356.

Harris, H. 1970. *The principles of human biochemical genetics.* American Elsevier, New York.

Harris, H. 1971. Protein polymorphism in man. *Can. J. Genet. Cytol.* **13**:381–396.

Harris, H., and D. A. Hopkinson. 1972. Average heterozygosity per locus in man: An estimate based on the incidence of enzyme polymorphisms. *Ann. Hum. Genet.* **36**:9–20.

Harris, K. F. 1979. Leafhoppers and aphids as biological vectors: Vector–virus relationships. In K. Maramorosch and K. F. Harris (eds.). *Leafhopper vectors and plant disease agents.* Academic Press, San Diego, Calif., pp. 217–308.

Harrison, G. W. 1979. Stability under environmental stress: Resistance, resilience, persistence, and variability. *Am. Nat.* **113**:659–669.

Harrison, S. 1994. Resources and dispersal as factors limiting a population of the tussock moth *(Orgyia vetusta),* a flightless defoliator. *Oecologia* **99**:27–34.

Harrison, S., and N. Cappuccino. 1995. Using density-manipulation experiments to study population regulation. In N. Cappuccino and P. W. Price (eds.). *Population dynamics: New approaches and synthesis.* Academic Press, San Diego, Calif., pp. 131–147.

Hartley, S. E., and J. H. Lawton. 1987. Effects of different types of damage on the chemistry of birch foliage, and the responses of birch feeding insects. *Oecologia* **74**:432–437.

Hartling, L. K., and R. C. Plowright. 1979. Foraging by bumblebees on patches of artificial flowers: A laboratory study. *Can. J. Zool.* **57**:1866–1870.

Harvey, P. H., and M. D. Pagel. 1991. *The comparative method in evolutionary biology.* Oxford Univ. Press, Oxford.

Harville, J. P. 1955. Ecology and population dynamics of the California oak moth *Phyrganidia californica* Packard (Lepidoptera: Dioptidae). *Microentomology* **20**:83–166.

Haskell, N., and E. P. Catts (eds.). 1990. *Entomology and death: A procedural guide.* Published by the authors.

Hassell, M. P. 1970. Parasite behaviour as a factor contributing to the stability of insect host–parasite interactions. In P. J. den Boer and G. R. Gradwell. *Dynamics of populations.* Centre for Agricultural Publications and Documentation, Wageningen, The Netherlands, pp. 366–378.

Hassell, M. P. 1978. *The dynamics of arthropod predator–prey systems.* Princeton Univ. Press, Princeton, N.J.

Hassell, M. P. 1980. Foraging strategies, population models and biological control: A case study. *J. Anim. Ecol.* **49**:603–628.

Hassell, M. P. 1981. Arthropod predator–prey systems. In R. M. May (ed.). *Theoretical ecology: Principles and applications.* Sinauer, Sunderland, Mass., pp. 105–131.

Hassell, M. P. 1986. Parasitoids and population regulation. In J. Waage and D. Greathead (eds.). *Insect parasitoids.* Academic Press, London, pp. 201–224.

Hassell, M. P., and C. B. Huffaker. 1969. The appraisal of delayed and direct density-dependence. *Can. Entomol.* **101**:353–361.

Hassell, M. P., and R. M. May. 1973. Stability in insect host–parasite models. *J. Anim. Ecol.* **42**:693–726.

Hassell, M. P., and D. J. Rogers. 1972. Insect parasite responses in the development of population models. *J. Anim. Ecol.* **41**:661–676.

Hassell, M. P., and G. C. Varley. 1969. New inductive population model for insect parasites and its bearing on biological control. *Nature* **223**: 1133–1137.

Hassell, M. P., J. H. Lawton, and J. R. Beddington. 1977. Sigmoid functional responses by invertebrate predators and parasitoids. *J. Anim. Ecol.* **46**: 249–262.

Hatchett, J. H., and R. L. Gallun. 1970. Genetics of the ability of the Hessian fly, *Mayetiola destructor,* to survive on wheats having different genes for resistance. *Ann. Entomol. Soc. Am.* **63**:1400–1407.

Haukioja, E. 1980. On the role of plant defenses in the fluctuation of herbivore populations. *Oikos* **35**:202–213.

Haukioja, E. 1982. Inducible defenses of white birch to a geometrid defoliator, *Epirrita autumnata.* In J. H. Visser and A. K. Minks (eds.). *Insect–plant relationships.* Centre for Agricultural Publishing and Documentation, Wageningen, The Netherlands, pp. 199–203.

Haukioja, E., and T. Hakala. 1975. Herbivore cycles and periodic outbreaks: Formulation of a general hypothesis. *Rep. Kevo Subarct. Res. Stn.* **12**:1–9.

Haukioja, E., and S. Neuvonen. 1987. Insect population dynamics and induction of plant resistance: The testing of hypotheses. In P. Barbosa and J. C. Schultz (eds.). *Insect outbreaks.* Academic Press, San Diego, Calif., pp. 411–432.

Haukioja, E., and P. Niemelä. 1976. Does birch defend itself actively against herbivores? *Rep. Kevo Subarct. Res. Stn.* **13**:44–47.

Haukioja, E., and P. Niemalä. 1977. Retarded growth of a geometrid larva after mechanical damage to leaves of its host tree. *Ann. Zool. Fenn.* **14**:48–52.

Hawkins, B. A. 1988. Species diversity in the third and fourth trophic levels: Patterns and mechanisms. *J. Anim. Ecol.* **57**:137–162.

Hawkins, B. A. 1994. *Pattern and process in host–parasitoid interactions.* Cambridge Univ. Press, Cambridge.

Hawkins, B. A., and R. D. Goeden. 1984. Organization of a parasitoid community associated with a complex of galls on *Atriplex* spp. in southern California. *Ecol. Entomol.* **9**:271–292.

Hawkins, B. A., and P. Gross. 1992. Species richness and population limitation in insect parasitoid–host systems. *Am. Nat.* **139**:417–423.

Hawkins, B. A., and W. Sheehan (eds.). 1994. *Parasitoid community ecology.* Oxford Univ. Press, Oxford.

Hawkins, B. A., R. R. Askew, and M. R. Shaw. 1990. Influences of host feeding-niche and foodplant type on generalist and specialist parasitoids. *Ecol. Entomol.* **15**:275–280.

Hay, M. E., and W. Fenical. 1988. Marine plant–herbivore interactions: The ecology of chemical defense. *Annu. Rev. Ecol. Syst.* **19**:111–145.

Hay, M. E., J. E. Duffy, C. A. Pfister, and W. Fenical. 1987. Chemical defense against different marine herbivores: Are amphipods insect equivalents? *Ecology* **68**:1567–1580.

Haynes, D. L., R. K. Brandenburg, and P. D. Fisher. 1973. Environmental monitoring network for pest management systems. *Environ. Entomol.* **2**:889–899.

Heatwole, H., and R. Levins. 1972a. Biogeography of the Puerto Rican Bank: Flotsam transport of terrestrial animals. *Ecology* **53**:112–117.

Heatwole, H., and R. Levins. 1972b. Trophic structure stability and faunal change during recolonization. *Ecology* **53**:531–534.

Heatwole, H., and R. Levins. 1973. Biogeography of Puerto Rican Bank: Species turnover on a small cay, Cayo Ahogado. *Ecology* **54**:1042–1055.

Heatwole, H., D. M. Davis, and A. M. Wenner. 1964. Detection of mates and hosts by parasitic insects of the genus *Megarhyssa* (Hymenoptera: Ichneumonidae). *Am. Midl. Nat.* **71**:374–381.

Hedin, P. A. (ed.). 1983. *Plant resistance to insects.* American Chemical Society, Washington, D.C.

Hedrick, P. W. 1983. *Genetics of populations.* Science Books International, Boston.

Hedrick, P. W. 1984. *Population biology: The evolution and ecology of populations.* Jones and Bartlett, Boston.

Hedrick, P. W., M. E. Ginevan, and E. P. Ewing. 1976. Genetic polymorphism in heterogeneous environments. *Annu. Rev. Ecol. Syst.* **7**:1–32.

Heinrich, B. 1971a. Temperature regulation of the sphinx moth, *Manduca sexta.* I. Flight energetics and body temperature during free and tethered flight. *J. Exp. Biol.* **54**:141–152.

Heinrich, B. 1971b. Temperature regulation of the sphinx moth, *Manduca sexta.* II. Regulation of heat loss by control of blood circulation. *J. Exp. Biol.* **54**:153–166.

Heinrich, B. 1973. The energetics of the bumblebee. *Sci. Am.* **288**(4): 96–102.

Heinrich, B. 1975a. Energetics of pollination. *Annu. Rev. Ecol. Syst.* **6**:139–170.

Heinrich, B. 1975b. Bee flowers: A hypothesis on flower variety and blooming times. *Evolution* **29**:325–334.

Heinrich, B. 1975c. The role of energetics in bumblebee–flower interrelationships. In L. E. Gilbert and P. H. Raven (eds.). *Coevolution of animals and plants.* Univ. Texas Press, Austin, Texas, pp. 141–158.

Heinrich, B. 1979a. *Bumble-bee economics*. Harvard Univ. Press, Cambridge, Mass.

Heinrich, B. 1979b. Foraging strategies of caterpillars. *Oecologia* **42**:325–337.

Heinrich, B. (ed.). 1981. *Insect thermoregulation*. Wiley, New York.

Heinrich, B. 1993. How avian predators constrain caterpillar foraging. In N. E. Stamp and T. M. Casey (eds.). *Caterpillars: Ecological and evolutionary constraints on foraging*. Chapman & Hall, New York, pp. 224–247.

Heinrich, B., and P. H. Raven. 1972. Energetics and pollination ecology. *Science* **176**:597–602.

Heinrich, G. H. 1977. *Ichneumoninae of Florida and neighboring states*. Florida Department of Agriculture and Consumer Services, Gainesville, Fla.

Heithaus, E. R., D. C. Culver, and A. J. Beattie. 1980. Models of some ant–plant mutualisms. *Am. Nat.* **116**:347–361.

Helgesen, R. G., and D. L. Haynes. 1972. Population dynamics of the cereal leaf beetle, *Oulema melanopus* (Coleoptera: Chrysomelidae): A model for age specific mortality. *Can. Entomol.* **104**:797–814.

Heller, R., and M. Milinski. 1979. Optimal foraging of sticklebacks on swarming prey. *Anim. Behav.* **27**:1127–1141.

Hellmers, H. 1964. An evaluation of the photosynthetic efficiency of forests. *Q. Rev. Biol.* **39**:249–257.

Hengeveld, R., and J. Haeck. 1982. The distribution of abundance. I. Measurements. *J. Biogeog.* **9**:303–316.

Hengeveld, R., S. A. L. M. Kooijman, and C. Taillie. 1979. A spatial model explaining species–abundance curves. In J. K. Ord, G. P. Patil, and C. Taillie (eds.). *Statistical distributions in ecological work*. International Cooperative Publishing House, Burtonsville, Md., pp. 333–347.

Hepburn, H. R. 1985. The integument. In M. S. Blum (ed.). *Fundamentals of insect physiology*. Wiley, New York, pp. 139–183.

Herbold, B., and P. B. Moyle. 1986. Introduced species and vacant niches. *Am. Nat.* **128**:751–760.

Hermann, H. R. (ed.). 1979–1982. *Social insects*. 4 vols. Academic Press, San Diego, Calif.

Herrebout, W. M. 1969. Some aspects of host selection in *Eucarcelia rutilla* Vill. (Diptera: Tachinidae). *Neth. J. Zool.* **19**:1–104.

Herrera, C. M. 1982. Grasses, grazers, mutualism, and coevolution: A comment. *Oikos* **38**:254–258.

Hespenheide, H. A. 1979. Are there fewer parasitoids in the tropics? *Am. Nat.* **113**:766–769.

Hespenheide, H. A. 1994. An overview of faunal studies. In L. A. McDade, K. S. Bawa, H. A. Hespenheide, and G. S. Hartshorn (eds.). *La Selva: Ecology and natural history of a neotropical rain forest*. Univ. Chicago Press, Chicago, pp. 238–243.

Hewitt, O. H. (ed.). 1954. A symposium on cycles in animal populations. *J. Wildl. Manage.* **18**:1–112.

Hickman, J. C. 1974. Pollination by ants: A low energy system. *Science* **184**:1290–1292.

Hicks, K. L., and J. O. Tahvanainen. 1974. Niche differentiation by crucifer feeding flea beetles (Coleoptera: Chrysomelidae). *Am. Midl. Nat.* **91**:406–423.

Hikino, H., and T. Takemoto. 1972. Arthropod moulting hormones from plants, *Achyranthes* and *Cyathula*. *Naturwissenschaften* **59**:91–98.

Hills, H. G., N. H. Williams, and C. H. Dodson. 1972. Floral fragrances and isolating mechanisms in the genus *Catasetum* (Orchidaceae). *Biotropica* **4**:61–76.

Hinde, R. A. 1956. The biological significance of the territories of birds. *Ibis* **98**:340–369.

Hinton, H. E. 1955. Protective devices of endopterygote pupae. *Trans. Soc. Br. Entomol.* **12**:49–92.

Hocking, B. 1975. Ant–plant mutualism: Evolution and energy. In L. E. Gilbert and P. H. Raven (eds.). *Coevolution of animals and plants*. Univ. Texas Press, Austin, Texas, pp. 78–90.

Holdridge, L. R. 1947. Determination of world plant formations from simple climatic data. *Science* **105**:367–368.

Holdridge, L. R. 1967. *Life zone ecology.* 2nd ed. Tropical Research Center, San José, Costa Rica.

Holdridge, L. R., W. C. Grenke, W. H. Hatheway, T. Liang, and J. A. Tosi, Jr. 1971. *Forest environments in tropical life zones: A pilot study.* Pergamon Press, Oxford.

Holland, W. J. 1903. *The moth book. A guide to the moths of North America.* Doubleday, Page, New York. Reprinted 1968, Dover, New York.

Hölldobler, B. 1976. Tournaments and slavery in a desert ant. *Science* **192**:912–914.

Hölldobler, B., and E. O. Wilson. 1990. *The ants.* Belknap Press, imprint of Harvard Univ. Press, Cambridge, Mass.

Holling, C. S. 1959a. Some characteristics of simple types of predation and parasitism. *Can. Entomol.* **91**:385–398.

Holling, C. S. 1959b. The components of predation as revealed by a study of small mammal predation of the European pine sawfly. *Can. Entomol.* **91**:293–320.

Holling, C. S. 1961. Principles of insect predation. *Annu. Rev. Entomol.* **6**:163–182.

Holling, C. S. 1965. The functional response of predators to prey density and its role in mimicry and population regulation. *Mem. Entomol. Soc. Can.* **45**:1–60.

Holling, C. S. 1966. The functional response of invertebrate predators to prey density. *Mem. Entomol. Soc. Can.* **48**:1–86.

Holling, C. S. 1968. The tactics of a predator. In T. R. E. Southwood (ed). *Insect abundance.* Symp. R. Entomol. Soc. London 4, pp. 47–58.

Holling, C. S. 1973. Resilience and stability of ecological systems. *Annu. Rev. Ecol. Syst.* **4**:1–23.

Holling, C. S., and S. Ewing. 1971. Blind man's buff: Exploring the response space generated by realistic ecological simulation models. In G. P. Patil, E. C. Pielou, and W. E. Waters (eds.). *Statistical ecology.* Vol. 2. Pennsylvania State Univ. Press, University Park, Pa., pp. 207–223.

Holmes, J. C. 1979. Parasite populations and host community structure. In B. B. Nickol (ed.). *Host–parasite interfaces.* Academic Press, San Diego, Calif., pp. 27–46.

Honigberg, B. M. 1967. Chemistry and parasitism among some protozoa. *Chem. Zool.* **1**:695–814.

Honigberg, B. M. 1970. Protozoa associated with termites and their role in digestion. In K. Krishna and F. M. Weesner (eds.). *Biology of termites.* Vol. 2. Academic Press, San Diego, Calif., pp. 1–36.

Hooker, J. D. 1847–1860. *The botany of the Antarctic voyage of H.M. Discovery ships Erebus and Terror in the years 1839–1843, under the command of Captain Sir James Clark Ross.* Reeve, London.

Horn, B. K. P. 1978. Dodo apocrypha. *Sci. News* **113**:19.

Howarth, F. G., and G. W. Ramsay. 1991. The conservation of island insects and their habitats. In N. M. Collins and J. A. Thomas (eds.). *The conservation of insects and their habits.* Academic Press, London, pp. 71–107.

Howe, H. F. 1985. Gomphothere fruits: a critique. *Am. Nat.* **125**:853–865.

Howe, H. F., and L. C. Westley. 1988. *Ecological relationships of plants and animals.* Oxford Univ. Press, New York.

Hsiao, T. H., and G. Fraenkel. 1968. Selection and specificity of the Colorado potato beetle for solanaceous and nonsolanaceous plants. *Ann. Entomol. Soc. Am.* **61**:493–503.

Hubbard, S. F., and R. M. Cook. 1978. Optimal foraging by parasitoid wasps. *J. Anim. Ecol.* **47**:593–604.

Huffaker, C. B. 1958. Experimental studies on predation: Dispersion factors and predatory–prey oscillations. *Hilgardia* **27**:343–383.

Huffaker, C. B. 1967. A comparison of the status of biological control of St. Johnswort in California and Australia. *Mushi* (Supl.) **39**:51–73.

Huffaker, C. B. (ed.). 1971. *Biological control.* Plenum Press, New York.

Huffaker, C. B., and C. E. Kennett. 1969. Some aspects of assessing efficiency of natural enemies. *Can. Entomol.* **101**:425–447.

Huffaker, C. B., and J. E. Laing. 1972. "Competitive displacement" without a shortage of resources? *Res. Popul. Ecol.* **14**:1–17.

Huffaker, C. B., and P. S. Messenger (eds.). 1976. *Theory and practice of biological control.* Academic Press, San Diego, Calif.

Humphries, D. A., and P. M. Driver. 1970. Protean defense by prey animals. *Oecologia* **5**:285–302.

Hungate, R. E. 1936. Studies on the nutrition of *Zootermopsis.* I. The role of bacteria and molds in cellulose decomposition. *Zentralbl. Bakteriol. Parasitenk. Abt.* II **95**:240–249.

Hungate, R. E. 1938. Studies on the nutrition of *Zootermopsis.* II. The relative importance of the termite and the protozoa in wood digestion. *Ecology* **19**:1–25.

Hungate, R. E. 1955. Mutualistic intestinal protozoa. In S. H. Hutner and A. Lwoff (eds.). *Biochemistry and physiology of protozoa.* Vol. 2. Academic Press, San Diego, Calif., pp. 159–199.

Hungate, R. E. 1966. *The rumen and its microbes.* Academic Press, San Diego, Calif.

Hunter, A. F. 1991. Traits that distinguish outbreaking and non-outbreaking Macrolepidoptera feeding on northern hardwood trees. *Oikos* **60**:275–282.

Hunter, A. F. 1995a. The ecology and evolution of reduced wings in forest macrolepidoptera. *Evol. Ecol.* **9**:275–287.

Hunter, A. F. 1995b. Ecology, life history, and phylogeny of outbreak and nonoutbreak species. In N. Cappuccino and P. W. Price (eds.). *Population dynamics: New approaches and synthesis.* Academic Press, San Diego, Calif., pp. 41–64.

Hunter, M. D. 1990. Differential susceptibility to variable plant phenology and its role in competition between two insect herbivores on oak. *Ecol. Entomol.* **15**:401–408.

Hunter, M. D. 1992a. A variable insect–plant interaction: The relationship between tree budburst phenology and population levels of insect herbivores among trees. *Ecol. Entomol.* **17**:91–95.

Hunter, M. D. 1992b. Interactions within herbivore communities mediated by the host plant: The keystone herbivore concept. In M. D. Hunter, T. Ohgushi, and P. W. Price (eds.). *Effects of resource distribution on animal–plant interactions.* Academic Press, San Diego, Calif., pp 287–325.

Hunter, M. D., and P. W. Price. 1992. Playing chutes and ladders: Heterogeneity and the relative roles of bottom-up and top-down forces in natural communities. *Ecology* **73**:724–732.

Hurd, L. E., M. V. Mellinger, L. L. Wolf, and S. J. McNaughton. 1971. Stability and diversity at three trophic levels in terrestrial successional ecosystems. *Science* **173**:1134–1136.

Hurd, L. E., M. V. Mellinger, L. L. Wolf, and S. J. McNaughton. 1972. *Science* **176**:545.

Hurley, P. M. 1968. The confirmation of continental drift. *Sci. Am.* **218**(4):52–64.

Hutchinson, G. E. 1953. The concept of pattern in ecology. *Proc. Acad. Nat. Sci. Phila.* **105**:1–12.

Hutchinson, G. E. 1957. Concluding remarks. *Cold Spring Harbor Symp. Quant. Biol.* **22**:415–427.

Hutchinson, G. E. 1965. *The ecological theater and the evolutionary play.* Yale Univ. Press, New Haven, Conn.

Hutchinson, G. E. 1971. Scale effects in ecology. In G. P. Patil, E. C. Pielou, and W. E. Waters (eds.). *Spatial patterns and statistical distributions.* Pennsylvania State Univ. Press, University Park, Pa., pp. xvii–xxvi.

Hutchinson, G. E., and E. S. Deevey. 1949. Ecological studies on populations. *Surv. Biol. Prog.* **1**:325–359.

Huxley, J. S. 1938. The present standing of the theory of sexual selection. In G. R. deBeer (ed.). *Evolution: Essays on aspects of evolutionary biology.* Clarendon Press, Oxford, pp. 11–42.

Huxley, C. R., and D. F. Cutler (eds.). 1991. *Ant–plant interactions.* Oxford Univ. Press, Oxford.

Illies, J. 1965. Phylogeny and zoogeography of the Plecoptera. *Annu. Rev. Entomol.* **10**:117–140.

Illies, J. 1969. Biogeography and ecology of neotropical freshwater insects, especially those from running waters. In E. J. Fittkau, J. Illies, H. Klinge, G. H. Schwabe, and H. Sioli (eds.). *Biogeography and ecology in South America.* Vol. 2. Monogr. Biol. 19, Dr. W. Junk, The Hague, The Netherlands, pp. 685–708.

Inoue, T., and M. Kato. 1992. Inter- and intraspecific morphological variation in bumblebee species, and competition in flower utilization. In M. D. Hunter, T. Ohgushi, and P. W. Price (eds.). *Effects of resource distribution on animal–plant interactions.* Academic Press, San Diego, Calif., pp. 393–427.

Istock, C. A. 1967. Transient competitive displacement in natural populations of whirligig beetles. *Ecology* **48**:929–937.

Istock, C. A. 1973. Population characteristics of a species ensemble of waterboatmen (Corixidae). *Behaviour* **46**:1–36.

Istock, C. A. 1981. Natural selection and life history variation: Theory plus lessons from a mosquito. In R. F. Denno and H. Dingle (eds.). *Insect life history patterns: Habitat and geographic variation.* Springer-Verlag, New York, pp. 113–127.

Itô, Y. 1959. *Hikaku Seitaigaku.* Iwanami, Tokyo.

Itô, Y. 1980. *Comparative ecology.* 2nd ed. Cambridge Univ. Press, Cambridge.

Itô, Y., and K. Miyashita. 1968. Biology of *Hyphantria cunea* Drury (Lepidoptera: Arctiidae) in Japan. V. Preliminary life tables and mortality data in urban areas. *Res. Popul. Ecol.* **10**:177–209.

Ives, W. G. H. 1976. The dynamics of larch sawfly (Hymenoptera: Tenthredinidae) populations in southwestern Manitoba. *Can. Entomol.* **108**: 701–730.

Iwasa, Y., M. Higashi, and N. Yamamura. 1981. Prey distribution as a factor determining the choice of optimal foraging strategy. *Am. Nat.* **117**: 710–723.

Jacobs, M. E. 1955. Studies on territorialism and sexual selection in dragonflies. *Ecology* **36**:566–586.

Jaksic, F. M. 1981. Abuse and misuse of the term "guild" in ecological studies. *Oikos* **37**:397–400.

Janetos, A. C. 1982. Foraging tactics of two guilds of web-spinning spiders. *Behav. Ecol. Sociobiol.* **10**:19–27.

Janzen, D. H. 1966. Coevolution of mutualism between ants and acacias in Central America. *Evolution* **20**:249–275.

Janzen, D. H. 1967a. Fire, vegetation structure, and the ant x acacia interaction in Central America. *Ecology* **48**:26–35.

Janzen, D. H. 1967b. Interaction of the bull's horn acacia (*Acacia cornigera* L.) with an ant inhabitant (*Pseudomyrmex ferruginea* F. Smith) in eastern Mexico. *Univ. Kans. Sci. Bull.* **47**:315–558.

Janzen, D. H. 1967c. Why mountain passes are higher in the tropics. *Am. Nat.* **101**:233–249.

Janzen, D. H. 1968. Host plants as islands in evolutionary and contemporary time. *Am. Nat.* **102**:592–595.

Janzen, D. H. 1969a. Seed-eaters versus seed size, number, toxicity and dispersal. *Evolution* **23**:1–27.

Janzen, D. H. 1969b. Allelopathy by myrmecophytes: The ant *Azteca* as an allelopathic agent of *Cecropia*. *Ecology* **50**:147–153.

Janzen, D. H. 1970. Herbivores and the number of tree species in tropical forests. *Am. Nat.* **104**:501–528.

Janzen, D. H. 1971. Euglossine bees as long-distance pollinators of tropical plants. *Science* **171**:203–205.

Janzen, D. H. 1973a. Host plants as islands, II. Competition in evolutionary and contemporary time. *Am. Nat.* **107**:786–790.

Janzen, D. H. 1973b. Comments on host-specificity of tropical herbivores and its relevance to species richness. In V. H. Heywood (ed.). *Taxonomy and ecology.* Academic Press, San Diego, Calif., pp. 201–211.

Janzen, D. H. 1973c. Evolution of polygynous obligate acacia-ants in western Mexico. *J. Anim. Ecol.* **42**:727–750.

Janzen, D. H. 1975. Interactions of seeds and their insect predators/parasitoids in a tropical deciduous forest. In P. W. Price (ed.). *Evolutionary strategies of parasitic insects and mites,* Plenum Press, New York, pp. 154–186.

Janzen, D. H. 1979. How to be a fig. *Annu. Rev. Ecol. Syst.* **10**:13–51.

Janzen, D. H. 1980. When is it coevolution? *Evolution* **34**:611–612.

Janzen, D. H. 1981. The peak in North American ichneumonid species richness lies between 38° and 42°N. *Ecology* **62**:532–537.

Janzen, D. H. 1986a. *Guanacaste National Park: Tropical ecological and cultural restoration.* Editorial Univ. Estatal a Distancia, San José, Costa Rica.

Janzen, D. H. 1986b. Mice, big mammals, and seeds: It matters who defecates what where. In A. Estrada and T. H. Fleming (eds.). *Frugivores and seed dispersal.* Dr. W. Junk, Dordrecht, The Netherlands, pp. 251–271.

Janzen, D. H. 1988. Tropical ecological and biocultural restoration. *Science* **239**:243–244.

Janzen, D. H., and P. S. Martin. 1982. Neotropical anachronisms: The fruits the gomphotheres ate. *Science* **215**:19–27.

Janzen, D. H., and C. M. Pond. 1975. A comparison, by sweep sampling, of the arthropod fauna of secondary vegetation in Michigan, England and Costa Rica. *Trans. R. Entomol. Soc. London* **127**:33–50.

Jeffords, M. R., J. G. Sternburg, and G. P. Waldbauer. 1979. Batesian mimicry: Field demonstration of the survival value of pipevine swallowtail and monarch color patterns. *Evolution* **33**:275–286.

Jeffords, M. R., G. P. Waldbauer, and J. G. Sternburg. 1980. Determination of the time of day at which diurnal moths painted to resemble butterflies are attacked by birds. *Evolution* **34**:1205–1211.

Jennings, D. H., and D. L. Lee (eds.). 1975. *Symbiosis*. Symp. Soc. Exp. Biol. 29. Cambridge Univ. Press, London.

Joern, A., and S. B. Gaines. 1990. Population dynamics and regulation in grasshoppers. In R. F. Chapman and A. Joern (eds.). *Biology of grasshoppers*. Wiley, New York, pp 415–482.

Johnson, C. D. 1981. Interaction between bruchid (Coleoptera) feeding guilds and behavioral patterns of pods of the Leguminosae. *Environ. Entomol.* **10**:249–253.

Johnson, C. D., and C. N. Slobodchikoff. 1979. Coevolution of *Cassia* (Leguminosae) and its seed beetle predators (Bruchidae). *Environ. Entomol.* **8**:1059–1064.

Johnson, C. G. 1966. A functional system of adaptive dispersal by flight. *Annu. Rev. Entomol.* **11**:233–260.

Johnson, C. G. 1969. *Migration and dispersal of insects by flight*. Methuen, London.

Johnson, C. G. 1971. Entomology department. In *Rothamsted Exp. Stn. Rep. 1970*, Part 1, pp. 183–200, 330–337.

Johnson, G. B. 1971. Metabolic implications of polymorphism as an adaptive strategy. *Nature* **232**:347–349.

Johnson, G. B. 1973. Enzyme polymorphism and biosystematics: The hypothesis of selective neutrality. *Annu. Rev. Ecol. Syst.* **4**:93–116.

Johnson, G. B. 1974. Enzyme polymorphism and metabolism. *Science* **184**:28–37.

Johnson, G. B. 1976. Polymorphism and predictability at the α-glycerophosphate dehydrogenase locus in *Colias* butterflies: Gradients in allele frequency with single populations. *Biochem. Genet.* **14**:403–426.

Johnson, W. T., and H. H. Lyon. 1976. *Insects that feed on trees and shrubs: An illustrated practical guide*. Cornell Univ. Press, Ithaca, N.Y.

Johnston, D. W., and E. P. Odum. 1956. Breeding bird populations in relation to plant succession on the Piedmont of Georgia. *Ecology* **37**:50–62.

Jones, C. E., and S. L. Buchman. 1974. Ultraviolet floral patterns as functional orientation cues in hymenopterous pollination systems. *Anim. Behav.* **22**:481–485.

Jones, C. E., and R. J. Little (eds.). 1983. *Handbook of experimental pollination biology*. Van Nostrand Reinhold, New York.

Jones, C. G., and R. D. Firn. 1978. The role of phytoecdysteroids in bracken fern, *Pteridium aquilinum* (L.) Kuhn as a defense against phytophagous insect attack. *J. Chem. Ecol.* **4**:117–138.

Jones, C. G., and J. H. Lawton. 1991. Plant chemistry and insect species richness of British umbellifers. *J. Anim. Ecol.* **60**:767–777.

Jones, C. G., and J. H. Lawton (eds.). 1995. *Linking species and ecosystems.* Chapman & Hall, New York.

Jones, C. G., J. R. Aldrich, and M. S. Blum. 1981a. 2-Furaldehyde from baldcypress: A chemical rationale for the demise of the Georgia silkworm industry. *J. Chem. Ecol.* **7**:89–101.

Jones, C. G., J. R. Aldrich, and M. S. Blum. 1981b. Baldcypress allelochemics and the inhibition of silkworm enteric microorganisms: Some ecological considerations. *J. Chem. Ecol.* **7**:103–114.

Jones, C. G., J. H. Lawton, and M. Shachak. 1994. Organisms as ecosystem engineers. *Oikos* **69**:373–386.

Jones, D. A. 1962. Selective eating of the acyanogenic form of the plant *Lotus corniculatus* L. by various animals. *Nature* **193**:1109–1110.

Jones, D. A. 1966. On the polymorphism of cyanogenesis in *Lotus corniculatus*. I. Selection by animals. *Can. J. Genet. Cytol.***8**:556–567.

Jones, D. A. 1971. Chemical defense mechanisms and genetic polymorphism. *Science* **173**:945.

Jones, D. A. 1972. Cyanogenic glycosides and their function. In J. B. Harborne (ed.). *Phytochemical ecology.* Academic Press, London, pp. 103–124.

Jones, D. A. 1979. Chemical defense: Primary or secondary function? *Am. Nat.* **113**:445–451.

Jones, D. A., J. Parsons, and M. Rothschild. 1962. Release of hydrocyanic acid from crushed tissues of all stages in the life-cycle of species of the Zygaeninae (Lepidoptera). *Nature* **193**:52–53.

Jones, D. A., R. J. Keymer, and W. M. Ellis. 1978. Cyanogenesis in plants and animal feeding. In J. B. Harborne (ed.). *Biochemical aspects of plant and animal coevolution.* Academic Press, London, pp. 21–34.

Kambysellis, M. P., and W. B. Heed. 1971. Studies of oogenesis in natural populations of Drosophilidae, I. Relation of ovarian development and ecological habitats of the Hawaiian species. *Am. Nat.* **105**:31–49.

Kamil, A. C., and T. D. Sargent (eds.). 1981. *Foraging behavior: Ecological, ethological and psychological approaches.* Garland Publishing, New York.

Kaneshiro, K. Y., H. L. Carson, F. E. Clayton, and W. B. Heed. 1973. Niche separation in a pair of homosequential *Drosophila* species from the island of Hawaii. *Am. Nat.* **107**:766–774.

Karban, R. 1986. Induced resistance against spider mites in cotton: Field verification. *Entomol. Exp. Appl.* **42**:239–242.

Karban, R., and J. R. Carey. 1984. Induced resistance of cotton seedlings to mites. *Science* **225**:53–54.

Karban, R., R. Adamchak, and W. C. Schnathorst. 1987. Induced resistance and interspecific competition between spider mites and a vascular plant wilt. *Science* **235**:678–680.

Kareiva, P. 1983. Influence of vegetation texture on herbivore populations: Resource concentration and herbivore movement. In R. F. Denno and M. S. McClure (eds.).*Variable plants and herbivores in natural and managed systems*. Academic Press, San Diego, Calif., pp. 259–289.

Kareiva, P. 1986. Trivial movement and foraging by crop colonizers. In M. Kogan (ed.). *Ecological theory and integrated pest management practice*. Wiley, New York, pp. 59–82.

Kareiva, P. 1989. Renewing the dialogue between theory and experiments in population ecology. In J. Roughgarden, R. M., May, and S. A. Levin (eds.). *Perspectives in ecological theory*. Princeton Univ. Press, Princeton, N.J., pp. 68–88.

Kaufmann, J. H. 1983. On the definitions and functions of dominance and territoriality. *Biol. Rev.* **58**:1–20.

Kay, Q. O. N. 1978. The role of preferential and assortative pollination in the maintenance of flower colour polymorphisms. In A. J. Richards (ed.). *The pollination of flowers by insects*. Academic Press, London, pp. 175–190.

Kearsley, M. C., and T. G. Whitham. 1989. Developmental changes in resistance to herbivory: Implications for individuals and populations. *Ecology* **70**:422–434.

Keast, A., R. L. Crocker, and C. S. Christian (eds.). 1959. Biogeography and ecology in Australia. *Monogr. Biol.* **8**:1–640.

Keeler, K. H. 1981. A model of selection for facultative nonsymbiotic mutualism. *Am. Nat.* **18**:488–498.

Keiper, R. R. 1969. Behavioral adaptations of cryptic moths. IV. Preliminary studies on species resembling dead leaves. *J. Lepid. Soc.* **23**:205–210.

Keiser, I., R. M. Kobayashi, D. H. Miyashita, E. J. Harris, E. L. Schneider, and D. L. Chambers. 1974. Suppression of Mediterranean fruit flies by Oriental fruit flies in mixed infestations in guava. *J. Econ. Entomol.* **67**:355–360.

Kellert, S. R. 1993. Values and perceptions of invertebrates. *Conserv. Biol.* **7**:845–855.

Kemp. G. A., and L. B. Keith. 1970. Dynamics and regulation of red squirrel (*Tamiasciurus hudsonicus*) populations. *Ecology* **51**:763–779.

Kendeigh, S. C. 1961. *Animal ecology*. Prentice Hall, Upper Saddle River, N.J.

Kendeigh, S. C. 1974. *Ecology with special reference to animals and man*. Prentice Hall, Upper Saddle River, N.J.

Kessel, E. L. 1955. The mating activities of balloon flies. *Syst. Zool.* **4**:97–104.

Kettlewell, H. B. D. 1955. Selection experiments on industrial melanism in the Lepidoptera. *Heredity* **9**:323–342.

Kettlewell, H. B. D. 1956. Further selection experiments on industrial melanism in the Lepidoptera. *Heredity* **10**:287–301.

Kettlewell, H. B. D. 1959. Darwin's missing evidence. *Sci. Am.* **200**(3):48–53.

Kettlewell, H. B. D. 1961. The phenomenon of industrial melanism in Lepidoptera. *Annu. Rev. Entomol.* **6**:245–262.

Kettlewell, H. B. D. 1973. *The evolution of melanism: The study of a recurring necessity with special reference to industrial melanism in the Lepidoptera.* Oxford Univ. Press, London.

Kevan, P. G. 1978. Floral coloration, its colorimetric analysis and significance in anthecology. In A. J. Richards (ed.). *The pollination of flowers by insects.* Academic Press, London, pp. 51–78.

Kevan, P. G., and H. G. Baker. 1983. Insects as flower visitors and pollinators. *Annu. Rev. Entomol.* **28**:407–453.

Kfir, R., and R. F. Luck. 1979. Effects of constant and variable temperature extremes on sex ratio and progeny production by *Aphytis melinus* and *A. lingnanensis* (Hymenoptera: Aphelinidae). *Ecol. Entomol* **4**: 335–344.

Kikkawa, J., and D. J. Anderson (eds.). 1986. *Community ecology: Pattern and process.* Blackwell Scientific, Melbourne, Australia.

Kimmins, J. P. 1971. Variations in the foliar amino acid composition of flowering and nonflowering balsam fir (*Abies balsamea* (L.) Mill.) and white spruce (*Picea glauca* (Moench) Voss) in relation to outbreaks of the spruce budworm (*Choristoneura fumiferana* (Clem.)). *Can. J. Zool.* **49**: 1005–1011.

Kimmins, J. P. 1972. Relative contributions of leaching, litter-fall and defoliation by *Neodiprion sertifer* (Hymenoptera) to the removal of cesium-134 from red pine. *Oikos* **23**:226–234.

Kimsey, L. S. 1980. The behaviour of male orchid bees (Apidae, Hymenoptera, Insecta) and the question of leks. *Anim. Behav.* **28**:996–1004.

Kimura, M. 1979. The neutral theory of molecular evolution. *Sci. Am.* **241**(5):98–126.

Kimura, M., and T. Ohta. 1971. *Theoretical aspects of population genetics.* Princeton Univ. Press, Princeton, N.J.

King, C. E., E. E. Gallaher, and D. A. Levin. 1975. Equilibrium diversity in plant–pollinator systems. *J. Theor. Biol.* **53**:263–275.

King, J. L., and T. H. Jukes. 1969. Non-Darwinian evolution. *Science* **164**:788–798.

King, P. E., R. R. Askew, and C. Sanger. 1969. The detection of parasitised hosts by males of *Nasonia vitripennis* (Walker) (Hymenoptera: Pteromalidae) and some possible implications. *Proc. R. Entomol. Soc. London A* **44**:85–90.

Kingsolver, J. G., and M. A. R. Koehl. 1985. Aerodynamics, thermoregulation, and the evolution of insect wings: Differential scaling and evolutionary change. *Evolution* **39**:488–504.

Kiritani, K., S. Kawahara, T. Sasaba, and F. Nakasuki. 1972. Quantitative evaluation of predation by spiders on the green rice leafhopper, *Nephotettix cincticeps* Uhler, by a sight-count method. *Res. Popul. Ecol.* **13:** 187–200.

Kirk, V. M. 1972. Seed-caching by larvae of two ground beetles, *Harpalus pennsylvanicus* and *H. erraticus. Ann. Entomol. Soc. Am.* **65:**1426–1428.

Kirk, V. M. 1973. Biology of a ground beetle, *Harpalus pennsylvanicus. Ann. Entomol. Soc. Am.* **66:**513–518.

Kirkpatrick, T. W. 1957. *Insect life in the tropics.* Longmans, Green, London.

Kitchell, J. F., R. V. O'Neill, D. Webb, G. W. Gallepp, S. M. Bartell, J. F. Koonce, and B. S. Ausmus. 1979. Consumer regulation of nutrient cycling. *BioScience* **29:**28–34.

Kloet, G. S., and W. D. Hincks. 1945. *A check list of British insects.* Kloet and Hincks, Stockport, Cheshire, England.

Kloet, G. S., and W. D. Hincks. 1964–1978. *A check list of British insects.* 2nd ed. rev. Handbooks for the identification of British insects. Vol. 11, Parts 1–5. Royal Entomological Society of London, London.

Klomp, H. 1966. The dynamics of a field population of the pine looper, *Bupalus piniarius* L. *Adv. Ecol. Res.* **3:**207–305.

Klopfer, P. H. 1959. Environmental determinants of faunal diversity. *Am. Nat.* **93:**337–342.

Klopfer, P. H., and R. H. MacArthur. 1960. Niche size and faunal diversity. *Am. Nat.* **94:**293–300.

Knipling, E. F. 1979. *The basic principles of insect population suppression and management.* U.S.D.A. Agric. Handb. 512. U.S. Department of Agriculture, Washington, D.C.

Kogan, M. 1977. The role of chemical factors in insect/plant relationships. *Proc. 15th Int. Contr. Entomol., Washington, D.C.,* pp. 211–227.

Kogan, M., and J. Paxton. 1983. Natural inducers of plant resistance to insects. In P. A. Hedin (ed.). *Plant resistance to insects.* American Chemical Society, Washington, D.C., pp. 153–171.

Kojima, K., J. H. Gillespie, and Y. N. Tobari. 1970. A profile of *Drosophila* species' enzymes assayed by electrophoresis. I. Number of alleles, heterozygosities, and linkage disequilibrium in glucose-metabolizing systems and some other enzymes. *Biochem. Genet.* **4:**627–637.

Kok, L. T. 1979. Lipids of ambrosia fungi and the life of mutualistic beetles. In L. R. Batra (ed.). *Insect-fungus symbiosis: Nutrition, mutualism, and commensalism.* Allanheld, Osmun, Montclair, N.J., pp. 33–52.

Kokubo, A. 1965. Population fluctuations and natural mortalities of the pine-moth, *Dendrolimus spectabilis. Res. Popul. Ecol.* 7:23–34.

Kozlovsky, D. G. 1968. A critical evaluation of the trophic level concept. I. Ecological efficiencies. *Ecology* 49:48–60.

Krebs, C. J. 1964. *The lemming cycle in Baker Lake, Northwest Territories, during 1959–1962.* Arct. Inst. N. Am. Tech. Paper 15. Arctic Institute of North America, Calgary, Alberta, Canada.

Krebs, C. J. 1970. *Microtus* population biology: Behavioral changes associated with the population cycle in *M. ochrogaster* and *M. pennsylvanicus*. *Ecology* 51:34–52.

Krebs, J. R. 1978. Optimal foraging: Decision rules for predators. In J. R. Krebs and N. B. Davies (eds.). *Behavioural ecology: An evolutionary approach*. Sinauer, Sunderland, Mass., pp. 23–63.

Krebs, J. R., and N. B. Davies (eds.). 1978. *Behavioural ecology: An evolutionary approach*. Sinauer, Sunderland, Mass.

Krebs, J. R., and N. B. Davies. 1981. *An introduction to behavioral ecology*. Sinauer, Sunderland, Mass.

Krebs, J. R., and N. B. Davies (eds.). 1991. *Behavioural ecology: An evolutionary approach*. 3rd ed. Blackwell Scientific, Oxford.

Krebs, J. R., J. C. Ryan, and E. L. Charnov. 1974. Hunting by expectation or optimal foraging? A study of patch use by chickadees. *Anim. Behav.* 22:953–964.

Krefting, L. W., and E. I. Roe. 1949. The role of some birds and mammals in seed germination. *Ecol. Monogr.* 19:269–286.

Krieger, R. I., P. P. Feeny, and C. F. Wilkinson. 1971. Detoxication enzymes in the guts of caterpillars: An evolutionary answer to plant defenses? *Science* 172:579–581.

Kritsky, G. 1991a. Beetle gods of ancient Egypt. *Am. Entomol.* 37:85–89.

Kritsky, G. 1991b. Darwin's Madagascan hawk moth prediction. *Am. Entomol.* 37:206–210.

Kukalová-Peck, J. 1978. Origin and evolution of insect wings and their relation to metamorphosis, as documented by the fossil record. *J. Morphol.* 156:53–126.

Kukalová-Peck, J. 1991. Fossil history and the evolution of the hexapod structures. In I. D. Naumann (ed.). *The insects of Australia*. Cornell Univ. Press, Ithaca, N.Y., pp. 141–179.

Kulesza, G. 1975. Comment on "Niche, habitat, and ecotope." *Am. Nat.* 109:476–479.

Kullenberg, B. 1950. Investigations on the pollination of *Ophrys* species. *Oikos* 2:1–19.

Kullenberg, B. 1956a. On the scents and colours of *Ophrys* flowers and their specific pollinators among the aculeate Hymenoptera. *Sven. Bot. Tidskr.* 50:25–46.

Kullenberg, B. 1956b. Field experiments with chemical sexual attractants on aculeate Hymenoptera males. I. *Zool. Bidr. Upps.* 31:253–354.

Kullenberg, B. 1961. Studies in *Ophrys* pollination. *Zool. Bidr. Upps.* 34:1–340.

Kunin, W. E., and K. J. Gaston. 1993. The biology of rarity: Patterns, causes, and consequences. *Trends Ecol. Evol.* 8:298–301.

Kuschel, G. 1962. The Curculionidae of Gough Island and the relationships of the weevil fauna of the Tristan de Cunha group. *Proc. Linn. Soc. London* 173:69–78.

Kuschel, G. 1969. Biogeography and ecology of South American Coleoptera. In E. J. Fittkau, J. Illies, H. Klinge, G. H. Schwabe, and H. Sioli (eds.). *Biogeography and ecology in South America.* Vol. 2. Monogr. Biol. 19, Dr. W. Junk, The Hague, The Netherlands, pp. 709–722.

Labandeira, C. C., and T. L. Phillips. 1996a. Insect fluid-feeding on Upper Pennsylvanian tree forms (Palaeodictyoptera, Marattiales) and the early history of the piercing-and-sucking functional feeding group. *Ann. Entomol. Soc. Am.* **89**:157–183.

Labandeira, C. C., and T. L. Phillips. 1996b. A carboniferous insect gall: Insight into early ecologic history of the Holometabola. *Proc. Natl. Acad. Sci. USA* **93**:8470–8474.

Labine, P. A. 1968. The population biology of the butterfly, *Euphydryas editha.* VIII. Oviposition and its relation to patterns of oviposition in other butterflies. *Evolution* **22**:799–805.

Lack, D. 1947a. The signficance of clutch size. Parts I and II. *Ibis* **89**:302–352.

Lack, D. 1947b. *Darwin's finches.* Cambridge Univ. Press, Cambridge.

Lack, D. 1949. Comments on Mr. Skutch's paper on clutch size. *Ibis* **91**:455–458.

Lack, D. 1954. *The natural regulation of animal numbers.* Clarendon Press, Oxford.

Lack, D. 1966. *Population studies of birds.* Oxford Univ. Press, London.

Lack, D. 1968. *Ecological adaptations for breeding in birds.* Methuen, London.

Laine, K., and H. Henttonen. 1983. The role of plant production in microtine cycles in northern Fennoscandia. *Oikos* **40**:407–418.

Laine, K. J., and P. Niemelä. 1980. The influence of ants on the survival of mountain birches during an *Oporinia autumnata* (Lep. Geometridae) outbreak. *Oecologia* **47**:39–42.

Landahl, J. T., and R. B. Root. 1969. Differences in the life tables of tropical and temperate milkweed bugs, genus *Oncopeltus* (Hemiptera: Lygaeidae). *Ecology* **50**:734–737.

Larsson, S. 1989. Stressful times for the plant stress–insect performance hypothesis. *Oikos* **56**:277–283.

Larsson, S., A. Wiren, L. Lundgren, and T. Ericsson. 1986. Effects of light and nutrient stress on leaf phenolic chemistry in *Salix dasyclados* and susceptibility to *Galerucella lineola* (Coleoptera). *Oikos* **47**:205–210.

Lauenroth, W. K., G. V. Skogerboe, and M. Flug (eds.). 1983. *Analysis of ecological systems: State-of-the-art in ecological modeling.* Developments in environmental modeling 5. Elsevier, New York.

Law, J. M., and F. E. Regnier. 1971. Pheromones. *Annu. Rev. Biochem.* **40**:533–548.

Lawton, J. H. 1978. Host-plant influences on insect diversity: The effects of space and time. In L. A. Mound and N. Waloff (eds.). *Diversity of insect faunas.* Symp. R. Entomol. Soc. London 9, Blackwell Scientific Publications, Oxford, pp. 105–125.

Lawton, J. H. 1982. Vacant niches and unsaturated communities: A comparison of bracken herbivores at sites on two continents. *J. Anim. Ecol.* **51**:573–595.

Lawton, J. H. 1983. Plant architecture and the diversity of phytophagous insects. *Annu. Rev. Entomol.* 23–29.

Lawton, J. H. 1984a. Herbivore community organization: General models and specific tests with phytophagous insects. In P. W. Price, C. N. Slobodchikoff, and W. S. Gaud (eds.). *A new ecology: Novel approaches to interactive systems.* Wiley, New York, pp. 329–352.

Lawton, J. H. 1984b. Non-competitive populations, non-convergent communities, and vacant niches: The herbivores of bracken. In D. R. Strong, D. Simberloff, L. G. Abele, and A. B. Thistle (eds.). *Ecological communities: Conceptual issues and the evidence.* Princeton Univ. Press, Princeton, N.J., pp. 67–100.

Lawton, J. H. 1990. Species richness and population dynamics of animal assemblages. Patterns in body size:abundance space. *Philos. Trans. R. Soc. London Ser. B* **330**:283–291.

Lawton, J. H. 1991. Species richness, population abundances, and body sizes in insect communities: Tropical versus temperate comparisons. In P. W. Price, T. M. Lewinsohn, G. W. Fernandes, and W. W. Benson (eds.). *Plant–animal interactions: Evolutionary ecology in tropical and temperate regions.* Wiley, New York, pp. 71–89.

Lawton, J. H., J. R. Beddington, and R. Bonser. 1974. Switching in invertebrate predators. In M. B. Usher and M. H. Williamson (eds.). *Ecological stability.* Chapman & Hall, London, pp. 141–158.

Lawton, J. H., and M. P. Hassell. 1981. Asymmetrical competition in insects. *Nature* **289**:793–795.

Lawton, J. H., and M. P. Hassell. 1984. Interspecific competition in insects. In C. B. Huffaker and R. L. Rabb (eds.). *Ecological entomology.* Wiley, New York.

Lawton, J. H., and M. MacGarvin. 1986. The organization of herbivore communities. In J. Kikkawa and D. J. Anderson (eds.). *Community ecology: Pattern and process.* Blackwell Scientific, Melbourne, Australia, pp. 163–186.

Lawton, J. H., T. M. Lewinsohn, and S. G. Compton. 1993. Patterns of diversity for the insect herbivores on bracken. In R. E. Ricklefs and D. Schluter (eds.). *Species diversity in ecological communities: Historical and geographical perspectives.* Univ. Chicago Press, Chicago, pp. 178–184.

Lawton, J. H., and S. McNeill. 1979. Between the devil and the deep blue sea: On the problem of being a herbivore. In R. M. Anderson, B. D. Turner, and L. R. Taylor (eds.). *Population dynamics.* Symp. Br. Ecol. Soc. 20, Blackwell Scientific Publications, Oxford, pp. 223–244.

Lawton, J. H., and P. W. Price. 1979. Species richness of parasites on hosts: Agromyzid flies on the British Umbelliferae. *J. Anim. Ecol.* **48**:619–637.

Lawton, J. H., and D. Schröder. 1977. Effects of plant type, size of geographical range and taxonomic isolation on number of insect species associated with British plants. *Nature* **265**:137–140.

Lawton, J. H., and D. R. Strong. 1981. Community patterns and competition in folivorous insects. *Am. Nat.* **118**:317–338.

Leather, S. R. 1986. Insect species richness of the British Rosaceae: The importance of host range, plant architecture, age of establishment, taxonomic isolation and species–area relationships. *J. Anim. Ecol.* **55**:841–860.

Leather, S. R. 1988. Size, reproductive potential and fecundity in insects: Things aren't as simple as they seem. *Oikos* **51**:386–389.

Leather, S. R. 1995. Factors affecting fecundity, fertility, oviposition, and larviposition in insects. In S. R. Leather and J. Hardie (eds.). *Insect reproduction.* CRC Press, Boca Raton, Fla., pp. 143–174.

Leather, S. R., and J. Hardie (eds.). 1995. *Insect reproduction.* CRC Press, Boca Raton, Fla.

LeCam, L. M., J. Neyman, and E. L. Scott (eds.). 1972. *Darwinian, neo-Darwinian and non-Darwinian evolution.* Proc. 6th Berkeley Symp. Math. Stat. Prob. Vol. 5. Univ. California Press, Berkeley, Calif.

Lee, J. J., and D. L. Inman. 1975. The ecological role of consumers: An aggregated systems view. *Ecology* **56**:1455–1458.

Lees, D. R., and E. R. Creed. 1975. Industrial melanism in *Biston betularia:* The role of selective predation. *J. Anim. Ecol.* **44**:67–83.

Lees, D. R., E. R. Creed, and J. G. Duckett. 1973. Atmospheric pollution and industrial melanism. *Heredity* **30**:227–232.

Legg, D. E., T. C. Schenk, and H. C. Chiang. 1986. European corn borer (Lepidoptera: Pyralidae) oviposition preference and survival on sunflower and corn. *Environ. Entomol.* **15**:631–634.

Leonard, D. E. 1970a. Intrinsic factors causing qualitative changes in populations of *Porthetria dispar* (Lepidoptera: Lymantriidae). *Can. Entomol.* **102**:239–249.

Leonard, D. E. 1970b. Intrinsic factors causing qualitative changes in populations of the gypsy moth. *Proc. Entomol. Soc. Ontario* **100**:195–199.

Leonard, J. E. 1950. *Flies, their origin, natural history, tying, hooks, patterns and selections of dry and wet flies, nymphs, streamers, salmon flies for fresh and salt water in North America and the British Isles, including a dictionary of 2200 patterns.* Barnes, New York.

Leppik, E. E. 1960. Early evolution of flower types. *Lloydia* **23**:72–92.

Leppik, E. E. 1972. Origin and evolution of bilateral symmetry in flowers. In T. Dobzhansky, M. K. Hecht, and W. C. Steere (eds.). *Evolutionary biology.* Vol. 5. Appleton-Century-Crofts, New York, pp. 49–85.

LeRoux, E. J., R. O. Paradis, and M. Hudon. 1963. Major mortality factors in the population dynamics of the eye-spotted bud moth, the pistol casebearer, the fruit-tree leaf roller, and the European corn borer in Quebec *Mem. Entomol. Soc. Can.* **32**:67–82.

LeRoux, E. J., et al. 1963. Population dynamics of agricultural and forest insect pests. *Mem. Entomol. Soc. Can.* **32**:1–103.

Leslie, P. H. 1959. The properties of a certain lag type of population growth and the influence of an external random factor on a number of such populations. *Physiol. Zool.* **32**:151–159.

Leslie, P. H., and J. C. Gower. 1960. The properties of a stochastic model for the predator–prey type of interaction between two species. *Biometrika* **47**:219–234.

Letourneau, D. K. 1986. Associational resistance in squash monocultures and polycultures in tropical Mexico. *Environ. Entomol.* **15**:285–292.

Letourneau, D. K. 1987. The enemies hypothesis: Tritrophic interactions and vegetational diversity in tropical agroecosystems. *Ecology* **68**: 1616–1622.

Levin, D. A., and H. Kerster. 1969a. Density-dependent gene dispersal in *Liatris. Am. Nat.* **103**:61–74.

Levin, D. A., and H. W. Kerster. 1969b. The dependence of bee-mediated pollen and gene dispersal upon plant density. *Evolution* **23**:560–571.

Levin, D. A., and A. C. Wilson. 1976. Rates of evolution in seed plants: Net increase in diversity of chromosome numbers and species numbers through time. *Proc. Natl. Acad. Sci. USA* **73**:2086–2090.

Levin, S. A. (ed.). 1975. *Ecosystem analysis and prediction.* Society for Industrial and Applied Mathematics, Philadelphia, Pa.

Levins, R. 1968. *Evolution in changing environments.* Princeton Univ. Press, Princeton, N.J.

Levins, R., and H. Heatwole. 1973. Biogeography of the Puerto Rican Bank: Introduction of species onto Palominitos Island. *Ecology* **54**:1056–1064.

Lewis, T. 1973. *Thrips: Their biology, ecology and economic importance.* Academic Press, London.

Lewis, W. H. P. 1971. Polymorphism of human enzyme proteins. *Nature* **230**:215–218.

Lewontin, R. C. 1965. Selection for colonizing ability. In H. G. Baker and G. L. Stebbins (eds.). *The genetics of colonizing species.* Academic Press, San Diego, Calif., pp. 77–91.

Lewontin, R. C. 1974. *The genetic basis of evolutionary change.* Columbia Univ. Press, New York.

Lewontin, R. C., and J. L. Hubby. 1966. A molecular approach to the study of genic heterozygosity in natural populations. II. Amount of variation and degree of heterozygosity in natural populations of *Drosophila pseudoobscura. Genetics* **54**:595–609.

Lidicker, W. Z. 1962. Emigration as a possible mechanism permitting the regulation of population density below carrying capacity. *Am. Nat.* **96**:29–33.

Liebherr, J. K. 1988. General patterns in West Indian insects, and graphical biogeographic analysis of some circum-Caribbean *Platynus* beetles (Carabidae). *Syst. Zool.* 37:385–409.

Liebherr, J. K. 1992. Phylogeny and revision of the *Platynus degallieri* species group (Coleoptera: Carabidae: Platynini). *Bull. Am. Mus. Nat. Hist.* 214:1–115.

Liebherr, J. K. 1994. Biogeographic patterns of montane Mexican and Central American Carabidae (Coleoptera). *Can. Entomol.* 126:841–901.

Liebig, J. 1840. *Chemistry in its application to agriculture and physiology.* Taylor and Walton, London.

Lieth, H. F. H. (ed.). 1978. *Patterns of primary production in the biosphere.* Dowden, Hutchinson & Ross, Stroudsburg, Pa.

Likens, G. E., and F. H. Bormann. 1974. Linkages between terrestrial and aquatic ecosystems. *BioScience* 24:447–456.

Lin, N. 1963. Territorial behaviour in the cicada killer wasp, *Sphecious spheciosus* (Drury) (Hymenoptera: Sphecidae). I. *Behaviour* 20:115–133.

Lindeman, R. L. 1942. The trophic–dynamic aspect of ecology. *Ecology* 23:399–418.

Lindroth, C. H. 1957. *The faunal connections between Europe and North America.* Wiley, New York.

Lindroth, C. H. 1963. The fauna history of Newfoundland, illustrated by carabid beetles. *Opusc. Entomol., Suppl.* 23:1–112.

Lindroth, C. H. 1971. On the occurrence of the continental element in the ground-beetle fauna of eastern Canada (Coleoptera: Carabidae). *Can. Entomol.* 103:1455–1462.

Linhart, Y. B. 1974. Intra-population differentiation in annual plants. I. *Veronica peregrina* L. raised under non-competitive conditions. *Evolution* 28:232–243.

Linsenmaier, W. 1972. *Insects of the world.* McGraw-Hill, New York.

Linsley, E. G., and J. W. MacSwain. 1947. Factors influencing the effectiveness of insect pollinators of alfalfa in California. *J. Econ. Entomol.* 40:349–357.

Linsley, E. G., J. W. MacSwain, and P. H. Raven. 1963a. Comparative behavior of bees and Onagraceae. I. *Oenothera* bees of the Colorado Desert. *Univ. Calif. Publ. Entomol.* 33:1–24.

Linsley, E. G., J. W. MacSwain, and P. H. Raven. 1963b. Comparative behavior of bees and Onagraceae. II. *Oenothera* bees of the Great Basin. *Univ. Calif. Publ. Entomol.* 33:25–58.

Linsley, E. G., J. W. MacSwain, and P. H. Raven. 1964. Comparative behavior of bees and Onagraceae. III. *Oenothera* bees of the Mohave Desert, California. *Univ. Calif. Publ. Entomol.* 33:59–98.

Llewellyn, M. 1972. The effects of the lime aphid, *Eucalypterus tiliae* L. (Aphididae) on the growth of the lime, *Tilia* x *vulgaris* Hayne. I. Energy requirements of the aphid population. *J. Appl. Ecol.* 9:261–282.

Lloyd, J. E. (ed.). 1980. Insect behavioral ecology. *Fla. Entomol.* 62:1–111.

Lloyd, J. E. (ed.). 1982. Insect behavioral ecology. *Fla. Entomol.* **65**:1–104.

Lloyd, M., and R. J. Ghelardi. 1964. A table for calculating the "equitability" component of species diversity. *J. Anim. Ecol.* **33**:217–225.

Lloyd, M., J. H. Zar, and J. R. Karr. 1968. On the calculation of information-theoretical measures of diversity. *Am. Midl. Nat.* **79**:257–272.

Loomis, W. D. 1967. Biosynthesis and metabolism of monoterpenes. In J. B. Pridham (ed.). *Terpenoids in plants.* Academic Press. London, pp. 59–82.

Lotka, A. J. 1924. *Elements of physical biology.* Williams & Wilkins, Baltimore. Reprinted as *Elements of mathematical biology.* 1956. Dover, New York.

Loucks, O. L. 1970. Evolution of diversity, efficiency, and community stability. *Am. Zool.* **10**:17–25.

Louda, S. M. 1982. Distribution ecology: Variation in plant recruitment over a gradient in relation to insect seed predation. *Ecol. Monogr.* **52**:25–41.

Louda, S. M. 1983. Seed predation and seedling mortality in the recruitment of a shrub, *Haplopappus venetus* (Asteraceae), along a climatic gradient. *Ecology* **64**:511–521.

Louda, S. M. 1989. Differential predation pressure: A general mechanism for structuring plant communities along complex environmental gradients? *Trends Ecol. Evol.* **4**:158–159.

Louda, S., and S. Mole. 1991. Glucosinolates: Chemistry and ecology. In G. A. Rosenthal and M. R. Berenbaum (eds.). *Herbivores: Their interactions with secondary plant metabolites.* Vol. 1. 2nd ed. Academic, San Diego, Calif., pp. 123–164.

Louda, S. M., P. M. Dixon, and N. J. Huntly. 1987a. Herbivory in sun and shade at a natural meadow–woodland ecotone in the Rocky Mountains. *Vegetatio* **72**:141–149.

Louda, S. M., M. A. Farris, and M. J. Blua. 1987b. Variation in methyl-glucosinolate and insect damage to *Cleome serrulata* (Capparaceae) along a natural soil moisture gradient. *J. Chem. Ecol.* **13**:569–581.

Loye, J. E., and M. Zuk (eds.). 1991. *Bird–parasite interactions: Ecology, evolution, and behavior.* Oxford Univ. Press, Oxford.

Lu, P.-Y., R. L. Metcalf, and E. M. Carlson. 1978. Environmental fate of five radiolabeled coal conversion by-products evaluated in a laboratory model ecosystem. *Environ. Health Perspect.* **24**:201–208.

Lu, P.-Y., R. L. Metcalf, N. Plummer, and D. Mandel. 1977. The environmental fate of three carcinogens: Benzo-(α)-pyrene, benzidine, and vinyl chloride evaluated in laboratory model ecosystems. *Arch. Environ. Contam. Toxicol.* **6**:129–142.

Luck, R. F. 1971. An appraisal of two methods of analyzing insect life tables. *Can. Entomol.* **103**:1261–1271.

Luck, R. F., and H. Podoler. 1985. Competitive exclusion of *Aphytis lingnanensis* by *A. melinus:* Potential role of host size. *Ecology* **66**: 904–913.

Luck, R. F., H. Podoler, and R. Kfir. 1982. Host selection and egg allocation behaviour by *Aphytis melinus* and *A. lingnanensis:* Comparison of two facultatively gregarious parasitoids. *Ecol. Entomol.* 7:397–408.

Luckinbill, L. S. 1973. Coexistence in laboratory populations of *Paramecium aurelia* and its predator *Didinium nasatum. Ecology* 54:1320–1327.

Lugo, A. E., E. G. Farnworth, D. Pool. P. Jerez, and G. Kaufman. 1973. The impact of the leaf cutter ant *Atta colombica* on the energy flow of a tropical wet forest. *Ecology* 54:1292–1301.

Lumley, J. S. P., and W. Benjamin. 1994. *Research: Some ground rules.* Oxford Univ. Press. Oxford.

Lyell, C. 1830. *Principles of geology, being an attempt to explain the former changes to the earth's surface by reference to causes now in operation.* Vol. 1. Murray, London.

MacArthur, R. H. 1955. Fluctuations of animal populations, and a measure of community stability. *Ecology* 36:533–536.

MacArthur, R. H. 1958. Population ecology of some warblers of northeastern coniferous forests. *Ecology* 39:599–619.

MacArthur, R. H. 1962. Some generalized theorems of natural selection. *Proc. Natl. Acad. Sci. USA* 48:1893–1897.

MacArthur, R. H. 1972. *Geographical ecology: Patterns in the distribution of species.* Harper & Row, New York.

MacArthur, R. H., and J. W. MacArthur. 1961. On bird species diversity. *Ecology* 42:594–598.

MacArthur, R. H., and E. R. Pianka. 1966. An optimal use of a patchy environment. *Am. Nat.* 100:603–609.

MacArthur, R. H., and E. O. Wilson. 1963. An equilibrium theory of insular zoogeography. *Evolution* 17:373–387.

MacArthur, R. H., and E. O. Wilson. 1967. *The theory of island biogeography.* Princeton Univ. Press, Princeton, N.J.

MacArthur, W. P. 1927. Old time typhus in Britain. *Trans. R. Soc. Trop. Med. Hyg.* 20:487.

Macauley, B. J., and L. R. Fox. 1980. Variation in total phenols and condensed tannins in *Eucalyptus:* Leaf phenology and insect grazing. *Aust. J. Ecol.* 5:31–35.

Macfadyen, A. 1957. *Animal ecology: Aims and methods.* Pitman, London.

MacGarvin, M. 1982. Species–area relationships of insects on host plants: Herbivores on rosebay willowherb. *J. Anim. Ecol.* 51:207–223.

MacKay, R. J., and J. Kalff. 1973. Ecology of two related species of caddis fly larvae in the organic substrates of a woodland stream. *Ecology* 54:499–511.

Mackerras, I. M. 1970. Composition and distribution of the fauna. In *The insects of Australia.* Melbourne Univ. Press, Melbourne, Australia, pp. 187–203.

MacLeod, J. 1937. The species of Diptera concerned in cutaneous myiasis of sheep in Britain. *Proc. R. Entomol. Soc. London A* **12**:127–133.

Madden, J. L., and M. P. Coutts. 1979. The role of fungi in the biology and ecology of woodwasps (Hymenoptera: Siricidae). In L. R. Batra (ed.). *Insect–fungus symbiosis: Nutrition, mutualism, and commensalism.* Allanheld, Osmun, Totowa, N.J., pp. 165–174.

Maelzer, D. A. 1970. The regression of log N_{n+1} on log N_n as a test of density dependence: An exercise with computer-constructed density-independent populations. *Ecology* **51**:810–822.

Maguire, B. 1971. Phytotelmata: Biota and community structure determination in plant-held waters. *Annu. Rev. Ecol. Syst.* **2**:439–466.

Maguire, L. A. 1983. Influence of collard patch size on population densities of lepidopteran pests (Lepidoptera: Pieridae, Plutellidae). *Environ. Entomol.* **12**:1415–1419.

Maier, C. T., and G. P. Waldbauer. 1979a. Dual mate-seeking strategies in male syrphid flies (Diptera: Syrphidae). *Ann. Entomol. Soc. Am.* **72**:54–61.

Maier, C. T., and G. P. Waldbauer. 1979b. Diurnal activity patterns of flower flies (Diptera: Syrphidae) in an Illinois sand area. *Ann. Entomol. Soc. Am.* **72**:237–245.

Malcolm, S. B. 1991. Cardenolide-mediated interactions between plants and herbivores. In G. A. Rosenthal and M. R. Berenbaum (eds.). *Herbivores: Their interactions with secondary plant metabolites.* Vol. 1. 2nd ed. Academic Press, San Diego, Calif., pp. 251–296.

Malotki, E. 1978. *Hopitutuwutsi—Hopi tales: A bilingual collection of Hopi Indian stories.* Museum of Northern Arizona, Flagstaff, Ariz.

Malthus, T. R. 1798. *An essay on the principle of population as it affects the future improvement of society.* Johnson, London.

Mani, M. S. 1968. *Ecology and biogeography of high altitude insects.* Ser. Entomol. Vol. 4. Dr. W. Junk, The Hague, The Netherlands.

Mani, M. S. (ed.) 1974. Ecology and biogeography in India. *Monogr. Biol.* **23**:1–773.

Manley, G. V. 1971. A seed-cacheing carabid (Coleoptera). *Ann. Entomol. Soc. Am.* **64**:1474–1475.

Maramorosch, K., and K. F. Harris. 1979. *Leafhopper vectors and plant disease agents.* Academic Press, San Diego, Calif.

Marden, J. H., and K. D. Waddington. 1981. Floral choices by honeybees in relation to the relative distances of flowers. *Physiol. Entomol.* **6**:431–435.

Margulis, L. 1981. *Symbiosis in cell evolution.* W.H. Freeman, San Francisco.

Margulis, L. 1993. *Symbiosis in cell evolution: Microbial communities in the Archean and Proterozoic eons.* W.H. Freeman, New York.

Margulis, L., and R. Fester (eds.). 1991. *Symbiosis as a source of evolutionary innovation: Speciation and morphogenesis.* MIT Press, Cambridge, Mass.

Marquis, R. J. 1991. Herbivore fauna of *Piper* (Piperaceae) in a Costa Rican wet forest: Diversity, specificity, and impact. In P. W. Price, T. M. Lewinsohn, G. W. Fernandes, and W. W. Benson (eds.). *Plant–animal interactions: Evolutionary ecology in tropical and temperate regions.* Wiley, New York, pp. 179–208.

Marquis, R. J., and H. E. Braker. 1994. Plant–herbivore interactions: Diversity, specificity, and impact. In L. A. McDade, K. S. Bawa, H. A. Hespenheide, and G. S. Hartshorn (eds.). *La Selva: Ecology and natural history of a neotropical rain forest.* Univ. Chicago Press, Chicago, pp. 261–281.

Marshall, A. G. 1977. Interrelationships between *Arixenia esau* (Dermaptera) and molossid bats and their ectoparasites in Malaysia. *Ecol. Entomol.* 2:285–291.

Marshall, A. G. 1981. *The ecology of ectoparasitic insects.* Academic Press, San Diego, Calif.

Marshall, D. R., and S. K. Jain. 1969. Interference in pure and mixed populations of *Avena fatua* and *A. barbata. J. Ecol.* 57:251–270.

Marshall. L. G., S. D. Webb, J. J. Sepkoski, and D. M. Raup. 1982. Mammalian evolution and the great American interchange. *Science* **215:** 1351–1357.

Martin, J. L. 1966. The insect ecology of red pine plantations in central Ontario. IV. The crown fauna. *Can. Entomol.* **98:**10–27.

Martins, R. P., T. M. Lewinsohn, and J. H. Lawton. 1995. First survey of insects feeding on *Pteridium aquilinum* in Brazil. *Rev. Bras. Entomol.* **39:**151–156.

Matsuda, R. 1979. Abnormal metamorphosis and arthropod evolution. In A. P. Gupta (ed.). *Arthropod phylogeny.* Van Nostrand Reinhold, New York, pp. 137–256.

Matthews, E. G. 1977. Signal-based frequency-dependent defense strategies and the evolution of mimicry. *Am. Nat.* **111:**213–222.

Matthews, J. V. 1976a. Insect fossils from the Beafort formation: Geological and biological significance. *Geol. Surv. Pap. Can.* **76-1B:**217–227.

Matthews, J. V. 1976b. Evolution of the subgenus *Cyphelophorus* (Genus *Helophorus,* Hydrophilidae, Coleoptera) description of two new fossil species and discussion of *Helophorus tuberculatus* Gyll. *Can. J. Zool.* **54:**652–673.

Matthews, R. W., and J. R. Matthews. 1978. *Insect behavior.* Wiley, New York.

Matthews, S. W. 1973. This changing earth. *Natl. Geog. Mag.* **143:**1–37.

Mattson, W. J. (ed.) 1977. *The role of arthropods in forest ecosystems.* Springer-Verlag, New York.

Mattson, W. J. 1980. Herbivory in relation to plant nitrogen content. *Annu. Rev. Ecol. Syst.* **11:**119–161.

Mattson, W. J., and N. D. Addy. 1975. Phytophagous insects as regulators of forest primary production. *Science* **190:**515–522.

Mattson, W. J., and R. A. Haack. 1987. The role of drought stress in provoking outbreaks of phytophagous insects. In P. Barbosa and J. C. Schultz (eds.). *Insect outbreaks*. Academic Press, San Diego, Calif., pp. 365–407.

Mattson, W. J., and J. M. Scriber. 1987. Nutritional ecology of insect folivores of woody plants: Nitrogen, water, fiber, and mineral considerations. In F. Slansky and J. G. Rodriguez (eds.). *Nutritional ecology of insects, mites, spiders, and related invertebrates*. Wiley, New York, pp. 105–146.

Mattson, W. J., P. Niemelä, I. Millers, and Y. Inguanzo. 1994. Immigrant phytophagous insects on woody plants in the United States and Canada: An annotated list. *USDA For. Serv. North Central For. Exp. Stn. Gen. Tech. Rep. NC-169*. U.S. Government Printing Office, Washington, D.C., pp. 1–27.

May, M. L. 1979. Insect thermoregulation. *Annu. Rev. Entomol.* **24:** 313–349.

May, R. M. 1973a. *Stability and complexity in model ecosystems*. Princeton Univ. Press, Princeton, N.J.

May, R. M. 1973b. Qualitative stability in model ecosystems. *Ecology* **54:**638–641.

May, R. M. 1975. Patterns of species abundance and diversity. In M. L. Cody and J. M. Diamond (eds.). *Ecology and evolution of communities*. Belknap Press, imprint of Harvard Univ. Press, Cambridge, Mass., pp. 81–120.

May, R. M. 1976. Models for two interacting populations. In R. M. May (ed.). *Theoretical ecology: Principles and applications*. W.B. Saunders, Philadelphia, pp. 49–70.

May, R. M. 1978a. Mathematical aspects of the dynamics of animal populations. In S. A. Levin (ed.). *Studies in mathematical biology. II. Populations and communities*. Mathematical Association of America, Washington, D.C., pp. 317–366.

May, R. M. 1978b. Host parasitoid systems in patchy environments: A phenomenological model. *J. Anim. Ecol.* **47:**833–844.

May, R. M. 1978c. The dynamics and diversity of insect faunas. *Symp. R. Entomol. Soc. London* **9:**188–204.

May, R. M. 1981. Models for two interacting populations. In R. M. May (ed.). *Theoretical ecology: Principles and applications*. 2nd ed. Sinauer, Sunderland, Mass., pp. 78–104.

May, R. M. 1986. How many species are there? *Nature* **324:**514–515.

May, R. M., M. P. Hassell, R. M. Anderson, and D. W. Tonkyn. 1981. Density dependence in host-parasitoid models. *J. Anim. Ecol.* **50:**855–865.

Mayhew, S. H., S. K. Kato, F. M. Ball, and C. Epling. 1966. Comparative studies of arrangements within and between populations of *Drosophila pseudoobscura*. *Evolution* **20:**646–662.

Maynard Smith, J. 1978. Optimization theory in evolution. *Annu. Rev. Ecol. Syst.* **9:**31–56.

Mayr, E. 1961. Cause and effect in biology. *Science* **134:**1501–1506.

Mayr, E. 1963. *Animal species and evolution*. Belknap Press, imprint of Harvard Univ. Press, Cambridge, Mass.

Mayr, E. 1982. *The growth of biological thought: Diversity, evolution, and inheritance.* Belknap Press, imprint of Harvard Univ. Press, Cambridge, Mass.

Mayr, E., and W. B. Provine (eds.). 1980. *The evolutionary synthesis: Perspectives on the unification of biology.* Harvard Univ. Press, Cambridge, Mass.

McClure, M. S. 1974. Biology of *Erythroneura lawsoni* (Homoptera: Cicadellidae) and coexistence in the sycamore leaf-feeding guild. *Environ. Entomol.* **3**:59–68.

McClure, M. S. 1975. Key to the eight species of coexisting *Erythroneura* leafhoppers (Homoptera: Cicadellidae) on American sycamore. *Ann. Entomol. Soc. Am.* **68**:1039–1043.

McClure, M. S. 1980. Foliar nitrogen: A basis for host suitability for elongate hemlock scale, *Fiorinia externa* (Homoptera: Diaspididae). *Ecology* **61**:72–79.

McClure, M. S., and P. W. Price. 1975. Competition among sympatric *Erythroneura* leafhoppers (Homoptera: Cicadellidae) on American sycamore. *Ecology* **56**:1388–1397.

McClure, M. S., and P. W. Price. 1976. Ecotope characteristics of coexisting *Erythroneura* leafhoppers (Homoptera: Cicadellidae) on sycamore. *Ecology* **57**:928–940.

McDade, L. A., K. S. Bawa, H. A. Hespenheide, and G. S. Hartshorn (eds.). 1994. *La Selva: Ecology and natural history of a neotropical rain forest.* Univ. Chicago Press, Chicago.

McDonald, J. F., and J. C. Avise. 1976. Evidence for the adaptive significance of enzyme activity levels: Interspecific variation in α-GPDP and ADH in *Drosophila. Biochem. Genet.* **14**:347–355.

McGuffin, W. C. 1977. Guide to the Geometridae of Canada (Lepidoptera) II. Subfamily Ennominae. 2. *Mem. Entomol. Soc. Can.* **101**:1–191.

McIntosh, R. P. 1985. *The background of ecology: Concept and theory.* Cambridge Univ. Press, Cambridge.

McIntosh, R. P. 1995. H. A. Gleason's "individualistic concept" and theory of animal communities: A continuing controversy. *Biol. Rev.* **70**:317–357.

McKey, D. 1974. Adaptive patterns in alkaloid physiology. *Am. Nat.* **108**:305–320.

McLeod, J. M. 1972. The Swaine jack pine sawfly, *Neodiprion swainei,* life system: Evaluating the long-term effects of insecticide applications in Quebec. *Environ. Entomol.* **1**:371–381.

McMahon, T. 1973. Size and shape in biology. *Science* **179**:1201–1204.

McNamee, R. J. 1987. The equilibrium structure and behavior of defoliating insect systems. Ph.D. thesis. Dept. Zoology, Univ. British Columbia, Vancouver, B.C., Canada.

McNaughton, S. J. 1976. Serengeti migratory wildebeest: Facilitation of energy flow by grazing. *Science* **191**:92–94.

McNaughton, S. J. 1979. Grazing as an optimization process: Grass–ungulate relationships in the Serengeti. *Am. Nat.* **113**:691–703.

McNaughton, S. J., M. Osterheld, D. A. Frank, and K. J. Williams. 1989. Ecosystem-level patterns of primary productivity and herbivory in terrestrial habitats. *Nature* **341**:142–144.

McNaughton, S. J., M. Osterheld, D. A. Frank, and K. J. Williams. 1991. Primary and secondary production in terrestrial ecosystems. In J. Cole, G. Lovett, and S. Findlay (eds.). *Comparative analyses of ecosystems.* Springer-Verlag, New York, pp. 120–139.

McNeill, S. 1971. The energetics of a population of *Leptopterna dolabrata* (Heteroptera: Miridae). *J. Anim. Ecol.* **40**:127–140.

McNeill, S. 1973. The dynamics of a population of *Leptopterna dolabrata* (Heteroptera: Miridae) in relation to its food resources. *J. Anim. Ecol.* **42**:495–507.

McNeill, S., and T. R. E. Sothwood. 1978. The role of nitrogen in the development of insect/plant relationships. In J. B. Harborne (ed.). *Biochemical aspects of plant and animal coevolution.* Academic Press, London, pp. 77–98.

Meeuse, A. D. J. 1973. Anthecology, floral morphology and angiosperm evolution. In V. H. Heywood (ed.). *Taxonomy and ecology.* Academic Press, San Diego, Calif., pp. 189–200.

Mégnin, P. 1894. La faune des cadavres. Encyclopédie scientifique des aidememoire. G. Masson, Gauthier-Villars, Paris.

Melber, A., and G. H. Schmidt. 1977. Sozial-phänomene bie Heteropteren. *Zoologica* **127**:19–53.

Menge, B. A. 1992. Community regulation: Under what conditions are bottom-up factors important on rocky shores? *Ecology* **73**:755–765.

Menge, B. A., and J. P. Sutherland. 1976. Species diversity gradients: Synthesis of the roles of predation, competition, and temporal heterogeneity. *Am. Nat.* **110**:351–369.

Mengel, R. M. 1964. The probable history of species formation in some northern wood warblers (Parulidae). *Living Bird* **3**:9–43.

Merriam, C. H. 1894. Laws of temperature control of the geographic distribution of terrestrial animals and plants. *Natl. Geog. Mag.* **6**:229–238.

Merriam, C. H. 1898. Life zones and crop zones. *U.S. Dept. Agric. Div. Biol. Surv. Bull.* **10**:9–79.

Mertz, D. B. 1970. Notes on methods used in life-history studies. In J. H. Connell, D. B. Mertz, and W. W. Murdoch (eds.). *Readings in ecology and ecological genetics.* Harper & Row, New York, pp. 4–17.

Mertz, D. B. 1972. The *Tribolium* model and the mathematics of population growth. *Annu. Rev. Ecol. Syst.* **3**:51–78.

Mertz, D. B. 1975. Senescent decline in flour beetle strains selected for early adult fitness. *Physiol. Zool.* **48**:1–23.

Messina, F. J. 1981. Plant protection as a consequence of an ant–membracid mutualism: Interactions on goldenrod (*Solidago* sp.). *Ecology* **62:** 1433–1440.

Metcalf, R. L. 1977. Model ecosystem approach to insecticide degradation: A critique. *Annu. Rev. Entomol.* **22:**241–261.

Metcalf, R. L. 1989. Insect resistance to insecticides. *Pesticide Sci.* **26:** 333–358.

Metcalf, R. L. 1994. Insecticides in pest management. In R. L. Metcalf and W. H. Luckmann (eds.). *Introduction to insect pest management.* 3rd ed. Wiley, New York, pp. 245–314.

Metcalf, R. L., and W. H. Luckmann. 1975. *Introduction to insect pest management.* Wiley, New York.

Metcalf, R. L., and W. H. Luckmann (eds.). 1982. *Introduction to insect pest management.* 2nd ed. Wiley, New York.

Metcalf, R. L., and W. H. Luckmann (eds.). 1994. *Introduction to insect pest management.* 3rd ed. Wiley, New York.

Metcalf, R. L., and E. R. Metcalf. 1992. *Plant kairomones in insect ecology and control.* Chapman & Hall, New York.

Metcalf, R. L., and R. A. Metcalf. 1993. *Destructive and useful insects: Their habits and control.* 5th ed. McGraw-Hill, New York.

Metcalfe, J. R. 1972. An analysis of the population dynamics of the Jamaican sugar-cane pest *Saccharosydne saccharivora* (Weston) (Hom., Delphacidae). *Bull. Entomol. Res.* **62:**73–85.

Mettler, L. E., and T. G. Gregg. 1969. *Population genetics and evolution.* Prentice Hall, Upper Saddle River, N.J.

Mettler, L. E., T. G. Gregg, and H. E. Schaffer. 1988. *Population genetics and evolution.* 2nd ed. Prentice Hall, Upper Saddle River, N.J.

Michener, C. D. 1947. A revision of the American species of *Hoplitis* (Hymenoptera, Megachilidae). *Bull. Am. Mus. Nat. Hist.* **89:** 257–318.

Michener, C. D. 1974. *The social behavior of the bees: A comparative study.* Belknap Press, imprint of Harvard Univ. Press, Cambridge, Mass.

Mikkola, K. 1984. On selective forces acting in the industrial melanism of *Biston* and *Oligia* moths (Lepidoptera: Geometridae and Noctuidae). *Biol. J. Linn. Soc.* **21:**409–421.

Miller, C. A. 1957. A technique for estimating the fecundity of natural populations of the spruce budworm. *Can. J. Zool.* **35:**1–13.

Miller, C. A. 1963. The spruce budworm. In R. F. Morris (ed.). The dynamics of epidemic spruce budworm populations. *Mem. Entomol. Soc. Can.* **31:**12–19.

Miller, C. A. 1966. The black-headed budworm in eastern Canada. *Can. Entomol.* **98:**592–613.

Miller, R. S. 1967. Pattern and process in competition. *Adv. Ecol. Res.* **4:**1–74.

Miller, W. E. 1967. The European pine shoot moth: Ecology and control in the lake states. *For. Sci. Monogr.* **14:**1–72.

Miller, W. E. 1996. Population behavior and adult feeding capacity in Lepidoptera. *Environ. Entomol.* **25**:213–226.

Miller, S., R. W. Pearcy, and E. Berger. 1975.Polymorphism at the α-glycerophosphate dehydrogenase locus in *Drosophila melanogaster.* I. Properties of adult allozymes. *Biochem. Genet.* **13**:175–188.

Milne, A. 1957a. The natural control of insect populations. *Can. Entomol.* **89**:193–213.

Milne, A. 1957b. Theories of natural control of insect populations. *Cold Spring Harbor Symp. Quant. Biol.* 22, 253–267.

Milne, A. 1962. On a theory of natural control of insect population. *J. Theor. Biol.* **3**:19–50.

Mitchell, R. 1968. Site selection by larval water mites parasitic on the damselfly *Cercion hieroglyphicum* Brauer. *Ecology* **49**:40–47.

Mitchell, R. 1970. An analysis of dispersal in mites. *Am. Nat.* **104**:425–431.

Mitchell, R. 1973. Growth and population dynamics of a spider mite (*Tetranychus urticae* K., Acarina: Tetranychidae). *Ecology* **54**:1349–1355.

Mitchell, R. 1977. Bruchid beetles and seed packaging by palo verde. *Ecology* **58**:644–651.

Mitter, C., B. Farrell, and D. J. Futuyma. 1991. Phyogenetic studies of insect–plant interactions: Insights into the genesis of diversity. *Trends Ecol. Evol.* **6**:290–293.

Mitter, C., B. Farrell, and B. Wiegmann. 1988. The phylogenetic study of adaptive zones: Has phytophagy promoted insect diversification? *Am. Nat.* **132**:107–128.

Mogford, D. J. 1978. Pollination and flower colour polymorphism, with special reference to *Cirsium palustre.* In A. J. Richards (ed.). *The pollination of flowers by insects.* Academic Press, London, pp. 191–199.

Monroe, D. D. 1974. The systematics, phylogeny, and zoogeography of *Symmerus* Walker and *Australosymmerus* Freeman (Diptera: Mycetophilidae: Ditomyiinae). *Mem. Entomol. Soc. Can.* **92**:1–183.

Mook, L. J. 1963. Birds and the spruce budworm. In R. F. Morris (ed.). *The dynamics of epidemic spruce budworm populations.* Mem. Entomol. Soc. Can. 31, Entomological Society of Canada, Ottawa, Ontario, pp. 268–271.

Moore, J. A. 1993. *Science as a way of knowing: The foundations of modern biology.* Harvard Univ. Press, Cambridge.

Moore, N. W. 1964. Intra- and interspecific competition among dragonflies (Odonata). *J. Anim. Ecol.* **33**:49–71.

Mopper, S., and T. G. Whitham. 1986. Natural bonsai at Sunset Crater. *Nat. Hist.* **95**(12):42–47.

Mopper, S., and T. G. Whitham. 1992. The plant stress paradox: Effects on pinyon sawfly sex ratios and fecundity. *Ecology* **73**:515–525.

Moran, P. A. P. 1954. The logic of the mathematical theory of animal populations. *J. Wildl. Manage.* **18**:60–66.

Moreton, B. D. 1969. *Beneficial insects and mites.* 6th ed. Ministry Agric. Fish. Food Bull. 20 H.M.S.O., London.

Morris, G. K, 1971. Aggression in male conocephaline grasshoppers (Tettigoniidae). *Anim. Behav.* **19**:132–137.

Morris, G. K. 1972. Phonotaxis of male meadow grasshoppers (Orthoptera: Tettigoniidae). *J. N.Y. Entomol. Soc.* **80**:5–6.

Morris, R. F. 1955. The development of sampling techniques for forest insect defoliators, with particular reference to the spruce budworm. *Can. J. Zool.* **33**:225–294.

Morris, R. F. 1957. The interpretation of mortality data in studies on population dynamics. *Can. Entomol.* **89**:49–69.

Morris, R. F. 1959. Single-factor analysis in population dynamics. *Ecology* **40**:580–588.

Morris, R. F. 1960. Sampling insect populations. *Annu. Rev. Entomol.* **5**:243–264.

Morris, R. F. 1963a. Predictive population equations based on key factors. *Mem. Entomol. Soc. Can.* **32**:16–21.

Morris, R. F. (ed.). 1963b. The dynamics of epidemic spruce budworm populations. *Mem. Entomol. Soc. Can.* **31**:1–332.

Morris, R. F. 1964. The value of historical data in population research, with particular reference to *Hyphantria cunea* Drury. *Can. Entomol.* **96**:356–368.

Morris, R. F. 1967. Influence of parental food quality on the survival of *Hyphantria cunea*. *Can. Entomol.* **99**:24–33.

Morris, R. F. 1969. Approaches to the study of population dynamics. In W. E. Waters (ed.). *Forest insect population dynamics.* U.S.D.A. For. Serv. Res. Paper NE-125, U.S. Government Printing Office, Washington, D.C., pp. 9–28.

Morris, R. F. 1971a. Observed and simulated changes in genetic quality in natural populations of *Hyphantria cunea*. *Can. Entomol.* **103**:893–906.

Morris, R. F. 1971b. The influence of land use and vegetation on the population density of *Hyphantria cunea*. *Can. Entomol.* **103**:1525–1536.

Morris, R. F. 1972. Predation by wasps, birds, and mammals on *Hyphantria cunea*. *Can. Entomol.* **104**:1581–1591.

Morris, R. F. 1976a. Influence of genetic changes and other variables on the encapsulation of parasites by *Hyphantria cunea*. *Can. Entomol.* **108**:673–684.

Morris, R. F. 1976b. Relation of parasite attack to the colonial habit of *Hyphantria cunea*. *Can. Entomol.* **108**:833–836.

Morris, R. F. 1976c. Relation of mortality caused by parasites to the population density of *Hyphantria cunea*. *Can. Entomol.* **108**:1291–1294.

Morris, R. F., and C. W. Bennett. 1967. Seasonal population trends and extensive census methods for *Hyphantria cunea*. *Can. Entomol* **99**:9–17.

Morris, R. F., and W. C. Fulton. 1970a. Models for the development and survival of *Hyphantria cunea* in relation to temperature and humidity. *Mem. Entomol. Soc. Can.* **70**:1–60.

Morris, R. F., and W. C. Fulton. 1970b. Heritability of diapause intensity in *Hyphantria cunea* and correlated fitness responses. *Can. Entomol.* **102:** 927–938.

Morris, R. F., and C. A. Miller, 1954. The development of life tables for the spruce budworm. *Can. J. Zool.* **32:**283–301.

Morris, R. F., and T. Royama. 1969. Logarithmic regression as an index of responses to population density. *Can. Entomol.* **101:**361–364.

Morrison, G., M. Auerbach, and E. D. McCoy. 1979. Anomalous diversity of tropical parasitoids: A general phenomenon? *Am. Nat.* **114:**303–307.

Morrow, P. A. 1977. The significance of phytophagous insects in the *Eucalyptus* forests of Australia. In W. J. Mattson (ed.). *The role of arthropods in forest ecosystems.* Springer-Verlag, New York, pp. 19–29.

Morrow, P. A., and V. C. La Marche. 1978. Tree ring evidence for chronic insect suppression of productivity in subalpine *Eucalyptus. Science* **201:** 1244–1246.

Morse, D. H. 1979. Prey capture by the crab spider *Misumena calycina* Araneae: Thomisidae). *Oecologia* **39:**309–319.

Morse, D. H. 1980. *Behavioral mechanisms in ecology.* Harvard Univ. Press, Cambridge, Mass.

Morse, D. H., and R. S. Fritz. 1982. Experimental and observational studies of patch choice at different scales by the crab spider *Misumena vatia. Ecology* **63:**172–182.

Mosquin, T. 1971. Competition for pollinators as a stimulus for the evolution of flowering time. *Oikos* **22:**398–402.

Mousseau, T. A., and H. Dingle. 1991. Maternal effects in insect life histories. *Annu. Rev. Entomol.* **36:**511–534.

Murdoch, W. W. 1966a. Aspects of the population dynamics of some marsh carabidae. *J. Anim. Ecol.* **35:**127–156.

Murdoch, W. W. 1966b. Population stability and life history phenomena. *Am. Nat.* **100:**5–11.

Murdoch, W. W. 1966c. "Community structure, population control, and competition" A critique. *Am. Nat.* **100:**219–226.

Murdoch, W. W. 1969. Switching in general predators: Experiments on predator specificity and stability of prey populations. *Ecol. Monogr.* **39:** 335–354.

Murdoch, W. W. 1970. Population regulation and population inertia. *Ecology* **51:**497–502.

Murdoch, W. W. 1971. The developmental response of predators to changes in prey density. *Ecology* **52:**132–137.

Murdoch, W. W. 1973. The functional response of predators. *J. Appl. Ecol.* **10:**335–342.

Murdoch, W. W. 1994. Population regulation in theory and practice. *Ecology* **75:**271–287.

Murdoch, W. W., and J. R. Marks. 1973. Predation by coccinellid beetles: Experiments on switching. *Ecology* **54:**160–167.

Murdoch, W. W., and A. Oaten. 1975. Predation and population stability. *Adv. Ecol. Res.* **9**:1–131.

Murdoch, W. W., F. C. Evans, and C. H. Peterson. 1972. Diversity and pattern in plants and insects. *Ecology* **53**:819–829.

Murie, A. 1944. *The wolves of Mount McKinley.* Fauna Natl. Parks U.S. Fauna Ser. 5. U.S. Government Printing Office, Washington, D.C.

Myers, J. H. 1988. Can a general hypothesis explain population cycles of forest Lepidoptera? *Adv. Ecol. Res.* **18**:179–242.

Myers, J. H. 1993. Population outbreaks of forest Lepidoptera. *Am. Sci.* **81**:240–251.

Myers, J. H., and C. J. Krebs 1971. Genetic, behavioral, and reproductive attributes of dispersing field voles *Microtus pennsylvanicus* and *Microtus ochrogaster. Ecol. Monogr.* **41**:53–78.

Myers, J. H., and L. D. Rothman. 1995. Field experiments to study regulation of fluctuating populations. In N. Cappuccino and P. W. Price (eds.). *Population dynamics: New approaches and synthesis.* Academic Press, San Diego, Calif., pp. 229–250.

Nachman, G. 1981. A simulation model of spatial heterogeneity and nonrandom search in an insect host–parasitoid system. *J. Anim. Ecol.* **50**: 27–47.

Nailand, P., and S. A. Hanrahan. 1993. Modelling brown locust, *Locustana pardalina* (Walker), outbreaks in the Karoo. *S. Afr. J. Sci.* **89**:420–424.

Nault, L. R., M. E. Montgomery, and W. S. Bowers. 1976. Ant–aphid asociation: Role of aphid alarm pheromone. *Science* **192**:1349–1351.

Naumann, I. D. (ed.). 1991. *The insects of Australia: A textbook for students and research workers.* 2nd ed. Cornell Univ. Press, Ithaca, N.Y.

Nei, M., and R. K. Koehn. 1983. *Evolution of genes and proteins.* Sinauer, Sunderland, Mass.

Neilson, M. M., and R. F. Morris. 1964. The regulation of European spruce sawfly numbers in the Maritime Provinces of Canada from 1937 to 1963. *Can. Entomol.* **96**:773–784.

Nelson, B. C., and M. D. Murray. 1971. The distribution of Mallophoga on the domestic pigeon *(Columba livia). Int. J. Parasitol.* **1**:21–29.

Nelson, B. C., and C. R. Smith. 1980. Ecology of sylvatic plague in lava caves at Lava Beds National Monument, California. In R. Traub and M. Starcke. (eds.). *Fleas.* A. A. Balkema, Rotterdam, The Netherlands, pp. 273–275.

Nelson, C. E., M. Walker-Simmons, D. Makus, G. Zuroske, J. Graham, and C. A. Ryan. 1983. Regulation of synthesis and accumulation of proteinase inhibitors in leaves of wounded tomato plants. In P. A. Hedin (ed.). *Plant resistance to insects.* American Chemical Society, Washington, D.C., pp. 103–122.

Neumann, U. 1971. Die Sukzession der Bodenfauna (Carabidae [Coleoptera], Diplopoda und Isopoda) in den forstlich rekultivierten Gebietendes Rheinischen Braunkohlenreviers. *Pedobiologia* **11**:193–226.

Neuvonen, S., and P. Niemelä. 1981. Species richness of Macrolepidoptera on Finnish deciduous trees and shrubs. *Oecologia* **51**:364–370.

Nevo, E. 1978. Genetic variation in natural populations: Patterns and theory. *Theor. Popul. Biol.* **13**:121–177.

Nevo, E. 1983. Population genetics and ecology: The interface. In D. S. Bendall (ed.). *Evolution from molecules to men.* Cambridge Univ. Press, Cambridge, pp. 287–321.

Nicholson, A. J. 1933. The balance of animal populations. *J. Anim. Ecol.* **2**(Suppl.):132–178.

Nicholson, A. J. 1954a. Compensatory reactions of populations to stress, and their evolutionary significance. *Aust. J. Zool.* **2**:1–8.

Nicholson, A. J. 1954b. An outline of the dynamics of animal populations. *Aust. J. Zool.* **2**:9–65.

Nicholson, A. J. 1957a. The self-adjustment of populations to change. *Cold Spring Harbor Symp. Quant. Biol.* **22**:153–172.

Nicholson, A. J. 1957b. Comments on paper of T. B. Reynoldson. *Cold Spring Harbor Symp. Quant. Biol.* **22**:326.

Nicholson, A. J. 1958. Dynamics of insect populations. *Annu. Rev. Entomol.* **3**:107–136.

Nicholson, A. J., and V. A. Bailey. 1935. The balance of animal populations. Part I. *Proc. Zool. Soc. London,* pp. 551–598.

Nickerson, R. P., and M. Druger. 1973. Maintenance of chromosomal polymorphism in a population of *Drosophila pseudoobscura.* II. Fecundity, longevity, viability and competitive fitness. *Evolution* **27**:125–133.

Nielsen, E. T. 1961. On the habits of the migratory butterfly *Ascia monuste* L. *Biol. Medd.* **23**:1–81.

Niemalä, P., and E. Haukioja. 1982. Seasonal patterns in species richness of herbivores: Macrolepidopteran larvae on Finnish deciduous trees. *Ecol. Entomol.* **7**:169–175.

Niemalä, P., and W. J. Mattson. 1996. Invasion of North American forests by European phytophagous insects: Legacy of the European crucible? *BioScience* **46**:741–753.

Niemalä, P., E.-M. Aro, and E. Haukioja. 1979. Birch leaves as a resource for herbivores: Damage-induced increase in leaf phenols with trypsin-inhibiting effects. *Rep. Kevo Subarct. Stn.* **15**:37–40.

Niemalä, P., S. Hanhimäki, and R. Mannila. 1981. The relationship of adult size in noctuid moths (Lepidoptera, Noctuidae) to breadth of diet and growth form of host plants. *Ann. Entomol. Fenn.* **47**:17–20.

Niklas, K. J. 1985. Wind pollination: A study in controlled chaos. *Am. Sci.* **73**:462–470.

Nitecki, M. H. (ed.). 1983. *Coevolution.* Univ. Chicago Press, Chicago.

Nordlund, D. A. 1981. Semiochemicals: A review of terminology. In D. A. Nordlund, R. L. Jones, and W. J. Lewis (eds.). *Semiochemicals: Their role in pest control.* Wiley, New York, pp. 13–28.

Nordlund, D. A., and W. J. Lewis. 1976. Terminology of chemical releasing stimuli in intraspecific and interspecific interactions. *J. Chem. Ecol.* 2:211–220.

Nordlund, D. A., R. L. Jones, and W. J. Lewis (eds.). 1981. *Semiochemicals: Their role in pest control*. Wiley, New York.

Norris, D. M. 1979. The mutualistic fungi of Xyleborini beetles. In L. R. Batra (ed.). *Insect-fungus symbiosis: Nutrition, mutualism, and commensalism*. Allanheld, Osmun, Totowa, N.J., pp. 53–63.

Nothnagle, P. J., and J. C. Schultz. 1987. What is a forest pest? In P. Barbosa and J. C. Schultz (eds.). *Insect outbreaks*. Academic Press, San Diego, Calif., pp. 59–80.

Novák, V., F. Hrozinka, and B. Starý. 1976. *Atlas of insects harmful to forest trees*. Vol. 1. Elsevier, Amsterdam.

Odhiambo, T. R. 1959. An account of parental care in *Rhinocoris albopilosus* (Signoret) (Hemiptera-Heteroptera: Reduviidae), with notes on its life history. *Proc. R. Entomol. Soc. London A* **34**:175–185.

Odum, E. P. 1950. Bird populations of the Highlands (North Carolina) Plateau in relation to plant succession and avian invasion. *Ecology* **31**:587–605.

Odum, E. P. 1953. *Fundamentals of ecology*. W.B. Saunders, Philadelphia, Pa.

Odum, E. P. 1959. *Fundamentals of ecology*. 2nd ed. W.B. Saunders, Philadelphia, Pa.

Odum, E. P. 1969. The strategy of ecosystem development. *Science* **164**:262–270.

Odum, E. P. 1971. *Fundamentals of ecology*. 3rd ed. W.B. Saunders, Philadelphia, Pa.

Odum, E. P., C. E. Connell, and L. B. Davenport. 1962. Population energy flow of three primary consumer components of old-field ecosystems. *Ecology* **43**:88–96.

Odum, H. T. 1957. Trophic structure and productivity of Silver Springs, Florida. *Ecol. Monogr.* **27**:55–112.

Ohgushi, T. 1986. Population dynamics of an herbivorous lady beetle, *Henosepilachna niponica,* in a seasonal environment. *J. Anim. Ecol.* **55**:861–879.

Ohgushi, T. 1992. Resource limitation of insect herbivore populations. In M. D. Hunter, T. Ohgushi, and P. W. Price (eds.). *Effects of resource distribution on animal–plant interactions*. Academic Press, San Diego, Calif., pp. 199–241.

Ohgushi, T. 1995. Adaptive behavior produces stability of herbivorous lady beetle populations. In N. Cappuccino and P. W. Price (eds.). *Population dynamics: New approaches and synthesis*. Academic Press, San Diego, Calif., pp. 303–319.

Ohgushi, T., and H. Sawada. 1981. The dynamics of natural populations of a phytophagous lady beetle, *Henosepilachna pustulosa* (Kôno) under differ-

ent habitat conditions. I. Comparison of adult population parameters among local populations in relation to habitat stability. *Res. Popul. Ecol.* **23**:94–115.

Ohgushi, T., and H. Sawada. 1985a. Population equilibrium with respect to available food resource and its behavioural basis in an herbivorous lady beetle, *Henosepilachna niponica. J. Anim. Ecol.* **54**:781–796.

Ohgushi, T., and H. Sawada. 1985b. Arthropod predation limits the population density of an herbivorous lady beetle, *Henosepilachna niponica* (Lewis). *Res. Popul. Ecol.* **27**:351–359.

Ohgushi, T., and H. Sawada. 1997. A shift toward early reproduction in an introduced herbivorous ladybird. *Ecol. Entomol.* **22**:90–96.

Ohno, S., C. Stenius, L. Christian, and G. Schipmann. 1969. *De novo* mutation-like events observed at the 6PGD locus of the Japanese quail, and the principle of polymorphism breeding more polymorphism. *Biochem. Genet.* **3**:417–428.

Oksanen, L. 1991. A century of community ecology: How much progress? *Trends Ecol. Evol.* **6**:294–296.

Oldfield, G. N. 1970. Mite transmission of plant viruses. *Anu. Rev. Entomol.* **15**:343–380.

Oldiges, H. 1958. Waldbodendüngung und Schädlingsfauna des Kronenraums. *Allg. Forst.* **13**:138–140.

O'Neill, R. V. 1967. Niche segregation in seven species of diplopods. *Ecology* **48**:983.

O'Neill, R. V. 1976. Ecosystem persistence and heterotrophic regulation. *Ecology* **57**:1244–1253.

O'Neill, R. V., and D. E. Reichle. 1980. Dimensions of ecosystem theory. In R. H. Waring (ed.). *Forests: Fresh perspectives from ecosystem analysis.* Oregon State Univ. Press, Corvallis, Oreg., pp. 11–26.

Opler, P. A. 1974. Oaks as evolutionary islands for leaf-mining insects. *Am. Sci.* **62**:67–73.

Orians, G. H. 1962. Natural selection and ecological theory. *Am. Nat.* **96**:257–263.

Orians, G. H. 1969. On the evolution of mating systems in birds and mammals. *Am. Nat.* **103**:589–603.

Orians, G. H. 1974. Diversity, stability, and maturity in natural ecosystems. In W. H. van Dobben and R. H. Lowe-McConnell (eds.). *Unifying concepts in ecology.* Dr. W. Junk, The Hague, The Netherlands, pp. 139–150.

Oshima, K., H. Honda, and I. Yamamoto. 1973. Isolation of an oviposition marker from azuki bean weevil, *Callosobruchus chinensis* (L.). *Agric. Biol. Chem.* **37**:2679–2680.

O'Toole, C., and A. Raw. 1991. *Bees of the world.* Blandford, London.

Otte, D., and A. Joern. 1975. Insect territoriality and its evolution: Population studies of desert grasshoppers on creosote bushes. *J. Anim. Ecol.* **44**:29–54.

Overal, W. L. 1980. Host-relations of the batfly *Megistopoda aranea* (Diptera: Streblidae) in Panama. *Univ. Kans. Sci. Bull.* **52**:1–20.

Ovington, J. D. 1962. Quantitative ecology and the woodland ecosystem concept. *Adv. Ecol. Res.* **1**:103–192.

Ovington, J. D., D. Heitkamp, and D. B. Lawrence. 1963. Plant biomass and productivity of prairie, savanna, oakwood, and maize field ecosystems in central Minnesota. *Ecology* **44**:52–63.

Owadally, A. W. 1979. The dodo and the tambalacoque tree. *Science* **203**:1363–1364.

Owen, D. 1980a. *Camouflage and mimicry.* Univ. Chicago Press, Chicago.

Owen, D. F. 1980b. How plants may benefit from the animals that eat them. *Oikos* **35**:230–235.

Owen, D. F., and D. O. Chanter. 1970. Species diversity and seasonal abundance in tropical Ichneumonidae. *Oikos* **21**:142–144.

Owen, D. F., and J. Owen. 1974. Species diversity in temperate and tropical Ichneumonidae. *Nature* **249**:583–584.

Owen, D. F., and R. G. Wiegert. 1976. Do consumers maximize plant fitness? *Oikos* **27**:488–492.

Owen, D. F., and R. G. Wiegert. 1981. Mutualism between grasses and grazers: An evolutionary hypothesis. *Oikos* **36**:376–378.

Owen, D. F., and R. G. Wiegert. 1982. Grasses and grazers: Is there a mutualism? *Oikos* **38**:258–259.

Paige, K. N., and T. G. Whitham. 1985. Individual and population shifts in flower color by scarlet gilia: a mechanism for pollinator tracking. *Science* **227**:315–317.

Paine, R. T. 1966. Food web complexity and species diversity. *Am. Nat.* **100**:65–75.

Paine, R. T. 1969a. A note on trophic complexity and community stability. *Am. Nat.* **103**:91–93.

Paine, R. T. 1969b. The *Pisaster–Tegula* interaction: Prey patches, predator food preference, and intertidal community structure. *Ecology* **50**:950–961.

Paine, R. T. 1971. The measurement and application of the calorie to ecological problems. *Annu. Rev. Ecol. Syst.* **2**:145–164.

Pajunen, V. I. 1966. The influence of population density on the territorial behaviour of *Leucorrhinia rubicunda* L. (Odon., Libellulidae). *Ann. Zool. Fenn.* **3**:40–52.

Papaj, D. R. 1994. Use and avoidance of occupied hosts as a dynamic process in tephritid flies. In E. A. Bernays (ed.). *Insect–plant interactions.* Vol. 5. CRC Press, Boca Raton, Fla., pp. 25–46.

Papavizas, G. C. (ed.). 1981. *Biological control in crop production.* Allenheld, Osmun, Totowa, N.J.

Paradis, R. O., and E. J. LeRoux. 1965. Recherches sur la biologie et la dynamique des populations naturelles d'*Archips argyrospilus* (Wlk.) dans le sud-ouest du Québec. *Mem. Entomol. Soc. Can.* **43**:1–77.

Park, T. 1932. Studies in population physiology: The relation of numbers to initial population growth in the flour beetle, *Tribolium confusum* Duval. *Ecology* **13**:172–181.

Park, T. 1948. Experimental studies of interspecies competition. I. Competition between populations of the flour beetles, *Tribolium confusum* Duval and *Tribolium castaneum* Herbst. *Ecol. Monogr.* **18**:265–308.

Park, T. 1954a. Experimental studies of interspecies competition. II. Temperature, humidity, and competition in two species of *Tribolium*. *Physiol. Zool.* **27**:177–238.

Park, T. 1954b. Competition: An experimental and statistical study. In O. Kempthorne, T. A. Bancroft, J. W. Gowen, and J. L. Lush (eds.). *Statistics and mathematics in biology*. State College Press, Ames, Iowa, pp. 175–195.

Park, T., P. H. Leslie, and D. B. Mertz. 1964. Genetic strains and competition in populations of *Tribolium*. *Physiol. Zool.* **37**:97–162.

Parker, G. A. 1970a. Sperm competition and its evolutionary consequences in the insects. *Biol. Rev.* **45**:525–567.

Parker, G. A. 1970b. The reproductive behavior and the nature of sexual selection in *Scatophaga stercoraria* L. (Diptera: Scatophagidae). VII. The origin and evolution of the passive phase. *Evolution* **24**:774–788.

Parker, G. A. 1970c. Sperm competition and its evolutionary effect on copula duration in the fly *Scatophaga stercoraria*. *J. Insect Physiol.* **16**:1301–1328.

Parker, G. A. 1971. The reproductive behaviour and the nature of sexual selection in *Scatophaga stercoraria* L. (Diptera: Scatophagidae). VI. The adaptive significance of emigration from the oviposition site during the phase of genital contact. *J. Anim. Ecol.* **40**:215–233.

Parker, G. A. 1974. The reproductive behaviour and the nature of sexual selection in *Scatophaga stercoraria* L. (Diptera; Scatophagidae). IX. Spatial distribution of fertilization rates and evolution of male search strategy within the reproductive area. *Evolution* **28**:93–108.

Parker, G. A. 1978a. Serching for mates. In J. R. Krebs and N. B. Davies (eds.). *Behavioral ecology: An evolutionary approach*. Sinauer, Sunderland, Mass., pp. 214–244.

Parker, G. A. 1978b. Evolution of competitive mate searching. *Annu. Rev. Entomol.* **23**:173–196.

Parker, G. A. 1979. Sexual selection and sexual conflict. In M. S. Blum and N. A. Blum (eds.). *Sexual selection and reproductive competition in insects*. Academic Press, San Diego, Calif., pp. 123–166.

Parker, M. A., and R. B. Root. 1981. Insect herbivores limit habitat distribution of a native composite, *Machaeranthera canescens*. *Ecology* **62**:1390–1392.

Parnell, J. R. 1966. Observations on the population fluctuations and life histories of the beetles *Bruchidius ater* (Bruchidae) and *Apion fuscirostre* (Curculionidae) on broom *(Sarothamnus scoparius)*. *J. Anim. Ecol.* **35**:157–188.

Parry, G. D. 1981. The meaning of *r*- and *K*-selection. *Oecologia* **48**:260–264.

Parsons, P. A., and J. A. McKenzie. 1972. The ecological genetics of *Drosophila. Evol. Biol.* **5**:87–132.

Pastorok, R. A. 1981. Prey vulnerability and size selection by *Chaoborus* larvae. *Ecology* **62**:1311–1324.

Patrick, R., and D. Strawbridge. 1963. Variation in the structure of natural diatom communities. *Am. Nat.* **97**:51–57.

Peakall, R., S. N. Handel, and A. J. Beattie. 1991. The evidence for, and importance of, ant pollination. In C. R. Huxley and D. F. Cutler (eds.). *Ant–plant interactions.* Oxford Univ. Press, Oxford, pp. 421–429.

Pearl, R., and J. R. Miner. 1935. Experimental studies on the duration of life. XIV. The comparative mortality of certain lower organisms. *Q. Rev. Biol.* **10**:60–79.

Pearl, R., and S. L. Parker. 1921. Experimental studies on the duration of life. I. Introductory discussion of the duration of life in *Drosophila. Am. Nat.* **55**:481–509.

Pearl, R., and L. J. Reed. 1920. On the rate of growth of the population of the United States since 1790 and its mathematical representation. *Proc. Natl. Acad. Sci. USA* **6**:275–288.

Pearson, G. A. 1996. Insect tattoos on humans: A "dermagraphic" study. *Am. Entomol.* **42**:99–105.

Pearson, O. P. 1964. Carnivore-mouse predation: An example of its intensity and bioenergetics. *J. Mammal.* **45**:177–188.

Pechuman, L. L. 1967. Observations on the behavior of the bee *Anthidium manicatum* (L.). *J. N.Y. Entomol. Soc.* **75**:68–73.

Pedley, T. J. (ed.). 1977. *Scale effects in animal locomotion.* Academic Press, London.

Pellmyr, O. 1992. The phylogeny of a mutualism: Evolution and coadaptation between *Trollius* and its seed-parasitic pollinators. *Biol. J. Linn. Soc.* **47**:337–365.

Pellmyr, O., and C. J. Huth. 1994. Evolutionary stability of mutualism between yuccas and yucca moths. *Nature* **372**:257–260.

Pellmyr, O., J. Leebens-Mack, and C. J. Huth. 1996. Non-mutualistic yucca moths and their evolutionary consequences. *Nature* **380**:155–156.

Pemberton, R. W., and T. Yamasaki. 1995. Insects: Old food in new Japan. *Am. Entomol.* **41**:227–229.

Pennycuick, C. J. 1972. *Animal flight.* Edward Arnold, London.

Peschken, D. P. 1972. *Chrysolina quadrigemina* (Coleoptera: Chrysomelidae) introduced from California to British Columbia against the weed *Hypericum perforatum:* Comparison of behaviour, physiology, and colour in association with post-colonization adaptation. *Can. Entomol.* **104:** 1689–1698.

Peterman, R. M. 1978. The ecological role of mountain pine beetle in lodgepole pine forests. In D. L. Kibbee, A. A. Berryman, G. D. Amman, and R. W. Stark (eds.). *Theory and practice of mountain pine beetle management in lodgepole pine forests.* Forest, Wildlife and Range Exp. Stn., Univ. Idaho, Moscow, Idaho, pp. 16–26.

Petras, M. L., J. D. Reimer, F. G. Biddle, J. E. Martin, and R. S. Linton. 1969. Studies of natural populations of *Mus*. V. A survey of nine loci for polymorphisms. *Can. J. Genet. Cytol.* **11**:497–513.

Petrusewicz, K., and W. L. Grodzinski. 1975. The role of herbivore consumers in various ecosystems. In D. E. Reichle, J. F. Franklin, and D. W. Goodall (eds.). *Productivity of world ecosystems*. National Academy of Sciences, Washington, D.C., pp. 64–70.

Philip, J. R. 1955. Note on the mathematical theory of population dynamics and a recent fallacy. *Aust. J. Zool.* **3**:287–294.

Philipson, J. 1966. *Ecological energetics*. Edward Arnold, London.

Pianka, E. R. 1966. Latitudinal gradients in species diversity: A review of concepts. *Am. Nat.* **100**:33–46.

Pianka, E. R. 1969. Sympatry of desert lizards *(Ctenotus)* in Western Australia. *Ecology* **50**:1012–1030.

Pianka, E. R. 1970. On *r*- and *K*-selection. *Am. Nat.* **104**:592–597.

Pianka, E. R. 1972. *r*- and *K*-selection or *b*- and *d*-selection? *Am. Nat.* **106**:581–588.

Pianka, E. R. 1974. *Evolutionary ecology*. Harper & Row, New York.

Pickett, S. T. A., and J. N. Thompson. 1978. Patch dynamics and the design of nature reserves. *Biol. Conserv.* **13**:27–37.

Pickett, S. T. A., J. Kolasa, and C. G. Jones. 1994. *Ecological understanding: The nature of theory and the theory of nature*. Academic press, San Diego, Calif.

Pielou, E. C. 1969. *An introduction to mathematical ecology*. Wiley-Interscience, New York.

Pielou, E. C. 1975. *Ecology diversity*. Wiley, New York.

Pielou, E. C. 1979. *Biogeography*. Wiley, New York.

Pietrewicz, A. T., and A. C. Kamil. 1979. Search image formation in the blue jay *(Cyanocitta cristata). Science* **204**:1332–1333.

Pimentel, D. 1961a. Species diversity and insect population outbreaks. *Ann. Entomol. Soc. Am.* **54**:76–86.

Pimentel, D. 1961b. Animal population regulation by the genetic feed-back mechanism. *Am. Nat.* **95**:65–79.

Pimentel, D. 1968. Population regulation and genetic feedback. *Science* **159**:1432–1437.

Pimentel, D. 1975. *Insects, science and society*. Academic Press, San Diego, Calif.

Pimm, S. L. 1979a. The structure of food webs. *Theor. Popul. Biol.* **16**: 144–158.

Pimm, S. L. 1979b. Complexity and stability: Another look at MacArthur's original hypothesis. *Oikos* **33**:351–357.

Pimm, S. L. 1980. Food web design and the effect of species deletion. *Oikos* **35**:139–149.

Pimm, S. L. 1982. *Food webs*. Chapman & Hall, New York.

Pimm, S. L., and J. H. Lawton. 1977. Number of trophic levels in ecological communities. *Nature* **268**:329–331.

Pimm, S. L., and J. H. Lawton. 1978. On feeding on more than one trophic level. *Nature* **275**:542–544.

Pitelka, F. A. 1941. Distribution of birds in relation to major biotic communities. *Am. Midl. Nat.* **25**:113–137.

Pitelka, F. A. 1959. Population studies of lemmings and lemming predators in northern Alaska. *Proc. Int. Congr. Zool.* **15**:757–75ɔ.

Pitelka, F. A. 1964. The nutrient-recovery hypothesis for arctic microtine cycles. I. Introduction. In D. Crisp (ed.). *Grazing in terrestrial and marine environments.* Blackwell Scientific, Oxford, pp. 55–56.

Pitelka, F. A., P. Q. Tomich, and G. W. Treichel. 1955. Ecological relations of jaegers and owls as lemming predators near Barrow, Alaska. *Ecol. Monogr.* **25**:85–117.

Plapp, F. W. 1976. Biochemical genetics of insecticide resistance. *Annu. Rev. Entomol.* **21**:179–197.

Platt, A. P., and L. P. Brower. 1968. Mimetic versus disruptive coloration in intergrading populations of *Limenitis arthemis* and *astyanax* butterflies. *Evolution* **22**:699–718.

Platt, A. P., R. P. Coppinger, and L. P. Brower. 1971. Demonstration of the selective advantage of mimetic *Limenitis* butterflies presented to caged avian predators. *Evolution* **25**:692–701.

Podoler, H., 1981. Effects of variable temperature on responses of *Aphytis melinus* and *A. lingnanensis* to host density. *Phytoparasitica* **9**:179–190.

Podoler, H., and D. Rogers. 1975. A new method for the identification of key factors from life-table data. *J. Anim. Ecol.* **44**:85–114.

Polis, G. A. 1991a. Food webs in desert communities: complexity via diversity and omnivory. In G. A. Polis (ed.). *The ecology of desert communities.* Univ. Arizona Press, Tucson, Ariz., pp. 383–437.

Polis, G. A. (ed.). 1991b. *The ecology of desert communities.* Univ. Arizona Press, Tucson, Ariz.

Polis, G. A., and D. R. Strong. 1996. Food web complexity and community dynamics. *Am. Nat.* **147**:813–846.

Pomeroy, L. R. 1974. *Cycles of essential elements.* Dowden, Hutchinson & ross, Stroudsburg, Pa.

Poole, R. W. 1974. *An introduction to quantitative ecology.* McGraw-Hill, New York.

Poorter, H., and H. Lambers. 1991. Is interspecific variation in growth rate positively correlated with biomass allocation to the leaves? *Am. Nat.* **138**:1264–1268.

Portmann, A. 1959. *Animal camouflage.* Univ. Michigan Press, Ann Arbor, Mich.

Pottinger, R. P., and E. J. LeRoux. 1971. The biology and dynamics of *Lithocolletis blancardella* (Lepidoptera: Gracillariidae) on apple in Quebec. *Mem. Entomol. Soc. Can.* **77**:1–437.

Powell, J. A., and R. A. Mackie. 1966. Biological interrelationships of moths and *Yucca whipplei. Univ. Calif. Publ. Entomol.* **42**:1–59.

Powell, J. R. 1975. Protein variation in natural populations of animals. *Evol. Biol.* **8**:79–119.

Power, M. E. 1992. Top-down and bottom-up forces in food webs: Do plants have primacy? *Ecology* **73**:733–746.

Prada, M., O. J. Marini-Filho, and P. W. Price. 1995. Insects in flower heads of *Aspilia foliacea* (Asteraceae) after a fire in a central Brazilian savanna: Evidence for the plant vigor hypothesis. *Biotropica* **27**:513–518.

Prakash, S. 1969. Genic variation in a natural population of *Drosophila persimilis. Proc. Natl. Acad. Sci. USA* **62**:778–784.

Prakash, S., R. C. Lewontin, and J. L. Hubby. 1969. A molecular approach to the study of genic heterozygosity in natural populations. IV. Patterns of genic variation in central, marginal and isolated populations of *Drosophila pseudoobscura. Genetics* **61**:841–858.

Pratt, T. K. 1982. Pleistocene seed dispersal. *Science* **216**:6.

Preszler, R. W., and P. W. Price. 1988. Host quality and sawfly populations: A new approach to life table analysis. *Ecology* **69**:2012–2020.

Price, M. V., and N. M. Waser. 1979. Pollen dispersal and optimal out-crossing in *Delphinium nelsoni. Nature* **277**:294–296.

Price, P. W. 1970a. A loosestrife sawfly, *Monostegia abdominalis* (Hymenoptera: Tenthredinidae). *Can. Entomol.* **102**:491–495.

Price, P. W. 1970b. Characteristics permitting coexistence among parasitoids of a sawfly in Quebec. *Ecology* **51**:445–454.

Price, P. W. 1970c. Trail odors: Recognition by insects parasitic on cocoons. *Science* **170**:546–547.

Price, P. W. 1971a. Niche breadth and dominance of parasitic insects sharing the same host species. *Ecology* **52**:587–596.

Price, P. W. 1971b. Toward a holistic approach to insect population studies. *Ann. Entomol. Soc. Am.* **64**:1399–1406.

Price, P. W. 1972. Behavior of the parasitoid *Pleolophus basizonus* (Hymenoptera: Ichneumonidae) in response to changes in host and parasitoid density. *Can. Entomol.* **104**:129–140.

Price, P. W. 1973a. Reproductive strategies in parasitoid wasps. *Am. Nat.* **107**:684–693.

Price, P. W. 1973b. Parasitoid strategies and community organization. *Environ. Entomol.* **2**:623–626.

Price, P. W. 1974a. Strategies for egg production. *Evolution* **28**:76–84.

Price, P. W. 1974b. Energy allocation in ephemeral adult insects. *Ohio J. Sci.* **74**:380–387.

Price, P. W. 1975a. Reproductive strategies of parasitoids. In P. W. Price (ed.). *Evolutionary strategies of parasitic insects and mites.* Plenum Press, New York, pp. 87–111.

Price, P. W. 1975b. Introduction: The parasitic way of life and its consequences. In P. W. Price (ed.). *Evolutionary strategies of parasitic insects and mites.* Plenum Press, New York, pp. 1–13.

Price, P. W. 1975c. *Insect ecology.* Wiley, New York.

Price, P. W. (ed.). 1975d. *Evolutionary strategies of parasitic insects and mites*. Plenum Press, New York.

Price, P. W. 1976. Colonization of crops by arthropods: Non-equilibrium communities in soybean fields. *Environ. Entomol.* 5:605–611.

Price, P. W. 1977. General concepts on the evolutionary biology of parasites. *Evolution* 31:405–420.

Price, P. W. 1980. *Evolutionary biology of parasites*. Princeton Univ. Press, Princeton, N.J.

Price, P. W. 1981a. Semiochemicals in evolutionary time. In D. A. Nordlund, R. L. Jones, and W. J. Lewis (eds.). *Semiochemicals: Their role in pest control*. Wiley, New York, pp. 251–279.

Price, P. W. 1981b. Relevance of ecological concepts to practical biological control. In G. C. Papavizas (ed.). *Biological control in crop production*. Allenheld, Osmun, Totowa, N.J., pp. 3–19.

Price, P. W. 1983. Hypotheses on organization and evolution in herbivorous insect communities. In R. F. Denno and M. S. McClure (eds.). *Variable plants and herbivores in natural and managed systems*. Academic Press, San Diego, Calif., pp. 559–596.

Price, P. W. 1984a. Communities of specialists: Vacant niches in ecological and evolutionary time. In D. Strong, D. Simberloff, L. G. Abele, and A. B. Thistle (eds.). *Ecological communities: Conceptual issues and the evidence*. Princeton Univ. Press, Princeton, N.J., pp. 510–523.

Price, P. W. 1984b. Alternative paradigms in community ecology. In P. W. Price, C. N. Slobodchikoff, and W. S. Gaud (eds.). *A new ecology: Novel approaches to interactive systems*. Wiley, New York, pp. 353–383.

Price, P. W. 1984c. The concept of the ecosystem: Organization, structure and dynamics. In C. B. Huffaker and R. L. Rabb (eds.). *Ecological entomology*. Wiley, New York, pp. 19–50.

Price, P. W. 1989. Clonal development of coyote willow, *Salix exigua* (Salicaceae), and attack by the shoot-galling sawfly, *Euura exiguae* (Hymenoptera: Tenthredinidae). *Environ. Entomol.* 18:61–68.

Price, P. W. 1990. Host populations as resources defining parasite community organization. In G. W. Esch, A. O. Bush, and J. M. Aho (eds.). *Parasite communities: Patterns and processes*. Chapman & Hall, London, pp. 21–40.

Price, P. W. 1991a. Darwinian methodology and the theory of insect herbivore population dynamics. *Ann. Entomol. Soc. Am.* 84:465–473.

Price, P. W. 1991b. Evolutionary theory of host and parasitoid interactions. *Biol. Control* 1:83–93.

Price, P. W. 1991c. Patterns in communities along latitudinal gradients. In P. W. Price, T. M. Lewinsohn, G. W. Fernandes, and W. W. Benson (eds.). *Plant–animal interactions: Evolutionary ecology in tropical and temperate regions*. Wiley, New York, pp. 51–69.

Price, P. W. 1991d. The plant vigor hypothesis and herbivore attack. *Oikos* 62:244–251.

Price, P. W. 1992a. Evolution and ecology of gall-inducing sawflies. In J. D.

Shorthouse and O. Rohfritsch (eds.). *Biology of insect-induced galls.* Oxford Univ. Press, New York, pp. 208–224.

Price, P. W. 1992b. Plant resources as the mechanistic basis for insect herbivore population dynamics. In M. D. Hunter, T. Ohgushi, and P. W. Price (eds.). *Effects of resource distribution on animal-plant interactions.* Academic Press, San Diego, Calif., pp. 139–173.

Price, P. W. 1992c. Evolutionary perspectives on host plants and their parasites. *Adv. Plant Pathol.* 8:1–30.

Price, P. W. 1993. Practical significance of tritrophic interactions for crop protection. In T. A. van Beek and H. Breteler (eds.). *Phytochemistry and agriculture.* Oxford Univ. Press, Oxford, pp. 87–106.

Price, P. W. 1994a. Phylogenetic constraints, adaptive syndromes, and emergent properties: from individuals to population dynamics. *Res. Popul. Ecol.* **36**:3–14.

Price, P. W. 1994b. Evolution of parasitoid communities. In B. A. Hawkins and W. Sheehan (eds.). *Parasitoid community ecology.* Oxford Univ. Press, Oxford, pp. 472–491.

Price, P. W. 1996a. *Biological evolution.* W.B. Saunders, Philadelphia, Pa.

Price, P. W. 1996b. Empirical research and factually based theory: What are their roles in entomology? *Am. Entomol.* **42**:209–214.

Price, P. W., and K. M. Clancy. 1986a. Interactions among three trophic levels: Gall size and parasitoid attack. *Ecology* **67**:1593–1600.

Price, P. W., and K. M. Clancy. 1986b. Multiple effects of precipitation on *Salix lasiolepis* and populations of the stem-galling sawfly, *Euura lasiolepis. Ecol. Res.* **1**:1–14.

Price, P. W., and M. D. Hunter. 1995. Novelty and synthesis in the development of population dynamics. In N. Cappuccino and P. W. Price (eds.). *Population dynamics: New approaches and synthesis.* Academic Press, San Diego, Calif., pp. 389–412.

Price, P. W., and H. Pschorn-Walcher. 1988. Are galling insects better protected against parasitoids than exposed feeders? A test using tenthredinid sawflies. *Ecol. Entomol.* **13**:195–205.

Price, P. W., and H. A. Tripp. 1972. Activity patterns of parasitoids on the Swaine jack pine sawfly, *Neodiprion swainei* (Hymenoptera: Diprionidae), and parasitoid impact on the host. *Can. Entomol.* **104**: 1003–1016.

Price, P. W., and G. P. Waldbauer. 1975. Ecological aspects of insect pest management. In R. L. Metcalf and W. H. Luckmann (eds.). *Introduction to insect pest management.* Wiley, New York, pp. 36–73.

Price, P. W., and G. P. Waldbauer. 1994. Ecological aspects of pest management. In R. L. Metcalf and W. H. Luckmann (eds.). *Introduction to insect pest management.* 3rd ed. Wiley, New York, pp. 35–72.

Price, P. W., C. E. Bouton, P. Gross, B. A. McPheron, J. N. Thompson, and A. E. Weis. 1980. Interactions among three trophic levels: Influence of plants on interactions between insect herbivores and natural enemies. *Annu. Rev. Ecol. Syst.* **11**:41–65.

Price, P. W., C. N. Slobodchikoff, and W. S. Gaud. 1984. *A new ecology: Novel approaches to interactive systems.* Wiley, New York.

Price, P. W., M. Westoby, B. Rice, P. R. Atsatt, R. S. Fritz, J. N. Thompson, and K. Mobley. 1986. Parasite mediation of ecological interactions. *Annu. Rev. Ecol. Syst.* **17**:487–505.

Price, P. W., H. Roininen, and J. Tahvanainen. 1987a. Plant age and attack by the bud galler, *Euura mucronata. Oecologia* **73**:334–337.

Price, P. W., H. Roininen, and J. Tahvanainen. 1987b. Why does the bud-galling sawfly, *Euura mucronata,* attack long shoots? *Oecologia* **74**: 1–6.

Price, P. W., G. L. Waring, R. Julkunen-Tiitto, J. Tahvanainen, H. A. Mooney, and T. P. Craig. 1989. The carbon-nutrient balance hypothesis in within-species phytochemical variation of *Salix lasiolepis. J. Chem. Ecol.* **15**: 1117–1131.

Price, P. W., N. Cobb, T. P. Craig, G. W. Fernandes, J. K. Itami, S. Mopper, and R. W. Preszler. 1990. Insect herbivore population dynamics on trees and shrubs: New approaches relevant to latent and eruptive species and life table development. In E. A. Bernays (ed.). *Insect–plant interactions.* Vol. 2. CRC Press, Boca Raton, Fla., pp. 1–38.

Price, P. W., I. Andrade, C. Pires, E. Sujii, and E. M. Vieira. 1995a. Gradient analysis using plant modular structure: Pattern in plant architecture and insect herbivore utilization. *Environ. Entomol.* **24**:497–505.

Price, P. W., T. P. Craig, and H. Roininen, 1995b. Working toward theory on galling sawfly population dynamics. In N. Cappuccino and P. W. Price (eds.). *Population dynamics: New approaches and synthesis.* Academic Press, San Diego, Calif., pp. 321–338.

Price, P. W., I. R. Diniz, H. C. Morais, and E. S. A. Marques. 1995c. The abundance of insect herbivore species in the tropics: The high local richness of rare species. *Biotropica* **27**:468–478.

Price, P. W., G. W. Fernandes, and R. DeClerck-Floate. 1996. Gall-inducing insect herbivores in multitrophic systems. In A. C. Gange and V. K. Brown (eds.). *Multitrophic interactions in terrestrial systems.* Blackwell Scientific, Oxford. pp. 239–255.

Price, P. W., G. W. Fernandes, A. C. F. Lara, J. Brawn, H. Barrios, M. G. Wright, S. P. Ribeiro, and N. Rothcliff. 1997. Global patterns in local number of insect galling species. *J. Biogeog.* In press.

Price, R. E. 1991. Oviposition by the African migratory locust, *Locusta migratoria migratoroides,* in a crop environment in South Africa. *Entomol. Exp. Appl.* **61**:169–177.

Price, R. E., and H. D. Brown. 1990. Reproductive performance of the African migratory locust, *Locusta migratoria migratoroides* (Orthoptera: Acrididae), in a cereal crop environment in South Africa. *Bull. Entomol. Res.* **80**:465–472.

Price Jones, D., and M. E. Solomon (eds.). 1974. *Biology in pest and disease control.* Symp. Brit. Ecol. Soc. 13, Blackwell Scientific Publications, Oxford.

Pritchard, G. 1969. The ecology of a natural population of Queensland fruit fly, *Dacus tryoni.* II. The distribution of eggs and its relation to behaviour. *Aust. J. Zool.* **17**:293–311.

Proctor, M., and P. Yeo. 1972. *The pollination of flowers.* Taplinger, New York.

Prokopy, R. J. 1968. Visual responses of apple maggot flies, *Rhagoletis pomonella:* Orchard studies. *Entomol. Exp. Appl.* **11**:403–422.

Prokopy, R. J. 1972. Evidence for a marking pheromone deterring repeated oviposition in apple maggot flies. *Environ. Entomol.* **1**:326–332.

Prokopy, R. J., and E. D. Owens. 1978. Visual generalist with visual specialist phytophagus insects: Host selection behavior and application to management. *Entomol. Exp. Appl.* **24**:409–420.

Prokopy, R. J., V. Moericke, and G. L. Bush. 1973. Attraction of apple maggot flies to odor of apples. *Environ. Entomol.* **2**:743–749.

Prop, N. 1960. Protection against birds and parasites in some species of tenthredinid larvae. *Arch. Neerl. Zool.* **13**:380–447.

Pukowski, E. 1933. Ökologische Untersuchungen an *Necrophorus* F. *Z. Morphol. Ökol. Tiere* **27**:518–586.

Pulliam, H. R. 1974. On the theory of optimal diets. *Am. Nat.* **108**:59–74.

Pyke, G. H. 1978a. Optimal foraging: Movement patterns of bumblebees between inflorescences. *Theor. Popul. Biol.* **13**:72–98.

Pyke, G. H. 1978b. Optimal foraging in bumblebees and coevolution with their plants. *Oecologia* **36**:281–293.

Pyke, G. H. 1981. Optimal foraging in nectar-feeding animals and coevolution with their plants. In A. C. Kamil and T. D. Sargent (eds.). *Foraging behavior: Ecological, ethological and psychological approaches.* Garland Publishing, New York, pp. 19–38.

Pyke, G. H., H. R. Pulliam, and E. L. Charnov. 1977. Optimal foraging: A selective review of theory and tests. *Q. Rev. Biol.* **52**:137–154.

Rabb, R. L., and F. E. Guthrie (eds.). 1970. *Concepts of pest management.* North Carolina State Univ., Raleigh, N.C.

Rabb, R. L., and G. G. Kennedy (eds.). 1979. *Movement of highly mobile insects: Concepts and methodology in research.* North Carolina State Univ., Raleigh, N.C.

Radtkey, R. R., and M. C. Singer. 1995. Repeated reversals of host-preference evolution in a specialist insect herbivore. *Evolution* **49**:351–359.

Raffa, K. F., and A. A. Berryman. 1980. Flight responses and host selection by bark beetles. In A. A. Berryman and L. Safranyik (eds.). *Dispersal of forest insects: Evaluation, theory and management implications.* Washington State Univ., Pullman, Wash., pp. 213–233.

Rainey, R. C. (ed.). 1976. *Insect flight.* Symp. R. Entomol. Soc. London 7, Royal Entomological Society of London, London.

Rainey, R. C. 1982. Putting insects on the map: Spatial inhomogeneity and the dynamics of insect populations. *Antenna* **6**:162–169.

Ramirez, B. W. 1969. Fig wasps: Mechanism of pollen transfer. *Science* **163**:580–581.

Ramirez, B. W. 1970. Host specificity of fig wasps (Agaonidae). *Evolution* 24:680–691.

Randolph, P. A. 1973. Influence of environmental variability on land snail population properties. *Ecology* 54:933–955.

Rathcke, B. J. 1976a. Competition and coexistence within a guild of herbivorous insects. *Ecology* 57:76–87.

Rathcke, B. J. 1976b. Insect-plant patterns and relationships in the stem-boring guild. *Am. Midl. Nat.* 96:98–117.

Rathcke, B. J. 1983. Competition and facilitation among plants for pollination. In L. Real (ed.). *Pollination biology.* Academic Press, San Diego, Calif., pp. 305–329.

Rathcke, B. J. 1992. Nectar distributions, pollinator behavior, and plant reproductive success. In M. D. Hunter, T. Ohgushi, and P. W. Price (eds.). *Effects of resource distribution on animal–plant interactions.* Academic Press, San Diego, Calif., pp. 113–138.

Rathcke, B. J., and P. W. Price. 1976. Anomalous diversity of tropical ichneumonid parasitoids: A predation hypothesis. *Am. Nat.* 110:889–893.

Rausher, M. D. 1979. Larval habitat suitability and oviposition preference in three related butterflies. *Ecology* 60:503–511.

Rausher, M. D. 1982. Population differentiation in *Euphydryas editha* butterflies: Larval adaptation to different hosts. *Evolution* 36:581–590.

Read, D. P., P. P. Feeny, and R. B. Root. 1970. Habitat selection by the aphid parasite *Diaeretiella rapae* (Hymenoptera: Braconidae) and hyperparasite *Charips brassicae* (Hymenoptera: Cynipidae). *Can. Entomol.* 102:1567–1578.

Reader, P. M., and T. R. E. Southwood. 1981. The relationship between palatability of invertebrates and the successional status of a plant. *Oecologia* 51:271–275.

Readshaw, J. L. 1965. A theory of phasmatid outbreak release. *Aust. J. Zool.* 13:475–490.

Real, L. (ed.). 1983. *Pollination biology.* Academic Press, San Diego, Calif.

Real, L. A. (ed.). 1994. *Ecological genetics.* Princeton Univ. Press, Princeton, N.J.

Reddingius, J. 1971. Gambling for existence. A discussion of some theoretical problems in animal population ecology. *Bibl. Biotheor.* 12:1–208.

Rees, C. J. C. 1969. Chemoreceptor specificity associated with choice of feeding site by the beetle, *Chrysolina brunsvicensis* on its foodplant, *Hypericum hirsutum.* In J. de Wilde and L. M. Schoonhoven (eds.). *Insect and host plant.* North-Holland, Amsterdam, pp. 565–583.

Regal, P. J. 1977. Ecology and evolution of flowering plant dominance. *Science* 196:622–629.

Rehr, S. S., P. P. Feeny, and D. H. Janzen. 1973. Chemical defense in Central American non-ant-acacias. *J. Anim. Ecol.* 42:405–416.

Reichle, D. E., R. A. Goldstein, R. I. Van Hook, Jr., and G. J. Dodson. 1973. Analysis of insect consumption in a forest canopy. *Ecology* 54:1076–1084.

Reichle, D. E., J. F. Franklin, and D. W. Goodall (eds.). 1975. *Productivity of world ecosystems.* National Academy of Sciences, Washington, D.C.

Reichstein, T., J. von Euw, J. A. Parsons, and M. Rothschild. 1968. Heart poisons in the monarch butterfly. *Science* **161**:861–866.

Remington, C. L. 1963. Historical backgrounds of mimicry. *Proc. XVI Int. Congr. Zool.* **4**:145–149.

Rensch, B. 1938. Einwirkung des Klimas bie der Ausprägung von Vogelrassen, mit besonderer Berücksichtigung der Flügelform und der Eizahl. *Proc. 8th Int. Orn. Congr. 1934*:305–311.

Rensch, B. 1959. *Evolution above the species level.* Columbia Univ. Press, New York.

Rettenmeyer, C. W. 1970. Insect mimicry. *Annu. Rev. Entomol.* **15**:43–74.

Reuter, O. M. 1913. *Lebensgewohnheiten und Instinkte der Insekten bis zum Erwachen der sozialen Instinkte.* Friedländer, Berlin.

Reynoldson, T. B. 1957. Population fluctuations in *Urceolaria mitra* (Peritricha) and *Enchytraeus albidus* (Oligochaeta) and their bearing on regulation. *Cold Spring Harbor Symp. Quant. Biol.* **22**:313–324.

Reynoldson, T. B., J. F. Gilliam, and R. M. Jaques. 1981. Competitive exclusion and coexistence in natural populations of *Polycelis nigra* and *P. tenuis* (Tricladida, Turbellaria). *Arch. Hydrobiol.* **92**:71–113.

Rhoades, D. F. 1979. Evolution of plant chemical defense against herbivores. In G. A. Rosenthal and D. H. Janzen (eds.). *Herbivores: Their interaction with secondary plant metabolites.* Academic Press, San Diego, Calif., pp. 3–54.

Rhoades, D. F. 1983. Responses of alder and willow to attack by tent caterpillar and webworms: Evidence for pheromonal sensitivity of willow. In P. A. Hedin (ed.). *Plant resistance to insects.* American Chemical Society, Washington, D.C., pp. 55–68.

Rhoades, D. F. 1985. Offensive-defensive interactions between herbivores and plants: Their relevance in herbivore population dynamics and ecological theory. *Am. Nat.* **125**:205–238.

Rhoades, D. F., and R. G. Cates. 1976. Toward a general theory of plant anti-herbivore chemistry. In J. W. Wallace and R. L. Mansell (eds.). *Biochemical interaction between plants and insects.* Planum Press, New York, pp. 168–213.

Rice, E. L. 1974. *Allelopathy.* Academic Press, San Diego, Calif.

Richards, A. J. (ed.). 1978. *The pollination of flowers by insects.* Academic Press, London.

Richards, O. W. 1927. Sexual selection and allied problems in the insects. *Biol. Rev.* **2**:298–364.

Richards, O. W. 1961. The theoretical and practical study of natural insect populations. *Annu. Rev. Entomol.* **6**:147–162.

Richards, O. W., and N. Waloff. 1961. A study of a natural population of *Phytodecta olivacea* (Forster) (Coleoptera: Chrysomelidae). *Philos. Trans. R. Soc. London B* **244**:205–257.

Richmond, R. C. 1972. Enzyme variability in the *Drosophila willistoni* group. III. Amounts of variability in the superspecies, *D. paulistorum. Genetics* **70**:87–112.

Ricklefs, R. E. 1970. Clutch size in birds: Outcome of opposing predator and prey adaptations. *Science* **168**:599–600.

Ricklefs, R. E., and D. Schluter (eds.). 1993. *Species diversity in ecological communities: Historical and geographical perspectives.* Univ. Chicago Press, Chicago.

Rickson, F. R. 1971. Glycogen plastids in Müllerian body cells of *Cecropia peltata:* A higher green plant. *Science* **173**:344–347.

Ridgway, R. L., and S. B. Vinson (eds.). 1977. *Biological control by augmentation of natural enemies.* Plenum Press, New York.

Ridley, M. 1978. Paternal care. *Anim. Behav.* **26**:904–934.

Riebesell, J. F. 1974. Paradox of enrichment in competitive systems. *Ecology* **55**:183–187.

Riley, C. V. 1892. The yucca moth and yucca pollination. *Mo. Bot. Gard. 3rd Annu. Rep.,* pp. 99–159.

Riley, C. V. 1893. Further notes on yucca insects and yucca pollination. *Proc. Biol. Soc. Wash.* **8**:41–54.

Ripper, W. E. 1956. Effect of pesticides on balance of arthropod populations. *Annu. Rev. Entomol.* **1**:403–438.

Risch, S. J. 1981. Insect herbivore abundance in tropical monocultures and polycultures: An experimental test of two hypotheses. *Ecology* **62**: 1325–1340.

Risch, S., and D. H. Boucher. 1976. What ecologists look for. *Bull. Ecol. Soc. Am.* **57**(3):8–9.

Risch, S. J., D. A. Andow, and M. Altieri. 1983. Agroecosystem diversity and pest control: Data, tentative conclusions and new research directions. *Environ. Entomol.* **12**:625–629.

Robinson, M. H. 1969. Defenses against visually hunting predators. *Evol. Biol.* **3**:225–259.

Robinson, T. 1974. Metabolism and function of alkaloids in plants. *Science* **184**:430–435.

Robinson, T. 1979. The evolutionary ecology of alkaloids. In G. A. Rosenthal and D. H. Janzen (eds.). *Herbivores: Their interaction with secondary plant metabolites.* Academic Press, San Diego, Calif., pp. 413–448.

Robinson, T. 1991. *The organic constituents of higher plants: Their chemistry and interrelationships.* Cordus Press, North Amherst, Mass.

Roeder, K. D. 1965. Moths and ultrasound. *Sci. Am.* **212**(4):94–102.

Roeder, K. D., and A. E. Treat. 1961. The detection and evasion of bats by moths. *Am. Sci.* **49**:135–148.

Roff, D. A. 1990. The evolution of flightlessness in insects. *Ecol. Monogr.* **60**:389–421.

Roff, D. A. 1992. *The evolution of life histories: Theory and analysis.* Chapman & Hall, New York.

Roff, D. A. 1994. The evolution of flightlessness: is history important? Evol. Ecol. **8**:639–657.

Rogers, D. J., and M. P. Hassell. 1974. General models for insect parasite and predator searching behaviour: Interference. *J. Anim. Ecol.* **43**:239–253.

Rohde, K. 1978a. Latitudinal gradients in species diversity and their causes. I. A review of the hypotheses explaining the gradients. *Biol. Zentralbl.* **97**:393–403.

Rohde, K. 1978b. Latitudinal gradients in species diversity and their causes. II. Marine parasitological evidence for a time hypothesis. *Biol. Zentralbl.* **94**:405–418.

Rohde, K. 1978c. Latitudinal differences in host-specificity of marine Monogenea and Digenea. *Marine Biol.* **47**:125–134.

Rohde, K. 1979. A critical evaluation of intrinsic and extrinsic factors responsible for niche restriction in parasites. *Am. Nat.* **114**:648–671.

Rohde, K. 1981. Niche width of parasites in species-rich and species-poor communities. *Experientia* **37**:359–361.

Rohde, K. 1992. Latitudinal gradients in species diversity: The search for the primary cause. *Oikos* **65**:514–527.

Rohfritsch, O., and J. D. Shorthouse. 1982. Insect galls. In G. Kahl and J. S. Schell (eds.). *Molecular biology of plant tumors.* Academic Press, San Diego, Calif., pp. 131–152.

Roininen, H., P. W. Price, and J. Tahvanainen. 1988. Field test of resource regulation by the bud-galling sawfly, *Euura mucronata,* on *Salix cinerea.* *Holarct. Ecol.* **11**:136–139.

Roininen, H., P. W. Price, R. Julkunen-Tiitto, and J. Tahvanainen. 1997. Oviposition stimulant for a galling sawfly, *Euura lasiolepis,* on willow is a phenolic glucoside. *J. Chem. Ecol.* In press.

Roitberg, B., and R. Prokopy. 1987. Insects that mark host plants. *BioScience* **37**:400–406.

Root, R. B. 1967. The niche exploitation pattern of the blue-gray gnatcatcher. *Ecol. Monogr.* **37**:317–350.

Root, R. B. 1973. Organization of a plant–arthropod association in simple and diverse habitats: The fauna of collards *(Brassica oleracea). Ecol. Monogr.* **43**:95–124.

Root, R. B. 1975. Some consequences of ecosystem texture. In S. A. Levin (ed.). *Ecosystem analysis and prediction.* Society for Industrial and Applied Mathematics, Philadelphia, Pa., pp. 83–97.

Root, R. B., and S. J. Chaplin. 1976. The life-styles of tropical milkweed bugs, *Oncopeltus* (Hemiptera: Lygaeidae) utilizing the same host. *Ecology* **57**:132–140.

Rose, M. R. 1984. Laboratory evolution of postponed senescence in *Drosophila melanogaster. Evolution* **38**:1004–1010.

Rosenthal, G. A. 1983. A seed-eating beetle's adaptations to a poisonous seed. *Sci. Am.* **249**(5):164–171.

Rosenthal, G. A., and M. R. Berenbaum (eds.). 1991. *Herbivores: Their in-*

teractions with secondary plant metabolites. Vol. 1. *The chemical partici-pants.* 2nd ed. Academic Press, San Diego, Calif.

Rosenthal, G. A., and M. R. Berenbaum (eds.). 1992. *Herbivores: Their in-teractions with secondary plant metabolites.* Vol. 2. *Ecological and evolu-tionary processes.* 2nd ed. Academic Press, San Diego, Calif.

Rosenthal, G. A., and D. H. Janzen (eds.). 1979. *Herbivores: Their interac-tion with secondary plant metabolites.* Academic Press, San Diego, Calif.

Rosenzweig, M. L. 1971. Paradox of enrichment: Destabilization of exploita-tion ecosystems in ecological time. *Science* 171:385–387.

Rosenzweig, M. L. 1995. *Species diversity in space and time.* Cambridge Univ. Press, Cambridge.

Rosenzweig, M. L., and Z. Abramsky. 1993. How are diversity and produc-tivity related? In R. E. Ricklefs and D. Schluter (eds.). *Species diversity in ecological communities: Historical and geographical perspectives.* Univ. Chicago Press, Chicago, pp. 52–65.

Ross, H. H. 1957. Principles of natural coexistence indicated by leafhopper populations. *Evolution* 11:113–129.

Ross, H. H. 1958. Further comments on niches and natural coexistence. *Evo-lution* 12:112–113.

Ross, H. H. 1965. Pleistocene events and insects. In H. E. Wright, Jr. and D. G. Frey (eds.). *The quaternary of the United States.* Princeton Univ. Press, Princeton, N.J., pp. 583–596.

Ross, H. H. 1967. The evolution and past dispersal of the Trichoptera. *Annu. Rev. Entomol.* 12:169–206.

Ross, H. H., and W. E. Ricker. 1971. The classification, evolution, and dis-persal of the winter stonefly genus *Allocapnia. Ill. Biol. Monogr.* 45:1–166.

Ross, H. H., and T. Yamamoto. 1967. Variations in the winter stonefly *Allo-capnia granulata* as indicators of Pleistocene faunal movements. *Ann. En-tomol. Soc. Am.* 60:447–458.

Rossiter, M. C. 1992. The impact of resource variation on population quality in herbivorous insects: A critical aspect of population dynamics. In M. D. Hunter, T. Ohgushi, and P. W. Price (eds.). *Effects of resource distribution on animal–plant interactions.* Academic Press, San Diego, Calif., pp. 13–42.

Rossiter, M. C. 1994. Maternal effects hypothesis of herbivore outbreak. *Bio-Science* 44:752–763.

Rossiter, M. C. 1995. Impact of life-history evolution on population dynam-ics: Predicting the presence of maternal effects. In N. Cappuccino and P. W. Price (eds.). *Population dynamics: New approaches and synthesis.* Aca-demic Press, San Diego, Calif., pp. 251–275.

Rothschild, M. 1965. The rabbit flea and hormones. *Endeavour* 24:162–168.

Rothschild, M., and T. Clay. 1952. *Fleas, flukes and cuckoos: A study of bird parasites.* Collins, London.

Rothschild, M., and B. Ford, 1964. Maturation and egg-laying of the rabbit flea (*Spilopsyllus cuniculi* Dale) induced by the external application of hydrocortisone. *Nature* **203**:210–211.

Rothschild, M., and B. Ford. 1972. Breeding cycle of the flea *Cediopsylla simplex* is controlled by breeding cycle of host. *Science* **178**:625–626.

Rothschild, M., and B. Ford. 1973. Factors influencing the breeding of the rabbit flea *(Spilopsyllus cuniculi):* A spring-time accelerator and a kairomone in nestling rabbit urine with notes on *Cediopsylla simplex,* another "hormone bound" species. *J. Zool. London* **170**:87–137.

Rothschild, M., and H. E. Hinton. 1968. Holding organs on the antennae of male fleas. *Proc. R. Entomol. Soc. London A* **43**:105–107.

Rothschild, M., and R. Traub. 1971. *A revised glossary of terms used in the taxonomy and morphology of fleas.* British Museum (Natural History), London.

Roubik, D. W. 1992. Loose niches in tropical communities: Why are there so few bees and so many trees? In M. D. Hunter, T. Ohgushi, and P. W. Price (eds.). *Effects of resource distribution on animal–plant interactions.* Academic Press, San Diego, Calif., pp. 327–354.

Roughgarden, J. 1975. Evolution of marine symbiosis: A simple cost-benefit model. *Ecology* **56**:1201–1208.

Roughgarden, J. 1979. *Theory of population genetics and evolutionary ecology: An introduction.* Macmillan, New York.

Roughgarden, J., R. M. May, and S. A. Levin (eds.). 1989. *Perspectives in ecological theory.* Princeton Univ. Press, Princeton, N.J.

Roush, R. T., and B. E. Tabashnik. 1990. *Pesticide resistance in arthropods.* Chapman & Hall, New York.

Rowell-Rahier, M., and J. M. Pasteels. 1992. Third trophic level influences of plant allelochemicals. In G. A. Rosenthal and M. R. Berenbaum (eds.). *Herbivores: Their interactions with secondary plant metabolites.* Vol. 2. 2nd ed. Academic Press, San Diego, Calif., pp. 243–277.

Royama, T. 1966. A re-interpretation of courtship feeding. *Bird Study* **13**:116–129.

Royama, T. 1970. Factors governing the hunting behaviour and selection of food by the great tit (*Parus major L.*). *J. Anim. Ecol.* **39**:619–668.

Royama, T. 1971. A comparative study of models for predation and parasitism. *Res. Popul. Ecol.* **13**(Suppl.1):1–91.

Royama, T. 1977. Population persistence and density dependence. *Ecol. Monogr.* **47**:1–35.

Royama, T. 1981a. Fundamental concepts and methodology for the analysis of animal population dynamics, with particular reference to univoltine species. *Ecol. Monogr.* **51**:473–493.

Royama, T. 1981b. Evaluation of mortality factors in insect life table analysis. *Ecol. Monogr.* **51**:495–505.

Royama, T. 1984. Population dynamics of the spruce budworm, *Choristoneura fumiferana. Ecol. Monogr.* **54**:429–462.

Royama, T. 1992. *Analytical population dynamics*. Chapman & Hall, London.

Royama, T. 1996. A fundamental problem in key factor analysis. *Ecology* 77:87–93.

Rudnew, D. F. 1963. Physiologischer Zustand der Wirtspflanze und Massenvermehrung von Forstschädlingen. *Z. Angew. Entomol.* 53:48–68.

Ruesink, W. G. 1975. Analysis and modeling in pest management. In R. L. Metcalf and W. H. Luckmann (eds.). *Introduction to insect pest management*. Wiley, New York, pp. 353–376.

Ruesink, W. G., and D. W. Onstad. 1994. Systems analysis and modeling in pest management. In R. L. Metcalf and W. H. Luckmann (eds.). *Introduction to insect pest management*. 3rd ed. Wiley, New York, pp. 393–419.

Russell, B. 1946. *History of Western philosophy*. Oxford Univ. Press, London.

Russell, E. P. 1989. Enemies hypothesis: A review of the effect of vegetational diversity on predatory insects and parasitoids. *Envir. Entomol.* **18:** 590–599.

Ryan, C. A. 1979. Proteinase inhibitors. In G. A. Rosenthal and D. H. Janzen (eds.). *Herbivores: Their interaction with secondary plant metabolites*. Academic Press, San Diego, Calif., pp. 599–618.

Salt, G. 1937. The sense used by *Trichogramma* to distinguish between parasitized and unparasitized hosts. *Proc. R. Soc. London Ser. B.* **122:**57–75.

Salt, G. 1970. *The cellular defense reactions of insects*. Cambridge Univ. Press, New York.

Salt, G. W. 1967. Predation in an experimental protozoan population (*Woodruffia–Paramecium*). *Ecol. Monogr.* 37:113–144.

Samarasinghe, S., and E. J. LeRoux. 1966. The biology and dynamics of the oystershell scale, *Lepidosaphes ulmi* (L.) (Homoptera: Coccidae) on apple in Quebec. *Ann. Entomol. Soc. Quebec* 11:206–292.

Samways, M. J. 1994. *Insect conservation biology*. Chapman & Hall, London.

Sanders, H. L. 1968. Marine benthic diversity: A comparative study. *Am. Nat.* **102:**243–282.

Sanders, H. L. 1969. Benthic marine diversity and the stability-time hypothesis. In G. M. Woodwell and H. H. Smith (eds.). *Diversity and stability in ecological systems*. Brookhaven Symp. Biol. 22, Brookhaven National Laboratory, Upton, N.Y., pp. 71–81.

Santos, P. F., and W. G. Whitford. 1981. The effects of microarthropods on litter decomposition in a Chihuahuan desert ecosystem. *Ecology* 62:654–663.

Santos, P. F., J. Phillips, and W. G. Whitford. 1981. The role of mites and nematodes in early stages of buried litter decomposition in a desert. *Ecology* 62:664–669.

Sapp, J. 1994. *Evolution by association: A history of symbiosis*. Oxford Univ. Press, New York.

Sargent, T. D. 1990. Startle as an anti-predator mechanism, with special reference to the underwing moths *(Catocala)*. In D. L. Evans and J. O. Schmidt (eds.). *Insect defenses: Adaptive mechanisms and strategies of prey and predators.* State Univ. New York Press, Albany, N. Y. pp. 229–249.

Sargent, T. D., and R. R. Keiper. 1969. Behavioral adaptations of cryptic moths. 1. Preliminary studies on bark-like species. *J. Lepid. Soc.* **23**:1–9.

Savage, J. M. 1958. The concept of ecologic niche, with reference to the theory of natural coexistence. *Evolution* **12**:111–112.

Schal, C., and W. J. Bell. 1982. Ecological correlates of paternal investment of urates in a tropical cockroach. *Science* **218**:170–173.

Schedl, K. E. 1958. Breeding habits of arboricole insects in Central Africa. *Proc. 10th Int. Congr. Entomol. Montreal* **1**:183–197.

Schell, J., M. Van Montagu, A. Depicker, D. De Waele, G. Engler, C. Genetello, J. P. Hernalsteens, M. Holsters, E. Messens, B. Silva, S. Van den Elsacker, N. Van Larebeke, and I. Zaenen. 1979. Crown-gall: Bacterial plasmids as oncogenic elements for eucaryotic cells. In I. Rubenstein, R. L. Phillips, C. E. Green, and B. G. Gengenbach (eds.). *Molecular biology of plants.* Academic Press, San Diego, Calif., pp. 315–337.

Schindler, D. W. 1978. Factors regulating phytoplankton production and standing crop in the world's fresh waters. *Limnol. Oceanogr.* **23**:478–486.

Schmidt, J. O. 1982. Biochemistry of insect venoms. *Annu. Rev. Entomol.* **27**:339–368.

Schmidt, D. J., and J. C. Reese. 1986. Sources of error in nutritional index studies of insects on artificial diet. *J. Insect. Physiol.* **32**:193–198.

Schmidt-Nielsen, K. 1972. Locomotion: Energy cost of swimming, flying, and running. *Science* **177**:222–228.

Schmidt-Nielsen, K. 1975. Scaling in biology: The consequences of size. *J. Exp. Zool.* **194**:287–308.

Schoener, T. W. 1982. The controversy over interspecific competition. *Am. Sci.* **70**:586–595.

Schoener, T. W. 1983. Field experiments on interspecific competition. *Am. Nat.* **122**:240–285.

Schoener, T. W. 1985. Some comments on Connell's and my reviews of field experiments on interspecific competition. *Am. Nat.* **125**:730–740.

Schoenly, K., and W. Reid. 1987. Dynamics of heterotrophic succession in carrion arthropod assemblages: Discrete seres or a continuum of change? *Oecologia* **73**:192–202.

Schoenly, K., R. A. Beaver, and T. A. Heumier. 1991. On the trophic relations of insects: A food-web approach. *Am. Nat.* **137**:597–638.

Scholtz, C. H., and S. L. Chown. 1995. The evolution of habitat use and diet in the Scarabaeoidea: A phylogenetic approach. In J. Pakaluk and S. A. Ślipiński (eds.). *Biology, phylogeny, and classification of Coleoptera: Papers celebrating the 80th birthday of Roy A. Crowson.* Muzeum i Instytut Zoologii PAN, Warsaw, pp. 355–374.

Scholtz, C. H. and E. Holm (eds.). 1985. *Insects of southern Africa.* Butterworth, Durban, South Africa.

Schultz, A. M. 1964. The nutrient-recovery hypothesis for arctic microtine cycles. II. Ecosystem variables in relation to the arctic microtine cycles. In D. J. Crisp (ed.). *Grazing in terrestrial and marine environments.* Blackwell Scientific, Oxford, pp. 57–68.

Schultz, J., D. Otte, and F. Enders. 1977. *Larrea* as a habitat component for desert arthropods. In T. Mabry, J. Hunziker, and D. Difeo (eds.). *Creosote bush: Biology and chemistry of* Larrea *in New World deserts.* U.S. Int. Biol. Prog. Syn. Ser. 6. Dowden, Hutchinson & Ross, Stroudsburg, Pa., pp. 176–208.

Schwenke, W. 1961. Walddüngung und Schadinsekten. *Anz. Schädlingsk.* **34**:129–134.

Schwerdtfeger, F. 1934. Studien über den Massenwechsel einiger Forstschädlinge. III. Untersuchungen über die Mortalität der Forleule (*Panolis flammea* Schiff.) im Krisenjahr einer Epidemie. *Mitt. Forstwirtsch. Forstwiss.* **5**:417–474.

Schwerdtfeger, F. 1968. *Ökologie der Tiere: Demökologie.* Vol. 2. Parey, Hamburg, Germany.

Scott, P. E., S. L. Buchmann, and M. K. O'Rourke. 1993. Evidence for mutualism between a flower-piercing carpenter bee and ocotillo: Use of pollen and nectar by nesting bees. *Ecol. Entomol.* **18**:234–240.

Scriber, J. M. 1996. Tiger tales: Natural history of native North American swallowtails. *Am. Entomol.* **42**:19–32.

Scriber, J. M., and F. Slansky. 1981. The nutritional ecology of immature insects. *Annu. Rev. Entomol.* **26**:183–211.

Scudder, G. G. E., and S. S. Duffey. 1972. Cardiac glycosides in the Lygaeinae (Hemiptera: Lygaeidae). *Can. J. Zool.* **50**:35–42.

Seifert, R. P. 1984. Does competition structure communities? Field studies on Neotropical *Heliconia* insect communities. In D. R. Strong, D. Simberloff, L. G. Abele, and A. B. Thistle (eds.). *Ecological communities: Conceptual issues and the evidence.* Princeton Univ. Press, Princeton, N.J., pp. 54–63.

Seifert, R. P., and F. H. Seifert. 1976. A community matrix analysis of *Heliconia* insect communities. *Am. Nat.* **110**:461–483.

Seifert, R. P., and F. H. Seifert. 1979. A *Heliconia* insect community in a Venezuelan cloud forest. *Ecology* **60**:462–467.

Seigler, D. S. 1977. Primary roles for secondary compounds. *Biochem. Syst. Ecol.* **5**:195–199.

Seigler, D. S., and P. W. Price. 1976. Secondary compounds in plants: Primary functions. *Am. Nat.* **110**:101–105.

Selander, R. K. 1976. Genic variation in natural populations. In F. J. Ayala (ed.). *Molecular evolution.* Sinauer, Sunderland, Mass., pp. 21–45.

Selander, R. K., and W. E. Johnson. 1973. Genetic variation among vertebrate species. *Annu. Rev. Ecol. Syst.* **4**:75–91.

Selander, R. K., S. Y. Yang, R. C. Lewontin, and W. E. Johnson. 1970. Genetic variation in the horseshoe crab *(Limulus polyphemus)*, a phylogenetic "relic." *Evolution* 24:402–414.

Sen-Sarma, P. K. 1974. Ecology and biogeography of the termites of India. In M. S. Mani (ed.). *Ecology and biogeography in India.* Monogr. Biol. 23, Dr. W. Junk, The Hague, The Netherlands, pp. 421–472.

Sernander, R. 1906. Entwurf einer Monographie der europaischen Myrmeco-choren. *Konigl. Svenska Vetenskap. Akad. Handlinger* 41:1–407.

Service, P. M. 1987. Physiological mechanisms of increased stress resistance in *Drosophila melanogaster* selected for postponed senescence. *Physiol. Zool.* 60:321–326.

Service, P. M. 1993. Laboratory evolution of longevity and reproductive fitness components in male fruit flies: Mating ability. *Evolution* 47:387–399.

Service, P. M., and A. J. Fales. 1993. Evolution of delayed reproductive senescence in male fruit flies: Sperm competition. *Genetica* 91:111–125.

Service, P. M., and R. E. Vossbrink. 1996. Genetic variation in "first" male effects on egg laying and remating by female *Drosophila melanogaster. Behav. Genet.* 26:39–48.

Service, P. M., E. W. Hutchinson, and M. R. Rose. 1988. Multiple genetic mechanisms for the evolution of senescence in *Drosophila melanogaster. Evolution* 42:708–716.

Shapiro, A. M. 1970. The role of sexual behavior in density-related dispersal of pierid butterflies. *Am. Nat.* 104:367–372.

Shaw, G. G., and C. H. A. Little. 1972. Effect of high urea fertilization of balsam fir trees on spruce budworm development. In J. G. Rodriguez (ed.). *Insect and mite nutrition.* North-Holland, Amsterdam, pp. 589–597.

Sheehan, W. 1986. Response by specialist and generalist natural enemies to agroecosystem diversification: A selective review. *Environ. Entomol.* 15:456–461.

Sheehan, W. 1991. Host range patterns of hymenopteran parasitoids of exophytic lepidopteran folivores. In E. A. Bernays (ed.). *Insect–plant interactions.* Vol. 3. CRC Press, Boca Raton, Fla., pp. 209–248.

Shelford, V. E. 1907. Preliminary note on the distribution of the tiger beetles *(Cicindela)* and its relation to plant succession. *Biol. Bull.* 14:9–14.

Shelford, V. E. 1913. *Animal communities in temperate America.* Univ. Chicago Press, Chicago.

Shelford, V. E. 1963. *The ecology of North America.* Univ. Illinois Press, Urbana, Ill.

Sheppard, P. M. 1952. Natural selection in two colonies of the polymorphic land snail *Cepaea nemoralis. Heredity* 6:233–238.

Sheppard, P. M. 1958. *Natural selection and heredity.* Hutchinson, London.

Sheppard, P. M. 1961. Recent genetical work on polymorphic mimetic Papilios. In J. S. Kennedy (ed.). *Insect polymorphism.* Symp. R. Entomol. Soc. London 1, Royal Entomological Society of London, London, pp. 20–29.

Sheppard, P. M. 1962. Some aspects of the geography, genetics and taxonomy of a butterfly. In D. Nichols (ed.). *Taxonomy and geography.* Syst. Assoc. Publ. 4, Systematics Association, London, pp. 135–152.

Shields, O. 1967. Hilltopping: An ecological study of summit congregation behavior of butterflies on a southern California hill. *J. Res. Lepid.* 6:69–178.

Shipley, B., and R. H. Peters. 1990. A test of the Tilman model of plant strategies: Relative growth rate and biomass partitioning. *Am. Nat.* 136:139–153.

Shipley, B., and R. H. Peters. 1991. The seduction by mechanism: A reply to Tilman. *Am. Nat.* 138:1276–1282.

Showler, A. T. 1995. Locust (Orthoptera: Acrididae) outbreak in Africa and Asia, 1992–1994: An overview. *Am. Entomol.* 41:179–185.

Shreeve, T. G., and C. F. Mason. 1980. The number of butterfly species in woodlands. *Oecologia* 45:414–418.

Shugart, H. H., and R. V. O'Neill (eds.). 1979. *Systems ecology.* Dowden, Hutchinson & Ross, Stroudsburg, Pa.

Shugart, H. H., and D. C. West. 1981. Long-term dynamics of forest ecosystems. *Am. Sci.* 69:647–652.

Sih, A. 1980. Optimal behavior: Can foragers balance the two conflicting demands? *Science* 210:1041–1043.

Silvertown, J. W. 1982. No evolved mutualism between grasses and grazers. *Oikos* 38:253–254.

Simberloff, D.S. 1969. Experimental zoogeography of islands: A model for insular colonization. *Ecology* 50:296–314.

Simberloff, D. 1976. Trophic structure determination and equilibrium in an arthropod community. *Ecology* 57:395–398.

Simberloff, D. S. 1978. Colonization of islands by insects: Immigration, extinction, and diversity. In L. A. Mound and N. Waloff (eds.). *Diversity of insect faunas.* Symp. R. Entomol. Soc. London 9, Royal Entomological Society of London, London, pp. 139–153.

Simberloff, D. 1981. Community effects of introduced species. In M. H. Nitecki (ed.). *Biotic crises in ecological and evolutionary time.* Academic Press, San Diego, Calif., pp. 53–81.

Simberloff, D. S. 1984. Morphological and taxonomic similarity and combinations of coexisting birds in two archipelagoes. In D. Strong, D. Simberloff, L. Abele, and A. B. Thistle (eds.). *Ecological communities: Conceptual issues and the evidence.* Princeton Univ. Press, Princeton, N.J.

Simberloff, D., and W. Boecklen. 1981. Santa Rosalia reconsidered: Size ratios and competition. *Evolution* 35:1206–1288.

Simberloff, D. S., and E. O. Wilson. 1969. Experimental zoogeography of islands: The colonization of empty islands. *Ecology* 50:278–296.

Simberloff, D. S., and E. O. Wilson. 1970. Experimental zoogeography of islands. A two-year record of colonization. *Ecology* 51:934–937.

Simpson, G. G. 1940. Mammals and land bridges. *J. Wash. Acad. Sci.* 30:137–163.

Simpson, G. G. 1964. Species density of North American recent mammals. *Syst. Zool.* **13:**57–73.

Simpson, B. B., and J. L. Neff. 1983. Evolution and diversity of floral rewards. In C. E. Jones and R. J. Little (eds.). *Handbook of experimental pollination biology.* Scientific and Academic Editions, New York, pp. 142–159.

Sinclair, A. R. E. 1970. Studies of the ecology of the East African buffalo. Ph.D. thesis, Oxford Univ.

Sinclair, A. R. E. 1973. Regulation, and population models for a tropical ruminant. *E. Afr. Wildl. J.* **11:**307–316.

Singer, M. C. 1971. Evolution of food-plant preference in the butterfly *Euphydryas editha. Evolution* **25:**383–389.

Singer, M. C. 1972. Complex components of habitat suitability within a butterfly colony. *Science* **176:**55–77.

Singer, M. C. 1983. Determinants of multiple host use by a phytophagous insect population. *Evolution* **37:**389–403.

Singer, M. C. 1986. The definition and measurement of oviposition preference in plant-feeding insects. In J. R. Miller and T. A. Miller (eds.). *Insect–plant interactions.* Springer-Verlag, New York, pp. 65–94.

Singer, M. C. 1994. Behavioral constraints on the evolutionary expansion of insect diet: A case history from checkerspot butterflies. In L. Real (ed.). *Behavioral mechanisms in evolutionary ecology.* Univ. Chicago Press, Chicago, pp. 279–296.

Singer, M. C., D. Ng, and C. D. Thomas. 1988. Heritability of oviposition preference and its relationship to offspring performance within a single insect population. *Evolution* **42:**977–985.

Singer, M. C., D. Ng, D. Vasco, and C. D. Thomas. 1992. Rapidly evolving associations among oviposition preferences fail to constrain evolution of insect diet. *Am. Nat.* **139:**9–20.

Singer, M. C., C. D. Thomas, and C. Parmesan. 1993. Rapid human-induced evolution of insect–host associations. *Nature* **366:**681–683.

Singer, M. C., C. D. Thomas, H. L. Billington, and C. Parmesan. 1994. Correlates of speed of evolution of host preference in a set of twelve populations of the butterfly *Euphydryas editha. Ecoscience* **1:**107–114.

Sinha, R. N. 1968. Adaptive significance of mycophagy in stored-product Arthropoda. *Evolution* **22:**785–798.

Skaife, S. H., J. Ledger, and A. Bannister. 1979. *African insect life.* Rev. ed. Struik, Cape Town, South Africa.

Skellam, J. G. 1951. Random dispersal in theoretical populations. *Biometrika* **38:**196–218.

Skutch, A. F. 1949. Do tropical birds rear as many young as they can nourish? *Ibis* **91:**430–455.

Skutch, A. F. 1967. Adaptive limitation of the reproductive rate of birds. *Ibis* **109:**579–599.

Skutch, A. F. 1971. *A naturalist in Costa Rica.* Univ. Florida Press, Gainesville, Fla.

Sláma, K. 1969. Plants as a source of materials with insect hormone activity. *Entomol. Exp. Appl.***12**:721–728.

Sláma, K. 1979. Insect hormones and antihormones in plants. In G. A. Rosenthal and D. H. Janzen (eds.). *Herbivores: Their interaction with secondary plant metabolites.* Academic press, San Diego, Calif., pp. 683–700.

Sláma, K., and C. M. Williams. 1965. Juvenile hormone activity for the bug *Pyrrhocoris apterus. Proc. Natl. Acad. Sci. USA* **54**:411–414.

Sláma, K., and C. M. Williams. 1966. The juvenile hormone. V. The sensitivity of the bug, *Pyrrhocoris apterus,* to a hormonally active factor in American paper-pulp. *Biol. Bull.* **130**:235–246.

Slansky, F. 1974. Energetic and nutritional interactions between larvae of the imported cabbage butterfly, *Pieris rapae* L., and cruciferous food plants. Ph.D. thesis, Cornell Univ., Ithaca, N.Y.

Slansky, F., and P. Feeny. 1977. Stabilization of the rate of nitrogen accumulation by larvae of the cabbage butterfly on wild and cultivated food plants. *Ecol. Monogr.* **47**:209–228.

Slansky, F., and J. G. Rodriguez (eds.). 1987a. *Nutritional ecology of insects, mites, spiders, and related invertebrates.* Wiley, New York.

Slansky, F., and J. G. Rodriguez. 1987b. Nutritional ecology of insects, mites, spiders, and related invertebrates: An overview. In F. Slansky and J. G. Rodriguez (eds.). *Nutritional ecology of insects, mites, spiders, and related invertebrates.* Wiley, New York, pp. 1–69.

Slansky, F., and J. M. Scriber. 1985. Food consumption and utilization. In G. A. Kerkut and L. I. Gilbert (eds.). *Comprehensive insect physiology, biochemistry and pharmacology.* Vol. 4. Pergamon Press, Oxford, pp. 87–163.

Slobodkin, L. B. 1962. Energy in animal ecology. *Adv. Ecol. Res.* **1**:69–101.

Slobodkin, L. B., F. E. Smith, and N. G. Hairston. 1967. Regulation in terrestrial ecosystems, and the implied balance of nature. *Am. Nat.* **101**: 109–124.

Smart, J., and N. F. Hughes. 1973. The insect and the plant: Progressive paleoecological integration. In H. F. van Emden (ed.). *Insect/plant relationships.* Symp. R. Entomol. Soc. London 6, Royal Entomological Society of London, London, pp. 143–155.

Smith, A. P. 1973. Stratification of temperate and tropical forests. *Am. Nat.* **107**:671–683.

Smith, F. E. 1954. Quantitative aspects of population growth. In E. J. Boell (ed.). *Dynamics of growth processes.* Princeton Univ. Press, Princeton, N.J., pp. 277–294.

Smith, F. E. 1961. Density dependence in the Australian thrips. *Ecology* **42**:403–407.

Smith, H. S. 1935. The role of biotic factors in the determination of population density. *J. Econ. Entomol.* **28**:873–898.

Smith, J. N. M. 1974. The food searching behaviour of two European thrushes. II: The adaptiveness of the search patterns. *Behaviour* **49**:1–61.

Smith, J. N. M., and R. Dawkins. 1971. The hunting behaviour of individual great tits in relation to spatial variations in their food density. *Anim. Behav.* **19**:695–706.

Smith, J. N. M., and H. P. A. Sweatman. 1974. Food-searching behavior of titmice in patchy environments. *Ecology* **55**:1216–1232.

Smith, K. G. V. 1986. *A manual of forensic entomology.* Cornell Univ. Press, Ithaca, N.Y.

Smith, P. H. 1972. The energy relations of defoliating insects in a hazel coppice. *J. Anim. Ecol.* **41**:567–587.

Smith, P. W. 1957. An analysis of post-Wisconsin biogeography of the prairie peninsula region based on distributional phenomena among terrestrial vertebrate populations. *Ecology* **38**:205–218.

Smith, R. L. 1979a. Paternity assurance and altered roles in the mating behaviour of a giant water bug, *Abedus herberti* (Heteroptera: Belostomatidae). *Anim. Behav.* **27**:716–725.

Smith, R. L. 1979b. Repeated copulation and sperm precedence: Paternity assurance for a male brooding water bug. *Science* **205**:1029–1031.

Smith, R. L. 1980. Evolution of exclusive postcopulatory paternal care in the insects. *Fla. Entomol.* **63**:65–78.

Soderstrom, T. R., and C. E. Calderon. 1971. Insect pollination in tropical rain forest grasses. *Biotropica* **3**:1–16.

Sokoloff, A. 1972. *The biology of* Tribolium *with special emphasis on genetic aspects.* Vol. I. Oxford Univ. Press, London.

Sokoloff, A. 1974. *The biology of* Tribolium *with special emphasis on genetic aspects.* Vol. II. Oxford Univ. Press, London.

Sokoloff, A. 1977. *The biology of* Tribolium *with special emphasis on genetic aspects.* Vol. III. Oxford Univ. Press, London.

Solomon, M. E. 1949. The natural control of animal populations. *J. Anim. Ecol.* **18**:1–35.

Solomon, M. E. 1964. Analysis of processes involved in the natural control of insects. *Adv. Ecol. Res.* **2**:1–58.

Solomon, M. E. 1969. *Population dynamics.* St. Martin's Press, New York. Reprinted 1973.

Somero, G. N. 1978. Temperature adaptation of enzymes: Biological optimization through structure–function compromises. *Annu. Rev. Ecol. Syst.* **9**:1–29.

Sondheimer, E., and J. B. Simeone (eds.). 1970. *Chemical ecology.* Academic Press, San Diego, Calif.

Southwood, T. R. E. 1960a. The abundance of the Hawaiian trees and the number of their associated insect species. *Proc. Hawaii. Entomol. Soc.* **17**:299–303.

Southwood, T. R. E. 1960b. The evolution of the insect–host tree relationship: A new approach. *Int. Congr. Entomol. 11, Vienna* **1**:651–655.

Southwood, T. R. E. 1961. The number of species of insect associated with various trees. *J. Anim. Ecol.* **30**:1–8.

Southwood, T. R. E. 1966. *Ecological methods with particular reference to the study of insect populations.* Methuen, London.

Southwood, T. R. E. (ed.). 1968. *Insect abundance.* Symp. R. Entomol. Soc. London 4, Royal Entomological Society of London, London.

Southwood, T. R. E. 1975. The dynamics of insect populations. In D. Pimentel (ed.). *Insects, science, and society.* Academic Press, San Diego, Calif., pp. 151–199.

Southwood, T. R. E. 1977a. The relevance of population dynamic theory to pest status. In J. M. Cherrett and G. R. Sagar (eds.). *Origins of pest, parasite, disease and weed problems.* Symp. Br. Ecol. Soc. 18, British Ecological Society, London, pp. 35–54.

Southwood, T. R. E. 1977b. Habitat, the templet for ecological strategies? *J. Anim. Ecol.* **46**:337–365.

Southwood, T. R. E. 1979. *Ecological methods with particular reference to the study of insect populations.* 2nd ed. Methuen, London.

Southwood, T. R. E. 1988. Tactics, strategies and templets. *Oikos* **52**:3–18.

Southwood, T. R. E., and H. N. Comins. 1976. A synoptic population model. *J. Anim. Ecol.* **45**:949–965.

Southwood, T. R. E., V. K. Brown, and P. M. Reader. 1979. The relationships of plant and insect diversities in succession. *Biol. J. Linn. Soc.* **12**:327–348.

Spanner, D. C. 1963. The green leaf as a heat engine. *Nature* **198**:934–937.

Spiess, E. B. 1977. *Genes in populations.* Wiley, New York.

Spieth, H. T. 1968. Evolutionary implications of sexual behavior in *Drosophila. Evol. Biol.* **2**:157–193.

Spieth, H. T. 1974. Courtship behavior in *Drosophila. Annu. Rev. Entomol.* **19**:385–405.

Springett, B. P. 1968. Aspects of the relationship between burying beetles, *Necrophorus* spp. and the mite, *Poecilochirus necrophori* Vitz. *J. Anim. Ecol.* **37**:417–424.

Spruce, R. 1908. *Notes of a botanist on the Amazon and Andes.* Vol. 2. Macmillan, London.

St. Amant, J. L. S. 1970. The detection of regulation in animal populations. *Ecology* **51**:823–828.

Stamp, N. E., and T. M. Casey (eds.). 1993. *Caterpillars: Ecological and evolutionary constraints on foraging.* Chapman & Hall, New York.

Stanton, M. L. 1983. Spatial patterns in the plant community and their effects upon insect search. In S. Ahmad (ed.). *Herbivorous insects: Host-seeking behavior and mechanisms.* Academic Press, San Diego, Calif., pp. 125–157.

Stark, R. W. 1959. Population dynamics of the lodgepole needle miner, *Recurvaria starki* Freeman, in Canadian Rocky Mountain parks. *Can. J. Zool.* **37**:917–943.

Stark, R. W. 1965. Recent trends in forest entomology. *Annu. Rev. Entomol.* **10**:303–324.

Starr, M. P. 1975. A generalized scheme for classifying organismic associations. *Symp. Soc. Exp. Biol.* **29**:1–20.

Stearns, S. C. 1976. Life-history tactics: A review of the ideas. *Q. Rev. Biol.* **51**:3–47.

Stearns, S. C. 1977. The evolution of life history traits: A critique of the theory and a review of the data. *Annu. Rev. Ecol. Syst.* **8**:145–171.

Stearns, S. C. 1981. On measuring fluctuating environments: Predictability, constancy, and contingency. *Ecology* **62**:185–199.

Stearns, S. C. 1992. *The evolution of life histories.* Oxford Univ. Press, Oxford.

Steiner, W. W. M. 1977. Niche width and genetic variation in Hawaiian *Drosophila. Am. Nat.* **111**:1037–1045.

Steiner, K. E., and V. B. Whitehead. 1988. The association between oil-producing flowers and oil-collecting bees in the Drakensberg of southern Africa. *Monogr. Syst. Bot. Mo. Bot. Gard.* **25**:259–277.

Steiner, K. E., and V. B. Whitehead. 1990. Pollinator adaptation to oil-secreting flowers—*Rediviva* and *Diascia. Evolution* **44**:1701–1707.

Steiner, K. E., and V. B. Whitehead. 1991. Oil flowers and oil bees: Further evidence for pollinator adaptation. *Evolution* **45**:1493–1501.

Steiner, K. E., V. B. Whitehead, and S. D. Johnson. 1994. Floral and pollinator divergence in two sexually deceptive South African orchids. *Am. J. Bot.* **81**:185–194.

Steinhaus, E. A. 1946. *Insect microbiology.* Comstock, Ithaca, N.Y.

Steinhaus, E. A. (ed.). 1963. *Insect pathology: An advanced treatise.* Vols. 1 and 2. Academic Press, San Diego, Calif.

Stephenson, S. N. 1973. Fertilization responses in six southern Michigan old-field communities. *Bull. Ecol. Soc. Am.* **54**(1):33 (abstract).

Stevens, G. C. 1989. The latitudinal gradient in geographical range: How so many species coexist in the tropics. *Am. Nat.* **133**:240–256.

Steward, R. C. 1977. Industrial and non-industrial melanism in the peppered moth, *Biston betularia* (L.). *Ecol. Entomol.* **2**:231–243.

Stiling, P. D. 1996. *Ecology: Theories and applications.* 2nd ed. Prentice Hall, Upper Saddle River, N.J.

Stockner, J. G. 1971. Ecological energetics and natural history of *Hedriodiscus truquii* (Diptera) in two thermal spring communities. *J. Fish. Res. Bd. Can.* **28**:73–94.

Stokes, A. W. (ed.). 1974. *Territory.* Dowden, Hutchinson & Ross, Stroudsburg, Pa.

Stoltz, D. B., and S. B. Vinson. 1979. Viruses and parasitism in insects. *Adv. Virus Res.* **24**:125–171.

Stoltz, D. B., S. B. Vinson, and E. A. MacKinnon. 1976. Baculovirus-like particles in the reproductive tracts of female parasitoid wasps. *Can. J. Microbiol.* **22**:1013–1023.

Stork, N. E. 1988. Insect diversity: Facts, fiction and speculation. *Biol. J. Linn. Soc.* **35**:321–337.

Stoutamire, W. P. 1974. Australian terrestrial orchids, thynnid wasps, and pseudocopulation. *Am. Orchid Soc. Bull.* **43**:13–18.

Straw, R. M. 1955. Hybridization, homogamy, and sympatric speciation. *Evolution* **9**:441–444.

Strong, D. R. 1974a. Nonasymptotic species richness models and the insects of British trees. *Proc. Natl. Acad. Sci. USA* **71**:2766–2769.

Strong, D. R. 1974b. The insects of British trees: Community equilibration in ecological time. *Ann. Mo. Bot. Gard.* **61**:692–701.

Strong, D. R. 1974c. Rapid asymptotic species accumulation in phytophagous insect communities: The pests of cacao. *Science* **185**:1064–1066.

Strong, D. R. 1979. Biogeographic dynamics of insect–host plant communities. *Annu. Rev. Entomol.* **24**:89–119.

Strong, D. R. 1981. The possibility of insect communities without competition: Hispine beetles on *Heliconia*. In R. F. Denno and H. Dingle (eds.). *Insect life history patterns: Habitat and geographic variation.* Springer-Verlag, New York, pp. 183–194.

Strong, D. R. 1982. Harmonious coexistence of hispine beetles on *Heliconia* in experimental and natural communities. *Ecology* **63**:1039–1049.

Strong, D. R. 1984a. The question of interspecies competition among phytophagous insects. In D. R. Strong, D. Simberloff, L. G. Abele, and A. B. Thistle (eds.). *Ecological communities: Conceptual issues and the evidence.* Princeton Univ. Press, Princeton, N.J.

Strong, D. R. 1984b. Banana's best friend. *Nat. Hist.* **93**(12):50–57.

Strong, D. R. 1992. Are trophic cascades all wet? Differentiation and donor-control in speciose ecosystems. *Ecology* **73**:747–754.

Strong, D. R., and D. A. Levin. 1975. Species richness of the parasitic fungi of British trees. *Proc. Natl. Acad. Sci. USA* **72**:2116–2119.

Strong, D. R., E. D. McCoy, and J. R. Rey. 1977. Time and the number of herbivore species: The pests of sugarcane. *Ecology* **58**:167–175.

Strong, D. R., L. A. Szyska, and D. S. Simberloff. 1979. Tests of community-wide character displacement against null hypotheses. *Evolution* **33**:897–913.

Strong, D. R., J. H. Lawton, and T. R. E. Southwood. 1984a. *Insects on plants: Community patterns and mechanisms.* Blackwell Scientific Oxford.

Strong, D. R., D. Simberloff, L. G. Abele, and A. B. Thistle (eds.). 1984b. *Ecological communities: Conceptual issues and the evidence.* Princeton Univ. Press, Princeton, N.J.

Stuart, A. M. 1972. Behavioral regulatory mechanisms in the social homeostasis of termites (Isoptera). *Am. Zool.* **12**:589–594.

Sudd, J. H. 1967. *An introduction to the behavior of ants.* St. Martin's Press, New York.

Sun, M. 1988. Costa Rica's campaign for conservation. *Science* **239**:1366–1369.

Swain, T. 1976. Secondary compounds: Primary products. In M. Luckner, K. Mothes, and L. Norer (eds.). *Secondary metabolism and coevolution.*

Deutsche Akademie Naturforscher Leopoldina, Halle, Germany, pp. 411–421.

Swain, T. 1979. Tannins and lignins. In G. A. Rosenthal and D. H. Janzen (eds.). *Herbivores: Their interaction with secondary plant metabolites.* Academic Press, San Diego, Calif., pp. 657–682.

Swaine, J. M. 1918. Canadian bark-beetles. Part II. A preliminary classification, with an account of the habits and means of control. *Dom. Can. Dept. Agric. Tech. Bull.* **14**:1–143.

Sweet, M. H. 1963. The biology and ecology of the Rhyparochromine of New England (Heteroptera: Lygaeidae). Part I. *Entomol. Am.* **43**:1–124.

Sweet, M. H. 1964. The biology and ecology of the Rhyparochromine of New England (Heteroptera: Lygaeidae). Part II. *Entomol. Am.* **44**:1–201.

Tabashnik, B. E., and F. Slansky. 1987. Nutritional ecology of forb foliage-chewing insects. In F. Slansky and J. G. Rodriguez (eds.). *Nutritional ecology of insects, mites, spiders, and related invertebrates.* Wiley, New York, pp. 71–103.

Tahvanainen, J. O. 1972. Phenology and microhabitat selection of some flea beetles (Coleoptera: Chrysomelidae) on wild and cultivated crucifers in central New York. *Entomol. Scand.* **3**:130–138.

Tahvanainen, J. O., and R. B. Root. 1970. The invasion and population outbreak of *Psylloides napi* (Coleoptera: Chrysomelidae) on yellow rocket *(Barbarea vulgaris)* in New York. *Ann. Entomol. Soc. Am.* **63**:1479–1480.

Tahvanainen, J. O., and R. B. Root. 1972. The influence of vegetational diversity on the population ecology of a specialized herbivore, *Phyllotreta cruciferae* (Coleoptera: Chrysomelidae). *Oecologia* **10**:321–346.

Takabayashi, J., M. Dicke, and M. A. Posthumus. 1991. Variation in composition of predator-attracting allelochemicals emitted by herbivore-infested plants: Relative influence of plant and herbivore. *Chemoecology* **2**:1–6.

Takahashi, F. 1968. Functional response to host density in a parasitic wasp, with reference to population regulation. *Res. Popul. Ecol.* **10**:54–68.

Tallamy, D. W., and M. J. Raupp (eds.). 1991. *Phytochemical induction by herbivores.* Wiley, New York.

Tamarin, R. M. 1978. *Population regulation.* Dowden, Hutchinson & Ross, Stroudsburg, Pa.

Tammaru, T., and E. Haukioja. 1997. Capital breeders and income breeders among Lepidoptera: Consequences to population dynamics. *Oikos.* In press.

Tansley, A. G. 1935. The use and abuse of vegetational concepts and terms. *Ecology* **16**:284–307.

Tansley, A. G. 1939. *The British islands and their vegetation.* Cambridge Univ. Press. London.

Taylor, L. R. 1979. The Rothamsted insect survey: An approach to the theory and practice of synoptic pest forecasting in agriculture. In R. L. Rabb and G. G. Kennedy (eds.). *Movement of highly mobile insects: Concepts and methodology in research*. North Carolina State Univ., Raleigh, N.C., pp. 148–185.

Taylor, R. A. J. 1981a. The behavioural basis of redistribution. I. The Δ-model concept. *J. Anim. Ecol.* 50:573–586.

Taylor, R. A. J. 1981b. The behavioural basis of redistribution. II. Simulations of the Δ-model. *J. Anim. Ecol.* 50:587–604.

Taylor, R. A. J., and L. R. Taylor. 1979. A behavioural model for the evolution of spatial dynamics. In R. M. Anderson, B. D. Turner, and L. R. Taylor (eds.). *Population dynamics.* Symp. Brit. Ecol. Soc. 20, British Ecological Society, London, pp. 1–27.

Taylor, R. J. 1974. Role of learning in insect parasitism. *Ecol. Monogr.* **44**:89–104.

Temple, S. A. 1977. Plant-animal mutualism: Coevolution with dodo leads to near extinction of plant. *Science* **197**:885–886.

Tepedino, V. J., and N. L. Stanton. 1976. Cushion plants as islands. *Oecologia* **25**:243–256.

Terborgh, J. 1973. On the notion of favorableness in plant ecology. *Am. Nat.* **107**:481–501.

Terborgh, J. W., and J. Faaborg. 1980. Saturation of bird communities in the West Indies. *Am. Nat.* **116**:178–195.

Tetley, J. 1947. Increased variability accompanying an increase in population in a colony of *Argynnis selene* (Lep. Nymphalidae). *Entomologist* **80**: 177–179.

Thayer, G. H. 1909. *Concealing-coloration in the animal kingdom: An exposition of the laws of disguise through color and pattern.* Macmillan, New York.

Thiery, T. G. 1982. Environmental instability and community diversity. *Biol. Rev.* **57**:671–710.

Thompson, D. W. 1942. *On growth and form.* 2nd ed. Cambridge Univ. Press, Cambridge.

Thompson, J. N. 1975. Reproductive strategies: A review of concepts with particular reference to the number of offspring produced per reproductive period. *Biologist* **57**:14–25.

Thompson, J. N. 1978. Within-patch structure and dynamics in *Pastinaca sativa* and resource availability to a specialized herbivore. *Ecology* **59**: 443–448.

Thompson, J. N. 1982. *Interaction and coevolution.* Wiley, New York.

Thompson, J. N. 1983. Selection pressures on phytophagous insects feeding on small host plants. *Oikos* **40**:438–444.

Thompson, J. N. 1988. Evolutionary ecology of the relationship between oviposition preference and performance of offspring in phytophagous insects. *Entomol. Exp. Appl.* **47**:3–14.

Thompson, J. N. 1994. *The coevolutionary process.* Univ. Chicago Press, Chicago.

Thompson, J. N., and O. Pellmyr. 1991. Evolution of oviposition behavior and host preference in Lepidoptera. *Annu. Rev. Entomol.* **36**:65–89.

Thompson, J. N., and P. W. Price. 1977. Plant plasticity, phenology and herbivore dispersion: Wild parsnip and the parsnip webworm. *Ecology* **58**: 1112–1119.

Thompson, W. R. 1939. Biological control and the theories of the interactions of populations. *Parasitology* **31**:299–388.

Thornhill, R. 1974. Evolutionary ecology of Mecoptera (Insecta). Ph.D. dissertation, Univ. Michigan, Ann Arbor, Mich.

Thornhill, R. 1976. Sexual selection and paternal investment in insects. *Am. Nat.* **110**:153–163.

Thornhill, R., and J. Alcock. 1983. *The evolution of insect mating systems.* Harvard Univ. Press, Cambridge, Mass.

Thorp, R. W. 1969. Systematics and ecology of bees of the subgenus *Diandrena* (Hymenoptera: Andrenidae). *Univ. Calif. Publ. Entomol.* **52**:1–146.

Thorp, R. W., D. L. Briggs, J. R. Estes, and E. H. Erickson. 1975. Nectar fluorescence under ultraviolet irradiation. *Science* **189**:476–478.

Tilman, D. 1978. Cherries, ants and tent caterpillars: Timing of nectar production in relation to susceptibility of caterpillars to ant predation. *Ecology* **59**:686–692.

Tilman, D. 1988. *Plant strategies and the dynamics and structure of plant communities.* Princeton Univ. Press, Princeton, N.J.

Tilman, D. 1989. Population dynamics and species interactions. In J. Roughgarden, R. M. May, and S. A. Levin (eds.). *Perspectives in ecological theory.* Princeton Univ. Press, Princeton, N.J., pp. 89–100.

Tilman, D. 1991a. Relative growth rates and plant allocation patterns. *Am. Nat.* **138**:1269–1275.

Tilman, D. 1991b. The schism between theory and ardent empiricism: A reply to Shipley and Peters. *Am. Nat.* **138**:1283–1286.

Tilman, D., and S. Pacala. 1993. The maintenance of species richness in plant communities. In R. E. Ricklefs and D. Schluter (eds.). *Species diversity in ecological communities: Historical and geographical perspectives.* Chicago Univ. Press, Chicago, pp. 13–25.

Tinbergen, L. 1960. The natural control of insects in pinewoods. I. Factors influencing the intensity of predation by songbirds. *Arch. Neerl. Zool.* **13**:265–343.

Tinbergen, L., and H. Klomp. 1960. The natural control of insects in pinewoods. II. Conditions for damping of Nicholson oscillations in parasite–host systems. *Arch. Neerl. Zool.* **13**:344–379.

Tinkle, D. W. 1969. The concept of reproductive effort and its relation to the evolution of life histories of lizards. *Am. Nat.* **103**:501–516.

Toft, C. A., and A. Aeschlimann. 1991. Introduction: Coexistence or conflict? In C. A. Toft, A. Aeschlimann, and L. Bolis (eds.). *Parasite–host associations: Coexistence or conflict?* Oxford Univ. Press, Oxford.

Toft, C. A., A. Aeschlimann, and L. Bolis (eds.). 1991. *Parasite–host associations: coexistence or conflict?* Oxford Univ. Press, Oxford.

Tonzetich, J., and C. L. Ward. 1973a. Adaptive chromosomal polymorphism in *Drosophila melanica. Evolution* **27**:486–494.

Tonzetich, J., and C. L. Ward. 1973b. Interaction effects of temperature and humidity on pupal survival in *Drosophila melanica. Evolution* **27**:495–504.

Tostowaryk, W. 1972. The effect of prey defense on the functional response of *Podisus modestus* (Hemiptera: Pentatomidae) to densities of the sawflies *Neodiprion swainei* and *N. pratti banksianae* (Hymenoptera: Neodiprionidae). *Can. Entomol.* **104**:61–69.

Tovar, D. C., J. T. M. Montiel, R. C. Bolaños, H. O. Yates, and J. E. F. Lara. 1995. *Insectos forestales de México.* Univ. Autónoma Chapingo, Chapingo, Mexico.

Townes, H. 1969. *The genera of Ichneumonidae.* Part I. Mem. Am. Entomol. Inst. 11, American Entomological Institute, Ann Arbor, Mich.

Townes, H. 1972. Ichneumonidae as biological control agents. *Proc. Tall Timbers Conf. Ecol. Anim. Control Habitat Manage.* **3**:235–248.

Traub, R. 1972. The Gunong Benom expedition 1967. 12. Notes on zoogeography, convergent evolution and taxonomy of fleas (Siphonaptera), based on collections from Gunong Benom and elsewhere in South-East Asia. II. Convergent evolution. *Bull. Br. Mus. (Nat. Hist,) Zool.* **23**:307–387.

Traub, R., and C. L. Wisseman. 1974. The ecology of chigger-borne rickettsiosis (scrub typhus). *J. Med. Entomol.* **11**:237–303.

Travis, C. C., and W. M. Post. 1979. Dynamics and comparative statics of mutualistic communities. *J. Theor. Biol.* **78**:553–571.

Trivers, R. L. 1972. Parental investment and sexual selection. In B. Campbell (ed.). *Sexual selection and the descent of man,* 1871–1971. Aldine, Chicago, pp. 136–179.

Tso, T. C., and R. N. Jeffrey. 1959. Biochemical studies on tobacco alkaloids. I. The fate of labeled tobacco alkaloids supplied to *Nicotiana* plants. *Arch. Biochem. Biophys.* **80**:46–56.

Tso, T. C., and R. N. Jeffrey. 1961. Biochemical studies on tobacco alkaloids. IV. The dynamic state of nicotine supplied to *N. rustica. Arch. Biochem. Biophys.* **92**:253–256.

Tucker, V. A. 1969. The energetics of bird flight. *Sci. Am.* **220**(5):70–78.

Tullock, G. 1970. The coal tit as a careful shopper. *Am. Nat.* **105**:77–80.

Tummala, R. L., W. G. Ruesink, and D. L. Haynes. 1975. A discrete component approach to the management of the cereal leaf beetle ecosystem. *Environ. Entomol.* **4**:175–186.

Tuomi, J., P. Niemalä, E. Haukioja, S. Siren, and S. Neuvonen. 1984. Nutrient stress: An explanation for plant anti-herbivore responses to defoliation. *Oecologia* **61**:208–210.

Turnbull, C. M. 1961. *The forest people: A study of the Pygmies of the Congo.* Simon & Schuster, New York.

Turnbull, C. M. 1962. *The lonely African*. Simon & Schuster, New York.

Turnbull, A. L. 1966. A population of spiders and their potential prey in an over-grazed pasture in eastern Ontario. *Can. J. Zool.* **44**:557–583.

Turesson, G. 1922. The genotypical response of the plant species to the habitat. *Hereditas* **3**:211–350.

Turnipseed, S. G., and M. Kogan. 1976. Soybean entomology. *Annu. Rev. Entomol.* **21**:247–282.

Tweedie, M. 1973. *All color book of insects*. Octopus Books, London.

Udvardy, M. D. F. 1969. *Dynamic zoogeography with special reference to land animals*. Van Nostrand Reinhold, New York.

Uetz, G. 1974. Species diversity: A review. *Biologist* **56**:111–129.

Ullrich, W. G., and G. R. Coope. 1974. Occurrence of the east palaearctic beetle *Tachinus jacutus* Poppius (Col. Staphylinidae) in deposits of the last glacial period in England. *J. Entomol.* (B) **42**:207–212.

Ulrich, H., A. Petalas, and R. Camenzind. 1972. Der Generationswechsel von *Mycophila speyeri* Barnes, einer Gallmücke mit paedogenetischer Fortpflanzung. *Rev. Suisse Zool.* **79**(Suppl.):75–83.

Usinger, R. L. (ed.). 1956. *Aquatic insects of California, with keys to North American genera and California species*. Univ. California Press, Berkeley, Calif.

Utida, S. 1952. Interspecific competition between two species of the bean weevil. *Res. Popul. Ecol.* **1**:166–172.

Utida, S. 1953. Interspecific competition between two species of bean weevil. *Ecology* **34**:301–307.

Utida, S. 1972. Density dependent polymorphism in the adult of *Callosobruchus maculatus* (Coleoptera, Bruchidae). *J. Stored Prod. Res.* **8**:111–126.

Uvarov, B. P. 1966. *Grasshoppers and locusts: A handbook of general acridology*. Vol. 1. Cambridge Univ. Press, Cambridge.

Uvarov, B. P. 1977. *Grasshoppers and locusts: A handbook of general acridology*. Vol. 2. Centre for Overseas Pest Research, London.

van Alphen, J. J. M., and F. Galis. 1982. Patch time allocation and parasitization efficiency of *Asobara tabida* Nees, a larval parasitoid of *Drosophila*. In J. J. M. van Alphen. *Foraging behaviour of* Asobara tabida, *a larval parasitoid of* Drosophila. Offsetdrukkerij Kanters, Alblasserdam, The Netherlands, pp. 133–158.

van Bronswijk, J. E. M. H., and R. N. Sinha. 1971. Interrelations among physical, biological, and chemical varieties in stored-grain ecosystems: A descriptive and multivariate study. *Ann. Entomol. Soc. Am.* **64**:789–803.

Vance, R. R. 1978. A mutualistic interaction between a sessile marine clam and its epibionts. *Ecology* **59**:679–685.

Van den Bosch, R., and P. S. Messenger. 1973. *Biological control*. Intext Press, New York.

Van den Bosch, R., E. I. Schlinger, E. J. Dietrick, J. C. Hall, and B. Puttler. 1964. Studies on succession, distribution, and phenology of imported parasites of *Therioaphis trifolii* (Monell) in southern California. *Ecology* **45**:602–621.

van der Aart, P. J. M. 1974. Spinnen zonder web. In K. Bakker et al. (eds.). *Meijendel: Duin-water-leven.* van Hoeve, The Hague, The Netherlands, pp. 178–182.

Van der Drift, J., and M. Witkamp. 1960. The significance of the break-down of oak litter by *Enoicyla pusilla* Burm. *Arch. Neerl. Zool.* **13**:486–492.

Vandermeer, J. H., and D. H. Boucher. 1978. Varieties of mutualistic interaction in population models. *J. Theor. Biol.* **74**:549–558.

Van der Pijl, L. 1972. *Principles of dispersal in higher plants.* 2nd ed. Springer-Verlag, New York.

Van der Pijl, L., and C. H. Dodson. 1966. *Orchid flowers, their pollination and evolution.* Univ. Miami Press, Coral Gables, Fla.

Van Emden, H. F. (ed.). 1972a. *Insect/plant relationships.* Symp. R. Entomol. Soc. London 6, Royal Entomological Society of London, London.

Van Emden, H. F. (ed.). 1972b. *Aphid technology.* Academic Press, London.

Van Hook, R. I., Jr., and G. L. Dodson. 1974. Food energy budget for the yellow-poplar weevil, *Odontopus calceatus* (Say). *Ecology* **55**:205–207.

Van Lenteren, J. C., R. W. van der Linden, and A. Gluvers. 1976. A "borderline detector" for recording locomotory activities of animals. *Oecologia* **26**:133–137.

Van Mieghem, J., P. van Oye, and J. Schell (eds.). 1965. Biogeography and ecology in Antarctica. *Monogr. Biol.* **15**:1–762.

Van Valen, L. 1973. Body size and numbers of plants and animals. *Evolution* **27**:27–35.

Varley, G. C. 1947. The natural control of population balance in the knap-weed gall-fly *(Urophora jaceana). J. Anim. Ecol.* **16**:139–187.

Varley, G. C. 1967. Estimation of secondary production in species with an annual life cycle. In K. Petrusewicz (ed.). *Secondary productivity in terrestrial ecosystems.* Institute of Ecology, Polish Academy of Sciences, Warsaw, pp. 447–457.

Varley, G. C., and G. R. Gradwell. 1960. Key factors in population studies. *J. Anim. Ecol.* **29**:399–401.

Varley, G. C., and G. R. Gradwell. 1968. Population models for the winter moth. In T. R. E. Southwood (ed.). *Insect abundance.* Symp. R. Entomol. Soc. London 4, Entomological Society of London, London, pp. 132–142.

Varley, G. C., and G. R. Gradwell. 1970. Recent advances in insect population dynamics. *Annu. Rev. Entomol.* **15**:1–24.

Varley, G. C., G. R. Gradwell, and M. P. Hassell. 1973. *Insect population ecology: An analytical approach.* Blackwell Scientific, Oxford.

Vaughan, T. A. 1978. *Mammalogy.* 2nd ed. W.B. Saunders, Philadelphia, Pa.

Vaurie, C. 1951. Adaptive differences between two sympatric species of nuthatches *(Sitta). Proc. 10th Int. Orn. Congr.,* pp. 163–166.

Vepsäläinen, K. 1978. Wing dimorphism and diapause in *Gerris:* Determination and adaptive significance. In H. Dingle (ed.). *Evolution of insect migration and diapause*. Springer-Verlag, New York, pp. 218–253.

Verhulst, P. F. 1938. Notice sur la loi que la population suit dans son accroissement. *Correspond. Math. Phys.* **10**:113–121.

Vieira, E. M., J. Andrade, and P. W. Price. 1996. Fire effects on a *Palicourea rigida* (Rubiaceae) gall midge: A test of the plant vigor hypothesis. *Biotropica* **28**:210–217.

Vinson, S. B. 1975. Biochemical coevolution between parasitoids and their hosts. In P. W. Price (ed.). *Evolutionary strategies of parasitic insects and mites*. Plenum Press, New York, pp. 14–48.

Vinson, S. B., and P. Barbosa. 1987. Interrelationships of nutritional ecology of parasitoids. In F. Slansky and J. G. Rodriguez (eds.). *Nutritional ecology of insects, mites, spiders, and related invertebrates*. Wiley, New York, pp. 673–695.

Vinson, S. B., and F. S. Guillot. 1972. Host marking: Source of a substance that results in host discrimination in insect parasitoids. *Entomophaga* **17**:241–245.

Vogel, S. 1978. Evolutionary shifts from reward to deception in pollen flowers. In A. J. Richards (ed.). *The pollination of flowers by insects*. Academic Press, London, pp. 89–96.

Volterra, V. 1926. Variazioni e fluttuazioni del numero d'individui in specie animali conviventi. *Mem. Acad. Lincei Roma* **2**:31–113.

von Frisch, K. 1967. *The dance language and orientation of bees*. Belknap Press, imprint of Harvard Univ. Press, Cambridge, Mass.

Vuilleumier, F. 1970. Insular biogeography in continental regions. I. The northern Andes of South America. *Am. Nat.* **104**:373–388.

Waage, Jeffrey K. 1978. Arrestment responses of the parasitoid *Nemeritis canescens* to a contact chemical produced by its host, *Plodia interpunctella*. *Physiol. Entomol.* **3**:135–146.

Waage, Jeffrey K. 1979a. The evolution of insect/vertebrate associations. *Biol. J. Linn. Soc.* **12**:187–224.

Waage, Jeffrey K. 1979b. Foraging for patchily-distributed hosts by the parasitoid, *Nemeritis canescens*. *J. Anim. Ecol.* **48**:353–371.

Waage, Jonathan K. 1974. Reproductive behavior and its relation to territoriality in *Calopteryx maculata* (Beauvois) (Odonata: Calopterygidae). *Behaviour* **47**:240–256.

Waage, Jonathan K. 1979. Dual function of the damselfly penis: Sperm removal and transfer. *Science* **203**:916–918.

Waage, J., and D. Greathead (eds.). 1986. *Insect parasitoids*. Academic Press, London.

Waddington, K. D. 1980. Flight patterns of foraging bees relative to density of artificial flowers and distribution of nectar. *Oecologia* **44**:199–204.

Waddington, K. D., and B. Heinrich. 1981. Patterns of movement and floral choice by foraging bees. In A. Kamil and T. Sargent (eds.). *Foraging behav-*

ior: Ecological, ethological, and psychological approaches. Garland Publishing, New York, pp. 215–230.

Waddington, K. D., and L. R. Holden. 1979. Optimal foraging: On flower selection by bees. *Am. Nat.* **114:**179–196.

Waddington, K. D., T. Allen, and B. Heinrich. 1981. Floral preferences of bumblebees *(Bombus edwardsii)* in relation to intermittent versus continuous rewards. *Anim. Behav.* **29:**779–784.

Wade, M. J., and S. J. Arnold. 1980. The intensity of sexual selection in relation to male sexual behaviour, female choice, and sperm precedence. *Anim. Behav.* **28:**446–461.

Waldbauer, G. P. 1968. The consumption and utilization of food by insects. *Adv. Insect Physiol.* **5:**229–288.

Waldbauer, G. P., and J. K. Sheldon. 1971. Phenological relationships of some aculeate Hymenoptera, their dipteran mimics, and insectivorous birds. *Evolution* **25:**371–382.

Waldbauer, G. P., and J. G. Sternburg. 1975. Saturniid moths as mimics: An alternative interpretation of attempts to demonstrate mimetic advantage in nature. *Evolution* **29:**650–658.

Walde, S. J. 1995. Internal dynamics and metapopulations: Experimental tests with predator-prey systems. In N. Cappuccino and P. W. Price (eds.). *Population dynamics: New approaches and synthesis.* Academic Press, San Diego, Calif., pp. 173–193.

Walker, W. F. 1980. Sperm utilization strategies in nonsocial insects. *Am. Nat.* **115:**780–799.

Wallace, A. R. 1867. *J. Proc. Entomol. Res.,* Feb. 4, p. lxxi.

Wallace, A. R. 1869. *The Malay archipelago.* Macmillan, London.

Wallace, A. R. 1876. *The geographical distribution of animals.* 2 vols. Harper, New York.

Wallace, J. W., and R. L. Mansell (eds.). 1976. *Biochemical interaction between plants and insects.* Plenum Press, New York.

Wallner, W. E. 1987. Factors affecting insect population dynamics: Differences between outbreak and non-outbreak species. *Annu. Rev. Entomol.* **32:**317–340.

Wallwork, J. A. 1958. Notes on the feeding behaviour of some forest soil Acarina. *Oikos* **9:**260–271.

Wallwork, J. A. 1970. *Ecology of soil animals.* McGraw-Hill, London.

Waloff, N. 1968a. A comparison of factors affecting different insect species on the same host plant. In T. R. E. Southwood (ed.). *Insect abundance.* Symp. R. Entomol. Soc. London 4, Royal Entomological Society of London, London, pp. 76–87.

Waloff, N. 1968b. Studies on the insect fauna on scotch broom *Sarothamnus scoparius* (L.) Wimmer. *Adv. Ecol. Res.* **5:**87–208.

Wang, Y. H., and A. P. Gutierrez. 1980. An assessment of the use of stability analyses in population ecology. *J. Anim. Ecol.* **49:**435–452.

Wangersky, P. J., and W. J. Cunningham. 1957. Time lag in prey–predator population models. *Ecology* **38**:136–139.

Ward, R. D., and P. D. N. Hebert. 1972. Variability of alcohol dehydrogenase activity in a natural population of *Drosophila melanogaster. Nat. New Biol.* **236**:243–244.

Waring, G. L., and N. S. Cobb. 1992. The impact of plant stress on herbivore population dynamics. In E. A. Bernays (ed.). *Insect–plant interactions.* Vol. 4. CRC Press, Boca Raton, Fla., pp. 167–226.

Warner, R. E. 1968. The role of introduced diseases in the extinction of the endemic Hawaiian avifauna. *Condor* **70**:101–120.

Waser, N. M., and M. V. Price. 1981. Pollinator choice and stabilizing selection for flower color in *Delphinium nelsonii. Evolution* **35**:376–390.

Waser, N. M., and M. V. Price. 1983. Optimal and actual outcrossing in plants, and the nature of plant–pollinator interaction. In C. E. Jones and R. J. Little (eds.). *Handbook of experimental pollination biology.* Van Nostrand Reinhold, New York, pp. 341–359.

Washburn, J. O. 1984. Mutualism between a cynipid gall wasp and ants. *Ecology* **65**:654–656.

Washburn, J. O., and H. V. Cornell. 1981. Parasitoids, patches, and phenology: Their possible role in the local extinction of a cynipid gall wasp population. *Ecology* **62**:1597–1607.

Watanabe, M. 1976. A preliminary study on population dynamics of the swallow-tail butterfly, *Papilio xuthus* L. in a deforested area. *Res. Popul. Ecol.* **17**:200–210.

Waterhouse, D. F. 1991. Insects and humans in Australia. In I. D. Naumann (ed.). *The insects of Australia: A textbook for students and research workers.* Cornell Univ. Press, Ithaca, N.Y., pp. 221–235.

Watson, A. (ed.). 1970. *Animal populations in relation to their food resources.* Br. Ecol. Soc. Symp. 10, British Ecological Society, London.

Watson, G. 1964. Ecology and evolution of passerine birds on the islands of the Aegean Sea. Ph.D. thesis, Dept. Biology, Yale Univ., New Haven, Conn.

Watt, K. E. F. 1963. Mathematical population models for five agricultural crop pests. *Mem. Entomol. Soc. Can.* **32**:83–91.

Watt, A. D., S. R. Leather, M. D. Hunter, and N. A. C. Kidd (eds.). 1990. *Population dynamics of forest insects.* Intercept, Andover, Hants, England.

Way, M. J., and M. Cammell. 1970. Aggregation behaviour in relation to food utilization by aphids. In A. Watson (ed.). *Animal populations in relation to their food resources.* Symp. Br. Ecol. Soc. 10, British Ecological Society, London, pp. 229–246.

Weber, N. A. 1979. Fungus-culturing by ants. In L. R. Batra (ed.). *Insect–fungus symbiosis: Nutrition, mutualism, and commensalism.* Allanheld, Osmun, Totowa, N.J., pp. 77–116.

Weis-Fogh, T. 1976. Energetics and aerodynamics of flapping flight: A synthesis. *Symp. R. Entomol. Soc. London* **7**:48–78.

Wellington, W. G. 1954a. Atmospheric circulation processes and insect ecology. *Can. Entomol.* **86:**312–333.

Wellington, W. G. 1954b. Weather and climate in forest entomology. *Meteorol. Monogr.* **2**(8):11–18.

Wellington, W. G. 1957a. Individual differences as a factor in population dynamics: The development of a problem. *Can. J. Zool.* **35:**293–323.

Wellington, W. G. 1957b. The synoptic approach to studies of insects and climate. *Annu. Rev. Entomol.* **2:**143–162.

Wellington, W. G. 1960. Qualitative changes in natural populations during changes in abundance. *Can. J. Zool.* **38:**289–314.

Wellington, W. G. 1962. Population quality and the maintenance of nuclear polyhedrosis between outbreaks of *Malacosoma pluviale* (Dyar). *J. Insect Path.* **4:**285–305.

Wellington, W. G. 1964. Qualitative changes in unstable environments. *Can. Entomol.* **96:**436–451.

Wellington, W. G. 1977. Returning the insect to insect ecology: Some consequences for pest management. *Environ. Entomol.* **6:**1–8.

Wellington, W. G. 1979. Insect dispersal: A biometeorological perspective. In R. L. Rabb and G. G. Kennedy (eds.). *Movement of highly mobile insects: Concepts and methodology in research.* North Carolina State Univ., Raleigh, N.C., pp. 104–108.

Wellington, W. G., P. J. Cameron, W. A. Thompson, I. B. Vertinsky, and A. S. Landsberg. 1975. A stochastic model for assessing the effects of external and internal heterogeneity of an insect population. *Res. Popul. Ecol.* **17:**1–28.

Went, F. W. 1968. The size of man. *Am. Sci.* **56:**400–413.

Werren, J. H. 1980. Sex ratio adaptations to local mate competition in a parasitic wasp. *Science* **208:**1157–1159.

Weseloh, R. M. 1981. Host location by parasitoids. In D. A. Nordlund, R. L. Jones, and W. J. Lewis (eds.). *Semiochemicals: Their role in pest control.* Wiley, New York, pp. 79–95.

Westlake, D. F. 1963. Comparisons of plant productivity. *Biol. Rev.* **38:** 385–425.

Westman, W. E. 1978. Measuring the inertia and resilience of ecosystems. *BioScience* **28:**705–710.

Weygoldt, P. 1969. *The biology of pseudoscorpions.* Harvard Univ. Press, Cambridge, Mass.

White, M. J. D. 1973. *Animal cytology and evolution.* Cambridge Univ. Press, London.

White, T. C. R. 1969. An index to measure weather-induced stress of trees associated with outbreaks of psyllids in Australia. *Ecology* **50:**905–909.

White, T. C. R. 1974. A hypothesis to explain outbreaks of looper caterpillars, with special reference to populations of *Selidosema suavis* in a plantation of *Pinus radiata* in New Zealand. *Oecologia* **16:**279–301.

White, T. C. R. 1976. Weather, food and plagues of locusts. *Oecologia* **22**:119–134.

White, T. C. R. 1978. The importance of a relative shortage of food in animal ecology. *Oecology* **33**:71–86.

White, T. C. R. 1984. The abundance of invertebrate herbivores in relation to the availability of nitrogen in stressed food plants. *Oecologia* **63**:90–105.

White, T. C. R. 1993. *The inadequate environment: Nitrogen and the abundance of animals.* Springer-Verlag, Berlin.

Whitehead, V. B., E. A. C. L. E. Schelpe, and N. C. Anthony. 1984. The bee, *Rediviva longimanus* Michener (Apoidea: Melittidae), collecting pollen and oil from *Diascia longicornis* (Thunb.) Druce (Scrophulariaceae). *S. Afr. J. Sci.* **80**:286.

Whitham, T. G. 1977. Coevolution of foraging in *Bombus* and nectar dispensing in *Chilopsis:* A last dreg theory. *Science* **197**:593–596.

Whitham, T. G. 1978. Habitat selection by *Pemphigus* aphids in response to resource limitation and competition. *Ecology* **59**:1164–1176.

Whitham, T. G. 1979. Territorial behaviour of *Pemphigus* gall aphids. *Nature* **279**:324–325.

Whitham, T. G. 1980. The theory of habitat selection: Examined and extended using *Pemphigus* aphids. *Am. Nat.* **115**:449–466.

Whitham, T. G. 1981. Individual trees as heterogeneous environments: Adaptation to herbivory or epigenetic noise? In R. F. Denno and H. Dingle (eds.). *Insect life history patterns: Habitat and geographic variation.* Springer-Verlag, New York, pp. 9–27.

Whitham, T. G., and S. Mopper. 1985. Chronic herbivory: Impacts on tree architecture and sex expression of pinyon pine. *Science* **227**:1089–1091.

Whitham, T. G., and C. N. Slobodchikoff. 1981. Evolution by individuals, plant–herbivore interactions, and mosaics of genetic variability: The adaptive significance of somatic mutations in plants. *Oecologia* **49**:287–292.

Whitney, H. S. 1982. Relationships between bark beetles and symbiotic organisms. In J. B. Mitton and K. B. Sturgeon (eds.). *Bark beetles in North American conifers: A system for the study of evolutionary biology.* University of Texas Press, Austin, pp. 183–211.

Whittaker, J. B. 1971. Population changes in *Neophilaenus lineatus* (L.) (Homoptera: Cercopidae) in different parts of its range. *J. Anim. Ecol.* **40**:425–443.

Whittaker, R. H. 1969. Evolution of diversity in plant communities. In G. M. Woodwell and H. H. Smith (eds.). *Diversity and stability in ecological systems.* Brookhaven Symp. Biol. 22, Brookhaven National Laboratory, Upton, N.Y., pp. 178–195.

Whittaker, R. H. 1970a. The biochemical ecology of higher plants. In E. Sondheimer and J. B. Simeone (eds.). *Chemical ecology.* Academic Press, San Diego, Calif., pp. 43–70.

Whittaker, R. H. 1970b. *Communities and ecosystems*. Macmillan, New York.

Whittaker, R. H. 1972. Evolution and the measurement of species diversity. *Taxon* **21**:213–251.

Whittaker, R. H. 1975. *Communities and ecosystems*. 2nd ed. Macmillan, New York.

Whittaker, R. H., and P. P. Feeny. 1971. Allelochemics: Chemical interactions between species. *Science* **171**:757–770.

Whittaker, R. H., and S. A. Levin (eds.). 1975. *Niche theory and application*. Dowden, Hutchinson & Ross, Stroudsburg, Pa.

Whittaker, R. H., and G. M. Woodwell. 1968. Dimension and production relations of trees and shrubs in the Brookhaven Forest, New York. *J. Ecol.* **56**:1–25.

Whittaker, R. H., and G. M. Woodwell. 1969. Structure, production and diversity of the oak–pine forest at Brookhaven, New York. *J. Ecol.* **57**: 155–174.

Whittaker, R. H., S. A. Levin, and R. B. Root. 1973. Niche, habitat, and ecotope. *Am. Nat.* **107**:321–338.

Whittaker, R. H., S. A. Levin, and R. B. Root. 1975. On the reasons for distinguishing "Niche, habitat, and ecotope." *Am. Nat.* **109**:479–482.

Wickler, W. 1968. *Mimicry in plants and animals*. McGraw-Hill, New York.

Wiebes, J. T. 1979. Co-evolution of figs and their insect pollinators. *Annu. Rev. Ecol. Syst.* **10**:1–12.

Wiegert, R. G. 1964a. The ingestion of xylem sap by meadow spittlebugs, *Philaenus spumarius* (L.). *Am. Midl. Nat.* **71**:422–428.

Wiegert, R. G. 1964b. Population energetics of meadow spittlebugs *(Philaenus spumarius* L.) as affected by migration and habitat. *Ecol. Monogr.* **34**:217–241.

Wiegert, R. G. 1965. Energy dynamics of the grasshopper populations in old field and alfalfa field ecosystems. *Oikos* **16**:161–176.

Wiegert, R. G. 1973. A general ecological model and its use in simulating algal–fly energetics in a thermal spring community. In P. W. Geier, L. R. Clark, D. J. Anderson, and H. A. Nix (eds.). *Insects: Studies in population management*. Mem. Ecol. Soc. Aust. 1, Ecological Society of Australia, Canberra City, Australia, pp. 85–102.

Wiegert, R. G. (ed.). 1976. *Ecological energetics*. Dowden, Hutchinson & Ross, Stroudsburg, Pa.

Wiegert, R. G., and F. C. Evans. 1967. Investigations of secondary productivity in grasslands. In K. Petrusewicz (ed.) *Secondary productivity in terrestrial ecosystems*. Institute of Ecology, Polish Academy of Sciences, Warsaw, pp. 499–518.

Wiegert, R. G., and D. F. Owen. 1971. Trophic structure, available resources and population density in terrestrial vs. aquatic ecosystems. *J. Theor. Biol.* **30**:69–81.

Wiegmann, B. M., C. Mitter, and B. Barrell. 1993. Diversification of carnivorous parasitic insects: Extraordinary radiation or specialized dead end? *Am. Nat.* **142**:737–754.

Wigglesworth, V. B. 1972. *The principles of insect physiology.* 7th ed. Chapman & Hall, London.

Wigglesworth, V. B. 1976. The evolution of insect flight. In R. C. Rainey (ed.). *Insect flight. Symp. R. Entomol. Soc. London* 7:255–269.

Wikel, S. K. 1982. Immune responses to arthropods and their products. *Annu. Rev. Entomol.* **27**:21–48.

Wiklund, C. 1974. Oviposition preferences in *Papilio machaon* in relation to the host plants of the larvae. *Entomol. Exp. Appl.* **17**:189–198.

Wiklund, C. 1975. The evolutionary relationship between adult oviposition preferences and larval host plant range in *Papilio machaon* L. *Oecologia* **18**:185–197.

Wiklund, C. 1981. Generalist vs. specialist oviposition behaviour in *Papilio machaon* (Lepidoptera) and functional aspects on the hierarchy of oviposition preferences. *Oikos* **36**:163–170.

Wiklund, C. 1984. Egg-laying patterns in butterflies in relation to their phenology and the visual apparency and abundance of their host plants. *Oecologia* **63**:23–29.

Wiklund, C. 1995. Protandry and mate acquisition. In S. R. Leather and J. Hardie (eds.). *Insect reproduction.* CRC Press, Boca Raton, Fla., pp. 175–197.

Wilding, N., N. M. Collins, P. M. Hammond, and J. F. Webber (eds.). 1989. *Insect–fungus interactions.* Academic Press, London.

Williams, C. B. 1964. *Patterns in the balance of nature and related problems in quantitative ecology.* Academic Press, London.

Williams, C. M., and W. E. Robbins. 1968. Conference on insect–plant interactions. *BioScience* **18**:791–799.

Williams, G. C. 1966a. *Adaptation and natural selection.* Princeton Univ. Press, Princeton, N.J.

Williams, G. C. 1966b. Natural selection, the costs of reproduction, and a refinement of Lack's principle. *Am. Nat.* **100**:687–690.

Williams, K. S. 1983. The coevolution of *Euphydryas chalcedona* butterflies and their larval host plants. III. Oviposition behavior and host plant quality. *Oecologia* **56**:336–340.

Williams, K. S., D. E. Lincoln, and P. R. Ehrlich. 1983. The coevolution of *Euphydryas chalcedona* butterflies and their larval host plants. II. Maternal and host plant effects on larval growth, development, and food-use efficiency. *Oecologia* **56**:330–335.

Williams, N. H. 1978. A preliminary bibliography on euglossine bees and their relationships with orchids and other plants. *Selbyana* **2**:345–355.

Williams, N. H. 1982. The biology of orchids and euglossine bees. In J. Arditti (ed.). *Orchid biology: Reviews and perspectives II.* Cornell Univ. Press, Ithaca, N.Y., pp. 119–171.

Williamson, M. 1972. *The analysis of biological populations.* Edward Arnold, London.

Willis, M. A., and M. C. Birch, 1982. Male lek formation and female calling in a population of the arctiid moth *Estigmene acrea. Science* **218**:168–170.

Willson, M. F. 1971. Life history consequences of death rates. *Biologist* **53**:49–56.

Willson, M. F. 1983. *Plant reproductive ecology.* Wiley, New York.

Willson, M. F., and R. I. Bertin. 1979. Flower-visitors, nectar production, and inflorescence size of *Asclepias syriaca. Can. J. Bot.* **57**:1380–1388.

Willson, M. F., and N. Burley. 1983. *Mate choice in plants: Tactics, mechanisms, and consequences.* Princeton Univ. Press, Princeton, N.J.

Willson, M. F., and P. W. Price. 1977. The evolution of inflorescence size in *Asclepias* (Asclepiadaceae). *Evolution* **31**:495–511.

Willson, M. F., and B. J. Rathcke. 1974. Adaptive design of the floral display in *Asclepias syriaca* L. *Am. Midl. Nat.* **92**:47–57.

Wilson, E. O. 1955. A monographic revision of the ant genus *Lasius. Bull. Mus. Comp. Zool. Harvard* **113**:1–201.

Wilson, E. O. 1961. The nature of the taxon cycle in a tropical ant fauna. *Evolution* **13**:122–144.

Wilson, E. O. 1969. The species equilibrium. In G. M. Woodwell and H. H. Smith (eds.). *Diversity and stability in ecological systems.* Brookhaven Symp. Biol. 22, Brookhaven National Laboratory, Upton, N.Y., pp. 38–47.

Wilson, E. O. 1971. *The insect societies.* Belknap Press, imprint of Harvard Univ. Press, Cambridge, Mass.

Wilson, E. O. 1975. *Sociobiology: The new synthesis.* Belknap Press, imprint of Harvard Univ. Press, Cambridge, Mass.

Wilson, E. O. (ed.). 1988. *Biodiversity.* National Academy Press, Washington, D.C.

Wilson, E. O. 1992. *The diversity of life.* W. W. Norton, New York.

Wilson, E. O., and W. H. Bossert. 1971. *A primer of population biology.* Sinauer, Stamford, Conn.

Wilson, E. O., and D. S. Simberloff. 1969. Experimental zoogeography of islands: Defaunation and monitoring techniques. *Ecology* **50**:267–278.

Wilson, F. 1961. Adult reproductive behaviour in *Asolcus basalis* (Hymenoptera: Scelionidae). *Aust. J. Zool.* **9**:739–751.

Wilson, J. W. 1974. Analytical zoogeography of North American mammals. *Evolution* **28**:124–140.

Wilson, R. D., P. H. Monaghan, A. Osanik, L. C. Price, and M. A. Rogers. 1974. Natural marine oil seepage. *Science* **184**:857–865.

Wise, D. H. 1981. A removal experiment with darkling beetles: Lack of evidence for interspecific competition. *Ecology* **62**:727–738.

Witkamp, M. 1971. Soils as components of ecosystems. *Annu. Rev. Ecol. Syst.* **2**:85–110.

Witmer, M. C., and A. S. Cheke. 1991. The dodo and the tambalacoque tree: An obligate mutualism reconsidered. *Oikos* **61**:133–137.

Wolda, H. 1983. Diversity, diversity indices and tropical cockroaches. *Oecologia* **58**:290–298.

Wolda, H. 1996. Between-site similarity in species composition of a number of Panamanian insect groups. *Misc. Zool. (Barcelona)*. In press.

Wolda, H., C. W. O'Brien, and H. P. Stockwell. 1996. Weevil diversity and seasonality in tropical Panama as deduced from light-trap catches (Coleoptera: Curculionidae). *Smithson. Contr. Zool.* In press.

Wood, T. K. 1976. Alarm behavior of brooding female *Umbonia crassicornis* (Homoptera: Membracidae). *Ann. Entomol. Soc. Am.* **69**:340–344.

Wood, T. K. 1978. Parental care in *Guayaquila compressa* Walker (Homoptera: Membracidae). *Psyche* **85**:135–145.

Wood, T. K. 1984. Life history patterns of tropical membracids (Homoptera: Membracidae). *Sociobiology* **8**:299–344.

Wood, T. K., and K. L. Olmstead. 1984. Latitudinal effects on treehopper species richness (Homoptera: Membracidae). *Ecol. Entomol.* **9**:109–115.

Woodwell, G. M. 1967. Toxic substances and ecological cycles. *Sci. Am.* **216**(3):24–31.

Woodwell, G. M. 1970. The energy cycle of the biosphere. *Sci. Am.* **223**(3):64–74.

Woodwell, G. M. 1974. Variation in the nutrient content of leaves of *Quercus alba*, *Quercus coccinea*, and *Pinus rigida* in the Brookhaven Forest from bud-break to abscission. *Am. J. Bot.* **61**:749–753.

Woodwell, G. M., C. F. Wurster, Jr., and P. A. Isaacson. 1967. DDT residues in an east coast estuary: A case of biological concentration of a persistent insecticide. *Science* **156**:821–824.

Woolridge, A. W., and F. P. Harrison. 1968. Effects of soil fertility on abundance of green peach aphids on Maryland tobacco. *J. Econ. Entomol.* **61**:387–391.

Wright, S. 1968. *Evolution and the genetics of populations.* Vol. 1. *Genetic and biometric foundations.* Univ. Chicago Press, Chicago.

Wright, S. 1969. *Evolution and the genetics of populations.* Vol. 2. *The theory of gene frequencies.* Univ. Chicago Press, Chicago.

Wright, S. 1977. *Evolution and the genetics of populations.* Vol. 3. *Experimental results and evolutionary deductions.* Univ. Chicago Press, Chicago.

Wright, S. 1978. *Evolution and the genetics of populations.* Vol. 4. *Variability within and among natural populations.* Univ. Chicago Press, Chicago.

Wright, S., and T. Dobzhansky. 1946. Genetics of natural populations. XII. Experimental reproduction of some of the changes caused by natural selection in certain populations of *Drosophila pseudoobscura*. *Genetics* **31**:125–156.

Wullems, G. J., L. Molendijk, G. Ooms, and R. A. Schnilperoort. 1981. Retention of tumor markers in F1 progeny plants from in vitro induced octopine and nopaline tumor tissues. *Cell* **24**:719–727.

Wyatt, I. J. 1964. Immature stages of Lestremiinae (Diptera: Cecidomyiidae) infesting cultivated mushrooms. *Trans. R. Entomol. Soc. London* **116**:15–27.

Wyatt, I. J. 1967. Pupal paedogenesis in the Cecidomyiidae (Diptera) 3: A reclassification of the Heteropezini. *Trans. R. Entomol. Soc. London* **119:** 71–98.

Wynne-Edwards, V. C. 1962. *Animal dispersion in relation to social behaviour.* Hafner Press, New York.

Wynne-Edwards, V. C. 1964. Population control in animals. *Sci. Am.* **211**(2):68–74.

Wynne-Edwards, V. C. 1965. Self-regulating systems in populations of animals. *Science* **147:**1543–1548.

Yamamoto, R. T. 1974. Induction of host plant specificity in the tobacco hornworm, *Manduca sexta. J. Insect Physiol.* **20:**641–650.

Yang, S., M. Wheeler, and I. Bock. 1972. Isozyme variations and phylogenetic relationships in the *Drosophila bipectinata* species complex. Studies in genetics. VII. *Texas Univ. Publ.* **7213:**213–227.

Yoda, K., T. Kira, H. Ogawa, and K. Hozumi. 1963. Self-thinning in overcrowded pure stands under cultivated and natural conditions (Intraspecific competition among higher plants XI). *J. Biol. Osaka City Univ.* **14:** 107–129.

Yoshida, T. 1957. Experimental studies on the interspecific competition among the granary beetles. 4. *Mem. Fac. Lib. Arts Educ. Miyazaki Univ.* **1:**55–80 (in Japanese with English summary).

Young, J. O., I. G. Morris, and T. B. Reynoldson. 1964. A serological study of *Asellus* in the diet of lake-dwelling triclads. *Arch. Hydrobiol.* **60:**366–373.

Zahavi, A., D. Eisikowitch, A. K. Zahavi, and A. Cohen. 1984. A new approach to flower constancy in honey bees. *5th Symp. Int. Poll., Versailles,* INRA, Versailles, France, pp. 89–95.

Zak, J., and D. W. Freckman. 1991. Soil communities in deserts: Microarthropods and nematodes. In G. A. Polis (ed.). *The ecology of desert communities.* Univ. Arizona Press, Tucson, Ariz., pp. 55–88.

Zak, J., and W. Whitford. 1988. Interactions among soil biota in desert ecosystems. *Agric. Ecosyst. Environ.* **24:**87–100.

Zimmerman, M. 1979. Optimal foraging: A case for random movement. *Oecologia* **43:**261–267.

Zinsser, H. 1935. *Rats, lice and history.* Little, Brown, Boston.

Zou, J., and R. G. Cates. 1994. Role of Douglas fir *(Pseudotsuga menziesii)* carbohydrates in resistance to budworm *(Choristoneura occidentalis). J. Chem. Ecol.* **20:**395–405.

Zucker, W. V. 1982. How aphids choose leaves: The roles of phenolics in host selection by a galling aphid. *Ecology* **63:**972–981.

Zucker, W. V. 1983. Tannins: Does structure determine function? An ecological perspective. *Am. Nat.* **121:**335–365.

Zumpt, F. 1965. *Myiasis in man and animals in the Old World: A textbook for physicians, veterinarians, and zoologists.* Butterworth, London.

Taxonomic Index

Note: only arthropods are indexed

Author Index

Note: articles with two or more authors have first author indexed.

861

Subject Index